Lecture Notes in Networks and Systems

Volume 333

Series Editor

Janusz Kacprzyk, Systems Research Institute, Polish Academy of Sciences, Warsaw, Poland

Advisory Editors

Fernando Gomide, Department of Computer Engineering and Automation—DCA, School of Electrical and Computer Engineering—FEEC, University of Campinas—UNICAMP, São Paulo, Brazil

Okyay Kaynak, Department of Electrical and Electronic Engineering, Bogazici University, Istanbul, Turkey

Derong Liu, Department of Electrical and Computer Engineering, University of Illinois at Chicago, Chicago, USA
Institute of Automation, Chinese Academy of Sciences, Beijing, China

Witold Pedrycz, Department of Electrical and Computer Engineering, University of Alberta, Alberta, Canada
Systems Research Institute, Polish Academy of Sciences, Warsaw, Poland

Marios M. Polycarpou, Department of Electrical and Computer Engineering, KIOS Research Center for Intelligent Systems and Networks, University of Cyprus, Nicosia, Cyprus

Imre J. Rudas, Óbuda University, Budapest, Hungary

Jun Wang, Department of Computer Science, City University of Hong Kong, Kowloon, Hong Kong

AF864743

The series "Lecture Notes in Networks and Systems" publishes the latest developments in Networks and Systems—quickly, informally and with high quality. Original research reported in proceedings and post-proceedings represents the core of LNNS.

Volumes published in LNNS embrace all aspects and subfields of, as well as new challenges in, Networks and Systems.

The series contains proceedings and edited volumes in systems and networks, spanning the areas of Cyber-Physical Systems, Autonomous Systems, Sensor Networks, Control Systems, Energy Systems, Automotive Systems, Biological Systems, Vehicular Networking and Connected Vehicles, Aerospace Systems, Automation, Manufacturing, Smart Grids, Nonlinear Systems, Power Systems, Robotics, Social Systems, Economic Systems and other. Of particular value to both the contributors and the readership are the short publication timeframe and the world-wide distribution and exposure which enable both a wide and rapid dissemination of research output.

The series covers the theory, applications, and perspectives on the state of the art and future developments relevant to systems and networks, decision making, control, complex processes and related areas, as embedded in the fields of interdisciplinary and applied sciences, engineering, computer science, physics, economics, social, and life sciences, as well as the paradigms and methodologies behind them.

Indexed by SCOPUS, INSPEC, WTI Frankfurt eG, zbMATH, SCImago.

All books published in the series are submitted for consideration in Web of Science.

More information about this series at https://link.springer.com/bookseries/15179

Atulya K. Nagar · Dharm Singh Jat ·
Gabriela Marín-Raventós · Durgesh Kumar Mishra
Editors

Intelligent Sustainable Systems

Selected Papers of WorldS4 2021, Volume 1

Springer

Editors
Atulya K. Nagar
School of Mathematics, Computer Science and Engineering
Liverpool Hope University
Liverpool, UK

Dharm Singh Jat
Department of Computer Science
Namibia University of Science and Technology
Windhoek, Namibia

Gabriela Marín-Raventós
University of Costa Rica
Curridabat, Costa Rica

Durgesh Kumar Mishra
Sri Aurobindo Institute of Technology
Indore, Madhya Pradesh, India

ISSN 2367-3370 ISSN 2367-3389 (electronic)
Lecture Notes in Networks and Systems
ISBN 978-981-16-6308-6 ISBN 978-981-16-6309-3 (eBook)
https://doi.org/10.1007/978-981-16-6309-3

© The Editor(s) (if applicable) and The Author(s), under exclusive license to Springer Nature Singapore Pte Ltd. 2022
This work is subject to copyright. All rights are solely and exclusively licensed by the Publisher, whether the whole or part of the material is concerned, specifically the rights of translation, reprinting, reuse of illustrations, recitation, broadcasting, reproduction on microfilms or in any other physical way, and transmission or information storage and retrieval, electronic adaptation, computer software, or by similar or dissimilar methodology now known or hereafter developed.
The use of general descriptive names, registered names, trademarks, service marks, etc. in this publication does not imply, even in the absence of a specific statement, that such names are exempt from the relevant protective laws and regulations and therefore free for general use.
The publisher, the authors and the editors are safe to assume that the advice and information in this book are believed to be true and accurate at the date of publication. Neither the publisher nor the authors or the editors give a warranty, expressed or implied, with respect to the material contained herein or for any errors or omissions that may have been made. The publisher remains neutral with regard to jurisdictional claims in published maps and institutional affiliations.

This Springer imprint is published by the registered company Springer Nature Singapore Pte Ltd.
The registered company address is: 152 Beach Road, #21-01/04 Gateway East, Singapore 189721, Singapore

Preface

The fifth edition of the WorldS4 2021—World Conference on Smart Trends in Systems Security and Sustainability—was held during 29–30 July 2021 digitally on Zoom and organised by Global Knowledge Research Foundation. The associated partners were Springer, InterYIT IFIP, and IEEE UK and Ireland Sections. The conference provided a useful and wide platform both for the display of the latest research and for the exchange of research results and thoughts. The participants of the conference were from almost every part of the world, with the background of either academia or industry, allowing a real multinational multicultural exchange of experiences and ideas.

A great pool of more than 870 papers were received for this conference from across 28 countries among which around 144 papers were accepted with this Springer Series and were presented through a digital platform during the two days. Due to the overwhelming response, we had to drop many papers in the hierarchy of the quality. Totally, 28 technical sessions were organised in parallel in two days along with few keynotes and panel discussions. The conference involved deep discussion and issues which were intended to solve at global levels. New technologies will be proposed, experiences will be shared, and future solutions for enhancement in systems and security will also be discussed. The final papers will be published in two volumes of proceedings by Springer LNNS Series.

Over the years, this conference has been organised and conceptualised with the collective efforts of a large number of individuals. We would like to thank each of the committee members and the reviewers for their excellent work in reviewing the papers. Grateful acknowledgements are extended to the team of Global Knowledge Research Foundation for their valuable efforts and support.

We look forward to welcome you to the sixth edition of this WorldS4 Conference in 2022.

Liverpool, UK — Atulya K. Nagar
Windhoek, Namibia — Dharm Singh Jat
Curridabat, Costa Rica — Gabriela Marín-Raventós
Indore, India — Durgesh Kumar Mishra

Contents

App OnViolet on UV Radiation Skin Care Habits 1
Patricia Janet Uceda Martos, Laura Sofía Bazán Díaz, and Elki Mimar Solano Cueva

Identifying Key Quality Features of mHealth Applications: Unsupervised Feature Selection Approach: MARS Case Study 13
Rolando Armas, Carlos Montenegro, Andrés Larco, and Cesar Yanez

Implementation of E-health in Burkina Faso: A Status Report 23
Sadouanouan Malo, Lydie Simone Kone/Tapsoba, and Yaya Traore

A Training Web Platform to Improve Cognitive Skills for Phishing Attacks Detection 33
María Fernanda Cazares, Diana Arévalo, Roberto O. Andrade, Walter Fuertes, and Manuel Sánchez-Rubio

Development of a Linear Model of Diabetes Mellitus with a PID Controller and a Method for Automatic Adjustment of the PID Controller 43
D. A. Andrikov and S. V. Kurbanov

Financial Decisions in Colombian Organisations in the Health Sector 51
Romel Ramón González-Díaz, Elena Cachicatari-Vargas, Santos Lucio Guanilo-Gómez, and Graciela Tristancho Cediel

Business Innovation Capabilities in the Road Infrastructure Sector in the Caribbean Region of Colombia 61
Romel Ramón González-Díaz, Santos Lucio Guanilo-Gómez, Elena Cachicatari-Vargas, Angel Acevedo-Duque, and Katiusca Cruz-Ayala

Smart Agricultural System Using IoT 73
Md. Faridul Islam Suny, Tania Khatun, Zahura Zaman, Md. Monjourur Roshed Fahim, Md. Ariful Islam, Rashida Jesmin, and Tajim Md. Niamat Ullah Akhund

IoT-Based Low-Cost Automated Irrigation System for Smart Farming 83
Tajim Md. Niamat Ullah Akhund, Nishat Tasnim Newaz, Zahura Zaman, Atia Sultana, Alistair Barros, and Md Whaiduzzaman

Simultaneous Measurements Reading from More Than One MiBand 3 Wristbands 93
Zuzanna Rodak, Krzysztof Tokarz, Pawel Mielnik, and Marcin Fojcik

When One Wireless Technology is Not Enough: A Network Architecture for Precision Agriculture Using LoRa, Wi-Fi, and LTE ... 103
Jose A. Brenes and Gabriela Marín-Raventós

Application of Generative Adversarial Network for Designer Assistance System 113
Bo Liu, You-Hsun Wu, and Yang-Cheng Lin

The Research of Macro-Control Diagram of ESP Well Based on Machine Learning 123
Zou Honglan, Yang Junzheng, Feng Gang, Wang Qianghua, and Ma Dan

Detection System Potholes on Roads based on Recurrent Neuronal Network 131
Younes Ed-Doughmi and Rachid El Ayachi

Eduinformatics: A New Academic Field Needed in the Age of Information and Communication Technology 139
Kunihiko Takamatsu, Katsuhiko Murakami, Yasuhiro Kozaki, Kenya Bannaka, Ikuhiro Noda, Kenichiro Mitsunari, Masato Omori, and Yasuo Nakata

Classification of Sagittal Lumbar Spine MRI for Lumbar Spinal Stenosis Detection Using Transfer Learning of a Deep Convolutional Neural Network 149
Friska Natalia and Sud Sudirman

Robust Facial-Based Inter-Beat Interval Estimation Through Spectral Signature Tracking and Periodic Filtering 161
Mostafa Salah, Loay Hassan, Somaya Abdel-khier, Ammar M. Hassan, and Osama A. Omer

Research and Application on Optimization of Drainage Gas Recovery Methods in Linxing Block 173
Shijie Liu, Zhenhua Cai, Wenrong Song, Tao Xuan, Chengzhen Liu, Song Wang, and Gang Feng

Can the Measurements of Albumin and Glucose Be Correlated Each Other? A Physics-Based Theory 183
Huber Nieto-Chaupis

Hybrid 2D Theory of Outdoor Infections in Times of Global Pandemic 197
Huber Nieto-Chaupis

Impacts of Blockchain on Digital Marketing 209
Sachin Garg, Sachin Gupta, and Bhoomi Gupta

A Review of Various Techniques Used in the Implementation of Lane Departure and Road Detection 219
Sagrika Chandra, Manav Gupta, Bhoomi Gupta, and Vandana Choudhary

Virtual Police Station System with Chat Bot Using FSM 229
Bhoomi Gupta, Parth Rustagi, Priyanshu Sinha, and Rohit Sroa

Khon Gesture Detection with Deep Learning Algorithms 237
Sathit Prasomphan, Pannawat Niyomrat, Burapa Tainsakukdej, Titiwat Kuarkamphun, and Noppawee Deejungwiphat

Contribution of Robotics in Medical Applications A Literary Survey 247
Abdul Hadi Himel, Farjana Abedin Boby, Sabrina Saba, Tajim Md. Niamat Ullah Akhund, and K. M. Akkas Ali

Development of Virtual Reality Training System Based on EEG Biofeedback 257
Pei-Jung Lin and Man-Lai Lam

Adapting Automatic Speech Recognition to the Radiology Domain for a Less-Resourced Language: The Case of Latvian 267
Normunds Gruzitis, Roberts Dargis, Viesturs Julijs Lasmanis, Ginta Garkaje, and Didzis Gosko

Contextualization of Personal Data Discovery and Anonymization Tools 277
Hassane Tahir and Patrick Brézillon

Early Lungs Tuberculosis Detection Using Deep Learning 287
Shiv Pranathi Kale, Jigisha Patil, Akshada Kshirsagar, and Varsha Bendre

Localising the Privacy Requirement to Improve the Scalability Issue of Blockchain 295
Ajay Singh and Preetida Vinayakray Jani

Performance Analysis of Wi-Fi and LTE-LAA/LTE-U 313
Prajwal S. Telkar, Deekshith Nayak, D. N. Rahul Raj, and V. Kiran

Blockchain-Based Decentralized Electronic Voting System 325
Mahima Pawar, Mansi Dabriwal, Raj Rathi, Shrey Kumar Sinha, and Sachin Suresh Pande

Human–Robot Intermittency and Invariants 335
Bernhard Heiden, Bianca Tonino-Heiden, and Volodymyr Alieksieiev

Breast Cancer Detection Using Machine Learning Algorithms 347
Kapil Tajane, Siddhesh Sheth, Rushikesh Satale, Tanmay Tumbare, and Omkar Panchal

A Machine Learning-Based Congenital Categorization Approach in Early Recognition of Breast Carcinoma 357
Sandeep Singh, Zunaid Aalam, Amritpal Kaur, and Prabhjot Singh Manocha

Review of Powering and Actuation in Bio-robotic Walking Orthoses ... 367
M. Shankar Lingam and A. Arun Kumar

Study of Device-to-Device Communication in the Next Generation of Wireless Communication 381
Sapana Dhanvijay and Rajneesh Karn

Atherosclerosis Disease Prediction Based on Feature Optimization and Ensemble Classifier 399
Brajesh Kumar and Harsh Mathur

Damage Estimation of Reed Fibers During Recycling 413
Ming Li, Kaitang Hu, and Suying Shao

Sentiment Analysis in Teachers Performance Rating Using Naïve Bayes Algorithm 421
Irish C. Juanatas, Ma. Corazon G. Fernando, Ace C. Lagman, and John Benedict C. Legaspi

The Lack of Focus Given to Ethical Training in the IT Industry 429
Pradnya Chitrao, Pravin Kumar Bhoyar, Rajiv Divekar, and Janhvi Pant

Heuristic Approach to Sustainable Agile Development with Secure Indicators 443
G. Priyalakshmi, R. Nithin, and N. M. Tharanitaran

Using Machine Learning Based CNN Architectural Models for Breast Ductal Carcinoma Recognition 453
Prashant Vats, Reenu Batra, Faraz Doja, Manu Phogat, Piyush Kumar Gupta, and Siddhartha Sankar Biswas

Improving Collaborative Filtering Based Recommender System with Season and Style Features 465
Raghav Modi and Ronakkumar Patel

Predicting Sentiment and Emotion on ASD Using Machine Learning and Deep Neural Networks Based on Twitter Data 473
T. Lakshmi Praveena and N. V. Muthu Lakshmi

Incidence of Dialectal Accents in the Students' Listening Understanding 489
Wilmer Patricio Collaguazo Vega, Gina Alexandra Albán Rocha, Paulina Maricela Jácome Santana, and Luis Gonzalo Maiza Vayas

Classification of Handwritten Indic Scripts Using Deep Learning 499
Aaditree Jaisswal, Prachi Ghalsasi, Aabha Jain, Misha Batra, and Sandhya Arora

Classification of Recyclable Waste Generated in Indian Households 507
Harshith MohanKumar, Mohammed Zayd Jamadar, S. Natarajan, and H. Nithin

A Hybrid Model on Deep Learning for the Diagnosis of Diabetic Retinopathy Using Image Cropping 515
A. Aruna Kumari and Santosh Kumar Henge

Automation of Translating Unit-Level Verification Scenarios for Test Vector Generation of SoC 527
Rahul Anilkumar, B. K. S. V. L. Varaprasad, and K. Padmapriya

Analysis of Firmware Security Mechanisms 537
Sandeep Gogineni Ravindrababu, Varsha Venugopal, and Jim Alves-Foss

Mutual Funds Categorization Using Ensemble Clustering 547
Anureet Bansal, Tushar Sharma, and Gaganmeet Kaur Awal

Load Balancing Synchronized Brownout Oriented Approach for Safeguarding Cloud Services Sustainability 557
Saumitra Vatsal and Shalini Agarwal

Blockchain-Based Location Sharing in 5G Open RAN Infrastructure for Sustainable Communities 571
Mojtaba Enayati, S. Sree Lekshmi, Tom Toby, Makarand Prabhu, K. P. Rahul, S. Parvathy, and Seshaiah Ponnekanti

Carbon Footprint Analysis Using Knowledge Graph 587
Sonam Sharma, Meghna Roy Chowdhury, and Sumukh Sirmokadam

Automatic Traffic Light Detection for Self-Driving Cars Using Transfer Learning 597
S. Gautam and A. Kumar

An Exploratory Study to Evaluate the Quality of Research Publication by the Indian Researchers 607
Saptasindhu Panda, Kajal Panda, Sharmistha Pradhan, Sanjib Kumar Mahanta, and Sunil Dhal

Data Center Simulation for Resource Utilization Analysis at Extended Loads 617
Mathew Koshy Karunattu, S. Sayanth, Akash Vijayan, Ansamma John, and Manu J Pillai

Detection of Alzheimer's Disease Through Speech Features and Machine Learning Classifiers 627
Ajay Sankar Gulapalli and Vinay Kumar Mittal

An Approach to Paperless Administration of Offices in India Using Blockchain 641
Atharva Paliwal, Chaitanya Kapre, Yash Roy, and Dilipkumar A. Borikar

Key Technology Adoption Indicators for Smart Universities: A Preliminary Proposal 651
Dewar Rico-Bautista, César D. Guerrero, César A. Collazos, Gina Maestre-Gongora, María Camila Sánchez-Velásquez, Yurley Medina-Cárdenas, Diana Teresa Parra-Sánchez, Antón Garcia Barreto, and Jose Swaminathan

Smart Farm Water Requirement Analysis Using Machine Learning 665
R. Akshay Sunny, R. Jyosthna, R. Melvin Raj, and M. Manoj

Abnormal Activity Detection Using Deep Learning 675
A. Dhanush Kumar, P. Shushruth Reddy, Kriti C. Parikh, C. Meghana Sarvani, and P. Loel Maansi

Noisy Dense Label Prediction with Noise-Robust Loss Functions 683
Redwan Hasif Alvi and Rashedur M. Rahman

A Systematic Review on Convolutional Neural Networks for Sentinel-2 Image Classification 695
Anjali R. Sharma and Suraj T. Sawant

Categorical Analysis of Code Smell Detection Using Machine Learning Algorithms 703
Ankita Bansal, Utkarsh Jayant, and Abha Jain

Foreign Accent Recognition Using a Combination of Native and Non-native Speech 713
Radha Krishna Guntur, Krishnan Ramakrishnan, and Vinay Kumar Mittal

A Novel Upgraded Uniform Embedding Technique for JPEG Steganography 723
Savina Bansal, R. K. Bansal, and Rakesh Kumar

Insights into Developers' Abandonment in FLOSS Projects 731
Rajdeep Kaur and Kuljit Kaur

A Machine Learning Approach for Heart Attack Prediction 741
Sanjay Gour, Punita Panwar, Divya Dwivedi, and Chetan Mali

Bio-Inspired Framework for Security in IoT Devices 749
Dion Mariyanayagam, Pancham Shukla, and Bal S. Virdee

Reaction Capacity Identification Problem: Is There Any Way to Formalize Scientific Knowledge and Automate Reasoning so that Intelligent Systems Can Solve It? 759
Karina A. Gulyaeva and Irina L. Artemieva

Author Index 769

Editors and Contributors

About the Editors

Atulya K. Nagar holds Foundation Chair as a professor of Mathematical Sciences and is pro-vice-chancellor for Research and Dean of the Faculty of Science at Liverpool Hope University, UK. He is also the head of the School of Mathematics, Computer Science, and Engineering, which he established at the University. He is an internationally respected scholar working at the cutting edge of theoretical computer science, applied mathematical analysis, operations research, and systems engineering. He received a prestigious Commonwealth Fellowship to pursue his doctorate (D.Phil.) in Applied Nonlinear Mathematics, which he earned from the University of York (UK) in 1996. He holds a B.Sc. (Hons.), a M.Sc., and M.Phil. (with distinction) in Mathematical Physics from the MDS University of Ajmer, India. His research expertise spans both applied mathematics and computational methods for nonlinear, complex, and intractable problems arising in science, engineering, and industry.

Dharm Singh Jat is a professor of Computer Science at Namibia University of Science and Technology (NUST). He has guided about eight Ph.D. and 24 master research scholars. He is the author of more than 150 peer-reviewed articles and the author or editor of more than 20 books. His interests span the areas of multimedia communications, wireless technologies, mobile communication systems, edge roof computing, software-defined networks, network security and Internet of things. He has given several guest lecturer/invited talks at various prestigious conferences. He is a fellow of the Institution of Engineers (I), Fellow of Computer Society of India, Chartered Engineer (I), Senior Member IEEE and Distinguished ACM Speaker. He also developed experiments based on simulation software for network experiments, research and project work for undergraduate and postgraduate students. His research in developing video communication platforms for solving QoS issues in video communications and also developed a framework for video transmission over wireless networks for undergraduate and postgraduate students.

Gabriela Marín-Raventós received a M.Sc. in Computer Science from Case Western Reserve University in 1985 and a Ph.D. in Business Analysis and Research from Texas A&M University, USA, in 1993. She has been a Computer Science faculty member at Universidad de Costa Rica (UCR) since 1980. She was the dean of Graduate Studies and the director of the Research Center for Communication and Information Technologies (CITIC), both at UCR. Currently, she is a director of the Graduate Program in Computer Science and Informatics. She has organized several international and national conferences, and has been, and actually is, a chair of several program and editorial committees. From 2012 to 2016, she was the president of the Latin American Center for Computer Studies (CLEI), becoming the first woman to occupy such a distinguished position. Since September 2016, she has been vice-president of the International Federation for Information Processing (IFIP), in charge of the Digital Equity Committee. Her research interests include smart cities, human–computer interaction, decision support systems, gender in IT and digital equity.

Durgesh Kumar Mishra has received M.Tech. degree in Computer Science from DAVV, Indore, in 1994 and Ph.D. degree in Computer Engineering in 2008. Presently, he has been working as a director and professor (CSE) at Sri Aurobindo Institute of Technology, Indore, MP, India. He is also a visiting faculty at IIT-Indore, MP, India. He has 24 years of teaching and 10 years of research experience. He has completed his Ph.D. on the area - Secure Multi-Party Computation for Preserving Privacy. He has published more than 90 papers in refereed international/national journals and conferences including IEEE, ACM conferences. His publications are listed in DBLP, Citeseer-x, Elsevier, and Scopus. He is a senior member of IEEE. He visited MIT Boston and presented his presentation on security and privacy. He also chaired a panel on "Digital Monozukuri" at "Norbert Winner in 21st century" at BOSTON. He became the member of Bureau of Indian standards (BIS), Government of India, for Information Security domain.

Contributors

Zunaid Aalam SGT University, Gurugram, Haryana, India

Somaya Abdel-khier Aswan University, Aswan, Egypt

Angel Acevedo-Duque Faculty of Business and Administration, Autonomous University of Chile, Providencia, Chile

Shalini Agarwal Shri Ramswaroop Memorial University, Barabanki, Uttar Pradesh, India

Tajim Md. Niamat Ullah Akhund Department of Computer Science and Engineering, Daffodil International University, Dhaka, Bangladesh;
Institute of Information Technology, Jahangirnagar University, Dhaka, Bangladesh

R. Akshay Sunny School of Electronics and Communication Engineering, REVA University, Bangalore, India

K. M. Akkas Ali Institute of Information Technology, Jahangirnagar University, Dhaka, Bangladesh

Volodymyr Alieksieiev Leibniz University Hannover, Garbsen, Germany

Jim Alves-Foss Center for Secure and Dependable Systems, University of Idaho, Moscow, Idaho, USA

Redwan Hasif Alvi Department of Electrical and Computer Engineering, North South University, Dhaka, Bangladesh

Roberto O. Andrade Faculty of Informatics, Escuela Politécnica Nacional, Quito, Ecuador

D. A. Andrikov Peoples' Friendship University of Russia, Moscow, Russian Federation

Rahul Anilkumar U R Rao Satellite Centre, Indian Space Research Organization, Bangalore, India

Rolando Armas International Associated Laboratory LIA-MODO, Nagano, Japan

Sandhya Arora M.K.S.S.S's Cummins College of Engineering for Women, Pune, India

Irina L. Artemieva Far Eastern Federal University, Vladivostok, Russia

A. Arun Kumar BITS, Warangal, India

A. Aruna Kumari Department of Computer Applications and Computer Science and engineering, Lovely Professional University, Jalandhar, Punjab, India

Diana Arévalo Computer Science Department, Universidad de Las Fuerzas Armadas ESPE, Sangolquí, Ecuador

Gaganmeet Kaur Awal USME, Delhi Technological University, Delhi, India

Kenya Bannaka Kobe Tokiwa University, Kobe, Japan;
Osaka Kyoiku University, Osaka, Japan

Ankita Bansal Netaji Subhas University of Technology, New Delhi, India

Anureet Bansal USME, Delhi Technological University, Delhi, India

R. K. Bansal Giani Zail Singh Campus College of Engineering and Technology, Maharaja Ranjit Singh Punjab Technical University Bathinda, Punjab, India

Savina Bansal Giani Zail Singh Campus College of Engineering and Technology, Maharaja Ranjit Singh Punjab Technical University Bathinda, Punjab, India

Antón Garcia Barreto Universidad Francisco de Paula Santander, Ocaña, Colombia

Alistair Barros School of Information Systems, Queensland University of Technology, Brisbane, Australia

Misha Batra M.K.S.S.S's Cummins College of Engineering for Women, Pune, India

Reenu Batra Department of Computer Science and Engineering, SGT University, Gurugram, India

Varsha Bendre Pimpri Chinchwad College of Engineering, Pune, India

Pravin Kumar Bhoyar Symbiosis Institute of Management Studies (SIMS), A Constituent of Symbiosis International Deemed University, Pune, India

Siddhartha Sankar Biswas Department of Computer Science and Engineering, SEST, New Delhi, India

Farjana Abedin Boby Daffodil International University, Dhaka, Bangladesh

Dilipkumar A. Borikar Shri Ramdeobaba College of Engineering and Management, Nagpur, Maharashtra, India

Jose A. Brenes Research Center for Communication and Information Technologies (CITIC), University of Costa Rica, San José, Costa Rica;
Computer Science Graduate Program (PPCI), University of Costa Rica, San José, Costa Rica

Patrick Brézillon LIP6, Sorbonne University, Paris, France

Elena Cachicatari-Vargas Universidad Nacional Jorge Basadre Grhomann, Tacna, Perú

Zhenhua Cai CNOOC Ener Tech-Drilling and Production Co., Tianjin, China

María Fernanda Cazares IDEIAGEOCA Research Group, Universidad Politécnica Salesiana, Quito, Ecuador

Graciela Tristancho Cediel Universidad Industrial de Santander, Bucaramanga, Colombia

Sagrika Chandra Maharaja Agrasen Institute of Technology, New Delhi, India

Pradnya Chitrao Symbiosis Institute of Management Studies (SIMS), A Constituent of Symbiosis International Deemed University, Pune, India;
Symbiosis International Deemed University, Pune, India

Vandana Choudhary Maharaja Agrasen Institute of Technology, New Delhi, India

César A. Collazos Universidad del Cauca, Popayán, Colombia

Katiusca Cruz-Ayala Centro Internacional de Investigación Y Desarrollo (CIID), Montería, Colombia

Mansi Dabriwal Department of Information Technology, Pune Institute of Computer Technology, Pune, India

Ma Dan Xi'an Zhongkong Tiandi Technology Development Co, Ltd, Xi'an Shaanxi, China

Roberts Dargis Institute of Mathematics and Computer Science, University of Latvia, Riga, Latvia

Noppawee Deejungwiphat Department of Computer and Information Science, Faculty of Applied Science, King Mongkut's University of Technology North Bangkok, Bangkok, Thailand

Sunil Dhal Sri Sri University, Cuttack, India

A. Dhanush Kumar Mahatma Gandhi Institute of Technology, Hyderabad, India

Sapana Dhanvijay Madhyanchal Professional University, Bhopal, Madhya Pradesh, India

Rajiv Divekar Symbiosis Institute of Management Studies (SIMS), A Constituent of Symbiosis International Deemed University, Pune, India

Faraz Doja Department of Computer Science and Engineering, SEST, New Delhi, India

Divya Dwivedi Department of Computer Science and Engineering, Bhupal Nobels University, Udaipur, India

Laura Sofía Bazán Díaz Faculty of Engineering, Universidad Privada del Norte, Cajamarca, Perú

Younes Ed-Doughmi Department of Computer Science, FST, University of Sultan Moulay Sliman, Beni Mellal, Morocco

Rachid El Ayachi Department of Computer Science, FST, University of Sultan Moulay Sliman, Beni Mellal, Morocco

Mojtaba Enayati School for Sustainable Development, Amrita Vishwa Vidyapeetham, Amritapuri, India

Md. Monjourur Roshed Fahim Department of Computer Science and Engineering, Daffodil International University, Dhaka, Bangladesh

Gang Feng Xi'an Zhongkong Tiandi Technology Development Co, Ltd, Xi'an, China

Ma. Corazon G. Fernando FEU Institute of Technology, Manila, Philippines

Marcin Fojcik Western Norway University of Applied Sciences, Bergen, Norway

Walter Fuertes Computer Science Department, Universidad de Las Fuerzas Armadas ESPE, Sangolquí, Ecuador

Feng Gang Xi'an Zhongkong Tiandi Technology Development Co, Ltd, Xi'an Shaanxi, China

Sachin Garg MVN University, Palwal, Haryana, India

Ginta Garkaje Institute of Mathematics and Computer Science, University of Latvia, Riga, Latvia

S. Gautam Department of Computer Science and Applications, Panjab University, Chandigarh, India

Prachi Ghalsasi M.K.S.S.S's Cummins College of Engineering for Women, Pune, India

Romel Ramón González-Díaz Centro Internacional de Investigación Y Desarrollo (CIID), Montería, Colombia

Didzis Gosko Riga East University Hospital, Riga, Latvia

Sanjay Gour Department of Computer Science and Engineering, Jaipur Engineering College and Research Centre, Jaipur, India

Normunds Gruzitis Institute of Mathematics and Computer Science, University of Latvia, Riga, Latvia

Santos Lucio Guanilo-Gómez Universidad Nacional Jorge Basadre Grhomann, Tacna, Perú

César D. Guerrero Universidad Autónoma de Bucaramanga, Bucaramanga, Colombia

Ajay Sankar Gulapalli Aditya Engineering College, Surampalem, India

Karina A. Gulyaeva Far Eastern Federal University, Vladivostok, Russia

Radha Krishna Guntur Department of ECE, VNRVJIET, Hyderabad, India

Bhoomi Gupta Maharaja Agrasen Institute of Technology, Rohini, Delhi, India

Manav Gupta Maharaja Agrasen Institute of Technology, New Delhi, India

Piyush Kumar Gupta Department of Computer Science and Engineering, SEST, New Delhi, India

Sachin Gupta MVN University, Palwal, Haryana, India

Ammar M. Hassan Arab Academy for Science, Technology and Maritime Transport, Aswan, Egypt

Loay Hassan Aswan University, Aswan, Egypt

Bernhard Heiden University of Applied Sciences, Villach, Austria;
University of Graz, Graz, Austria

Santosh Kumar Henge Department of Computer Applications and Computer Science and engineering, Lovely Professional University, Jalandhar, Punjab, India

Abdul Hadi Himel Daffodil International University, Dhaka, Bangladesh

Zou Honglan PetroChina Research Institute of Petroleum Exploration and Development, Haidian, Beijing, China

Kaitang Hu Yuncheng Polytechnic College, Yuncheng, China

Md. Ariful Islam Department of Computer Science and Engineering, Daffodil International University, Dhaka, Bangladesh

Aabha Jain M.K.S.S.S's Cummins College of Engineering for Women, Pune, India;
Deen Dayal Upadhyaya College, University of Delhi, Delhi, India

Abha Jain Deen Dayal Upadhyaya College, University of Delhi, Delhi, India

Aaditree Jaisswal M.K.S.S.S's Cummins College of Engineering for Women, Pune, India

Mohammed Zayd Jamadar PES University, Bengaluru, Karnataka, India

Preetida Vinayakray Jani Sardar Patel Institute of Technology, Mumbai, India

Utkarsh Jayant Netaji Subhas Institute of Technology, New Delhi, India

Rashida Jesmin Department of Zoology, Government Hajee Korop Ali Memorial College, Sirajganj, Bangladesh

Ansamma John TKM College of Engineering, Kollam Kerala, India

Irish C. Juanatas FEU Diliman, Quezon City, Philippines

Yang Junzheng PetroChina Research Institute of Petroleum Exploration and Development, Haidian, Beijing, China

R. Jyosthna School of Electronics and Communication Engineering, REVA University, Bangalore, India

Shiv Pranathi Kale Pimpri Chinchwad College of Engineering, Pune, India

Chaitanya Kapre Shri Ramdeobaba College of Engineering and Management, Nagpur, Maharashtra, India

Rajneesh Karn Madhyanchal Professional University, Bhopal, Madhya Pradesh, India

Mathew Koshy Karunattu TKM College of Engineering, Kollam Kerala, India

Amritpal Kaur SGT University, Gurugram, Haryana, India

Kuljit Kaur Guru Nanak Dev University, Amritsar, India

Rajdeep Kaur Guru Nanak Dev University, Amritsar, India

Tania Khatun Department of Computer Science and Engineering, Daffodil International University, Dhaka, Bangladesh

V. Kiran R V College of Engineering, Bengaluru, India

Lydie Simone Kone/Tapsoba Algebra, Discrete Mathematics and Informatics Laboratory (LAMDI), Bobo-Dioulasso, Burkina Faso

Yasuhiro Kozaki The University of Tokyo, Tokyo, Japan;
Osaka Kyoiku University, Osaka, Japan

Akshada Kshirsagar Pimpri Chinchwad College of Engineering, Pune, India

Titiwat Kuarkamphun Department of Computer and Information Science, Faculty of Applied Science, King Mongkut's University of Technology North Bangkok, Bangkok, Thailand

A. Kumar Department of Computer Science and Applications, Panjab University, Chandigarh, India

Brajesh Kumar Madhyanchal Professional University, Bhopal, Madhya Pradesh, India

Rakesh Kumar Giani Zail Singh Campus College of Engineering and Technology, Maharaja Ranjit Singh Punjab Technical University Bathinda, Punjab, India

S. V. Kurbanov Peoples' Friendship University of Russia, Moscow, Russian Federation

Ace C. Lagman FEU Institute of Technology, Manila, Philippines

T. Lakshmi Praveena Sri Padmavati Mahila Visvavidyalayam, Tirupati, India

Man-Lai Lam Department of Computer Science and Information Engineering, Hungkuang University, Taichung, Taiwan

Andrés Larco Escuela Politécnica Nacional, Quito, Ecuador

Viesturs Julijs Lasmanis Institute of Mathematics and Computer Science, University of Latvia, Riga, Latvia

John Benedict C. Legaspi FEU Institute of Technology, Manila, Philippines

S. Sree Lekshmi Center for Wireless Networks and Applications (Amrita WNA), Amritapuri, India

Ming Li University of Michigan-Flint, Flint, MI, USA

Pei-Jung Lin Department of Computer Science and Information Engineering, Hungkuang University, Taichung, Taiwan

Yang-Cheng Lin Department of Industrial Design, National Cheng Kung University, Tainan, Taiwan

Bo Liu Department of Industrial Design, National Cheng Kung University, Tainan, Taiwan

Chengzhen Liu CNOOC Ener Tech-Drilling and Production Co., Tianjin, China

Shijie Liu CNOOC Ener Tech-Drilling and Production Co., Tianjin, China

P. Loel Maansi Mahatma Gandhi Institute of Technology, Hyderabad, India

Gina Maestre-Gongora Universidad Cooperativa de Colombia, Medellín, Colombia

Sanjib Kumar Mahanta Sri Sri University, Cuttack, India

Chetan Mali Guest Faculty, Department of Computer Science& IT Mohanlal Sukhadia University, Udaipur, India

Sadouanouan Malo Algebra, Discrete Mathematics and Informatics Laboratory (LAMDI), Bobo-Dioulasso, Burkina Faso

Prabhjot Singh Manocha Chandigarh University, Chandigarh, India

M. Manoj School of Electronics and Communication Engineering, REVA University, Bangalore, India

Dion Mariyanayagam Centre for Communications Technology, London Metropolitan University London, London, UK

Patricia Janet Uceda Martos Faculty of Engineering, Universidad Privada del Norte, Cajamarca, Perú

Gabriela Marín-Raventós Computer Science Graduate Program (PPCI), University of Costa Rica, San José, Costa Rica

Harsh Mathur Madhyanchal Professional University, Bhopal, Madhya Pradesh, India

Yurley Medina-Cárdenas Universidad Francisco de Paula Santander, Ocaña, Colombia

C. Meghana Sarvani Mahatma Gandhi Institute of Technology, Hyderabad, India

R. Melvin Raj School of Electronics and Communication Engineering, REVA University, Bangalore, India

Pawel Mielnik Section for Rheumatology, Department for Neurology, Rheumatology and Physical Medicine, Helse Førde, Førde, Norway

Kenichiro Mitsunari Kobe Tokiwa University, Kobe, Japan;
Osaka Kyoiku University, Osaka, Japan

Vinay Kumar Mittal KL Deemed To Be University, Vijayawada, India; Department of ECE, KLEF, Vijayawada, India

Raghav Modi U & P U. Patel Department of Computer Engineering, Chandubhai S. Patel Institute of Technology (CSPIT), Charotar University of Science and Technology (CHARUSAT), Anand, India

Harshith MohanKumar PES University, Bengaluru, Karnataka, India

Carlos Montenegro Escuela Politécnica Nacional, Quito, Ecuador

Katsuhiko Murakami The University of Tokyo, Tokyo, Japan; Osaka Kyoiku University, Osaka, Japan

N. V. Muthu Lakshmi Sri Padmavati Mahila Visvavidyalayam, Tirupati, India

Yasuo Nakata Kobe Tokiwa University, Kobe, Japan; Osaka Kyoiku University, Osaka, Japan

Friska Natalia Department of Information System, Universitas Multimedia Nusantara, Tangerang, Indonesia

S. Natarajan PES University, Bengaluru, Karnataka, India

Deekshith Nayak R V College of Engineering, Bengaluru, India

Nishat Tasnim Newaz Institute of Information Technology, Jahangirnagar University, Dhaka, Bangladesh

Huber Nieto-Chaupis Universidad Autonoma del Peru, Lima, Peru

H. Nithin PES University, Bengaluru, Karnataka, India

R. Nithin PSG College of Technology, Coimbatore, Tamil Nadu, India

Pannawat Niyomrat Department of Computer and Information Science, Faculty of Applied Science, King Mongkut's University of Technology North Bangkok, Bangkok, Thailand

Ikuhiro Noda Kobe Tokiwa University, Kobe, Japan; Osaka Kyoiku University, Osaka, Japan

Osama A. Omer Aswan University, Aswan, Egypt

Masato Omori Kobe Tokiwa University, Kobe, Japan; Osaka Kyoiku University, Osaka, Japan

K. Padmapriya U R Rao Satellite Centre, Indian Space Research Organization, Bangalore, India

Atharva Paliwal Shri Ramdeobaba College of Engineering and Management, Nagpur, Maharashtra, India

Omkar Panchal Department of Computer Engineering, Pimpri Chinchwad College of Engineering, Nigdi, Maharashtra, India

Kajal Panda Sri Sri University, Cuttack, India

Saptasindhu Panda Sri Sri University, Cuttack, India

Sachin Suresh Pande Department of Information Technology, Pune Institute of Computer Technology, Pune, India

Janhvi Pant Symbiosis Institute of Management Studies (SIMS), A Constituent of Symbiosis International Deemed University, Pune, India

Punita Panwar Department of Computer Science and Engineering, Jaipur Engineering College and Research Centre, Jaipur, India

Kriti C. Parikh Mahatma Gandhi Institute of Technology, Hyderabad, India

Diana Teresa Parra-Sánchez Universidad Autónoma de Bucaramanga, Bucaramanga, Colombia

S. Parvathy Center for Wireless Networks and Applications (Amrita WNA), Amritapuri, India

Ronakkumar Patel U & P U. Patel Department of Computer Engineering, Chandubhai S. Patel Institute of Technology (CSPIT), Charotar University of Science and Technology (CHARUSAT), Anand, India

Jigisha Patil Pimpri Chinchwad College of Engineering, Pune, India

Mahima Pawar Department of Information Technology, Pune Institute of Computer Technology, Pune, India

Manu Phogat Department of Computer Science and Engineering, SGT University, Gurugram, India

Manu J Pillai TKM College of Engineering, Kollam Kerala, India

Seshaiah Ponnekanti Center for Wireless Networks and Applications (Amrita WNA), Amritapuri, India

Makarand Prabhu Center for Wireless Networks and Applications (Amrita WNA), Amritapuri, India

Sharmistha Pradhan Sri Sri University, Cuttack, India

Sathit Prasomphan Department of Computer and Information Science, Faculty of Applied Science, King Mongkut's University of Technology North Bangkok, Bangkok, Thailand

G. Priyalakshmi PSG College of Technology, Coimbatore, Tamil Nadu, India

Wang Qianghua PetroChina Research Institute of Petroleum Exploration and Development, Haidian, Beijing, China

Rashedur M. Rahman Department of Electrical and Computer Engineering, North South University, Dhaka, Bangladesh

K. P. Rahul Center for Wireless Networks and Applications (Amrita WNA), Amritapuri, India

D. N. Rahul Raj R V College of Engineering, Bengaluru, India

Krishnan Ramakrishnan Centre for CYS, Amritha University, Coimbatore, India

Raj Rathi Department of Information Technology, Pune Institute of Computer Technology, Pune, India

Sandeep Gogineni Ravindrababu Center for Secure and Dependable Systems, University of Idaho, Moscow, Idaho, USA

Dewar Rico-Bautista Universidad Francisco de Paula Santander, Ocaña, Colombia

Gina Alexandra Albán Rocha Departamento de Seguridad Y Defensa, Universidad de Las Fuerzas Armadas ESPE, Av. General Rumiñahui Y Ambato, Sangolquí, Ecuador

Zuzanna Rodak Department of Graphics, Computer Vision and Digital Systems, Faculty of Automatic Control, Electronics and Computer Science, Silesian University of Technology, Gliwice, Poland

Yash Roy Shri Ramdeobaba College of Engineering and Management, Nagpur, Maharashtra, India

Meghna Roy Chowdhury Tata Consultancy Services, Mumbai, India

Parth Rustagi Maharaja Agrasen Institute of Technology, Delhi, India

Sabrina Saba Daffodil International University, Dhaka, Bangladesh

Mostafa Salah Sohag University, Sohag, Egypt

Manuel Sánchez-Rubio Faculty of Engineering, Universidad Internacional de La Rioja, Logroño, Spain

María Camila Sánchez-Velásquez Universidad Francisco de Paula Santander, Ocaña, Colombia

Paulina Maricela Jácome Santana Escuela de Educación Básica General Córdova, Ambato, Ecuador

Rushikesh Satale Department of Computer Engineering, Pimpri Chinchwad College of Engineering, Nigdi, Maharashtra, India

Suraj T. Sawant Department of Computer Science and IT, College of Engineering Pune, Pune, India

S. Sayanth TKM College of Engineering, Kollam Kerala, India

M. Shankar Lingam NIRDPR MoRD, Government of India, Hyderabad, India

Suying Shao Tianjin University of Science & Technology, Tianjin, China

Anjali R. Sharma Department of Computer Science and IT, College of Engineering Pune, Pune, India

Sonam Sharma Tata Consultancy Services, Mumbai, India

Tushar Sharma USME, Delhi Technological University, Delhi, India

Siddhesh Sheth Department of Computer Engineering, Pimpri Chinchwad College of Engineering, Nigdi, Maharashtra, India

Pancham Shukla Centre for Communications Technology, London Metropolitan University London, London, UK

P. Shushruth Reddy Mahatma Gandhi Institute of Technology, Hyderabad, India

Ajay Singh Sardar Patel Institute of Technology, Mumbai, India

Sandeep Singh SGT University, Gurugram, Haryana, India

Priyanshu Sinha Maharaja Agrasen Institute of Technology, Delhi, India

Shrey Kumar Sinha Department of Information Technology, Pune Institute of Computer Technology, Pune, India

Sumukh Sirmokadam Tata Consultancy Services, Mumbai, India

Elki Mimar Solano Cueva Faculty of Engineering, Universidad Privada del Norte, Cajamarca, Perú

Wenrong Song Beijing Yadan Petroleum Technology Development Co. Ltd, Beijing, China

Rohit Sroa Maharaja Agrasen Institute of Technology, Delhi, India

Sud Sudirman Department of Computer Science, Liverpool John Moores University, Liverpool, UK

Atia Sultana Department of CSE, Daffodil International University, Dhaka, Bangladesh

Md. Faridul Islam Suny Department of Computer Science and Engineering, Daffodil International University, Dhaka, Bangladesh

Jose Swaminathan Vellore Institute of Technology, Vellore, India

Hassane Tahir Links Consultants, Paris, France

Burapa Tainsakukdej Department of Computer and Information Science, Faculty of Applied Science, King Mongkut's University of Technology North Bangkok, Bangkok, Thailand

Kapil Tajane Department of Computer Engineering, Pimpri Chinchwad College of Engineering, Nigdi, Maharashtra, India

Kunihiko Takamatsu Kobe Tokiwa University, Kobe, Japan;
Osaka Kyoiku University, Osaka, Japan

Prajwal S. Telkar R V College of Engineering, Bengaluru, India

N. M. Tharanitaran PSG College of Technology, Coimbatore, Tamil Nadu, India

Tom Toby Center for Wireless Networks and Applications (Amrita WNA), Amritapuri, India

Krzysztof Tokarz Department of Graphics, Computer Vision and Digital Systems, Faculty of Automatic Control, Electronics and Computer Science, Silesian University of Technology, Gliwice, Poland

Bianca Tonino-Heiden University of Graz, Graz, Austria

Yaya Traore Algebra, Discrete Mathematics and Informatics Laboratory (LAMDI), Bobo-Dioulasso, Burkina Faso

Tanmay Tumbare Department of Computer Engineering, Pimpri Chinchwad College of Engineering, Nigdi, Maharashtra, India

B. K. S. V. L. Varaprasad U R Rao Satellite Centre, Indian Space Research Organization, Bangalore, India

Prashant Vats Department of Information Technology, Fairfield Institute of Management and Technology, GGSIP University, New Delhi, India

Saumitra Vatsal Shri Ramswaroop Memorial University, Barabanki, Uttar Pradesh, India

Luis Gonzalo Maiza Vayas Universidad Técnica de Ambato, Ambato, Ecuador

Wilmer Patricio Collaguazo Vega Universidad Técnica de Cotopaxi, Latacunga, Ecuador

Varsha Venugopal Center for Secure and Dependable Systems, University of Idaho, Moscow, Idaho, USA

Akash Vijayan TKM College of Engineering, Kollam Kerala, India

Bal S. Virdee Centre for Communications Technology, London Metropolitan University London, London, UK

Song Wang College of Petroleum Engineering, China University of Petroleum, Beijing, China

Md Whaiduzzaman Institute of Information Technology, Jahangirnagar University, Dhaka, Bangladesh;
School of Information Systems, Queensland University of Technology, Brisbane, Australia

You-Hsun Wu Department of Industrial Design, National Cheng Kung University, Tainan, Taiwan

Tao Xuan CNOOC Ener Tech-Drilling and Production Co., Tianjin, China

Cesar Yanez Escuela Politécnica Nacional, Quito, Ecuador

Zahura Zaman Department of Computer Science and Engineering, Daffodil International University, Dhaka, Bangladesh

App OnViolet on UV Radiation Skin Care Habits

Patricia Janet Uceda Martos, Laura Sofía Bazán Díaz, and Elki Mimar Solano Cueva

Abstract With the development of the OnViolet application, it was sought to determine its impact on people's habits when faced with exposure to the sun's rays. For the development of the system, OpenUp was used as a framework. ISO 9126 was used in the technical guide for evaluating the software, obtaining a rating from good to excellent. The research was of an applied type and of a quasi-experimental design. We worked with a sample of 370 young university volunteers, divided into an experimentation group and a control group. The survey was used as a technique and as an instrument a 20 items questionnaire, built from the theoretical framework, validated by experts and with an acceptable reliability by Cronbach's alpha statistic equal to 0.720, in a pilot test, to be applied through of a pretest and a posttest. It was concluded in the identification of a positive influence of the use of OnViolet for the improvement of the habits of skin care of the people, obtaining in the Chi-square test of Pearson, the items with 0.0, less than 0.05 of significance: use protection against UV radiation; knowledge of the maximum time of exposure to the sun without protection; knowledge of the accepted UV radiation index for unprotected exposure; use of means for the care and protection of the skin against UV radiation; knowledge of applications that report on the levels of UV radiation and recommendations on skin care; and visit the dermatologist to receive recommendations on skin care against solar radiation.

Keywords Software · UV radiation · Sun protection

1 Introduction

In recent years, environmental pollution has increased dramatically, generating strong damage to the ozone layer, which caused changes in people's lifestyle due to increased

P. J. U. Martos · L. S. B. Díaz · E. M. Solano Cueva (✉)
Faculty of Engineering, Universidad Privada del Norte, Cajamarca, Perú
e-mail: N00034677@upn.pe

P. J. U. Martos
e-mail: patricia.uceda@upn.edu.pe

© The Author(s), under exclusive license to Springer Nature Singapore Pte Ltd. 2022
A. K. Nagar et al. (eds.), *Intelligent Sustainable Systems*, Lecture Notes in Networks and Systems 333, https://doi.org/10.1007/978-981-16-6309-3_1

exposure to the sun during the day. These common changes have led to an increase in dermatological and skin diseases, even reaching skin cancer, which in many cases leads to the death of people.

For the Spanish Association Against Cancer [1], skin cancer has increased by 5% in recent years. In 2019, American Cancer Society estimates estimated that there were 96,480 new cases of melanoma in the United States (about 57,220 in men and 39,260 in women) and about 7230 people (4740 men and 2490 women) who a cause of melanoma died; these cases increased massively during the last 30 years, generating concern and alert in the public [2].

In Peru during 2013, the Government enacted Law N° 30102 that provides preventive measures for health due to prolonged exposure to solar radiation, which requires public and private institutions to take measures [3]. Even though the mortality rate due to skin cancer is 20% of the cases detected and that Peru ranks first in the world for high radiation (solar index of 14 points), they have not yet been developed mobile applications, Web platforms or technology that allow the prevention and change of habits during exposure to sunlight [4].

Likewise, the General Directorate of Epidemiology (DGE) in Peru carried out an analysis of the cancer situation at the national level based on epidemiological care, finding during the period 2006 and 2010 a total registry of 5975 cases of skin cancer (2744 in men and 3231 in women) that represent 6.6% of all cancers [5]. On the other hand, in the Regional Hospital of Cajamarca during the years 2012 to 2016, data was collected on the care of 89 cancer patients, with 42.70% skin cancer, and 75.29% of 60-year-old patients with greater frequency, turning the most affected decade of life between 60 and 69 years of age [6].

Blumthaler investigated UV monitoring for public health that was provided through a Web site, where he considered it important to emphasize that the index varies greatly in time (of exposure and duration) and in space and is important at the time of issue recommendations to users [7].

Llerena-Izquierdo, Viera-Sanchez and Rodríguez-Moreira developed and implemented a kit with a portable device and a mobile application that allowed the detection and prevention of natural or artificial UV radiation. The prototype consisted of an Arduino with low cost and portability components. The Android mobile application transmitted data from the portable device during the wireless connection with a friendly interface [8].

Nicholson, Murphy, Walker, Tinker and Dobbinson investigated UV routines and forecasting through the SunSmart app. The qualitative research aimed to assess the understanding of the UV index among Australian users who regularly used the app as a UV forecasting tool in daily life, rather than just displaying the UV index. Misperceptions about ultraviolet radiation were found to be common; adults who used the application for the first time rarely transformed the information into new behaviors, but old users recognized the excellent ability of the application to provide UV forecast results and therefore incorporated this information into their daily routine that allowed improvement in its sun protection behavior [9].

Toro-Huamanchumo, Burgos-Muñoz, Vargas-Tineo and Burgos-Muñoz y Bada described the awareness, behavior and attitudes toward sun exposure among bathers on the north coast of Peru; for this, in the cross-sectional study, they used a questionnaire that was applied to 410 bathers, where only 13.66% of the respondents correctly answered the seven questions related to sun exposure and skin cancer, regarding sun protection practices; women used an umbrella and sunscreen more frequently than the group of men. There was a greater presence of negative attitude toward skin care and UV rays in men, while women presented better practices [10].

Tellez, Nieto-Gutiérrez and Taype-Rondán investigated about the availability of mobile sun protection applications through a descriptive observational study, searching the AppStore and Google Play Store; mobile applications that aimed to improve the use of sunscreens were registered, finding 134 of which 88 were available only in English and 92 were free. Regarding free applications: 5 included videos or animations, 2 included games, 28 had institutional certification, 64 indicated the UV index according to the current location and 16 provided information on the proper use of sunscreen. Most of the applications were downloaded more than 1000 times [11].

Julian, Thorburn and Geldhof investigated the beliefs of 334 college students regarding UV exposure, sunscreen use and its associations with skin cancer risk and protective behaviors in cloudy weather. Of the use of sunscreen, 54% almost never use it and often 34%. The use of a wide-brimmed hat: 74% almost never use it and 8% often. The use of sunglasses: 18% almost never use them while 60% use them often. In the use of a long-sleeved shirt, 20% almost never do it and 50% often. The users who visit the specialist: 31% disagree, and 38% responded to agree [12].

With the development of the OnViolet application, it was sought to determine the impact on people's habits in the face of exposure to the sun's rays, as a prevention and awareness measure of the population.

2 Method

This research was of a quantitative approach, of an applied type and with a quasi-experimental design, emphasizing the solution of problems for the improvement of skin care habits in people. The population was made up of 10,000 young university students, taking random accessibility to both genders as a criterion. The students selected for the research were obtained through a probabilistic sample, using the statistical formula for known populations, using a margin of error of 5%, for which 370 people were obtained that make up the sample divided into two groups: control and experimental.

The data collection technique was the survey; the instrument used to measure skin care habits was a 20 items questionnaire, with closed and multiple-choice questions, built from the theoretical framework. The instrument was subjected to an evaluation by experts for its validation, obtaining a 90% validity on average, and through a pilot test, an acceptable reliability was obtained by the Cronbach's Alpha statistic equal to 0.720. To measure the functionality of the system, the ISO 9126 standard was

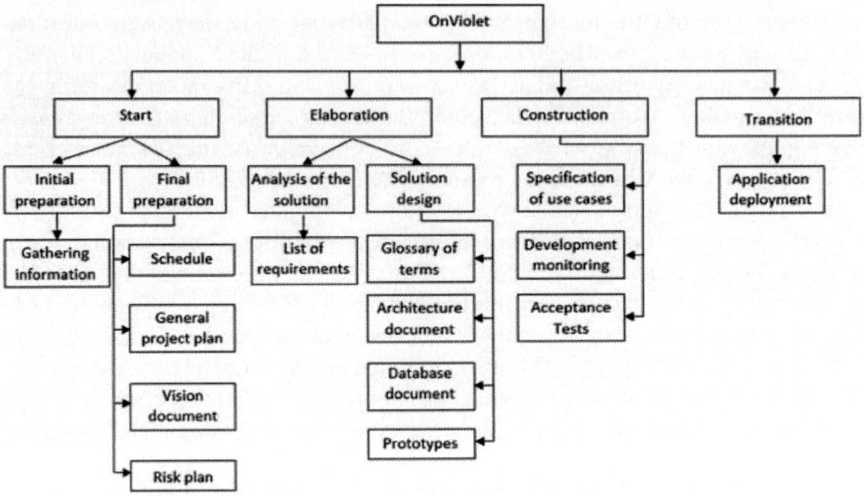

Fig. 1 Phases of implementation of the APP OnViolet [13]

used, based on the technical guide for software evaluation, evaluating criteria such as suitability, accuracy, interoperability, security and functionality compliance.

Likewise, for the development and adequate implementation of the OnViolet system, the four phases of the life cycle provided by Open Up were used, among them phases: (a) Initiation phase: Scope, initial evaluation of the cost and planning scheme; (b) Preparation phase: Analysis and definition of the system architecture; (c) Construction phase: Implementation of various components of the system and functionality and (d) Transition phase: Functional verification of the software product to be distributed to users.

For the development of the OnViolet system, the Open Up framework was used and included both the mobile application and the Web application. For the development of the mobile application, Java, the Android Studio IDE, XML with layouts, Material Design and Retrofit were used. In the development of the Web application, JavaScript, Node.Js environment, HTML, JQuery, the CSS BoosTrap Framework, SMTP.Js and Firebase were used (Figs. 1, 2 and 3).

3 Results

The OnViolet system was developed, at the mobile application and Web application level (Figs. 4 and 5). ISO 9126 was used in the technical guide for evaluating the software, obtaining a rating from good to excellent. In the construction of the prototype hardware, an Arduino uno, an ML8511 UV radiation sensor, an 8266 Wi-Fi module

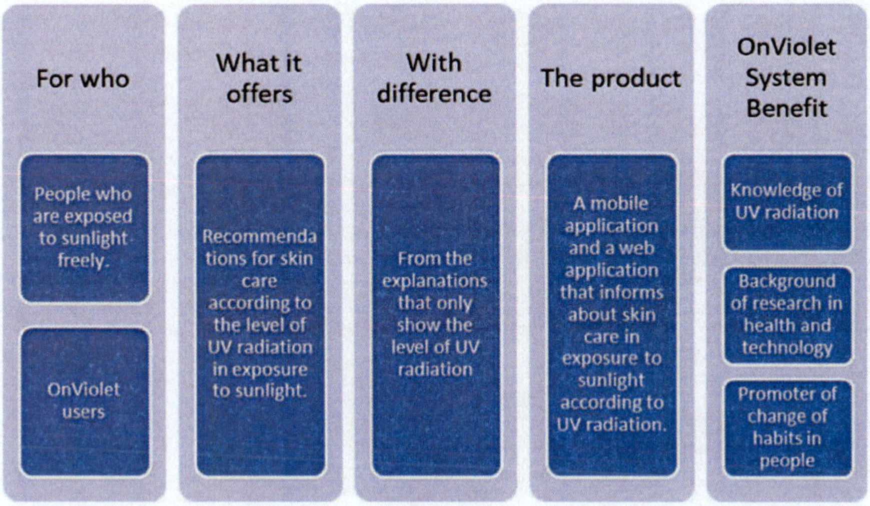

Fig. 2 Technological alternative of OnViolet solution [13]

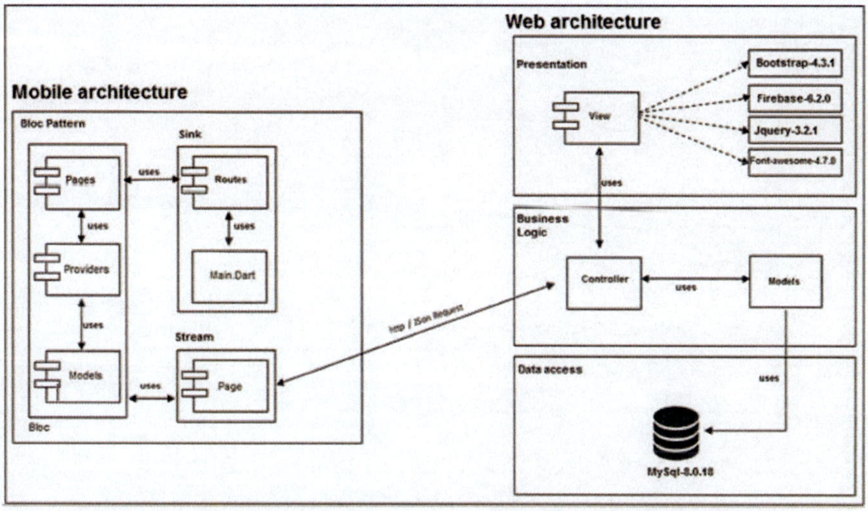

Fig. 3 OnViolet system architecture [13]

and a breadboard were used; these elements were interconnected for the reception and subsequent output of UV radiation data in real time (Fig. 6).

The skin care habits that were observed taking into account the following aspects: use of protection against UV radiation; knowledge of the maximum time of exposure to the sun without protection; knowledge of the accepted UV radiation index for unprotected exposure; use of means for the care and protection of the skin against

Fig. 4 OnViolet interface of the mobile application [13]

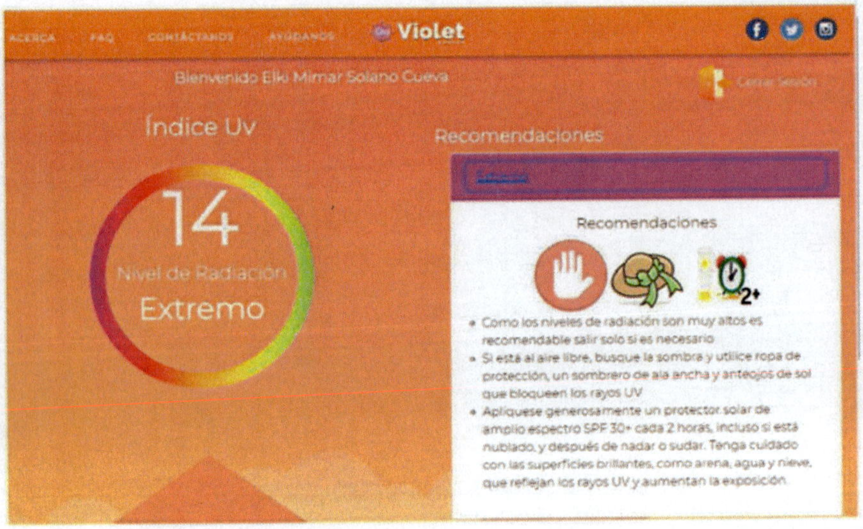

Fig. 5 OnViolet interface of the Web application [13]

UV radiation; knowledge of applications that report on UV radiation levels and skin care recommendations and visit the dermatologist to receive recommendations on skin care against solar radiation.

Data on habits were taken in four observations, in the pretest and posttest, in the control group and in the experimental group. Pearson's chi-square test was used,

Fig. 6 Prototype hardware for real-time UV radiation detection [13]

Table 1 Use of protection against UV radiation

Group indicators		Pretest	Posttest
Control	Does not use protection	73% (135)	75.1% (139)
	Protection once a day	26.5% (49)	24.3% (45)
	Protection 2 times a day	0% (0)	0% (0)
	Protection 3 times a day	0.5% (1)	0.5% (1)
Experimental	Does not use protection	63.2% (117)	1.6% (3)
	Protection once a day	33% (61)	91.4% (169)
	Protection 2 times a day	2.2% (4)	7% (13)
	Protection 3 times a day	1.6% (3)	0% (0)

and those items where 0.0 of significance or less than 0.05 were obtained are shown (Tables 1, 2, 3, 4, 5 and 6).

Table 2 Knowledge of the maximum time of exposure to the sun without protection

Group Indicators		Pretest	Posttest
Control	Does not know	47% (87)	44.3% (82)
	20 min	0.5% (1)	0.5% (1)
	30 min	27.6% (51)	25.9% (48)
	40 min	24.9% (46)	29.2% (54)
Experimental	Does not know	32.4% (60)	0% (0)
	20 min	3.2% (6)	100% (85)
	30 min	18.4% (34)	0% (0)
	40 min	45.9% (85)	0% (0)

Table 3 Knowledge of the accepted UV index for unprotected exposure

Group indicators		Pretest	Posttest
Control	Low	7% (13)	5.9% (11)
	Moderate	60% (111)	62.7% (116)
	High or tall	31.9% (59)	30.3% (56)
	Very high	0.5% (1)	0.5% (1)
	Extreme	0.5% (1)	0.5% (1)
Experimental	Low	7.6% (14)	100% (185)
	Moderate	58.4% (108)	0% (0)
	High or tall	33.5% (62)	0% (0)
	Very high	0.5% (1)	0% (0)
	Extreme	0.5% (1)	0% (0)

Table 4 Use of means for the care and protection of the skin against UV radiation

Group indicators		Pretest	Posttest
Control	Does not use	55.1% (102)	55.1% (102)
	Cap	14.1% (26)	20% (37)
	Sunscreen	10.8% (20)	10.8% (20)
	Glasses	20% (37)	14.1% (26)
Experimental	Does not use	36.8% (68)	0% (0)
	Cap	19.5% (36)	10.8% (20)
	Sunscreen	23.8% (44)	89.2% (165)
	Glasses	20% (37)	0% (0)

Although there is general information on sun exposure and UV radiation in different media, the mobile application that developed on a Web site provides recommendations on skin care in sun exposure, according to the index of UV radiation that is provided by a hardware component that receives and emits in real time.

The use of the OnViolet software shows that people are more conscientious in the use of sun protection, they improved their knowledge about the areas of the skin that

Table 5 Knowledge of applications that report UV radiation levels and skin care recommendations

Group indicators		Pretest	Posttest
Control	Does not know	99.5% (184)	99.5% (184)
	App en Android	0% (0)	0% (0)
	App en IOS	0% (0)	0% (0)
	Página web	0.5% (1)	0.5% (1)
Experimental	Does not know	97.3% (180)	0% (0)
	App en Android	0% (0)	94.6% (175)
	App en IOS	0% (0)	0% (0)
	Página web	2.7% (5)	5.4% (10)

Table 6 Visit the dermatologist to receive recommendations on skin care against solar radiation

Group indicators		Pretest	Posttest
Control	Once a month	1.1% (2)	1.1% (2)
	Once a year	5.9% (11)	5.9% (11)
	Once at two years	12.4% (23)	12.4% (23)
	Did not come	80.5% (149)	80.5% (149)
Experimental	Once a month	1.1% (2)	2.2% (4)
	Once a year	12.4% (23)	93% (172)
	Once at two years	14.1% (26)	4.9% (9)
	Did not come	72.4% (134)	0% (0)

are more prone to be affected by UV radiation, they better identify the maximum time that can be exposed to the sun without protection, they learned about the UV radiation index accepted for exposure without protection, as well as the means for the care and protection of the skin, using various types of applications or Web sites that report on the levels of UV radiation and on skin care, and finally, they became aware of the visit to the dermatologist to receive recommendations on skin care.

In Pearson's chi-square test, the items were obtained with 0.0, less than 0.05 of significance (Table 7).

4 Discussion

It is necessary to consider prevention measures [3] using mobile applications and Web platforms such as OnViolet as support, which promote the change of habits of people during their exposure to sunlight [4]. It is important to consider prevention from an early age since people between 60 and 19 years of age are the cancer patients with the highest frequency [6].

Like Blumthaler, it was possible to investigate UV monitoring through a Web site, emphasizing the exposure time and the duration time when issuing recommendations

Table 7 OnViolet influence results obtained from chi-square

Aspect of the evaluated item	Level of significance
1. Use of protection against UV radiation	0.000000
2. Knowledge of the maximum time of exposure to the sun without protection	0.000000
3. Knowledge of the accepted UV index for unprotected exposure	0.000000
4. Use of means for the care and protection of the skin against UV radiation	0.000000
5. Knowledge of applications that report UV radiation levels and skin care recommendations	0.000000
6. Visit the dermatologist to receive recommendations on skin care against solar radiation	0.000000

Source [13]

to users [7]. The application developed was implemented using a hardware device that detects UV radiation, like Llerena-Izquierdo, Viera-Sánchez and Rodríguez-Moreira, also using an Arduino [8].

With the OnViolet app, skin care habits were investigated as well as Nicholson, Murphy, Walker, Tinker and Dobbinson, and Toro-Huamanchumo, Burgos-Muñoz, Vargas-Tineo and Burgos-Muñoz y Bada with their SunSmart application, having similar results to the knowledge about UV radiation [9, 10]. The usefulness of the OnViolet application has common aspects with Tellez et al. [11] in their research on the availability of mobile applications on sun protection, to improve the use of sunscreens including videos and games to inform and show the UV index according to the location of the user, finding a great acceptance due to the concern about the consequences of UV radiation [11].

Julian et al.'s [12] research also worked with university students and UV exposure, use of sunscreen and risk of skin cancer. Similar results were obtained regarding the use of sunscreen, but they also detailed the type of protection such as the use of a hat, the use of sunglasses and the use of a long-sleeved shirt.

The present research demonstrates the impact of the influence of the OnViolet application on skin care habits, being a global concern that international associations address to minimize the risk of skin cancer that has increased in recent years and has generated human losses [1, 2, 5].

Research from national and international repositories had to be investigated and analyzed, resulting in few studies that provide recommendations for skin care using technology, but information was found on the consequences and effects of radiation exposure. UV, given the scarce habits and insufficient knowledge and awareness that people have [13].

The added value of the present research is that it works in conjunction with a mobile application and a Web site, providing multiple and varied forms of skin care in front of exposure to sunlight, and especially according to the index of UV radiation that presents the user's location area at a given time. This type of application, because it is hand in hand with the mobile device, becomes an easy way of monitoring and orientation, which influences and regulates the behavior and habits of people in the face of sun exposure.

Likewise, for the full functionality of the Web site and mobile, the construction of the prototype hardware fulfills the purpose of receiving and emitting the UV radiation information in real time, contributing with the correct location information and radiation indices to guide the recommendations of improving people's skin care habits, by including new care alternatives in the face of the silent enemy of skin health. The validation and standardization of the hardware for later use are pending.

5 Conclusions

It was concluded in the identification of a positive influence of the use of OnViolet for the improvement of the skin care habits of the people, obtaining in the Pearson's chi-square test, the items with 0.0, less than 0.05 of significance: use of protection against UV radiation; knowledge of the maximum time of exposure to the sun without protection; knowledge of the accepted UV radiation index for unprotected exposure; use of means for the care and protection of the skin against UV radiation; knowledge of applications that report on UV radiation levels and skin care recommendations; and visit the dermatologist to receive recommendations on skin care against solar radiation.

It was determined that the use of the developed software (mobile and Web application) "OnViolet" positively influenced people's skin care habits, improving knowledge about skin care, UV radiation index and time maximum exposure.

The quality of the "OnViolet" software was verified according to ISO 9126, in terms of functionality, with a high degree of acceptance, for the benefit of users.

It is important to continue implementing skin care alternatives, and since by changing your protection habits against UV radiation, you will help prevent skin cancer, which progresses silently.

References

1. Asociación Española contra el Cáncer, Asociación Española contra el Cáncer, [En 6línea]. Available: https://www.aecc.es/es/todo-sobre-cancer/prevencion/evita-exposicion-sol/sol-cancer-piel
2. American Cancer Society, American Cancer Society, American Cancer Society, 9 Enero 2018. [En línea]. Available: https://www.cancer.org/es/cancer/cancer-de-piel-tipo-melanoma/acerca/estadisticas-clave.html. [Último acceso: 2020 Enero 2019]
3. Diario El Correo, Diario El Correo, Grupo Epensa, 7 Diciembre 2017. [En línea]. Available: https://diariocorreo.pe/salud/estadisticas-cancer-de-piel-peru-790644/. [Último acceso: 10 Enero 2019]

4. RPP,¿Por qué el Perú tuvo la radiación solar más alta del mundo?, 2017. [En línea]. Available: https://rpp.pe/lima/actualidad/por-que-el-peru-tiene-la-radiacion-mas-alta-del-mundo-noticia-927749
5. Sordo, C., Gutiérrez, C.: Cáncer de piel y radiación solar: experiencia peruana en la prevención y detección temprana del cáncer de piel y melanoma, Revista Peruana de Medicina Experimental y Salud Pública, pp. 3–4 (2015)
6. Vásquez, H.: Frecuencia del Cáncer de Piel y sus Características Anatopatológicas, Universidad Nacional de Cajamarca (2018)
7. Blumthaler, M.: UV monitoring for public health. Int. J. Environ. Res. Public Health **15**(8) (2018)
8. Llerena-Izquierdo, J., Viera-Sanchez, N., Rodriguez-Moreira, B.: Portable device and mobile application for the detection of ultraviolet radiation in real time with a low cost sensor in arduino, ICAT 2019. Comunicaciones en informática y ciencias de la información, vol. 1193 (2020)
9. Nicholson, A., Murphy, M., Walker, M., Tinker H., Dobbinson, S.: Not part of my routine: a qualitative study of use and understanding of UV forecast information and the SunSmart app. BMC Public Health 19(1) (2019)
10. Toro-Huamanchumo, C.J., Burgos-Muñoz, S.J., Vargas-Tineo, L.M., Burgos-Muñoz, R.M., Bada, C.: Awareness, behavior and attitudes concerning sun exposure among beachgoers in the northern coast of Peru. PeerJ **2019**(1) (2019)
11. Tellez, W.A., Nieto-Gutierrez, W., Taype-Rondan, A.: Sunscreen mobile apps: a content analysis. [Applications mobiles de protection solaire: une analyse de contenu], Eur. Res. Telemed. **6**(3–4), 157–163 (2017)
12. Julian, A., Thorburn S., Geldhof, G.J.: Health beliefs about UV and skin cancer risk behaviors. Cancer Control: J. Moffitt Cancer Center **27**(4) (2020)
13. Solano Cueva, E.M.: Influencia del uso del sistema de medición de radiación UV—OnViolet—con estándares de funcionalidad, en los hábitos de cuidado de la piel de las personas. Cajamarca 2019, Universidad Privada del Norte, Cajamarca (2020)

Identifying Key Quality Features of mHealth Applications

Unsupervised Feature Selection Approach: MARS Case Study

Rolando Armas, Carlos Montenegro, Andrés Larco, and Cesar Yanez

Abstract In this work, the influential features to improve mHealth apps quality are identified, using a wrapper structure that integrates multi-objective evolutionary algorithm (MOEA) as a search method and clustering algorithm and index clustering performance as evaluation criteria. The study uses a dataset that comes from the employment of the mobile app rating scale (MARS) model. The dataset comprehends 565 Spanish language apps for people with special skills, published during the last twenty years. The results show that it is feasible to find specific characteristics that could lead to improving the apps, and the process used is a practical contribution directed to the stakeholders that promote the enhancement of the quality of life through the mHealth approach.

Keywords mHealths apps quality · Unsupervised feature selection · MARS · Multi-objective evolutionary algorithms

1 Introduction

Unsupervised feature selection (UFS) is a relatively new research field where several techniques and methods have been developed to build models for tasks such as regression, classification, or clustering. Given a case of study where some features

R. Armas (✉)
International Associated Laboratory LIA-MODO, Nagano, Japan
e-mail: rolandoarmas@gmail.com
URL: https://sites.google.com/view/lia-modo/members

C. Montenegro · A. Larco · C. Yanez
Escuela Politécnica Nacional, Quito, Ecuador
e-mail: carlos.montenegro@epn.edu.ec

A. Larco
e-mail: andres.larco@epn.edu.ec

C. Yanez
e-mail: cesaryanezv@outlook.com

© The Author(s), under exclusive license to Springer Nature Singapore Pte Ltd. 2022
A. K. Nagar et al. (eds.), *Intelligent Sustainable Systems*, Lecture Notes in Networks and Systems 333, https://doi.org/10.1007/978-981-16-6309-3_2

could be redundant or irrelevant, UFS supports the selection of the most useful features for a better understanding or a compact model development [1].

In this work, the specific case that we deal with is software development under quality standards. Quality under this context implies the needs of users combining software features. For medical software, quality assurance and measurements are indispensable elements to provide users satisfaction [2, 3]. Published by Stoyanov et al. [4, 5], mobile app rating scale (MARS) is a specific multidimensional measure for mobile health apps and provides a health domain evaluation criteria. Those studies propitiated multiple applications [6–8], and some adaptations [9–12]. For more papers related to MARS see "Read by QxMD" [13].

MARS quality rating criteria comprehend 23 quality characteristics arranged into five categories: engagement, functionality, aesthetics, information, and subjective quality. MARS quality rating determines values for the apps according to a Likert scale and a mean value representing the quality; then, MARS does not support developers in identifying the key aspects of app quality. Given a dataset with many features, a combinatorial problem arises to select the important ones, becoming a challenging task to find them to reduce the dimensionality. Proposed UFS methods such as filter and wrapper have been developed to select the most exciting subspace [1]. The wrapper approach enables us to group the data as suitably we can see in each candidate feature subspace according to "feasible" groups and select the most "feasible subspace" with the minimum number of features [14]. We search for "feasible" groups, meaning "feasible" as a subset of MARS features that lead developers to find the most relevant characteristics to build applications.

This work explores the unsupervised feature selection problem through K-modes clustering and the DBindex performance criteria for evaluating candidate feature subsets. For searching the optimal feature combination, we use a bi-objective evolutionary algorithm where DB index and the number of features are the objective functions that are minimized simultaneously.

2 Method

The wrapper method proposed in [14] defines the integration of a search method, clustering algorithm, and a feature evaluation criterion under an iteration way. In this work, we propose an evolutive iterative approach where: as a search method, we use a multi-objective evolutionary algorithm (MOEA), K-modes for categorical data clustering, and Davies–Bouldin (DB) index as feature evaluation criteria. We follow the feature selection approach illustrated in Fig. 1.

A solution is a subset or combination of selected features from MARS dataset. Each proposed solution is evaluated by K-modes, and after that, the clustering quality is evaluated by DB index. A normalized version of DB index (NDBI) and the number of features (nVar) are the objective functions. MOEA finds the best combination of

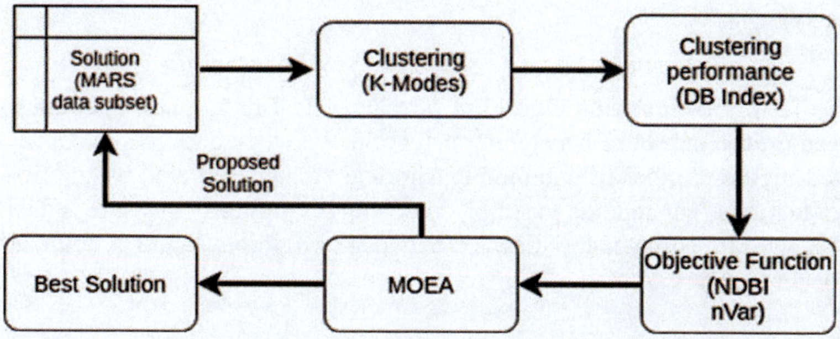

Fig. 1 Method

features that minimize simultaneously those two objectives. MOEA applies genetic operators and proposes new solutions to be evaluated, starting a new process iteration. Finally, after a specific number of generations, the optimal solutions emerge which are followed by a brief description of the components.

2.1 MARS Dataset

As said before, MARS identifies 23 quality characteristics (model variables) grouped into five categories [15]:

- Engagement: Entertainment (V1), Interest (V2), Customization (V3), Interactivity (V4), and Target group (V5).
- Functionality: Performance (V6), Easy to use (V7), Navigation (V8), and Gestural design (V9).
- Aesthetics: Layout (V10), Graphics (V11), and Visual appeal (V12).
- Information: Accuracy (V13), Goals (V14), Quality of information (V15), Quantity of information (V16), Visual information (V17), and Credibility (V18).
- Subjective points to user satisfaction.

Here, we do not consider the subjective quality category for this study due to the overall mean quality score supports the objectivity of MARS [5]. MARS items are evaluated, for each app, using a Likert scale (1—Inadequate, 2—Poor, 3—Acceptable, 4—Good, and 5—Excellent). Apps were collected for Android, iOS, Desktop, and Web platforms [16, 17], using the search chain "desktop applications for (autism OR down syndrome OR cerebral palsy)". Particularly, Web applications for people with autism were examined in developer and promoter websites [18]. The MARS dataset that we employ contains 565 evaluation records of applications for people with disabilities collected through 20 years (2000–2020).

2.2 K-Modes

K-modes [19] is a clustering algorithm for categorical data types. It defines clusters based on the number of matching categories between data points. Distance is measured by the number of common categorical attributes shared by the two data points or features. The method to simple matching dissimilarity measure is the Hamming distance. It estimates the distance between two points X and Y defined by m categorical attributes defined by Eq. 1.

$$d(X, Y) = \sum_{i=1}^{m} \delta(x_i, y_i) \tag{1}$$

where: $\delta(x_i, y_i) = 0$ if $x_i = y_i$ or $\delta(x_i, y_i) = 1$ if $x_i \neq y_i$, x_i and y_i are the categorical values of feature i in X and Y. The bigger the dissimilarity between the two objects, the more mismatches there are between X and Y and therefore the bigger the Hamming distance.

A cluster centroid clustering is created by the vector of modes of categorical attributes, which minimizes the sum of distances between each object and the centre of the corresponding cluster and therefore minimizes their dissimilarities. The k-modes algorithm randomly selects k different objects out of the dataset as the initial cluster centroids. After that, it calculates the distance between each object and centroid and assigns each object to the cluster containing its closest centroid. Finally, for each cluster, a new mode centroid is computed and compared with the previous one, and if the two modes differ, the modes and clusters are updated iteratively.

2.3 DB Index

The quality of clusters is measured by the within-cluster scatter and between-cluster separation. The Davies–Bouldin (DB) index [20] relates to a model with better separation between the clusters and the data points density in the clusters.

The index is defined as the average similarity between each cluster and determined by $R_{i,j} = \frac{s_i + s_j}{d_{i,j}}$, where: s_i is the average distance between each point of cluster i and the centroid of that cluster, and $d_{i,j}$ is the distance between cluster centroids i and j. Given a k number of clusters, the Davies–Bouldin index is defined by Eq. 2.

$$\mathrm{DB} = \frac{1}{k} \sum_{i=1}^{k} \max_{i \neq j} R_{i,j} \tag{2}$$

2.4 MOEA

Multi-objective evolutionary algorithms (MOEAs) have been used as an effective search heuristic for combinatorial problems where search space is huge and complex and is not feasible to do an exhaustive search. We propose a bi-objective EA approach for searching and select an optimal combination of MARS features. For that, we use the NSGA-II algorithm [21] and the next configuration:

Representation: We use a binary representation. V_j represents the genetic expression where each j_{th} is a model variable ($V_j = \{0, 1\}$, $j = \{1, ..., P\}$). If $V_j = 1$, that variable will be selected, and the opposite, it is not used when $V_j = 0$. P is the number of model variables.

Operators: We use bit-flip mutation where one point is randomly selected from genes, and a change is made ($1 \rightarrow 0$ or $0 \rightarrow 1$) to the selected one based on a preset probability. For crossover, we use the single-point operator as a recombination method. Randomly select crossover point in two parents, and exchange genes with crossover point as a reference.

Objective Functions: We evaluate each solution according to two objective functions which are defined following:

f1: Normalized DB Index (NDBI): To qualify each combination of features by k-modes clustering, we normalized dividing by the number of variables due to the following reasons: (1) Experimentally, we observed that MOEA converges better providing a set of high diversity solutions, (2) DB index is based on distance metrics, and it is influenced by the dimensionality of the space, which is changeable according to the number of features [14], (3) As the number of features increases, index decreases, and normalization correlates this effect [22], and (4) A normalized value allows a general metric to qualify the solution fitness. Therefore, we compute f1 by Eq. 3.

$$f1 = \text{NDBI} = \frac{\text{DB}}{\text{nVar}} \tag{3}$$

f2: Number of features (nVar): Represents the number of active variables ($V_j = 1$) in each solution and is given by Eq. 4.

$$f2 = \text{nVar} = \sum_{j=1}^{P} V_j \tag{4}$$

2.5 Experimental Set-up

A first task previous to unsupervised classification by clustering is to estimate the k number of clusters. Taking the complete MARS dataset, we run the K-modes algorithm [23] with a different number of clusters ($k \in [2, 9]$). As a metric, we use the cost function. Clustering cost is defined as the sum distance of all points to their respective cluster centroids. Complementarily, we use the silhouette score function [24]. It computes silhouette coefficients of each point that estimate how much a point is similar to its cluster than others; a higher silhouette coefficient score $[-1, 1]$ relates to a model with better-defined clusters. After analysing those parameters, we choose $k = 5$ for our experiments. After that, we configure the experiment running the genetic algorithm 15 times with a different random seed each. The number of generations is set to 25, and the population size is 25. Crossover rate is set to $P_c = 1.0$, and the mutation rate is $P_m = 1/n$, where $n = 18$ is the number of model variables. To find feasible solutions with at least one variable for each MARS category, we define a constraint to search for solutions with nVar > 4.

3 Results

We verify the trade-off between objectives through the correlation metric. We found a negative correlation of -0.88 between them which is expressed as a Pareto set in Fig. 2. The left side of Fig. 2 shows all solutions that are gathering from all runs in the last generation. Note the diversity of the solutions into a range from 5 to 18 features. Also, note that for the same number of features, different solutions arise.

Those solutions are different combinations of features with different NDBI values. That reveals the combinatorial problem that emerges for searching the best combination of features and additionally proves that the evolutionary algorithm is working properly for searching the solutions that minimize both objectives (NDBI and nVar). Note that some solutions with a specific number of features do not converge as others, for example, solutions with 11, 16, 17, and 18 features. The right side of Fig. 2 displays the Pareto optimal set (POS), the optimal best solutions after all runs. Note that the difference between the two extremes of the POS is approximately 0.1. That scale enables developers to discriminate solutions in a diverse set without to lose too much "quality" (cluster consistency).

The genetic expression of POS is shown in the left side of Fig. 3. Each column identifies a variable. A row corresponds to the settings of all variables in a specific

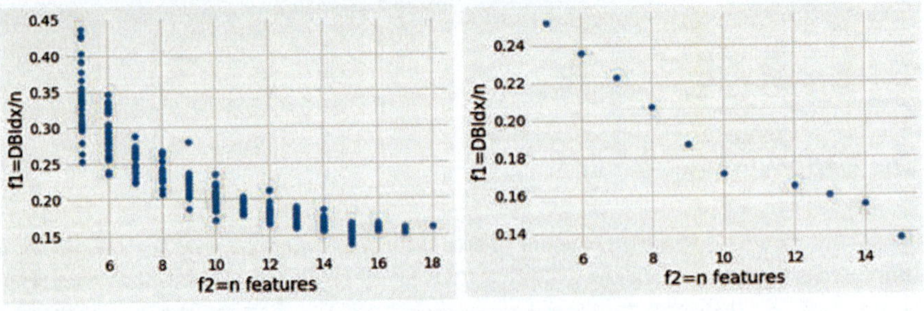

Fig. 2 Solutions at last generation (left) and Pareto optimal set (right)

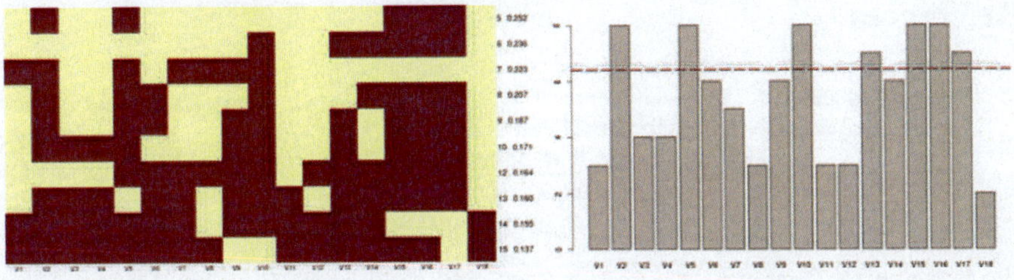

Fig. 3 POS genetic expression (left) and POS variable frequency distribution (right)

solution. The dark colour indicates that the variable is active, and pale colour indicates that the variable is not considered for that solution. The labels for each row report the name of the corresponding solution where the first number indicates the number of selected or active variables (nVar), and the second represents the NDBI value. The genetic expression of POS is displayed in ascending sort with the solution with the fewer features at the top. Observe that there is a common pattern from the first five solutions (nVar: 5, 6, 7, 8, 9), features: V1, V3, V4, V11, V12, and V18 are not selected. Variables that belong to category information appear in the majority of solutions: V15, V16, V17. Other common variables are V2, V5, V10. Also, notice that for almost all solutions, variables V1, V11, and V18 are not selected.

The right side of Fig. 3 shows the variable frequency distribution for all solutions in the POS. To select the variables whose frequency is above 80%, we define a limit represented by a dashed line. The most frequent variables are V2, V5, V10, V13, V15, V16, and V17.

4 Conclusions

In this work, we proposed an unsupervised feature selection for the MARS quality criteria dataset to identify essential features in the case study that could lead developers to create better applications for people with special skills. We implemented a wrapper method that integrates a MOEA as a search method and clustering algorithm and index clustering performance as evaluation criteria. Given a MARS feature dataset, our bi-objective approach, minimizing DB index and the number of features (nVar), revealed that the evolutionary approach converges correctly, exploring optimal solutions where the essential features arise.

The results show that seven features that belong to Engagement (V2, V5), Aesthetics (V10), and Information category (V13, V15, V16, V17) are the most frequent. Consequently, for the case study, MARS' information category variables become an essential group to consider to improve mHealth apps' quality.

Finally, this research's approach and case study describe a practical and reusable process whose application generates results that increase people's quality of life with unique skills.

References

1. Solorio-Fernández, S., Carrasco-Ochoa, J.A., Martínez-Trinidad, J.F.: A review of unsupervised feature selection methods. Artif. Intell. Rev. **53**(2), 907–948 (2020)
2. Ahamed, N.U., Sundaraj, K., Ahmad, R.B., Rahman, M., Ali, A.: A framework for the development of measurement and quality assurance in software-based medical rehabilitation systems. Procedia Engineering **41**, 53–60 (2012). International Symposium on Robotics and Intelligent Sensors 2012 (IRIS 2012)
3. Holzinger, A., Searle, G., Kleinberger, T., Seffah, A., Javahery, H.: Investigating usability metrics for the design and development of applications for the elderly. In: Miesenberger, K., Klaus, J., Zagler, W., Karshmer, A. (eds.) Computers Helping People with Special Needs, pp. 98–105. Springer, Berlin, Heidelberg (2008)
4. Stoyanov, S.R., Hides, L., Kavanagh, D.J., Wilson, H.: Development and Validation of the User Version of the Mobile Application Rating Scale (uMARS). JMIR mHealth and uHealth **6** (2016)
5. Stoyanov, S.R., Hides, L., Kavanagh, D.J., Zelenko, O., Tjondronegoro, D., Mani, M.: Mobile app rating scale: a new tool for assessing the quality of health mobile apps. JMIR mHealth and uHealth **3**(1) (2015)
6. Choi, Y., Demiris, G., Lin, S., Iribarren, S., Landis, C., Thompson, H., McCurry, S., Heitkemper, M., TM, W.: Smartphone applications to support sleep self-management: review and evaluation. J. Clin. Sleep Med. **10** (2018)
7. Davalbhakta, S., Advani, S., Kumar, S.: A systematic review of smartphone applications available for corona virus disease 2019 (COVID19) and the assessment of their quality using the mobile application rating scale (MARS). J. Med. Syst. (2020)
8. Salehinejad, S., Niakan, K.S., Hajesmaeel, G.S., Bahaadinbeigy, K., Fatehi, F.: A review and content analysis of national apps for COVID-19 management using Mobile Application Rating Scale (MARS). Inform. Health. Soc. Care **3** (2021)
9. Bardus, M., Awada, N., Ghandour, L.A., Fares, E.J., Gherbal, T., Al-Zanati, T., Stoyanov, S.R.: The Arabic version of the mobile app rating scale: Development and validation study. JMIR Mhealth Uhealth **8**(3), e16956 (2020)
10. Domnich, A., Arata, L., Amicizia, D., Signori, A., Patrick, B., Stoyanov, S., Hides, L., Gasparini, R., Panatto, D.: Development and validation of the Italian version of the Mobile Application Rating Scale and its generalisability to apps targeting primary prevention. BMC Med. Inform. Decis. Mak. **7** (2016)
11. Martin Payo, R., Fernandez Álvarez, M., Blanco Díaz, M., Cuesta Izquierdo, M., Stoyanov, S., Llaneza Suárez, E.: Spanish adaptation and validation of the mobile application rating scale questionnaire. Int. J. Med. Inform. **129**, 95–99 (2019)
12. Messner, E.M., Terhorst, Y., Barke, A., Baumeister, H., Stoyanov, S., Hides, L., Kavanagh, D., Pryss, R., Sander, L., Probst, T.: The German version of the mobile app rating scale (MARS-G): development and validation study. JMIR Mhealth Uhealth **8**(3), e14479 (2020)
13. Read by qxmd. https://read.qxmd.com/keyword/156406 (2021)
14. Dy, J.G., Brodley, C.: Feature selection for unsupervised learning. J. Mach. Learn. Res. **5**, 845–889 (2004)
15. Usabilitynet: usability resources for practitioners and managers. http://www.usabilitynet.org/home.htm (2021)
16. Fundación orange. http://www.fundacionorange.es/ (2021)
17. Portal aragonés de la comunicación aumentativa y alternativa. http://www.arasaac.org/ (2021)
18. Pictoaplicaciones. https://www.pictoaplicaciones.com/ (2021)
19. Huang, J.Z.: Clustering Categorical Data with k-Modes, 2nd edn. IGI Global
20. Davies, D.L., Bouldin, D.W.: A cluster separation measure. IEEE Trans. Pattern Anal. Mach. Intell. **PAMI-1**(2), 224–227 (1979)
21. Srinivas, N., Deb, K.: Muiltiobjective optimization using nondominated sorting in genetic algorithms. Evol. Comput. **2**(3), 221–248 (1994)

22. Morita, M., Sabourin, R., Bortolozzi, F., Suen, C.Y.: Unsupervised feature selection using multi-objective genetic algorithms for handwritten word recognition. In: Proceedings of the International Conference on Document Analysis and Recognition, ICDAR **2003-January**, 666–670 (2003)
23. de Vos, N.J.: kmodes categorical clustering library. https://github.com/nicodv/kmodes (2015–2021)
24. Rousseeuw, P.J.: Silhouettes: a graphical aid to the interpretation and validation of cluster analysis. J. Comput. Appl. Math. **20**, 53–65 (1987)

Implementation of E-health in Burkina Faso: A Status Report

Sadouanouan Malo, Lydie Simone Kone/Tapsoba, and Yaya Traore

Abstract The term e-health refers to all areas where information and communication technologies (ICT) are used for health. E-health appears more and more as a relevant solution to meet the challenges health systems are facing. However, despite its interest, e-health has long remained underexploited in Burkina Faso even if many softwares have been the subject of investments in hospitals as part of a modernization of administrative operations, which does not directly contribute to the quality of care. Hospital information systems (HIS) are the foundation on which e-health is based. They organize, at the IT level, the exchange of information between departments within the same hospital. It is on those systems that the shared medical record (SMR) is based. In this paper, we present an inventory of the implementation of e-health in Burkina Faso. We collected data on HIS deployed in major hospitals in Burkina Faso. Then we analyzed the level of interoperability of those HIS, and we finally proposed an interoperability approach adapted to Burkina Faso.

Keywords Interoperability · Hospital information system · Ontology · Web service

1 Introduction

According to the World Health Organization (WHO), e-health is defined as "digital services for the well-being of the person." E-health is therefore the application of information and communication technologies (ICT) to the field of health and well-being. An attempt to inventory the disciplines and concepts that claim to be e-health allows e-health to include:—regional and national health information networks and distributed electronic record systems, including information systems for health professionals and hospitals, online services such as electronic prescribing, databases, portals, and online health promotion systems;—telemedicine systems and associated services (teleconsultation, teleradiology, telesurveillance, etc.);—specialized tools for health professionals and researchers. Three technological bases are

S. Malo (✉) · L. S. Kone/Tapsoba · Y. Traore
Algebra, Discrete Mathematics and Informatics Laboratory (LAMDI), Bobo-Dioulasso, Burkina Faso

© The Author(s), under exclusive license to Springer Nature Singapore Pte Ltd. 2022

A. K. Nagar et al. (eds.), *Intelligent Sustainable Systems*, Lecture Notes in Networks and Systems 333, https://doi.org/10.1007/978-981-16-6309-3_3

essential to develop e-health:—information and communication technologies (information systems for hospitals and health professionals),—e-health supports (health network infrastructures) and—e-health products and services (software and equipment). The first base is a prerequisite for the establishment of the second, itself being an essential prerequisite for the establishment of the third. None of the three pillars is strong enough today in Burkina Faso. This technological obstacle gives a virtual character to e-health in Burkina Faso. Significant efforts must therefore be made to ensure that Burkina Faso's health system makes optimum use of the enormous potential offered by ICTs. To this end, on the recommendation of several international organizations (resolution 58/28 of the World Health Assembly, resolution 65 of the World Telecommunication Development Conference),[1] regional (Recommendation RC60/5 taken by the African ministers of health in 2010 on e-health)[2] and sub-regional (AU, ECOWAS, UEMOA),[3] Burkina Faso implemented an e-health strategy for the improvement of the health system. The national e-health strategy is a strategy that supports the implementation of the orientations of the national health development plan 2011–2020. It is the operational translation of the national health policy, and the vision of which is to achieve "the best state of health for the entire population through an efficient national health system." To date, there is no mapping of health facilities with a hospital information systems (HIS), network infrastructure and a permanent, and stable internet connection.

Our study presents the situation of the deployment of Hospital Information Systems (HIS) and interconnection networks in the main hospitals and the most important health structures in Burkina Faso. We then present a deployment approach of these infrastructure for better implementation of e-health in Burkina.

The rest of the paper is organized as follows: in the second section, we will present e-health practices and challenges. In the third section, we present the situation of E-health in Burkina Faso, and we will end with a conclusion and perspectives in the fifth section.

2 E-health Practices and Challenges

Based on a review of the literature, this synthesis brings together the knowledge elements of the field of e-health in its practices and uses.

The concept of e-health dates from the early 2000s and represents an emerging field at the crossroads of medical informatics, public health and the corporate world. It refers to health services provided through the internet or digital technologies [1].

---

[1] Resolution 58/28 of the World Health Assembly, resolution 65 of the World Telecommunication Development Conference.

[2] Recommendation RC60 / 5 taken by the African ministers of health in 2010 on e-health.

[3] (AU: African Union, ECOWAS: Economic Community of West African States, UEMOA: Union Economique et Monétaire Ouest Africaine - West African Economic and Monetary Union).

The perimeters of e-health are very broad and understanding the uses requires a clarification of its components. Its definitions are multiple and not harmonized between countries [2]. The majority of definitions characterize e-health as the networked use of digital technologies dedicated to health [3]. Digital health involves a systemic approach comprising technological tools and actors able to understand the functioning of the system (uses and functions of the tools) as well as the information transmitted [4, 5].

Scope overlaps exist between e-health, tele-health, m-health, and telemedicine. The term "E-health" has a wide acceptance, covers all fields of digital health [6] and identifies two areas with permeable borders [7, 8]:—Hospital Information Systems (HIS) and—Tele-health which includes telemedicine and mobile health or m-health (this component represents all electronic devices, applications for mobiles (Apps), connected objects (OCs) related to health).

E-health devices, and in particular telemedicine, are technological platforms that combine terminals and IT solutions that strongly depend on their location context. They cannot function without integration with already existing information systems, which induces a strong "organizational embeddedness" with the structures and organizations using these technologies [9, 10].

3 The Situation of E-health in Burkina Faso

3.1 Organization of the Health System in Burkina Faso

To meet the health needs of its populations, Burkina Faso has a health system based on both the public and private sub-sectors as well as traditional medicine. The administration of the health sector is provided by the Ministry of Health through three levels:—the central level made up of the Minister's office, the general secretariat, central departments and attached structures;—the intermediate level made up of 13 regional health directorates and—the peripheral level with 70 health districts.

In terms of care, there is also a three-level pyramid:—the first level made up of health and social promotion centers (CSPS), medical centers (CM) and medical centers with a surgical branch (CMA),—the second level made up of regional hospital centers (RHC),—the third level made up of university hospital centers (UHC). In the context of this paper, the study was carried out in the UHRs and RHCs.

3.2 Status of HIS and Tele-Health in Burkina Faso

The completion of this study involved collecting data in all UHCs, RHCs, in the department in charge of ICT of the Ministry of Health of Burkina Faso and in Non-Governmental Organizations (NGOs) involved in health in Burkina Faso during the

Table 1 Description of HIS in Burkina Faso

Applications	The scope	Local/RESINA	Database	Field of application
ENDOS-BF	All UHCS, RHCS	No/Yes	PostgreSQL	Health data warehouse
ONE HEALTH	All UHCS, RHCS	No/Yes	PostgreSQL	Health data warehouse
TELIA SANTE	UHC Yalagado Ouedraogo	Yes/No	Oracle database 10G	Patient record management software
DOOR	UHC Tengandogo	Yes/No	DB2	This software has several functionalities: administrative management and patient file management
GARDMI	UHC Charles De Gaule	Yes/No	Microsoft access	Patient record management and billing
SIHDPI	UHC Bogodogo	Yes/No	PostgreSQL	This HIS is composed of several sub-modules: administrative management, fund management and clinical data management

period from October 6 to 24, 2020. Table 1 illustrates the information collected according to the scope, the type of computer network (local/RESINA[4]), the database and the functionalities covered.

We note that all UHCs and RHCs have a hospital information system that takes into account or not the management of the shared medical record (SMR). We note that only the UHCs of Bogodogo, Tengandogo, and the Charles De Gaule pediatric hospital have an HIS taking into account the management of the SMR. This support is made from the entry of the patient into the health structure upon discharge. However, in each HIS, the applications used are different from one UHC or RHC to another. In general, in all the HIS of the UHC and RHC, the applications are diverse and varied and operate autonomously. There are therefore no interactions between the applications.

Some NGOs and private health structures that support the state of Burkina Faso in the health sector also use applications dedicated to improving the health of populations. The data from these health structures and NGOs are presented in Table 2.

[4] RESINA: Réseau Informatique National de l'Administration (National Administration Network).

Table 2 Description of HIS of NGOs' health structures in Burkina Faso

Structures	Field of application	The scope	Applications
INDIAN PAN e-NETWORK Project, RAFT Project	Tele-health and continuing medical education	UHC Yalagado Ouédraogo, UHC Sanou Souro, RHC Tenkodogo, CMA Nouna, CMA Pouytenga	Ipath, Bogou, Dudal, VIPS, eCourse platform by videoconference
Nouna Health Research Centre	Epidemiological surveillance	Nouna	Configured tablets embedding forms
The NGO Terre des Hommes, Nouna Health Research Center	Mother–child health and disease monitoring	CSPS (Northern and Boucle du Mouhoun Regions)	mHealth, Mos@n, integrated patient management platform, electronic consultation registry, diagnostic assistance platform
Informatics and Tele-health Services Directorate	Computerized data management	In all health districts	Health data warehouse

The data in this table show that tele-health and m-health applications are well used for improving the health of populations.

As in the case of applications deployed in UHCs and RHCs, the applications used by NGOs, and private health structures are very heterogeneous, making their interoperability difficult.

Burkina Faso Ministry of Health has an electronic health information management system based on a data warehouse for the collection of epidemiological data called "ENDOS-BF." The "ENDOS-BF" client is installed in all the UHCs and RHCs which are all connected to the Ministry of Health by the national interconnection network called "RESINA." In addition, the Ministry of Health has an electronic platform called "One Health," integrated with the "ENDOS-BF" system, which provides real-time data on unusual health events and suspected cases related to these health events. The data are provided to the central directorates of the Ministry of Health, to the laboratories of the same ministry and to the Center for Health Emergency Response Operations called "CORUS." Since March 2020, this "One Health" platform has played a key role in the management of the COVID-19 pandemic.

3.3 *Interoperability of Information Systems in Burkina Faso*

Burkina Faso has started a vast project to modernize its administration with electronic administration projects. As part of this modernization project, several national interconnection networks were deployed and a private cloud set up as shown in Table 3.

Table 3 Electronic administration projects

Infrastructures	The scope	Remark
National Backbone	National territory	Broadband fiber optic data transport network
Government Intranet	National territory	National administration network RESINA
G-Cloud	National territory	IaaS; PaaS; AaaS
PARICOM	Project	connectivity; e-services; education and training; e-Skills; tele-health, Health System Management;

As can be seen in Table 3, Burkina Faso has a national interconnection network called "RESINA." RESINA is a national intranet set up to interconnect the country's public administration. In addition to RESINA, a national backbone based on optical fiber has been set up as part of the G-Cloud project. G-Cloud is a national cloud set up to pool IT resources at the national level.

In addition, a program to support the strengthening of communication infrastructures called "PARICOM" has been set up to improve the availability, efficiency, and accessibility of the offer of electronic communication services.

Finally, on March 9, 2021, the ministry in charge of ICTs in Burkina Faso adopted a national standard for computer cabling. Once adopted, this standard should provide an efficient, effective and a practical guide to ensure compliance with current standards in networking infrastructure construction projects.

In order to provide the confidence necessary for the development of electronic governance within the public administration and between the administration and all actors of the digital economy, Burkina Faso has adopted a General Interoperability Repository (GIR[5]) regulating electronic exchanges between different actors. This repository deals with the interoperability of information systems which will make it possible to adopt a common format (syntactic and semantic) for electronic exchanges. In this repository, an interoperability protocol in a manual, semi-automatic and automatic process scheme has been proposed. According to the GIR, the manual interoperability process scheme relates to manual exchanges during the execution of an interoperability process. The semi-automatic and automatic interoperability process scheme relates to the exchange of files during the execution of an interoperability process. In all these interoperability process schemes, the GIR proposes three levels of interoperability: semantic, syntactic, and technical interoperability. The levels of interoperability proposed in the GIR protocol are well included in those proposed in the literature. Indeed, in the literature, some authors mention the existence of two levels of interoperability: the semantic level and the technical level [11]. In [12], the authors go further by presenting five levels of interoperability that they consider important to ensure communication between distributed information systems: the technical level, the structural level, the syntactic level, the semantic level, and the organizational level. In the semantic level of the GIR, it is proposed an approach based on a glossary. The interoperability issue is more complicated when it comes

[5] https://anssi.bf/fileadmin/user_upload/Document_RGI_version_1.0_version_adoptee_Mai_2018.pdf.

Table 4 Synthesis of the state of the art on semantic interoperability in HIS

Level of interoperability	Approaches	Limits
Syntactic	Web services: SOAP, WSDL [14–17]	Web services do not take into account the semantics of information Variety of medical vocabularies is the major obstacle to the integration of independently developed applications
Semantic and Syntactic	Semantic Web: XML, WSDL and UDDI, SOA [18]; SOA: XML, WSDL, UDDI [17, 19]; Standards: HPRIM, HL7, CEN TC251, DICOM l'IHE, LOINC, SNOMED, UMLS [20–22]	Difficulties in aligning web service standards with those of medical field Diversity of norms and standards is a serious problem for data integration
Semantic	Ontology [23–28]	Diversity of ontologies is a problem

to HIS given the heterogeneity of the applications used. In this context, some applications may not be technically or semantically interoperable. It is this difficulty that the authors of [13] have highlighted in their work. Several research projects have explored the notion of HIS interoperability. These works are summarized in Table 4. The approaches proposed in the GIR are therefore not suited to our context.

3.4 Review of the Implementation of E-health in Burkina Faso

As a result, we can see that in almost all HIS in Burkina Faso, the software offer is plethora and diversified, not only by specialty, but also according to the health structure. This situation is obviously an obstacle to the use of this software and therefore implies the absence of interoperability, namely the absence of data sharing between the various existing software based on norms and standards. In addition, despite the implementation of telemedicine solutions in certain health structures, very few health professionals have experimented with teleconsultation or telesurveillance. Finally, despite the existence of national connection networks, the health structures are not interconnected and the Internet connection in these health structures is not reliable and efficient.

We can then say that despite the political will and the adoption of ICTs in health structures, hopes for an advance in e-health in Burkina Faso are still struggling to materialize.

4 HIS Implementation Scenario for Successful E-health in Burkina Faso: Our Approach

In order to successfully implement e-health in Burkina Faso, it is necessary to develop a rigorous HIS architecture, which will allow information to be shared correctly. It is also necessary to ensure the quality of the data entered or captured. In our work [29], we propose an architecture for the interoperability of Hospital Information Systems in Burkina Faso. In this paper, we tried to use the approaches proposed in the GIR but they were not appropriate in our context. In the semantic level of the GIR, we propose to use health domain ontologies repository instead of the glossary. In the technical level of the GIR, we focus on web services. Figure 1 illustrates our proposal for interoperability between two hospitals A and B.

For each application of Hospital, A for example, a local ontology is created from the database. This ontology makes it possible to reason on the data of the hospital A and to establish a correspondence with the data of hospital B. The functionalities of the applications of each hospital are implemented thanks to web services. Some functionalities will be implemented by a composition of web services. This approach therefore requires that a repository of ontologies in the health field is used at the semantic level instead of the glossary as proposed in the GIR. The web services of each hospital are used at the technical level in the GIR, and a semantic annotation of these services is necessary for the composition of services. The strategy used to create the ontologies local and the web services with our approach is the following:

1. For each hospital database used
2. Generate an local ontology
3. Add the local ontology to the referential of ontology
4. If (the local Ontology is not aligned with the referential of ontology) then
5. Align the local ontology with the referential ontology
6. The referential ontology is updated
7. Create a web service from hospital application used

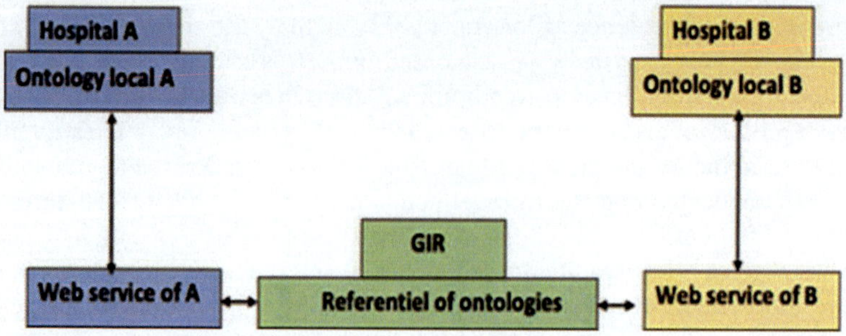

Fig. 1 Interoperability of HIS in Burkina Faso

8. Annotate the web service with local ontology
9. Connect the web service annotated with the composition of web services
10. The composition of web services is updated.

5 Conclusion

In this paper, we have presented e-health practices and challenges and the results of the study on HIS in Burkina Faso. We can conclude that despite the interest of political stakeholders and adoption of ICTs by health structures, there is not yet a real materialization of e-health in Burkina Faso. Then we made a critical analysis of HISs in this same country, and then we have proposed an approach for the implementation of interoperable HIS adapted to the case in this country. Our methodology is based on a set of domain ontologies and web services.

In the future work, first we propose an approach to generate an ontology form a database. Next, we describe an approach to create and annotate semantically web service with an ontology in order to allow the composition of services.

References

1. Eysenbach, G.: What is eHealth? J. Med. Internet Res. **3**(2): e20 (2001)
2. Oh, H., et al.: What is eHealth: a systematic review of published definitions. J. Med. Internet Res. **7**(1), e1 (2005)
3. Pagliari, C., et al.: What is eHealth: a scoping exercise to map the field. J. Med. Internet Res. **7**(1), e9 (2005)
4. Dumez, H., Marrauld, L., Minvielle, E.: États des lieux de l'innovation en santé numérique. In: Working Paper 15-CRG-01, 66 p (2015)
5. Dumez, H., Minvielle, E.: L'e-santé rend-elle la démocratie sanitaire pleinement performative? Systèmes d'Information et Management, Eska, special issue. Health IT **22**(1), 9–37 (2017)
6. Pon, D., Coury, A.: Stratégie de transformation du système de santé - Rapport final: Accélérer le virage numérique. Editeur : Ministère des Solidarités et de la Santé. Septembre, 33 p (2018)
7. Scala B.: E-santé: la médecine à l'ère du numérique. Sci. Santé **29**, 33 (2016)
8. Cnom. Santé connectée. De l'e-santé à la santé connectée. Livre Blanc du Conseil national de l'Ordre des médecins. Janvier. 36 p (2015)
9. Fernandez, V., Gille, L., Houy, T.: Les technologies numériques de santé. Examen prospectif et critique. Presse des Mines. Février. 112 p (2015)
10. Minvielle, E.: Le patient et le système. En quête d'une organisation sur mesure - Approches innovantes du parcours de santé. Editions Seli Arslan. Avril. 288 p (2018)
11. Salzano, G., et Bourret, C.: Identification and composition of metadata for cooperative information systems illustration on the health information systems (2005)
12. Lopez, D.M., Blobel, B.G.D.M.E.: A development framework for semantically interoperable health information systems. Int. J. Med. Inf. **78**(2), 83–103 (2009)
13. Della Valle, E., Cerizza, D., Celino, I., Dogac, A., Gokce, B., Kabak, Y., Okcan, A., Gulderen, O., Namli, T., Bicer. V.: An eHealth case study applying semantic web service technology in the healthcare environment (2007)
14. Elridge, D., Jerzy, R.: Design for a patient-centric medical information system using XML web services. In: International Conference on Information Technology (ITNG'07)

15. Boniface, M., et Wilken, P.: ARTEMIS: towards a secure interoperability infrastructure for healthcare information systems. Stud. Health Technol. Inf. **112**, 181–189 (2005)
16. D'Mello, E., et Rozenblit, J.: Design for a patient-centric medical information system using XML web services. In: Fourth International Conference on Information Technology (ITNG'07), pp. 562–567. IEEE (2007)
17. Da Silva, C.F., Da Cunha, P.R., Ghodous, P., et al.: The semantic side of service-oriented architectures. In: Semantic Enterprise Application Integration for Business Processes: Service-Oriented Frameworks, pp. 90–104. IGI Global (2010)
18. Goldschmidt, P.G.: HIT and MIS: implications of health information technology and medical information systems. Commun. ACM **48**(10), 68–74 (2005)
19. Kart, F., Miao, G., Moser, L.E., Melliar-Smith, P.M.: A distributed e-healthcare system based on the service oriented architecture. In: IEEE International Conference on Services Computing (SCC 2007), pp. 652–659
20. Semantic Interoperability for better health and safer healthcare. Semantic Health Report (2009)
21. Touzi, J.; Aide à la conception de Système d'Information Collaboratif support de l'interopérabilité des entreprises, thèse de doctoraz (2007)
22. Jérôme velo, Jacques la biche, Y.S.: Intégration et interopérabilité des systèmes d'information : un système de médiation de données basé sur l'architecture orientée services SOA (2010)
23. Rolf, E., Josef, I., Jörg, R.: Relevance of terminological standards and services in telemedicine. In: Information Technology Solutions for Healthcare. Springer (2006)
24. Khan, W.A., Khattak, A.M., Hussain, M., et al.: An adaptive semantic based mediation system for data interoperability among health information systems. J. Med. Syst. **38**(8), 28 (2014)
25. Salma, G., El Mehdi, S., et al.: System of mediation based on ontology to solve the semantic heterogeneity between information systems. Int. J. Comput. Appl. **69** (2013)
26. Mezghani, E., Exposito, E., Drira, K., et al.: A semantic big data platform for integrating heterogeneous wearable data in healthcare. J. Med. Syst. **39**(12), 185 (2015)
27. Hadji, A., Kholladi, M.K.: Traitement et Fusion de Données dans le Cadre de l'Interopérabilité Sémantique des Systèmes d'Information Géographiques (2012)
28. Ali, S., et Chong, I.: Semantic mediation model to promote improved data sharing using representation learning in heterogeneous healthcare service environments. Appl. Sci. **9**(19), 4175 (2019)
29. Tapsoba, L.S., Traore, Y., et Malo, S.: Towards an architecture for the interoperability of hospital information systems in Burkina Faso. Stud. Health Technol. Inf. **272**, 159–162 (2020)

A Training Web Platform to Improve Cognitive Skills for Phishing Attacks Detection

María Fernanda Cazares, Diana Arévalo, Roberto O. Andrade, Walter Fuertes, and Manuel Sánchez-Rubio

Abstract Attackers adapt their attacks to be more effective based on exploiting technical and human vulnerabilities. Social engineering attacks take advantage of human vulnerabilities considering that psychological factors could extend these vulnerabilities due to anxiety, stress, or high credibility. The development of technological solutions must be accompanied by training for people. However, training should not focus only on showing the technical details. In addition, it should consider strengthening the cognitive processes used in decision-making processes. This work aims to propose a methodology and architecture to develop a phishing training Web platform based on cognitive processes.

Keywords Cognitive security · Cyberattacks · Decision-making · Phishing

1 Introduction

Cybersecurity attacks have presented significant growth during the last years. During 2020, the technological inclusion in daily activities increased due to the context of the COVID-19 pandemic [1]. Attacks such as ransomware, cross-site scripting,

M. F. Cazares (✉)
IDEIAGEOCA Research Group, Universidad Politécnica Salesiana, Quito, Ecuador
e-mail: mcazares@ups.edu.ec

D. Arévalo · W. Fuertes
Computer Science Department, Universidad de Las Fuerzas Armadas ESPE, Sangolquí, Ecuador
e-mail: daarevalo3@espe.edu.ec

W. Fuertes
e-mail: wmfuertes@espe.edu.ec

R. O. Andrade
Faculty of Informatics, Escuela Politécnica Nacional, Quito, Ecuador
e-mail: roberto.andrade@epn.edu.ec

M. Sánchez-Rubio
Faculty of Engineering, Universidad Internacional de La Rioja, Logroño, Spain
e-mail: manuel.sanchezrubio@unir.net

© The Author(s), under exclusive license to Springer Nature Singapore Pte Ltd. 2022

A. K. Nagar et al. (eds.), *Intelligent Sustainable Systems*, Lecture Notes in Networks and Systems 333, https://doi.org/10.1007/978-981-16-6309-3_4

and social engineering (SE) were more relevant. Cybercriminals took advantage of teleworking, tele-education, and the massive use of electronic services and social networks to carry out banking fraud, information theft, online scams, and cyberstalking [2]. SE attacks have been among the most used by attackers in recent years because they allow them to take advantage of human vulnerabilities generated by psychological factors (emotional and cognitive) to carry out attacks. The pandemic context affected the psychological factors due to anxiety, stress, and people's need for information [3].

We consider that the training design should be focused not only on instructing people on the theoretical aspects of phishing but also on improving the decision-making processes that will improve phishing message detection [4]. Understanding the factors associated with the decision-making process in low real-world environments will allow for better training programs to be defined. This work aims to define a Web-training prototype that considers mental models in decision making based on analysis task cognitive for the improvement of phishing detection processes.

The remainder of this article is structured as follows. Section 2 presents background about phishing performs an analysis of the cognitive factors. Section 3 explains the methodology to develop the prototype. Section 4 shows the model of the proposed training platform. Finally, Sect. 5 presents the conclusions of this work.

2 Background

Attackers have taken advantage of social context to adapt and increase their attacks based on SE, specifically phishing and fake news. Malicious codes were introduced to users' devices through applications that indicate coronaviruses' location in real time, such as Corona Virus Map. In agree with [5], more than 600 malicious applications, including mobile banking Trojans and spyware, posing as applications offering a COVID-19 related service. Phishing attacks during the COVID-19 pandemic use the name of agencies such as the WHO, FBI, and OMS to try to persuade victims to fall for the frauds. Security portals and government sites worldwide have shown the relevance of this security issue and have put forward alternatives for this fight. Table 1 indicates several projects by European Commission to combat COVID-19 misinformation [6].

To deal with SE attacks, organizations could use specialized technological tools to detect and block the attacks. However, it also requires that people make a critical judgment before performing an action that makes them a potential victim of frauds or information theft [7]. With the more human-related aspect, the importance of training and coaching has been defined in academic and industry sectors [8].

Naturalistic Decision Making.

The naturalistic decision making (NDM) expanded beyond choosing between options and is understood as a process that includes the perception and recognition of situations that identify the essence of the human experience by describing

Table 1 Existing projects related to digital security

Project	Description
HERoS	Focus on the efficiency of decision response during public health emergencies
Co-Inform	Based on tools to foster critical thinking and digital literacy related with COVID-19 for maintaining information to society
TRESCA	Working on developing trust in science and innovation by communication practices between scientific researchers, journalists, and policymakers who have analyzed basic digital sanitary practices to stop misinformation
FANDANGO	Based on detecting fake news by analyzing and verifying different typologies of news data, media sources, and social media
COMPROP	Based on methods to detect possible solutions in social and computer science
Misinformation Classifier	Focus on identifying patterns in the language for identifying fake news. The Commission and European Parliament are currently using the tool
Social Rumors	Based on identified accounts on Twitter that are posting links to known dodgy sources. Then, monitors other links posted by these accounts to pinpoint new narratives emerging on social media

concepts of representation of scripts, schemes, and mental models. This is part of the advantage of advances in cognitive psychology [9]. Some of the main theories that explain these mental models are cognitive continuum theory, cognitive control, and the recognition primed decision model. A mental model method focuses on cognitive task analysis (CTA), allow to know the cognitive skills needed in a task. It also facilitates understanding the complex problem and captures cognitive strategies used by experienced personnel to develop formal training [10]. According to Militello Laura, "A streamlined method of CTA, stands for Applied Cognitive Task Analysis (ACTA) consists of three interview methods that help the practitioner to extract information about the cognitive demands and skills required for a task" [11].

Cognitive Model applied to the decision-making process.

The cognitive processes involved in two-choice decision making have been studied in psychology, neuroscience, and neuro-economics. The studies have been based on determining the factors that drive people to choose between one decisions or another when choosing between two specific options [12]. These concepts can be applied within this work context to understand why a person takes a specific decision, e.g., when they can choose between two possible options, such as whether or not to click on the attachment, visit the Web page or not, or forward or not forward the possible suspicious email. Following, we mention in Table 2 some diffusion models proposed a long time.

The Ratcliff diffusion model [15] is of interest in the field of psychology and neuroscience. The model is mathematically defined and addresses cognitive processes such as information accumulation speed and decision bias in binary decision tasks. As

Table 2 Some diffusion models proposed a long time

Cognitive Process Model	Cognitive process	Resource
Diffusion [12]	Response time two-choices	Speed of information accumulation decision bias
Ballistic [13]	Evidence accumulation	Drift signal decision
Random walk [14]	Response time	Amount of evidence obtained

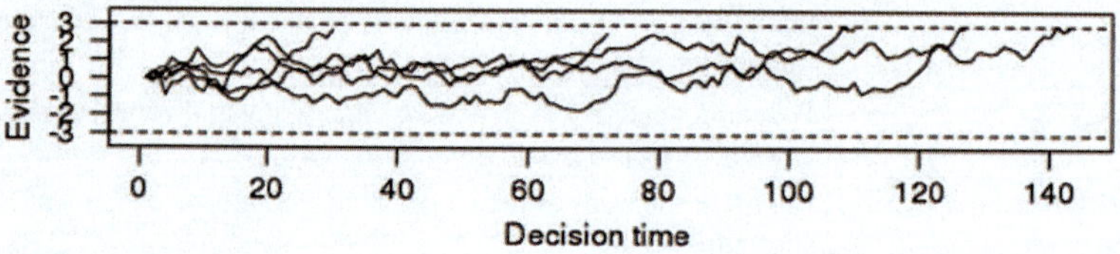

Fig. 1 Increase of evidence reduces the response time in a random walk model

stated Lerche [16], the parameters measured in the initial Ratcliff diffusion model are the following:

- Drift rate involves the speed and direction of the accumulated information. It can be considered positive if there is a rapid acquisition of information and harmful otherwise;
- Threshold separation represents the decision criterion that allows users to decide with minor errors;
- A starting point is the turning point at which one can define the amount of preliminary information necessary to make between one decision or another;
- Non-decision time specifies the time between information processing and decision making.

The ballistic diffusion models consider the amount of information accumulated. Donkin et al. [17], in a recognition memory task, one accumulator gathers evidence in favor of the "old" response, and another gathers evidence in favor of the "new" response. The random walk model establishes that the number of trials can influence the amount of information accumulated and the time taken to decide [18]. Figure 1 shows that a higher number of proofs supports the reduction in the decision time.

3 Design a Web Platform Based on the Decision-Making Model

The present work aims to develop a Web-based training platform for understanding natural decision-making processes in cybersecurity and is founded on the following research question: How to apply decision-making models in the field of SE attacks?

The methodology used for the development of the Web-based training platform is based on the following aspects: (1) the cognitive processes involved in decision-making. (2) the features used in phishing attack processes, and the TAM processes to validate technology-based prototypes. To develop these three steps, we applied the following resources:

- Applied cognitive task analysis (ACTA);
- Interviews to extract information about the cognitive demands and skills required for a task;
- Interview using simulation scenarios to extract information related to the cognitive process to face against phishing events.

From the cybersecurity perspective, specifically in cybersecurity incident management processes, the detection time of an attack MTTD (Mean Time To Detect) and the accuracy of detection (accuracy) are relevant to reduce the impact of cyberattacks. From the cognitivist approach, the response time is related to the amount of previous information (starting point) and the speed at which new information can be acquired (threshold separation), and the influence that this information has to make a decision (drift). Whether the interest of having a relatively low MTTD is associated with Response Time in the decision-making process will depend on the speed at which new information can be acquired and processed. Additionally, if more information (evidence) could be processed, the decision time will be slower. At this point, two perspectives related to response time we present: First, from the technological perspective, i.e., the development of security solutions for phishing detection requires the capacity to process a large amount of evidence in a short time, so the incorporation of big data and artificial intelligence can be an excellent support for this objective [19], [20]. On the other hand, from the human perspective, processing a large amount of evidence to improve detection's cognitive processes required many training hours; this approach could not be adequate for all organizations [21].

3.1 Cognitive Requirements for Decision-Driven Design

The following parameters have been established to establish a measurement process:

- Response time (RT). The time it takes for people to select between two options. In the case of this work: phishing or legitimate;
- Identification accuracy. Measures the person's ability to identify phishing components;
- Hit rate. Measures the person's ability to identify phishing.

For the platform's design, we identify four sub-tasks that the experts described during the interviews (see, Fig. 2). We identified the ability to solve problems, and the elaboration of judgments are the most used cognitive skills.

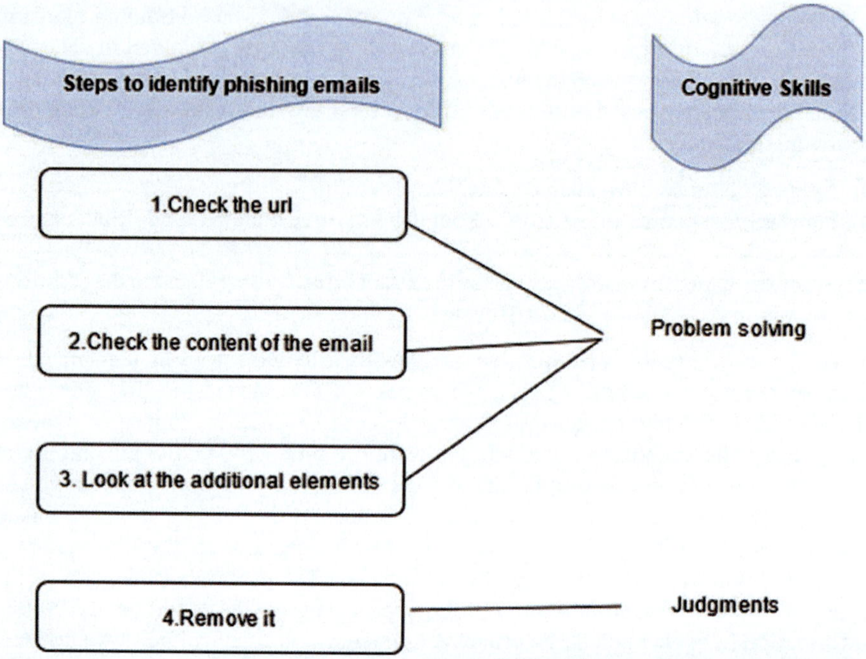

Fig. 2 Results of cognitive task analysis CTA

3.2 *TAM Process*

At this stage, we selected the tools to be used in the Web platform for training to improve cognitive skills for phishing detection. To do so, HTML and JavaScript allow the development of iterative Web applications and PHP, allowing us to connect with the MongoDB Atlas database, which is where we will host our information. The proposed platform takes as a starting point the user's predisposition to test his knowledge against a Phishing attack. The proposed architecture (see Fig. 1) is divided into three sections: (i) the scenario of user interaction with the platform and their answers to the exercises, (ii) data storage in the MongoDB Atlas database, and (iii) the application of data mining and decision trees to analyze the data and obtain the behaviors as explained in Fig. 3. For the first scenario, the Web platform was developed using JavaScript and PHP programming languages for the Back-end and Bootstrap for the Front-end. In the second phase, MongoDB Atlas was used to store the data of each user. In the last phase, it is planned that with the data collection and the application of the KDD methodology, apply data mining algorithms using decision trees to obtain users' behavior to a phishing attack.

To make the coding, we used an integrated development environment for three aspects: quick debugging, rigorous testing, and team collaboration. Through HTML code (HyperText Markup Language) and Bootstrap, we developed a platform's start

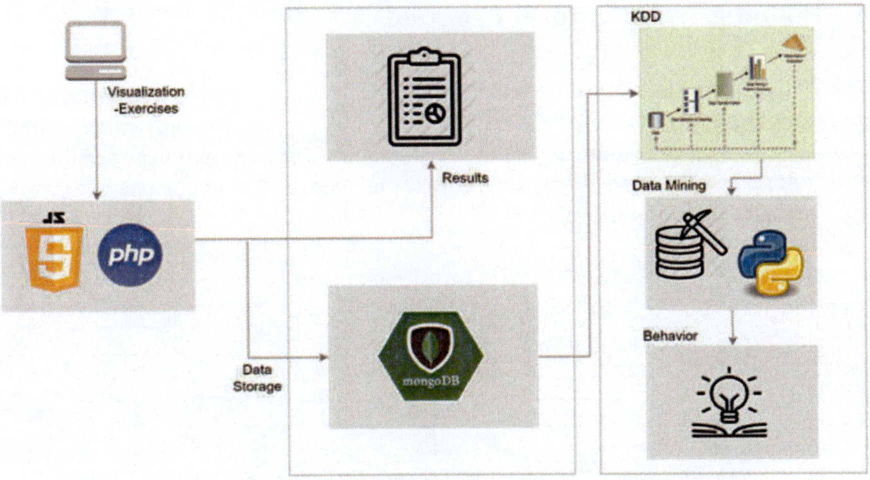

Fig. 3 Proposed of architecture for Web-training platform of phishing attacks

window to detail the objective. In the next window, a registration form was developed. The user must enter his name, surname, current activity (teacher or student), educational career, email, and years of experience, considering that these data will help us initially identify the type of user. Finally, we show the interface where it is shown one by one the simulated exercises. The development of the exercises was conducted with the characteristics found in the literature search. Each exercise shows a real-life case, and the user will be able to interact with the URLs of the links, either hyperlinks or attachments shown in the exercises. The time of each exercise is timed with JavaScript from the time the exercise is shown until the user presses the option's button. After the user chooses an option from the exercise presented, either phishing or legitimate, a pop-up window is displayed to write the reason for choosing this answer. This parameter will help in the analysis of the cognitive skills for phishing detection. When the user completes the ten exercises, the report window is displayed, including the user's name, the average time to complete the exercises, and the success rate. It also shows the error rate in a descriptive statistical graph. MongoDB Atlas database is a secure cloud database with high availability and performance, works through clusters, which are sets of databases, and gives the option of adding the computer's IP address from which are working. This manager helps us to save for each of the users a document that will contain the registration data, the response of each exercise, the comment, the time of completion of each exercise, and the coordinates by which the user navigated in the application for each exercise.

4 Training Web Platform Proposal

The Web platform is focused on improving cognitive processes in people during the process of detecting phishing attacks. Figure 4 shows the mockup proposed for Web-training based on the following variables: Response time in each exercise; hit and error rate; response latency (the time it takes to respond); exercise characteristics; user interaction (Help information).

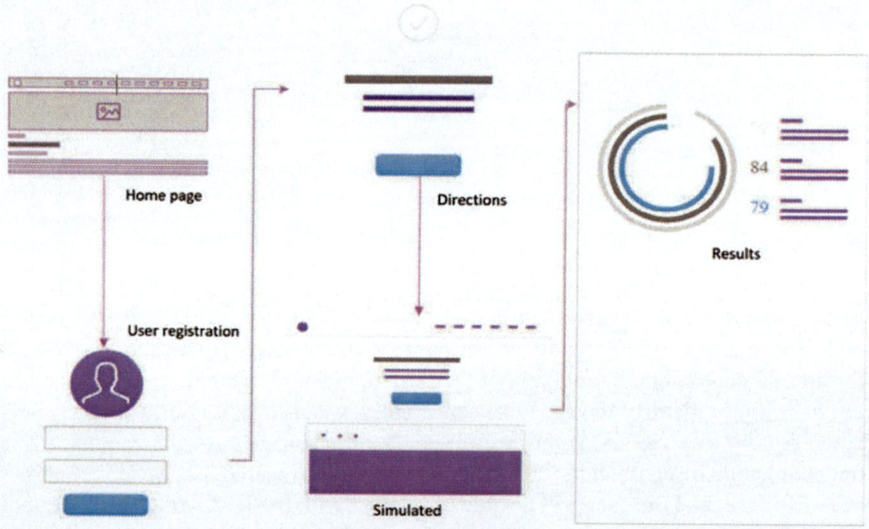

Fig. 4 Modules: user registration, simulated phishing exercises, and results

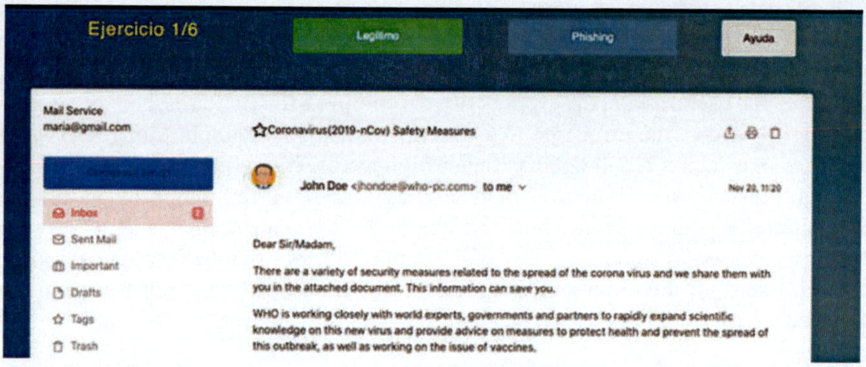

Fig. 5 Simulated phishing exercise on the web-training platform

Figure 5 shows a simulated phishing exercise that includes the most common components used by attackers in a phishing message that includes a logo that personifies a person or organization, the content of interest to the victim, and documents or links that simulated the injected malicious code.

5 Conclusions

In the design of the Web-training platform, the characteristics of phishing attacks were considered, such as persuasion through words that arouse people's interest, such as COVID-19 or vaccines, job offers, change of passwords, and photographs to achieve that each one of the exercises is the closest to reality. It is essential to consider that the more the simulation resembles reality, the greater the probability of achieving success in the training objective or becoming aware of each person's response patterns to face cyberattacks. The variables as response time, errors, and successes concerning years of experience help in comparative research on how experts or novices solve and face cyberattacks. Based on these aspects stored in the platform's database, it will be possible to perform mathematical analysis of human behavior. The analysis of cognitive skills in the different tasks to solve phishing is a guideline to generate a mental model that explains the actions and evaluations that individuals perform when facing possible cybersecurity attacks.

References

1. FBI PSA: A COVID 19-Driven Increase in Telework from Hotels Could Pose a Cyber Security Risk for Guests | AHA. (2021). Retrieved 12 March 2021, from https://www.aha.org/other-cybersecurity-reports/2020-10-06-fbi-psa-covid-19-driven-increase-telework-hotels-could-pose
2. COVID-19 cyberthreats. (2021). Retrieved 12 March 2021, from https://www.interpol.int/en/Crimes/Cybercrime/COVID-19-cyberthreats
3. Rodríguez-Hidalgo, A.J., Pantaleón, Y., Dios, I., Falla, D.: Fear of COVID-19, stress, and anxiety in university undergraduate students: a predictive model for depression. Front. Psychol. **11**, 591797 (2020). https://doi.org/10.3389/fpsyg.2020.591797
4. SANS (2021). Robust Phishing Awareness Simulation Training that Changes Behavior Retrieved 12 March 2021, from https://www.sans.org/security-awareness-training/products/phishing
5. Avast. (2021). 2020 Year In Review. Retrieved 12 March 2021, from https://blog.avast.com/2020-year-in-review-avast
6. EU-funded projects on Digital security—Shaping Europe's digital future—European Commission. (2021). Retrieved 12 March 2021, from https://ec.europa.eu/digital-single-market/en/programme-and-projects/eu-funded-projects-digital-security
7. Cyber Security Hub.: Phishing Attacks Work Because... Humans (2021). Retrieved from https://www.cshub.com/attacks/articles/phishing-attacks-work-because-humans
8. Basit, A., Zafar, M., Liu, X., et al.: A comprehensive survey of AI-enabled phishing attacks detection techniques. Telecommun Syst **76**, 139–154 (2021). https://doi.org/10.1007/s11235-020-00733-2

9. Klein, G.: Naturalistic decision making. Human Factors: J. Hum. Factors Ergon. Soc. **50**(3), 456–460 (2008). https://doi.org/10.1518/001872008X288385
10. Militello, Laura G.: J. Cognitive Eng. Decis. Making 2016 **10**(4), 327 (2016). https://doi.org/10.1177/1555343416669318
11. Militello, L.G., Hutton, R.J.: Applied cognitive task analysis (ACTA): a practitioner's toolkit for understanding cognitive task demands. Ergonomics **41**(11), 1618–1641 (1998). https://doi.org/10.1080/001401398186108. PMID: 9819578
12. Ratcliff, R., Smith, P.L., Brown, S.D., McKoon, G.: Diffusion decision model: current issues and history. Trends Cogn. Sci. **20**(4), 260–281 (2016). https://doi.org/10.1016/j.tics.2016.01.007
13. Goldfarb, S., Leonard, N.E., Simen, P., Caicedo-Núñez, C.H., Holmes, P.: A comparative study of drift diffusion and linear ballistic accumulator models in a reward maximization perceptual choice task. Front. Neurosci. **8**, 148 (2014). https://doi.org/10.3389/fnins.2014.00148
14. Blurton, S.P., Kyllingsbæk, S., Nielsen, C.S., Bundesen, C.: A poisson random walk model of response times. Psychol. Rev. **127**(3), 362–411 (2020). https://doi.org/10.1037/rev0000179
15. Ratcliff, R., McKoon, G.: The diffusion decision model: theory and data for two-choice decision tasks. Neural Comput. **20**(4), 873–922 (2008). https://doi.org/10.1162/neco.2008.12-06-420
16. Lerche, V., Voss, A.: Retest reliability of the parameters of the Ratcliff diffusion model. Psychol Res. **81**(3), 629–652 (2017 May). https://doi.org/10.1007/s00426-016-0770-5. Epub 2016 Apr 23 PMID: 27107855
17. Donkin, C., Brown, S., Heathcote, A., Wagenmakers, E.J.: Diffusion versus linear ballistic accumulation: different models but the same conclusions about psychological processes? Psychon. Bull. Rev. **18**(1), 61–69 (2011). https://doi.org/10.3758/s13423-010-0022-4
18. Farrell, S., Lewandowsky, S.: Computational models as aids to better reasoning in psychology. Curr. Dir. Psychol. Sci. **19**(5), 329–335 (2010). https://doi.org/10.1177/0963721410386677
19. Andrade, R., Torres, J., Cadena, S.: Cognitive security for incident management process. In: Rocha, Á., Ferrás, C., Paredes, M. (eds.) Information Technology and Systems. ICITS 2019. Advances in Intelligent Systems and Computing, vol. 918. Springer, Cham (2019). https://doi.org/10.1007/978-3-030-11890-7_59
20. Andrade, R., Torres, J., Flores, P.: "Management of information security indicators under a cognitive security model." 2018 IEEE 8th Annual Computing and Communication Workshop and Conference (CCWC), Las Vegas, NV, USA, pp. 478-483 (2018). https://doi.org/10.1109/CCWC.2018.8301745
21. Abdelhamid, N., Ayesh, A., Thabtah, F.A.: Phishing detection based associative classification data mining. Expert Syst. Appl. **41**, 5948–5959 (2014)

Development of a Linear Model of Diabetes Mellitus with a PID Controller and a Method for Automatic Adjustment of the PID Controller

D. A. Andrikov and S. V. Kurbanov

Abstract The design of a linear control system of blood sugar level is considered. Currently, the trend of using individual portable regulatory systems to maintain optimal blood sugar levels prevails. We analyze differences in the quality of life of a patient with diabetes and a healthy person and control and normalize the blood sugar level of the patient. A review of mathematical models of diabetes mellitus is presented, and the choice of a mathematical model for the development of an automatic control system for an insulin pump is justified. The linearization of the existing nonlinear model was performed, and the adequacy of the linearized model was verified and confirmed. The choice of the PID controller and the CHR method of its adjustment is justified. The model of the patient with the controller is constructed, and the algorithm of automatic adjustment of parameters of the PID controller is developed and implemented. The directions of continuing work on the problem of regulation in the system under study are proposed.

Keywords Diabetes mellitus · Insulin pump · Automatic control system · Insulin–glucose interaction models · PID controller · Linearization · Automatic control system · Parameter setting

1 Introduction

Diabetes mellitus is dangerous primarily for its complications. Unfortunately, to date, there is no unambiguous generally accepted understanding of the causes of DM and methods of their elimination, so the treatment of DM is symptomatic. Normalization of sugar levels practically eliminates the difference in the quality of life of a patient with diabetes and a healthy person.

The development of medical science and technology has made possible the outpatient treatment of DM, and the prevailing trend is the use of individual regulatory

D. A. Andrikov (✉) · S. V. Kurbanov
Peoples' Friendship University of Russia, 6 Miklukho-Maklaya St, Moscow 117198, Russian Federation
e-mail: andrikov-da@rudn.ru

© The Author(s), under exclusive license to Springer Nature Singapore Pte Ltd. 2022

A. K. Nagar et al. (eds.), *Intelligent Sustainable Systems*, Lecture Notes in Networks and Systems 333, https://doi.org/10.1007/978-981-16-6309-3_5

systems to maintain optimal blood sugar levels. Such systems "glucose sensor—controller—dosing device for insulin delivery—patient's body" are classic automatic control systems and include a glucometer and a dispenser for drug injection—an insulin pump, as well as algorithms for calculating the dose and rate of insulin injection. These algorithms can be implemented using a controller.

Since diabetes mellitus is one of the most widespread endocrine diseases in the Russian Federation [1], the topic of developing new and improving existing automatic control systems for regulating blood sugar levels in patients with diabetes mellitus is relevant.

2 Goals and Objectives

The aim of the work is to develop a technique for regulating blood sugar levels using an insulin pump, with the prospect of using this technique in the implementation of compact medical devices, in particular an insulin pump.

1. To achieve this goal, the following tasks were solved:
2. Analysis of numerous existing mathematical models in order to select the most suitable for linearization;
3. Modification of the selected model [2] and its linearization;
4. Determination of the parameters of the linearized model and its transfer functions for food intake and insulin administration;
5. Checking the adequacy of the linearized model, comparing its reaction with the reaction of the original model; the test method is modeling the processes of changes in blood sugar levels;
6. Checking the possibility of achieving the main goal of setting the PID controller normalization of blood sugar levels using an insulin pump, taking into account the features of the object of regulation;
7. Analysis of PID controller tuning methods; selection of the simplest and most reliable method that does not require complex and time-consuming calculations and has passed a long practical test on other regulatory objects [3]; Checking the possibility of achieving the main goal of setting the PID controller normalization of blood sugar levels using an insulin pump, taking into account the features of the object of regulation;
8. Adaptation of the method of setting the PID controller by the selected CHR method, in relation to the task of correcting the sugar level;
9. Setting up the PID controller according to the adapted method, determining the parameters of the controller;
10. Experimental verification of the functioning of the patient with the regulator, comparison of the obtained transition curves with the transition curve of a healthy person; if necessary, adjustment of the regulator.

3 The Main Stages of Solving Problems

The concept of artificial pancreas required the development of a mathematical model of the interaction of insulin and glucose in the human body. Mathematical models of the interaction of glucose and insulin (virtual patient models) are conventionally divided into pharmacokinetic and physiological models. Pharmacokinetic models describe the kinetics of the glucose–insulin interaction, and they are easier to deduce from experimental data than physiological ones. In physiological models, organs are considered as components of the system, and the balance of substances is determined by metabolic processes and diffusion between blood and organ cells. A common disadvantage of physiological models compared to pharmacokinetic models is the large number of parameters, and, as a result, low accuracy. This circumstance makes the physiological model unsuitable for controlling a medical device for regulating blood glucose levels. Therefore, only some pharmacokinetic models were considered in detail: the Bergman model, the Hovorka model, and the simplest basic differential mathematical model of the form

$$\begin{aligned} \frac{dx}{dt} &= -a_1 \cdot x \cdot y + a_2(x_0 - x) \cdot H(x_0 - x) + a_3 \cdot P(t) \\ \frac{dy}{dt} &= b_1 \cdot (x - x_0) \cdot H(x - x_0) - b_2 \cdot y + b_3 \cdot u(t) \end{aligned} \quad (1)$$

where

$x(t)$ is the blood sugar level;
$y(t)$ is the level of insulin in the blood;
x_0 fasting blood sugar level.

The constants a_1, a_2, a_3 and b_1, b_2, b_3 are positive and are, respectively, the sensitivities of the sugar and insulin gradients; the function $H(x)$ is a single step; $P(t)$ describes changes in sugar levels from food intake; $u(t)$ describes changes in insulin levels.

The undoubted advantages of this model are its simplicity and compactness, consisting in a minimum number of equations and parameters. The adequacy of this model is proved in practice [4].

The model was linearized by entering a new variable—the deviation of the sugar level from «zero», corresponding to «hunger»:

$$z = x - x_0$$

and the following matrices of the linear system in the state space were obtained

$$A = \begin{pmatrix} -a_2 \cdot H(-z) & -a_1 \cdot x_0 \\ b_1 \cdot H(z) & -b_2 \end{pmatrix}; \; B = \begin{pmatrix} a_3 & 0 \\ 0 & b_3 \end{pmatrix}; \; C = (1 \; 0); \; D(x_0)$$

The transfer function with a second-order characteristic polynomial was derived from the representation in the state space

$$Z(s) = -\frac{a_1 \cdot x_0 \cdot b_3}{s^2 + b_2 \cdot s + a_1 \cdot x_0 \cdot b_1} \cdot U(s) + a_3 \cdot \frac{s + b_2}{s^2 + b_2 \cdot s + a_1 \cdot x_0 \cdot b_1} \cdot P(s)$$

as well as transfer functions:

- at $P(s) = 0$ for insulin input

$$W_u(s) = \frac{a_1 \cdot x_0 \cdot b_3}{s^2 + b_2 \cdot s + a_1 \cdot x_0 \cdot b_1}$$

- at $U(s) = 0$ for nutrition

$$W_p(s) = a_3 \cdot \frac{s + b_2}{s^2 + b_2 \cdot s + a_1 \cdot x_0 \cdot b_1}$$

The simulation showed a practical coincidence of the transition functions of the linearized and the original system within two to three hours from the moment of eating (Fig. 1), so the linear model was accepted for further research, provided that the maximum of the transition function of the system with the controller is reached at the specified time.

When choosing a method of setting up the controller, a fundamental preference was given to engineering methods that allow calculating the parameters of the controller based on the characteristics of the transient process graph. A comparative analysis of the methods was carried out, and the CHR method was adopted as providing a better quality of the transient process and a greater margin of stability of the system with a regulator compared to the Ziegler–Nichols method. A method was developed for determining the parameters of the PID controller based on a single input variable—the angle of inclination of the tangent to the transient graph. The method was based on the approximation of the transition process graph by a polynomial, followed by analytical differentiation of this polynomial, in order to accurately determine the angle of the tangent.

In the case of a projected insulin input control system, the adjustment process is complicated by the following factors, which are inherent properties of the control object:

1. The subsystem of nutrition (food intake) in the patient is practically not connected with the subsystem of regulation (production or administration of insulin);
2. The PID controller cannot be embedded into the control object in the sense that the controller is an external device to this object;
3. The parameters of the model of a particular patient with DM are individual and vary quite widely.

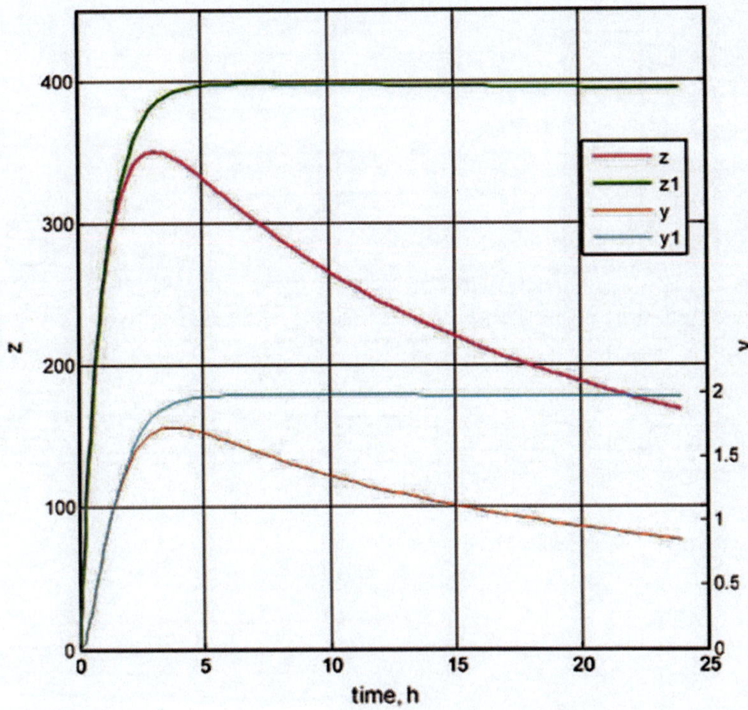

Fig. 1 Transients in the patient's body during a single meal in the initial (z, y) and linearized (z_1, y_1) models

Therefore, an experimental study of a linearized system with a controller was carried out.

The MATLAB/Simulink software package was selected to study the linearized insulin input control system and test the PID controller tuning algorithm. The model is shown in Fig. 2.

A number of computational experiments were carried out on the model, as a result of which the limits of changing the parameters of the PID controller were determined:

- gain $K = 12 \ldots 24$;
- time constant of the integrator $T_\mathrm{i} = 0.01 \ldots 0.05$ h;
- derivation time constant $T_\mathrm{d} = 0.004 \ldots 0.008$ h.

A transient graph with regulator settings.

$K = 18$; $T_\mathrm{i} = 0.0333$ h; $T_\mathrm{d} = 0.0058$ h.

For a patient with an insulin pump (IP) is shown in Fig. 3. For comparison, the same graph shows the transition process in the "healthy" model.

Thus, as a result of the work done, the prospects of applying the considered approach to the development of controllers for insulin pumps can be considered proven.

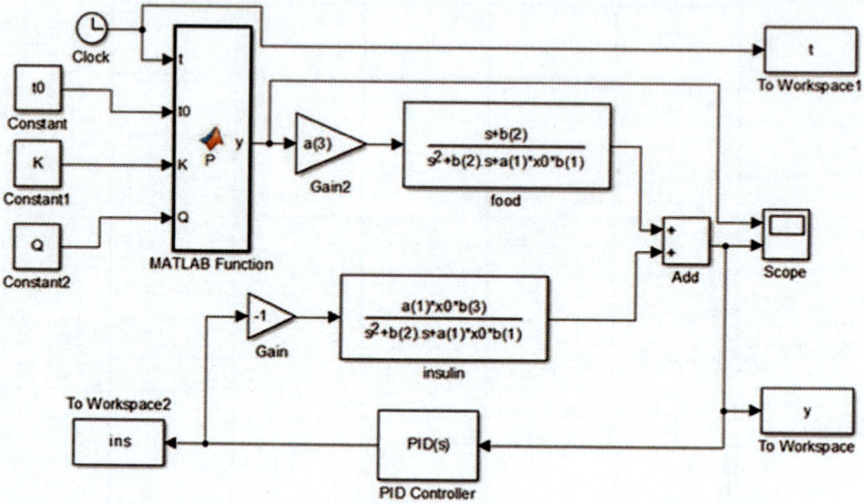

Fig. 2 Model of the object with a PID controller

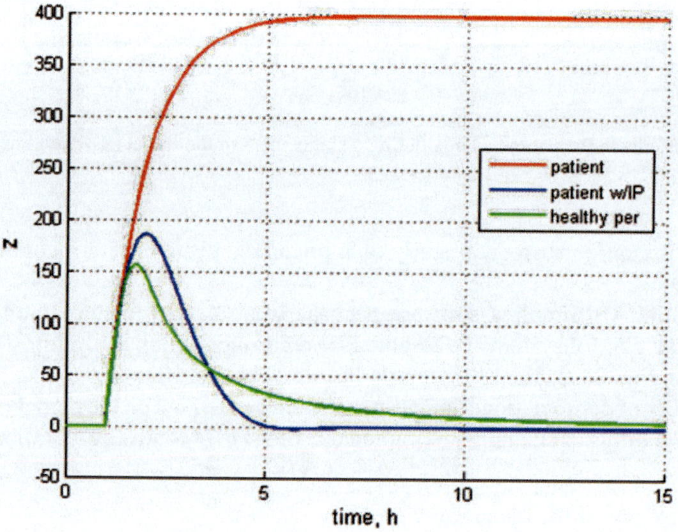

Fig. 3 Transients on a control object with a PID controller

The advantage of using a linear model over a nonlinear one is that a well-developed theory applies to the linear model. This simplifies the controller configuration procedure and allows you to automate this procedure. Thus, it is possible to achieve the dynamics of the level of sugar in the blood of the "patient" characteristic of the "healthy."

4 Conclusion

To summarize, we can say that the goal of regulation set in the work was achieved as a result of the following stages:

1. analysis of existing mathematical models of the interaction of sugar and insulin in the human body, modification of the selected model and its linearization;
2. determination of the transfer functions of the linearized model for food intake and insulin injection;
3. checking the adequacy of the linearized model;
4. checking the possibility of normalizing blood sugar levels with an insulin pump using a PID controller;
5. development of a method for setting the PID controller by the CHR method in relation to the task of correcting the sugar level and determining the parameters of the controller using this method;
6. experimental verification of the functioning of a digital model of a patient with DM with a controller.

A transition process was obtained, and the dynamics of which slightly differs from the dynamics of the digital model of a healthy person, namely the sugar level is normalized to a "healthy" state within 3 h, with the maximum excess over the "healthy" level being 20%. The development of the work could be to improve regulation by adding nonlinear links to the regulator and clarifying the parameters of insulin input (e.g., the glucose concentration (sugar level) at the time when the insulin input should be stopped) in order to further bring the transition process to the level corresponding to the norm [5].

References

1. Health Care in Russia, p. 170. Rosstat. M. (2019)
2. Kozeko, L.G., Tsirulyova, V.M., Shapovalova, I.A.: The study and numerical implementation of the mathematical model of the task of correcting blood sugar levels. Study and implementation of numerical mathematical model the task of adjusting the level of sugar in the blood. TVSU. Series: Chemistry. No. 1, pp. 169–178 (2018)
3. Chien, K.L., Hrones, J.A., Reswick, J.B.: On automatic control of generalized passive systems. Trans ASME **74**, 175–185 (1952)
4. Davis, M.J.: Differential model of diabetes mellitus. In: Andrews, J.G., McLone, R.R. (eds.) Mathematical Modelling, vol. 260 pp. 128–139. Butterworths, London (1976)
5. Gusev, A.V., Gavrilov, D.V., Korsakov, I.N., et al.: Prospects for the use of machine learning methods for predicting cardiovascular disease. Doctor Inf. Technol. **2019**(3) C. 41–47 (2019) ISSN 1811-0193

Financial Decisions in Colombian Organisations in the Health Sector

Romel Ramón González-Díaz, Elena Cachicatari-Vargas, Santos Lucio Guanilo-Gómez, and Graciela Tristancho Cediel

Abstract The purpose of this investigation was to analyse the current financial situation of the Colombian IPS about decision-making. It was developed under a quantitative methodology; the sample was 364 financial managers from the Colombian IPS, an instrument with 13 items with a Likert scale was applied, validated by expert judgement and discriminant validity, the reliability was 0.91. The significant results show that financial managers consider making the right decisions (Good 41.7% and Excellent 55.3%), and 71.2% think they have an excellent financial situation. This allows a relationship between the variables, so the alternative hypothesis is accepted if there is a relationship between decision-making and the financial status of the IPS in Colombia. Likewise, according to the contingency coefficient, the intensity of the association is 0.616, considered vital.

Keywords Colombian IPS · Financial situation · Decision-making

1 Introduction

The processes of globalisation and internationalisation have introduced in organisations a dizzying series of changes related to the way they are managed; all this to adapt to the economic and social dynamics that these changes bring with it, to

R. R. González-Díaz (✉)
Centro Internacional de Investigación Y Desarrollo (CIID), 230001 Montería, Colombia
e-mail: director@ciid.com.co

E. Cachicatari-Vargas · S. L. Guanilo-Gómez
Universidad Nacional Jorge Basadre Grhomann, 23000 Tacna, Perú
e-mail: ecachicatariv@unjbg.edu.pe

S. L. Guanilo-Gómez
e-mail: lguanilog@unjbg.edu.pe

G. T. Cediel
Universidad Industrial de Santander, 680001 Bucaramanga, Colombia

© The Author(s), under exclusive license to Springer Nature Singapore Pte Ltd. 2022

A. K. Nagar et al. (eds.), *Intelligent Sustainable Systems*, Lecture Notes in Networks and Systems 333, https://doi.org/10.1007/978-981-16-6309-3_6

successfully conduct a company, that is why timely and accurate financial information is needed; reliable and accessible to serve management to make decisions on a particular topic.

In Latin America, there has been remarkable growth in small and medium-sized enterprises (SMEs) due to its expansion of more than 3% in Latin American GDP growth; in the case of Colombia, growth of 3.8% of its GDP is forecast for 2019 [1] of his PIB. This economic growth is due to 80% of the contributions presented by SMEs. Pymes in their various sectors [2, 3].

In the health sector, Colombia is experiencing a crisis in its system that exceeds the economy and is affecting the society because it involves all the people in the order, who frequently suffer the deterioration of the quality of the Health Ndering Institutions (IPS); according to Ortiz-Campillo et al. [4] in recent years the growth in the delivery of health services has increased by 40%. However, this growth is not accompanied by a timely payment by the Health Promoters (EPS), and a change in portfolios can be observed, but with the closure of other institutions due to lack of liquidity [5].

According to the Colombian Association of Clinics and Hospitals [6], in a survey of more than 50% of the Health Institutions affiliated to the national health system, a portfolio of more than 500 million dollars was obtained, 500 million de dollars 64% of which corresponds to portfolio over 180 days. The worrying situation, because these institutions have short-term concurrent expenses (supplier payment, payroll, among others) if the company does not maintain sufficient working capital, it could fall into a process of illiquidity, where recourse to financial instruments (bank loans, factoring, cash discount) would lead to an imminent decrease in its profitability [2, 7].

Therefore, this situation is represented in a high percentage of Colombia's IPS; they have an average payment period of 180 days, a situation that may be unsustainable to any organisation, who must honour their commitments to third parties in less than 90 days, generating negative cash flow, causing the search for financing in credit institutions, to maintain their operations, that is why the need to carry out a constant analysis of the financial statements is created to make sound decisions and minimise the risks of bankruptcy.

A financial analysis, according to Refs. [8, 9] considers that an analysis of the financial statements is a systematic process of reflection of the financial statements to diagnose the economic and financial position of the production unit, be able to generate considerations for decision-making according to the objectives established by the organisation, this process of containing ethical criteria by the analyst says [10–12].

Likewise, financial analysis provides quantitative data about the company's situation, making comparisons with the past and projections of the future with the sole objective of generating criteria of certainty in management decisions. Ignoring the review of the Company's Economic and Financial Situation in the decision-making of the Board of Directors; may result in a financial catastrophe; therefore, it is necessary to know the reasons why wrong decisions are made, according to Silva are several reasons why bad decisions are made, however, the leading cause is the effects on the cognitive bias that lead us to a state of apparent normality, but one is not acting rationally.

In addition to the comment, other reasons influence decision-making, lack of information, feelings, and moods, which we somehow resort to as a kind of mental shortcut when deciding. According to a problem in assigning importance to tasks, all this has consequences, generating crazy decisions. Similarly, the retribution of erroneous resources can interfere with the flow of work activities and create emergencies. Mistakes are caused by poor decision-making and take valuable time and resources to correct them. Companies are affected by the decision because decisions are made prematurely, which increases the likelihood of costly errors. The decision's preference can come from the overload of information, pressure of deadlines, excessive choices, or the lack of judgement of decision-makers.

This paper aims to describe the current financial situation of Colombian service providers and their management decision-making process. This provides relevant information for the Colombian health sector because it allows us to understand these institutions' wide-ranging financial difficulties. To this end, 364 managers of Colombian IPS were surveyed to study decision-making and financial situation variables. The results show the existence of a statistically significant relationship between decision-making and financial situation.

1.1 Decision-Making

Decision-making, by Financiera [13] an activity planned and based on the real needs of a company or organisation and oriented towards a change in the knowledge, skills, and attitudes of the collaborator. Operationally, it is defined as the total score achieved through the application of a similar scale questionnaire prepared by the researcher, aimed at measuring dimensions, type of decision-making, and stages of the decision-making process with its indicators: scheduled decisions, unscheduled decisions, problem identification, decision alternatives and decision effectiveness assessment.

1.2 Financial Situation

On the other hand, as raised by authors [13–15], those who agree that the financial situation relates to the final result of any accounting and financial process translated into the financial statements, which are drawn up based on the balances faithfully taken from the ledgers, which may well be from the general ledger and balance sheets, or from the auxiliaries, which must be consistent. For consequently shows the company's history during an accounting period and of all the decisions made by its managers.

2 Materials and Methods

This research focused on quantitative research through a field design, non-experimental cross-sectional design, where a questionnaire of 13 items and Likert-type scale was applied, validated by three experts, with a Cronbach's Alpha coefficient of 0.91 (excellent) to a population of 8000 IPS according to data from the Association of Clinics and Hospitals (2019), a sampling calculator was used for the sampling technique (https://www.netquest.com/) with a confidence level of 95% and a margin of error of 5%.

The data were analysed with descriptive statistical tools to understand their behaviour; then, the chi-square statistic was applied to test the hypotheses referred to. The hypothesis system of the research was: null hypothesis (H0) "there is no relationship between decision-making and the financial situation of the IPS". In contrast, the alternative hypothesis (H1) "there is a relationship between decision-making and the financial situation of the IPS". The asymptotic (bilateral) significance considered is 0.005, and if H1 is accepted, the contingency coefficient is considered to measure the strength of the relationship between the variables. In particular, most of these investigations need to be approved. Subsequently, we constructed the 3PR system and approved it. This structure was effectively used as an establishment for the improvement of an SPM efficiency metric.

3 Analysis and Discussion of Results

Once the results have been obtained, they are synthesised by each of the variables under study. On the one hand, the variable decision-making is based on an organisation's actual needs, resulting in financial managers considering making sound decisions (Good 41.7% and Excellent 55.3%) (see Table 1). Likewise, two types of decisions are generated, on the one hand, scheduled decisions such as those routine and frequent, and on the other hand, those that occur with a lower frequency called

Table 1 Descriptive statistics of the variable: decision-making

		Frequency	Percentage	Percentage valid	Cumulative percentage
Valid	Very bad decisions	1	0.3	0.8	0.8
	Regular decisions	3	0.8	2.3	3.0
	Good decisions	55	15.1	41.7	44.7
	Excellent decisions	73	20.1	55.3	100.0
	Total	132	36.3	100.0	
Lost	System	232	63.7		
Total		364	100.0		

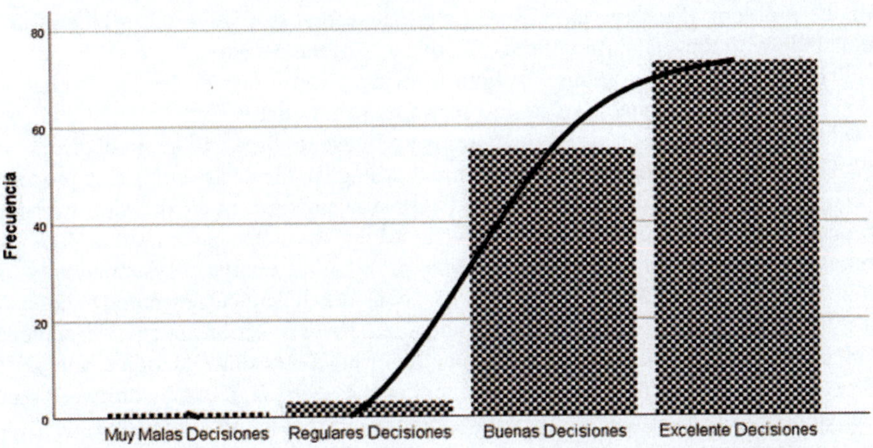

Fig. 1 Histogram of decision-making behaviour

unscheduled, therefore, after the processing of the data and analysed through the scale of interpretation, it can be determined that there are the IPS Colombians a deficient presence of the decisions both programmed and unscheduled, which implies that this organisation shows evident disorientation in terms of decision-making.

As for the scheduled decisions, 40% of managers say that routine problems in the organisation are never solved through the participation of those involved but are, on the contrary, tough decisions emanated unilaterally (see Fig. 1). Likewise, 60% of the population believes that neither policies nor rules in the institution guide decision-making. All of this can generate misguided decisions regarding financial decisions (financing and investment) and reflected in the same financial statements.

In this regard, authors such as Refs. [16–18] propose that scheduled decisions are made by policies, procedures, or rules, written or unwritten, facilitating decision-making in recurring situations because they limit or exclude other options. So there must be routine procedures to treat usual problems.

About the indicator concerning unscheduled decisions, it could be shown that 40% of the total respondent stated that decisions are never made to solve the rare problems in Colombian IPS, as well as decision-making under a specific model, and 80% believe that decisions are never taken to improve the relationship with the community. All this is with the approaches of Marín [19], which refers that unscheduled decisions, also called unstructured, are decisions that are made in the face of problems or situations presented infrequently, or those where a specific model or process of solution is needed. It is necessary to follow a model of decision-making to generate specific solutions for specific problems.

About the stages of the decision-making process of IPS managers, they show an average of indicators with a deficient presence in the monitoring of the stages of the decision-making process, so it contrasts with what is stated by Regaliza et al. [20] establish that a series of steps should be followed that begins with the identification of the problem, the steps to select an alternative that can solve the problem and conclude with the evaluation of the effectiveness of the decision.

This process can be applied to both individual and group decisions. Decision-making matters the choice of a way forward, so alternatives of action should be evaluated in a previous state. If the latter is not present, there will be no decision.

The identification of the problem as the first stage in the decision-making process is considered by the population surveyed as the low presence in the decision-making process used by the entity under study; however, the items indicate that 40% of managers state that rarely the causes of the problem are analysed systematically to make decisions also, 60% of managers never consider the logical reasoning to choose a solution to problems in the institution and alternatives of action are never accepted, since they are imposed solutions, without structuring a coherent and logical analysis.

As for the evaluation of the effectiveness of the decision, it can be observed that deficient presence in the IPS of Colombia represents that the strategies of evaluation of the effectiveness of decisions bring observable deficiencies since the achievements achieved by the institution are never judged and evaluated, as well as, rarely used models to understand the reality of the decision taken, a situation contrary to what was raised by, who adds that to make some types of decisions, models are often used. These can be defined as a simplified representation of a part of a complex reality, which requires analysis and understanding to decide how to solve the problems presented in organisations, promoting change. This gives you an idea of the consequences that each of the alternative actions that have been defined and that can serve to choose the most appropriate behaviour as the course of action that will solve the threat.

Moreover, according to the unstable financial situation approaches, which coincide, the analysis of the financial statements allows us to visualise and interpret the facts, situations, or states that are desired to examine the company. In this research, the questionnaire's perception was known through the questionnaire; their answers are summarised in Table 2.

Table 2 shows that 71.2% of financial managers consider an excellent financial situation, which contrasts with the approaches. They manifest in their study a temporary financial situation of economic surplus due to late payments by EPS.

Figure 2 refers to the low IPS considering a poor financial situation. However, authors like Velasco López et al. [21] and González Briceño and Acosta Aldana [22]

Table 2 Descriptive statistics of the variable: financial situation

		Freq	%	% valid	Cumulat. %
Valid	Regular financial situation	3	0.8	2.3	2.3
	Good financial situation	35	9.6	26.5	28.8
	Excellent financial situation	94	25.8	71.2	100.0
	Total	132	36.3	100.0	
Lost	System	232	63.7		
Total		364	100.0		

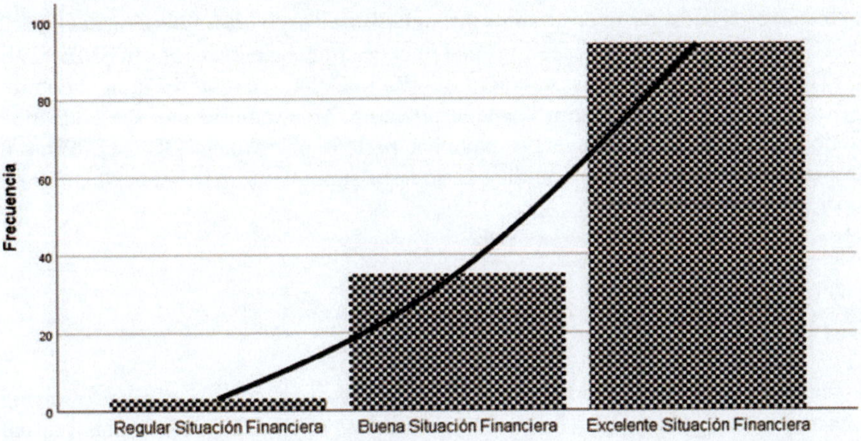

Fig. 2 Histogram of financial situation behaviour

argue that more than 50% of IPS makes terrible decisions in their financing and fall into severe problems of liquidity.

Once Pearson's chi-square hypothesis test statistic was applied, it could be determined that the asymptotic (bilateral) signification level is 0.000 lower than the significance level set by study 0.05. Therefore, the null hypothesis is rejected, and the alternative hypothesis is accepted if there is a relationship between decision-making and the financial situation of IPS in Colombia. Also, the intensity of the ratio according to the contingency coefficient is 0.616, considered vital (see Table 3).

These results are in line with the approaches of Geiger and Kjellberg [23], Iheanachor et al. [24], Toorajipour et al. [25], who in their studies address financial management as the source of information for decision-making. However, this is not an isolated situation because many external (governmental measures, diseases, legal problems, suppliers, among others) and internal (marketing strategies, innovation processes and internal processes) components determine the decision-making process of managers of service providers Colombia. On the other hand, studies related

Table 3 Chi-square tests between variables

	Value	N	Asymptotic significance (bilateral)
Pearson Chi-square	81,405^{to}	6	000
Reason for plausibility	40,610	6	000
Linear association by linear	39,011	1	000
Contingency coefficient	618		
N of valid cases	364		

(a) Eight boxes (66.7%) have expected a count of less than 5. The expected minimum count is 0.02

to managerial decision-making show through Bijmolt et al. [26], Lobschat et al. [27], and Shrestha et al. [28] that feelings and hunches influence decisions in SMEs[29].

Therefore, the financial situation of the IPS in Colombia requires greater control of financial resources through financial planning. This requires average collection periods of 240 days and average payment periods of 58 days [30, 31]. Thus, it is necessary to use ample liquidity periods to support the payments of different operating expenses.

4 Conclusion

Once the analysis and interpretation of the results were completed, the following conclusion was generated. The Board of Directors of the IPS in Colombia present a coherent and coordinated decision-making process with their work team that has benefited their fair financial situation. However, these organisations show a clear orientation in their decision-making, which despite being primarily intuitive, generate good results in the financial situation of the IPS. Likewise, the decision-making processes must be rational because it is considered a stroke of luck and the sound profitability of the health business that has made the organisations dedicated to this area financially sustainable in the Colombian market.

References

1. Cepal, N.: Agenda 2030 y los Objetivos de Desarrollo Sostenible: una oportunidad para América Latina y el Caribe (2018)
2. González-Díaz, R.R., Acevedo-Duque, Á.E., Guanilo-Gómez, S.L., Cachicatari-Vargas, E.: Business counterintelligence as a protection strategy for SMEs. Entrepreneurship Sustain. Issues **8**(3), 340–352 (2021)
3. González-Díaz, R.R., Acosta-Moltó, E., Flores-Ledesma, K., Vargas, E.C., Menacho-Rivera, A.: Marketing experiencial en organizaciones sin fines de Lucro: una mirada a los proveedores de experiencias. Rev. Ibérica Sistemas Tecnologias Informação **E36**, 186–202 (2020)
4. Ortiz-Campillo, L., Ortiz-Ospino, L.E., Coronell-Cuadrado, R.D., Hamburger-Madrid, K., Orozco-Acosta, E.: Incidencia del clima organizacional en la productividad laboral en instituciones prestadoras de servicios de salud (IPS): un estudio correlacional (2019)
5. Ramírez Carvajal, Z., Gutiérrez Sosa, J.: Modelo de optimización de flujos de efectivo mediante la diversificación de los métodos de contratación en salud–Aplicado a la IPS Universitaria (2019)
6. Clínicas, A.C.D.: Hospitales (01 de 11 de 2017). Asociación Colombiana de Clínicas y Hospitales (2017)
7. González-Díaz, R.R., Perez, L.A.B.: Análisis financiero empresarial del sector comercio como factor de competitividad través de la lógica difusa. Estrategia **1**(1), 1–10 (2015)
8. Sáenz, L., Sáenz, L.: Razones financieras de liquidez: un indicador tradicional del estado financiero de las empresas. Orbis Cognita **3**(1), 81–90 (2019)
9. Revuelta Villorejo, J.P.: Análisis Financiero y Estratégico de las principales Ganaderías de Toros de Lidia en España (2019)

10. Ledesma, K.N.F., González-Díaz, R.R.: Etica sindical en la negociación de contrato colectivo de construcción de obras públicas. Caso Ducolsa. CIID J. **1**(1), 01–14 (2020)
11. Noroño Sánchez, J., Nuñez Villavicencio, M., González Díaz, R.: Ética sindical como mecanismo impulsor de competitividad en las pequeñas y medianas empresas. Union ethics as a mechanism driving competitiveness in small and medium-sized enterprises. Utopía Praxis Latinoamericana **25**(1), 154–173 (2020)
12. Gonzalez-Díaz, R.R., Becerra-Peréz, L.A., Acevedo-Duque, Á.E.: Narco-marketing as a strategy for local tourism development. Rev. Ibérica Sistemas Tecnologias Informação **36**(E36), 71–85 (2020)
13. Financiera, I.D.S.: Estados financieros. Impuestos **1**, 731 (2020)
14. Alvarado, M.J., Cruz, C.G.M.: Hacia el Estado de Situación Financiera de acuerdo a las Normas de Información Financiera. Boletín Científico Ciencias Económico Administrativas ICEA **8**(16), 37–41 (2020)
15. Rodríguez Ariza, L., Gómez Miranda, M.E.: Actividad:¿ Condicionante de la situación económico-financiera empresarial? (2020)
16. González, J., Salazar, F., Ortiz, R., Verdugo, D.: Gerencia estratégica: herramienta para la toma de decisiones en las organizaciones. Telos **21**(1), 242–267 (2019)
17. Freire, A.G.H., Gonzaga, V.A.B., Freire, A.H.H., Rodríguez, S.R.V., Granda, E.C.V.: Razones financieras de liquidez en la gestión empresarial para toma de decisiones. Quipukamayoc **24**(46), 153–162 (2016)
18. Esteban Salvador, L.: La contabilidad de gestión como herramienta para la toma de decisions (2019)
19. Marín, L.E.: Auditoria financiera como proceso para la toma de decisiones gerenciales. Observatorio de la Economía Latinoamericana, no. octubre (2018)
20. Regaliza, J.C.P., Gual, J.C., Val, P.A.: Simulación como herramienta de ayuda para la toma de decisiones empresariales. Un caso práctico. Rev. Métodos Cuantitativos Economía Empresa **21**, 188–204 (2016)
21. Velasco López, L.A., Rivera Urbano, A.C., Granados, Y.A., Solarte, P.A., Portilla, O.: Estrategias para restablecer el equilibrio financiero de la IPS Comfacauca (2019)
22. González Briceño, E.A., Acosta Aldana, L.A.: Efecto de implementación de NIIF en la cartera de las IPS (2018)
23. Geiger, S., Kjellberg, H.: Market mash ups: the process of combinatorial market innovation. J. Bus. Res. **124**, 445–457, (2021). https://doi.org/10.1016/j.jbusres.2020.11.010
24. Iheanachor, N., David-West, Y., Umukoro, I.O.: Business model innovation at the bottom of the pyramid—a case of mobile money agents. J. Bus. Res. **127**, 96–107 (2021). https://doi.org/10.1016/j.jbusres.2021.01.029
25. Toorajipour, R., Sohrabpour, V., Nazarpour, A., Oghazi, P., Fischl, M.: Artificial intelligence in supply chain management: a systematic literature review. J. Bus. Res. **122**, 502–517 (2021). https://doi.org/10.1016/j.jbusres.2020.09.009
26. Bijmolt, T.H.A., et al.: Challenges at the marketing–operations interface in omni-channel retail environments. J. Bus. Res. **122**, 864–874 (2021). https://doi.org/10.1016/j.jbusres.2019.11.034
27. Lobschat, L., et al.: Corporate digital responsibility. J. Bus. Res. **122**, 875–888 (2021). https://doi.org/10.1016/j.jbusres.2019.10.006
28. Shrestha, Y.R., Krishna, V., von Krogh, G.: Augmenting organisational decision-making with deep learning algorithms: principles, promises, and challenges. J. Bus. Res. **123**, 588–603 (2021). https://doi.org/10.1016/j.jbusres.2020.09.068
29. González-Díaz, R.R., Becerra-Perez, L.A.: Stimulating components for business development in Latin American SMEs. In: Trends and Applications in Information Systems and Technologies, pp. 366–374. Springer International Publishing, Cham (2021). https://doi.org/10.1007/978-3-030-72651-5_35
30. Bustamante-Cabrera, G.Í., González-Díaz, R.R.: Times of pandemic in Bolivia: role of political and social organization in health policy. Eur. J. Mol. Clin. Med. **8**(02), 2021 (2021)

31. González-Díaz, R., Acevedo-Duque, Á., Salazar-Sepúlveda, G., Castillo, D.: Contributions of subjective well-being and good living to the contemporary development of the notion of sustainable human development. Sustainability **13**(6), (2021). https://doi.org/10.3390/su13063298
32. Hernández-Julio, Y.F., Meriño-Fuentes, I., González-Díaz, R.R., Guerrero-Avendaño, A., Toledo, L.V.O., Bernal, W.N.: Fuzzy knowledge discovery and decision-making through clustering and dynamic tables: application in Colombian business finance. In: 2020 15th Iberian Conference on Information Systems and Technologies (CISTI), 24–27 June 2020, pp. 1–5 (2020). https://doi.org/10.23919/CISTI49556.2020.9141117
33. González-Díaz, R.R., Vásquez Llamo, C.E., Hurtado Tiza, D.R., Menacho Rivera, A.S.: Plataformas interactivas y estrategias de gestión del conocimiento durante el Covid-19. Rev. Venezolana Gerencia, **25**(4), 68–81 (2020). [Online]. Available: https://produccioncientificaluz.org/index.php/rvg/article/view/35177
34. Silva Díaz, I.: Análisis financiero de los estados financieros para la toma de decisiones financieras (2019)
35. Huacchillo Pardo, L.A., Ramos Farroñan, E.V., Pulache Lozada, J.E.: La gestión financiera y su incidencia en la toma de decisiones financieras. Rev. Univ. Soc. **12**(2), 356–362 (2020)
36. Castalo Suescún, V., Gallego Álvarez, A.: Análisis de la incidencia de las finanzas conductuales en la toma de decisiones empresariales (2020)
37. González, Y.R., Arceo, A.C., Sánchez, N.M., Domínguez, A.H.: Agrupamiento conceptual lógico combinatorio: una alternativa para la toma de decisiones. Intel. Artif. **19**(57), 82–96 (2016)
38. Ivancevich, J.M., Konopaske, R., Matteson, M.T.: Comportamiento organizacional. McGraw Hill (2006)
39. Weston, J.F., Brigham, E.F., Mont, J.G.: Fundamentos de administración financiera. Nueva Editorial Interamericana (1987)
40. Brealey, R.A., Myers, S.C., Marcus, A.J.: Fundamentals of Corporate Finance. McGraw-Hill/Irwin (2012)
41. González-Díaz, R.R., Ledesma, K.N.F.: Cultura organizacional y Sustentabilidad empresarial en las Pymes durante crisis periodos de confinamiento social. CIID J. **1**(1), 28–41 (2020)
42. Cuevas Rendón, L.F., Restrepo Agudelo, J.: Análisis financiero de la IPS REMY de la ciudad de Cali del periodo 2016–2017 con el fin de determinar estrategias que mejoren sus procesos (2018)

Business Innovation Capabilities in the Road Infrastructure Sector in the Caribbean Region of Colombia

Romel Ramón González-Díaz, Santos Lucio Guanilo-Gómez, Elena Cachicatari-Vargas, Angel Acevedo-Duque, and Katiusca Cruz-Ayala

Abstract Business innovation capabilities have become the most critical driver of transformation and growth in companies in the road construction sector. To this must be added the multiple benefits it brings to companies: increasing profit. This article aims to analyse the innovation capabilities of companies in the infrastructure sector in the Colombian Caribbean region. The sample consisted of 331 managers/directors of construction companies dedicated to road infrastructure located in the Colombian Caribbean region, to whom a Likert scale questionnaire was applied. The current results allude to the existence of significant gaps between the articulation of Colombian state entities, the university and the company, evidenced in the low levels of business innovation capacity.

Keywords Innovation capabilities · Road infrastructure · Business innovation · Sustainable development

R. R. González-Díaz (✉) · K. Cruz-Ayala
Centro Internacional de Investigación Y Desarrollo (CIID), 230001 Montería, Colombia
e-mail: director@ciid.com.co

S. L. Guanilo-Gómez · E. Cachicatari-Vargas
Universidad Nacional Jorge Basadre Grhomann, 23000 Tacna, Perú
e-mail: lguanilog@unjbg.edu.pe

E. Cachicatari-Vargas
e-mail: ecachicatariv@unjbg.edu.pe

A. Acevedo-Duque
Faculty of Business and Administration, Autonomous University of Chile, 9170124 Providencia, Chile
e-mail: angel.acevedo@uautonoma.cl

© The Author(s), under exclusive license to Springer Nature Singapore Pte Ltd. 2022

A. K. Nagar et al. (eds.), *Intelligent Sustainable Systems*, Lecture Notes in Networks and Systems 333, https://doi.org/10.1007/978-981-16-6309-3_7

1 Introduction

The processes of internationalisation and globalisation have removed space barriers in the commercialisation of goods and services, resulting in high levels of competition and the emergence of new efficiency paradigms, which have forced organisations to rethink their business operations and remain competitive in an increasingly demanding market. Las Information and Communication Technologies (ICT), as a trend, has caused a real revolution, with a more significant impact than those originated by riots in the history of humanity (industrial, agricultural, and technological revolution) [1–4].

In the knowledge economy, various authors such as [5–7] agree that innovation capabilities are the vital sources in increasing productivity, competitiveness, and growth, justifying that the growth and competitiveness of enterprises and the economy as a whole will depend heavily on the building of innovation capabilities, as well as on the sustained efforts of companies nestled in a territory to generate them [4, 8, 9].

In Latin America, the government is aware of the relationship between innovation and local development levels in Colombia. It has adjusted its public policy on innovation aligned to the Sustainable Development Goals (SDGs) to the Colombian context through international cooperation with the global and national development agenda of the United Nations Development Programme (UNDP), which brings together practical tools for the design and implementation of public policies aimed at meeting the SDG goals and their incorporation into local government plans within new periods of government.

One of the challenges of Colombia is to land this agenda in the national context and to be able to generate synergies between the different sectors of the State, which allows enhancing the achievement of the objectives in the field of sustainable development, highlighting among them, objective 9 on industry, innovation, infrastructure, building resilient infrastructures, promoting inclusive and sustainable industrialisation, and fostering innovation, given the diverse dispersion in infrastructure that permeates local development, being part of the strategic diagnosis reflected in the National Development Plan (NDS) (2014–2018), where the backhand in the provision of logistics and transport infrastructure [10].

The transport infrastructure is one of the fundamental pillars to generate competitiveness and forms one of the engines of economic growth. In Colombia, it was observed through the Global Competitiveness Report 2014–2019 that cities develop in isolated territories with no relationship or complementarity between them [11]. It should be noted that only 11.8% of the tracks are paved, highlighting that more than half of the paved roads are in poor condition, representing an infrastructure deficit that increases logistical costs and exceeds 23% of GDP.

Investments in the transport infrastructure sector have been relatively low at an average of less than 1% of GDP. However, in recent years the national government has concentrated its efforts on bringing investment levels to 3% of GDP. Thus, the GDP of the transport infrastructure sector rose from a growth of 6% in 2012 to 9.8%

in 2013, consolidating, and the industry that most boosted investment and national growth during 2013.

In this sense, it has been noticed in the Contractors of this sector to have certain deficiencies in the Innovation indicators that directly impact the performance parameters of each company, demonstrated in the number of improvements in processes, productivity, and employment. These processes directly traverse the innovation capabilities applied directly to the field of Road Infrastructure, so a picture can be established through the investments implemented in the Departmental Development Plan of the governorate of Córdoba that reached a total of $494,565 million during the years 2015–2019) [2, 8, 12–14].

1.1 Business Innovation Capabilities

It is widely documented that innovative business capacity can mobilise the knowledge possessed by employees and the combination of these to create a new experience, resulting in products or processes of innovation [4, 5, 14–16]. In the infrastructure sector of the Colombian Caribbean region, reference is made to the following Table 1.

1.1.1 Environment

It is considered the space or environment outside the company as an organisation where there is no control and has difficulties in determining with certainty or accuracy where the borders or the correct limits of the company are concerning what is known as the external environment [17–19].

1.1.2 Business Innovation

In the economic literature, innovation is defined as the introduction of a new or higher quality good (product) for consumers than the previous ones, the introduction of new production methods for an industry sector, the opening of new markets, the use of new sources of supply, or the introduction of new ways of competing that lead to a redefinition of industry [4, 13–15, 20].

1.1.3 Innovation Activities

Innovation activities vary greatly depending on the company, so some companies are engaged in well-defined innovation projects, such as developing and introducing a new product. In contrast, others are dedicated to making continuous improvements to their products, processes, or operation. It is also conceived as a broad notion,

Table 1 Table of variable operationalisation

Variable	Dimensions	Indicators
Business innovation capabilities	Environment	Infrastructure conditions
		Labor cost
		Clustering
		Higher education institutions
		Environmental conditions
		Economic aspects
		Legislation
		Government
	Company innovation	Services innovation
		Product innovation
	Impacts and innovation activities	Licences or copyrights
		Spare parts, maintenance and purchase programme
		Modernisation of machinery, hardware and software
		Standardisation of processes in standards and manuals: operational processes and technological processes
		Organisational structure
		Organisational culture and managers

expressing all the dynamism of human nature, with work being only one of them, alongside the activities carried out in the domestic, educational, cultural, sports, trade union, political spheres, among others [21–23].

2 Methodology

This study is part of a quantitative methodology under a transactional, non-experimental field design. The data collection instrument was the Likert-type scale questionnaire with 20 reagents, piloted (10 managers), subsequently validated at the opinion of 5 experts and valued under the reliability criteria with Cronbach's Alpha coefficient 0.83 (appropriate). The study universe is 2352 builders, for which sample techniques were applied, with a confidence level of 95% and a margin of error of 5%, for a sample of 331 managers/directors. The questionnaire was applied to 331 managers in the Colombian Caribbean region with 20 reagents, according to the following data interpretation Table 2.

Table 2 Criteria for interpretation of results

Variable	Ordinal Scale	Cut Points
Business innovation capabilities	Very bad business innovation capabilities	1.00–1.80
	Poor business innovation capabilities	1.81–2.60
	Regular business innovation capabilities	2.61–3.40
	Good business innovation capabilities	3.41–4.20
	Excellent business innovation capabilities	4.20–5.00

3 Analysis and Discussion of Results

Once the data collection instrument was applied to the analysis unit, the descriptive statistic was performed through SPSS25, resulting in the following.

3.1 Business Innovation Capabilities

The study variable analysed was framed in the following statistical data (Table 3).

Table 3 Descriptive statistics of the variable innovation capabilities of entrepreneurship

		Frequency	Percentage	Percentage valid	Cumulative percentage
Valid	Poor business innovation capabilities	22	6.4	6.7	6.7
	Regular business innovation capabilities	88	25.5	26.7	33.3
	Good business innovation capabilities	198	57.4	60.0	93.3
	Excellent business innovation capabilities	22	6.4	6.7	100.0
	Total	330	95.7	100.0	
Lost	System	15	4.3		
Total		345	100.0		

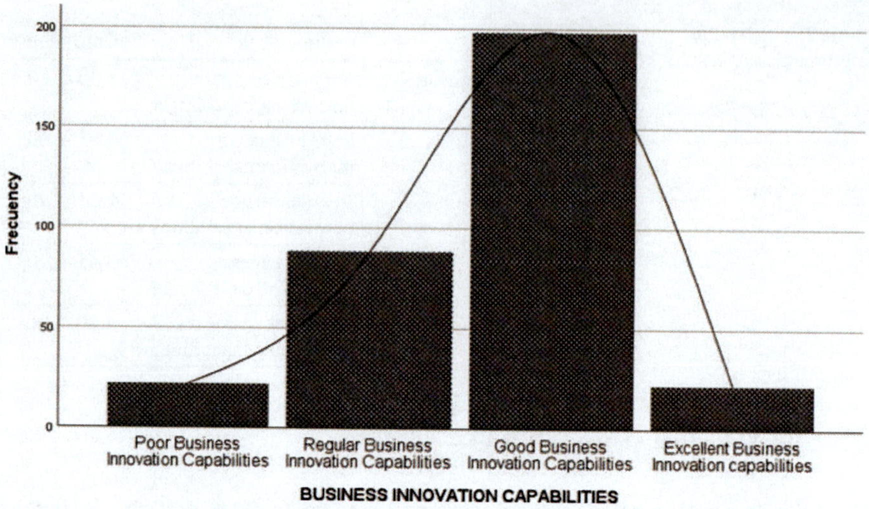

Fig. 1 Business innovation capabilities

Figure 1 shows that construction companies are 60.0% in excellent business innovative capabilities. These results coincide with the studies of Martínez and Maldonado [16], Velez and Afcha [20].

Figure 1 presents the behaviour as the business innovation capabilities of companies in the road infrastructure sector, which demonstrate the right business environment in the construction companies located in the main cities of the Colombian Caribbean region, reaffirmed in Fig. 2.

Specifically, the transport infrastructure as an indicator of the environment shows that managers consider 80% to agree on the minimum conditions for the operations of construction companies. This situation is consistent with the approaches of Valeriano Cárdenas, Ticona [18], who expresses that the requirements of the infrastructure contribute to the remuneration of the factors taking into account a timely allocation of resources.

As for the conditions of the essential public services offered in the areas where the offices of the construction companies are operating, and it was observed that 86.7% of the directors surveyed have a positive experience in terms of public services in the Colombian Caribbean industry.

On the other hand, the cost of the specialised labour in the road infrastructure sector, it is observed that 73.3% of managers consider that the prices are adequate to operate in the construction of road works, all this is consistent with the approaches of Torres González [24], Ortiz and Restrepo [25], who is considered that the allocation of labour costs is indispensable in determining the values of activities, processes, and products, thereby facilitating decision-making, planning, and administrative control. Therefore, we see the importance of costs to carry out the planning, development, and control of all those expenses or investments made in road construction companies.

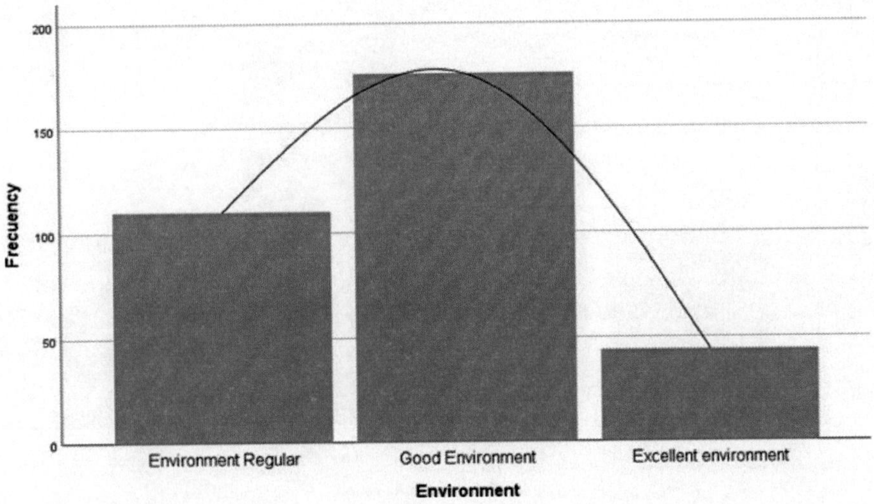

Fig. 2 Environment

It is to itself, and evidence that 93.3% of the managers of the construction companies stated that the road infrastructure sector and the companies that converge in it have the necessary technology derived from the different local universities or failing that the technical support. Environmental restrictions in the road infrastructure sector, it was observed that 64.3% of managers of construction companies say that there are sufficient restrictions to delay operational and administrative activities regarding permission to start construction work.

Therefore, the economic environment for the development of commercial operations is a wide discrepancy between the managers of companies engaged in road infrastructure. The articulation of government entities, 71.4% of the directors of the construction companies consider that the road infrastructure sector is covered by the strategic plan of the local, regional and national government, all in line with the National Development Plan (2018–2022), which tries to articulate the different entities, focusing on planning, organisation, and territorial development.

3.1.1 Business Innovation

In terms of business innovation, Fig. 3 shows an exemplary business innovation, which is a critical element of the innovation capabilities of road infrastructure companies, highlighting the advances in services, products, licences, or copyright and purchase of new equipment according to the modernisation of the environment.

Inside, it is evident in Fig. 3 that the directors of the companies of road infrastructure companies show in 66.7% that they frequently present innovations in terms of service, all in line with the approaches of INVÍAS, states that the sector, in general,

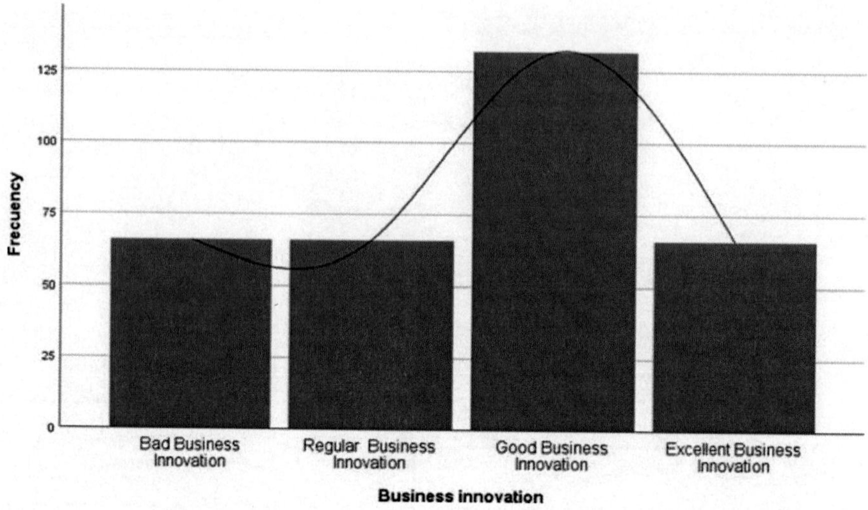

Fig. 3 Business innovation

is undergoing a process of innovation through the orange economy, these aspects continue to be discussed through the Innovation and Sustainability Wheels of INVS.

About product innovation, it is evident that the directors of the construction companies manifest in 73.4% that they have participated in the processes of improvement of the products generated from their commercial activities, all in line with the strategies of INVÍAS, who highlight the importance of launching new technologies according to the needs of the country in the different modes of transport and its infrastructure.

As well, the innovation of the products, it is evident that the directors of the construction companies manifest in 73.4% that they have participated in the processes of improvement of the products generated from their commercial activities, all in line with the strategies of INVÍAS, who highlight the importance of launching new technologies according to the needs of the country in the different modes of transport and its infrastructure.

3.1.2 Impacts and Innovation Activities

The impact of innovation within organisations is conceived, such as the game rules, including laws and regulations, socio-cultural patterns, technical standards, and routing patterns. Consequently, the product of these relationships between organisations and institutions directly impacts the company as a waste agent of the system, influencing its internal innovation capabilities. In this sense, the following results of this study are shown in Fig. 4.

In Fig. 4, we can observe the high participation of the activities and impacts of innovation in the companies consulted, detailed in the following sections. In detail,

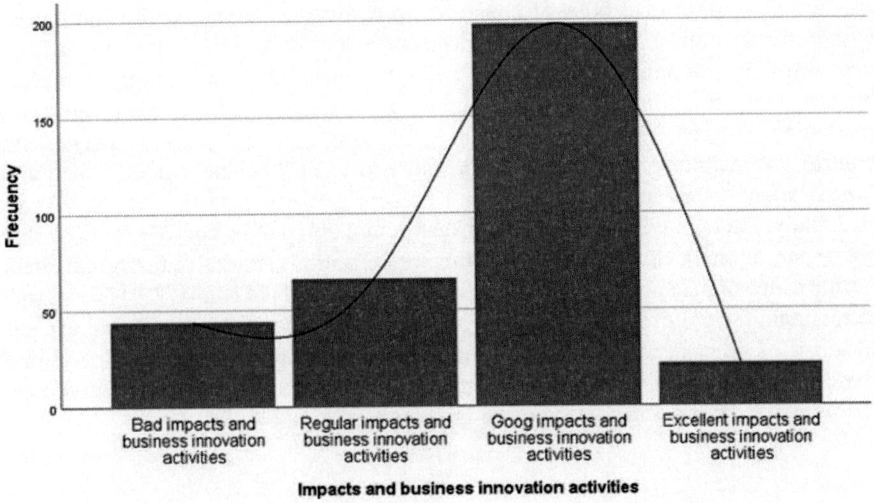

Fig. 4 Impacts and business innovation activities

when designing and implementing a new method of productive development from an innovative approach, each company or institution requires patenting it to prevent competition from stealing its copyright. Thus, registering its trademark protects the initiative of its activities. Likewise, only expressions, but not ideas, procedures, methods of operation, or mathematical concepts themselves may or may not cover elements such as titles, slogans, or logos, depending on whether the paternity of the work is sufficient.

On the other hand, it is noted that there are discrepancies between the directors of the construction companies, who concentrate their opinions at 35.7% on being undecided about the submission of licences or copyright. However, it is necessary to comment that national copyright law (DNDA) reports considering a slight increase in in-licences or copyright, arriving in 2018 with 78,347 new registrations.

It is also noted that the directors of the construction companies present a disjunction regarding the replacement programmes maintenance of purchase of new equipment according to the modernisation. These results coincide with the studies of Duarte Cabarique [26], who states that Latin American organisations present in 85% of problems of updating machinery and other technological components, meriting as the primary cause, the mismanagement of resources for amortisation and costs for obsolescence inherent in operations and the wear and tear of them.

It is also noted that 60% of the directors of construction companies frequently modernise machinery, hardware, and computer programmes to improve organisational and operational processes, all in line with the studies of Munar [27]. They argue that organisations must remain in a continuous change and review of the different organisational processes to detect failures and minimise response times that improve business performance.

Currently, more than 60% of construction companies are using the centralised system for the control of construction projects called SINCO ERP (Integrated Software for Construction Management), which has enabled them to improve performance in construction project management, accounting, and finance management, marketing and real estate management, marketing and real estate management, control and maintenance of machinery and equipment, business intelligence and human talent management.

Finally, the managers of the road infrastructure companies consider at 60% that the organisational culture that reigns in their organisations is suitable for the excellent performance of its commercial activities, all by the studies of Munar [27] is essential to maintain the culture organisation, which not only generates motivational aspects in terms of identification, appropriation, and sense of belonging, but also allows to detect problems in the working groups, or form teams with a positive and clear ideology for the proper functioning of the company.

4 Conclusion

Once the analysis and interpretation of the results have been completed, the following conclusions were generated.

About the environment for business innovation capabilities in companies in the road infrastructure sector in the Colombian Caribbean, it is concluded that there is a significant influence of the environment in the face of business innovation capabilities, highlighting the existence of adequate transport infrastructure, the labour costs are fair, the presence of at least one Higher Education Institutions with academic programmes related to the construction sector is recognised, highlighting the active participation of architecture and engineering programmes in the processes of improvement of the routine operations of construction companies. It is important to note that the national, regional, and local government have clearly defined within their strategic plans the improvement of road infrastructure, which favours the market of that construction sector. However, there is a clear disarticulation among government authorities regarding following a single route to concentrate decentralised resources in road infrastructure.

On the other hand, there is no evidence of clustering processes related to construction. However, consortia practices conformation as strategic alliances to generate competitiveness factors that allow participation in public works procurement processes before the Colombian State. It should also be noted that there were economic aspects that create uncertainty and instability in the development of the sector, which has led to a slow pace of GDP growth. Regarding the legal framework in which the road infrastructure construction sector is developed, there is a lack of support from state institutions that stimulate commercial activities, highlighting the lack of encouragement by the government in terms of local taxation, research, and development.

About the innovation processes of the goods and services of companies in the road infrastructure sector in the department of Córdoba, it is concluded that there is a high presence of the elements of innovation of goods and services in the capacities of business innovation, highlighting that companies in the road infrastructure sector frequently present innovations in their products and services. On the other hand, there are a series of needs in developing licences or copyright and deficiencies in the replacement programme maintenance and purchase of new equipment according to the modernisation of the environment.

Concerning the description of companies' impacts and innovation activities in the road infrastructure sector in the Colombian Caribbean region, it is concluded that there is a high presence of innovation impacts and activities on business innovation capabilities. Therefore, companies in the road infrastructure sector are frequently modernised machinery, hardware, and software and have an adequate organisational structure to meet the demand for road infrastructure construction. On the other hand, certain nonconformities are in terms of administrative processes, marketing, human and operational talent, which are not standardised through manuals of norms and procedures, which often hinders the fluidity of managerial procedures indirectly affecting the factors of organisational culture for organisational development.

Finally, about the analysis of the innovation capabilities of companies in the road infrastructure sector in the Colombian Caribbean region, it is concluded that, despite describing appropriate situations regarding business innovation capabilities for the development of the commercial operations of road infrastructure construction companies, there are gaps between the articulation of Colombian state entities, the university, and the company.

References

1. González-Díaz, R.R., Ledesma, K.N.F.: Cultura organizacional y Sustentabilidad empresarial en las Pymes durante crisis periodos de confinamiento social. CIID J. **1**(1), 28–41 (2020)
2. Ledesma, K.N.F., González-Díaz, R.R., Contreras, E.E.L.: Estrategia alternativa para la resolución de conflictos en las redes sociales digitales. CIID J. **1**(1), 103–113 (2020)
3. Hernández-Julio, Y.F., Hernández, H.M., Guzmán, J.D.C., Nieto-Bernal, W., Díaz, R.R.G., Ferraz, P.P., (eds.): Fuzzy knowledge discovery and decision-making through clustering and dynamic tables: application in medicine. In: International Conference on Information Technology & Systems. Springer (2019)
4. Regalado, V.B., López, N.T.M., González-Díaz, R.R., Polo, E.A.S.: Innovación y desempeño económico en México. CIID J. **1**(1), 80–102 (2020)
5. Leonard, Y.G., Rodríguez, R.A.S.: La teoría de los recursos y capacidades como fundamento metodológico para el estudio de la gestión de la innovación empresarial. Rev. Científica Ecociencia **7**, 1–15 (2020)
6. Morales Sánchez, M.A., Díaz Rodríguez, H.E.: Determinantes de las capacidades de innovación en el sector biotecnológico en México. Investigacióneconómica **78**(307), 90–118 (2019)
7. Ortiz, A.B., Salgado, M.C., Espinosa, N.V.: Gerencia de la innovación empresarial: U. Externado de Colombia (2020)
8. Ledesma, K.N.F., González-Díaz, R.R.: Etica sindical en la negociación de contrato colectivo de construcción de obras públicas. Caso Ducolsa. CIID J. **1**(1), 01–14 (2020)

9. Díaz, R.G.: EXPRO as a Schizoaffective Stimulator through Experiential Marketing in Nonprofit Organisations (2020)
10. Raynaud, N.C., Mogrovejo, J.M.: El desarrollo y financiación de la infraestructura en Colombia. Gestión Desarrollo Libre **3**(5), (2020)
11. Sánchez Castiblanco, K.N., Cáceres Correa, E.N., Rodríguez Mancipe, C.R.: ¿Qué le aporta Colombia a la OCDE y viceversa? (2020)
12. Arboleda, M.L.O., Díaz, R.R.G.: El carácter axiológico del proceso educativo en Colombia. Centros: Rev. Científica Univ. **6**(2), 1–17 (2017)
13. González-Díaz, R.R., Lara, R.J.V., López, R.O., Hernández-Royett, J.: Tax on advertising and commercial advertising: an analysis from municipal tax management. Globalciencia **2**(1), 20–34 (2016)
14. Díaz, R.R.G., Gutiérrez, P.M.H., Araujo, E.J.R.S., (eds.): Zonas francas de la región Caribe colombiana, ventaja competitiva y costos logísticos. Memorias del III Congreso internacional en administración de negocios internacionales (CIANI): Retos y oportunidades del desarrollo sostenible en los negocios internacionales. Universidad Pontificia Bolivariana (2019)
15. Pérez, C.M.: Innovación empresarial al servicio de la micro y pequeña empresa nortesantandereana: por la competitividad regional. Económicas Cuc **40**(1), (2019)
16. Martínez, S.V.H., Maldonado, L.G., Casallas, Md.P.J., Ojeda, R.M.G.: Capacidad dinámica de innovación como apuesta competitiva de valor empresarial. Gestión Desarrollo Libre **3**(5), (2020)
17. Celis Olave, W.D., Duarte Velásquez, S.J., Villabona Santos, J.P.: Estudio para la Mejora del Sector educativo en su entorno empresarial del Municipio de Barrancabermeja (2020)
18. Valeriano Cárdenas, J.A., Ticona, V., Muruxia, R.: La planeación estratégica y su impacto en el entorno empresarial de las MYPES exportadores de orégano-provincia Tacna, región Tacna año 2017 (2019)
19. González-Díaz, R.R., Cruz-Ayala, K.: Contraloría financiera en la contratación pública. Una revisión de los contratos de obras públicas del estado Venezolano. Inquietud Empresarial **20**(1), 43–58 (2020)
20. Velez, C.I., Afcha, S.M., Bustamante, M.A.: Cooperación universidad-empresa y su efecto sobre el desempeño innovador empresarial. Informacióntecnológica **30**(1), 159–168 (2019)
21. Wang, C., Hu, Q.: Knowledge sharing in supply chain networks: effects of collaborative innovation activities and capability on innovation performance. Technovation **94**, 102010 (2020)
22. Khotamov, I., Avazov, N.I.: Digital innovation activities today and their types. Архивнаучныхисследований 11 (2020)
23. Grama-Vigouroux, S., Saidi, S., Berthinier-Poncet, A., Vanhaverbeke, W., Madanamoothoo, A.: From closed to open: a comparative stakeholder approach for developing open innovation activities in SMEs. J. Bus. Res. (2019)
24. Torres González, J.: Colombia: logros en dos décadas de modelo de desarrollo aperturista-Analisis según resultados de balanza comercial (2020)
25. Ortiz, C.H., Restrepo, D.M.J., Cruz, G.N.: El impacto de la infraestructura en el crecimiento económico colombiano: un enfoque smithiano. Lecturas Economía **90**, 97–126 (2019)
26. Duarte Cabarique, W.: Límites de la propiedad intelectual frente a la monopolización de medicamentos en Colombia. Universidad de la Costa (2020)
27. Munar, L.M.: Análisis de la propiedad industrial en Colombia: un análisis de la decisión andina 486 del año 2000. Inciso **22**(1), 102–114 (2020)

Smart Agricultural System Using IoT

Md. Faridul Islam Suny, Tania Khatun, Zahura Zaman, Md. Monjourur Roshed Fahim, Md. Ariful Islam, Rashida Jesmin, and Tajim Md. Niamat Ullah Akhund

Abstract In today's people's existence, the perception of IoT is soaring. No exception is agriculture. Their pretty substantial excellence is already evident. The relevance of IoT in agriculture is also beginning to emerge. We have displayed a framework of auto irrigation, pest detection, and notification system incorporating the utilization of IoT in agriculture in this paper. The implementation of this framework by using different types of sensors and IoT devices is pretty optimistic and would make a major contribution to agricultural growth.

Keywords Internet of things (IoT) · Determining the land condition · Auto irrigation · Pest identification · Remote monitoring

1 Introduction

The Internet of Things (IoT) encompasses a range of intertwined, Internet-connected abstractions which, without user intercession, are able to retrieve and transmit data over a wireless network [1]. The intensifying adverse impact of IoT is visible in different fields. No exception is agriculture. Agriculture is the root of economic extension for our nation. It is the agriculture which assists the developed countries to survive on. Science has made the unfeasible to become feasible where IoT has been a massive part of it from fewer decades. IoT has given uncountable consolation to living being with its application in different sectors where the implementation in agricultural sector is one of the most dominant one. In the context of today's world, the stronger the economy of a country, the more influential it is to influence others and being able to develop quickly in the other fields. According to our agriculture-dependent economy, if we can involve the implementation of IoT in our cultivation

Md. F. I. Suny · T. Khatun · Z. Zaman · Md. M. R. Fahim · Md. A. Islam ·
T. Md. N. U. Akhund (✉)
Department of Computer Science and Engineering, Daffodil International University, Dhaka, Bangladesh

R. Jesmin
Department of Zoology, Government Hajee Korop Ali Memorial College, Sirajganj, Bangladesh

© The Author(s), under exclusive license to Springer Nature Singapore Pte Ltd. 2022

A. K. Nagar et al. (eds.), *Intelligent Sustainable Systems*, Lecture Notes in Networks and Systems 333, https://doi.org/10.1007/978-981-16-6309-3_8

or simply if we can combine the IoT and farming sector, then surely our economy will develop in a rushing manner. Building a smart agricultural system based on IoT will be able to act as a major regulator in the economy. A smart agricultural system is a system that builds an automated and transparent agricultural system through a networking system using various sensors, software, and smart devices. Involvement of IoT in our husbandry can be an aid to prevent any kind of harmful effect over our agriculture even before it occurs. So, if we can detect destruction earlier than it occurs, then we will be able to take the production of our crops to great extent. In this manner, the production cost will be less, and so IOT application over farming will be really methodical. In addition, agriculture is a means of meeting the growing needs of the growing population. In addition, agriculture is a means of meeting the growing food needs of the growing population. So, from now on the development of sustainable agricultural system has become very foremost. Smartly advanced data collection, analysis and processing through IoT integration in agriculture are able to make farmer decision making much uncomplicated. In this way, if the farmers are up to make more precise determination about their crop or cultivable land, then it is also achievable to enlarge the productivity remarkably.

2 First Literature Review

By 2050, that means in next 30 years, it is predicted that the number of world's population will reach from 7.8 to 9.7 billion [2]. The food demand of this growing population will increase exponentially. According to the survey of United Nations—Food and Agriculture Organizations, agricultural production will need to increase by 70% from now to meet the needs of the population by 2050 [3]. Besides, agriculture is the center of the global economy. 60% of the world's population is directly or indirectly dependent on agriculture for their livelihood [4]. In addition, considering Bangladesh as one of the most agricultural countries in the world, it is seen that 48.4% of the people here are engaged in agriculture, 70.1% of the total land area of the country is used for agriculture, and the contribution of agriculture is 16.5% of the total GDP of Bangladesh [5]. The maximum nitrate ratio in drinking water should not exceed 44.2 ppm according to the US Environmental Protection Agency (EPA) [6]. We can control this amount through proper irrigation with the help of various sensors, IoT technology and wireless sensor network (WSN) under auto irrigation system [6], while increasing the fertility of arable land, another problem that is agriculture production is inversely tormented by pest infestation and plant diseases. In this case, using smart technology and devices, using photograph processing techniques, if the crop image can be processed and divided into pixel by pixel of pest and leaves, the disease can be identified and taken action before it reaches a critical stage, and the damage can be greatly reduced. So, a process named support vector machine (SVM) is used where shade function is needed to teach the process to classify pest pixels and the leaf pixels also [7]. Properly using robots in agriculture, the care system is able to save 99.9% of the cost of fertilizer on land, which is an advantage of IoT [8]. Big data, cloud, mobile computing, and Wi-Fi/3G/4G are currently playing an important role in the advancement of IoT technology [9, 10]. Therefore, if smart IoT technology, i.e., robots, sensors, or WSN, and remote control devices are being used in agriculture, mobile applications will be integrated and notification will be provided to the farmers so that they will be able to build a strong agricultural system with virtual and real combination [2, 11]. Some substantial information that we have accumulated from our existing literature is shown in the following table (Table 1). Electronic voting [12] and biometric EVM [13] are using IoT-based systems. IoT-based systems are

Table 1 Some recent works and their lacking

Recent works	Lacking
Utilizing Internet of Things from the data analysis, e.g., sprinkler water flow, humidity, and temperature, and soil moisture, evaluation, and preventive movement can be taken [22]	There is rhetoric of land surveillance and notifying farmers or those engaged in agriculture, but no automated solution is assigned
By observing the temperature and moisture of the soil, to arrange auto irrigation and to notify the farmer via the GSM module [2]	Although it was said that by detecting the temperature and humidity of the land and informing the farmer via the GSM module, the auto irrigation system was provided, and nothing was said to display the measurement of the current humidity, temperature, and humidity of the land
Facilitating automatic irrigation by detecting temperature, humidity, and soil moisture and notifying the movement of potentially harmful animals via SMS [10]	Direct current temperature, humidity, or soil moisture is not specified. Notifications with pictures of dangerous insects or animals are also not listed
Ascertaining the soil's temperature and humidity and procuring the farmer with MMS by capturing images [3]	Soil moisture of the land has not been determined, and the issue of the auto irrigation system in the land has not been observed as required
Temperature, soil level, and humidity are checked in graphical portrayal in the IoT ThingSpeak web page [23]	Use of site page or web page is here in this paper where farmers need to login with email ID, but according to Bangladesh's context, rural farmers might not have proper knowledge of using an application and also might not have email address
Using PIR and acoustic sensors, pest identification in the field [24]	While its pest identification method is explained here, but how the farmer will be informed after identification was not clear
Remote monitoring system [23]	Only the temperature, soil level, and humidity are found out and informed about in the application, but no pest infection is determined
Ascertaining the soil's temperature and humidity [3]	Soil moisture of the land has not been determined, and the issue of the auto irrigation system in the land has not been observed as required

helping Alzheimer's patients also [14]. IoT-based robots are serving in restaurants [15]. Smart farms are doing well with the help of IoT [16]. COVID-19 virus-affected patients [17] are getting help from IoT [18]. Paddy cultivations [19], nurse robots [20], and remote sensing [21] robots are working with IoT.

3 Contribution

We have summarized some of the works in Table 1, and they are lacking in light of our existing literature. Within that scenario, via the work of this study, we are contributing to inspecting the soil across the measurement of temperature, humidity, and moisture, while at the same point, we have illuminated the auto irrigation system as a solvent to land droughts. Not just that, we have used LCD to simulate the current temperature, humidity, and moisture in the soil and have made arrangements to provide the farmer with timely updates by SMS from time to time using the GSM module to provide a proper land update. Not just that, we include labeled system to identify the pests in the farmland and urgently alerting the farmers via SMS.

4 Proposed Methodology

Different types of sensors, and perhaps other internals, have been used to construct a strategically functioning and effective smart agriculture system in the equivalent circuit (Fig. 1) underneath portraying our IOT-based smart agriculture system. Soil moisture sensor, DHT11 sensor/temperature, humidity detection sensor, sound sensor (acoustic sensor), and PIR sensor are critiques of sensors that can be used to strengthen our framework. Aside from the sensors, the DC motor, GSM module (900A), Arduino UNO, breadboard, and LCD screen are all crucial parts in the overall system.

Soil Moisture Sensor: Soil moisture sensors are extraordinarily beneficial in determining water degrees, considerably facilitating farmers' efforts and lowering costs. A soil moisture sensor empowers agriculturalists to estimate the water degrees or levels without the want to be physically present in the discipline.

Soil Moisture Sensor Functionality: The soil moisture sensor is extremely easy to use. The fork-shaped probe, which has two exposed conductors, serves as a resistor (analogous to a potentiometer) whose resistance varies in relation to the amount of water in the soil. It is more of an inversely proportional relationship than anything else.

1. As much water in the soil implies greater conductivity, which can lead to a reduction in resistance.

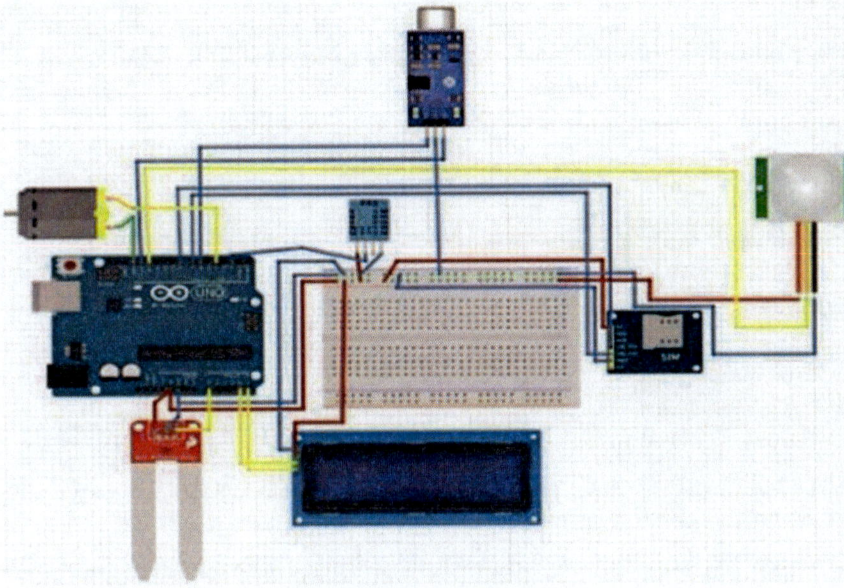

Fig. 1 Circuit diagram of IOT-based smart agriculture system

2. Even if there is little water in the soil, it has low conductivity which can result in higher resistance.

The soil moisture sensor produces an output voltage per resistance that by mensuration we are able to verify the moisture or dryness level.

DHT11 Sensor/Temperature and Humidity Detection Sensor: Humidity refers to the volume of water vapor released into the air. The proportion of humidity in the air has an influence on the amount of chemical and organic practices. Quite apart from agriculturists, it is important to keep up to date on temperature shifts. The DHT11 is a low-cost digital sensor that assesses temperature and humidity. A haptic humidity sensing module and a thermistor for temperature sensing are included in the DHT11 sensor.

Sound Sensor: A sound sensor is a module that detects seismic waves based on their amplitude and incorporates them to electric impulses. In the sense of our smart agriculture device, farmers use this sensor to grasp pest noise, as well as insects and other animals' moving noises, so that immediate action can be taken. The electric signal is sent to the farmer as a caution of pest invasion within the agricultural land after the noise is intercepted.

PIR Sensor: PIR sensors sense activity and can be used to assess if an animal, pest, or even a pedestrian has reached or escaped the human range. It sends a warning or notification to the farmer about the danger as soon as it senses motion.

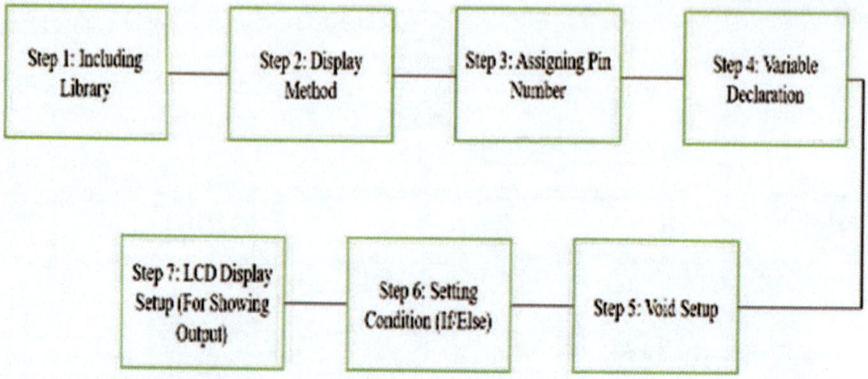

Fig. 2 Steps to implement smart agricultural system using IoT

GSM Module (SIM 900A): The SIM900A is a GSM module that is pretty widely used in mobile phones. This module alerts users or messages to farmers.

DC Motor: DC motors are the most basic motors that can be used as a water pump. This type of machine is used in household appliances as well as industrial settings. This motor was used as a water pump with 5 V power in this smart agricultural device.

LCD Screen: The LCD monitor is a flat panel display that shows the current temperature, humidity, and water level.

To establish the interface, and perhaps even read the input and output, a breadboard and an Arduino UNO are used. These are the total ingredients we used to expand our smart agriculture device circuit diagram (Fig. 1). Afterward we initiated a few steps (Fig. 2) to recite the Arduino code we anticipated. Temperature, moisture, and humidity sensors work to find out the amount of soil temperature, moisture, and humidity and to send that data to Arduino and let it know if the land needs water. Accordingly, Arduino water pumps off/on, and the LCD shows the amount of temperature, humidity, soil moisture and acoustic value. To monitor temperature and moisture, we utilized DHT11 sensor. All of this work requires the combination of the Arduino with the sensors and the LCD with the Arduino. GSM module will be connected also. It will also be able to trace pest mobility thru passive infrared (PIR) sensors in colloidal conditions and to detect insect infestation using acoustic sensor by monitoring noise level and to alert farmer through SMS with the aid of the GSM module. And to reach the destination adequately, the Arduino code must be in existence. Via the flowchart (Fig. 3), we are revealing the entire system. In the following diagram (Fig. 4), the execution of the entire system is shown. Here the power supply has acted as a source of energy. In other words, the power supply has supplied energy to the system. And the temperature sensor, humidity sensor, and the moisture sensor provide information to Arduino by measuring the temperature and humidity of the land, respectively. Arduino provides instructions accordingly.

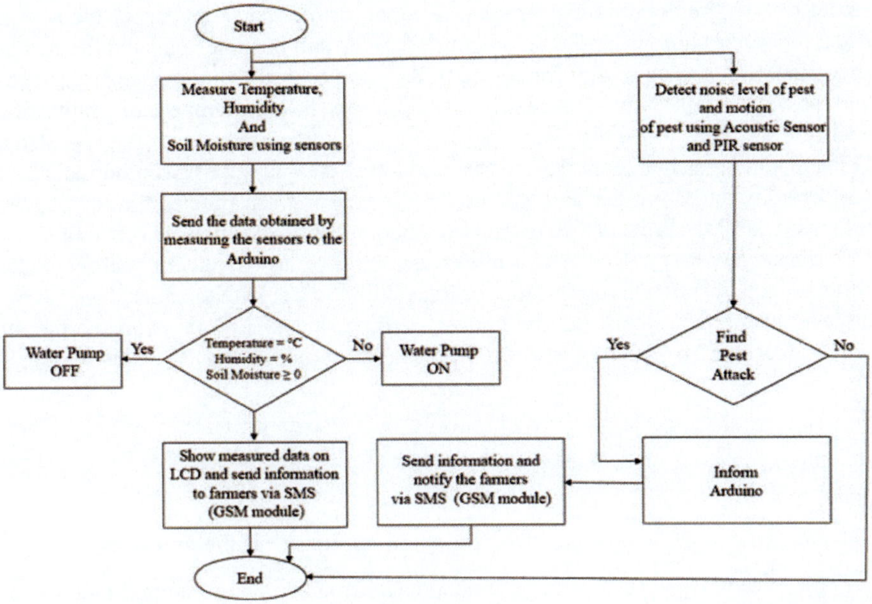

Fig. 3 Flowchart for smart agricultural system using IoT

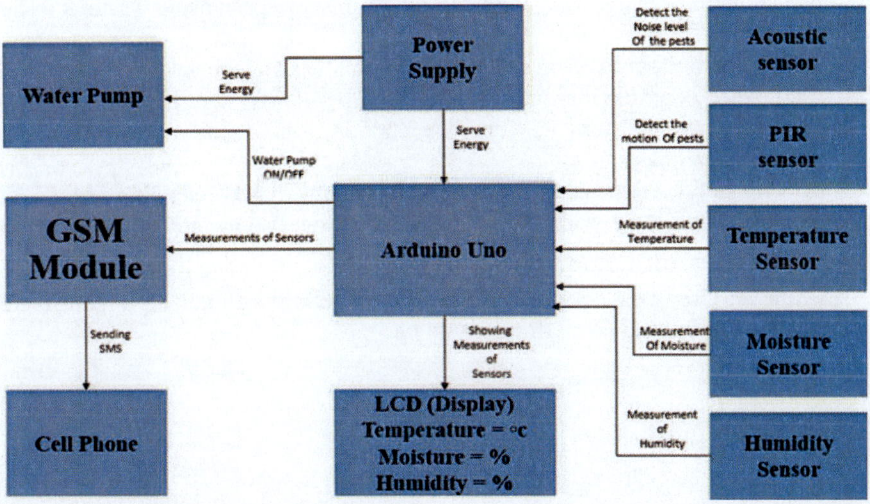

Fig. 4 Block diagram for smart agricultural system using IoT

According to the condition of temperature, moisture, and humidity, if the soil is drier, the water pump is turned on. Again, when the soil becomes moist at the end of irrigation, the sensors send information to Arduino according to their measurement and the water pump shuts off. And the LCD's job is to show the temperature, humidity, and humidity sensors on the display according to their measurement. There is also a GSM module attached to the device that will know the amount of Arduino moisture, temperature, and humidity and will notify the farmer from time to time about the intensity via SMS. We had also listed two others sensors: the acoustic sensor and the PIR sensor. Where the PIR sensor tracks the activity and movement of insects, the acoustic sensor is able to discern insect infestation by tracking the noise level. Even after pest outbreak is noticed, the updates will be sent to Arduino, and the farmer will be reassured by sending an SMS with the aid of the GSM module from there.

5 Advantages

Several of the notable benefits of our proposed approach are the following:

1. Farmers may get the privilege to obtain accustomed to new emerging technologies through our simulation.
2. A large amount of waste can be resisted by this smart farming mechanism.
3. Monopolizing the auto irrigation system, for instance, can avoid waste of water and reduce the use of excess energy.
4. This will empower the firm remotely to be analyzed adequately.
5. And, since coordination and an adequate irrigation system are the prerequisites for decent farming systems, we hope that this method will boost crop productivity.
6. As they notice the pests in the crop or field immediately, enabling farmers to respond accordingly instantaneously, we hope that this method will boost crop productivity.

Beyond all else, we agree that digital agriculture will be inclined to contribute significantly since it is a cost-effective system.

6 Conclusion

In IoT-based smart farming, a system is typically fitted in the paper to track the crop field with the aid of different sensors (soil moisture, humidity, temperature, PIR, and acoustic) and to automate the mechanism of irrigation. Using advanced technology, farmers can display the conditions of the sector remotely and with the guidance of the GSM module, via SMS, enabling farmers to carry out the responsibilities of the land. We have set out to establish the IoT-based farming method in the form of gear via the proposed scheme in this paper. But one of the drawbacks of the proposed model in this paper, however, is that no work on weather forecasting or weather warning has been done. For instance, while the auto irrigation system is structurally sound, the elevated water will cause damage to the ecosystem if it rains heavily after proper water irrigation. Ergo, work can be done in the future for this model to ensure that the irrigation system is accomplished as per the weather forecast.

References

1. Ali, Z.H., Ali, H.A., Badawy, M.M.: Internet of things (IoT): definitions, challenges and recent research directions. Int. J. Comput. Appl. **128**(1), 37–47 (2015)
2. Suresh, P., Koteeswaran, S.: An effective novel IOT framework for water irrigation system in smart precision agriculture. Int. J. Innov. Technol. Exploring Eng. (IJITEE) **8**(6), (2019). ISSN: 2278-3075
3. Prathibha, S.R., Anupama, H., Jyothi, M.P.: IOT based monitoring system in smart agriculture. In: 2017 International Conference on Recent Advances in Electronics and Communication Technology (2017)
4. Economy/Gianluca Zavatta—27 October 2014, visited on 2 February 2021, 10:40 am
5. Food and Agriculture Organization of United Nations, visited on 30 January 2021, 11:35 am
6. Alahi, Md.E.E., Pereira-Ishak, N., Mukhopadhyay, S.C., Burkitt, L.: An internet-of-things enabled smart sensing system for nitrate monitoring. IEEE Internet of Things J. https://doi.org/10.1109/JIOT.2018.2809669
7. Rajan, P., Radhakrishnan, B., Padma Suresh, L.: Detection and classification of pests from crop images using support vector machine. In: 2016 International Conference on Emerging Technological Trends [ICETT]
8. Mat, I., Kassim, M.R.M., Harun, A.N., Yusoff, I.M.: Smart agriculture using internet of things. In: 2018 IEEE Conference on Open Systems (ICOS)
9. Nagaraja, G.S., Soppimath, A.B., Soumya, T., Abhinith, A.: IoT Based Smart Agriculture Management System. IEEE 978-1-7281-2619-7/19/$31.00
10. Sushanth, G., Sujatha, S.: IOT based smart agriculture system 978-1-5386-3624-4/18/$31.00
11. Bauer, J., Aschenbruck, N.: Design and implementation of an agricultural monitoring system for smart farming. In: 2018 IoT Vertical and Topical Summit on Agriculture—Tuscany (IOT Tuscany)
12. Sarker, M.M., Shah, M.A.I., Akhund, T.M.N.U., Uddin, M.S.: An approach of automated electronic voting management system for Bangladesh using biometric fingerprint. Int. J. Adv. Eng. Res. Sci. **3**(11), 236907 (2016)
13. Sarker, M.M., Akhund, T.M.N.U.: The roadmap to the electronic voting system development: a literature review. Int. J. Adv. Eng. Manage. Sci. **2**(5), 239465 (2016)
14. Akhund, T.M.N.U., Mahi, M.J.N., Tanvir, A.H., Mahmud, M., Kaiser, M.S.: ADEPTNESS: Alzheimer's disease patient management system using pervasive sensors-early prototype and preliminary results. In: International Conference on Brain Informatics, pp. 413–422. Springer, Cham (2018)
15. Akhund, T.M.N.U., Siddik, M.A.B., Hossain, M.R., Rahman, M.M., Newaz, N.T., Saifuzzaman, M.: IoT Waiter Bot: a low cost IoT based multi functioned robot for restaurants. In: 2020 8th International Conference on Reliability, Infocom Technologies and Optimization (Trends and Future Directions) (ICRITO), pp. 1174–1178. IEEE (2020)
16. Akhund, T.M.N.U., Snigdha, S.R., Reza, M.S., Newaz, N.T., Saifuzzaman, M., Rashel, M.R.: Self-powered IoT-based design for multi-purpose smart poultry farm. In International Conference on Information and Communication Technology for Intelligent Systems, pp. 43–51. Springer, Singapore (2020)

17. Akhund, T.M.N.U., Jyoty, W.B., Siddik, M.A.B., Newaz, N.T., Al Wahid, S.A., Sarker, M.M.: IoT based low-cost robotic agent design for disabled and Covid-19 virus affected people. In: 2020 Fourth World Conference on Smart Trends in Systems, Security and Sustainability (WorldS4), pp. 23–26. IEEE (2020)
18. Akhund, T.M.N.U., Newaz, N.T., Hossain, M.R.: Low-Cost Remote Sensing IoT based Smartphone Controlled Robot for Virus Affected People (2020)
19. Hasan, M.A., Akhund, T.M.N.U.: An approach to Create IOT based Automated Smart Farming System for Paddy Cultivation
20. Akhund, T.M.N.U.: Study and Implementation of Multi-Purpose IoT Nurse-BoT
21. Akhund, T.M.N.U., Sagar, I.A., Sarker, M.M.: Remote Temperature Sensing Line Following Robot with Bluetooth Data Sending Capability
22. Patokar, A.M., Gohokar, V.V.: Precision agriculture system design using wireless sensor network. In: Information and Communication Technology, pp. 169–177. Springer, Singapore (2018)
23. Anusha, A., et al.: A model for smart agriculture using IOT. Int. J. Innov. Technol. Exploring Eng. (IJITEE) 8(6), (2019)
24. Saranya, K., et al.: IoT based pest controlling system for smart agriculture. In: 2019 International Conference on Communication and Electronics Systems (ICCES). IEEE (2019)

IoT-Based Low-Cost Automated Irrigation System for Smart Farming

Tajim Md. Niamat Ullah Akhund, Nishat Tasnim Newaz, Zahura Zaman, Atia Sultana, Alistair Barros, and Md Whaiduzzaman

Abstract In this research, we present a low-cost intelligent irrigation system for farming. Nowadays, farming is shifted to automated and remote monitoring and management systems integrated with Cloud, Fog, and IoT networks. Our developed prototype can measure water level, temperature, and humidity with a hardware sensor and micro-controller unit. We use different sensors to take different readings and values to decide to turn on or off the motor. We provide essential algorithm and flowchart to explain the system. Our IoT-application-based automated developed system makes an automated system for irrigation and provides notification via mobile SMS to inform us of the irrigation field's details remotely.

Keywords IoT · Smart farming · Smart irrigation · SMS notification · SIM Module

1 Introduction

Agriculture is still one of the main GDP growth and national number one activities in many countries. Nowadays significant amount of people depend on farming and agriculture worldwide Therefore, farming and agriculture have a strong impact on our food chain and economic growth and activities nationally and internationally. Thus, we are still largely depend on the farming and agriculture for essential food supply and trade. This work will result a system which will help farmers is agriculture. However, in many countries, worldwide agriculture and farming related tasks are not fully automated. Our farmers are suffering from various problem. Although they

Tajim Md. N. Akhund (✉) · Z. Zaman · A. Sultana
Department of CSE, Daffodil International University, Dhaka, Bangladesh

Tajim Md. N. Akhund · N. T. Newaz · M. Whaiduzzaman
Institute of Information Technology, Jahangirnagar University, Dhaka, Bangladesh

A. Barros · M. Whaiduzzaman
School of Information Systems, Queensland University of Technology, Brisbane, Australia

© The Author(s), under exclusive license to Springer Nature Singapore Pte Ltd. 2022

A. K. Nagar et al. (eds.), *Intelligent Sustainable Systems*, Lecture Notes in Networks and Systems 333, https://doi.org/10.1007/978-981-16-6309-3_9

are working hard but not getting desired outcome without automation. This scenario motivated us to make this research project. This work wants to meet the following objectives:

1. Collecting data of temperature, humidity and water level from farm and show those data to a monitor.
2. Giving notification via Mobile SMS.
3. Making intelligent decision and controlling the irrigation system based on the collected data.

This work will result in an IoT based system which will prevent electricity waste and water wastage. Farmers will be notified via SMS too. But Sensors only work within a specific area and sending SMS notification is not free of cost. Sensors and MCUs life time are also not very large. These are some challenges to implement this project. In the paper organization, first section is containing introduction and objectives of this work. The second section is containing the background study. Methodology and system model are mentioned in section three. The results and output are shown in the fourth section. The last section is containing the conclusion and future works.

2 Literature Review

Smart cultivation is the most important harvest in the sector of agriculture and with the help of this. Nowadays, thanks to the modernity of agriculture, farmers are able to grow rice thrice annually. Poultry farm automation [7], paddy cultivation [11], fruits cultivation [4], and hotel management [6] are doing well with the help of IoT. Remote sensing [3], remote Monitoring [5], virus affected people management [1], neuro-patient management [2] are blessed with IoT nowadays. Electronic voting [17] and bio-metric security [18] are also using IoT. We can collect data with robot from remote place by using IoT [18]. Farmers would be motivated to grow jute in order to expand the demand for these jute crops, revitalize the jute mills, raise prices, boost the advertising of jute [14]. In agricultural development, many new technologies are there in the cycle and with the improvement of these new technologies the smart farming system gone beyond description enhancement [6]. Agriculture and agricultural science are progressing at a rapid rate thanks to this IoT method [16]. To the point that there is no simpler way or adopt better of a product at a reduced amount, people still sought to do further experiments [20]. Around 75-77% of the population in Bangladesh are reliant on farming [20]. The new web creation is the IoT that assemble the usefulness of a machine like camera sensors [15], RFID smartphones, and wearable devices, etc., used by so many companies for their own need [22]. Agribusiness adds to the natural maintainability needed to blast crop yields and reduce costs [12]. Agri-cloud that increments rural creation and the accessibility of data related to bombed research activities, the effect of doing this could spare time and worth creation discussion less confounded and quicker [9]. The higher the populace in that nation, the higher the need to utilize the IoT framework [8]. Cloud based systems can make a secure [10] way to monitor various data and e-governance application [21]. IoT-based education can help people [19] with mobile cloud computing [23]. Cloud based can also trace virus community transmission [24].

3 Methodology and System Model

3.1 Requirements

We several hardware, and software part C++ programming language and Arduino IDE. We have used low-cost micro-controller unit Arduino UNO Revision 3. Moisture sensor and Water level sensors are used to sense the moister and Water level of the soil and send it to Arduino. DHT11 temperature and humidity sensor is used to sense the temperature and humidity. A LCD monitor, a SIM Module sim 800 is used to send the collected data and motor state to a mobile as notification. A relay module is used to control the high-voltage motor with only 5 volts provided from Arduino.

3.2 Algorithm

The system sense the data about temperature, humidity and water level from field. With the value, the system will automatically turn on and off the water pump. The system can show the data and notifications in a LCD monitor. If a condition fulfilled then the system can send SMS notification to a selected mobile phone. The algorithm is as follows:

Step 1: Power on.
Step 2: Temperature, humidity, water level data sensed.
Step 3: Show the data to a LCD monitor.
Step 4: Control the relay and motor by following the sensor data.
Step 5: if (water $<$ threshold value) Motor on.
else Motor off.
Step 6: Send SMS to a mobile number if provided condition is met.
Step 7: Stop.

3.3 System Model

The flow of system is discussed in Fig. 1. The system sense the data about temperature, humidity and water level from field. The system can show the data and notifications in a LCD monitor. With the value the system will automatically turn on and off the

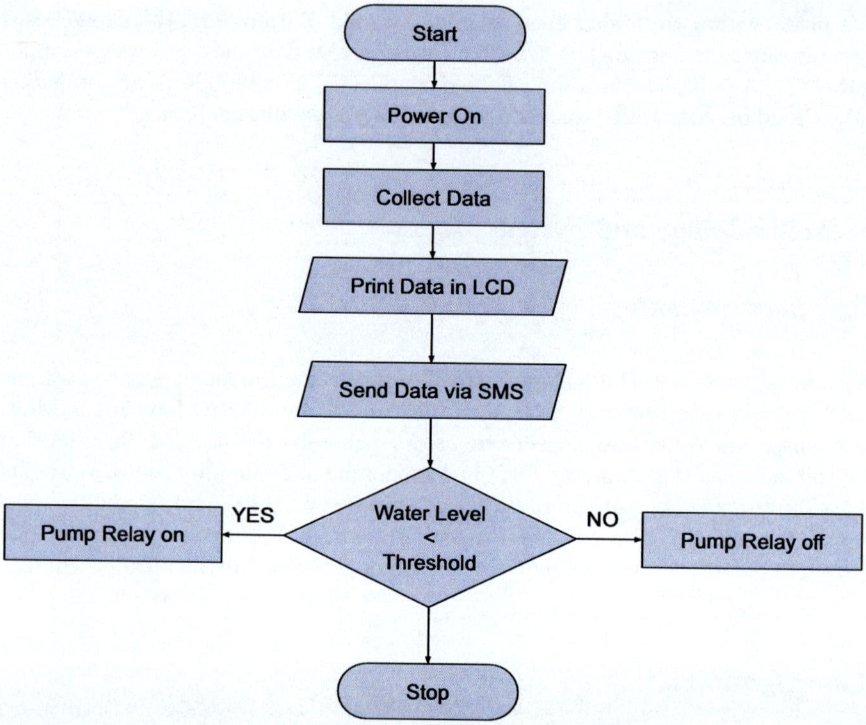

Fig. 1 System flow chart

water pump. The system sense the data with sensor for temperature, humidity and water level from field. With the value of the sensor the system will automatically turn of and off the water pump by following the algorithm. In a LCD monitor, the system can show the data and notifications.

3.4 Conceptual Circuit Diagram

The prototype 1 is made without SIM module is illustrated in Fig. 2. In this module, the project is made with SIM module SIM800L, Water Level Sensor, DHT11 Temperature and Humidity Sensor, Relay module, Arduino Micro-controller, LCD Module, and Motor. SIM module SIM800L, Water Level Sensor, DHT11 Temperature and Humidity Sensor, Relay module are connected with Arduino Micro-controller. Arduino sense the data and show the collected data and motor state in LCD Module. Arduino sends the data to relay module to on and off it. Relay module turn on and off the motor.

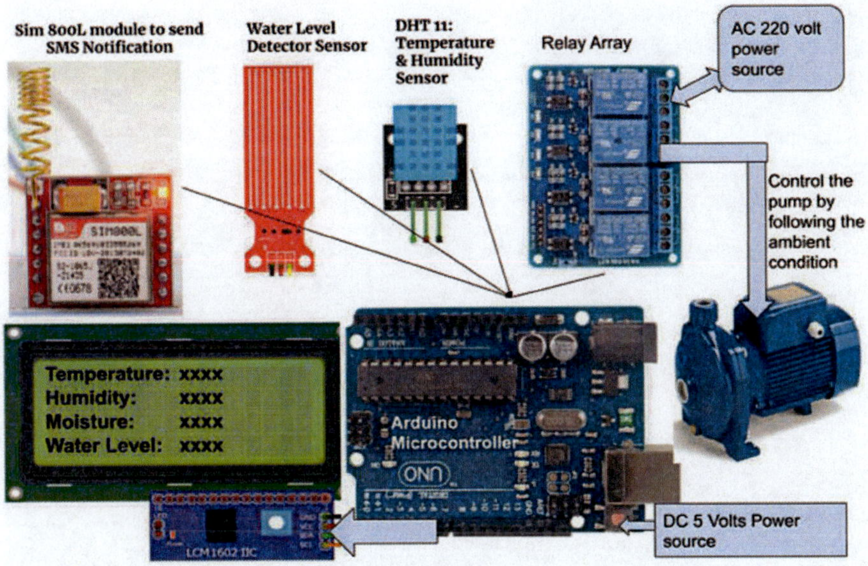

Fig. 2 Conceptual circuit diagram with SIM module

4 Results

In this section, we discuss the output of our designed prototype and its significant.

4.1 Prototype Output

The final prototype made for this project is illustrated in Fig. 3. Herc, we can see the 2 power supply were used. One is for MCU unit, and another is for running the Water pump. The SIM800L module is situated in the bottom side with a SIM. Water level sensor is inserted in a small bowl to test. The 20 × 4 LCD monitor is showing data and situated on the top. Relay module is situated beside the small pump. In prototype, we used small pump but in final real-time project we can use big field pumps. The MCU and Power supply and power converter is situated in the middle. An analog switch is also set upped to start on the condition for sending the message. Finally, it worked nicely. We got the SMS notification also. The SMS notification is shown in Fig. 4. The SMS notification notifies about the temperature, water level and humidity of field and also notify about the Pump state On or Off. It worked successfully.

Fig. 3 Final prototype with SIM module

4.2 Preliminary Results

From the result, we received successful output which are mentioned is Fig. 5. In this research project, we have tested our implemented prototype around 1000 times. The features obtained from this project's final prototype are as follows:

1. System can make Automatically motor start and off by following the threshold value of sensor data.
2. System can measure Moisture, Water level, Temperature, and Humidity.
3. Can Show the measured data on LCD.
4. The system can send Mobile SMS Notification about all state.

Our developed system costs us only around 20 USD with all of these features where as in marketplace amazon and others developed systems costs more than 50 USD without having the similar facilities such as SMS notification [13].

5 Conclusion

This work results a practical system for irrigation which is better than some other systems available in market. The developed system can monitor the water level, humidity and temperature of the field and not only show the data in a monitor but also sends the data to a mobile phone with SMS. Farmers do not need to use any smart

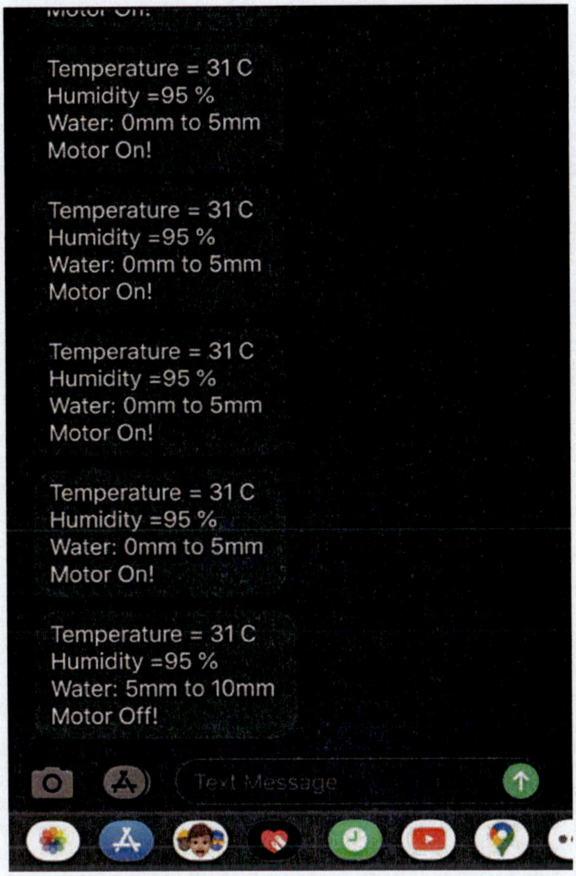

Fig. 4 SMS notification in mobile

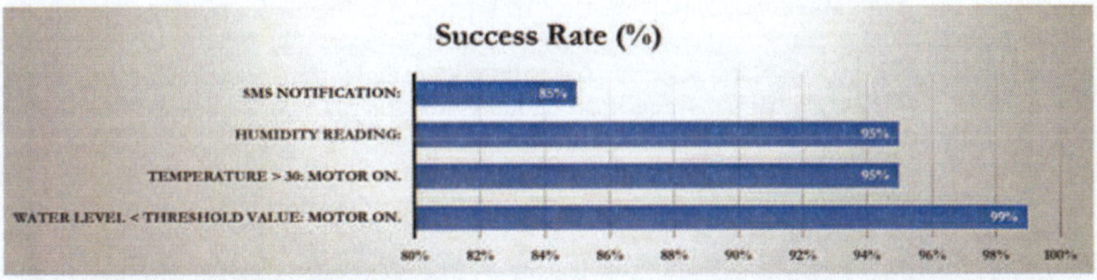

Fig. 5 Success rate of various features

phone or internet. They can monitor their field with any feature phone via mobile SMS. The system can control the water pump turning on and off automatically by considering the collected data and ambient environment. In future, this system will detect the problem in the irrigation and agricultural field with the help of machine learning and will be able to send data for predictive edge analytic to an integrated cloud IoT echo system.

Acknowledgements This work is partly supported through the Australian Research Council Discovery Project: DP190100314.

References

1. Akhund, T.M.N.U., Jyoty, W.B., Siddik, M.A.B., Newaz, N.T., Al Wahid, S.A., Sarker, M.M.: Iot based low-cost robotic agent design for disabled and covid-19 virus affected people. In: 2020 Fourth World Conference on Smart Trends in Systems, Security and Sustainability (WorldS4), pp. 23–26. IEEE (2020)
2. Akhund, T.M.N.U., Mahi, M.J.N., Tanvir, A.H., Mahmud, M., Kaiser, M.S.: Adeptness: Alzheimer's disease patient management system using pervasive sensors-early prototype and preliminary results. In: International Conference on Brain Informatics, pp. 413–422. Springer (2018)
3. Akhund, T.M.N.U., Newaz, N.T., Hossain, M.R.: Low-cost remote sensing IoT based smartphone controlled robot for virus affected people (2020)
4. Akhund, T.M.N.U., Rahman, M.H.: Bat banisher: an approach to create a high frequency ultrasound system to protect agricultural field from bats
5. Akhund, T.M.N.U., Sagar, I.A., Sarker, M.M.: Remote temperature sensing line following robot with bluetooth data sending capability
6. Akhund, T.M.N.U., Siddik, M.A.B., Hossain, M.R., Rahman, M.M., Newaz, N.T., Saifuzzaman, M.: IoT waiter bot: a low cost IoT based multi functioned robot for restaurants. In: 2020 8th International Conference on Reliability, Infocom Technologies and Optimization (Trends and Future Directions) (ICRITO), pp. 1174–1178. IEEE (2020)
7. Akhund, T.M.N.U., Snigdha, S.R., Reza, M.S., Newaz, N.T., Saifuzzaman, M., Rashel, M.R.: Self-powered IoT-based design for multi-purpose smart poultry farm. In: International Conference on Information and Communication Technology for Intelligent Systems, pp. 43–51. Springer (2020)
8. Ashifuddin Mondal, M., Rehena, Z.: IoT based intelligent agriculture field monitoring system. In: 2018 8th International Conference on Cloud Computing, Data Science & Engineering (Confluence), pp. 625–629. IEEE (2018)
9. Balamurugan, S., Divyabharathi, N., Jayashruthi, K., Bowiya, M., Shermy, R., Shanker, R.: Internet of agriculture: applying IoT to improve food and farming technology. In. Res. J. Eng. Technol. (IRJET) **3**(10), 713–719 (2016)
10. Farjana, N., Roy, S., Mahi, M.J.N., Whaiduzzaman, M.: An identity-based encryption scheme for data security in fog computing. In: Proceedings of International Joint Conference on computational intelligence, pp. 215–226. Springer (2020)
11. Hasan, M.A., Akhund, T.M.N.U.: An approach to create IoT based automated smart farming system for paddy cultivation
12. Kamienski, C., Soininen, J.P., Taumberger, M., Dantas, R., Toscano, A., Salmon Cinotti, T., Filev Maia, R., Torre Neto, A.: Smart water management platform: IoT-based precision irrigation for agriculture. Sensors **19**(2), 276 (2019)

13. Ortiz, D., Litvin, A.G., Salas Fernandez, M.G.: A cost-effective and customizable automated irrigation system for precise high-throughput phenotyping in drought stress studies. PloS one **13**(6), e0198546 (2018)
14. Raj, S., Sehrawet, S., Patwari, N., Sathiya, K.C.: IoT based model of automated agricultural system in india. In: 2019 3rd International Conference on Trends in Electronics and Informatics (ICOEI), pp. 88–93. IEEE (2019)
15. Rukhmode, S., Vyavhare, G., Banot, S., Narad, A., Tugnayat, R.: IoT based agriculture monitoring system using wemos. no. March, pp. 14–19 (2017)
16. Sarkar, P.J., Chanagala, S.: A survey on IoT based digital agriculture monitoring system and their impact on optimal utilization of resources. J. Electron. Commun. Eng. (IOSR-JECE) **11**(1), 1–4 (2016)
17. Sarker, M.M., Akhund, T.M.N.U.: The roadmap to the electronic voting system development: a literature review. Int. J. Adv. Eng. Manage. Sci. **2**(5), 239465
18. Sarker, M.M., Shah, M.A.I., Akhund, T.M.N.U., Uddin, M.S.: An approach of automated electronic voting management system for Bangladesh using biometric fingerprint. Int. J. Adv. Eng. Res. Sci. **3**(11), 236907 (2016)
19. Satu, M.S., Roy, S., Akhter, F., Whaiduzzaman, M.: Iolt: an IoT based collaborative blended learning platform in higher education. In: 2018 International Conference on Innovation in Engineering and Technology (ICIET), pp. 1–6. IEEE (2018)
20. Shahzadi, R., Tausif, M., Ferzund, J., Suryani, M.A.: Internet of things based expert system for smart agriculture. Int. J. Adv. Comput. Sci. Appl. **7**(9), 341–350 (2016)
21. Shovon, A.R., Roy, S., Sharma, T., Whaiduzzaman, M.: A restful e-governance application framework for people identity verification in cloud. In: International Conference on Cloud Computing, pp. 281–294. Springer (2018)
22. Stojkoska, B.L.R., Trivodaliev, K.V.: A review of internet of things for smart home: challenges and solutions. J. Clean. Prod. **140**, 1454–1464 (2017)
23. Whaiduzzaman, M., Gani, A., Naveed, A.: Towards enhancing resource scarce cloudlet performance in mobile cloud computing. Comput. Sci. Inf. Technol. 1 (2015)
24. Whaiduzzaman, M., Hossain, M.R., Shovon, A.R., Roy, S., Laszka, A., Buyya, R., Barros, A.: A privacy-preserving mobile and fog computing framework to trace and prevent covid-19 community transmission. IEEE J. Biomed. Health Inform. **24**(12), 3564–3575 (2020)

Simultaneous Measurements Reading from More Than One MiBand 3 Wristbands

Zuzanna Rodak, Krzysztof Tokarz, Pawel Mielnik, and Marcin Fojcik

Abstract This paper presents the research done on simultaneous data acquisition from multiple smart wristbands. The wristbands, besides monitoring fitness features, can be used to monitor health parameters like heart rate, blood pressure, and speed and dynamics of movement. The movement analysis may be used to monitor progress in rehabilitation. In order to analyze human movement more accurate, it is useful to gather data from several accelerometers simultaneously. The studies were based on raw data gathered from wristbands to collect and analyze the time between the readings depending on the number of connected devices. Additionally, the analysis was made of the difference in readings of the accelerometers from several devices to estimate the mean error of measurements.

Keywords Wristband · Sport band · Smart wearables · Internet of Things · Rehabilitation · Bluetooth low energy · Wireless data transmission

1 Introduction

Nowadays, wearable IoT devices are getting more popular. On a daily basis, we meet people with accessories such as smartwatches or fitness wristbands. They are widely used not only for their intended purpose, to monitor fitness activities, but also as

Z. Rodak (✉) · K. Tokarz
Department of Graphics, Computer Vision and Digital Systems, Faculty of Automatic Control, Electronics and Computer Science, Silesian University of Technology, Gliwice, Poland
e-mail: krzysztof.tokarz@polsl.pl

P. Mielnik
Section for Rheumatology, Department for Neurology, Rheumatology and Physical Medicine, Helse Førde, Førde, Norway
e-mail: pawel.franciszek.mielnik@helse-forde.no

M. Fojcik
Western Norway University of Applied Sciences, Bergen, Norway
e-mail: marcin.fojcik@hvl.no

© The Author(s), under exclusive license to Springer Nature Singapore Pte Ltd. 2022
A. K. Nagar et al. (eds.), *Intelligent Sustainable Systems*, Lecture Notes in Networks and Systems 333, https://doi.org/10.1007/978-981-16-6309-3_10

simple watches. The number of those devices increases rapidly every year [1]; they are getting smaller, more accurate, and equipped with new enhanced sensors. Above-mentioned factors will lead to the increased usage of smart wearable devices in the future [2]. In a wide variety of smart wearable technologies, the most comfortable for the users are single wrist-worn devices [3] which also provides good quality of experience [4].

The most common application of smartwatches and smart bands is to monitor heart rate and a number of steps. The variety of research done on smart wearable devices shows that besides monitoring fitness features, wearable devices may be a good solution to monitor health. One of the usages of smart wearable devices is to record blood pressure or heart rate in order to detect heart attacks [5]. Besides monitoring physical health such devices, by the use of the same sensors, may monitor behavioral health like bipolar disorder, depression, or stress [6].

The problem arises when there exists a need to monitor other parameters that are not supported directly by the device nor the application connected to it. In that case, it is required to read the raw data from the devices' built-in sensors. There may also appear a limitation caused by the construction of the device or the method used for the transmission of the data.

For each of the aforementioned purposes, fitness, and healthcare, the wearable devices require different accuracies and features, e.g., the possibility of reading proper measurements from quick and slow movements. Based on those factors, the authors of this paper will take into consideration different parameters of the Xiaomi MiBand 3 fitness tracker. The most focus will be put into conducting a research on the possibility of the simultaneous connection of several smart bands to one receiving device. The general idea is to analyze the influence of a number of devices on the time intervals of sending the data and their accuracy.

In the second part of this paper, a review of the uses of smart wearables is presented. In the next part, the measuring process is briefly described. Later, the analysis of time data and corresponding readings of the accelerometer is performed. In the end, the conclusions are drawn.

2 Literature Review

In the professional literature, there is a wide variety of research on the usage of smart wearable devices applied to monitor physical health. The most common application of wearable sensors is fall detection. In the research on fall risk assessment and prediction done by Montesinos et al. [7], the authors reviewed the placement of the sensor to best predict the risk of a fall. An example of that is a study done by Hsieh et al. [8] who proposes a scheme of fall risk detection based on a wristband device and a smartphone. The detection method contains three phases: first one is to detect the movement of the sensing device, the second check if after the fall the movement is still for some time, and the third phase detects persons slants in comparison to the ground. Another example is a system of monitoring of chronic arthritis patients proposed by authors in [9]. The system is based on a Xiaomi Mi Band that sends data to any Android device, e.g., smartphone, and then sends the data to a database in order to collect and analyze it. In another research conducted by Grym et al. [10], the authors consider the feasibility of smart wristbands to monitor women, during and one month after pregnancy. During the study on 20 pregnant women, the researchers concluded that chosen commercial wristbands are generally a good solution for that purpose. Another publication by Mantua et al. [11] compares commercially available personal health monitoring devices to polysomnography in sleep analysis. In the outcome, they concluded that some specific categories of data gathered by wrist-worn devices did not differ from readings from professional equipment.

Another application of wearable devices may be found in monitoring behavioral health. In a publication by Rincon et al. [12], the authors present their own solution for detecting emotional states for the elderly. It was achieved with the use of a self-designed wrist-worn device containing heart rate and skin resistance sensors. Additionally, the authors used a mini-camera to acquire photos in order to analyze the emotional states. The images from the camera are processed by the ESP-32 chip and an inertial measurement unit. Another research [13] shows that with the use of a heart rate sensor and galvanic skin response sensor, which were put either on the wrist or the ankle, it is possible to mitigate meltdowns in people with autism. In other studies, van Lier et al. [14] shows that usage of E4 wristband in a study of physiological and psychological measures is sufficient for real-time monitoring of alcohol craving.

All the above solutions require only one wearable device to provide feasible results. The usage of several wearable IoT devices appears in the scope of rehabilitation and fitness. In the study of computer-assisted systems for group classes of exercised-based rehabilitation [15, 16], the authors proposed a system design containing Kinect motion sensors and wristband heart rate monitors. The idea of this system is to maximize health benefits for chronic patients by monitoring the effectiveness of their exercises. Another application of the usage of several wearable IoT devices is proposed in a survey [17] within the description of Personal Area Network with several smart devices collecting data during exercises. In the paper by Liu [18], the author describes ideas of using wearable devices that are applied in athlete muscle and physical fitness monitoring. In another publication [19], the author proposed a system of several wearable sensors connected together to monitor sports rehabilitation.

The literature shows that the subject of simultaneous data acquisition from several IoT devices, including wristbands, is being discussed. In a publication [20], the authors describe the method for comparison of a number of commercially available sport bands including MiBand 3. In another paper done by Weghorn [21], the author analyzes the accuracy of inexpensive wearable devices in comparison to professional medical equipment.

To the best knowledge of the authors of this paper, there has been no other publication describing the parameters of collected data from several wearable IoT devices connected at the same time. Therefore, the authors planned and conducted a series of research in order to inquiry the parameters of data transmission from several sport bands simultaneously. The research consists of inspecting the frequency of sent data with different numbers of bands connected, the difference of reading times from several wristbands and the occurring differences in the data sent by devices.

3 Research Process

For the purpose of this study, the authors chose to use five Xiaomi MiBand 3 devices. The choice has been made based on the accessibility of mentioned devices, the possibility of raw data acquisition, and other features like price, size of the device, comfort of wearing, etc. The raw data from MiBands is sent via Bluetooth Low Energy to the computer application. The application collects the data about, e.g., heart rate or coordinates along three axes. The data collected in such a way is given a timestamp, and it is written into text files. Then, the data from the text files are processed in the MATLAB environment.

For the purpose of this research, the data was gathered in two different ways. In the first case, the devices were laying still and the connections were made randomly to 1, 2, 3, 4, and 5 wristbands at a time. The measurement process took about 5 min each. The data collected that way was used to establish the difference of time intervals in which the measurements are received by the computer. In the second case, all five sport bands were connected simultaneously. They were put on a rod, as shown in Fig. 1 and moved up and down with different acceleration and rotated with constant velocity. The data collected that way was used to compare the accuracy of readings of the accelerometer for each MiBand.

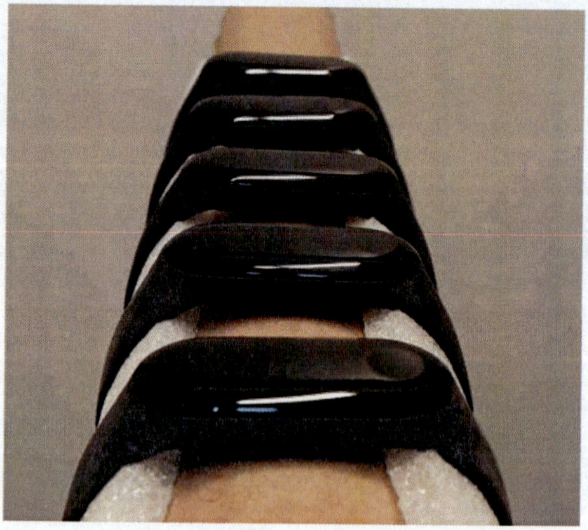

Fig. 1 Five wristbands fastened on a common rod

4 Analysis of the Results

4.1 Time Analysis

In the first part of the measurements, the collected data were analyzed in terms of the time of their acquisition. The timestamps of the measurements were compared, and the results of minimum, maximum, and mean time between the measurements are presented in Table 1.

From the presented data, it can be seen that when connected to only one device at a time, the data is received faster than when connected to multiple devices. While there is no great difference in time intervals for 2, 3, and 4 wristbands, we can notice that in the case of five devices, the mean value differs by 9 milliseconds in comparison to other cases of multiple connections. After repeating the measurements, it was observed that in the case of the connection of five sport bands, the time between the measurements from one random device is always two times greater than from the other four devices.

4.2 Data Accuracy

The data gathered during the second part of the measurements were processed in order to find out if the readings of the position of MiBands are similar to each other. To obtain the result, the maximum values from each peak from the measurements, shown in, e.g., Fig. 2, were gathered and analyzed. For each peak, the mean was calculated and compared to the highest reading of every MiBand. Then for every device, the mean of their accuracy was calculated. The results of that process are presented in Table 2.

From Table 2, we can see that the mean error of the readings for all wristbands is equal to 2.8%.

Table 1 Time between measurements

No. of MiBands	Time (ms)		
	Minimum	Maximum	Mean
1	40.39	42.74	42.41
2	40.54	86.55	45.19
3	29.47	101.48	45.77
4	40.55	90.44	45.18
5	40.83	121.51	54.49

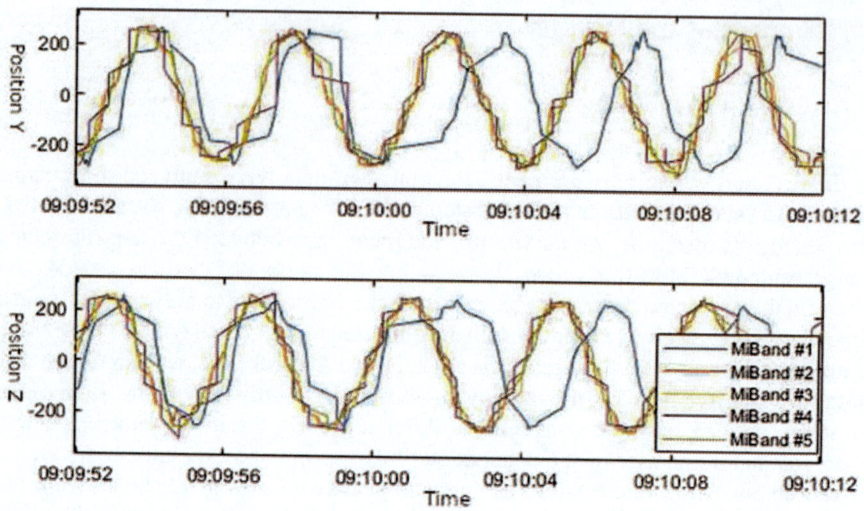

Fig. 2 Plot of the measurements for slow movement

Table 2 Mean error of measurements

MiBand number	Mean error	
	Y axis (%)	*Z* axis (%)
1	3.55	2.17
2	2.92	1.81
3	3.10	3.04
4	3.78	2.17
5	3.12	2.29
Mean	3.30	2.30

4.3 Additional Observations

During the analysis, it was discovered that when there are five wristbands connected simultaneously, some of the reading may occur lag. This phenomena is visible in Figs. 2 and 3. From the taken measurements, we can conclude that the lag occurs for a random number of devices. The wristband for which the lag occurs is also random. Further, we can notice that the lag increases over time. In Fig. 3, we can observe that for slower movements, we can identify which reading corresponds to which peak, while as the movement is getting faster, it makes it difficult to identify correctly the readings. What was also discovered during the research is that sometimes the measurements are not received in a long time. After that, the readings are received more frequently in a short period. It can be seen in a comparison of data from Table 1,

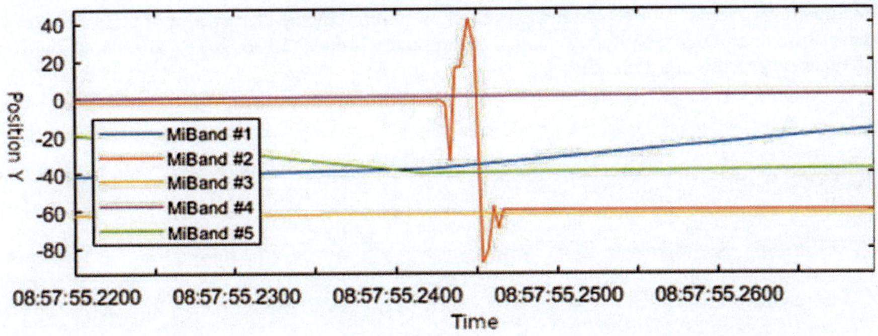

Fig. 3 Plot of the measurements for fast movement

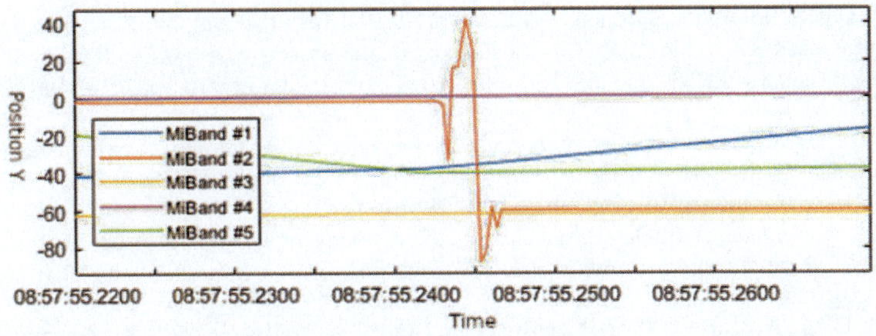

Fig. 4 The phenomena of frequent data receiving after the gap

stating that the mean of time of receiving data from five wristbands is about 54.49 ms, and analysis of Fig. 4, which shows that in a time interval of about 50 ms a number of 12 measurements were gathered.

5 Conclusions

From the performed research, the conclusion can be reached that the readings from five Xiaomi MiBands 3 wristbands show a high similarity of the values, which may mean good accuracy and repeatability of the measurements. The main disadvantage is the appearance of the lags that reduce the maximal frequency of cyclic movement that can be measured. Although the lags usually are followed by accumulated readings that can be used for improving the accuracy of the calculations of the average value, the data obtained is suitable to be post-processed rather than processed in real-time. It is necessary to conduct more studies to find the reason for the appearance of the measurement lags. The research will focus on checking whether the cause of the delays is the specificity of the computer system receiving the data or whether they result from the limitations of the Bluetooth Low Energy protocol or if it originates from possible interference from other wireless transmission systems.

Acknowledgements This research was partially supported by the Department of Graphics, Computer Vision and Digital Systems, under statue research project (Rau6, 2021), Silesian University of Technology (Gliwice, Poland).

References

1. Henriksen, A., Mikalsen, M.H., Woldaregay, A.Z., Muzny, M., Hartvigsen, G., Hopstock, L.A., Grimsgaard, S.: Using fitness trackers and smartwatches to measure physical activity in research: analysis of consumer wrist-worn wearables. J. Med. Internet Res. **20**(3) (2018). https://doi.org/10.2196/jmir.9157
2. Windasari, N.A., Lin, F.R., Kato-Lin, Y. C.: Continued use of wearable fitness technology: a value co-creation perspective. Int. J. Inf. Manage. **57**, 102292 (2021). https://doi.org/10.1016/j.ijinfomgt.2020.102292
3. Mehrang, S., Pietila, J., Tolonen, J., Helander, E., Jimison, H., Pavel, M., Korhonen, I.: Human activity recognition using a single optical heart rate monitoring wristband equipped with triaxial accelerometer. IFMBE Proc. **65**, 587–590 (2017). https://doi.org/10.1007/978-981-10-5122-7_147
4. Pal, D., Tassanaviboon, A., Arpnikanondt, C., Papasratorn, B.: Quality of experience of smart-wearables: from fitness-bands to smartwatches. IEEE Consumer Electron. Mag. **9**(1), 49–53 (2020). https://doi.org/10.1109/MCE.2019.2941462
5. Surantha, N., Atmaja, P., David, Wicaksono, M.: A review of wearable internet-of-things device for healthcare. Procedia Comput. Sci. **179**, 939–943 (2021). https://doi.org/10.1016/j.procs.2021.01.083
6. Hilty, D.M., Armstrong, C.M., Luxton, D.D., Gentry, M.T., Krupinski, E.A.: A scoping review of sensors, wearables, and remote monitoring for behavioral health: uses, outcomes, clinical competencies, and research directions. J. Technol. Behav. Sci. 1–36 (2021). https://doi.org/10.1007/s41347-021-00199-2
7. Montesinos, L., Castaldo, R., Pecchia, L.: Wearable inertial sensors for fall risk assessment and prediction in older adults: a systematic review and meta-analysis. In: Transactions on Neural Systems and Rehabilitation Engineering, vol. 26, Issue 3, pp. 573–582. Institute of Electrical and Electronics Engineers Inc. (2018). https://doi.org/10.1109/TNSRE.2017.2771383
8. Hsieh, S.L., Yang, C.T., Li, H.J.: Combining wristband-type devices and smartphones to detect falls. In: 2017 IEEE International Conference on Systems, Man, and Cybernetics, SMC 2017, Jan 2017, pp. 2373–2377 (2017). https://doi.org/10.1109/SMC.2017.8122977
9. Mielnik, P., Tokarz, K., Mrozek, D., Czekalski, P., Fojcik, M., Hjelle, A.M., Milik, M.: Monitoring of chronic arthritis patients with wearables–A report from the concept phase. In: Lecture Notes in Computer Science (Including Subseries Lecture Notes in Artificial Intelligence and Lecture Notes in Bioinformatics), 11684 LNAI, pp. 229–238 (2019). https://doi.org/10.1007/978-3-030-28374-2_20
10. Grym, K., Niela-Vilén, H., Ekholm, E., Hamari, L., Azimi, I., Rahmani, A., Liljeberg, P., Löyttyniemi, E., Axelin, A.: Feasibility of smart wristbands for continuous monitoring during pregnancy and one month after birth 11 Medical and Health Sciences 1117 Public Health and Health Services. BMC Pregnancy Childbirth **19**(1), 1–9 (2019). https://doi.org/10.1186/s12884-019-2187-9
11. Mantua, J., Gravel, N., Spencer, R.M.C.: Reliability of sleep measures from four personal health monitoring devices compared to research-based actigraphy and polysomnography. Sensors (Switzerland) **16**(5) (2016). https://doi.org/10.3390/s16050646
12. Rincon, J.A., Costa, A., Novais, P., Julian, V., Carrascosa, C.: Intelligent wristbands for the automatic detection of emotional states for the elderly. In: Lecture Notes in Computer Science (Including Subseries Lecture Notes in Artificial Intelligence and Lecture Notes in Bioinformatics), 11314 LNCS, pp. 520–530 (2018). https://doi.org/10.1007/978-3-030-03493-1_54

13. Cabibihan, J.J., Chellali, R., So, C.W.C., Aldosari, M., Connor, O., Alhaddad, A.Y., Javed, H.: Social robots and wearable sensors for mitigating meltdowns in autism—A pilot test. In: Lecture Notes in Computer Science (Including Subseries Lecture Notes in Artificial Intelligence and Lecture Notes in Bioinformatics), 11357 LNAI, pp. 103–114 (2018) https://doi.org/10.1007/978-3-030-05204-1_11
14. van Lier, H.G., Oberhagemann, M., Stroes, J.D., Enewoldsen, N.M., Pieterse, M.E., Schraagen, J.M. C., Postel, M.G., Vollenbroek-Hutten, M.M.R., de Haan, H.A., Noordzij, M.L.: Design decisions for a real time, alcohol craving study using physio- and psychological measures. In: Lecture Notes in Computer Science (Including Subseries Lecture Notes in Artificial Intelligence and Lecture Notes in Bioinformatics), 10171 LNCS, pp. 3–15 (2017). https://doi.org/10.1007/978-3-319-55134-0_1
15. Filos, D., Triantafyllidisa, A., Chouvardaa, I., Buys, R., Cornelissen, V., Budts, W., Walsh, D., Woods, C., Moran, K., Maglaverasa, N.: PATHway: decision support in exercise programmes for cardiac rehabilitation. Stud. Health Technol. Inform. **224**, 40–45 (2016). https://doi.org/10.3233/978-1-61499-653-8-40
16. Triantafyllidis, A., Filos, D., Buys, R., Claes, J., Cornelissen, V., Kouidi, E., Chatzitofis, A., Zarpalas, D., Daras, P., Chouvarda, I., Maglaveras, N.: A computer-assisted system with kinect sensors and wristband heart rate monitors for group classes of exercise-based rehabilitation. IFMBE Proc. **66**, 237–241 (2018). https://doi.org/10.1007/978-981-10-7419-6_39
17. Qiu, H., Wang, X., Xie, F.: A survey on smart wearables in the application of fitness. In: Proceedings —2017 IEEE 15th International Conference on Dependable, Autonomic and Secure Computing, 2017 IEEE 3rd International Conference on Big Data Intelligence and Computing, pp. 303–307 (2018)
18. Liu, S.: Mobile sensor network and wearable devices application in athlete muscle and physical fitness image monitoring. Microprocess. Microsyst. **82**, 103856 (2021). https://doi.org/10.1016/j.micpro.2021.103856
19. Jiang, Y.: Combination of wearable sensors and internet of things and its application in sports rehabilitation. Comput. Commun. **150**, 167–176 (2020). https://doi.org/10.1016/j.comcom.2019.11.021
20. Pardamean, B., Soeparno, H., Mahesworo, B., Budiarto, A., Baurley, J.: Comparing the accuracy of multiple commercial wearable devices: a method. Procedia Comput. Sci. **157**, 567–572 (2019). https://doi.org/10.1016/j.procs.2019.09.015
21. Weghorn, H.: Unsubstantial health and sports monitoring reliability of commercial fitness tracker bracelets induced by their all-in-one sensing unit approach—Experimental evaluation of measurement accuracy in dynamic and in steady physical effort scenarios. Commun. Comput. Inf. Sci. **975**, 55–74 (2019). https://doi.org/10.1007/978-3-030-14526-2_4

When One Wireless Technology is Not Enough: A Network Architecture for Precision Agriculture Using LoRa, Wi-Fi, and LTE

Jose A. Brenes and Gabriela Marín-Raventós

Abstract The world population will reach nearly 10 billion people by 2050, according to the United Nations. Therefore, more food to supply the world's demand will be required in the following years. Precision agriculture emerges as an option to satisfy the growing demand. In smart farming, wireless sensor networks (WSNs) are crucial in the deployment of sensors in crop fields. Precision agriculture includes crop monitoring and fertigation control. Monitoring and control have distinct network requirements. While monitoring stations deployment requires long-range networks, control stations have other requirements like low latency. For that reason, the use of a combination of WSN is necessary. In this paper, we present an option of network architecture for precision agriculture projects. The architecture includes the use of LoRa for monitoring stations and Wi-Fi/LTE for control stations. Currently, we are working on smart fertigation in greenhouses. For the architecture, we consider the typical requirements for smart farming projects, but also our project's requirements.

Keywords Wireless sensor networks · Precision agriculture · Smart farming · Network architecture · LoRa · Wi-Fi · LTE

1 Introduction

The world population prospects, published by the United Nations in 2019, state that the world population is projected to reach about 8.5 billion people in 2030, 9.7 billion in 2050, and 10.9 billion in 2100 [1]. A study carried out by the OECD and FAO determines that population growth will be the main driver of total food intake in the current decade [2]. The study also states that consumers in low-income countries get

J. A. Brenes (✉)
Research Center for Communication and Information Technologies (CITIC), University of Costa Rica, San José, Costa Rica
e-mail: joseantonio.brenes@ucr.ac.cr

J. A. Brenes · G. Marín-Raventós
Computer Science Graduate Program (PPCI), University of Costa Rica, San José, Costa Rica
e-mail: gabriela.marin@ucr.ac.cr

© The Author(s), under exclusive license to Springer Nature Singapore Pte Ltd. 2022
A. K. Nagar et al. (eds.), *Intelligent Sustainable Systems*, Lecture Notes in Networks and Systems 333, https://doi.org/10.1007/978-981-16-6309-3_11

the bulk of their daily calories from vegetable sources. Therefore, agriculture activity will be crucial in the following decades, but especially for these countries.

An analysis made by FAO in 2016 pointed out that about 84% of farms around the world are smaller than 2 ha. These farms cover approximately 12% of global farmland and are owned by small-scale farmers [3]. According to their analysis, thanks to technology adoption, small farmers in high-income countries operate farms with lower land extensions but reaching high yields and efficiency [4]. For this reason, it is crucial to improve farming practices and to provide technological support to small-scale farmers in low-income countries.

Precision agriculture refers to a farming management method based on constantly observing, measuring, and responding to variability in agricultural crops [5]. These actions are performed to increase crop yield and production profits but also to achieve more sustainable farming processes. By implementing precision agriculture, it is possible to make small-scale farmers more competitive, improving their economic situation.

In precision agriculture supported by technology, Internet of things (IoT) is used to carry out the sensing of physical variables from crops. For it, at least one wireless sensor network is required to deploy the sensors in the field and to send all the collected data to a storage and processing computational system. Even in the context of greenhouses, wireless sensor networks are needed since implementing wired networks implies hard work. In other contexts, like extensive crops, using wired networks is almost impossible, complicating the deployment of sensor stations.

There are a few different types of wireless sensor networks that can be implemented in IoT projects. Each option has advantages and disadvantages related to implementation cost (including components and network usage costs), power consumption, and connectivity range, among other characteristics.

There are many examples of smart agriculture applications in which wireless sensor networks are present in the literature. However, these applications are difficult to implement or useless in a real scenario. In real smart agriculture projects, it is necessary to use a combination of networks to accomplish the different requirements. Reliable connection between sensors, stations, and devices must be guaranteed. For this reason, it is crucial to select the correct wireless technology and when needed, a combination of them.

In this paper, we analyze the different wireless sensor networks that can be useful for smart agriculture projects. We also present the implementation of the network architecture, which supports our smart farming platform. In the next section, we introduce some concepts and popular wireless sensor networks. Next, we compare the main characteristics of different wireless sensor networks that can be used in smart agriculture projects. In the next section, we specify the different requirements that are usually present in this type of projects, by making emphasis on our requirements. Also, we present the specification of the architecture we implemented. Finally, in the last section, we present some conclusions and future work.

2 Theoretical Framework

In this paper, we focus on analyzing different wireless sensor networks available for IoT projects, specifically such networks which are helpful for precision agriculture.

2.1 *Internet of Things (IoT)*

According to the International Telecommunication Union (ITU), IoT can be defined as a global infrastructure for the information society, enabling advanced services by interconnecting real world and virtual objects through information and communication technologies [6]. In precision agriculture supported by technology, these interconnected objects refer to the monitoring and control stations deployed in crop fields, and the computational platforms are used to manage stations and process collected data.

The IoT architecture commonly includes three layers: application, transport, and sensing. In this paper, we focus on the network layer. The network layer oversees transferring data collected from devices to the computational platforms [7]. This transferring takes two stages: to get the data from each node or station located in the crop field and to transfer the collected data to the computational system that stores and processes it. The first stage is achieved by using wireless sensor networks, while the second stage is carried out by using a transport network.

2.2 *Wireless Sensor Networks (WSNs)*

The most popular wireless sensor networks used in the context of precision agriculture are introduced.

LPWAN: LPWAN is the abbreviation of low-power wide-area network. LPWANs support long-range device deployment and consume low power [8]. These two features make LPWANs very suitable for contexts in which the sensors or measurement stations must be deployed in the field or at least far from a place with Internet connection facilities. LPWANs sacrificed network data rates and latency to achieve long-range connectivity.

In smart agriculture applications, LPWANs are especially useful. It is effortless to deploy sensors in crop fields with them. Besides, in this context, low data rates and high latency do not represent a significant problem due to the importance of getting data directly from crops commonly geographically distributed.

There are several types of LPWANs. Here we detail the two more popular; LORA and Sigfox.

LoRa: LoRa is a proprietary modulation technique for long-range and low-power wireless communications which use the chirp spread spectrum (CSS) technology.

Since 2012, this technology has been owned and patented by Semtec, a company that founded the LoRa Alliance.[1] The LoRa Alliance is a non-profit association with the objective of enabling the large-scale deployment of LPWAN IoT through the development and promotion of the LoRaWAN open standard.[2] The LoRaWAN standard is a networking protocol designed to connect low-powered devices (commonly battery-assisted) to the Internet or local networks. The protocol enables bidirectional communication.

Sigfox: Sigfox is a global LPWAN. It was founded in 2010 as a company dedicated to connecting every physical world object to the "digital universe".[3] The company oversees the maintenance of the network and constantly expands its coverage. Since the network in Sigfox is operated and maintained by a company (or network operator), it is not possible to deploy private networks. Instead of that, a contract with a network operator must be subscribed to use the ground stations to send and receive data packets. That implies some advantages and disadvantages. Having a company behind the network operation makes the service very stable and enables to get support directly from the company. However, network usage costs could be high. Also, the duty cycle is incredibly low, making it difficult to use the network when it is needed to send and receive data during the day.

WLAN: Today, wireless local area networks are one of the most used networks in multiple contexts. We can find WLAN everywhere. The communication is based on the standard IEEE 802.11. There are several versions (b/g/n/ac/ax/ah) that vary in terms of range, bandwidth, and power consumption. For IoT projects, Wi-Fi HaLow will offer some benefits. Wi-Fi HaLow corresponds to a WLAN version that incorporates 802.11ah technology by operating the network in frequencies below 1 GHz, offering a more extended range and lower power consumption.[4] One of the main advantages behind the use of Wi-Fi is that it makes it feasible to interoperate the IoT and transport networks thanks to the support of TCP/IP stack. Nevertheless, this technology is not yet available commercially. According to a recent article published by a Wi-Fi Alliance member, it could be available soon [9].

Mobile Networks: In the last decades, the use of mobile networks has increased dramatically. These networks include technologies known as 3G, 4G, and more recently 5G. In this case, even though the network is not always exclusively offered by a company, costs associated with network usage exist. Besides, due to government restrictions on frequencies, it is not possible to deploy a private mobile network. The main advantage of mobile networks is their low latency. One of the main goals of 4G LTE was to reduce the latency to a maximum of 10 ms [10]. It is expected that by implementing 5G networks, latency and bandwidth can be improved. In IoT projects, the used version is LTE-M.

---

1 https://www.semtech.com/

2 https://lora-alliance.org/

3 https://www.sigfox.com/en.

4 https://www.wi-fi.org/discover-wi-fi/wi-fi-halow.

2.3 *Transport Networks*

Once data have been acquired from sensor nodes in crop fields, it is necessary to transport it to computational platforms, which are located far away. Transport networks are used to send the collected data. These refer to traditional data communication networks used to access the Internet.

In crop fields, wired connections are commonly unavailable. If available, the collected data can be sent using it. In other cases, transport networks like cellular (LTE-A) or satellite (VSAT) can be used thanks to their extended coverage. In reduced scenarios, Wi-Fi networks can be present as an option to connect WSN to the Internet.

3 Wireless Sensor Network Comparison

In the previous section, the most common WSN used in smart farming projects was described. In this section, we present a comparison between these distinct networks. The analysis was made considering advantages and disadvantages in several aspects. Table 1 shows the technical specifications for each of the considered wireless sensors

Table 1 Wireless sensor networks' technical specs

Item	LoRa	Sigfox	Wi-Fi HaLow	LTE-M
Wireless bands	ISM unlicensed bands Depends on the region, see [11]	ISM unlicensed bands Depends on the region, see [11]	ISM unlicensed bands. 902–928 MHz	Licensed bands Depends on region and government grants
Bandwidth	125/250/500 MHz channels	100 MHz channels	1 MHz to 16 MHz wide channels	1.4/3/5/10/15/20 MHz channels
Data rates	Up to 50 kbps	Uplink: 100 bps. Downlink: 600 bps	Up to 86.7 Mbps	Up to 100 Mbps
Range	Up to 10 km in rural areas. Up to 5 km in dense urban areas	Up to 40 km in rural areas. Up to 10 km in urban areas	Depends on the channel used Up to 1 km	Depends on network coverage offered by operator
Duty cycle	No limits in private networks. Network operator can specify restrictions in public networks	Uplink: up to 12 bytes of payload and up to 140 messages/day Downlink: up to 8 bytes of payload and up to 4 messages/day	No limits due to private deployment	No limits Service providers can set limits to the amount of data sent and received

network. The data were extracted from the Web sites of distinct providers and organizations behind the network's operation or standardization, all of them mentioned in the previous section.

In addition to each network's technical specification, it is crucial to consider the deployment and network usage costs when designing an IoT solution. In this respect, LTE-M and Sigfox have lower deployment costs than LoRa networks. This is because the network operator offers access to the network, eliminating the need to deploy devices like routers, access points, or gateways. However, LoRa and Wi-Fi HaLow have lower network use costs than Sigfox and LTE-M.

LTE-M, LoRa, and Sigfox are more suitable for mobility. Sensor stations communicating with Wi-Fi HaLow send data packets to access points. However, although Wi-Fi HaLow allows mobility, LTE-M, LoRa, and Sigfox-enabled stations can be moved over longer distances.

4 Network Requirements for Precision Agriculture Projects

So far, we have included the general characteristics of wireless sensor networks. In this section, we provide a list of requirements that we have identified for precision agriculture projects. The network requirements will be analyzed from a general perspective considering solution features, types of stations, and deployment location. It is crucial to consider that precision agriculture projects can include not only monitoring physical variables but also making decisions and taking actions according to the data collected from crops. For this reason, requirements can vary between projects.

4.1 Based on the Features of the Solution

Usually, there is no need to transfer large amounts of data. Sensor stations generate data packets in the order of tens of bytes every determined time period. Thus, data rates are not determinant. However, in the case of network latency (delay in communication over the network), the needs can be different between projects. When the solution includes actuators (in control stations), low latency is required to receive instructions and act on time. As an example, we can mention the case of a water pump. That actuator must be activated/deactivated at the precise moment to guarantee the crop is correctly irrigated.

4.2 Based on the Station Type

As it was mentioned before, precision farming projects may have different types of stations. Monitoring stations usually are deployed directly in crop fields. For that reason, WSN is helpful to transfer the data collected and consumes as little power as possible. Meanwhile, control stations manage actuators which require direct communication with the decision-making systems. Usually, control stations are located near electrical outlets, making it possible to use other networks with higher power consumption. Moreover, control stations can connect directly to the Internet.

For precision farming projects, all types of stations are required. Unless the solution only involves crop monitoring, we must use control stations to manage irrigation and nutrition, among other tasks. Hence, in most projects, a combination of networks is needed.

4.3 Based on the Deployment Location

There is no doubt that monitoring stations must be deployed in the crop field. Wireless sensor networks are required for that. However, other stations could be deployed in different locations, not only near the crops. As an example, stations can be used to monitor and control water tanks and pumps, water reservoirs, or fertilizers deposits. These items are usually located near workers' offices and not in crop fields. A mid-range network can be used to connect them.

Next, we present the combination of networks that we use for our smart farming project.

5 Network Architecture and Implementation

As stated at the beginning, we are currently working on a precision agriculture project for greenhouse farming. Our team includes experts from areas like computing, physics, and agronomy. We are trying to combine the knowledge from these distinct areas to create a low-cost fertigation platform useful for small-scale farmers in Costa Rica. It includes the use of monitoring and control stations to manage the fertigation automatization, but also to make better decisions based on collected data [12]. These decisions may be deployed as automatic actions requested to control stations. We also include deploying alerts and suggestions for farmers to improve their manual activities.

Analyzing the different requirements stated in the previous section and the objectives followed in our project, we selected a combination of networks for our smart farming solution. LoRa, Wi-Fi, and LTE-M were depicted as the more suitable combination for our purposes.

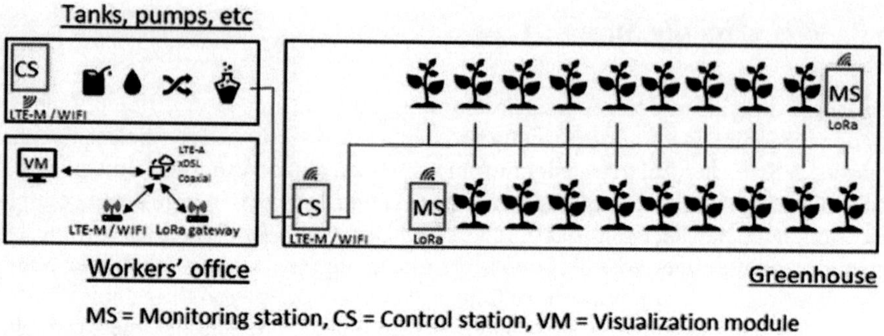

Fig. 1 Selected networks and platform components

Our precision agriculture project is for greenhouse fertigation. The platform must use different monitoring and control stations in constant communication with a decision-making platform. Although we are working with greenhouse crops, thanks to LoRa, we can deploy monitoring stations easily in all types of crops. The monitoring stations are constantly sending data to the system. That data can be accessed on-site through visualization modules located in the greenhouse or elsewhere, through the Internet by using a Web application.

Several control stations prepare the nutritional solutions and manage the fertigation. All control stations have implemented in them the minimum logic needed to do the corresponding task. Still, the decision-making platform can overwrite control stations' instructions to improve the irrigation or the fertilizers' application. The control stations use Wi-Fi when it is available and use LTE-A when Wi-Fi is not available.

To send data to processing platforms, we use wired connections (coaxial/fiber optic/xDSL) when they are available, or LTE-A, as a transport network. Figure 1 shows the selected networks and the platform components. The figure shows the interaction between monitoring and control station with networks.

It is essential to point out that the design can be used for all types of small-scale farmers. The architecture shown in Fig. 1 is advantageous because it permits deployment of monitoring stations at remote locations easily. Besides, it is helpful to have the visualization module, distinct gateways, and wireless access points in the same place. That enables the use of only one transport network to connect the platform to the Internet.

It is crucial to consider that precision agriculture implies not only monitoring. Control of fertigation is as important as monitoring to achieve the automation of productive processes. The proposed architecture provides support for both types of stations.

6 Discussion and Conclusions

In this paper, we analyzed the main features of distinct WSN useful for smart farming projects. Also, we detailed the different network requirements for these projects. As the result of in-depth analysis and technical trials for the precision agriculture project, which involves monitoring and control stations in greenhouses, we present a selection of WSN that we have implemented for our smart farming solution. The stations require a combination of WSN to enable their deployment in crop fields easily and to do effective management of fertigation.

There are many WSN used in IoT projects. Smart farming solutions typically use LPWANs due to their advantages. By using the proposed combination of networks, we can effectively collect data from crops and manage the fertigation in the exact amount at the precise time. Although we are currently working on precision farming in greenhouses, we think the architecture proposed can be used for extensive crops.

The proposed architecture considers the use of several wireless technologies. Currently, there is no platform to manage all our communication requirements. This can be considered a limitation. Nevertheless, in future, a tool can be developed to manage all the traffic and networks in a single place.

As future work, we also plan to test Wi-Fi HaLow technology. By using it, it can be possible to create a simple network architecture in which sensors and actuators can be directly integrated into decision-making systems through the Internet. However, the Wi-Fi HaLow technology is not yet available commercially.

Acknowledgements This work was partially supported by the Research Center for Communication and Information Technologies (CITIC) and the School of Computer Science and Informatics (ECCI) at the University of Costa Rica. Research Project No. 834-B9-189.

References

1. United Nations, Department of Economic and Social Affairs, P.D.: World Population Prospects 2019: Highlights (ST/ESA/SER.A/423). United Nations (2019)
2. Organisation for Economic Co-operation and Development OECD: Food and agriculture organization of the United Nations: OECD-FAO agricultural outlook 2020–2029. OECD (2020). https://doi.org/10.1787/1112c23b-en
3. Lowder, S.K., Skoet, J., Raney, T.: The number, size, and distribution of farms, smallholder farms, and family farms worldwide. World Dev. **87**, 16–29 (2016). https://doi.org/10.1016/j.worlddev.2015.10.041
4. Fuglie, K., Gautam, M., Goyal, A., Maloney, W.F.: Harvesting prosperity: technology and productivity growth in agriculture. World Bank (2019). https://doi.org/10.1596/978-1-4648-1393-1
5. of the European Commission, J.R.C. (JRC), H04, M.A.R. (MARS) U., Zarco-Tejada, P., Hubbard, N., Loudjani, P.: Precision agriculture:an opportunity for EU farmers potential support with the CAP 2014 2020: in depth analysis. Publications Office (2014). https://doi.org/10.2861/58758

6. ITU, I.T.U.: Recommendation ITU-T Y.2060 Overview of the Internet of things. International Telecommunication Union (2012)
7. Triantafyllou, A., Sarigiannidis, P., Lagkas, T.D.: Network protocols, schemes, and mechanisms for internet of things (IoT): features, open challenges, and trends, (2018). https://doi.org/10.1155/2018/5349894
8. Farrell, S.: Low-Power Wide Area Network (LPWAN) Overview. RFC Editor (2018)
9. Walsh, K.: Wi-Fi HaLow™—Worth the wait | Wi-Fi Alliance, https://www.wi-fi.org/beacon/kevin-walsh/wi-fi-halow-worth-the-wait. Last accessed 15 Apr 2021
10. Huang, J., Qian, F., Gerber, A., Mao, Z.M., Sen, S., Spatscheck, O.: A close examination of performance and power characteristics of 4G LTE networks. In: Proceedings of the 10th International Conference on Mobile Systems, Applications, and Services. pp. 225–238. Association for Computing Machinery, New York, NY, USA (2012). https://doi.org/10.1145/2307636.2307658
11. LoRa Alliance Technical Committee Regional Parameters Workgroup: RP002–1.0.1 LoRaWAN Regional Parameters. (2020)
12. Brenes, J.A., Soto, F., Marín-Raventós, G.: A cost-efficient smart solution for small-scale farmers : a multidisciplinary approach. In: Albornoz, V.M., Mac Cawley, A. (eds.) Proceedings of the III International Conference on Agro BigData and Decision Support Systems in Agriculture. pp. 101–104. Universidad Técnica Federico Santa María, Chile, Santiago, Chile (2019)

Application of Generative Adversarial Network for Designer Assistance System

Bo Liu, You-Hsun Wu, and Yang-Cheng Lin

Abstract With the continual innovation of technology and product design methods, consumer needs change, and there is a gradual shift in product development from a function-oriented to consumer-oriented approach. The transformation of consumer perceptions into product design elements has become the focus of designers. This study develops an automated design generation system for product forms, which assists product designers to rapidly incorporate the consumer perceptions on the product form into the design process. It combines the Kansei engineering theory with the deep convolutional generative adversarial network (DCGAN), using the side-view contouring of cars as a case study. A questionnaire is given to the participants for selecting the top six subjective terms perceived by consumers in car styling. Further, 1006 car side views are scored on the six Kansei characteristics using a semantic differential scale. The results are clustered and analysed using fuzzy c-means. Subsequently, the clusters are applied as training datasets for the DCGAN, which generates antagonistic data to support the design process. The results of this study show that the data generated through the proposed method can facilitate designers in adjusting the product shape and realising an efficiently product shape design that matches the consumer's desired imagery.

Keywords Deep Learning · Kansei Engineering · Generative Adversarial Networks · Shape Design · Automatic Design

1 Introduction

The trend in product development is shifting towards a consumer-oriented approach. Consumer perceptions and needs are being considered valuable in product development by manufacturers [1]. From being purely functional, the product design trend has shifted to incorporating product semantics for creating products that are more in tune with consumer perceptions [2]. Various factors influence consumer purchase

B. Liu (✉) · Y.-H. Wu · Y.-C. Lin
Department of Industrial Design, National Cheng Kung University, No.1, University Road, Tainan, Taiwan

© The Author(s), under exclusive license to Springer Nature Singapore Pte Ltd. 2022
A. K. Nagar et al. (eds.), *Intelligent Sustainable Systems*, Lecture Notes in Networks and Systems 333, https://doi.org/10.1007/978-981-16-6309-3_12

decisions, which depend not only on the functionality and suitability for use of the product but also on the psychological response evoked by its appearance [3]. However, in traditional product design, the product shape is often determined by the subjective wishes of the designer or the previous creative perceptions, resulting in a gap between the traditional design process and consumer perceptions. Therefore, businesses need to understand the perceived needs of the consumer [4] in order to create a deep-seated desire in the consumer to purchase. Many researchers have proposed a consumer-oriented design approach to assist designers in product development [5]. The most extensively used approach is Kansei Engineering (KE) [1], which is a technique for translating the sensory imagery of the consumer into product design elements. It was first proposed by Japanese scholar Yamamoto in 1986 [6]. Nagamachi later proposed six different KE methods and applied them in the automotive, cosmetic and housing design areas. Of late, KE is being extensively used in various fields. Some studies have developed a design support model or system for product designers based on the perceptual engineering theory [7] in order to determine the best design for a product form. Thus, KE has been recognised as a consumer-centred and viable technology, and most research is concerned with the more precise mapping of the user needs to the design elements. However, most studies examine the existing data samples, and the generation of new data is limited, for research and analysis.

In view of the above, we combine KE and a generative adversarial network (GAN) to investigate product modelling and introduce KE in the design process. This study aims to develop a product form generation system to assist product designers in incorporating consumer perceptions and feelings on the product form into the design process. We focus on data generation after mapping the relationship between user perception and product styling. Figure 1 depicts the research framework. Furthermore, the practicality of the proposed methodology is demonstrated through a case study. We first apply KE theory [1] and a semantic difference scale [8] to obtain the relationship between consumer perception and car styling. We then use fuzzy c-means (FCM) [9] to cluster the data for analysis. Finally, we employ a GAN [10] to train the clustered data separately for generating styling that represents different consumer perceptions. The specific implementation steps are described in detail below.

2 Kansei Engineering for Extracting Experimental Data

In this study, we use car side views as a case study because a car is a typical example of a modern and sophisticated product. Aesthetic aspects of car styling influence 70% of customers' purchase intentions [11]. This section describes the application of the perceptual engineering theory to define the process of experimenting with samples and perceptual vocabulary. It includes three phases: (1) sample collection and preprocessing, (2) selection and extraction of Kansei words and (3) definition of the relationship between the samples and Kansei words.

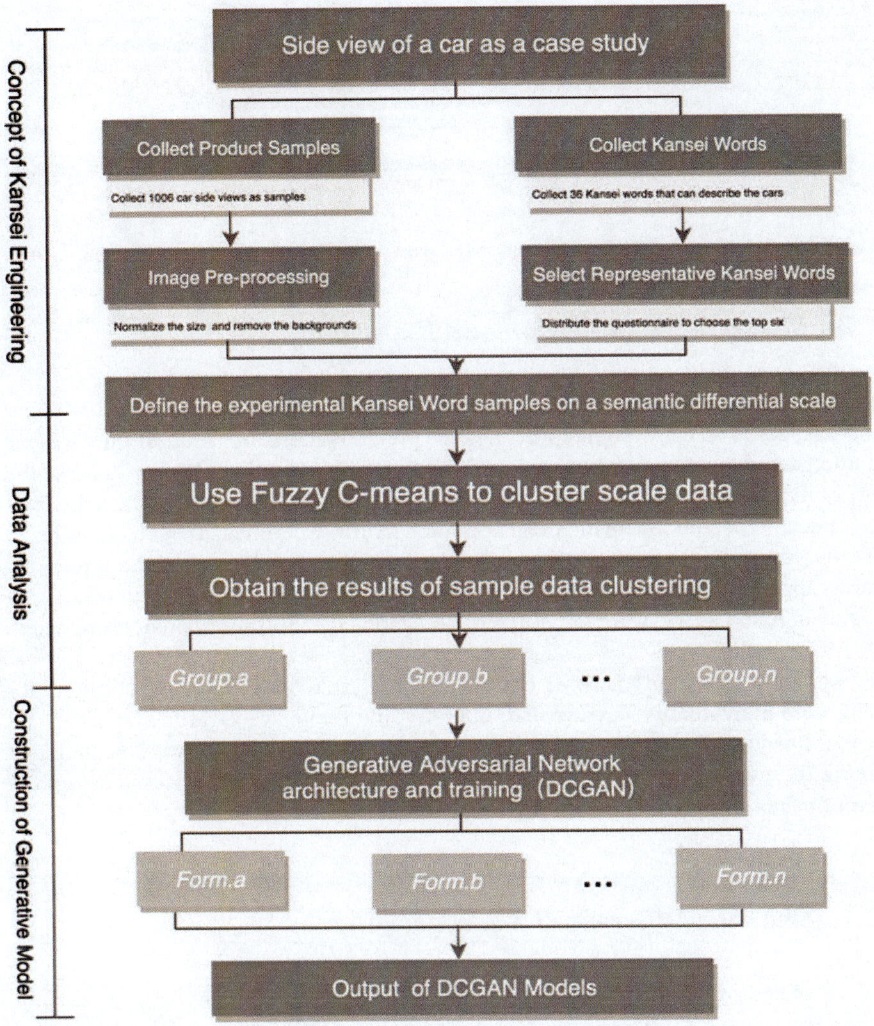

Fig. 1 Flow chart of the research framework

2.1 Sample Collection and Preprocessing

We collected a total of 1006 2D samples of car side views from various sources such as Websites and advertisements as experimental samples and preprocessed them. As this case study focuses only on car side views, we used Adobe Photoshop to remove the background in the collected car side-view samples and reset all the sample sizes to 256*192. Finally, we numbered the samples on car side views.

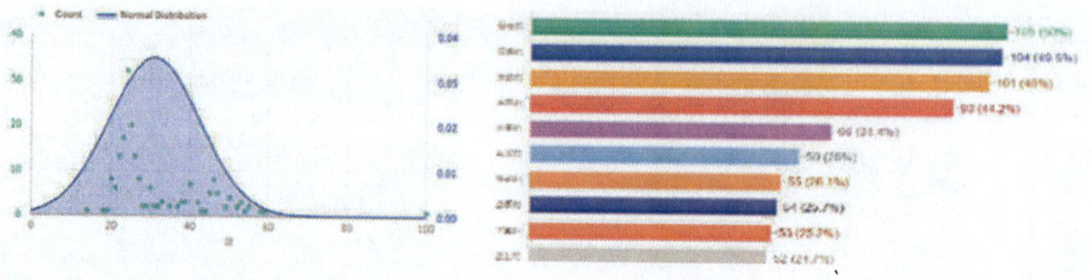

Fig. 2 Age distribution of the participants and top ten Kansei words for experimentation

2.2 *Kansei Word Collection and Extraction Experiment*

We selected a total of 36 Kansei words and consumer perceptions applicable to car models, based on car design reports and the previous literature. A questionnaire was administered to understand consumer perceptions on car styling. In this phase of the experiment, we aimed to obtain a more recognised perceptual image of the vehicle as an indicator for subsequent data segmentation. In this experiment, the selected Kansei words were provided in a questionnaire to the participants for selecting the perceptual terms applicable to a car model. The participants' average age was 30.89 years, 70.3% were male, and 29.7% were female (Fig. 2). Among the 209 questionnaires collected, 205 were valid.

Among the 36 Kansei words options, the top six most suitable descriptions for cars were individuality, streamlined, liquidity, futuristic, avant-garde and grand. As streamlined and liquidity were similar, a seventh word "powerful" was also included in the follow-up study category in order to differentiate the categories. The top ten experimental results are displayed in Fig. 2.

2.3 *Semantic Differential Scale Design*

We defined the relationship between the Kansei words and samples using a semantic differential scale [8]. We recruited a total of eight participants, including four males and four females, to fill the questionnaire. As shown in Fig. 3, the results of the previous phase of the experiment (six Kansei words) were utilised in the questionnaire to set up a ten-point semantic differential scale.

The eight participants were asked to evaluate the side styling of each sample car on a scale of 1–10 against the six perceptual terms, as depicted in Fig. 7. As the large data quantity in this study could induce product fatigue and affect the perceptual classification accuracy of the participants, the experiment was conducted in five sessions, with 200 randomised questions each. After completing the scale, we obtained the consumer ratings for each sample on the six different perceptual terms. For each subject, 1006*6 = 6036 data were obtained. In this experiment, we obtained perceptual data from the eight participants (6036*8 = 48,288). Due to space

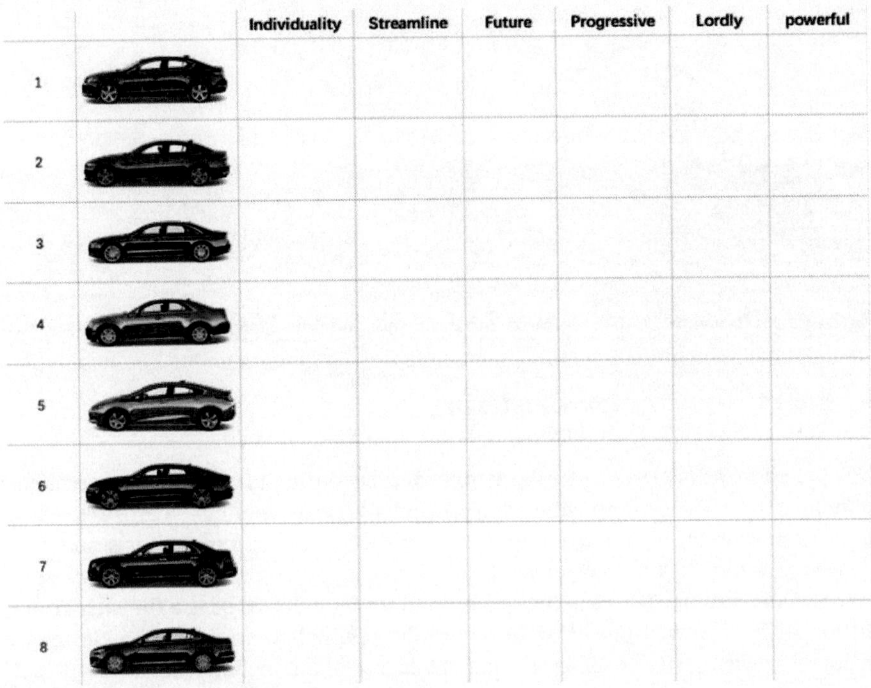

Fig. 3 Semantic differential scale

constraints, we only present the results of the first ten samples of one of the subjects and the final averages (Fig. 4).

We averaged the perceptual scores obtained from the respondents. We quantified the consumer perceptions and used them as a quantitative data source for the FCM. We describe the application of the FCM technique to process the experimental data in the next section.

	Individuality	Streamline	Future	Progressive	Lordly	powerful
1	7	4	2	4	6	8
2	7	3	2	4	6	7
3	7	3	2	3	5	8
4	7	4	1	3	6	8
5	7	4	2	4	5	8
6	2	2	1	1	4	4
7	4	3	3	4	6	6
8	7	3	5	5	5	8
9	2	1	1	1	8	6
10	6	4	4	6	8	4
11	5	3	5	5	7	3
12	6	4	3	3	7	4
13	5	4	3	3	7	4
14	6	4	2	2	8	3
15	3	3	2	2	5	7
16	3	6	1	2	7	4
17	5	5	3	3	7	4
18	8	1	5	7	3	8
19	7	2	6	5	4	7
20	6	6	2	2	7	8

	Individuality	Streamline	Future	Progressive	Lordly	powerful
1	3.25	3.625	2.125	2	4.625	5.75
2	3.875	3.625	2.375	2	4.875	5.125
3	3.25	3.125	2	2.375	4.125	5.875
4	3.875	3.25	2.25	2.875	5.125	5.375
5	3.375	3.25	2.375	2.625	3.375	4.75
6	3.875	3.625	3.25	3.25	5	4.875
7	3.875	3	3	3.75	4.75	5.25
8	7.125	5.75	3.75	5	5.125	7
9	4.75	2.5	2.875	3.875	6.25	6.125
10	3.75	3.625	3.125	3.625	4.125	4.375
11	3.75	3	3	3.25	4.25	3.625
12	3.625	3.25	2.75	2.75	4.5	4.5
13	3.375	3.125	2.625	2.625	4.125	4
14	3.75	3	2.125	2.25	4.625	4.375
15	3.75	3.125	2.5	2.875	4.875	4.875
16	4.125	4.125	2.625	3.25	5.375	5.625
17	5.375	4.25	3.25	3.75	5.5	5.5
18	7.375	3.75	5.375	7.375	5.875	6.875
19	6.625	3.875	6.75	7.5	6.125	6.5
20	4.875	4	3.75	4.25	6	6

Fig. 4 Scale scores of the first ten questions from one participant and mean scale scores for the first ten questions from eight participants

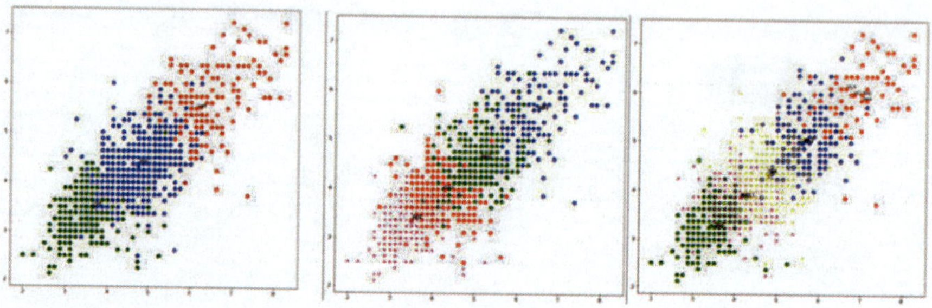

Fig. 5 Point plots with the data clustered into three, four and five categories, respectively

3 Fuzzy C-means for Clustering

We applied FCM [12] to cluster the sample data according to consumer perceptions, including the 1006 experimental samples and six perceptual characteristics. FCM is an unsupervised clustering algorithm often used for feature analysis and data clustering, which has been extensively applied in many fields [13]. It forms clusters based on the distance between data points and forms a cluster centre for each group. In this study, we used the FCM to obtain the optimal solution for the cluster by updating the centroid of each group and the degree of the membership value of each group data.

We set fuzzy parameter $M = 2$ and attempted to divide the data in this study into three, four and five clusters. As shown in Fig. 5, the best binning result was for three clusters, but the difference in data between each class after binning was not conducive for subsequent GAN training. Therefore, we finally divided the results into four clusters. The four categories were used as datasets for GAN training. The data quantity in the four clusters was 337, 284, 130 and 250, respectively. The cluster centroid samples for the four clusters were No. 67, No.163, No. 871 and No. 248, respectively.

4 Generative Adversarial Network (GAN)

After obtaining the clustering results, we categorised the training sets of the four clusters of the GAN according to the data sample numbers. The GAN is an unsupervised learning task in machine learning, proposed in 2014 [10]. The basic idea of the GAN involves the training of two opposing neural networks, the generator and discriminator. The generator takes random noises from latent space as the input, and the output needs to mimic the real samples in the training set as closely as possible. The discriminator discriminates the data generated by the generator and outputs it if eligible. Otherwise, the data are reintroduced into the generator and regenerated. The optimisation process of the GAN is a minima and maxima game for achieving the

Nash equilibrium [14]. During the training process, the generated sample quality and identifier discrimination are interactively improved [15]. In GAN architecture, the generator can be any algorithm, which can learn the training data distribution, and the discriminator can extract the features and perform training [16]. With the successful development of artificial intelligence, GANs are already satisfying the requirements of many applications. The GAN has brought new ideas into developing different fields [17]. In this study, we used the Deep Convolutional Generative Adversarial Network (DCGAN) [18] for training and generation, which combines the GAN and CNN.

DCGAN improves the original GAN's training stability and the quality of the generated results as well as the original GAN network architecture. Figure 6 illustrates the neural network architecture adopted in this study. We utilised the results of the previous FCM binning stage as the dataset for the GAN. After separate binning, we ranked the data and selected the top 100 of each group as the dataset for GAN training such that there are significant differences in each group of the dataset. In the generator, we first applied random noise as the input, decoded them into artefacts and then combined them with the actual image. Further, we flattened the image. We added a stride =2 convolution layer instead of a pooling layer and used the Conv2DTranspose layer for upsampling instead of the original Conv2D layer. We applied the LeakyReLu instead of ReLu as the activation function according to the original structure of the discriminator [18]. After testing, we used the Adam optimizer in both the generator and discriminator and set the learning rate to 0.0001 for training generation.

Figure 7 shows the entire process from training only one generation up to training 5,000 generations. Through the generation process, we determined that the generated data also differed with respect to features such as the number of training sessions,

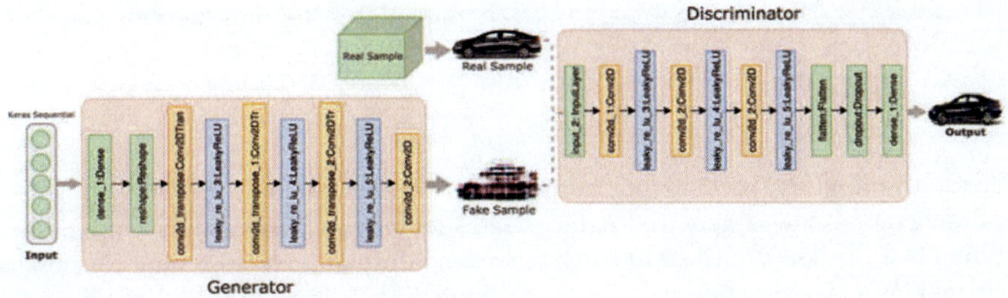

Fig. 6 Neural network architecture adopted in this study (we refer to the DCGAN architecture for mapping)

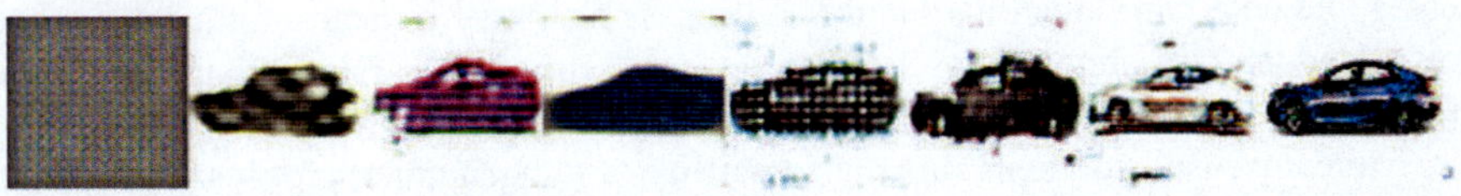

Fig. 7 Training process using DCGAN (1–5000 times)

Fig. 8 Results of each category generated after 10,000 times of cluster training

depending on the GAN input data. After 10,000 training sessions, the loss functions of the discriminator and generator were stable between 0.4 and 0.6, and 0.9 and 1.2, respectively. We exported the generated data after 10,000 training sessions for each dataset.

Figure 8 depicts the comparison of the generated results. The training results show that the generated results continue to improve and become more distinctive as the number of training sessions on the dataset increases. The above-generated results fully validate the feasibility of this research concept. The DCGAN selected as the generative network in this study has a stable training process and generated samples that are richly diverse. The generated results show that the scores of each group car side profiles differ significantly on different Kansei words and displayed in their appearance. These data are vital for the designer because they represent the feelings of consumers.

5 Conclusion

This study proposes a method for incorporating user perceptions in the design process and generating new shapes and demonstrates it through a case study on the side styling of a car. The proposed method generates information representing consumer perceptions and assists designers in determining the relationship between consumer perceptions and the product form more accurately. In this study, the concept of KE was applied to define the relationship between consumer perception and the side styling of a car. Consumer perceptual data were then obtained applying a semantic differential scale. The data were quantified and analysed using fuzzy c-means, and the results were used as the training dataset for an adversarial network. For realising training stability and quality output, a DCGAN was adopted to generate the car profiles. The experimental results demonstrate the versatility of the generated profiles. Designers can use this generative data to determine the best combination of the product form and consumer experience. However, this paper is a pilot study of the design of a car side view only. In practice, we do not design products from one perspective. In the future, we will extend GAN to generate product shapes from different angles. Secondly, we will expand our Kansei words database to allow more words to be grouped and selected. Finally, the systematic approach proposed in this study can not only be applied to car profiles, but also for the design of other consumer products.

References

1. Nagamachi, M.: Kansei engineering: a new ergonomic consumer-oriented technology for product development. Int. J. Ind. Ergon. **15**(1), 3–11 (1995)
2. Krippendorff, K.: On the essential contexts of artifacts or on the proposition that" design is making sense (of things)". Des. Issues **5**(2), 9–39 (1989)
3. Aoussat, A., Christofol, H., Le Coq, M.: The new product design-a transverse approach. J. Eng. Des. **11**(4), 399–417 (2000)
4. Lai, H.H., Lin, Y.C., Yeh, C.H., Wei, C.H.: User-oriented design for the optimal combination on product design. Int. J. Prod. Econ. **100**(2), 253–267 (2006)
5. Karnes, C.L., Sridharan, S.V., Kanet, J.J.: Measuring quality from the consumer's perspective: a methodology and its application. Int. J. Prod. Econ. **39**(3), 215–225 (1995)
6. Yamamoto, K.: Kansei engineering: The art of automotive development at Mazda. Special Lecture at the university of Michigan (1986)
7. Kengpol, A.: Design of a decision support system to evaluate the investment in a new distribution centre. Int. J. Prod. Econ. **90**(1), 59–70 (2004)
8. Osgood, C.E., Suci, G. J., Tannenbaum, P.H.: The measurement of meaning (No. 47). University of Illinois press (1957).
9. Bezdek, J.C., Ehrlich, R., Full, W.: FCM: the fuzzy c-means clustering algorithm. Comput. Geosci. **10**(2–3), 191–203 (1984)
10. Goodfellow, I.J., Pouget-Abadie, J., Mirza, M., Xu, B., Warde-Farley, D., Ozair, S., Bengio, Y., et al.: Generative adversarial networks. arXiv preprint arXiv:1406.2661(2014)
11. Cheutet, V.: 2D semantic sketcher for a car aesthetic design. Proc. CPI (2007)
12. Dunn, J.C.: A fuzzy relative of the ISODATA process and its use in detecting compact well-separated clusters (1973)
13. Yong, Y., Chongxun, Z., Pan, L.: A novel fuzzy c-means clustering algorithm for image thresholding. Measurement Sci. Rev. **4**(1), 11–19 (2004)
14. Goodfellow, I.: Nips 2016 tutorial: Generative adversarial networks. arXiv preprint arXiv:1701.00160(2016)
15. Creswell, A., White, T., Dumoulin, V., Arulkumaran, K., Sengupta, B., Bharath, A.A.: Generative adversarial networks: an overview. IEEE Signal Process. Mag. **35**(1), 53–65 (2018)
16. Cao, Y.J., Jia, L.L., Chen, Y.X., Lin, N., Yang, C., Zhang, B., Dai, H.H.: Recent advances of generative adversarial networks in computer vision. IEEE Access **7**, 14985–15006 (2018)
17. Wang, K., Gou, C., Duan, Y., Lin, Y., Zheng, X., Wang, F.Y.: Generative adversarial networks: introduction and outlook. IEEE/CAA J. Automatica Sinica **4**(4), 588–598 (2017)
18. Radford, A., Metz, L., Chintala, S.: Unsupervised representation learning with deep convolutional generative adversarial networks (2015). arXiv preprint arXiv:1511.06434

The Research of Macro-Control Diagram of ESP Well Based on Machine Learning

Zou Honglan, Yang Junzheng, Feng Gang, Wang Qianghua, and Ma Dan

Abstract The macro-control diagram plays an important role in the management and optimization of electric submersible pump wells (ESP). The judgment of ESP well working condition is determined by boundary value of macro-control diagram. However, in the actual process, the standards and theory of building macro-control diagram are not uniform, which cannot meet the needs of optimization of wells under different reservoir conditions and different working conditions. In this article, K-means method was used to perform clustering quantitative analysis and then combined with support vector machine to determine the best value of each decision boundary of the macro-control diagram, which avoids the influence of human factors on judgment results of ESP wells. It is shown that the macro-control diagram based on machine learning is more accurate in judging the working condition of ESP wells, which realizes the dynamic monitoring and judgment. It is of great significance and economic value for guiding the well stimulation.

Keywords ESP wells · Macro-control diagram · Machine learning · Cluster analysis · Support vector machine · Dynamic evaluation

1 Introduction

The daily management of electric submersible pump wells directly affects the pumping effect and the service life of electric pump, and these two indicators directly reflect the economic benefits of ESP. In order to understand the production status of oil wells and reduce the occurrence of faults, it is necessary to timely and accurately collect various production performance parameters of oil wells and analyze and master the production conditions of oil wells. The macro-control diagram of ESP

Z. Honglan · Y. Junzheng · W. Qianghua
PetroChina Research Institute of Petroleum Exploration and Development, Haidian, Beijing 100083, China

F. Gang (✉) · M. Dan
Xi'an Zhongkong Tiandi Technology Development Co, Ltd, Xi'an Shaanxi 710018, China
e-mail: feng9587@foxmail.com

© The Author(s), under exclusive license to Springer Nature Singapore Pte Ltd. 2022
A. K. Nagar et al. (eds.), *Intelligent Sustainable Systems*, Lecture Notes in Networks and Systems 333, https://doi.org/10.1007/978-981-16-6309-3_13

wells is drawn according to the needs of production and management of ESP wells. It can reflect the coordination of formation fluid supply capacity and drainage capacity of the ESP system, the reasonable utilization of electric pump equipment, and the management level of oilfield blocks. It can timely and accurately grasp the production status of oil wells, and take certain measures for abnormal production wells, further improve the working efficiency of electric pump wells, tap the production potential of oil wells, comprehensively guide the optimization of ESP system, and improve the management level of oil wells [1, 2]. Based on the analysis of the flow pressure and pump efficiency which affect the working condition of the ESP wells and their correlation, the sixth oil production plant of Daqing Oilfield summarized the macro-control chart of ESP wells with the bottom hole flow pressure as the ordinate and the displacement efficiency as the abscissa [3]. Sun Huanhe et al. [4] deduced and established the special working condition analytical diagram and comprehensive working condition analytical diagram of three-dimensional Cartesian coordinate system to analyze the working condition of ESP wells based on the characteristics of the relationship among head, displacement, and power of electric pump well unit and some specific intervals and specific points on the characteristic curve. Li Guoqing et al. [5] referred to the evaluation model of pumping wells according to the situation of electric pump wells in Zhongyuan Oilfield, combining with the characteristics of ESP wells to establish the evaluation model of ESP wells, and use its developed software and hardware system to test and evaluate the system efficiency of electric pump wells. Zeng Yongfeng et al. [6] established the macro-control diagram by using the relationship between relative flow pressure and displacement efficiency to analyze the working conditions of ESP wells. Liu Yigang et al. [7] in order to reduce the deviation of macro-control diagram caused by crude oil viscosity, water content, production, bottom hole pressure, emulsification, and other factors in actual heavy oil reservoir and dimensionless macro-dynamic control chart were established with dimensionless pressure (bottom hole flowing pressure/maximum hydrostatic pressure) and dimensionless displacement (actual liquid production/rated pump displacement) as the coordinate axis for electric pump well evaluation.

According to the above research and analysis, the current macro-control diagram of ESP wells usually adopts the method of expert experience or industry standard division to determine the boundary values of the macro-control diagram. The diagram is relatively ordinary, and the area division depends on human experience knowledge which has no effective data support. At present, big data technology has been used to realize the optimization of artificial lifting method, preventive maintenance of electric pump well, and reliability identification of oil production data [8–11]. This paper applies the big data analysis method of K-means clustering analysis and support vector machine (SVM) to optimize the boundary calculation method, which can comprehensively and accurately calculate the boundary of the macro-control diagram. The dynamic macro-control diagram is analyzed systematically, which not only represents the liquid supply capacity of oil reservoir, but also directly reflects the actual working status of electric submersible pump, so as to improve the working efficiency of electric pump well and tap the production potential of oil well.

2 Evaluation Index of Production Condition of ESP

The production status of ESP wells is determined by two aspects: One is the liquid supply capacity of the well, and the second is the discharge capacity of ESP [12]. The fluid supply capacity of oil well is mainly determined by formation pressure, reservoir thickness, permeability, flow pressure, and crude oil physical properties. The pump intake pressure can indirectly measure the formation fluid supply capacity of ESP wells and further reflect the working state of pump. Generally, the pump intake pressure is calculated by the driven liquid level of annular multiphase flow. With the progress of electric pump downhole monitoring technology, the pump intake pressure is calculated by the driven liquid level of annular multiphase flow, and now, the pump intake pressure can be directly obtained by downhole monitoring device, which avoids the influence of multiphase fluid properties on pump intake pressure calculation and improves the accuracy of pump intake pressure acquisition.

The type of pump directly determines the discharge capacity of the ESP, and the actual displacement of ESP wells is related to the discharge capacity and submergence degree of the reservoir, so the discharge capacity of different pump types cannot be simply compared with the actual displacement. At the same time, in order to show the coordination degree of the oil well, the displacement efficiency (as shown in Eq. 1) is selected as the evaluation index of the discharge capacity of the ESP, which not only reflects the discharge capacity of the oil well, but also makes the oil wells with different pump types comparable.

$$E = \frac{Q_f}{Q_c} \times 100\% \tag{1}$$

where Q_f is actual displacement of pump, m^3/d, and Q_c is rated displacement of pump, m^3/d.

According to the analysis of the evaluation index of the production condition of ESP wells, the displacement efficiency and the pump intake pressure reflect the production condition of ESP wells. Therefore, the displacement efficiency and the pump intake pressure are selected as the abscissa and ordinate of the macro-control diagram.

3 Establishment of Macro-Control Diagram

The decision boundary of conventional macro-control diagram is determined by empirical method and industry standard [10–14]. The theoretical basis for the selection of boundary value is closely related to the technical experience of researchers. There is an uncertainty which leads to inaccurate judgment of production conditions of electric pump wells and seriously affects the effect of pump inspection. In order to reduce the interference of human factors and obtain the most suitable macro-control diagram for studying the actual conditions of the block reservoir, K-means clustering and support vector machine (SVM) method were used in this paper to obtain the optimal value of each decision boundary of the macro-control diagram.

K-means clustering method is used to cluster all well data after removing the abnormal data in the study block under the coordinate of pump intake pressure displacement efficiency to obtain a clear data set. K-means clustering algorithm is a method based on centroid to divide known clustering categories. Its purpose is to obtain the clusters (the set of data objects) with minimum distance from centroid. As shown in Fig. 1a and b, the basic idea is to randomly select k clustering centers (centroids), divide all data objects into clusters according to the nearest principle, recalculate the centroids of all clusters, redistribute all data to each cluster, and iterate until the centroids do not change or move in a large range [15, 16] (as shown in Fig. 1c). Then, the support vector machine (SVM) method (Fig. 1d) is used to find the separation hyperplane with the largest geometric interval between data sets after K-means clustering, that is, the optimal decision boundary. The samples are

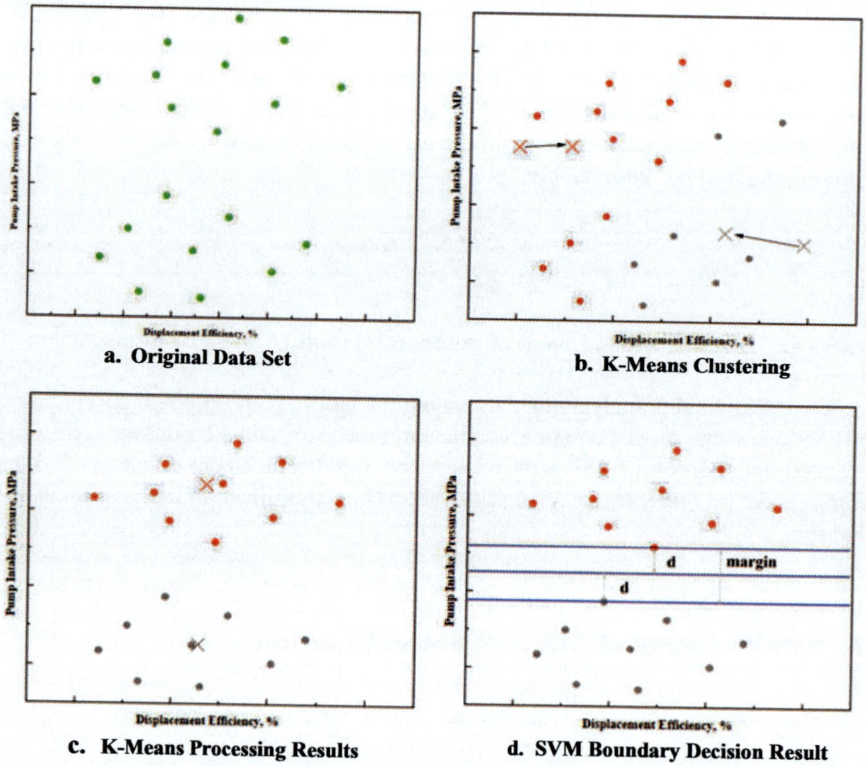

Fig. 1 Schematic diagram of the process of determining the best value of each decision boundary of the macro-control chart based on machine learning

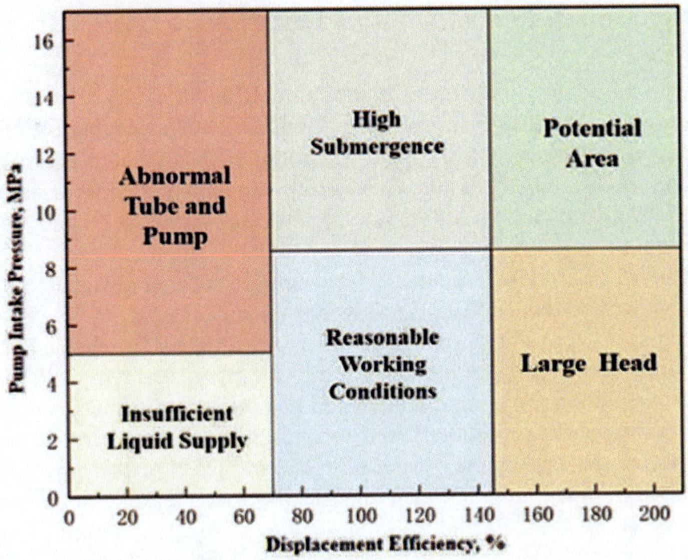

Fig. 2 Macro-control chart based on machine learning

segmented on both sides of the hyperplane, and the distance between the hyperplane and the nearest sample point on both sides is the largest [17].

By determining the upper and lower limits of pump intake pressure and displacement efficiency, the macro-control diagram of ESP wells can be divided into six different areas, as shown in Fig. 2. Firstly, according to the displacement efficiency of ESP wells operation engineering, the working conditions are divided into three types: normal range (displacement efficiency is between 70 and 145%), low displacement efficiency (displacement efficiency is less than 70%), and high displacement efficiency (displacement efficiency is more than 145%); by using k-means clustering algorithm to get different classification regions of pump intake pressure and displacement efficiency of electric pump well and combining with support vector machine to determine the decision boundary, the upper and lower limits of pump intake pressure are 8.5 MPa and 5.0 MPa, respectively, and the working conditions are divided into six categories: reasonable working conditions (displacement efficiency and pump intake pressure are in the normal range), abnormal area of pipe pump (low displacement efficiency and high pump intake pressure), insufficient liquid supply area (low displacement efficiency, outflow capacity of oil well is greater than inflow, and pump intake pressure is small), high submergence area (displacement efficiency is within normal and reasonable range, and pump intake pressure is greater), potential area (high displacement efficiency, shallow dynamic liquid level, and inflow capacity of oil well is greater than outflow capacity, and low pump intake pressure, the pump intake pressure is too high), and large head (the displacement efficiency is high, and the pump intake pressure is in a reasonable range).

4 Application of Macro-Control Diagram of ESP

According to the actual production situation of field and the data of all ESP wells in nearly three years, the macro-control diagram of ESP wells suitable for this oilfield operation area is established by using the algorithm model of macro-control diagram. The working condition analysis and adjustment evaluation of 15 ESP wells are carried out, and the working condition dynamic monitoring of one well is also taken into consideration.

A total of 15 ESP wells in an oilfield are comprehensively evaluated by macro-control diagram, and the evaluation results are shown in Fig. 3.

It can be seen from Fig. 3 that there are nine wells in the reasonable working condition area, one well in the insufficient liquid supply area, two wells in the high head area, one well in the potential area, and two wells in the abnormal pipe pump area. The corresponding treatment measures are taken for the electric pump well which is not in the reasonable working condition area. For the two wells in the abnormal area of pipe pump, it was found that there was leakage in the pipe string, so the pipe was replaced; for the wells in the area of insufficient liquid supply, the small displacement ESP should be replaced by a large displacement ESP or the injection volume of the surrounding water injection wells should be increased to make them enter the reasonable area; for the electric pump wells located in the area with large head, the measures of replacing small nozzles are taken to reach full formation potential; for the electric pump well in the potential area, it shows that the liquid production of the well is greater than the reasonable displacement range of the

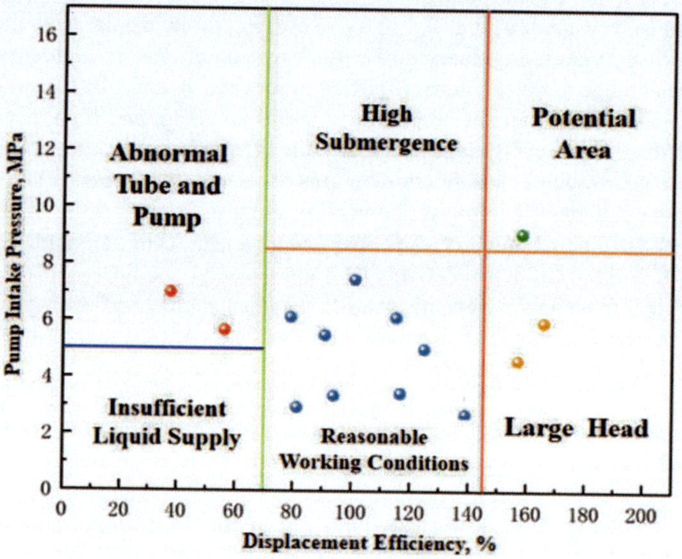

Fig. 3 Comprehensive evaluation result of ESP wells in an oil field

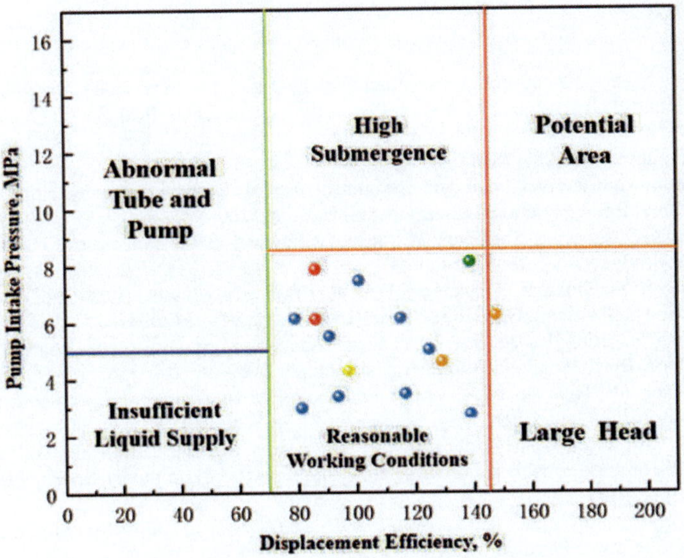

Fig. 4 Comprehensive evaluation diagram of ESP wells in an oil field after adjustment

electric submersible pump. Increasing the opening of the oil nozzle makes it enter the reasonable area.

The macro-control diagram is used to re-evaluate the six ESP wells after adjustments. The evaluation results are shown in Fig. 4. It can be seen from Fig. 4 that the production conditions of the six wells have been improved after the adjustment. Among them, five wells are in the reasonable working condition area after adjustment, and one well is close to the reasonable working condition area after adjustment.

5 Conclusions and Understandings

(1) Taking the pump intake pressure and displacement efficiency as the evaluation index, the decision boundary of macro-control diagram is determined based on K-means clustering and support vector machine method in machine learning. The macro-control diagram is more in line with the actual needs of the oilfield, and the evaluation results are more accurate.

(2) The macro-control diagram is used to judge the working condition of 15 ESP wells in an oilfield and dynamic monitoring of single well oil production condition. According to the judgment results of working conditions, the oil wells not in the reasonable working condition area are adjusted. After adjustment, the oil production conditions of the problem wells are improved, which shows that the macro-control diagram based on machine learning is accurate to judge the production condition of oil wells; it is of great significance to guide the production of ESP wells.

References

1. Liangyu, X.: Energy consumption evaluation and working condition analysis method based on macro-control charts. Energy Conserv. Petrol. Petrochem. Indus. **9**(3), 38–40 (2019)
2. Guojian, Z.: ESP characteristic curve revision and application of macro-control diagram. Oil Field Equipment **43**(08), 78–81 (2014)
3. Compilation and use method of dynamic control chart of electric submersible pump well: SY/T 5746–2009. Beijing, National Energy Administration (2009)
4. Huanhe, S., Yuren, L., Shicheng, W., et al.: Analytical method of electric pump well. Oil Drilling Prod. Technol. **05**, 43–54 (1990)
5. Guoqing, L., Youhong, X., Qiancheng, F., et al.: Efficiency test and evaluation of electric pump production well system. Oil Drilling Prod. Technol. **10**, 91–92 (2003)
6. Yongfeng, Z., Sijun, L., Yuming, L., et al.: Development of macro-control chart of ESP Well Production. Inner Mongolia Petrochemical Industry **2**, 85–86 (2006)
7. Liu Yigang, W., Huaxiao, M.X., et al.: Application of dimensionless macro-control chart of ESP in Bohai Oilfield. Chem. Eng. Equipment **11**, 265–267 (2018)
8. Wasfi, B., Alqasimi, M., Al Kadem, M., et al.: Innovative machine learning method to locate the root cause of the unreliable data coming from intelligent field equipment. Society of Petroleum Engineers (2019)
9. Ounsakul, T., Rittirong, A., Kreethapon, T., et al.: Data-driven diagnosis for artificial lift pump's failures. Society of Petroleum Engineers (2019)
10. Jansen Van Rensburg, N., Kamin, L., Davis, S.: Using machine learning-based predictive models to enable preventative maintenance and prevent ESP downtime. Society of Petroleum Engineers (2019)
11. Shi, J., Chen, S., Zhang, X., et al.: Artificial lift methods optimising and selecting based on big data analysis technology. Int. Petrol. Technol. Conf. (2019)
12. Kuang, B.: Research and application on the analysis of the working conditions of ESP well. China University of Petroleum (East China) (2010)
13. Mei, S., Shao, Y., Liu, J., et al.: Technology of electric submersible pump. Beijing, Petroleum Industry Press **7**, 45–56 (2004)
14. Li, J.: Oil well pumping conditions analysis and energy consumption evaluation method based on marco-control charts. Yanshan University (2016)
15. Mac Queen, J.: Some methods for classification and analysis of multivariate observations. Proceedings of the Fifth Berkeley Symposium on Mathematical Statistics and Probability, University of California Press, Berkeley, 281–297 (1967)
16. Jain, A.K.: Data clustering: 50 years beyond K-means. Pattern Recogn. Lett. **31**(8), 651–666 (2009)
17. Tian Z.D., Gao, X.W., Shi, T.: Combination kernel function least squares support vector machine for chaotic time series prediction. Acta Physica Sinica, **63**(16), 160508 (2014)

Detection System Potholes on Roads based on Recurrent Neuronal Network

Younes Ed-Doughmi and Rachid El Ayachi

Abstract The supervision of road quality is among the main lines of research in road security, exploiting the various techniques of machine learning. Our goal is to propose an efficient and accurate system to provide a complete and detailed report on the state of the road and anticipate the maintenance to be done to road or decision support if the system uses real-time in the context of an autonomous car example. For this, we propose a deep learning architecture in two steps. In the first step, we start a segmentation to determine the sides of the road and limiting the area detection of potholes. The second step to detecting potholes in this area reduces the number of false positives detected by the model. It reduces the detections of potholes that are not relevant for this detection, including potholes at the side of the road or the sidewalk. For this, each model was training on images from a public database. The result of our approach showed its effectiveness by reaching accuracy **93%**.

Keywords Pothole detection · Road damage · Road quality · Segmentation · Computer vision · Deep learning · Convolutional neural network

1 Introduction

Road safety is among the world's issues trying to improve with investments, laws, and controls changing with the rapid evolution of global transportation.

In addition, the detection of damaged and cracked parts of the infrastructure depended only on the engineering team's expertise to experiment. However, with the growing demand for road condition analysis, the analysis reports of the engineering team consume more time and financial resources. Therefore, automatic road analysis

Y. Ed-Doughmi (✉) · R. El Ayachi
Department of Computer Science, FST, University of Sultan Moulay Sliman, Beni Mellal, Morocco
e-mail: younes.ed-doughmi@usms.ma

R. El Ayachi
e-mail: rachid.elayachi@usms.ma

© The Author(s), under exclusive license to Springer Nature Singapore Pte Ltd. 2022

A. K. Nagar et al. (eds.), *Intelligent Sustainable Systems*, Lecture Notes in Networks and Systems 333, https://doi.org/10.1007/978-981-16-6309-3_14

systems are emerging as an alternative or helpful tool to speed up the road inspection process and provide more relevant analysis reports in record time.

For this, the research community has focused on this scourge to reduce the number of incidents caused, either by human error, car performance, or the quality of roads. In our work, we are interested in analyzing the quality of the road and the solutions that can improve road maintenance and give an objective estimate of the budget allocated.

This paper presents a hybrid approach to pothole detection based on two types of deep learning architectures, the first architecture is based on convolutional neural network (CNN), and the second architecture is based on deep convolutional encoder-decoder. Our solution is easy to implement, such as a low-cost camera that can be easily installed inside or outside a car, to retrieve sequences of images recorded via this camera. Then, all these images will be used by our system, to analyze the road condition. Our system which is based on two deep learning architectures can detect pothole areas in various road conditions, such as dry, wet, and shaded lanes, as well as several sizes and shapes of potholes. To evaluate the performance, we compare our results with other methods in the same field. The performance obtained indicates that our system can detect potholes with an optimal prediction rate.

2 Related Works

The number of papers published on road quality analysis continues to increase, which shows the scientific community's interest in the road quality problem.

Among the first and most standard approaches, we find machine learning for pothole detection based on support vector machine (SVM) [1], this approach is based on the histogram and training of the set of pavement images on the svn method. As a result, this study allowed detection of potholes more easily.

On the other hand, other works based on 3D data [2, 3], this work uses a cheap Kinect sensor, which allowed us to provide measurements of the depth of the detected potholes and give an estimation on the level of danger related to this detected pothole. Still, this type of sensor is limiting by the environmental bills and the intensity of the LED light (light emitting diode).

Other works include the implementation of captures to automatically detect cracks with a conventional neural network CNN [4], or using a stereo vision system [5, 6], to make a better analysis of any road surface, which can give more information in the case of driver-assisted driving or also for autonomous cars.

The works based on deep learning have shown their effectiveness on more complex processing, such as the detection of objects like the pothole [7, 8], the detection is base on architectures based on deep learning, which can classify and detect cracks and potholes on the road. Also, works use semantic segmentation architectures as U-Net [9]. Also, detection models based on YOLOV4 [10, 11], such as this type of model, have shown their effectiveness on the same problem our work.

3 Background

3.1 Classification Using Deep Learning

Convolutional Neural Networks [12] (ConvNets or CNNs) are a class of neural networks that have proven to be very effective in areas such as image recognition and classification. ConvNets have successfully identified faces, objects, and traffic signs, outside of the power supply of vision in robots and self-propelled cars. In our work, we will see the classification of born hens using deep learning (CNN).

Figure 1 represents the basic architecture of ConvNet, which contains basic functionalities such as Pooling, FC Layer. Many other architectures have developed based on the basic architecture of convolution, which has allowed spectacular advances in artificial intelligence.

3.2 Deep Convolutional Encoder–Decoder

Encoder–decoder neural network architecture combines with a new strategy to improve the overall consistency of labels, which an improved image segmentation model accompanies. Label distribution predictions extracted from the SegNet [13] network are studied and used in the project for image labeling Fig. 2. For our work, this type of model allows us to isolate the road area and target the areas to classify and increase the accuracy rate.

The above figure shows the Segnet architecture, and the network is only convolutional because there are no tightly connected layers. A decoder upsamples its

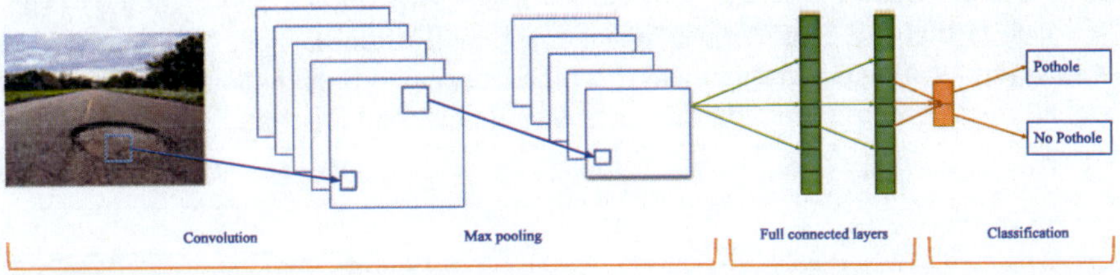

Fig. 1 Global view of the cnn architecture

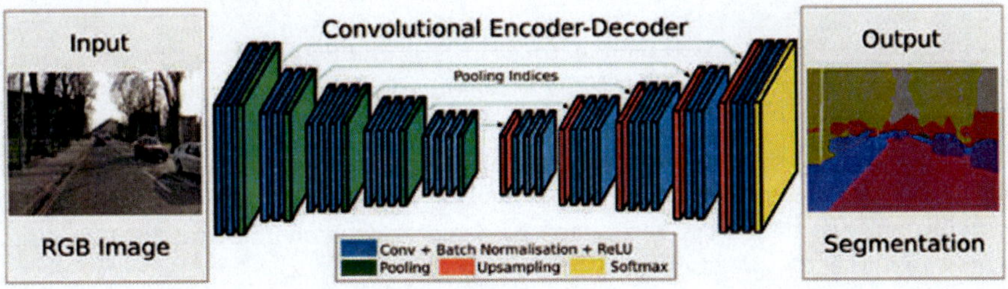

Fig. 2 Convolution encoder–decoder

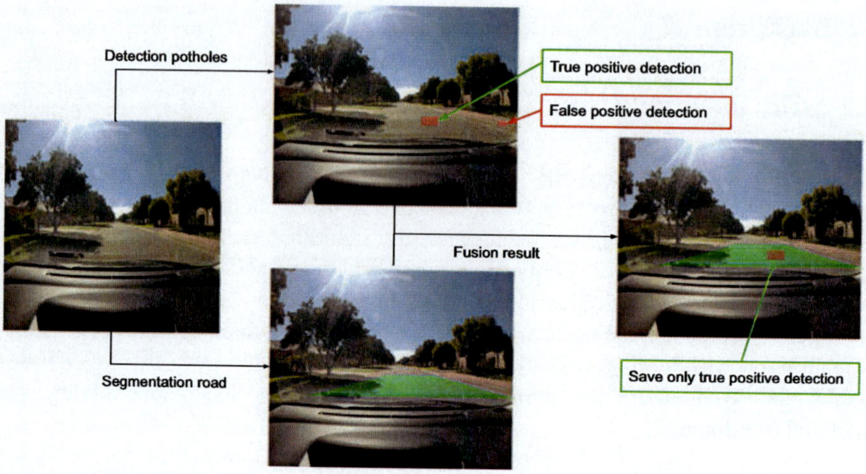

Fig. 3 Proposed architecture potholes detection

input using the transferred pool indices from its encoder to produce a sparse feature map(s). It then performs convolution with a trainable filter bank to densify the feature map, a final decoder output feature map fed to a softmax classifier for pixel-wise classification.

4 Proposed Approach

The proposed architecture used in delayed time Fig. 3, the combination of segmentation and detection increases the accuracy rate. However, the disadvantage of this approach is that the data storage management will often be bulky. This proposed architecture design uses fragmented machine learning to reduce the load on massive computing resources regarding data processing power and memory size.

5 Experiments

5.1 Dataset

To train our model, we used the dataset [14, 15]. The dataset images used for this study were captured by a GoPro Hero 3+ camera with a resolution of 3680×2760. The dataset is composed of two different sets: one considered simple and the other more complex. This quality allows the data is splitting into two groups, data used for training which presents 70% of the dataset. This division enables the model to validate better its training on the part of the dataset on which it has not done training, which significantly increases the model's accuracy.

5.2 *Training*

The training of the dataset is splitting into two steps. The first step is the training on the segmentation of the road, with a division of an image into two classes, such as the first class that surrounds the road, and the second class is the background that represents the rest of the image after the step labeling is necessary because the dataset not provide this labeling.

The second step of the training is the detection of potholes. This step of training is basing on the labeling already done in the work of the dataset used.

At the end of the training, the validation is performed by combining two segmentation and detection models to measure each pair's performance.

6 Result Analysis

In this section, we present the training results of several pairs of architecture studied in Table 1, with the accuracy rates and the validation obtained, as well as the rate of testing performed on images not used in the learning phase.

Our approach indicates that the combination of the two models increases the accuracy and reduces the false positives related to detecting potholes that are not in the road area.

Table 1 Score experimental results

Segmentation	Method	Accuracy (%)	Validation (%)	Test (%)
DenseASPP [17]	EfficientNet [16]	**95.40**	**94.51**	**93.3**
	YOLOv3 [18]	92.32	92.9	92.7
	GPipe [19]	84.82	79.0	83.0
	SENet [20]	82.7	85.0	83.04
	GoogleNet [21]	74.8	71.45	73.38
	AlexNet [22]	73.55	71.8	70.64
Unet [23]-ResNet [24]	EfficientNet	94.13	91.8	91.3
	YOLOv3	94.57	90.47	90.64
	GPipe	84.6	78.04	80.55
	SENet	82.0	81.79	81.04
	GoogleNet	73.47	71.17	70.15
	AlexNet	70.75	70.64	69.53
Unet-VGG16 [25]	EfficientNet	94.06	92.21	90.32
	YOLOv3	94.15	90.88	91.13
	GPipe	85.47	77.51	80.65
	SENet	81.02	80.78	79.13
	GoogleNet	72.5	70.94	68.65
	AlexNet	71.45	70.48	70.25

7 Conclusions

In this paper, we presented pothole detection by combining two deep learning architectures, segmentation, and detection. The two computational models used in our architecture and training on a public dataset containing images collected from different urban and rural road areas. The experimental results confirmed that using our approach; the potholes are well detected and achieved high accuracy.

This result also encourages government authorities to use our system to monitor and evaluate the condition of roads in all territories of the country.

Finally, for every solution, there is always a limitation. It is the case of our work, namely that our solution is not optimal for real-time use. For this reason, in the continuation of our work, we will try to merge the two models, segmentation and detection, in a single model that can make both approaches simultaneously, so we can upgrade to real-time detection and segmentation without losing the overall accuracy.

References

1. Lin, J., Liu, Y.: Potholes detection based on SVM in the pavement distress image. In:2010 Ninth International Symposium on Distributed Computing and Applications to Business, Engineering and Science, pp. 544–547. IEEE (2010)
2. Ahmed, A., Ashfaque, M., Ulhaq, M.U., Mathavan, S., Kamal, K., Rahman, M.: Pothole 3d reconstruction with a novel imaging system and structure from motion techniques. IEEE Trans. Intell. Transport. Syst. (2021)
3. Moazzam, I., Kamal, K., Mathavan, S., Usman, S., Rahman, M.: Metrology and visualization of potholes using the microsoft kinect sensor. In: 16th International IEEE Conference on Intelligent Transportation Systems (ITSC 2013), pp. 1284–1291. IEEE (2013)
4. Fan, R., Bocus, M.J., Zhu, Y., Jiao, J., Wang, L., Ma, F., Cheng, S., Liu, M.: Road crack detection using deep convolutional neural network and adaptive thresholding. In: 2019 IEEE Intelligent Vehicles Symposium (IV), pp. 474–479. IEEE (2019)
5. Fan, R., Liu, Y., Yang, X., Bocus, M.J., Dahnoun, N., Tancock, S.: Real-time stereo vision for road surface 3-d reconstruction. In: 2018 IEEE International Conference on Imaging Systems and Techniques (IST), pp. 1–6. IEEE (2018)
6. Fan, R., Bocus, M.J., Dahnoun, N.: A novel disparity transformation algorithm for road segmentation. Inf. Process. Lett. **140**, 18–24 (2018)
7. Maeda, H., Sekimoto, Y., Seto, T., Kashiyama, T., Omata, H.: Road damage detection and classification using deep neural networks with smartphone images. Comput.-Aided Civil Infrastruct. Eng. **33**(12), 1127–1141 (2018)
8. Cao, M.-T., Tran, Q.-V., Nguyen, N.-M., Chang, K.-T.: Survey on performance of deep learning models for detecting road damages using multiple dashcam image resources. Adv. Eng. Inform. **46**, 101182 (2020)
9. Pereira, V., Tamura, S., Hayamizu, S., Fukai, H.: Semantic segmentation of paved road and pothole image using u-net architecture. In: 2019 International Conference of Advanced Informatics: Concepts, Theory and Applications (ICAICTA), pp. 1–4. IEEE (2019)
10. Sarmiento, J.-A.: Pavement distress detection and segmentation using YOLOv4 and DeepLabv3 on pavements in the Philippines. arXiv:2103.06467 (2021)
11. Bochkovskiy, A., Wang, C.-Y., Liao, H.-Y.M.: YOLOv4: Optimal speed and accuracy of object detection. arXiv:2004.10934 (2020)

12. Krizhevsky, A., Sutskever, I., Hinton, G.E.: Imagenet classification with deep convolutional neural networks. In: Advances in Neural iInformation Processing Systems, pp. 1097–1105 (2012)
13. Badrinarayanan, V., Kendall, A., Cipolla, R.: Segnet: a deep convolutional encoder-decoder architecture for image segmentation. IEEE Trans. Pattern Anal. Mach. Intell. **39**(12), 2481–2495 (2017)
14. Nienaber, S., Booysen, M.J., Kroon, R.S.: Detecting potholes using simple image processing techniques and real-world footage (2015)
15. Nienaber, S., Kroon, R.S., Booysen, M.J.: A comparison of low-cost monocular vision techniques for pothole distance estimation. In: 2015 IEEE Symposium Series on Computational Intelligence, pp. 419–426. IEEE (2015)
16. Tan, M., Le, Q.: Efficientnet: Rethinking model scaling for convolutional neural networks. In: International Conference on Machine Learning, pp. 6105–6114. PMLR (2019)
17. Yang, M., Yu, K., Zhang, C., Li, Z., Yang, K.: Denseaspp for semantic segmentation in street scenes. In: Proceedings of the IEEE Conference on Computer Vision and Pattern Recognition, pp. 3684–3692 (2018)
18. Redmon, J., Farhadi, A.: YOLO3: an incremental improvement. arXiv:1804.02767 (2018)
19. Huang, Y., Cheng, Y., Bapna, A., Firat, O., Chen, M.X., Chen, D., Lee, H.J., Ngiam, J., Le, Q.V., Wu, Y., et al.: GPipe: Efficient training of giant neural networks using pipeline parallelism. arXiv:1811.06965, 2018
20. Hu, J., Shen, L., Sun, G.: Squeeze-and-excitation networks. In: Proceedings of the IEEE Conference on Computer Vision and Pattern Recognition, pp. 7132–7141 (2018)
21. Al-Qizwini, M., Barjasteh, I., Al-Qassab, H., Radha, H.: Deep learning algorithm for autonomous driving using googlenet. In: 2017 IEEE Intelligent Vehicles Symposium (IV), pp. 89–96. IEEE (2017)
22. Zifeng, W., Shen, C., Van Den Hengel, A.: Wider or deeper: revisiting the resnet model for visual recognition. Pattern Recogn. **90**, 119–133 (2019)
23. Ronneberger, O., Fischer, P., Brox, T.: U-net: convolutional networks for biomedical image segmentation. In: International Conference on Medical Image Computing and Computer-Assisted Intervention, pp. 234–241. Springer (2015)
24. He, K., Zhang, X., Ren, S., Sun, J.: Deep residual learning for image recognition. In: Proceedings of the IEEE Conference on Computer Vision and Pattern Recognition, pp. 770–778 (2016)
25. Simonyan, K., Zisserman, A.: Very deep convolutional networks for large-scale image recognition. arXiv:1409.1556 (2014)

Eduinformatics: A New Academic Field Needed in the Age of Information and Communication Technology

Kunihiko Takamatsu, Katsuhiko Murakami, Yasuhiro Kozaki, Kenya Bannaka, Ikuhiro Noda, Kenichiro Mitsunari, Masato Omori, and Yasuo Nakata

Abstract Worldwide, evidence-based education, institutional research, and learning analytics have become key terms in higher education in the twenty-first century. In order to realize these key terms, a great deal of data is necessary. With the development of information and communication technology (ICT) and the accompanying evolution of learning management systems, large amounts of educational data can be accumulated more easily and quickly than ever before. In this paper, we discuss how ICT can be used effectively and efficiently to improve the quality of higher education. We believe that one of the answers is the immediate establishment of a new concept called "eduinformatics," which we proposed in 2018. Eduinformatics is a novel field of education that combines both education and informatics. Eduinformatics not only deals with students' data but also provides new analytical methods and concepts to handle data in education, similar to bioinformatics. To investigate the significance of eduinformatics, we introduce some practical examples from Kobe Tokiwa University in Japan, and we present results of recent eduinformatics. From these examples, we show that the construction of ICT-based eduinformatics will lead to the improvement of the quality of higher education and higher education reform.

Keywords Eduinformatics · Higher education · Evidence-based education

1 Introduction

Everything in the world is about to be digitized and converted into data. The rapid development of information and communication technology (ICT) is playing a role in this. Digitalizing and dataifying is expected to enhance our abilities to manage,

K. Takamatsu (✉) · K. Bannaka · I. Noda · K. Mitsunari · M. Omori · Y. Nakata
Kobe Tokiwa University, Kobe 6530838, Japan

K. Murakami · Y. Kozaki
The University of Tokyo, Tokyo 108-8639, Japan

K. Takamatsu · K. Murakami · Y. Kozaki · K. Bannaka · I. Noda · K. Mitsunari · M. Omori · Y. Nakata
Osaka Kyoiku University, Osaka 582-8582, Japan

© The Author(s), under exclusive license to Springer Nature Singapore Pte Ltd. 2022

A. K. Nagar et al. (eds.), *Intelligent Sustainable Systems*, Lecture Notes in Networks and Systems 333, https://doi.org/10.1007/978-981-16-6309-3_15

analyze, and predict what is happening in society. This wave is also sweeping into higher education. For example, evidence-based education (EBE), institutional research (IR), and learning analytics (LA) are attracting attention in many countries as the centerpiece of higher education reform.

In 2008, Japan's Ministry of Education, Culture, Sports, Science and Technology (MEXT) clearly stated that it would promote IR in higher education [1]. Further, in 2017, it advised that it would promote educational policies that emphasize objective evidence [2] and, in 2018, that it would promote the visualization of learning outcomes [3]. Most recently, in 2021, it added a budget to promote the digital transformation of higher education [4].

Realizing EBE, IR, and learning analytics requires a great deal of data. Of late, data have been accumulating at an accelerated rate with the development and spread of ICT. Furthermore, the rapid spread of remote classes and distance learning using learning management systems (LMS) during the COVID-19 pandemic notably accelerated the accumulation of data on student learning in 2020, including in Japan. However, while LMS automatically accumulate data, Japanese universities remain responsible for its analysis and use.

Many educational researchers and IR staff are working to determine best practices for managing these data. However, this line of inquiry cannot be effectively pursued using conventional pedagogical techniques alone. If researchers can adequately use informatics, such as data science, machine learning, statistics, and so on, then data analysis and prediction will be promoted, which can dramatically contribute to the development of higher education, just as bioinformatics has significantly developed biology [5].

To quicken this research, we proposed the concept of "eduinformatics," a novel field of education that combines both education and informatics, in 2018 [6]. The term is a portmanteau of "education" and "informatics." Eduinformatics deals with students' data and provides novel analytical methods and concepts for using data in education, similar to bioinformatics [7] (Fig. 1).

Please note that the first paragraph of a section or subsection is not indented. The first paragraphs that follows a table, figure, equation, etc., does not have an indent, either.

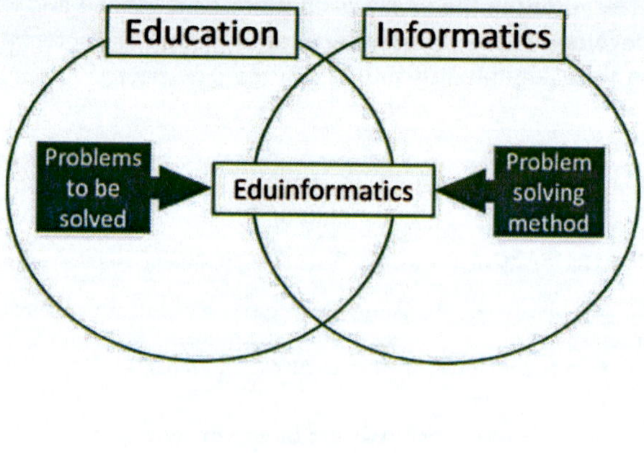

Fig. 1 Concept of eduinformatics (from [7])

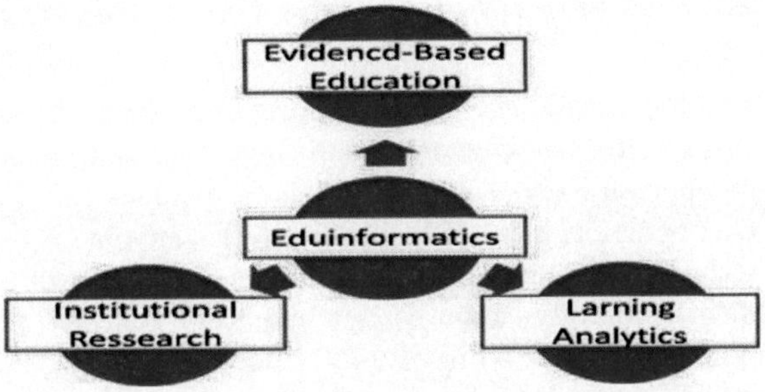

Fig. 2 Effectiveness of eduinformatics

Eduinformatics is a necessary concept in this age of the rapid development of ICT. Today, a large number of educational data points can be easily accumulated using LMS. Establishing eduinformatics can further accelerate EBE, IR, and LA which in turn will improve the quality of higher education (Fig. 2).

This paper discusses how eduinformatics can fully apply ICT to improve the quality of higher education by thinking with some practical examples of our own efforts at Kobe Tokiwa University. Using these examples, we argue that ICT is crucial for connecting education and student success.

2 Some Practical Examples

2.1 Established Research Unit of Mathematical and Data Science Education in a Center for the Promotion of Excellence in Research and Development in Higher Education within Kobe Tokiwa University

In 2016, the Cabinet Office in Japan proposed "Society 5.0," a human-centered society, in the fifth phase of the Basic Plan for Science and Technology as the future society that should be pursued, and the government is currently working to realize it. [8]. As a part of the plan, MEXT has explicitly stated the need to strengthen mathematical data science education in universities [9]. In response to this trend, we established a research unit of mathematical and data science education in the existing center for the promotion of excellence in research and development in higher education within Kobe Tokiwa University. This unit supports not only mathematical and data science education for students, but also the promotion of IR and statistical support for researchers within our university. Notably, we theorized eduinformatics as part of our work in this unit. In the following, we introduce some examples of the use of eduinformatics at Kobe Tokiwa University.

2.2 Some Examples of the Use of Eduinformatics

Ensuring equal evaluation among teachers in first-year experience course

Kobe Tokiwa University and Kobe Tokiwa College are both located in the same compound. Kobe Tokiwa University includes two faculties and four departments. One is the Faculty of Health Science, which has a department of nursing, medical technology, and radiology. Another is the Faculty of Education, which has a department of child education. Meanwhile, Kobe Tokiwa College is home to a department of oral health. Approximately, 450 students were registered at Kobe Tokiwa University in 2020.

In Japan, the importance of the first-year experience at university is increasing, as the country has advanced to the universal stage of university (i.e., enrollment rate of more than 50%). In 2017, Kobe Tokiwa University began a new first-year experience course named Manaburu I [10]. Here, we developed the word "Manaburu" as a portmanteau of the Japanese words for "learning" (manabu) and "able." The term "Manaburu" thus indicates that a "student is able to learn by themselves." Nineteen teachers teach this course, with approximately 350 students in total and 16–17 students per class. In this class, students gather into groups of approximately six. It is difficult for 19 different teachers to evaluate students equally. In a previous study that used the Steel–Dwass test, we found that the teachers of this course did not evaluate students in a standard way, even though rubrics (Fig. 3) [11].

Based on our above finding, we created a standard evaluation method—a revised rubric—for the teachers to follow. Moreover, we explained the rubric in detail to both the teachers and students of the Manaburu II after Manaburu I [13]. A multiple comparison analysis of the results showed no significant differences in the teachers' evaluations of students.

Development of web-based visualization methods for learning outcomes

As mentioned above, MEXT promotes the visualization of learning outcomes [3]. In step with MEXT's affirmation of the visualization of student outcomes, we developed a web-based radar chart system of Tokiwa competencies, which consists of 19 abilities to help students independently measure their competency level (Fig. 4) [14]. We designed this web-based radar chart system to be compatible with both

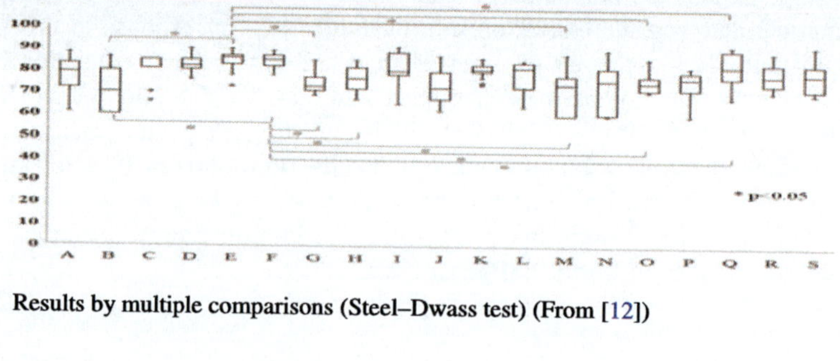

Fig. 3 Results by multiple comparisons (Steel–Dwass test) (From [12])

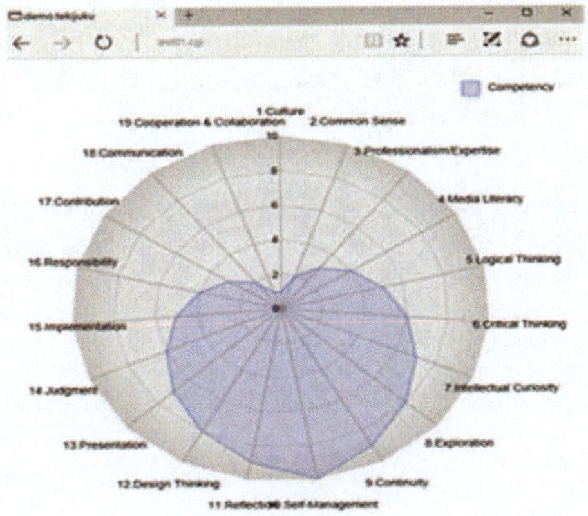

Fig. 4 Example of a radar chart (From [14])

computers and smartphones because most students now use a smartphone instead of a computer.

Predicting the probability of student dropout through EMIR using data from current and graduated students

In 2016, an office for IR promotion opened at Kobe Tokiwa University. The aim of this office is to propose, manage, arrange, and collect information on students at the university not only as a general management strategy, but also as a support for enrollment management. An IR promotional unit was also newly established in 2017. In contrast to this office, which consists only of administrative staff, this unit comprises a collaborative team of administrative staff and faculty/researchers. As part of this unit, we analyzed the nuances of "student dropout" by using data from current, graduated, and dropout students [15]. Dropping out has become a major problem in Japanese higher education.

At the time of our analysis, we had 3495 data points (i.e., headcounts) in our database containing 1246 items of numerical value. We obtained results with these data by processing them using a machine learning technique and two methods. A logistic regression yielded a test correction rate exceeding 95% and a random forest yielded a correction rate of about 90% [15].

Best graph to enable clear visualization of the data

If a significant difference is observed through analysis of student data in IR and/or LA, we must check the distribution of the data. However, when comparing and analyzing the data, IR practitioners should avoid using bar graphs in the form of a mean.

It is commonly known that scatter plots (Fig. 5a), beeswarm plots (Fig. 5b), and violin plots (Fig. 5c) are preferable to bar graphs or box plots [16]. We confirmed which graph was the best fit for our study by using data from our own university. Ultimately, we found that a scatter plot would best illustrate the original distribution of the data [17].

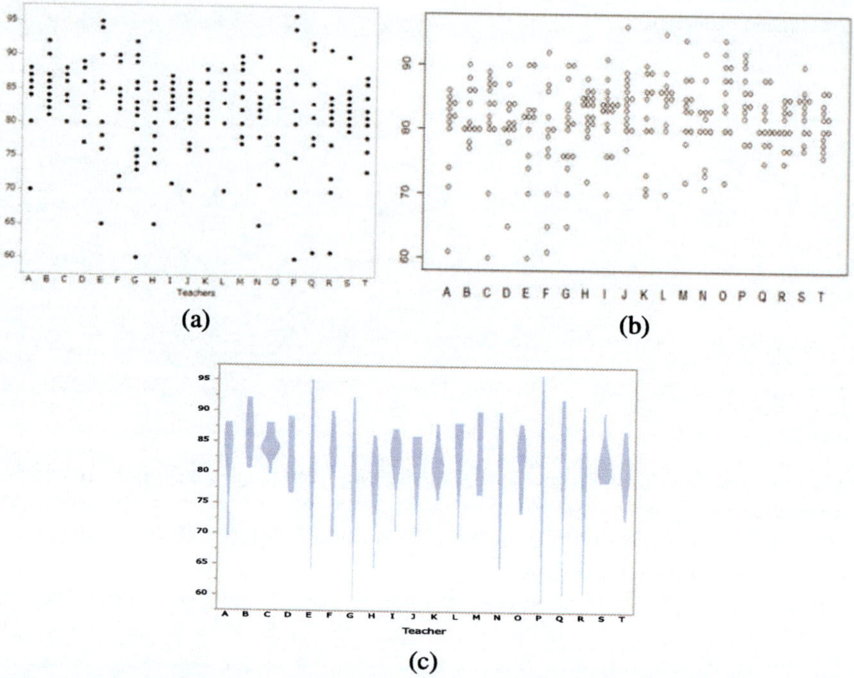

Fig. 5 a Scatter plot (From [17]) **b** Beeswarm plot (From [17]), **c** Violin plot (From [17])

3 Discussion

Today, ICT is becoming indispensable for the improvement of educational practices and quality. In the field of higher education, societal demands are increasing for EBE, IR, and the visualization of learning outcomes. Responding to this growing demand first and foremost requires data; however, data collection and accumulation were difficult before the spread of ICT. The development of ICT and the accompanying evolution of LMS have made it easier and faster than ever before to accumulate large amounts of educational data. For example, digitizing students' responses on paper questionnaires requires typing these data points into a computer; given the large amounts of data necessary to achieve significant and valid results, this process is usually time-consuming. Using ICT, such as Google Forms, to collect survey responses quickens the production of a digital summary [5]; moreover, such a use of ICT is further facilitated today by the fact that most students in Japanese universities now have smartphones.

However, there are many cases in Japan where the accumulated data are not fully used for educational improvement. This is because a systematic methodology for handling data has not been established. Eduinformatics, as a new academic field, works to fill this gap (Fig. 6).

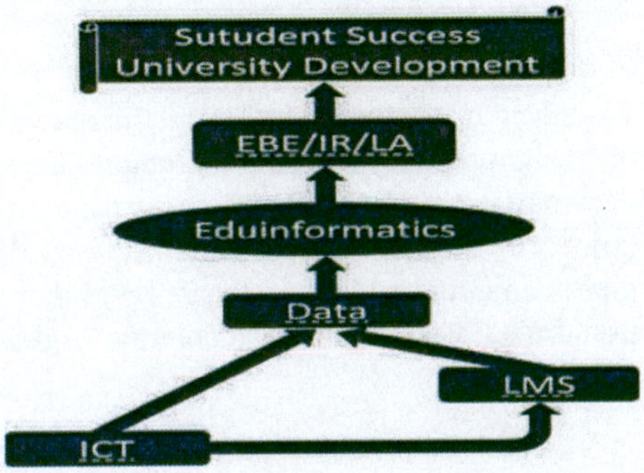

Fig. 6 Using of Eduinformatics in higher education

Eduinformatics, illustrated by the above examples, is not just statistics but a new interdisciplinary field that integrates pedagogy with statistics, mathematics, and informatics. Eduinformatics crucially helps prevent individual disciplines getting caught up with trivial matters, losing sight of the big picture, and settling for temporary solutions when handling educational data.

Using ICT enables us to develop new ways of teaching and learning; however, the development of teaching and learning methods alone is not enough. For student success and university development, thorough analyses of student data must form the basis of the development of new teaching and learning methods. To be sure, all fields experience such flows of innovation. For example, the p-value controversy continues in the field of statistics [18] and other new methods for data visualization, such as the recently proposed new beeswarm plot (Fig. 7), are constantly emerging. Notably, our particular contribution, eduinformatics, incorporates the latest developments of various disciplines to develop comprehensive data analysis and prediction methods appropriate for the current educational situation.

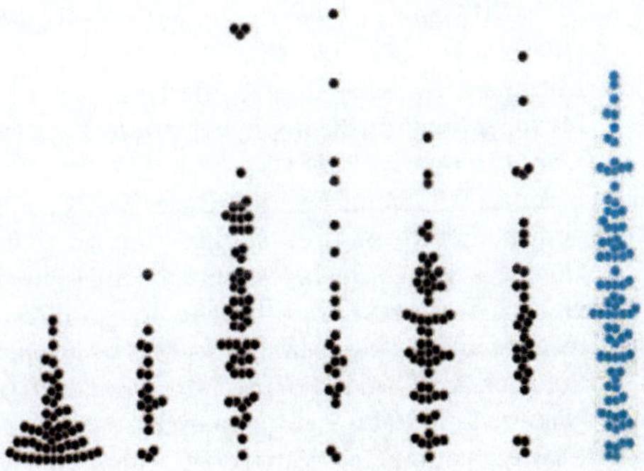

Fig. 7 New Beeswarm plot (From [15])

4 Conclusions

With the development of ICT and the accompanying evolution of LMS, large amounts of educational data can now be accumulated more easily and quickly than ever before. This paper presents a new line of academic inquiry interested in harnessing these powers for student success and university development: "eduinformatics." Eduinformatics enables us to make local and global analyses and predictions based on educational data and stands to accelerate higher education reform at an unprecedented rate.

References

1. MEXT: Construction of education for university (2008). http://www.mext.go.jp/b_menu/shingi/chukyo/chukyo4/houkoku/080410.htm
2. MEXT: Promoting Evidence-Based Education Policy (2017). https://www.mext.go.jp/content/20200522-mxt_koktou01-000006607_2-3.pdf
3. MEXT: Grand Design for Higher Education Toward 2040 (Report) (2018). https://www.mext.go.jp/b_menu/shingi/chukyo/chukyo0/toushin/1411360.htm
4. MEXT: Plan for Enhancing Education in Universities and Colleges of Technology Using Digital Technology (2021). https://www.mext.go.jp/content/20201224-mxt_senmon01-000011618_1.pdf
5. Takamatsu, K., Noda, I., Bannaka, K. et al.: Concept of ICT on Eduinformatics in Higher Education. In: 6th International Congress on Information and Communication Technology. p in press (2021)
6. Takamatsu, K., Murakami, K., Kirimura, T., et al.: "Eduinformatics": a new education field promotion. Bull kobe Tokiwa Univ **11**, 27–44 (2018). https://doi.org/10.20608/00000958
7. Takamatsu, K., Kozaki, Y., Murakami, K., et al.: Review of recent Eduinformatics research. In: IEEE/IIAI International Congress on Applied Information Technology 2019. pp. 27–32 (2019)
8. Cabinet Office: Society 5.0 (2016). https://www8.cao.go.jp/cstp/society5_0/
9. MEXT: Enhancement Measures to Mathematical and Data Science Education in Universities (2016). https://www.mext.go.jp/b_menu/shingi/chousa/koutou/080/gaiyou/__icsFiles/afieldfile/2016/12/21/1380788_01.pdf
10. Mitsunari, K., Kirimura, T., Kunisaki, T., et al. Paradigm shift in education from teaching to learning: focus on the implementation of academic skills and deep learning I. Bull kobe Tokiwa Univ 11:in press (2018)
11. Nakata, Y., Takamatsu, K., Bannaka, K., et al. Assessment of rubric-based evaluation by nonparametric multiple comparisons in first-year education in a Japanese University. In: Congress on 9th Quality Assurance in Higher Education. pp 39–40 (2019)
12. Nakata, Y., Kozaki, Y., Kunisaki, T., et al.: Assessment of rubric-based evaluation by nonparametric multiple comparisons in first-year education in a Japanese university. PEOPLE Int. J. Soc. Sci. **4**, 631–641 (2018). https://doi.org/10.20319/pijss.2018.41.631641
13. Nakata, Y., Kozaki, Y., Mitsunari, K., et al.:Ensuring Equal Evaluation among Teachers in First-Year Education Courses through Rubrics: A Multiple Comparison Analysis. In: International Conference on Education, Psychology, and Learning (ICEPL2018). pp 40–46 (2018)
14. Matsumoto, E., Takamatsu, K., Bohgaki, M., et al.: Development of course guidance support method for "Competency-Based Education (CBE)" in university and its Issues. Bull Kobe Tokiwa Univ. **12**, 17–28 (2019). https://doi.org/doi/10.20608/00001037

15. Murakami, K., Takamatsu, K., Kozaki, Y., et al.: Predicting the probability of student dropout through emir using data from current and graduate students. In: Advanced Applied Informatics (IIAI-AAI), 2018 7th International Institute of Applied Informatics (IIAI) International Congress on. Institute of Electrical and Electronics Engineers (IEEE). pp. 478–481 (2018)
16. Tracey, L.W., Milic, N.M., Winham, S.J., Vesna, D.G.: Beyond bar and line graphs: time for a new data presentation paradigm. Plos Biol **13**, e1002128 (2015). https://doi.org/10.1371/journal.pbio.1002128
17. Takamatsu, K., Murakami, K., Kozaki, Y., et al.: Using student data analysis based on Eduinformatics for criteria in institutional research. Int. J. Inst. Res. Manag. **4**, 15–29 (2020)
18. Wasserstein RL, Schirm AL, Lazar NA (2019) Moving to a world beyond "p<0.05"

Classification of Sagittal Lumbar Spine MRI for Lumbar Spinal Stenosis Detection Using Transfer Learning of a Deep Convolutional Neural Network

Friska Natalia and **Sud Sudirman**

Abstract Analysis of sagittal lumbar spine MRI images remains an important step in automated detection and diagnosis of lumbar spinal stenosis. There are numerous algorithms proposed in the literature that can measure the condition of lumbar intervertebral discs through analysis of the lumbar spine in the sagittal view. However, these algorithms rely on using suitable sagittal images as their inputs. Since an MRI data repository contains more than just these specific images, it is, therefore, necessary to employ an algorithm that can automatically select such images from the entire repository. In this paper, we demonstrate the application of an image classification method using deep convolutional neural networks for this purpose. Specifically, we use a pre-trained Inception-ResNet-v2 model and retrain it using two sets of T1-weighted and T2-weighted images. Through our experiment, we can conclude that this method can reach a performance level of 0.91 and 0.93 on the T1 and T2 datasets, respectively, when measured using the accuracy, precision, recall and f1-score metrics. We also show that the difference in performance between using the two modalities is statistically significant and using T2-weighted images is preferred over using T1-weighted images.

Keywords Medical image classification · Magnetic resonance imaging · Lumbar spinal stenosis · Transfer learning · Deep convolutional neural networks

1 Introduction

Millions of people around the world suffer from chronic lower back pain. It is a chronic disease that is detrimental to the health, social life and employment of its sufferers. Lumbar spinal stenosis (LSS), a narrowing of the lumbar spinal canal

F. Natalia (✉)
Department of Information System, Universitas Multimedia Nusantara, Tangerang, Indonesia
e-mail: friska.natalia@umn.ac.id

S. Sudirman
Department of Computer Science, Liverpool John Moores University, Liverpool, UK
e-mail: s.sudirman@ljmu.ac.uk

© The Author(s), under exclusive license to Springer Nature Singapore Pte Ltd. 2022
A. K. Nagar et al. (eds.), *Intelligent Sustainable Systems*, Lecture Notes in Networks and Systems 333, https://doi.org/10.1007/978-981-16-6309-3_16

that is resulted from bone or soft tissue inflammation, is one of the most common causes of chronic lower back pain. The pressure on the spinal nerve roots caused by this inflammation is responsible for the pain that is experienced by patients with LSS. A diagnosis of LSS in these patients is often carried out through an inspection of Magnetic Resonance Imaging (MRI) of the patients' lumbar spine by an expert radiologist. Recent advances in medical image processing allow the application of computer algorithms to help radiologists carried out this procedure. Some of these algorithms work only on midsagittal MRI images [1–3], whereas some others work only on traverse images that cut through the mid-height of an intervertebral disc (IVD) [4–8]. Since a patient's data repository contains more than just these specific images, the process to select suitable images as inputs to these algorithms is often done manually. To make the whole process more automated, it is, therefore, essential that a reliable algorithm to select such images is applied as well.

The objective of this study is to design a suitable solution for selecting all suitable sagittal images from a database of sagittal images that can be used as inputs to other algorithms that diagnose LSS. Our solution is using a pre-trained deep convolutional neural network (DCNN) model that has been developed for general image classification and retraining it to make the model suitable for medical image classification.

2 Literature Review

The task of selecting medical images that possess certain characteristics from a collection of medical images falls into the category of image classification, which is a fundamental task in computer vision that categorizes images into one of several predefined classes. The traditional approach in image classification involves two stages, with the first being the extraction of relevant information from the images via the calculation of low-level handcrafted features [9–11]. This is then, followed by a classification of the calculated features using trainable classifiers. Despite the success of this approach, it has a significant drawback when used in a wider image classification problem since the features are often task-dependent. In other words, the handcrafted image features that are optimized for a particular task often perform poorly when used in a different task, and the accuracy of the classification is very dependent on the design of these features.

The first DCNN model was originally proposed to overcome the problems associated with the traditional approach of image classification by allowing learning of such features through forward and back propagation of information in a series of convolutional and nonlinear neural network layers [12, 13]. This approach, however, has a significant practical problem due to its high-computational cost and the amount of data it needs to create a general set of features applicable for typical images. But only recently, in the advent of huge computational power from using graphics processing units and the large-scale acquisition and availability of image data resulting from the proliferation of the internet and social media that this approach has gained renewed attention from the research community which generated faster and better algorithms [14].

The popularity of DCNN compared to the more traditional approach of image classification is that the features are no longer manually handcrafted but instead are automatically learnable. These features are sufficiently general that they can be used in many different types of image classification tasks through transfer learning. Transfer learning is a widely accepted method in machine learning where a model developed for a task, often by training using a very large dataset, is used as a starting point for developing another model to solve a different task. This approach is less data and label dependent than other more traditional machine learning approaches and gains popularity recently, especially when a deep learning model is concerned because developing one from scratch requires a vast amount of computational and time resources [15]. One example application of a bespoke DCNN for medical image classification is CheXNet [16]. It is a 121-layer DCNN trained on a dataset with more than 100,000 frontal-view chest X-rays and is claimed to achieve a better performance than the average performance of four radiologists.

3 Material and Method

The material used in this research is taken from our Lumbar Spine MRI Dataset which is available publicly [6, 17]. This dataset contains anonymized clinical MRI studies of 515 patients with symptomatic back pains. The dataset consists of 48,345 T1-weighted and T2-weighted traverse and sagittal images of the patients' lumbar spine. The images were taken using a 1.5 T Siemens Magnetom Essenza MRI scanner mostly when the patients were in Head-First-Supine position. From the entire dataset, we took 19,176 sagittal images for this study. This consists of 9903 T1-weighted and 9273 T2-weighted images. The summary of the technical information of the scanning parameters carried out when recording these images is provided in Table 1.

Based on the advice from an expert radiologist, we categorize the sagittal images into two groups, namely *Mid* and *Lateral* groups, corresponding to their suitability for the analysis of the IVD. The first group consists of midsagittal slices that divide the left and right sides of an IVD into two symmetrical parts. It also contains neighbouring sagittal slices to the midsagittal slice which still show clear cross-sectional views of the lumbar vertebrae and discs. The second group consists of other sagittal slices that do not meet this requirement. An example of this grouping is shown in Figs. 1 and 2. The first figure shows the intersection lines of fifteen sagittal images with the shown traverse image, whereas the second figure shows a collage of the fifteen sagittal images. The midsagittal slice is identified as slice number 8 in the figures. That slice and four nearest adjacent slices, highlighted in yellow in the figures, are then put in the mid group. The other ten sagittal slices, highlighted in red, are then put in the lateral group. The population distribution of these classes for both T1-weighted and T2-weighted datasets is shown in Table 2.

Table 1 Sagittal MRI scanning parameters

Sequence types	T1-weighted	T2-weighted
Number of echoes (ETL)	3	15–18
Repetition time (ms)	330–926	3190–4000
Echo Time (ms)	9.2–12.0	67.0–96.0
Slice thickness (mm)	3.0– to 4.0	3.0–5.0
Spacing between slices (mm)	3.3–4.8	3.3–6.5
Field of view (mm)	280	280
Matrix (Freq. x Phase)	100%	100%
Imaging frequency (MHz)	63.7	63.7
Number of phase encoding steps	288–540	408–544
Scanning sequence	SE	SE
Sequence Variant	SK\SP\OSP	SK\SP\OSP
Scan options	SAT1\FS	SAT1
Number of averages	1, 2, 3 or 4	2
Echo train length	3	15–17
Per cent sampling	50–70	66–90
Per cent PHASE FIELD OF VIEW	100	100
Pixel bandwidth	150 or 235	150, 160 or 195
Flip angle	150	150

We performed validation of this dataset by checking the *Slice Location* information that is stored as part of the DICOM metadata. The value of the slice location attribute of a DICOM image metadata is the relative position, expressed in mm, of the image plane in the patient 3D axis system. In the absence of the slice location attribute in the DICOM image metadata, we calculated its value from the *Image Position* and *Image Orientation* attributes of the DICOM image metadata using the following technique. Let us denote p as the 3D coordinate of the top-left point on the plane specified in the Image Position attribute. Also, let u and v be two orthogonal unit vectors that lie on the sagittal plane as specified in the Image Orientation attribute. The slice location s of the plane is defined as the shortest distance from the origin point of the patient coordinate axes to the plane and can be calculated as the dot product $n \bullet p$, where n is the unit vector perpendicular to the plane which can be calculated as the cross product $u \times v$. The directions of the patient coordinate axes themselves are defined by the patient's orientation. The positive direction of the x-axis goes from the right-hand side of the patient to the left-hand side. The positive direction of the y-axis goes from the anterior side of the patient to the posterior side, whereas the z-axis is increasing from the feet towards the head of the patient. The position and orientation of the elements used in this calculation with respect to the patient's coordinate axes are illustrated in Fig. 3.

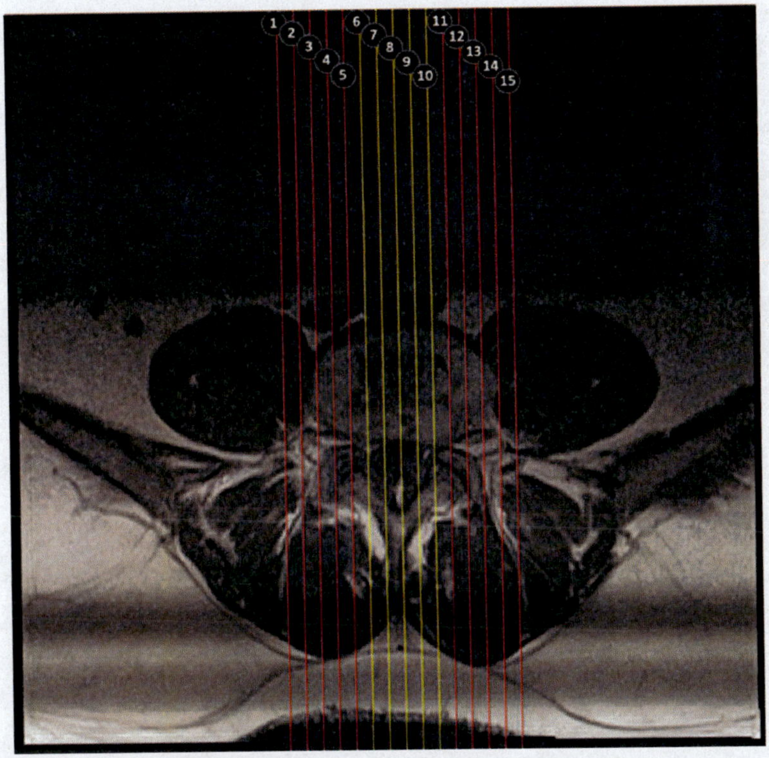

Fig. 1 A traverse image of a lumbar spine showing its intersection lines with fifteen sagittal images that are shown in Fig. 2. The yellow lines mark the sagittal images that are categorized in the *Mid* group and the red lines mark those categorized in the *Lateral* group

Using either the stored or the calculated slice location information, we could confirm that the sagittal planes in the mid groups of the dataset are within the acceptable range of normal IVD width [18].

Once the dataset has been developed, we then use it to train an Inception-ResNet-v2 model. Inception-ResNet-v2 [19] is a convolutional neural architecture that improves on the Inception family of architectures by incorporating the ResNet approach of using residual connections to replace the filter concatenation stage of the Inception architecture. Inception-Resnet-v2 is one of the newer generations of deep convolutional neural networks that is gaining popularity in the classification and annotation of medical images [20]. We adopted a methodology called Transfer Learning that transfers the learned network parameters of the model that was pre-trained using the ImageNet database [21] and then retrain the whole model after replacing its classification layers using the new dataset. The flowchart of the process is illustrated in Fig. 4.

To evaluate the suitability of the method, we use four performance metrics namely overall accuracy ($A.$), precision (P), recall (R), and f1-score (F). Using the standard

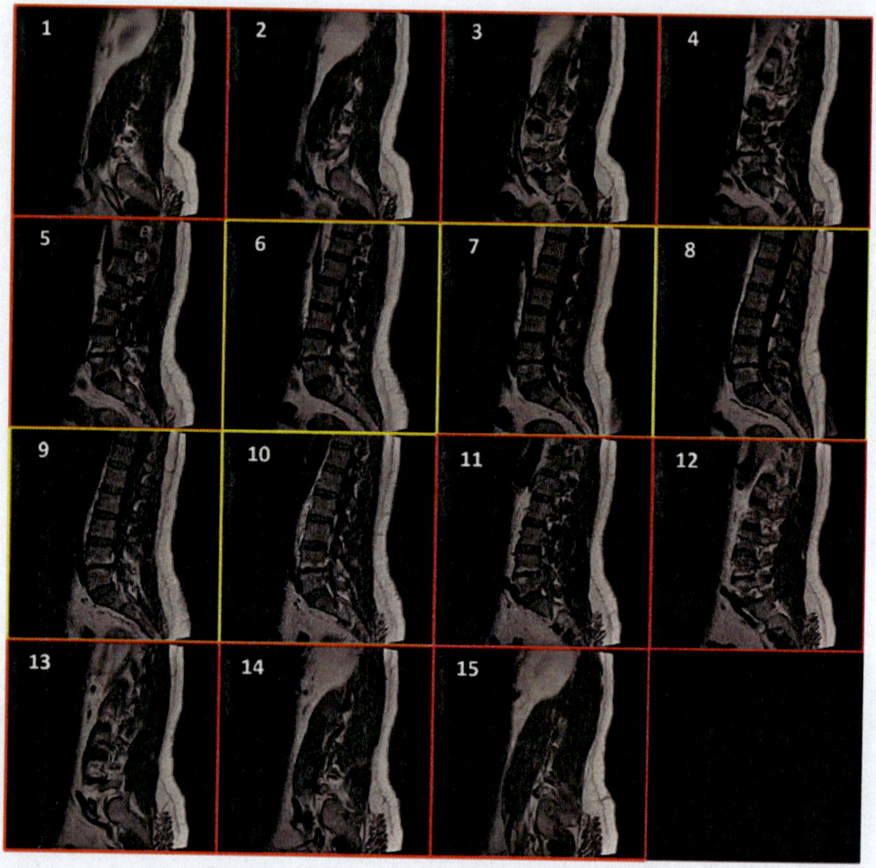

Fig. 2 An example of fifteen sagittal images of a patient's lumbar spine which intersections with a traverse image are shown in Fig. 1. The yellow rectangles highlight the sagittal images that are put in the *Mid* group and the red rectangles highlight those put in the *Lateral* group

Table 2 Class distribution in the dataset

Class\sequence types	T1-weighted	T2-weighted
Mid	4667	5236
Lateral	4316	4957
Total	9903	9273

notations of true positive (tp), true negative (tn), false positive (fp) and false negative (fn), the metrics are calculated as:

$$A = \frac{tp + tn}{tp + tn + fp + fn} \tag{1}$$

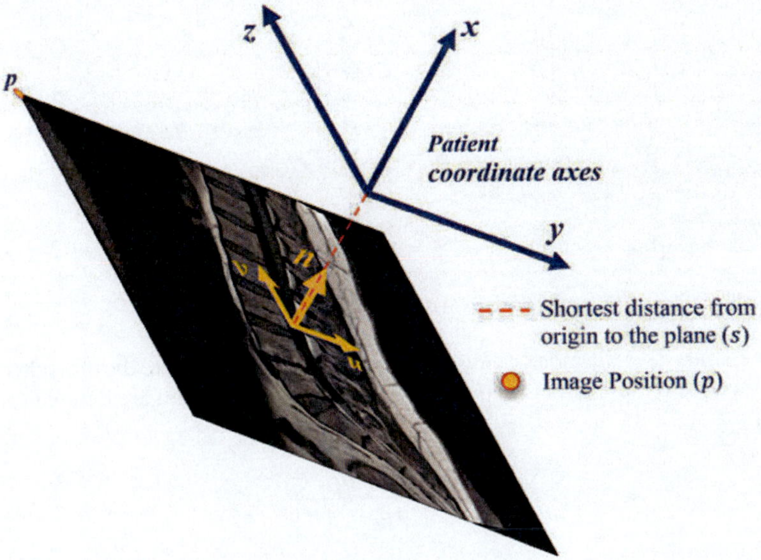

Fig. 3 An illustration of the position and orientation of the elements used to determine the slice location s with respect to the patient's coordinate axes

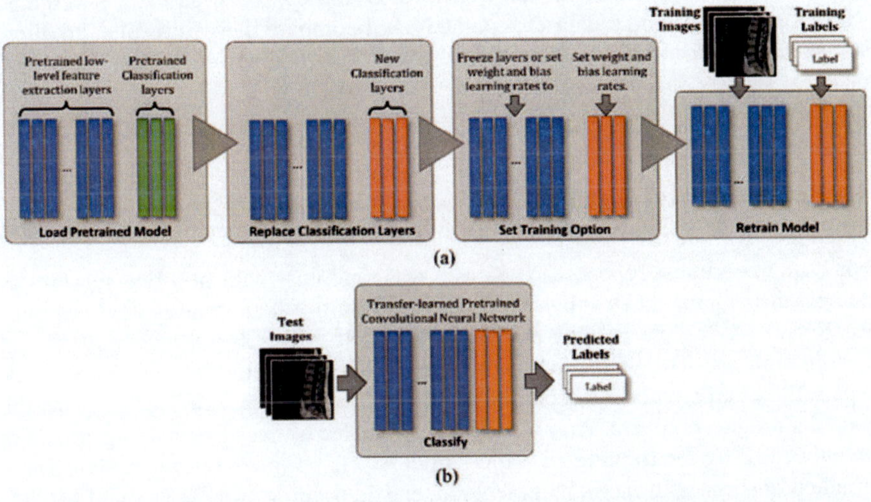

Fig. 4 A flowchart illustrating **a** the retraining process of a pre-trained deep convolutional neural network and **b** the classification of the images in the test dataset using the retrained model

$$P = \frac{1}{C}\sum_{i}^{C} P_i \tag{2}$$

$$R = \frac{1}{C}\sum_{i}^{C} R_i \tag{3}$$

$$F = \frac{2}{C}\sum_{i}^{C} \frac{P_i \cdot R_i}{P_i + R_i} \tag{4}$$

where $i \in \{1, 2\}$ is the index of the ith. class and P_i, R_i and F_i are the class precision, class recall and class f1-score of the ith class, respectively, and are defined as:
(5)

$$R_i = \frac{tp_i}{tp_i + fn_i} \tag{6}$$

$$F_i = 2 \times \frac{P_i \cdot R_i}{P_i + R_i} \tag{7}$$

We applied the aforementioned transfer learning approach on the T1-weighted dataset and T2-weighted dataset separately and compare the two results. To allow statistical analysis of the results, we carried out the process twenty times, each with different subsets of training and test data.

4 Experimental Results, Discussion, and Analysis

For each experiment repeat, each dataset was randomly split into two sub-groups namely the training and test dataset by an 80:20 ratio. When training the Inception-ResNet-v2 model, the training set was further split into a smaller training set and a validation set. The validation set was used solely to provide an unbiased estimate of the trained model's accuracy during training and was not used to adjust the model's weights or biases. They were instead adjusted by back propagating the errors calculated using the training set which process was optimized using the stochastic gradient descent with momentum technique. The training is done in small batches of ten samples for up to four epochs. The order of the samples in the training set was shuffled in every epoch to prevent the model from learning the order of the samples. The initial learning rate of the transferred (feature extraction) layers was set to 10^{-4} and that of the new classification layer was set 20 times higher. The training duration for the twenty repeats ranges from 750 to 765 min when run on a Windows 10 PC with an Intel i9-7900X CPU @ 3.30 GHz with 128 GB RAM and four NVIDIA Titan XP GPUs. A plot showing the validation accuracy during the training process is shown in Fig. 5. As can be seen from this figure, the accuracy values for both datasets plateau roughly before the final epoch. The final recorded validation accuracy for the T1 and T2 datasets is 0.92 and 0.91, respectively.

Once the training process is completed, it is used to predict the classes of each sample in the test set. These predicted classes are then used to calculate the four performance metrics defined in Eq. (1) to (4). The results are shown as box plots in Figs. 6 and 7. The figures show the minimum, the maximum, the sample median, and the first and third quartiles of each metric for ease of visual inspection of the classification performance.

We performed a statistical test to show that the difference in the classification results produced by using the T1 and T2 datasets is statistically significant. We use

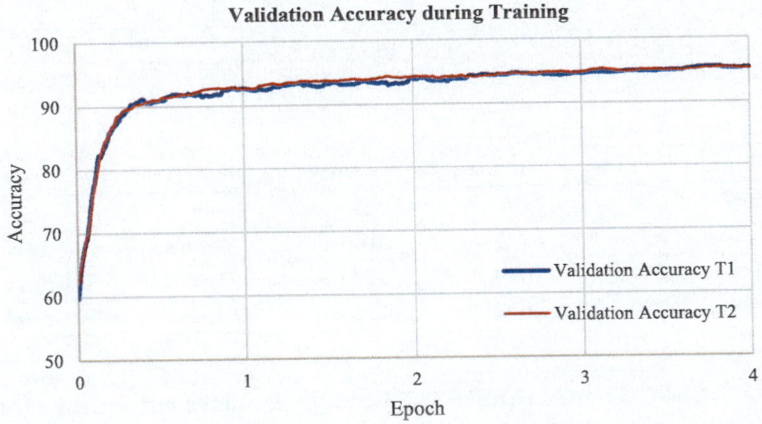

Fig. 5 A history of the validation accuracy during the training process

Fig. 6 A box plot showing the statistics of classification performance using the T1 dataset

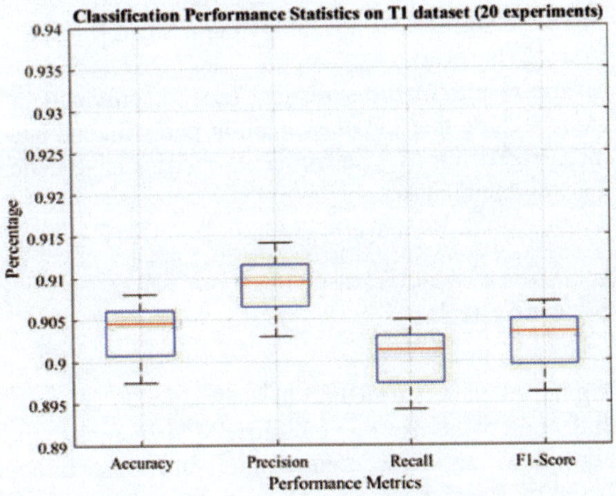

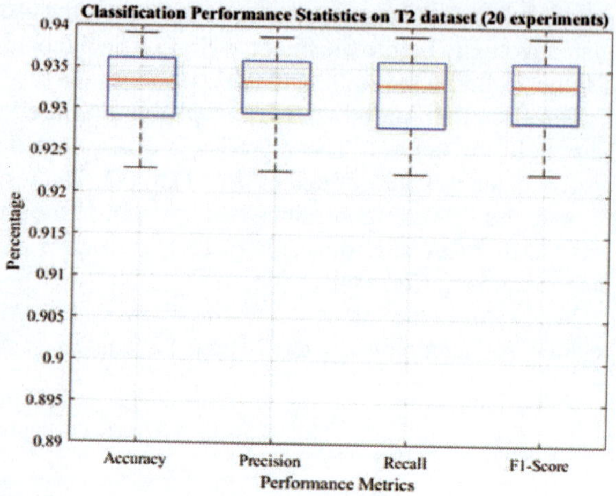

Fig. 7 A box plot showing the statistics of classification performance using the T2 dataset

Table 3 Summary of the statistical tests

Metrics	T1 Mean	T2 Mean	Difference	t-test Type	*p*-value
Accuracy	0.90	0.93	0.03	t-test	<0.05
Precision	0.91	0.93	0.02	Welch's t-test	<0.05
Recall	0.90	0.93	0.03	t-test	<0.05
F1-Score	0.90	0.93	0.03	t-test	<0.05

the Kolmogorov–Smirnov (KS) test [22] and the Bartlett's test for Homogeneity of Variance [23] to check if the populations follow a Normal distribution and if they have identical variances, respectively. Satisfying both tests allows us to use the standard t-test hypothesis testing to prove or disprove the null hypothesis. The null hypothesis here being there is no difference between the two results. If any one of the two populations does not satisfy the KS and Bartlett's tests then we use the Welch's t-test [24] instead.

The results of the statistical test are provided in Table 3. From the table, we can conclude that the classification performance using the T2-weighted dataset as measured using four performance metrics is statistically better than that using the T1-weighted dataset.

5 Conclusion

We have detailed a method to automatically select suitable sagittal images from a collection of lumbar spine sagittal MRI images that can be used as inputs to other algorithms that detect abnormalities in lumbar intervertebral discs. The method is based on transfer learning of a pre-trained Inception-Resnetv2 deep convolutional neural network model. We experimented with this method on a subset of a publicly available lumbar spine MRI dataset that consists of 9,903 T1-weighted and 9,273 T2-weighted MRI images. Our experimental results show that the average classification performance is 0.90 and 0.93 on the T1 dataset and the T2 dataset, respectively. We have also shown through statistical analysis of our experimental results that the classification performance using the T2 dataset is statistically better than that using the T1 dataset. We plan in the future to expand the method to include a wider range of deep convolutional neural network models as well as investigate the effect of applying the dimensionality reduction method on the generalization performance of this approach. We will also explore the possibility of applying the method to classify lumbar spine traverse MRI images.

References

1. Davarpanah, S.H., Liew, A.W.C.: Spatial possibilistic fuzzy C-mean segmentation algorithm integrated with brain mid-sagittal surface information. Int. J. Fuzzy Syst. **19**, 591–605 (2017). https://doi.org/10.1007/s40815-016-0247-0
2. Alomari, R.S., Ghosh, S., Koh, J., Chaudhary, V.: Vertebral Column Localization, Labeling, and Segmentation. In: Li, S. and Yao, J. (eds.) Spinal Imaging and Image Analysis. pp. 193–229. Springer International Publishing, Cham (2015). https://doi.org/10.1007/978-3-319-12508-4_7.
3. Ghosh, S., Chaudhary, V.: Supervised methods for detection and segmentation of tissues in clinical lumbar MRI. Comput. Med. Imaging Graph. **38**, 639–649 (2014). https://doi.org/10.1016/j.compmedimag.2014.03.005.
4. Natalia, F., Meidia, H., Afriliana, N., Young, J.C., Yunus, R.E., Al-Jumaily, M., Al-Kafri, A., Sudirman, S.: Automated measurement of anteroposterior diameter and foraminal widths in MRI images for lumbar spinal stenosis diagnosis. PLoS ONE **15**, 1–27 (2020). https://doi.org/10.1371/journal.pone.0241309
5. Paul, C.P.L., Smit, T.H., de Graaf, M., Holewijn, R.M., Bisschop, A., van de Ven, P.M., Mullender, M.G., Helder, M.N., Strijkers, G.J.: Quantitative MRI in early intervertebral disc degeneration: T1rho correlates better than T2 and ADC with biomechanics, histology and matrix content. PLoS One. **13**, e0191442 (2018)
6. Al-Kafri, A.S., Sudirman, S., Hussain, A., Al-Jumeily, D., Natalia, F., Meidia, H., Afriliana, N., Al-Rashdan, W., Bashtawi, M., Al-Jumaily, M.: Boundary delineation of MRI images for lumbar spinal stenosis detection through semantic segmentation using deep neural networks. IEEE Access. **7**, 43487–43501 (2019). https://doi.org/10.1109/ACCESS.2019.2908002
7. Zhang, Q., Bhalerao, A., Hutchinson, C.: Weakly-supervised evidence pinpointing and description. In: International Conference on Information Processing in Medical Imaging. pp. 210–222 (2017)
8. Al Kafri, A.S., Sudirman, S., Hussain, A.J., Al-Jumeily, D., Fergus, P., Natalia, F., Meidia, H., Afriliana, N., Sophian, A., Al-Jumaily, M., others, Al-Kafri, A.S., Sudirman, S., Hussain, A.J., Al-Jumeily, D., Fergus, P., Natalia, F., Meidia, H., Afriliana, N., Sophian, A., Al-Jumaily, M., Bashtawi, M., Al-Rashdan, W.: Segmentation of lumbar spine MRI images for stenosis detection using patch-based pixel classification neural network. In: 2018 IEEE Congress on Evolutionary Computation (CEC). pp. 1–8. , Rio de Janeiro (2018)
9. Baloch, S.H., Krim, H.: Flexible skew-symmetric shape model for shape representation, classification, and sampling. IEEE Trans. image Process. **16**, 317–328 (2007)
10. Song, Y., Cai, W., Zhou, Y., Feng, D.D.: Feature-based image patch approximation for lung tissue classification. IEEE Trans. Med. Imaging. **32**, 797–808 (2013)

11. Koitka, S., Friedrich, C.M.: Traditional Feature Engineering and Deep Learning Approaches at Medical Classification Task of ImageCLEF 2016. In: CLEF (Working Notes). pp. 304–317 (2016)
12. LeCun, Y., Boser, B.E., Denker, J.S., Henderson, D., Howard, R.E., Hubbard, W.E., Jackel, L.D.: Handwritten digit recognition with a back-propagation network. In: Advances in neural information processing systems. pp. 396–404 (1990)
13. LeCun, Y., Boser, B., Denker, J.S., Henderson, D., Howard, R.E., Hubbard, W., Jackel, L.D.: Backpropagation applied to handwritten zip code recognition. Neural Comput. **1**, 541–551 (1989)
14. Rawat, W., Wang, Z.: Deep convolutional neural networks for image classification: a comprehensive review. Neural Comput. **29**, 2352–2449 (2017)
15. Zhuang, F., Qi, Z., Duan, K., Xi, D., Zhu, Y., Zhu, H., Xiong, H., He, Q.: A comprehensive survey on transfer learning. Proc. IEEE. (2020). https://doi.org/10.1109/JPROC.2020.3004555
16. Rajpurkar, P., Irvin, J., Zhu, K., Yang, B., Mehta, H., Duan, T., Ding, D., Bagul, A., Langlotz, C., Shpanskaya, K., others: Chexnet: Radiologist-level pneumonia detection on chest x-rays with deep learning. arXiv Prepr. arXiv1711.05225 (2017)
17. Sudirman, S., Kafri, A. Al, Natalia, F., Meidia, H., Afriliana, N., Al-Rashdan, W., Bashtawi, M., Al-Jumaily, M.: Lumbar Spine MRI Dataset, https://data.mendeley.com/datasets/k57fr854j2/2, Last accessed 13 May 2019. https://doi.org/10.17632/k57fr854j2.2.
18. Zhou, S.H., McCarthy, I.D., McGregor, A.H., Coombs, R.R.H., Hughes, S.P.F.: Geometrical dimensions, of the lumbar vertebrae—analysis of data from digitised CT images. Eur. Spine J. **9**, 242–248 (2000). https://doi.org/10.1007/s005860000140
19. Szegedy, C., Ioffe, S., Vanhoucke, V., Alemi, A.: Inception-v4, inception-resnet and the impact of residual connections on learning. arXiv Prepr. arXiv1602.07261 (2016)
20. Pelka, O., Nensa, F., Friedrich, C.M.: Annotation of enhanced radiographs for medical image retrieval with deep convolutional neural networks. PLoS ONE **13**, 1–18 (2018). https://doi.org/10.1371/journal.pone.0206229
21. Deng, J., Dong, W., Socher, R., Li, L.-J., Li, K., Fei-Fei, L.: Imagenet: a large-scale hierarchical image database. In: Computer Vision and Pattern Recognition, 2009. CVPR 2009. IEEE Conference on. pp. 248–255 (2009)
22. Massey, F.J., Jr.: The Kolmogorov-Smirnov test for goodness of fit. J. Am. Stat. Assoc. **46**, 68–78 (1951)
23. Snedecor, G.W., Cochran, W.G.: Statistical Methods. Wiley India (2014)
24. Mendenhall, W.M., Sincich, T.L.: Statistics for Engineering and the Sciences. CRC Press (2016)

Robust Facial-Based Inter-Beat Interval Estimation Through Spectral Signature Tracking and Periodic Filtering

Mostafa Salah, Loay Hassan, Somaya Abdel-khier, Ammar M. Hassan, and Osama A. Omer

Abstract Recently, there is an increasing interest in approximating the contact PPG measurements through a remote camera-based sensing. However, extracting biomedical signs such as cardiac activities from facial videos is more challenging than contact-based measurements. In this paper, the spectral signature of the contact PPG (cPPG) is investigated against the corresponding remote-PPG (rPPG). In spite of the noisy nature of the video extracted signal, the semi-periodic nature of the pulsating blood volume is properly employed in tracking the spectral signature by periodic filtering around fundamental frequency and the first two harmonics. That salient spectral signature is reflected in better estimating inter-beat interval (IBI) and heart rate variability (HRV). The essential tradeoff resides in estimating the high (featuring detail preserving) beat wave quality against the main signal approximation (timing preserving) quality. Fortunately, IBI and HRV are less sensitive to the quality of the estimated beat wave. They only need sharp timing information of the wave. Our claims are assessed using videos from public database with the corresponding ground truth (GT) PPG. The results reflect robust IBI estimation with high accuracy compared to the state-of-the-art methods.

Keywords PPG estimation · IBI · Heart rate · Biomedical signs

This work is supported in part by the Information Technology In-dustry Development Agency (ITIDA) under number ARP2019.R27.4 and in part by Advanced Computer Systems (ACS).

M. Salah
Sohag University, Sohag, Egypt

L. Hassan · S. Abdel-khier · O. A. Omer (✉)
Aswan University, Aswan 81452, Egypt
e-mail: omer.osama@aswu.edu.eg

A. M. Hassan
Arab Academy for Science, Technology and Maritime Transport, Aswan, Egypt

© The Author(s), under exclusive license to Springer Nature Singapore Pte Ltd. 2022
A. K. Nagar et al. (eds.), *Intelligent Sustainable Systems*, Lecture Notes in Networks and Systems 333, https://doi.org/10.1007/978-981-16-6309-3_17

1 Introduction

Continuous and flexible monitoring of vital signs such as heart rate (HR), blood pressure (BP) and heart rate variability (HRV) is increasingly demanding nowadays. Traditional contact-based photoplethysmography (cPPG) monitoring is no longer proper for many scenarios. Therefore, the remote/imaging PPG monitoring (rPPG/iPPG) gains increasing concern in both the biomedical research community and industry as well. The iPPG detects the changes in the blood volume pulse (BVP) by sensing the ambient light intensity variation reflected from the individual's face by simple camera with reasonable resolution [1, 2]. Its principal of operation resembles the reflection mode of cPPG, however, the light source is replaced by ambient light and the light detector is replaced by camera sensor.

Really, the detected raw data is too noisy to extract the corresponding PPG approximation from rPPG signals. However, the exact PPG approximation is not required in some monitoring applications. Consequentially, we may classify the rPPG tasks into three main classes according to its difficulty in terms of the algorithmic signal processing:

1. Heart rate (HR) monitoring [3]: HR is the average number of beats per time interval (minute). It can be simply measured by detecting the fundamental (most salient) frequency component in the cardiac spectral range [0.7–5] Hz.
2. Inter-beat interval (IBI) and heart rate variability (HRV) [4]: IBI measures the time interval between beats. It is a measure of instantaneous frequency of the HR. IBI estimation can be regarded as a beat segmentation problem.
3. Detail-preserving pulse wave extraction [5]: Clean PPG approximation is needed for further estimating BP. It represents the highest challenging problem related to monitoring vital sign from facial videos.

2 The Proposed IBI Estimation Framework

The proposed framework for estimating IBI is presented in Fig. 1, and it is explained in detail in the following subsections.

2.1 Video Processing Stages

Face Detection and Tracking

As shown in Fig. 1, the first steps in the proposed framework are face detection followed by face tracking. Face detection algorithms can be divided into two groups, traditional face detection and deep learning-based face detection. In this framework, we rely on the deep learning-based algorithm [14] since deep learning models achieve the state of the art at almost of all computer vision tasks.

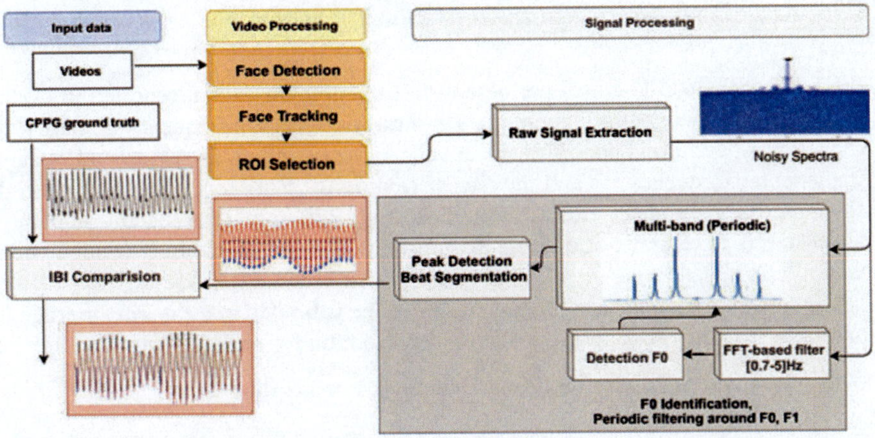

Fig. 1 Proposed IBI estimation framework

Face tracking algorithm tracks the presence of a human face detected by face detection algorithm in the first frame for every subsequent frame. In this work, we used the KCF tracking algorithm [15] which is high speed and accurate tracker.

Region-Of-Interest (ROI) Selection

The ROI is the area within a video frame that contains pixels providing the raw signal. ROI selection is a critical process, and its necessity includes as many skin pixels as likely. In previous studies, many rPPG methods were suggested to select ROI. Recent approach [4] involves defining multiple ROIs and generating one RGB signal each for subsequent analysis. We focus on refining the quality of the extracted cardiovascular wavc using skin segmentation for defining the ROI. Only the green channel is regarded for further processing based on the reported higher signal power in the green channel compared to other colors.

2.2 *Signal Processing Stages*

The traditional processing involves a long processing chain [3] including smoothing and de-trending, band-pass filtering and BSS through PCA or ICA. In the proposed framework, these stages are replaced by simple two-stage periodic filtering.

2.3 PPG Periodicity Nature and Quality Metrics

The fact of PPG/rPPG signal periodicity in time domain is well reported in [4, 7, 8]. The time domain periodicity implies spectra sparseness in frequency domain [9]. The whole spectrum is concentrated in very narrow bandwidth Δf around certain fundamental frequency f_0 and the first harmonic f_1. This represents the spectral signature of PPG signals. However, by regarding rPPG signal, its spectral signature is masked by excessive noise levels that prevent its simple extraction through wide band filtering.

In almost previous works, the periodicity usage is limited in measuring the signal quality by defining goodness and signal-to-noise ratio for each ROI signal.

- Goodness [8] metric for each ROI signal is given as follows:

$$G(\text{ROI}, \text{Frame}_j) = \frac{\int_{f_0-\Delta f}^{f_0+\Delta f} \hat{P}_f df}{\int_{\Omega} \hat{P}_f \text{df} - \left(\int_{f_0-\Delta f}^{f_0+\Delta f} \hat{P}_f \text{df}\right)} \tag{1}$$

where $\hat{P}_f$ represents spectral power density measured over the cardiac band and $\Omega = [0.5 - 5]$Hz over frame time interval T_{frame}. It measures the spectral power ratio between the very narrow band around the fundamental frequency ($f_0$band) to the other signal power in the whole cardiac band [0.5–5] Hz except the $f_0$band.

- Signal-to-noise (SNR) reliability [10] metric

$$Q_{\text{SNR}}(\text{ROI, Frame}_j) = \frac{\int_{f_0-\Delta f}^{f_0+\Delta f} \hat{P}_f \text{df} + \int_{2f_0-2\Delta f}^{2f_0+2\Delta f} \hat{P}_f \text{df}}{\int_{\Omega} \hat{P}_f \text{df} - \left(\int_{f_0\ -\Delta f}^{f_0+\Delta f} \hat{P}_f \text{df} + \int_{2f_0\ -2\Delta f}^{2f_0+2\Delta f} \hat{P}_f \text{df}\right)} \tag{2}$$

The bandwidth (Δf) may be set empirically to 0.2 Hz or 0.1 Hz as in [5].

This quality metric is used for selecting best ROIs' signal to be assigned for further processing.

2.4 Time Series-Based Analysis and Ideal Filtering

Real-time monitoring of HR and HRV can be simplified by permitting small processing intervals, about 5 or 10 s, delay between the picked signal and displayed values updates. Also, it is accepted as a proper updating rate from biomedical perspectives. $\text{Displayed}_{\text{Frame}}(i) = \text{picked}_{\text{Frame}}(i-1)$.

This leads to a great simplification in signal treatment to be regarded as a full-known time series. Consequently, it enables an essential filtering relaxation. The filtering can be performed by zero-phase non-causal forward–backward band-pass filter in the range [0.5–5] Hz similar to [6]. It may be defined as ideal filter because it provides the best filtering performance provided by corresponding causal counterpart.

2.5 *Periodic Filtering and Spectral Signature*

The main idea resides in exploiting the fact of spectral concentration around the fundamental and first harmonic components, in extracting the main building features of PPG with truthful timing information. So, we do not have to pass the whole cardiac band [0.5–5] Hz that contains unrelated signal components in addition to the required PPG information.

Further than employing SNR metric for identifying signal quality, we believe that a multi-band periodic filtering [10] can be very effective in extracting the required clear timing signal by extracting its spectral signature. Also, the nature of quasi-periodicity represents an additional degree of freedom that should be imposed in signal extraction [11, 12].

In this paper, we employ simple multi-band filtering based on the aforementioned ideal filter. The proposed approach involves the following stages:

1. Fundamental **frequency identification**: the traditional ideal filter [12] of the cardiac signal in range [0.5–5] Hz is applied on the green channel raw data. IBI estimation cannot be estimated from this signal accurately. There are many unrelated oscillations that give rise to false beat identification. This problem is alleviated by dropping the instances of signals corresponding to high peak variance.
2. **Two Band filtering**: the same non-causal filter is applied again on $f_0$band and $f_1 band$ with $\Delta f = 0.1\text{Hz}$. Hence, the pass range can be defined as passrange $=$ $([f_0 - \Delta f : f_0 - \Delta f]; [f_1 - \Delta f : f_1 - \Delta f])$
3. The filtered signal has better timing of heart beating through providing essential building features approximation of the PPG signal. This point has great impact on minimizing false IBI identification.

By remarking this fundamental frequency, the PPG spectral signature can be recovered by masking all the spectra except the very narrow pass band.

So, if $X(f)$ represents the spectrum of the raw data, then the spectral mask $X_{\text{mask}}(f)$ sets all elements to zeros except for the indices of the intended pass range will be set as ones.

$$X_{\text{mask}}(f) = \left[00\cdots000\underbrace{11\cdots11}_{f_0\text{band}}00\cdots000\underbrace{11\cdots11}_{f_1\text{band}}00\cdots000 \right]$$

By setting $\Delta f = +/-0.1Hz$, around f_0 and f_1 frequency bins. Each band will have about 0.2 Hz width.

The recovered (filtered) signal becomes

$$x_r(t) = \text{IFFT}(X(f).*X_{\text{mask}}(f))$$

This has the same effect of the time series ideal filter.

2.6 *Peak Finder and Beat Segmentation*

Thanks to the sharp timing PPG signal approximation, beats can be segmented based on the peak finder algorithm [14] with satisfactory accuracy from rPPG signal compared to the corresponding cPPG signal. It estimates local maxima (peaks) or minima (valleys) in the signal.

3 Experimental Evaluation

In this section, the proposed approach of periodic filtering will be evaluated through exploring its impact on approximating the essential features of both cPPG signals and rPPG signals extracted from different ROIs of the corresponding videos of public dataset. Also, the overall impact can be assessed by clarifying the experimental worst case errors in the estimated IBI.

3.1 *Data Set*

The framework will be tested under public data set such as Tokyo Tech. data [4]. However, all our results are performed on the lower rate videos (30 fps). Also, the quality of the estimated signals and IBIs is validated by comparing it against the ground truth contact cPPG signal that is measured in time synchronization with recorded videos.

3.2 *CPPG-Based Validation*

The two-stage selective filtering will be operated under $\Delta f = 0.1Hz$ (empirically assigned) as shown in Fig. 2. As noted from Fig. 3, applying the periodic selective filtering on the ground truth cPPG signal preserves the main building features of the signal including timing accuracy. Moreover, the selective filtering provides more

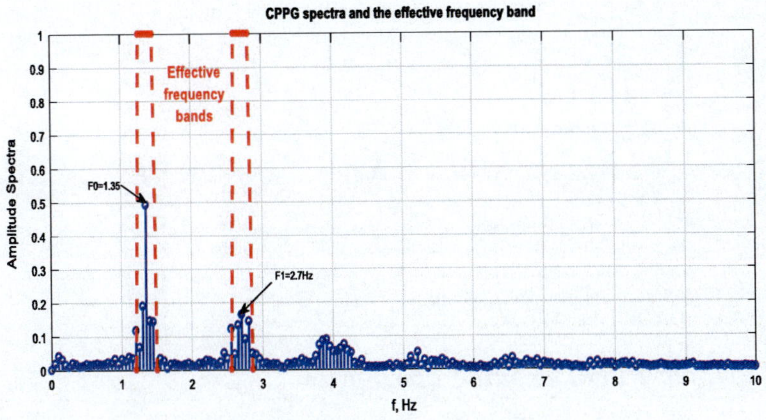

Fig. 2 Applying selective filter in frequency domain

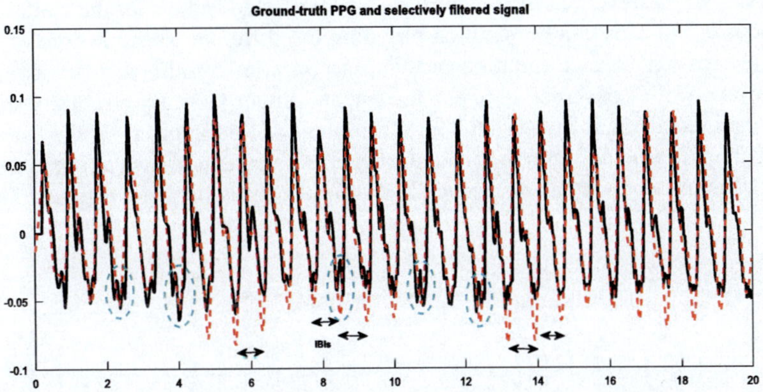

Fig. 3 cPPG signal (with circles) against two band selectively filtered signal (dotted

sharp timing at minima points than unfiltered signal. It agrees with spectral periodicity assumption that the narrow band selective filtering does not defect the essential signal construction (except diastolic notch).

3.3 Salient Spectra Signature

The raw data extracted from video sequence has noisy spectra (high noise floor) as shown in Fig. 4. However, the fundamental frequency still can be identified easily in the [0.5–5] Hz band.

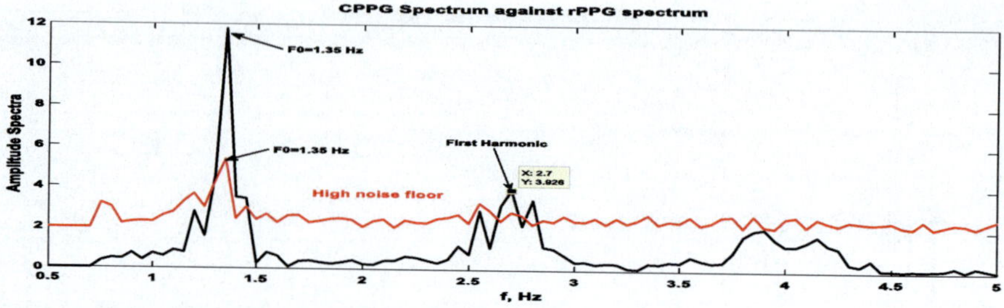

Fig. 4 cPPG spectrum against rPPG spectrum

3.4 Remote-PPG Periodic Filtering

By applying the periodic filtering on the raw data extracted from the video, we have an excellent (timing) approximation of the ground truth PPG signal as shown in Fig. 5. However, we observe that the facial rPPG signal is in advance over the contact PPG by about 0.2 s. This can be justified based on the different transmit time of blood volume between the face and the hand. This seems similar to the difference in pulse transit time (PTT) observed between the toy and finger [13]. By circularly shifting if the estimated signal by o.2 s (about 6 samples), the beat timing of the estimated signal almost coincides with the ground truth cPPG signal as shown in Fig. 6 we note no more than two samples difference in the beat lengths between estimated signal and the GT cPPG signal.

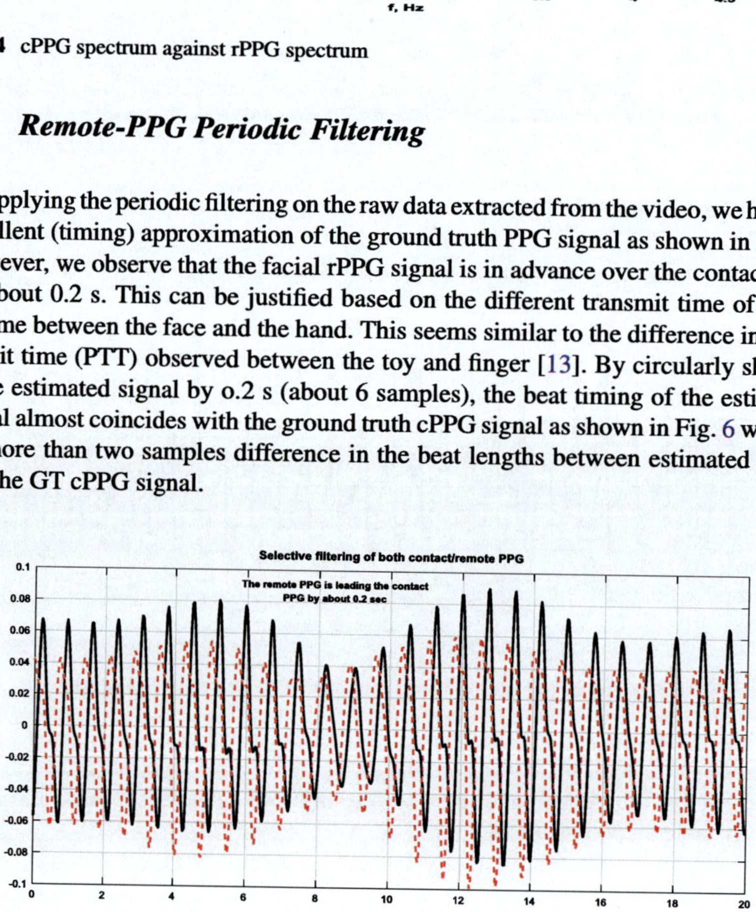

Fig. 5 Excellent approximation of PPG through rPPG periodic filtering

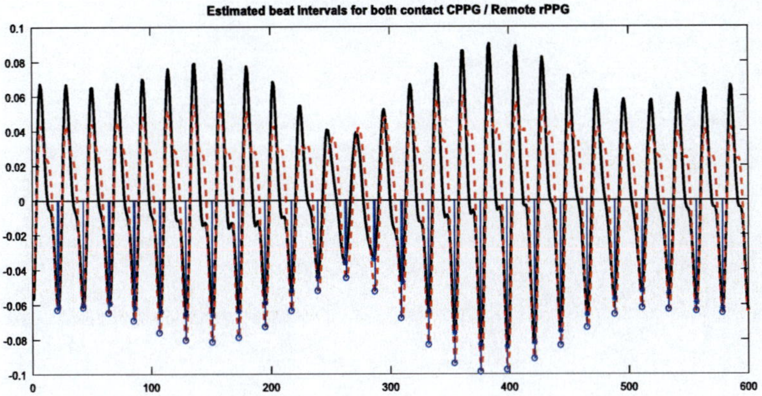

Fig. 6 Shifted rPPG periodic filtering against cPPG

3.5 *IBI Estimation*

Here, the timing accuracy of the periodically filtered signal will be indicated. The PPG spectral signature of multiple ROIs (whole face, skin face, and forehead) can be emphasized. Through testing over many videos from public dataset, the absolute IBI error may be upper-bounded by two samples difference between the estimated IBIs compared to GT IBIs. More specifically, the two samples time corresponds to about 0.0667 s, where we are operating on 30fps videos. The raw red, green, and blue signals are shown in Fig. 7. The corresponding spectral signature is shown in Fig. 8. This outperforms the corresponding upper bound absolute IBI error reported in [4] as 0.15 s as shown in Fig. 9.

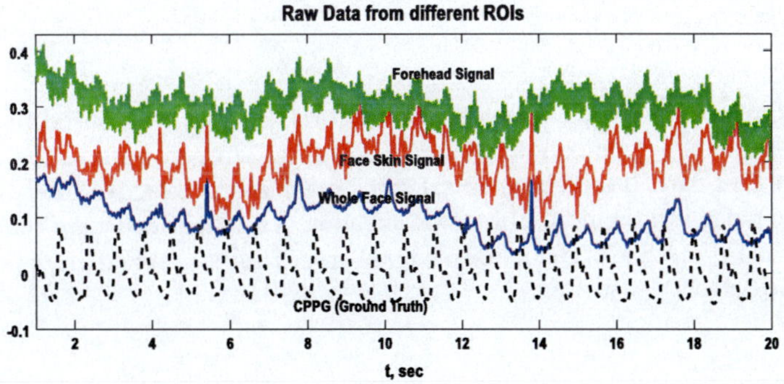

Fig. 7 Raw data from different ROIs against GT CPPG signal

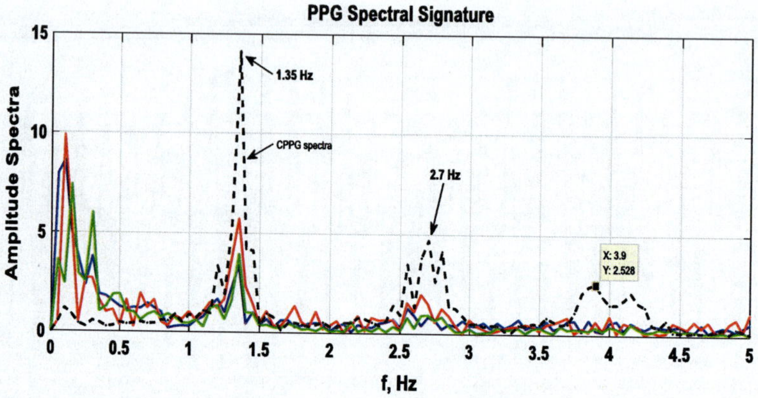

Fig. 8 Corresponding spectral signature

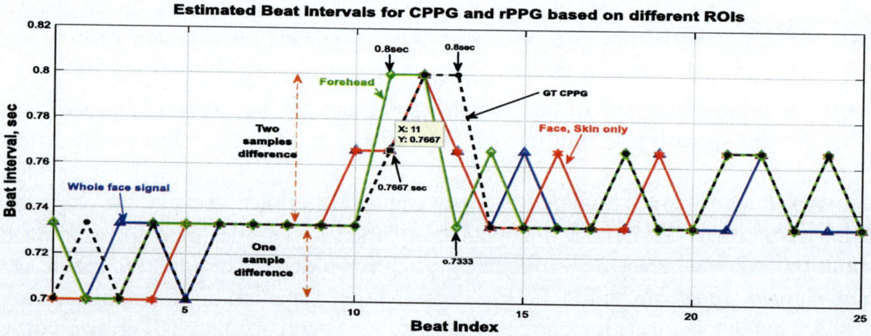

Fig. 9 Corresponding IBI estimation

4 Conclusion

In this paper, we proposed a new approach for IBI estimation based on spectral analysis and skin segmentation. Skin segmentation aims to remove the effect of facial non-skin parts on the biomedical signs estimation. Selective filter is used to extract the pulse information. The proposed approach outperforms the state-of-the-art approach [4] in sense of reduction of signal processing stages and the upper-bound absolute IBI error.

References

1. Fouad, R., Omer, O.A., Aly, M.H.: Optimizing remote photoplethysmography using adaptive skin segmentation for real-time heart rate monitoring. IEEE Access **7**, 76513–76528 (2019)

2. Boccignone, G., Conte, D., Cuculo, V., D'Amelio, A., Grossi, G., Lanzarotti, R.: An open framework for remote-PPG methods and their assessment. IEEE Access **8**, 216083–216103 (2020)
3. Poh, M.-Z., McDuff, D.J., Picard, R.W.: Non-contact, automated cardiac pulse measurements using video imaging and blind source separation. Opt. Express **18**, 10762–10774 (2010)
4. Maki, Y., Monno, Y., Yoshizaki, K., Tanaka, M., Okutomi, M.: "Inter-beat interval estimation from facial video based on reliability of BVP signals." In 2019 41st Annual International Conference of the IEEE Engineering in Medicine and Biology Society (EMBC), pp. 6525–6528 (2019)
5. Argüello-Prada, E.J.: "The mountaineer's method for peak detection in photoplethysmographic signals." Revista Facultad de Ingeniería Univ. de Antioquia, pp. 42–50 (2019)
6. Kumar, M., Veeraraghavan, A., Sabharwal, A.: DistancePPG: robust non-contact vital signs monitoring using a camera. Biomed. Opt. Express **6**, 1565–1588 (2015)
7. Macwan, R., Bobbia, S., Benezeth, Y., Dubois, J., Mansouri, A.: "Periodic variance maximization using generalized eigenvalue decomposition applied to remote photoplethysmography estimation." In Proceedings of the IEEE Conference on Computer Vision and Pattern Recognition Workshops, pp. 1332–1340 (2018)
8. Pai, A., Veeraraghavan, A., Sabharwal, A.: "CameraHRV: robust measurement of heart rate variability using a camera," in Optical Diagnostics and Sensing XVIII: Toward Point-of-Care Diagnostics, p. 105010S (2018)
9. Magdalena Nowara, E., Marks, T.K., Mansour, H., Veeraraghavan, A.: "SparsePPG: towards driver monitoring using camera-based vital signs estimation in near-infrared." In Proceedings of the IEEE conference on computer vision and pattern recognition workshops, pp. 1272–1281 (2018)
10. Tan, L., Jiang, J.: Novel adaptive IIR filter for frequency estimation and tracking [DSP Tips and Tricks]. IEEE Signal Process. Mag. **26**, 186–189 (2009)
11. Elvander, F., Ding, J., Jakobsson, A.: "On harmonic approximations of inharmonic signals," in ICASSP 2020–2020 IEEE International Conference on Acoustics, Speech and Signal Processing (ICASSP), pp. 5360–5364 (2020)
12. Nathanael Yoder, "peakfinder(x0, sel, thresh, extrema, includeEndpoints, interpolate)," MATLAB Central File Exchange (2020)
13. Nitzan, M., Khanokh, B., Slovik, Y.: The difference in pulse transit time to the toe and finger measured by photoplethysmography. Physio. measurement **23**, 85 (2001)
14. Redmon, J., Divvala, S., Girshick, R., Farhadi, A.: You only look once: unified, real-time object detection. IEEE Conference on Computer Vision and Pattern Recognition (CVPR) **2016**, 779–788 (2016). https://doi.org/10.1109/CVPR.2016.91
15. Henriques, J.F., Caseiro, R., Martins, P., Batista, J.: "High-speed tracking with kernelized correlation filters." Trans. Pattern Anal. Mach. Intell. **37**(3), :583–596 (2015). https://doi.org/10.1109/TPAMI.2014.2345390

Research and Application on Optimization of Drainage Gas Recovery Methods in Linxing Block

Shijie Liu, Zhenhua Cai, Wenrong Song, Tao Xuan, Chengzhen Liu, Song Wang, and Gang Feng

Abstract Linxing block is a tight gas reservoir with low pressure and production. Water can be seen at the initial stage of production, so drainage gas recovery (DGR) measures are needed. In Linxing block, the selection of DGR technology mainly depends on field experience, and the low recovery rate is caused by strong subjective factors and unreasonable DGR. This paper summarizes the production evaluation table of each DGR, obtains the influence evaluation table of each DGR influence factor combined with the actual production situation, and obtains the weight of each influence factor by using the analytic hierarchy process. Finally, according to the membership function, the membership matrix between each DGR and the influencing factors is established, and the weight of each DGR is obtained by using the fuzzy comprehensive evaluation method. The larger the value is, the more suitable the DGR is. Thus, a set of DGR optimization model is established. The actual data of a well in Linxing block are verified. The example shows that the model can optimize the DGR according to the weight. Foam DGR in Linxing block is the best way, and optimizing tubing and screw pump are the second, and the gas lift is the most unsuitable. The results are consistent with the actual production in Linxing block.

Keywords Linxing block · Optimization of DGR technology · Analytic hierarchy process · Fuzzy comprehensive evaluation · Membership function

S. Liu (✉) · Z. Cai · T. Xuan · C. Liu
CNOOC Ener Tech-Drilling and Production Co., Tianjin 300459, China
e-mail: liushj23@cnooc.com.cn

W. Song
Beijing Yadan Petroleum Technology Development Co. Ltd, Beijing 102299, China

S. Wang
College of Petroleum Engineering, China University of Petroleum, Beijing 102249, China

G. Feng
Xi'an Zhongkong Tiandi Technology Development Co, Ltd, Xi'an 710000, China

© The Author(s), under exclusive license to Springer Nature Singapore Pte Ltd. 2022
A. K. Nagar et al. (eds.), *Intelligent Sustainable Systems*, Lecture Notes in Networks and Systems 333, https://doi.org/10.1007/978-981-16-6309-3_18

Nomenclature

DGR	Drainage gas recovery
GL	Gas lift
PGL	Plunger gas lift
OT	Optimized tubing
SP	Screw pump
FDGR	Foam drainage gas recovery
RWA	Rod wear adaptability
VDGR	Vortex drainage gas recovery
DG	Discharge rate
FS	Fracturing sand
HW	Horizontal well
DD	Design difficulty
IL	Intelligent level
IC	Investment cost
CR	Corrosion resistance
MF	Membership function

1 Introduction

Linxing block is located in the east margin of Ordos Basin, which is a typical low-pressure and low permeability tight sandstone gas reservoir. Most wells in this block are generally below 2000 m deep, and the oil pressure is below 6Mpa. Water is generally produced. The output water is generally less than 10m^3/d, and the average daily production is 5m^3/d. Water is seen in the early stage of exploitation, and the stable production period is short. Most of the gas wells in this block will begin to accumulate water after one year of production. And, the intermittent production has seriously affected the gas production [1–3]. Therefore, DGR technology is imminent. At present, scholars have done lots of research on the optimization of DGR methods. . Yang Chuandong [4] introduced the development process of DGR technology in Sichuan gas field, summarized the advantages, disadvantages and applicable conditions of six DGR methods, put forward corresponding DGR measures for different types of effluent and production capacity, and achieved remarkable results in field application. He Zuny [5] also considered the applicability of DGR measures and economic factors, determined the technical economic discriminant and discriminant analysis method of gas lift and mechanical DGR technology, and focused on the influence of economic factors on the optimization of DGR technology. Qu Lin et al. [6] summarized the characteristics and application scope of seven DGR technologies, proposed the geological conditions for the implementation of DGR technology, and comprehensively considered the influence of gas well dynamic factors, process operation cost, operation efficiency, complexity and simplicity of process matching, management difficulty, etc. On the premise of determining the applicability of DGR technology, Gong Yanfen et al. [7] considered the economic factors, established the economic index matrix of DGR technology, and adopted TOPSIS method to quantitatively analyze and select the optimal scheme; Li Wenkui et al. [8] drew control charts of DGR technology selection for different layers by analyzing daily gas production, water–gas ratio, bottom hole pressure, critical liquid carrying flow, tubing size, and other indicators and determined corresponding DGR technology in the control chart according to the actual index value of each well. By establishing the relationship between gas–water ratio and monthly gas production, the best time to implement DGR technology is determined; Jiang Beibei et al. [9] classified the gas wells through the production decline analysis of gas wells, defined the comprehensive productivity index of gas wells, characterized the controlled water bearing area, formation pressure, and reservoir physical properties of single well, made the optimization chart combined with the comprehensive productivity index, formation pressure, and water production law of gas wells, and determined the DGR technology schemes of different types of gas wells; most of the above studies are based on the mechanism, which has a certain guiding role for the optimization of DGR technology. This paper proposes a selection of DGR technology based on the combination of artificial intelligence algorithm and the actual production situation, which provides technical guidance and experience for DGR technology.

2 Evaluation of DGR Technology

Through the investigation of the relevant literature of DGR technology of tight gas wells and combining with the actual production of tight gas wells in Linxing area[, the DGR technology is evaluated as follows (Table 1):

3 Establishment of Optimal Model for DGR

It is not objective and scientific to optimize the DGR technology of gas wells only through empirical analysis. AHP and fuzzy comprehensive evaluation can systematically and quantitatively evaluate the influence of various factors such as IC, system efficiency, IL, and well type, making the final optimization result more reasonable and reliable.

3.1 Establishment of MF of Each Factor

At present, there are several main methods to determine the MF: fuzzy statistical test method, function segmentation method, binary comparison sorting method, and

Table 1 Evaluation of various DGR methods

DGR	Technical advantages	Disadvantage
DGR by pump	Mature technology and durable equipment; low investment and operation cost; head and displacement are suitable for most wells	The rod and tube are susceptible to friction and corrosion
DGR by screw pump	Affected by solid particles, the displacement is easy to adjust, the system efficiency is high, and the floor area is easy to maintain	The stator wear affects the pump efficiency, the rod and tube are subject to friction and corrosion, and it may be difficult to start pumping after shutdown
Gas lift	The wellhead and downhole equipment are simple; affected by solid particles; the displacement can be adjusted flexibly; suitable for highly deviated wells	High pressure gas pipeline is needed; the ground equipment system is complex; large one-time investment; limited by gas source
Optimized tubing	The wellhead and downhole equipment are simple, belonging to one-time investment, without subsequent maintenance, and the process design is simple	When the small tubing is no longer suitable for DGR, it is more complex to take other drainage measures
Plunger gas lift	Suitable for low pressure and production tight gas reservoir with high intelligence, convenient operation, and labor saving	The effect of increasing yield is general, and it mainly plays the role of stabilizing yield
Foam DGR	The process management and operation are very convenient, with less investment, high benefit, and easy promotion	Defoaming agent may cause corrosion to pipe network and compressor and shorten its service life
Vortex DGR	Easy to operate without using workover equipment, shutting in, and killing well; it can slow down the wax and scale formation; the cost is low; no moving parts, reliable operation, the service cycle can be as long as more than half a year	The technology is relatively new, and the gas well adapts to harsh conditions

calculating the membership function of fuzzy set through the operation between fuzzy sets. In this paper, the function segmentation method is used.

(1) Membership function of discharge

DGR by pump, screw pump, and vortex are suitable for gas wells with liquid production; plunger gas lift, foam DGR, and optimized tubing are suitable for low liquid production gas wells, while gas lift is suitable for high liquid production gas wells. When the liquid production of gas well is $Q > 150\ m^3/d$, plunger gas lift, foam DGR and optimized tubing are not used. When the liquid production of gas well is

Q >300 m^3/d, rod pump, screw pump, and vortex are not used; When Q <50 m^3/d, gas lift is not used.

① Pump, screw pump, vortex

$$u_{n1} = \begin{cases} \frac{1}{1+(x_L/120)^{3.5}} & x_L < 300 \\ 0 & x_L \geq 300 \end{cases} \quad (1)$$

② Plunger gas lift, foam DGR, and optimized tubing

$$u_{n1} = \begin{cases} 1 & x_L < 50 \\ \frac{50-x_L}{150-50} & 50 \leq x_L \leq 150 \\ 0 & x_L > 150 \end{cases} \quad (2)$$

③ Gas lift

$$u_{n1} = \begin{cases} 0 & x_L < 50 \\ \frac{1}{1+((x_L-50)/60)^{-3}} & 50 \leq x_L < 300 \\ 0.6 & x_L \geq 300 \end{cases} \quad (3)$$

where x_L is the maximum liquid production.

(2) Membership function of fractured sand

Because of its low permeability and low porosity, tight gas wells have large-scale production only after fracturing. The influence of pumping unit, plunger gas lift, and eddy current drainage on pressure sand cracking is great.

① Pump, plunger gas lift, vortex

$$u_{n2} = \begin{cases} 1 & x_{n2} < 1 \\ \frac{4-x_{n2}}{4-1} & 1 \leq x_{n2} < 4 \\ 0 & x_{n2} \geq 4 \end{cases} \quad (4)$$

② Gas lift, optimized tubing, foam DGR, screw pump

$$u_{n2} = \begin{cases} 0 & x_{n2} < 1 \\ \frac{x_{n2}-1}{4-1} & 1 \leq x_{n2} < 4 \\ 1 & x_{n2} \geq 4 \end{cases} \quad (5)$$

where x_{n2} is the suitable capacity requirement for fractured sand.

(3) Membership function of inclined well and horizontal well

For the pumping unit and plunger gas lift, because there are moving parts, the movement in inclined wells and horizontal wells is limited. Gas lift, optimized tubing, bubble discharge, and screw pump are more suitable for production in inclined and horizontal wells.

$$u_{n3} = \begin{cases} 1 - \frac{5-x_{n3}}{5} x_{n3} < 5 \\ 1 \quad x_{n3} \geq 5 \end{cases} \tag{6}$$

where x_{n3} is the adaptability of each DGR equipment.

(4) Membership function of design difficulty

If the design is more complex, it will not be easy to popularize this method, among which the design of pumping unit, optimizing tubing and foam DGR is less difficult.

$$u_{n4} = \begin{cases} 1 - \frac{4-x_{n4}}{4} \; x_{n4} < 4 \\ 1 \;\; x_{n4} \geq 4 \end{cases} \tag{7}$$

where x_{n4} is the design difficulty of each DGR equipment.

(5) Membership function of rod wear adaptability

① Pump, screw pump, vortex

$$u_{n5} = \begin{cases} 1 x_{n5} < 1 \\ \frac{4-x_{n5}}{4-1} 1 \leq x_{n5} < 4 \\ 0 x_{n5} \geq 4 \end{cases} \tag{8}$$

② Gas lift, optimized tubing, plunger gas lift, foam DGR

$$u_{n5} = \begin{cases} 0 \; x_{n5} < 1 \\ \frac{x_{n5}-1}{4-1} \; 1 \leq x_{n5} < 4 \\ 1 x_{n5} \geq 4 \end{cases} \tag{9}$$

where x_{n5} is the field requirement for the adaptability of the pole.

(6) Corrosion membership function

$$u_{n6} = \begin{cases} 0 x_{n6} \leq 2 \\ \frac{x_{n6}}{5} 2 < x_{n6} < 5 \\ 1 \; x_{n6} \geq 5 \end{cases} \tag{10}$$

where x_{n6} is the CR of each DGR equipment.

(7) Intelligent-level membership function

$$u_{n7} = \begin{cases} \frac{x_{n7}}{5} \; x_{n7} < 4 \\ 1 \quad x_{n7} \geq 4 \end{cases} \tag{11}$$

where x_{n7} is the intelligent level of each DGR equipment.

(8) Membership function of investment cost

$$u_{n8} = \begin{cases} \frac{x_{n8}}{5} x_{n8} < 4 \\ 1 \quad x_{n8} \geq 4 \end{cases} \tag{12}$$

where x_{n8} is the investment cost of each DGR equipment.

3.2 Calculation of Weight of DGR

Based on AHP, the weight value of feature vector of DGR technology is obtained.

$$\mathrm{W} = [w_1, w_2, \ldots, w_n] \tag{13}$$

Based on membership function of each influence factor, the membership matrix is obtained as follows.

$$\mathrm{R} = \begin{bmatrix} R_{11} & R_{12} & \cdots & R_{1,n} \\ R_{21} & R_{22} & \cdots & R_{2,n} \\ \vdots & \vdots & \cdots & \vdots \\ R_{m1} & R_{m2} & \cdots & R_{m,n} \end{bmatrix} \tag{14}$$

where m is the number of types of DGR methods and N is the type number of influencing factors for the selection of production pattern in the block.

The objective function B is obtained by multiplying the formulas (13) and (14), which is the weight of each row and production method.

$$B = W \cdot R = [w_1, w_2, \ldots, w_n] \cdot \begin{bmatrix} R_{11} & R_{12} & \cdots & R_{1,n} \\ R_{21} & R_{22} & \cdots & R_{2,n} \\ \vdots & \vdots & \cdots & \vdots \\ R_{m1} & R_{m2} & \cdots & R_{m,n} \end{bmatrix} = [b_1, b_2, \ldots, b_n] \tag{15}$$

where b_i is the weight of the ith DGR method. The greater the weight is, the more suitable DGR method is for the block.

4 Case Study

Linxing block has been put into production since about 2015. The block is mainly composed of low-pressure and low-yield gas wells with buried depth between 1200 and 1800 m, formation pressure of 12–18 MPa formation temperature of 41–51 °C, formation pressure gradient of 0.92–0.99 MPa per 100 m, pressure coefficient of 0.96–1.02 and temperature gradient of 3.2 °C per 100 m. It belongs to normal pressure temperature system. The oil pressure is 3~5 MPa, and the average gas production is about $0.9 \times 10^4 m^3/d$.

Combined with the actual production situation, the comprehensive analysis table of each factor on the DGR technology is established, as shown in Table 2.

The design objectives and the design target parameters are shown in Table 3.

Using the MF of each factor, the membership matrix R is obtained.

$$R = \begin{bmatrix} 0.999 & 0 & 0.6 & 0.5 & 0.67 & 0.6 & 0.6 & 1 \\ 0.999 & 1 & 0.2 & 0.75 & 0.67 & 0.6 & 0.6 & 1 \\ 0 & 1 & 1 & 0.75 & 1 & 0.8 & 0.4 & 0.4 \\ 1 & 1 & 0.6 & 0.5 & 1 & 0.6 & 0.4 & 1 \\ 1 & 0 & 0.6 & 0.75 & 1 & 0.8 & 1 & 0.6 \\ 1 & 1 & 0.8 & 0.5 & 1 & 1 & 0.6 & 1 \\ 0.999 & 0 & 0.6 & 1 & 0 & 0.8 & 0.6 & 1 \end{bmatrix}$$

The weight of each factor is analyzed and calculated, and the weight comparison of each factor is shown in Table 4.

Table 2 Evaluation data of each factor

DGR	Pump	SP	GL	OT	PGL	FDGR	VDGR
DR	100	100	500	100	50	150	100
FS	2	4	4	5	2	4	3
HW	3	1	5	3	3	4	3
DD	2	3	3	2	3	2	4
RWA	2	2	5	5	4	5	4
CR	3	3	4	3	4	5	4
IL	3	3	2	2	5	3	3
IC	4	4	2	4	3	5	4

Table 3 Design objectives

	DR	FS	HW	DD	RWA	CR	IL	IC
Design objectives	10	high	high	More suitable	suitable	CR	high	low
target data	10	4	5	4	3	5	5	5

Table 4 Comparative data of the importance of influence factors

	DR	FS	HW	DD	RWA	CR	IL	IC
DR	1	1/3	1/5	3	1/3	1/2	1/3	1/4
FS	3	1	1/3	5	1	1	1	1/3
HW	5	3	1	6	3	3	1	1/3
DD	1/3	1/5	1/6	1	1/3	1/2	1/6	1/7
RWA	3	1	1/3	3	1	3	1/2	1/3
CR	2	1	1/3	2	1/3	1	1/4	1/5
IL	3	1	1	6	2	4	1	1/4
IC	4	3	3	7	3	5	4	1

Table 5 Evaluation of DGR methods

DGR	Pump	SP	GL	OP	PGL	FDGR	VDGR
Target value	0.6936	0.7247	0.6481	0.7976	0.6735	0.8892	0.6521

The analytic hierarchy process is used to calculate, and the parameters in Table 4 are input into the computer program as the judgment matrix to obtain the weight value of each factor.

$$W = \begin{bmatrix} 0.048 \; 0.101 \; 0.191 \; 0.026 \; 0.099 \; 0.059 \; 0.149 \; 0.327 \end{bmatrix}$$

The final objective function can be obtained from Eq. (15).

$$B = R \times W = \begin{bmatrix} 0.6936 \; 0.7247 \; 0.6481 \; 0.7976 \; 0.6735 \; 0.8892 \; 0.6521 \end{bmatrix}$$

Then the target values are shown in Table 5.

It can be seen from Table 5 that foam DGR is the best, followed by the optimized tubing and screw pump, and gas lift is the least suitable. In the actual production, there are many wells in Linxing block that choose the foam DGR, which shows that the equipment selected by the fuzzy evaluation method is more in line with the reality and also reflects the application prospect of foam production in tight gas wells.

5 Conclusion

(1) Through consulting the domestic and foreign literatures, the production evaluation tables of common DGR of tight gas wells are summarized, and the main influencing factors that need to be considered in the selection of DGR of tight gas wells in Linxing block are proposed from both technical and economic aspects.

(2) By using the analytic hierarchy process, the construction matrix of each influence factor is established, the weight of each influence factor vector is obtained, and the fuzzy relation matrix is established by using the MF, and thus, the fuzzy evaluation model of DGR is established. On this basis, the most suitable DGR technology of tight gas wells is optimized, which effectively avoids the evaluation deviation caused by subjective factors.
(3) Using AHP and multi-level fuzzy comprehensive evaluation method to optimize tight gas wells in Linxing block has good applicability, and foam DGR is the best DGR technology in this block, which has good reference significance for the selection of DGR technology in other gas fields.

References

1. Zhu, C., Wang, X., Dong, S., Li, X., Fu, S., Zhou, C.: Study on tubing diameter optimization of tight gas wells in Linxing block [J]. Petroleum Mach. **45**(04), 74–78 (2017)
2. Huan, C., Zihan, L., Cao Yanfeng, Y., Jifei, L.M.: Dynamic simulation analysis of wellbore effusion in linxingzhi dense gas well [J]. Lithologic Reservoir **30**(02), 154–160 (2018)
3. Qi, Y., Ju, Y. ,Meng, S.,Yu, K., Li, W., Jia, T., Wu, J., Chen, W., Luo, L.: Geological controls on high production of tight sandstone gas in Linxing Block, Eastern Ordos Basin, China[J]. Acta Geologica Sinica(English Edition) **94**(2) (2020)
4. Yang, C.: The supporting technology and application of drainage and gas production in Sichuan gas field [j]. Natural Gas Industry (03), 37–41+107 (1995)
5. He, Z.: Discussion on calculation method for feasibility demonstration of DGR technology [J]. Natural Gas Industry (02), 68–71 + 10–11 (1997)
6. Qu, L., Qu, J.: Discussion on selection of DGR technology [J]. Drilling Prod. Technol. (02), 49–51+113–114 (2005)
7. Gong, Y., Wang, L., Qu, L., Dong, G., Li, X.: Application of TOPSIS method in optimization of DGR technology [J]. J. Daqing Petrol. Inst. (05), 58–60+64+122 (2007)
8. Li, W., Zhou, G., Bi, G., Cao, G., Wang, Y., L, X., Li, J., Liu, Y.: Optimization model of DGR in Sebei gas field [J]. Natural Gas Industry. **29**(09), 60–63+137–138 (2009)
9. Beibei, J., Haitao, L., Zhouzhou, Y., Yusheng, J., Tianqi, Z.: A new method for optimization of drainage and production measures of water well in su59 block [J]. Oil Drilling Prod. Technol. **35**(05), 79–83 (2013)

Can the Measurements of Albumin and Glucose Be Correlated Each Other? A Physics-Based Theory

Huber Nieto-Chaupis

Abstract As it is well known, the confirmation of albuminuria in diabetic patients requires of a test entirely different to blood-based tests. However, from the idea that the glucose dipoles can cancel the negative charges along the renal glomerulus, the measurement of glucose in blood might be strongly related to the exit of albumin through the Bowman space. In this paper, a theoretical treatment based at electrical interactions focused on bunches of glucose in the renal glomerulus is presented. The purpose of this theoretical work is toward the exact estimation of the relation between glucose and albumin that allows to measure in only one test both glucose and albumin. A few scenarios by using computational simulation are presented. The results suggest the importance of the geometry of flux of albumin and glucose in the theoretical assumptions.

1 Introduction

It is well known that Diabetes Type-II is one of the main causes of mortality to date [1]. Diabetes in a nutshell is a disease that essentially consists in the lack of capabilities of pancreas to segregate efficiently granules of glucose [2]. A lot of efforts in the pharmacological side have been done at the past. In present, diabetes is already part of prospective programs inside the prospective Internet technologies [3] such as the Internet of Bio-Nano things invented by Akyildiz [4].

Previous to the acquisition of disease, potential candidates pass through the phase of pre-diabetes [5] in which the values of glucose are ranging between 90 and 135 $\pm$ 10 mg/dL. Commonly, this phase is fully unknown in such candidates. It actually can be called the phase-0 in the sense that people do not exhibit any symptomatology. It is an important phase because it is possible to anticipate subsequent denouement that opens the door to worst scenarios as cause of diabetes that were not experienced at the very beginning [6]. Once the patient has a concrete diagnostic of diabetes, the endocrinologist has as target to put down the high values of glucose through common pharmacology that mainly (and in most cases) is based at (1) Metformin,

H. Nieto-Chaupis (✉)
Universidad Autonoma del Peru, Panamericana Sur Km. 16.3, Villa el Salvador, Lima, Peru

© The Author(s), under exclusive license to Springer Nature Singapore Pte Ltd. 2022

A. K. Nagar et al. (eds.), *Intelligent Sustainable Systems*, Lecture Notes in Networks and Systems 333, https://doi.org/10.1007/978-981-16-6309-3_19

and (2) Vidagliptin [7]. Once the patient has crossed the line that separates the pre-diabetes and diabetes states, the values of glucose are very sensitive to all those foods rich in carbohydrates and sugar. In this manner, the patient must guarantee a healthy lifestyle that manifests security as to do not compromise others organs. In according to world-wide data, diabetes disease can trigger (1) kidney disease, (2) alteration of nervous system, and others types [8].

In this paper, the problem of dual measurement of glucose and albumin is treated. Because the diabetes patients would initialize an episode of albuminuria, it is expected that in most cases, there is a necessity as to take blood and urine tests in order to align the recovery of the normal values. In this manner, it is desired to find a kind of **universal relationship** in the sense that by knowing the value of glucose then the value of albuminuria is also known inside the systematic errors. In fact, one can arrive in a straightforward manner that

$$3M_{\mathrm{AL}} = 10^3 M_{\mathrm{GL}} \tag{1}$$

with M_{AL} and M_{GL} the masses of molecules albumin and glucose. Clearly, the electric composition of glucose is perceived as a cause to push out albumin from blood vessels. Thus, the interaction of both inside the inner layers of renal glomerulus gives rise to the expulsion of albumin through the Bowman space.

The accurate knowledge of these molecular interactions would lead to potential applications such as:

- Dual glucometer (measures glucose and albumin),
- Fast diagnosis of level of kidney degradation,
- Establishment of scenarios that contributes to a best management of diabetes.

The ionic composition of glucose would produce an electric force to almbuin so that one expects that there is well-defined electrodynamics inside the renal glomerulus. Mathematically, this can be expresses as follows [9]:

$$\mathbf{E} = -\nabla\Phi(\mathbf{r}), \tag{2}$$

with $\Phi(\mathbf{r})$ the electric potential the exerts the glucose molecule onto the albumin one. Therefore, one can establish the main equation that follows from the well-known Coulomb law that is given by:

$$\mathbf{F} = \pm Q\nabla\Phi(\mathbf{r}) \tag{3}$$

where Q the charge that "perceives" the electric field $\mathbf{E}$ that can be the ionic component either albumin or glucose. The presence of $\pm$ is due to the cases of attraction or repulsion among the charged compounds. The rest is this paper is structured a follows: in second section, we provide all classical electrodynamics machinery that sustains the present proposal. In third section, simulations and results are presented. Finally, in last section, the conclusion of paper is drawn.

2 Physics of Albumin–Glucose Interaction

2.1 Basic Relations

The master equation that governs the electrical interaction between glucose and albumin proteins can be written as:

$$N_{\mathrm{GL}} = \gamma N_{\mathrm{AL}} \tag{4}$$

that states that the number of dipoles N_{GL} of glucose can exert an electrical force on a number γN_{AL} of albumin proteins. The proportion is dictated by the constant γ that depends on their masses. For instance, the mass relation between glucose and albumin is given by Eq. 1. In this manner, depending on the total mass of charged bunching of glucose ions, all of them can exert a net electric force on the albumin ones.

Throughout the paper is assumed that albumin and glucose are transported along the micro vascular systems inside the renal glomerulus. In this manner, we will work out in function of the master equation written down as:

$$\rho_{\mathrm{GL}}(t) = \gamma(t)\rho_{\mathrm{AL}}(t), \tag{5}$$

indicating that the density of released proteins such as albumin is proportional to the one of glucose molecules entering into the area of renal glomerulus. From Eq. 5, one can state that:

$$\rho_{\mathrm{GL}}(t) = \gamma(t)\mathrm{Log}\left\{\mathrm{Exp}\left[\rho_{\mathrm{AL}}(t)\right]\right\} = \mathrm{Log}\left\{\left[\mathrm{Exp}\rho_{\mathrm{AL}}(t)\right]^{\gamma(t)}\right\} \tag{6}$$

fact that would indicate for example a type of universal relation given by:

$$\mathrm{Exp}\rho_{\mathrm{GL}}(t) = \left[\mathrm{Exp}\rho_{\mathrm{AL}}(t)\right]^{\gamma(t)}. \tag{7}$$

Now, the right side is written as follows:

$$\mathrm{Exp}\rho_{\mathrm{AL}}(t) = \left[1 + \rho_{\mathrm{AL}}(t) + \frac{1}{2!}(\rho_{\mathrm{AL}}(t))^2 + \cdots\right]^{\gamma(t)}. \tag{8}$$

It is feasible to apply same procedure on left side of Eq. 7 so that one has that:

$$1 + \rho_{\mathrm{GL}}(t) + \cdots = \left[1 + \rho_{\mathrm{AL}}(t) + \frac{1}{2!}(\rho_{\mathrm{AL}}(t))^2 + \cdots\right]^{\gamma(t)}. \tag{9}$$

In order to build a phenomenological model, one can set $\gamma(t) = 1$ so that one arrives to an equation given by: $x^2 + 2x - 2b = 0$ whose trivial solution is given by $x = -2 \pm \sqrt{4 + 2b}$ yielding for the albumin:

$$\rho_{\mathrm{AL}}(t) = -2 \pm \sqrt{4 + 2\rho_{\mathrm{GL}}(t)}. \tag{10}$$

The case of $\rho_{\mathrm{AL}}(t) = -2 + \sqrt{4 + 2\rho_{\mathrm{GL}}(t)}$ deserves a special attention in the sense that while $\sqrt{4 + 2\rho_{\mathrm{GL}}(t)} \approx 2$, then one gets that $\rho_{\mathrm{AL}}(t) \approx 0$. For example, $\sqrt{4 + 2\rho_{\mathrm{GL}}(t)} \approx 2.1$ for instance the one has that $\rho_{\mathrm{GL}}(t) \approx 0.205$. Clearly, this result would need of a depth analysis that must incorporate facts from human physiology, and under what circumstances, these numerical replacements can be achieved in realistic cases. It should be noted that one can make more specific with respect to the meaning of this toy model. It is to employ different values to the proportionality factor given by $\gamma(t)$.

2.2 *Electrodynamics of Interactions*

Turning now to the electrodynamics treatment, the densities as written in Eq. 5 can be explicitly defined in terms of charge and volume as:

$$\frac{\mathrm{d}Q_{\mathrm{GL}}(t)}{\mathrm{d}V_{\mathrm{GL}}(t)} = \frac{\gamma(t)\mathrm{d}Q_{\mathrm{AL}}(t)}{\mathrm{d}V_{\mathrm{AL}}(t)} \tag{11}$$

from the fundamental relationship $Q = \int \rho \mathrm{d}V$ then the Coulomb-like force between glucose and albumin (or also perceived as protein-protein interactions [10]) can be written below as:

$$\mathcal{F} = \frac{1}{4\pi\epsilon}\frac{Q_{\mathrm{GL}}(t)\mathrm{Q}_{\mathrm{AL}}(t)}{|\mathbf{R}(t) - \mathbf{R}'(t)|^2} = \frac{1}{4\pi\epsilon}\frac{Q_{\mathrm{GL}}(t)\int \rho_{\mathrm{AL}}(t)\mathrm{d}V}{|\mathbf{R}(t) - \mathbf{R}'(t)|^2}. \tag{12}$$

With this, one can define below the electric force per unit of charge. In this case, per unit of glucose charge while is under interaction to albumin. In this way one has that:

$$\frac{\mathcal{F}(t)}{Q_{\mathrm{GL}}(t)} = \frac{1}{4\pi\epsilon}\frac{\int \rho_{\mathrm{AL}}(t)\mathrm{d}V}{|\mathbf{R}(t) - \mathbf{R}'(t)|^2}, \tag{13}$$

that defines the left side as the electric field in a genuine manner. Therefore, this turns out to be dependent on the integration of charge density of albumin compound. With this, then one gets that:

$$\mathcal{E}(t) = \frac{1}{4\pi\epsilon}\frac{\int \rho_{\mathrm{AL}}(t)\mathrm{d}V}{|\mathbf{R}(t) - \mathbf{R}'(t)|^2}. \tag{14}$$

Commonly, in electrostatics one extracts the electric field from the electric potential through the well-known relationship given by $\mathcal{E}(t) = -\nabla\Phi(t)$. Because the present study does not exhibit electrostatics phenomenology because we are dealing with fluids that are constantly varying in time, then one can proceed through the mathematical definition of gradient to arrive to a physical meaning with respect to the glucose–albumin interaction, then:

$$\nabla\Phi = \frac{(-4\pi\epsilon)^{-1}\int\rho_{AL}(t)}{|\mathbf{r}(t)-\mathbf{r}'(t)|^2}$$
$$\Rightarrow \frac{d\Phi}{d\mathbf{r}(t)} = \frac{(-4\pi\epsilon)^{-1}\int\rho_{AL}(t)}{|\mathbf{r}(t)-\mathbf{r}'(t)|^2}. \tag{15}$$

For the sake of simplicity, the usage of units where $4\pi\epsilon = 1$ shall be assumed along the definitions and derivation of equations. From Eq. 15 solving for Φ one has:

$$\Phi(t) = -\int d\mathbf{r}(t)\left[\frac{\int\rho_{AL}(t)dV}{|\mathbf{r}(t)-\mathbf{r}'(t)|^2}\right]. \tag{16}$$

The integration on the volume clearly is not affected by the one over $d\mathbf{r}(t)$

$$\Phi(\mathbf{r}(t)) = -\int\rho_{AL}(t)dV\int\frac{d\mathbf{r}(t)}{|\mathbf{r}(t)-\mathbf{r}'(t)|^2} \tag{17}$$

with the last integration approximated as

$$\int\frac{d\mathbf{r}(t)}{|\mathbf{r}(t)-\mathbf{r}'(t)|^2} \approx \frac{-1}{|\mathbf{r}(t)-\mathbf{r}'(t)|} \tag{18}$$

that leads to the well-known electric potential given by:

$$\Phi(\mathbf{r}(t)) = \frac{\int\rho_{AL}(t)dV}{|\mathbf{r}(t)-\mathbf{r}'(t)|} = \frac{Q_{AL}}{|\mathbf{r}(t)-\mathbf{r}'(t)|}, \tag{19}$$

as expected. Of course, one can apply approximations that emphasize the role of the charge of albumin.

2.3 Criterion of Oscillating Forces

In praxis, one might expect electric oscillations because repulsion as well as attraction forces might be governing the electrodynamics [11]. It is actually a potential scenario since large amount of glucose concentration dictates the repulsion of albumin when both species have same sign. Thus, the oscillatory model suggests that:

$$\mathbf{r}(t) = r\mathrm{Sin}(\omega t)$$
$$\mathbf{r}'(t) = r'\mathrm{Cos}(\omega t) \tag{20}$$

that leads us to write down the electric potential as:

$$\begin{aligned}\Phi(r,r',t) &\approx \frac{\mathcal{Q}_{\mathrm{AL}}(t)}{|r\mathrm{Sin}(\omega t)-r'\mathrm{Cos}(\omega t)|}\\ &\approx \frac{\mathcal{Q}_{\mathrm{AL}}(t)}{\mathrm{Cos}(\omega t)|r\mathrm{Tan}(\omega t)-r'|}\\ &\approx \frac{\mathcal{Q}_{\mathrm{AL}}(t)}{\mathrm{Cos}(\omega t)|r\mathrm{Tan}(\omega t)-r'|}\end{aligned} \quad (21)$$

indicating that the electric potential that the ionic compound of albumin exerts on adjacent charged compounds. Clearly, the point where it exerts the field is r so that one can carry out a subsequent approximation written by:

$$\Phi(r,r',t) \approx \frac{\mathcal{Q}_{\mathrm{AL}}(t)}{r\mathrm{Cos}(\omega t)|\frac{r'}{r}-\mathrm{Tan}(\omega t)|} \quad (22)$$

with $\frac{r'}{r} << 1$ one finally arrives to

$$\Phi(r,r',t) \approx \frac{\mathcal{Q}_{\mathrm{AL}}(t)}{r\mathrm{Sin}(\omega t)} = \frac{\mathcal{Q}_{\mathrm{AL}}(t)}{r}\mathrm{Csc}(\omega t) \quad (23)$$

that seems to be an exact separation of variables in the form: $\Phi(r,r',t) = R(r)U(t)$.

2.4 Implications of Proteins Mass

The Coulomb force can also be written as:

$$\mathcal{F} = \frac{\mathcal{Q}_{\mathrm{GL}}\mathcal{Q}_{\mathrm{AL}}}{4\pi\epsilon_M}\frac{\mathbf{R}-\mathbf{R}'}{|\mathbf{R}-\mathbf{R}'|^3}. \quad (24)$$

Under the assumption that is the bunch of albumin that is fluxing in an anomalous manner, then it is plausible to write down the second Newton's law as:

$$M_{\mathrm{AL}}\frac{d^2\mathbf{R}(t)}{dt^2} = \frac{\mathcal{Q}_{\mathrm{GL}}\mathcal{Q}_{\mathrm{AL}}}{4\pi\epsilon_M}\frac{\mathbf{R}-\mathbf{R}'}{|\mathbf{R}-\mathbf{R}'|^3}. \quad (25)$$

Even though a more precise description of the flux of albumin the usage of diffusion equation in its more general representation [12] and with its time derivative in both sides:

$$\begin{aligned} d\mathbf{R}/dt &= D\nabla^2\mathbf{R}\\ d^2\mathbf{R}^2/dt &= D\frac{\nabla^2\mathbf{R}}{dt}\end{aligned} \quad (26)$$

with **R** denoting the center of gravity of ionic compound. With this one can write down:

$$M_{\mathrm{AL}}\frac{Dd\nabla^2\mathbf{R}(t)}{dt}=\frac{\mathcal{Q}_{\mathrm{GL}}\mathcal{Q}_{\mathrm{AL}}}{4\pi\epsilon_M}\frac{\mathbf{R}-\mathbf{R}'}{|\mathbf{R}-\mathbf{R}'|^3} \tag{27}$$

whose analysis is beyond the scope of this paper. Because from Eq. (25), one can read that ions of albumin molecules are under dynamics then it is of interest to extract a dynamical expression of it. Therefore, it is feasible to couple the diffusion equation and second Newton's law as:

$$\mathcal{Q}_{\mathrm{AL}}=\frac{M_{\mathrm{AL}}4\pi\epsilon_M|\mathbf{R}-\mathbf{R}'|^3\frac{d^2|\mathbf{R}|(t)}{dt^2}}{\mathcal{Q}_{\mathrm{GL}}|\mathbf{R}-\mathbf{R}'|}, \tag{28}$$

that is telling that the albumin charged compound is depending on the spatial locations of glucose. In terms of density, one arrives to:

$$\mathcal{Q}_{\mathrm{AL}}=\rho_{\mathrm{AL}}V_{\mathrm{AL}}=\frac{M_{\mathrm{AL}}4\pi\epsilon_M|\mathbf{R}-\mathbf{R}'|^3\frac{d^2|\mathbf{R}|(t)}{dt^2}}{\mathcal{Q}_{\mathrm{GL}}|\mathbf{R}-\mathbf{R}'|} \tag{29}$$

and the volume of albumin flux moving to a constant acceleration $\frac{\mathrm{d}^2|\mathbf{R}|(t)}{\mathrm{d}t^2}\approx 1$ thus one has that:

$$V_{\mathrm{AL}}=\frac{M_{\mathrm{AL}}4\pi\epsilon_M|\mathbf{D}|^2}{\rho_{\mathrm{AL}}\rho_{\mathrm{GL}}V_{\mathrm{GL}}}. \tag{30}$$

Under the assumption that albumin ions pass through the different zones of renal glomerulus with an approximate cylindric like geometry, then the radius of volume that fluxes is:

$$r=\sqrt{\frac{1}{\pi L}\frac{M_{\mathrm{AL}}4\pi\epsilon_M|\mathbf{D}|^2}{\rho_{\mathrm{AL}}\rho_{\mathrm{GL}}V_{\mathrm{GL}}}}\approx\sqrt{\frac{1}{L}\frac{M_{\mathrm{AL}}4\epsilon_M|\mathbf{D}|^2}{\rho_{\mathrm{AL}}\rho_{\mathrm{GL}}V_{\mathrm{GL}}}}. \tag{31}$$

A crude estimation is that of setting both glucose and albumin to have same densities approximately the value of

$$\rho=\rho_{\mathrm{GL}}=\rho_{\mathrm{AL}}. \tag{32}$$

In this sense one arrives to:

$$r\approx\frac{|\mathbf{D}|}{\rho}\sqrt{\frac{1}{L}\frac{M_{\mathrm{AL}}4\epsilon_M}{V_{\mathrm{GL}}}}. \tag{33}$$

2.5 Classical Electrodynamics at the Kidneys

Now, one can go back to Eq. 3 by the which the interaction between albumin and glucose, the aim of this paper, is desribed by the Coulomb force, that to some extent this formulation might be enough to describe molecular interactions either in the classical [13] or quantum territories [14]. For instance the full formulation can be done through the usage of $\mathbf{E} = -\nabla\Phi$ and $\mathbf{E} = \mathbf{F}/Q$ as well as $\mathbf{F} = -Q\nabla\Phi$. When the gradient is applied in both sides one gets:

$$\nabla \mathbf{F} = -Q\nabla^2\Phi. \tag{34}$$

On the other side, due to the Poisson's equation $\nabla^2\Phi = -\rho/\epsilon_0$, one can arrive to:

$$\nabla \mathbf{F} = \frac{Q\rho}{\epsilon_0}. \tag{35}$$

By taking advantage of the divergence of electric force, one can carry out the volume integration in both sides so that one can apply the divergence's theorem that is interpreted as the flux of albumin passing through the areas of the geometry that corresponds to the microscopic areas of kidneys. Thus one has that:

$$\int \nabla \mathbf{F} dV = \frac{Q\int \rho dV}{\epsilon_0} = \int \mathbf{F} d\mathbf{A}_{\mathrm{KIDNEY}} = \frac{Q_{\mathrm{AL}} Q_{\mathrm{GL}}}{\epsilon_0}, \tag{36}$$

again the fundamental Coulomb-like has been obtained and now it describes the interaction as the single product of the total charges.
The left side of Eq. (36) can be exploited in a territory that emphasizes the masses of the ones that are subjected to electrical interaction. For a general formulation, one can consider $dV = 4\pi r^2 dr$ then one gets:

$$\int \nabla \mathbf{F} r^2 dr = \frac{Q_{\mathrm{AL}} Q_{\mathrm{GL}}}{4\pi\epsilon_0}, \tag{37}$$

with the electric force explicitly written as function of mass $\mathbf{F} = M_{\mathrm{AL}} \frac{d^2 R(r,t)\hat{e_{\mathbf{R}}}}{dt^2}$ with $\hat{e}_{\mathbf{R}}$ along the radial direction. Therefore, one obtains:

$$M_{\mathrm{AL}} \nabla \int \frac{d^2 R(r,t)\hat{e_{\mathbf{R}}}}{dt^2} r^2 dr = \frac{Q_{\mathrm{AL}} Q_{\mathrm{GL}}}{4\pi\epsilon_0}, \tag{38}$$

thus by inserting $1/V_{\mathrm{AL}}$ in both sides then one obtains:

$$\frac{M_{\mathrm{AL}}}{V_{\mathrm{AL}}} \nabla \int \frac{d^2 \mathrm{R}(r,t)\hat{e_{\mathbf{R}}}}{dt^2} r^2 dr = \frac{Q_{\mathrm{AL}} Q_{\mathrm{GL}}}{4\pi\epsilon_0 V_{\mathrm{AL}}}, \tag{39}$$

and $\rho_{GL} = Q_{GL}/V_{GL}$ the volumetric charge density of glucose then one can arrive to:

$$\frac{M_{AL}}{V_{AL}}\nabla\int\frac{d^2\mathrm{R}(r,t)\hat{e_R}}{dt^2}r^2 dr = \frac{Q_{AL}}{4\pi\epsilon_0}\frac{\rho_{GL}V_{GL}}{\mathrm{V}_{AL}}. \tag{40}$$

It should be noted that this equation can be perceived as the divergence of electric force being directly proportional to Q^2/V. Aside with the radial component of the divergence then one arrives to:

$$\frac{M_{AL}}{V_{AL}}\int\frac{1}{r^2}\frac{\partial}{\partial r}\left[\frac{r^2 d^2\mathrm{R}(r,t)\hat{e_R}\cdot\hat{e_R}}{dt^2}\right]r^2 dr = \frac{Q_{AL}}{4\pi\epsilon_0}\frac{\rho_{GL}V_{GL}}{V_{AL}}. \tag{41}$$

In order to arrive a general formulation, one can separate this integration in two parts to go through each one in a closed form:

$$\begin{aligned}\frac{M_{AL}}{V_{AL}}\int\left[\frac{d^2\partial R(r,t)}{dt^2\partial r}\right]dr + \int\left[\frac{2rd^2R(r,t)}{dt^2}\right]r^2 dr \\ = \frac{Q_{AL}}{4\pi\epsilon_0}\frac{\rho_{GL}\mathrm{V}_{GL}}{V_{AL}}.\end{aligned} \tag{42}$$

In left side, one can see that the first integration is operated in a straightforward manner, resulting in:

$$\begin{aligned}\frac{M_{AL}}{V_{AL}}\left[\frac{d^2R(r,t)}{dt^2}\right] + \frac{M_{AL}}{V_{AL}}\int\frac{d^2}{dt^2}\left[2r^3R(r,t)\right]dr \\ = \frac{M_{AL}}{V_{AL}}\left[\frac{d^2R(r,t)}{dt^2}\right] \\ +\frac{d^2}{dt^2}R(t)\int dr 2r^2 R(r) = \frac{Q_{AL}}{4\pi\epsilon_0}\frac{\rho_{GL}\mathrm{V}_{GL}}{V_{AL}},\end{aligned} \tag{43}$$

where in the second integration was assumed the separation of variable for the radial function in the form $R(r,t) = \mathrm{R}(r)R(t)$.

2.6 *Bessel Functions as Radial Solutions*

From the idea that repulsive forces are producing electrical oscillations beween glucose and albumin ions, then along the radial part, one can propose the integer-order Bessel functions. In addition, the time-dependent part is proposed to be driven by a sinusoidal function. In this manner, one has that:

$$\mathbf{R}(r,t) = R(r)R(t) = \sum_q J_q(\beta r)\mathrm{Cos}\left[\frac{\omega t}{\gamma}\right] \tag{44}$$

where γ and β are free parameters and ω the frequency of oscillation. From it, one might suggest that the order of Bessel function might be associated to those spatial zones where the probability of repulsion acquires its maximum values. Therefore, the insertion of Eq. (44) into Eq. (43) yields:

$$\frac{M_{\rm AL}}{V_{\rm AL}} J_q(\beta r)\left(\frac{\omega}{\gamma}\right)^2 \mathrm{Sin}\left(\frac{\omega t}{\gamma}\right) - \frac{M_{\rm AL}}{V_{\rm AL}}\left(\frac{\omega}{\gamma}\right)^2 \mathrm{Sin}\left(\frac{\omega t}{\gamma}\right)\int \mathrm{d}r 2r^3 J_q(\beta r)\mathrm{d}r = \frac{Q_{\rm AL}}{4\pi\epsilon_0}\frac{\rho_{\rm GL}}{V_{\rm GL}}. \quad (45)$$

The factorization of the squared frequency and the "Sin" function leads us to a compact expression by the which again the time-dependent and radial part appear to be separated each other. Subsequently, one arrives to:

$$\frac{M_{\rm AL}}{V_{\rm AL}}\left(\frac{\omega}{\gamma}\right)^2 \mathrm{Sin}\left(\frac{\omega t}{\gamma}\right)\left(J_q(\beta r) - \int \mathrm{d}r 2r^3 J_q(\beta r)\mathrm{d}r\right) = \frac{Q_{\rm AL}}{4\pi\epsilon_0}\frac{\rho_{\rm GL}}{V_{\rm GL}} \quad (46)$$

that leads to an universal formulation between the quantity of albumin that fluxes through the glomerulus

$$\frac{M_{\rm AL}}{V_{\rm AL}} = \frac{Q_{\rm AL}}{4\pi\epsilon_0\left(\frac{\omega}{\gamma}\right)^2 \mathrm{Sin}\left(\frac{\omega t}{\gamma}\right)\left[J_q(\beta r) + \int 2r^3 J_q(\beta r)\mathrm{d}r\right]}\frac{\rho_{\rm GL}}{V_{\rm GL}} \quad (47)$$

$$\frac{M_{\rm AL}}{M_{\rm GL}} = \frac{Q_{\rm AL}}{4\pi\epsilon_0\left(\frac{\omega}{\gamma}\right)^2 \mathrm{Sin}\left(\frac{\omega t}{\gamma}\right)\left[J_q(\beta r) + \int 2r^3 J_q(\beta r)\mathrm{d}r\right]}\frac{V_{\rm AL}}{V^2_{\rm GL}} \quad (48)$$

3 Results and Simulations

As seen in Eqs. 47 and 48, one can see that in essence, the expression depends on up to three independent variables: frequency ω, radius r and free parameters γ and β. The pass of albumin through the layers of renal glomerulus would affect the basement membrane and defenestrate area previous to the reach the surface where albumin interacts with podocytes. In this manner, one can ask about the electrical force done by the repulsion with glucose dipoles that establishes a flux with an approximate geometry that characterizes the bunching of albumin. As states above, the assumption of a cylindric shape is a tentative choice that allows us to make estimates. In fact, as seen in Eq. 48 the denominator dominates the rate mass/volume of albumin so that in Fig. 1 (top) the green line denotes the mass/volume distribution of albumin with

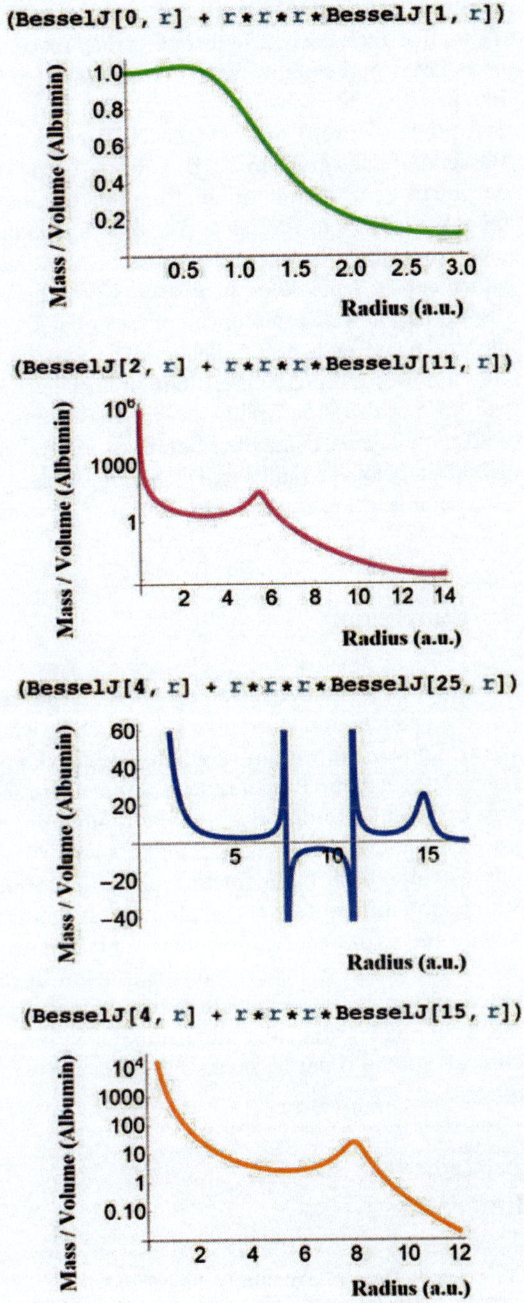

Fig. 1 The mass/volume rate of expulsed albumin by repulsion electric forces by glucose dipole as function of radius of cylindrical bunching for four different scenarios of radial part of denominator of Eq. 47

respect to radius. In this case, the combinations of the Bessel functions $J_0(\beta r)$ and $J_1(\beta, r)$ are considered. In this case, the flux of albumin keeps constant over a range of values of radius; however, beyond 0.9, the density of albumin decays fact that is interpreted as the repulsion forces among them so that there is not an effective flux. In Fig. 1 (below of top), the magenta line denotes same distribution for a combination of Bessel functions given by $J_2(\beta r)$ and $J_{11}(\beta r)$. The peak on 5.9 exhibits the fact that despite of a misconfiguration of the flux of albumin, at that radius equals 5.9 again the flux recovers its direction. The blue line at the distribution plotted above bottom plot shows the albumin density characterized as a complicated distribution by the which can be understood as phases. Here, the combination of $J_4(\beta r)$ and $J_{25}(\beta r)$ displays up to well-pronounced phases related entirely to the value of radius. While the first phase ends at 7.5, between 7.5 and 10.5 one would expect no any flux of albumin. Beyond this range, albumin initializes its transportation. Even though one can see a peak at 15. Finally, at Fig. 1 (Bottom), one can see a similar morphology with the magenta color distribution of above. In fact, in this case, the orange line exhibits a peak at a radius equals 8. For all four plots, the Wolfram `Mathematica` [14] command the denominator of Eq. 48 is written up in each one of them.

4 Conclusion

In this paper, it was derived an equation that contemplates the close relation between glucose and albumin as done in Eq. 48. Thus, while the density and volumen of glucose is known, the volume of albumin is also known. The resulting relation depends on the integer-order Bessel functions that are in accordance with the cylindric geometry of vascularity that encloses both albumin and glucose. From all simulations presented above, one can conclude that while for large values of radius of cylindric one would expect that a fraction of the initial number of albumin might reach to surpass the different layers of glomerulus, going to the Bowman's space. The fact that all curves of mass/volume albumin falls down with the radius, it clearly demonstrates the action-to-distance of the ion–ion interaction of glucose and albumin in vascularity while is passing along the kidneys. At next work, the quantitative side of this study shall be done in order to compare the theory developed so far with realistic clinical data, in order to assess the correctness of all assumptions provided in this paper.

References

1. Lakey, W.C., et al.: Are current clinical trials in diabetes addressing important issues in diabetes care? Diabetologia. https://doi.org/10.1007/s00125-013-2890-4
2. Pugliese, G., et al.: Diabetic kidney disease: new clinical and therapeutic issues. Springer Link, Review Open Access, Published: 02 Oct 2019

3. Nature Reviews Disease Primers, Diabetic Kidney Disease, Nature Reviews Disease Primers, vol. 1, Article number: 15038 (2015)
4. Nieto-Chaupis, H.: The Feynman-Kac formula to estimate the very beginning of the diabetic nephropathy. In: 53rd Annual Conference on Information Sciences and Systems (CISS) (2019)
5. Nieto-Chaupis, H., Unluturk, B.D.: Can the nano level electrodynamics explain the very early diagnosis of diabetic nephropathy in type-2 diabetes patients? IEEE 17th International Conference on Nanotechnology (IEEE-NANO) (2017)
6. Kling, J.: Diabetes treatments near and far. Nat Biotechnol **14**, 428 (1996). https://doi.org/10.1038/nbt0496-428b
7. Nieto-Chaupis, H.: Prospects for anticipating kidney damage in type-2 diabetes patients through the sensing of albumin passing through the renal glomerulus. In: IEEE EMBS International Conference on Biomedical and Health Informatics (BHI) (2017)
8. Yang, Y., et al.: Optimization of protein–Protein interaction measurements for drug discovery using AFM force spectroscopy. In: IEEE Trans. Nanotechnol. **18** (2019)
9. Jackson, John David: Classical Electrodynamics, 2nd edn. Wiley, New York, NY (1975)
10. Liu, W., Liu, L., Xu, K., Yu, Q.: Determination of glucose and albumin in urine using mid-infrared attenuated total reflection system. In: IEEE International Conference on Information Acquisition (2006)
11. Nieto-Chaupis, H.: Modeling the dynamics of failure of the electric shield at the renal glomerulus. In: IEEE XXIV International Conference on Electronics, Electrical Engineering and Computing (2017)
12. Coalson, R.D., Kurnikova, M.G.: Poisson-Nernst-Planck theory approach to the calculation of current through biological ion channels. IEEE Trans. NanoBiosci. **4**(1) (2005). Nieto-Chaupis, H.: Closed-form solutions of the diffusion equation to model prospective nanodevice to anticipate diabetes kidney disease through electric forces. In: IEEE 13th Nanotechnology Materials and Devices Conference (NMDC) (2018)
13. Nieto-Chaupis, H.: Understanding the beginning of diabetic nephropathy with electric potentials and repulsion forces around the renal glomerulus in the human kidney. In: IEEE URUCON (2017)
14. Wolfram Mathematica. www.wolfram.com

Hybrid 2D Theory of Outdoor Infections in Times of Global Pandemic

Huber Nieto-Chaupis

Abstract This paper targets to study the concrete scenario of aerosols transporting virus fact that is seen as a potential cause of massive infections in open spaces through the usage of a two dimension theory. For this end, the conjunction of a deterministic model and the equation of Weiss with a probabilistic meaning is compactified in one single equation. With this approach, different distributions of fractions of infections as function of distance are presented. Although in most cases, this hybrid theory yields distances of 1.5 and 2.0 m. The distributions also suggest the role of wind as a strong factor that might keep the aerosols beyond the established social distancing as cause of delaying of dehydration. Therefore, outdoor characteristics are factors that might not be seen at first sight but would constitute a risk factor in epochs of peaks of pandemic.

Keywords Covid-19 · Epidemiology · Aerosol · Outdoor infection

1 Introduction

Your Corona Virus Disease (Covid-19 in short) has arrived in most countries at the beginning of 2020 [1, 2]. This has created a world-wide pandemic reaching to unseen peaks of infections when it is compared to an epidemiological disaster in world history [3]. Although to date in a few countries massive vaccine's plan has been implemented, still a large fraction of world is still facing the pandemic aiming to minimize the increasing of infections, particularly in all those human group composed by vulnerable population [4]. In this manner, after the public health has acquired the focus of attention as to a robust implementation of solid policies, on the other hand emerges the restoration of social and economic activities that because quarantine and curfew were minimized in order to contain the first wave of infections [5]. In fact, a large number of economies might have been strongly affected causing damage in their objectives to surpass or maintain the PIB being this a strong indicator of the whole economy activity of a country [6]. Furthermore, with the arrival of a first

H. Nieto-Chaupis (✉)
Universidad Autónoma del Perú, Panamericana Sur Km. 16.3 Villa el Salvador, Lima, Perú

© The Author(s), under exclusive license to Springer Nature Singapore Pte Ltd. 2022

A. K. Nagar et al. (eds.), *Intelligent Sustainable Systems*, Lecture Notes in Networks and Systems 333, https://doi.org/10.1007/978-981-16-6309-3_20

wave, most countries are still experiencing the imminent arrival of a second wave that might to deteriorate even more the structures of the countries, so that this demands to impose a firm policy of social distancing that to some extent would guarantee the decreasing the infections along the time that the wave takes place [7]. In fact, global recommendations appeal to keep a distance between people in a range of 1.5–2.0 m or more. In addition, it is also suggested natural ventilation in indoor spaces since the stability of infected aerosols [8, 9] that can enter in physical contact with healthy people. Although this can be also applied to outdoor spaces, still is to some extent unclear the knowledge about the role of the different variables such as the velocity of wind, for example. Therefore, it appears the necessity to apply robust policies of care that minimize the probability of getting infected in outdoor areas. In this manner, emerges also the question about if the so-called "social distancing" suggested to be from 1.5 m might to have its upper limit more than 2.0 m due to the presence of others factors that cannot be easily recognized at a first sight and that can be potential causes to produce infection.

2 Deterministic Model of Outdoor Infection

In order to construct a deterministic model of infection in open areas, some ingredients in both tangible and non-tangible characters must be considered. In this manner for the present analysis, the following criteria are considered: (i) Wind velocity, (ii) the probability for finding infected people outdoor in a random time t, (iii) periods of accumulation of people, (iv) the aerodynamics of aerosol, although this in a first instance this can be considered negligible since its volume is perceived as being too small that would accelerate its dehydration at the shortest distances once were expelled by the source, and (v) source as all those infected ones that are expelling huge amount of aerosols such as the scenarios of loud speech.

In this way, once the source is expelling aerosol one can define a distribution of aerosols given by $n(r, t)$ with "r" and "t" distance and time, respectively. Under the assumption that basic kinematics dictates the mobility and displacements.

of bodies, the one can derive in a straightforward manner that the number of aerosols ΔN per unit of distance and unit of time is defined as $n(r, t)$:

$$n(r, t) = \frac{\Delta N}{\Delta r \Delta t}. \tag{1}$$

In this way from this one can estimate the net number of aerosols that reaches a distance "R" in a time "T" and it is given by:

$$N = \int_0^T \mathrm{d}t \int_0^R \mathrm{d}r n(r, t). \tag{2}$$

Of course, Eq. 1 as well as Eq. 2 do not tell nothing about the problem of aerosols in a scenario of outdoor spaces. Thus, going through the central problem that concerns this paper, it is feasible to introduce the wind velocity with the usage of a trivial chain rule that is applied in the following manner:

$$N = \int_0^T \int_0^R n(r, t) \frac{\mathrm{d}r}{\mathrm{d}t} \mathrm{d}t = \int_0^T n(r, t) v(t)\, \mathrm{d}t, \tag{3}$$

that is calculated per a single source. Clearly experience suggests that this can be generalized for a large number of sources (or infected people) in the form written below as:

$$\begin{aligned} \overline{N} = \sum_{q=1}^{Q} N_q = \sum_{q=1}^{Q} \int_0^{T_q} n_q\left(r_q, t_q\right) \frac{\mathrm{d}r_q}{\mathrm{d}t_q} \mathrm{d}t_q \\ = \sum_{q=1}^{Q} \int_0^{T_q} n_q\left(r, t_q\right) v_q\left(t_q\right) \mathrm{d}t_q, \end{aligned} \tag{4}$$

and from this one can arrive to a general formulation of the total number of aerosols that can reach the distance "r". To accomplish this, one can rewrite Eq. 4 as follows:

$$\frac{\mathrm{d}\overline{N}}{\mathrm{d}t} = \sum_{q=1}^{Q} n_q(r, t) v_q(t). \tag{5}$$

Clearly one can see that the velocity increases, the number of aerosols also can increase as well. This sounds in agreement with experience. In fact, it is assuming that only all those microdroplets that would have a negligible mass are vanishing because the pressure of wind. This suggest to propose the following question: What is the influence of the wind's pressure on the dehydration time of aerosol? A wide analysis is needed, but this is beyond the scope of this paper. From Eq. 5, one important point in this debate is the fraction of number of infected aerosols that can reach people in a certain distance. Thus, one is interested in the fraction $\eta \overline{N}/n_q$ with "η" a random number with unknown sign. Thus, from Eq. 5 one gets that:

$$\frac{\mathrm{d}\overline{N}}{\mathrm{d}t} = \sum_{q=1}^{Q} \frac{\eta \overline{N}}{n_q(r, t)} v_q(t). \tag{6}$$

One can see that the derivative $\mathrm{d}\overline{N}/\overline{N}$ can be formed at the left side of Eq. 6. Also, now the right side passed to be multiplied by $\mathrm{d}t$ so that one gets below that:

$$\frac{d\overline{N}}{\overline{N}} = \sum_{q=1}^{Q} \frac{\eta v_q(t)}{n_q(r,t)} dt. \tag{7}$$

That allows to carry out a straightforward integration in both sides yielding the exponential function at the left side with $\overline{N}_0$ the initial value at t_0. Indeed because $v_q(t) = dr_q/dt$ then Eq. 7 turns out to be written below with $\eta \Rightarrow \pm\eta$ (the sign can be random) as follows:

$$\overline{N} = \overline{N}_0 \text{Exp}\left[\eta \sum_{q=1}^{Q} \int_0^T \frac{v_q(t)}{n_q(r,t)} dt \right] = \overline{N}_0 \text{Exp}\left[\frac{\pm\eta \sum_{q=1}^{Q} \int_0^{R_q} \frac{dr_q}{n_q(r,t)}}{Q} \right]$$
$$= \overline{N}_0 \text{Exp}\left[\frac{\pm\eta}{Q} \sum_{q=1}^{Q} \int_0^{R_q} \frac{dr_q}{n_q(r,t)} \right]. \tag{8}$$

It should be noted that has been defined the average of a source of "Q" expellers. With the change $\int_0^{R_q} \frac{dr_q}{n_q(r,t)} = F_q(R_q, t)$ then Eq. 8 acquires a simplified form given by:

$$\overline{N}(t) = \text{Exp}\left[\frac{\pm\eta}{Q} \sum_{q=1}^{Q} F_q(R_q, t) \right]. \tag{9}$$

Actually, a more precise meaning of R_Q would be that describe the average distance $\langle R \rangle$ in the which one finds the net number of aerosols $\overline{N}(t)$. From Eq. 9 then the formulation deterministic yields the following equation that is inverse to the velocity of wind.

$$\frac{d\overline{N}(t)}{dR} = \frac{d\overline{N}(t)}{dR} \frac{dR}{dt} = v(t) \frac{d\overline{N}(t)}{dR} \Rightarrow \overline{N}(t)$$
$$= \frac{1}{v(t)} \int_0^{\langle R \rangle} dR \, \text{Exp}\left[\pm \frac{\eta}{Q} \sum_{q=1}^{Q} \int_0^{R} \frac{dr}{n_q(r,t)} \right]. \tag{10}$$

3 The Weiss Equation

Despite that Eq. 10 can to some extent model the social distancing as found to be ranging between 1.5 m [11] and beyond but inside a deterministic framework, the imminent presence of various probabilistic variables, it leads to usage the equation of Weiss [12, 13] based on the following event that can be described as follows:

When N objects are sent in a set of Q cells with no any hierarchy between Q and N (neither Q nor N are large with respect each other). Under the assumption that the Q cells are not any under of dependence of being closer or far each other. Then, when this repeated M times so that N and NM acquires a large number $\ll \infty$, then the probability to find an unoccupied cell or occupancy [14, 15] is given by:

$$P = \frac{Q\,\mathrm{Log\,Q}}{NM}. \tag{11}$$

With the interpretation: N objects $\rightarrow$ N infected microdroplets, Q cells $\rightarrow$ Q people, M the times by which is expected the arrived of N microdroplets, P = Probability that people that are not receiving not any aerosols. An important assumption for the application of Eq. 11 is the fact that the action of infection is carry out with a limited number of aerosols or microdroplets [16]. Thus, one can establish below:

$$P(\omega) = \frac{\omega Q\,\mathrm{Log\,Q}}{N}. \tag{12}$$

In contrast to the deterministic case, one can see that the number of aerosols "N" that in the theory of Weiss are defined as the objects appear now at the denominator. Interestingly, the exact probability that someone is not receiving not any aerosols is the integration over all known frequency yields:

$$P = \int_0^W \mathrm{d}\omega P(\omega) = \frac{W^2 Q\,\mathrm{Log}\,Q}{2N}, \tag{13}$$

with this it is plausible to define the probability that the cell is occupied, that means the case the probability "p" someone receives an infected aerosol before that this reaches the time of dehydration. In this manner one has that:

$$p = 1 - P = 1 - \int_0^W \mathrm{d}\omega P(\omega) = 1 - \frac{W^2 Q\,\mathrm{Log\,Q}}{2N}, \tag{14}$$

thus, it constitutes of great importance what are those values by which the quantity $\frac{W^2 Q\,\mathrm{Log\,Q}}{N}$ cannot reach the value of 1. It is analogue to expect that $p \ll 1$. In praxis, it is desired to know the values that $\frac{W^2 Q\,\mathrm{Log\,Q}}{N}$ acquires the value of 0.99 for example. With this the probability of cell occupation "or infection" is small for example of order of 0.01. In this manner one should focus on the Eq. 14. When Eq. 9 is introduced Eq. 14 has the form:

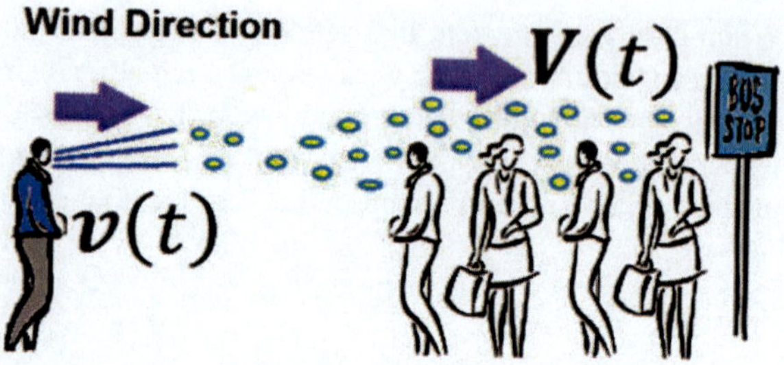

Fig. 1 Sketch of the potential scenario of coexistence of up to well-defined aerosol velocities as dictated by Eq. 27

$$P = \frac{W^2 Q \,\mathrm{Log}\, Q}{2\overline{N}_0 \mathrm{Exp}\left[\pm\eta \sum_{q=1}^{Q} \int_0^T \frac{v_q(t)}{n_q(r,t)} \mathrm{d}t\right]}. \quad (15)$$

As seen above, Fig. 2 suggests to employ the sign "+" so that this probability acquires the dependence of a negative exponential (obviously it happens because the exponential passed to numerator) (Fig. 1):

$$P = \frac{W^2 Q \,\mathrm{Log}\, Q}{2\overline{N}_0} \mathrm{Exp}\left[-\eta \sum_{k=1}^{K} \int_0^T \frac{v_k(t)}{n_k(r,t)} \mathrm{d}t\right]. \quad (16)$$

It would be a hybrid version of deterministic model for a generalization for "K" sources (or "K" different people). Nevertheless, a pedagogical analysis contemplates the case of a single source. Thus, by expanding the exponential and with the assumption of $\eta^L \approx 0$, for $L \geq 2$ one gets for $\eta \to -\eta$ since this has a random sign (as seen above previous to write Eq. 8), so that one gets:

$$P = \frac{+\eta W^2 Q \,\mathrm{Log}\, Q}{2\overline{N}_0} \int_0^T \frac{v(t)}{n(r,t)} \mathrm{d}t. \quad (17)$$

That justifies now the positive sign in Eq. 17. That can also be seen as a derivative in the sense as:

$$\frac{\mathrm{d}P}{\mathrm{d}t} = \frac{\eta W^2 Q \,\mathrm{Log}\, Q}{2\overline{N}_0} \frac{v(t)}{n(r,t)}, \quad (18)$$

and one can appeal once more to the chain rule:

$$\frac{\mathrm{d}P}{\mathrm{d}t}\frac{\mathrm{d}r}{\mathrm{d}t} = \frac{\mathrm{d}P}{\mathrm{d}r} V(t) = \frac{\eta W^2 Q \,\mathrm{Log}\, Q}{\overline{N}_0} \frac{v(t)}{n(r,t)}, \quad (19)$$

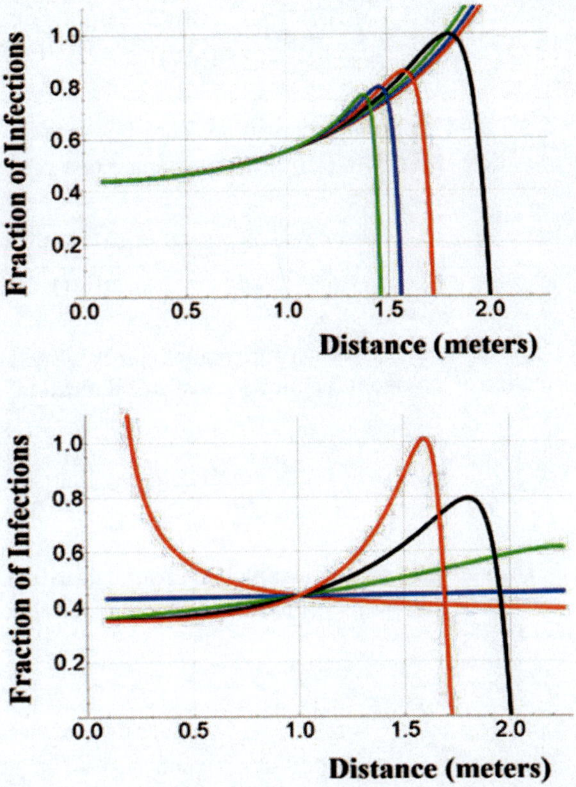

Fig. 2 Manifestations of Eq. 28 up to for 2 scenarios with polynomial forms done with Wolfram package [17]

by the which it is established the velocity $V(t)$ that denotes the velocity of all those aerosols reaching the people or cells defined by the integer number "Q".

Therefore, it was defined so far two well-defined velocities: the deterministic given by $v(t)$ that is directly proportional to the velocity of wind. On the other hand, $V(t)$ appears in a spontaneous manner as the probabilistic velocity as result of the fusion of both approaches. Under this manner one can arrives to:

$$\frac{\mathrm{d}P}{\mathrm{d}r} = \frac{\eta W^2 Q \,\mathrm{Log}\, Q}{n(r,t)\overline{N}_0}\frac{v(t)}{V(t)}. \tag{20}$$

Now it is plausible to carry out an integration over all possible values of Q in both sides of Eq. 20, over the lowest value namely "0", up to the upper limit $Q(r)$ denoting the fact that the expelled aerosols might to reach to "$Q(r)$ "people at the distance "r". Therefore, one gets:

$$\int_0^{Q(r)} \frac{\mathrm{d}P}{\mathrm{d}r}\mathrm{d}Q = \frac{\eta W^2}{n(r,t)\overline{N}_0}\frac{v(t)}{V(t)}\int_0^{Q(r)} Q\,\mathrm{Log}\,Q\,\mathrm{d}Q. \quad (21)$$

A valid approximation (that can be verified by the reader) is that of $\int_0^{Q(r)} \mathrm{d}Q\,Q\,\mathrm{Log}\,Q \approx Q^2(r)$. In this manner one gets:

$$\int_0^{Q(r)} \frac{\mathrm{d}P}{\mathrm{d}r}\mathrm{d}Q \approx \frac{\eta W^2 Q^2(r)}{n(r,t)}\frac{v(t)}{V(t)}. \quad (22)$$

Because P constitutes a pure random probability then one expects that this is also function of a random function for instance defined by F, therefore, one expects that:

$$\int_0^{Q(r)} \frac{\mathrm{d}P(F,Q,r)}{\mathrm{d}r}\mathrm{d}Q \approx \frac{\eta W^2 Q^2(r)}{n(r,t)}\frac{v(t)}{V(t)}. \quad (23)$$

The introduction of F is exploited in order to arrive to a phenomenological expression for the number of people that might to receive the infected aerosols. Therefore, one arrives to:

$$\int_0^{Q(r)} \frac{\mathrm{d}P(r)}{\mathrm{d}F}\frac{\mathrm{d}F}{\mathrm{d}r}\mathrm{d}Q = \frac{\mathrm{d}P(r)}{\mathrm{d}F}\frac{\mathrm{d}F}{\mathrm{d}r}\int_0^{Q(r)} \mathrm{d}Q$$

$$= \frac{\mathrm{d}P(r)}{\mathrm{d}F}\frac{\mathrm{d}F}{\mathrm{d}r}\int_0^{Q(r)} \mathrm{d}Q = \frac{\mathrm{d}P(r)}{\mathrm{d}F}\frac{\mathrm{d}F}{\mathrm{d}r}Q(r) \approx Q^{F(r)} \quad (24)$$

By which it was assumed that $\mathrm{d}P/\mathrm{d}F \approx 1$ (both probabilities would exhibit similar behaviour). In this way from Eq. 24 one gets

$$\frac{\mathrm{d}F}{\mathrm{d}r}Q^{F(r)} = \frac{\eta W^2 Q^2(r)}{n(r,t)}\frac{v(t)}{V(t)}, \quad (25)$$

And solving for Q one can see that that it has a well-defined random exponential because $F(r)$. So that one arrives to:

$$Q^{F(r)-2} = \frac{\eta W^2}{\frac{\mathrm{d}F(r)}{\mathrm{d}r}n(r,t)}\frac{v(t)}{V(t)}. \quad (26)$$

In this manner, in order to compactify Eq. 26, a direct relation alone for $Q(r)$ is needed. Thus, one gets finally that the number of people (or cells under the Weiss's model) that are receiving aerosols can be written in this hybrid approach as:

$$Q(r) \approx \left(\frac{\eta\, W^2}{\frac{\mathrm{d}F(r)}{\mathrm{d}r} n(r,t)} \frac{v(r,t)}{V(r,t)} \right)^{\frac{1}{F(r)-2}} \tag{27}$$

That is translated as the number of people that can be infected at the distance "*r*" because the coexistence of two well velocities: one $v(r, t)$ the wind's velocity and the velocity of arriving $V(r, t)$.

4 Results and Discussion

As seen above Eq. 27 denotes a hybrid formulation of the Weiss's equation. As one can see it exhibits a dependence on the rate $v(r, t)/V(r, t)$ that indicates the importance of the final velocity of the aerosol arriving on the people. In Fig. 1, a sketch about this point is displayed. One can see the clear existence of two well-defined velocities of aerosol. The physical conditions on the final velocity $V(r, t)$ might to depend on the humidity and heat so that the dehydration would be faster than other climate conditions. Although these points cannot be totally neglected, it is beyond the scope of this paper. In this manner the main equation Eq. 27 will be numerically approximated. For this a phenomenological model derived from this is proposal without the loss of the generality. In this way, the following approximation given by:

$$Q(r) \approx \left(\frac{\eta W^2}{\frac{\mathrm{d}F(r)}{\mathrm{d}r} n(r,t)} \frac{v(r,t)}{V(r,t)} \right)^{\frac{1}{F(r)-2}} \Rightarrow \left(\frac{\beta}{A(r)} \right)^{\frac{1}{C(r)}} \tag{28}$$

In Fig. 2 (Up) the case of $1/2.27\left(\frac{1}{-r\wedge n+r*r+12}\right)^{\frac{r*r}{r\wedge(n-1)-12}}$ for $n = 1,\ldots,7$ is displayed. Interestingly, one can see those lines green and black falls down in 1.5 m and 2 m, respectively. A more explicit view of hybrid model dictated by Eq. 21 is seen in Fig. 2 (down) where is plotted the case given by $1/2.85\left(\frac{1}{-r\wedge n+r*r+12}\right)^{\frac{r\wedge(n-2)}{r\wedge(n-1)-12}}$ $n = 1,\ldots,5$. One can see that lines red and black fall down on 1.7 m and 2 m, respectively. Here is also seen that lines with colours orange, blue and green keep their value that means that still a 40% up to a 50% of people away from the source 2.0 m or more can be sensitive to receive aerosols. The why the lines red and black exhibit a parabolic shape is because to the random composition of Eq. 21. In fact, why these lines back again to show a clear up is due to an eventual apparition of wind that elevates the aerosols. However, although not explicitly defined, the fall of lines is due to the expected dehydration. Thus, these results would suggest the importance of wind that operates as vector for transmission of virus through the expelled aerosols.

5 Conclusion

In this paper, it was theoretically proposed in a hybrid manner the idea that the wind velocity might be a strong factor for outdoor infections of virus in epochs of pandemic. For this, it was introduced in one hand a deterministic model based on elementary kinematics and on the other hand, the equation of Weiss defined inside a purely probabilistic scenario. The hybridization has yielded distributions of fraction of infections as function of distance with some congruence and coherence with the approved social distancing and established to be ranging between 1.5 and 2.0 m. However, these distances are not absolute in all those cases where wind velocities can displace the aerosols as well as to delay their intrinsic dehydration. However, experimental tests are needed to check the results of this study.

References

1. Wu, F., et al.: A new coronavirus associated with human respiratory disease in China. Nature **579**, 265269 (2020)
2. Shang, J., et al.: Structural basis of receptor recognition by SARS-CoV2. Nature **581**, 221224 (2020)
3. Westgarth, D.: Coronavirus pandemic: a time for reflection. Cell. Mol. Immunol. **17**, 889891 (2020)
4. Yang, J., et al.: Disease burden and clinical severity of the first pandemic wave of COVID-19 in Wuhan, China. Nat. Commun. **11**, 110 (2020)
5. Chauhan, S.: Comprehensive review of coronavirus disease 2019 (COVID-19). Biomed. J. **43**(4), 334–340 (2020)
6. Nicola, M.: The socio-economic implications of the coronavirus pandemic (COVID-19): a review. Int. J. Surg. **78**, 185193 (2020)
7. Duffey, R.B., Zio, E.: Analysing recovery from pandemics by learning theory: the case of CoVid-19. IEEE Access **8** (2020)
8. Domingo, J.L., Marques, M., Rovira, J.: Influence of airborne transmission of SARS-CoV-2 on COVID-19 pandemic. A review. Environ. Res. **188,** Article 109861 (2020)
9. van Doremalen, N., et al.: Aerosol and surface stability of SARSCoV-2 as compared with SARS-CoV-1. N Engl J Med **382**, 1564–1567 (2020). https://doi.org/10.1056/NEJMc2004973
10. Stadnytskyi, V., et al.: The airborne lifetime of small speech droplets and their potential importance in SARS-CoV-2 transmission. Proc. Natl. Acad. Sci. **117**(22), 11875–11877 (2020). https://doi.org/10.1073/pnas.2006874117
11. Morawskaa, L., et al.: How can airborne transmission of COVID-19 indoors be minimised? Environ. Int. **142**, 105832 (2020)
12. Nieto-Chaupis, H.: Theory of virus public infection through the weiss approach. In: 20th IEEE BIBE, 26–28 October 2020, USA
13. Weiss, I.: Limiting distributions in some occupancy problems. Ann. Math. Stat. **29**, 878–884 (1958)
14. Johansson, A.: Crowd and environmental management during mass gatherings. Lancet. Infect. Dis **12**(2), 150–156 (2012)

15. Nieto-Chaupis, H.: How effective are teleconsults to persuade patients of pulmonary tuberculosis of avoiding to use public transport at Lima City? In: IEEE 29th International Symposium on Computer-Based Medical Systems (CBMS), 2016
16. Flock, A.K., et al.: Experimental statistics of droplet trajectory and air flow during aerodynamic fragmentation of liquid drops. Int. J. Multiph. Flow **47**, 37–49 (2012)
17. Wolfram: Computation Meets Knowledge: https://www.wolfram.com/

Impacts of Blockchain on Digital Marketing

Sachin Garg, Sachin Gupta, and Bhoomi Gupta

Abstract Blockchain adoption is covering the technological landscape at a rapid pace and its industrial uses are evolving across the entire industrial spectrum. Several technology giants and startups have initiated novel use cases of this technology amidst significant rise in investments in blockchain projects. Blockchain is perceived as an enabler to transform systems across multiple domains. Recent trends in digital marketing have witnessed extensive use of big data analytics and artificial intelligence in reshaping digital marketing. It is believed that the next big transformation in digital marketing shall be due to blockchain while proving to be a disruptive technology that will change the digital marketing process as we know it. Digital Marketing is all set to be significantly benefited with the use of blockchain in a similar manner like finance and banking industry did. This paper is an attempt to explore the evolution of blockchain and its potential impacts on digital marketing. Digital marketing appears to be evolving at a pace of information technology, but blockchain holds the potential to transform digital marketing in a riotous, perhaps even irrevocable manner.

Keywords Digital marketing · Blockchain · Smart contracts

1 Introduction

Marketing has remained one of the most critical foundations for any business. It has been an ever growing domain driven by industrial revolution and technological transformations. Marketing activities have evolved over the past few decades across several different formats including print and television. However, the usual formats

S. Garg (✉) · S. Gupta
MVN University, Palwal, Haryana, India
e-mail: 20cs9003@mvn.edu.in

S. Gupta
e-mail: sachin.gupta@mvn.edu.in

B. Gupta
Maharaja Agrasen Institute of Technology, Rohini, Delhi, India
e-mail: bhoomigupta@mait.ac.in

© The Author(s), under exclusive license to Springer Nature Singapore Pte Ltd. 2022

A. K. Nagar et al. (eds.), *Intelligent Sustainable Systems*, Lecture Notes in Networks and Systems 333, https://doi.org/10.1007/978-981-16-6309-3_21

of marketing are losing sheen in today's world. Not astonishingly, almost everyone in this digital world has already been privy to the magical world of Digital Marketing in one way or the other. As per a definition by the Digital Marketing Institute, "Digital Marketing is the use of digital channels to promote or market the products and services to targeted consumers and businesses." Continuous growth in consumption of digital content are signs that traditional marketing approaches will be disrupted and digital marketing will be the primary mode of marketing in future. The reason digital marketing is more effective is because of its affordability, scalability and reachability, helping organizations manage a quicker reach to a larger customer base [1].

In 1993, the first clickable banner went active and this initiated the evolution of the digital marketplace. Buoyed by the success of this, big players like Yahoo boosted digital marketing and achieved 1 million hits on their website in the very first year. This triggered large-scale changes in the digital marketing space and introduced the concept of Search Engine Optimization, wherein companies started customizing their websites to achieve higher ranking in popular search engines like Yahoo and Google. With the advent of Web 2.0, users had the ability to access and contribute content that made them active participants rather than staying a passive user. It further permitted users to directly connect with other users and businesses. As a result, information flow volumes—including channels deployed by digital marketers-increased manifold, and by 2004, web advertising and marketing in the US alone brought in around $2.9 billion. The online marketing world saw its first rapid jump in 2006 when search engine traffic was reported to have risen to about 6.4 billion in a single month.

The biggest impact was experienced with the emergence of social networking websites. MySpace was the first social networking website to arrive, soon followed by Facebook and Orkut. Various companies noticed that the newly developed websites can offer a global platform with enormous opportunities, to facilitate digital marketing of their products and brands. This phenomenon unlocked fresh paths for businesses and ushered business marketing on a faster growth trajectory. With availability of these new resources, the companies started exploring new techniques and mechanisms to publicize their brands and products over various social networking platforms [2]. The cookie was another important breakthrough in the digital marketing industry. It is being used to trace common browsing habits and usage patterns of recurring users of the internet so as to customize promotions and marketing content to their tastes. Now digital market products are available to customers at every time. Three biggest social networking websites used by marketers are LinkedIn, Twitter, and FB. Digital marketing has been evolving and has taken many forms. Next section focuses on various types of digital marketing [3, 4].

2 Digital Marketing Types, Issues and Challenges

Digital marketing has become a multi-billion dollar industry that spreads to local businesses and major corporations worldwide. Digital marketing involves promotion of products on the internet using various electronic media. We are now living in a pervasive digital world and as of January 2021, and the internet has witnessed over 4.66 billion users out of which 3.8 billion use social media on a routine basis. Technologies like mobile computing and high-speed internet have resulted in significant attrition of the customer base of traditional marketing agencies and departments. Customers have descended on to computers, smart phones, and smart TVs that have forced digital marketers to explore these platforms for marketing. Ever since the evolution of World Wide Web and web applications, there has been significant growth in innovative delivery capabilities in reaching out to the customers through different techniques of digital marketing. Digital marketing can be done in many ways and various types are listed in Table 1.

As the annual amount of investment in digital marketing efforts increases, so do the expenses related to fraudulent activities at this place [5]. In fact, statistical data predicts an exponential growth of costs caused by digital advertising fraud. And given how various digital marketing strategies are routinely developed, it is becoming increasingly difficult to quickly detect fraudulent people from legitimate ones. Ad

Table 1 Types of digital marketing

Fraud	Details
Search Engine Optimization (SEO)	Website management and data analytics to gets top listing against searches by customers
Pay per Click Advertising (PPC)	Bidding for advertisements for certain keywords related to your business on search engine
Social Media Marketing	Listing advertisements on social media platforms
Influencer Marketing	Hiring famous celebrity to promote you product
Public Relations (PR) Content	Includes reviews, feedbacks and discussions
Email Marketing	Attractive info graphic content and newsletters in email
Mobile Marketing	Marketing and advertising using mobile apps
Brand Journalism	Creating attractive content that can pulls customers
Video Marketing	Using short videos/films to promote your product
Audio Marketing	Podcasts, talk shows on radio
Affiliate Marketing	Hire others to sell your products
Native Advertising	Advertisements are organized within articles and websites
Live Chat	AI based chat bots to address queries of customers
Augmented Reality Marketing	Virtual malls and shops
FAQs	Presenting answers of frequently asked questions
Instant Messaging Marketing	Using instant messaging apps like WhatsApp and Facebook Messenger for marketing

Table 2 Techniques of digital marketing frauds

Fraud	Details
Impression fraud	Generating fake ad view count using smartly designed bots
Click fraud	Fake clicks using transparent layers on content or duplicate clicks
Affiliate fraud	Fake affiliates to generate leads, impressions, sales or clicks
Lead fraud	Using people to fill fake lead generation forms
E-commerce fraud	Generating sales and earn commissions using orders canceled later
Sourced traffic	Buying fake traffic to boost the ranking of a website
Fake website	Fake website sites to sell advertising space

fraud is a serious problem [6]. According to Juniper Research, losses from ad fraud reached $42 billion in 2019 [7]. That is up 21% from the previous year, and they expect losses to hit $100 billion a year by 2023. This type of cybercrime leverages several techniques to execute digital marketing frauds shown in Table 2.

The hackers and fraudsters rely on botnets and human fraud farms to commit these crimes.

- Botnets are collections of computers that are infected with malware. These computers can exist anywhere in schools, businesses, hospitals, or even homes. They do what the malware that infects them tells them to do.
- Human fraud farms are a bit more expensive, but that is because they are better at avoiding most fraud detection solutions. Human fraud farms are collections of real people whose job it is to commit acts of ad fraud listed above on behalf of the criminals that pay them. Most of us never know how much we lose in time and other resources when securing, tracking, and creating ads that wind up falling victim to fraudulent activity.

3 Blockchain Introduction

Blockchain is an exciting computing infrastructure to facilitate distributed applications and can ensure robust transaction processing and secure information storage mechanism with complete transparency to all the participating nodes [8]. Each node maintains copy of the ledger that holds all previous transactions.

A node initiates a transaction and batch of transactions are verified and bundled to make a block. After successful verification and the block is sent to all the nodes to update their respective ledgers [9] as shown in Fig. 1.

Blockchain is a distributed ledger that keeps records of all the transactions with the capabilities like resiliency, trustworthiness, security and immutability [10, 11]. Following are some exciting blockchain features [12, 13].

- Immutability: This means transaction data cannot be changed once processed as valid transaction.

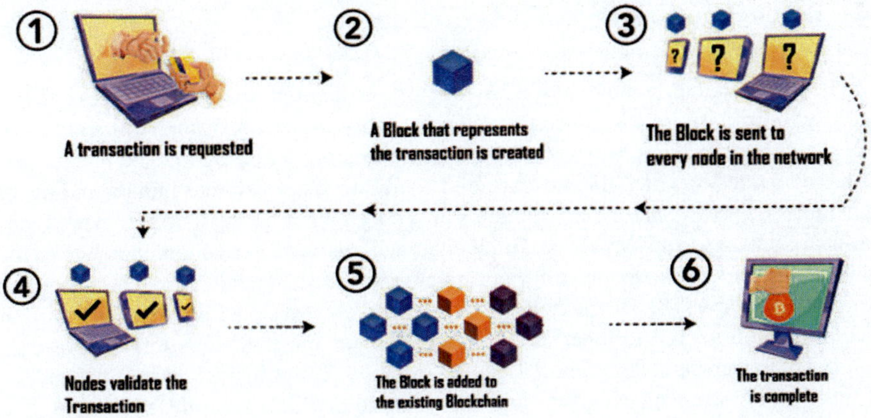

Fig. 1 Transaction cycle in a blockchain

- Distributed ledger: This means elimination of single point of failure and ensuring high availability and less likely to failure. This makes the system transparent and every change on the blockchain is observable and makes it more concrete.
- Enhanced Security: The blockchain based uses encryption to add another layer of security. Authentication and identities are managed by public key infrastructure that involve use of key pairs (public key and private key).
- Consensus algorithms make blockchain system more trustworthy and independent of relying on people for trust. The consensus is a decision-making process that facilitates transaction verification.

A typical blockchain has various components as listed in Table 3.

Table 3 Components of a blockchain

Component	Details
Node	A computing device that can help user to initiate a transaction
Transaction	An activity that involves data related to transfer of value or asset
Block	A bundle of transactions to be sent to all the nodes in the network
Chain	Blocks connected to each other in a well-defined order
Network node	These nodes perform the block verification process
Consensus	A set of rules and computations to facilitate block verification
Smart contract	This includes terms, conditions and other business rules that govern business transactions, programmed into a software and gets invoked automatically when transaction is initiated, and all conditions are executed as per the terms of all the stakeholders to control and document legally relevant facts, events and actions according to the specifications of the contract

4 Blockchain for Digital Marketing

As highlighted in previous section blockchain carries significant capabilities and has ability to bring more assurance in system's ability to deliver business objectives. It has been transforming every sector and marketing is no exception. The technology enables ownership of personal data and facilitate more reliable understanding of customer behaviors in relation to brands [14]. Digital marketing involves many types of data exchanges and lacks trust. Blockchain will disrupt this phenomenon within the advertising environment, and all frauds related to online advertising can be addressed using blockchain [15]. Blockchain can create trust and transparency within marketing and has the potential to meet the many challenges that marketers today have with digital marketing and e-commerce. Following section highlights how blockchain can transform digital marketing [16] and add value to digital marketing ecosystem.

- *Ensuring Data Privacy and fair use*: Blockchain can help in controlling what data goes into the system and ensures that it cannot be abused.
- *Countering Ad Fraud*: Blockchain can prevent duplicate and fake clicks.
- *Eliminating fake intermediaries*: Blockchain with strong marketing policies based smart contract can ensured affiliate identity to avoid redundancy or abuse.
- *Accurate Analytics and performance measurement*: Blockchain can facilitates audit process and verification of details related to advertising activities.
- *Credibility of Content*: Blockchain can be used to identify fake content.
- *Transparency and Traceability in Digital Marketing*: The fact that blockchain offers the characteristics of transparency and traceability enables customers to keep track of the brands, and if the brands are following their promises [17, 18].
- *Smart Contract based operations*: Using blockchain and smart contracts can allow a company to take the role of a trusted third party. If the contract is not fulfilled by one party, then preset declarations related to the contract will take place. As a consequence, the opposite party can have his or her money rolled back automatically if a product is not delivered as promised.

5 Use Case–Loyalty Program

Loyalty program is a marketing trick to attract and retain customers. For example a retailer starts a reward system in the form of points and customer earns some points on every purchase. The points earned can be redeemed on future purchases. Many retailers and e-commerce players are using loyalty programs for their business development. But it has been observed that there is scope of further improvements in these programs as most of the loyalty programs are brand specific and lacks usability outside that brand. One major limitation in most of these loyalty programs is the ability to offer integrated services between online and offline channels [19]. An integrated system connecting various vendors and facilitating tracking of entire consumer behavior can be an effective marketing strategy. Customers will have more options, while redeeming their earned points.

Developing such an integrated system is a serious challenge and following pain points needs to be addressed.

- Integrating various vendors across multiple business segments under FMCG, BFSI, etc.
- Transaction processing based on vendor and product specific rules to award points to customers.
- Standardizing reward points system to facilitate uniform redemption.
- Ensuring security and immutability of processed transactions.

This forms a strong case for use of blockchain to develop an effective loyalty program to address all the above mentioned pain points [20–22], which highlights the key objectives that such loyalty program, should address. Figure 2 demonstrates a basic flow diagram for such a system. Blockchain based system can ensure security and immutability along with compliance to all vendor specific rules using a smart contract. Figure 2 shows flow of a transaction initiated by a user and after successful processing of transaction, and the reward points are computed by Reward Points Management System based on rules defined by the vendor stored in Vendor Rule

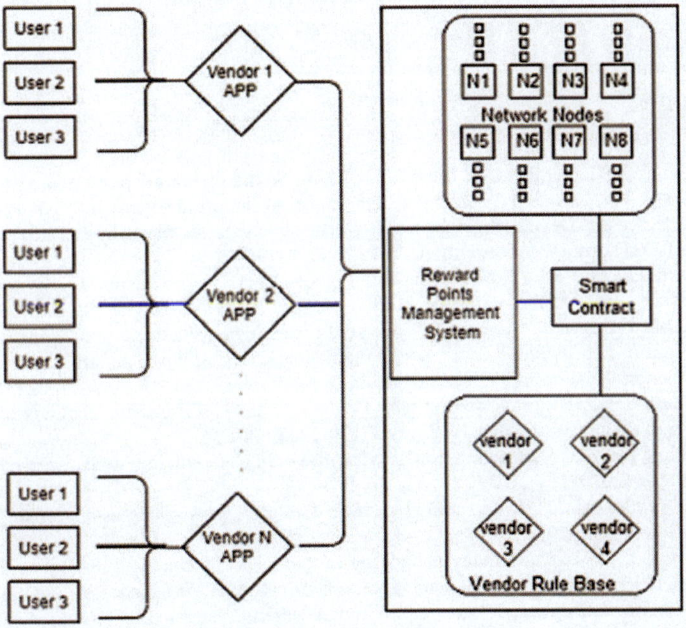

Fig. 2 Transaction flow in blockchain based loyalty program

Base. After computation the transaction is send to network nodes for verification and after successful verification the transaction is added to the blockchain maintained by all network nodes.

6 Conclusion

Blockchain has a potential to transform digital marketing by addressing existing digital marketing challenges. Blockchain can facilitate more consumer-friendly system with focus on Data privacy compliance. Blockchain can eliminate dependency on third parties, which means it will also be simple for marketing units to calculate the return on investment of their advertisements. Blockchain is going to modify marketing in terms of generating performance by slicing the agent and being clear and delivering trust within the system rather than relying on people involved. Blockchain is modifying digital marketing in a riotous way, probably erasing out an entire generation of traditional companies. Certainly, the positive affect of blockchain in digital marketing is not just in the new use cases that are emerging. It is in how those use cases will influence whole systems that are part of the digital marketplace. Digital marketing appears to be evolving with information technology, but blockchain is modifying digital marketing in riotous, perhaps even irrevocable manner.

References

1. Dwivedi, Y.K., Ismagilova, E., Rana, N.P. et al.: Social media adoption, usage & impact in business-to-business (B2B) context: a state-of-the-art literature review. Inf. Sys. Front. (2021)
2. Mosca, F., et al.: Developing successful global strategies for marketing luxury brands. IGI Global (2021). https://doi.org/10.4018/978-1-7998-5882-9
3. Fondevila-Gascón, J.F., Rom-Rodríguez, J., Polo-López, M., Crespo, J.L.: Marketing and advertising: trends of the sector. In: Martínez-López, F., D'Alessandro, S. (eds.) Advances in Digital Marketing and eCommerce. Springer Proceedings in Business and Economics (2020)
4. Li, F., Larimo, J., Leonidou, L.C.: Social media marketing strategy: definition, conceptualization, taxonomy, validation, & future agenda. J. Acad. Mark. Sci. **49**, 51–70 (2021)
5. Diez-Martin, F., Blanco-Gonzalez, A., Prado-Roman, C.: Research challenges in digital marketing: sustainability. Sustainability **11**(10), 2839 (2019)
6. Dörnyei, K.R.: Marketing professionals' views on online advertising fraud. J. Curr. Issues Res. Advertising (2020)
7. Juniper Research: https://www.juniperresearch.com/press/press-releases/advertising-fraud-losses-to-reach-42-bn-2019
8. Thomason, J., et al.: Blockchain technology for global social change. IGI Global (2019)
9. Krishna, G.J., et al.: Smart blockchain: a formidable combination of analytics and blockchain—a survey. In: Rodrigues, J.M.F., et al. (eds.) Smart Systems Design, Applications, and Challenges, pp. 43–62. IGI Global (2020)
10. Saberi, S., Kouhizadeh, M., Sarkis, J., Shen, L.: Blockchain technology and its relationships to sustainable supply chain management. Int. J. Prod. Res. **57**(7), 2117–2135 (2019)
11. Saini, V., Gupta, S.: Blockchain in supply chain: journey from disruptive to sustainable. J. Mech. Continua Mathe. Sci. (2019)

12. Ma, Y., Sun, Y., Lei, Y., et al.: A survey of blockchain technology on security, privacy, and trust in crowdsourcing services. World Wide Web (2019)
13. Syed, T.A., Alzahrani, A., Jan, S., Siddiqui, M.S., Nadeem, A., Alghamdi, T.: A comparative analysis of blockchain architecture and its applications: problems and recommendations. IEEE Access (2019)
14. Rejeb, A., Keogh, J.G., Treiblmaier, H.: How blockchain technology can benefit marketing: six pending research areas. Front. Blockchain **3** (2020)
15. Sihi, D.: Impacts of blockchain technology in marketing. In: Martínez-López, F., D'Alessandro, S. (eds.) Advances in Digital Marketing and eCommerce. Springer Proceedings in Business and Economics. Springer, Cham (2020)
16. Agrawal, M., Amin, D., Dalvi, H., Gala, R.: Blockchain-based universal loyalty platform. In: 2019 International Conference on Advances in Computing, Communication and Control (ICAC3), Mumbai, India, 2019, pp. 1–6
17. Hader, M., Elmhamedi, A., Abouabdellah, A.:Blockchain technology in supply chain management and loyalty programs: toward blockchain implementation in retail market. In: 2020 IEEE 13th International Colloquium of Logistics and Supply Chain Management (LOGISTIQUA), Fez, Morocco, 2020, pp. 1–6
18. Gheitanchi, S.: An autonomous loyalty program based on blockchains for IoT solution providers. In: 2020 IEEE Global Conference on Artificial Intelligence and Internet of Things (GCAIoT), Dubai, United Arab Emirates, 2020, pp. 1–6
19. Sönmeztürk, O., Ayav, T., Erten, Y.M.: Loyalty program using blockchain. In: 2020 IEEE International Conference on Blockchain (Blockchain), Rhodes, Greece, 2020, pp. 509–516
20. Pradana, I.G.M.T., Djatna, T.: A design of traceability system in coffee supply chain based on hierarchical cluster analysis approach. In: 2020 International Conference on Computer Science and Application in Agriculture (ICOSICA), Bogor, Indonesia, 2020, pp.1–5
21. Pramanik, B.K., Rahman, A.Z.M.S., Li, M.: Blockchain-based reward point exchange systems. Multimed Tools Appl. **79**, 9785–9798 (2020)
22. Chen, J., Ying, W., Chen, Y., Wang, Z.: Design principles for blockchain-enabled point exchange systems: an action design research on a polycentric collaborative network for loyalty programs. In: IFIP Advances in Information and Communication Technology, vol. 598. Springer, Cham (2020)

A Review of Various Techniques Used in the Implementation of Lane Departure and Road Detection

Sagrika Chandra, Manav Gupta, Bhoomi Gupta, and Vandana Choudhary

Abstract In this paper, we talk about the various ways through which driver steering assistance along with lane detection systems can be implemented. With the influx of various technological advancements, it is of absolute necessity to work on ideas and execute them. The driver steering assistance system helps an inattentive, ill, or fatigued driver to drive well ensuring the safety of them self and others. Similarly, a lane detection system works on preventing the driver from violating lane driving. Staying in lanes, while driving is recommended at all times. An extensive number of accidents are caused by reckless driving habits and seem to be a huge safety concern, the systems mentioned above help in reducing these accidents significantly. Once these are deployed, we can combat the problems faced, while driving and avoid losing lives to fatal road accidents as a result save lives. Both the concepts not only prove to be great measures to ensure safety but also are used in autonomous vehicles and other intelligent automobile and journey applications.

Keywords Computer vision · Autonomous systems · Driver safety · Lane detection · Steer monitoring · Edge detection

1 Introduction

Traffic and road safety have become concerns prominent in nature because of the ever-increasing number of vehicles-both personal and commercial. All over the world, there is a spike in the number of deaths happening due to road accidents [1]. A global report produced by the World Health Organization (WHO) in 2018 establishes this harsh truth. According to this report, 1.35 million individuals die every year

S. Chandra (✉) · M. Gupta · B. Gupta · V. Choudhary
Maharaja Agrasen Institute of Technology, New Delhi 110086, India

B. Gupta
e-mail: bhoomigupta@mait.ac.in

V. Choudhary
e-mail: vandanachoudhary@mait.ac.in

© The Author(s), under exclusive license to Springer Nature Singapore Pte Ltd. 2022
A. K. Nagar et al. (eds.), *Intelligent Sustainable Systems*, Lecture Notes in Networks and Systems 333, https://doi.org/10.1007/978-981-16-6309-3_22

because of road traffic mishaps. These reports also suggest that people aged 5–29 years are highly susceptible to road traffic fatalities and injuries which can even lead to death [2]. Furthermore, driving any vehicle is a complicated, multifaceted, and potentially risky activity to be performed as it involves the entire implementation of cognitive, physiological, and intuitive resources. Any depreciation in the above-mentioned factors can compromise the lives of people involved (such as drivers, passengers, and other people in other vehicles). Any drivers' operational state for driving a vehicle involves a demanding and complex set of physical, physiological, and psychological parameters. As a result, it is of utmost importance to have safety measures in check.

With the advent of the field of autonomous automation, the concept of an autonomous vehicle has been in the picture ever since the start of this century. It is believed that the era of autonomous vehicles will be comfortable as well as safe transportation [3]. Now that we have established that driving is complex in nature and that the probability of road accidents occurring is not negligible, we want to talk about the mishaps that happen due to driver fatigue and drowsiness. The table displayed shows the fatality rates due to road accidents across different countries [4]. For the nation of India, it is not clear how many fatal road accidents are caused by driver drowsiness. However, we know that there exists no substitute for sleep, and it is the need of the hour to make sure that road and vehicle accidents happening due to driver drowsiness and fatigue are lowered significantly, to the point where it does not exist altogether. The reason for us undertaking this project will be to come up with safety measures to help reduce and ultimately finish the problem of driver drowsiness (Table 1).

Table 1 Comparison of death tolls caused due to road accidents

Country name	Motorization rate/1000 people	No of road deaths/100,000 people	No of road deaths/100,000 motor vehicles
Africa (2016)	38.9	26.6	574
Brazil (2016)	209.3	23.4	57.5
Canada (2016)	686.3	5.8	8.9
China (2016)	141.2	18.8	104.5
India (2016)	36.3	16.6	130.1
Indonesia (2016)	87.2	15.3	36.7
United States (2016)	831.9	12.4	14.2

2 Related Work

In this section, we mean to talk about the existing literature, this would help us explore various perspectives, algorithms that have been proposed for lane disclosure, and gain insights.

In the year 1996 D. Pomerleau et al. came up with a RALPH system [5]. This system can be implemented to gauge the sidewise position of a self-governing vehicle. This is done through a matching algorithm method that adaptively aligns and adjusts to a predefined specified template to the averaged-out scan line intensity profile to limit the lateral offsets as well as the curvature of the lane. This system performed well under expansive and thorough tests, inclusive of a coast-to-coast, 2,850-mile drive. A successor to the RALPH system came in the year 1998, documented by Broggi [6], known as the GOLD (generic obstacle and lane detection) system. This uses an algorithm constructed on the edge and ends-based boundary detection. The image hence acquired is restored into a different set of a new image which shows the top view of the roadway in which the differential of the lane is straight standing clear lines against a dark backdrop. To extract the supposedly vertical vivid lines which are grouped onto larger portions. Specific adaptive filtering is used. This system is stereo-based hardware and software to increment and increase road safety. A deformable template approach was established in the LOIS algorithm [7]. All the feasible and attainable methods that the lane edges could emerge in a picture are described by a parametric group of shapes. An operation is then defined which proportionally measures how perfectly the pixel data of a supplied image is matched by a set of lane form characteristics. A 2004 paper written by Wang et al. [8] suggests that a B-Snake spline can be used as a progressive geometric model to showcase the lane and vehicles. The features needed by the mentioned geometric model are extracted from these processed pictures using Canny/Hough Estimation of Vanishing Points (CHEVP). Outputs that are obtained are promising, as they also included shadows on the road into consideration, such as trees and telegraph poles. Chen et al. [9] developed the AURORA (a vision-based roadway departure warning system) system. This marks the lanes drawn out on roads using a color video camera that faces downwards onto the road. A wide-angle lens, a portable Sun Sparc workstation, and a digitizer is included with these cameras. In 2008 M. Aly suggested an approach for recognizing road boundaries that is real-time, efficient, and resilient. The method uses a bird's-eye view of the input roadway picture, filtering through Gaussian kernels, and then implement the RANSAC spline fitting technique along with line detection to detect lanes. This algorithm delivered great results except in the cases when where it encountered crosswalks, zebra crossings, and writing with paint on the roads. To deal with challenging possibilities such as faded out lane markings, lane changes, emerging, ending, lanes merging, lane curvature, and splitting lanes. Arbitrary sample collective unison and particle filtering were used to create the algorithm, this particular algorithm produced several workable hypotheses in real-time in comparison with the other existing methods from that time. This technique was proposed by Z. Kim in 2008 in their publication robust lane detection and tracking in challenging scenarios [10]. Teng et al. [11] suggested a real-time algorithm for the identification of a vehicle driving on a highway utilizing video sequence. This algorithm was also able to perform well in a change of light intensity and shadows. The road boundaries were identified using Hough transformation with a limited region for carrying search. This algorithm could also be used for slightly curved roads, roads with different weather conditions, painted, and unpainted roads as well. In the same year, Zhou et al. [12] gave a new method to detect lanes using the motion video camera approach. This identification algorithm was fused with a tracking method that applied two Extended Kalman filters using the Interacting Multiple Models (IMM) algorithm. In the presence of noise, weak and faded markers, the method was linear in time and resilient. It could at the same time be implemented to determine line segments, positions, and slopes. Khalifa and Hashim [13] offered a boundary detection method on the streets based on the Gabor filter along with the Geometrical model. This algorithm finds successful implementation in lane departure warning systems as well as for various other auxiliary driving systems. The starting position, lane original orientation, width, and curvature were all features taken into consideration in the lane geometrical model because of its very few errors in edge detection, showcased by heightened accuracy in passengers on the road and tree shadows, and this method has the potential to defeat universal lane detection difficulties, such as the shadows and crosswalks. In 2010, Ghazali et al. [14] proposed a system that would be real-time and vision-based to find the placement and category of each lane in every input video frame. The continuous progression of the lane is approximated using a Bayesian probability model. A specialty of the above-mentioned flow of instructions is that it does not require special parameters for the camera, background models, or any other similar things, therefore, making this a more adaptive approach. A lane departure system was proposed by Phaneendra et al. [15]. The primary objective of this model was to detect lanes and give caution on departure from the lane by implementing an image processing algorithm. The lane fleeting conclusion is established on the distance between lanes and center of the lowest point in the acquired image coordinate, which required fewer parameters. The mentioned technique demonstrated more efficiency and viability when contrasted with existing systems.

3 Method

To apply lane detection, various techniques have come up over time and many ways exist through which lane detection can be implemented. The flowchart below shows a clear division for the same. The three broad categories are feature-based on boundary detection, region-based features, and a combination of both boundary-based and region-based (Fig. 1).

First, we will cover feature-based methods, starting with the edge detection-based approach.

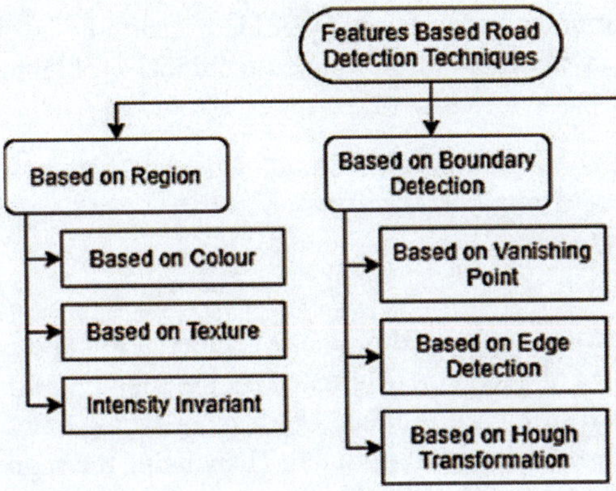

Fig. 1 Classification of detection techniques

3.1 Feature Based on Boundary Detection

Edge Detection based Approach. Based on mathematical formulae, the edge detection technique can be further divided into two groups, derivatives of second-order and first-order [16]. In 1965, Lawrence Roberts proposed a technique for edge detection within an image. The approach is simple mathematics yet it is efficient [17]. It will measure the spatial gradient for the picture and the value of a pixel in the resultant picture. The input image is taken as greyscale, and then the output image contains the edges in it. It was given by John Canny hence was named the Canny Approach [18]. This approach undergoes the successive steps:

1. To reduce noise and smoothen the image, a determined value of sigma is used with the Gaussian filter.
2. At every point, the local gradient and edge-direction are calculated. The edge point is a point where dominance in the gradient-direction is locally maximal.
3. The number of ridges in the gradient-image magnitude is intensified by the edge point. A narrow line is then produced as an output. This method is called non-maximum suppression.

Vanishing Point-based Boundary Detection. Most images contain some sort of perspective distortion. A vanishing point is where a given group of parallel lines in a 3D plane converge into one exclusive point on the image plane. A vanishing point approach helps us identify various parameters like object depth, object distance. However, in certain cases, the approach is not the best as there maybe be object irregularities that might not produce accurate results.

Hough Transform based Approach. Hough proposed a computationally efficient yet interesting procedure for spotting lines in a picture. The computations were based on the slope-intercept structure of mathematics. It was pointed in the 1970s that the same calculations can be carried out through the angle-radius parameters which can make the computations much more efficient [19]. Identification of straight lines in processed images is one of the most often noticed technical problems in image processing.

3.2 *Region Based Features*

Texture based Features. The texture is used in image segmentation and processing for detecting changes. The idea of a method originated for the segmentation of pavement boundaries from the part of a photo-induced by a front-facing camera placed on the top of a moving vehicle uses the said approach. Then, using the segmentation-algorithm, texture descriptors having a statistical nature and texture descriptors which are built on a gray-level co-occurrence matrix are combined [20]. Mostly, if a person wants to split up the perimeter into several portions in a picture that is speaking for the visuals of the vehicle-front then there can be two possible propositions:

1. Edge Detection: The boundaries around the regions are marked by substantial light intensity variations and to uproot these border pixels, some high-pass filtration applied beforehand is recommended after which the relevant segmentation approach divides the points where the gradient values surpass a margin value, commonly known as threshold.
2. Region Detection: Regions of a roadway can be distinguished utilizing a value or a set of values of some descriptors, such that the dissimilar portions of the image are well differentiable from the roadway and the boundaries of the roadway are extractable in this manner. These descriptors are associated with texture, color, and light-intensity, etc.

Color based Features. In content-based retrieval systems, color histograms are popularly used and are known to give good results. Color is an important value parameter in any image or video. It is an essential cue used for extracting information from an image [21]. Color is a point property, therefore, it helps in knowing the underlying details and these details have a high chance of being missed out when using a more expansive texture-based approach. The phenomena of over-segmentation are observed if, after segmentation, an image has few small regions that correspond to analogous items existing in the original image.

Intensity Invariant Features. Feature detection and description are common tasks used while image processing and in computer vision. However, the existing methods mostly do not take into consideration the factors of luminance, noise degradation, and order of the intensity values [22]. The human visual system can recognize a lot based on illumination conditions.

Results

The above-mentioned techniques for the edge, boundary-based detection methods, and intelligent vehicle systems development methods have their disadvantages and advantages. The advantage of the lane departure and road detection method is precise and inherent. We can more specifically define the degree of driver fatigue. Other than that, it discerns the feature of boundary, which does not take into account the various parameters. The various disadvantages involved in implementing this technique are that of unable to predict steep curvatures, recognition complexity, difficulty in the precise extraction of boundary, moreover, the final results can be easily affected by factors such as illumination and unpredictability. However, in the recent past few years, newer implementations have come forward which can be put into use to achieve better and more accurate results. This will mark greater advancements for safe driving. Machine learning technology has made exceptional progress in this particular domain. The application of these systems is extremely high. In any way, through which operational and equipment costs can be cut down will multiple times increase the viability of the system and could be implemented to larger portions of drivers. They are not only for better safety but also a futuristic idea that needs to be developed in an efficient manner. Conclusively, the ever so steep spike in the number of cars specifically in the urban areas yields a higher probability for an accident to occur. As a result, it is of high significance to study and implement lane driving, steer assist, and boundary detection systems. This is vital research work discussed which can provide great insights.

4 Conclusion

In this review, a detailed analysis of the techniques that exist for implementing lane detection systems has been carried out. So far, we have talked about the various methods through which we can test road boundaries. Lane detection, intervention, and monitoring are an important concept and these are essential for the successful development of intelligent vehicles and transport systems. It is also seen that the developing and implementing processes for lane driving are extremely crucial for safety reasons as well. Modern vehicles should have features that decrease the number of vehicle accidents and ensures passenger's safety. Various performance evaluation metrics have also been discussed. It is established that lane departure, intervention, and detection is a fundamental and inevitable module in the upcoming advanced driver assistance systems. These also have widespread and important applications in autonomous vehicles. From the beginning of this concept up until now, considerable advancements have been achieved. However, due to the extremely variable and dynamic environments of a road with traffic, we should work toward the enhancement of the development and implementation of driver assistance systems.

References

1. Global status reports on road safety 2018: World Health Organization, Geneva 2018. Licence: CC BY- NC-SA 3.0 IGO
2. Global Status Report on Road Safety 2018 by World Health Organisation Team: Social Determinants of Health, ISBN 9789241565684
3. Bimbraw, K.: Autonomous cars: past, present and future—a review of the developments in the last century, the present scenario and the expected future of autonomous vehicle technology. In: Proceedings of ICINCO 2015—12th International Conference on Informatics in Control, Automation and Robotics, vol. 1, pp. 191–198 (2015). https://doi.org/10.5220/0005540501910198
4. Mohammed, A.A., Ambak, K., Mosa, A.M., Syamsunur, D.: A review of traffic accidents and related practices worldwide. TOTJ 13–65. https://doi.org/10.2174/1874447801913010065
5. Pomerleau, D., Jochem.: Rapidly Adapting Machine Vision for Automated Vehicle Steering. IEEE (1996)
6. Broggi, B.M.: GOLD: a parallel real-time stereo vision system for generic obstacle and lane detection. IEEE Trans. Image Process. 4–6 (1998)
7. Kreucher, C., Lakshmanan, S.K.: A driver warning System based on the LOIS lane detection algorithm. In: Proceeding of IEEE International Conference on Intelligent Vehicles, pp. 17–22, 1998
8. Wang, Y., Teoha, E.K., Shen, D.: Lane detection and tracking using B-Snake. Image Vis. Comput. **22**, 269–28 (2004)
9. Chen, M., Jochem, T., Pomerleau, D.T.: AURORA: a vision-based roadway departure warning system. In: Proceeding of IEEE Conference on Intelligent Robots and Systems, 2004
10. Jung, C.R., Kelber, C.R.: Lane following and lane departure using a linear-parabolic model. Image Vis. Comput. 1192–1202 (2005)
11. Teng, Z., Kin, J.H., Kang, D.J.: Real-time lane detection by using multiple cues. In: IEEE International Conference on Control Automation and Systems, pp. 2334–2337, 2010
12. Zhou, S., Ziang, Y., Xi, J., Gong, J., Xiong, G., Chen, H.: A novel lane detection based on the geometrical model and Gabor filter. In: IEEE Intelligent Vehicles Symposium, pp. 59–64, 2010
13. Khalifa, O.O., Hashim, A.H.A.: Vision-based lane detection for autonomous artificial intelligent vehicles. In: IEEE International Conference on Semantic Computing, pp. 636–641, 2009
14. Ghazali, K., Xiao, R., Ma, J.: Road lane detection using h-maxima and improved hough transform. In: Fourth International Conference, 2012
15. Phaneendra, N., Goud, G., Padmaja, V.: Accident avoiding system using lane detection. Int. J. Res. Electron. Commun. Eng. **1**(1), 1–4 (2013)
16. Ansari, M., Kurchaniya, D., Dixit, M.: A comprehensive analysis of image edge detection techniques. Int. J. Multimedia Ubiquitous Eng. **12**, 1–12 (2017). https://doi.org/10.14257/ijmue.2017.12.11.01
17. Das, S.: Comparison of various edge detection technique. Int. J. Signal Process. Image Process. Pattern Recogn. **9**(2), 143–158 (2016). [2] Nadernejad, E., Sharifzadeh, S., Hassanpour, H.: Edge detection
18. Juneja, M., Singh Sandhu, P.: Performance evaluation of edge detection techniques for images in spatial domain. Int. J. Comput. Theory Eng. **1**(5), 614–621 (2009)
19. Duda, R.O., Hart, P.E.: Use of the Hough Transformation To Detect Lines and Curves in Pictures. Communications of the ACM, January 1972, vol. 15. Stanford Research Institute, Menlo Park, California (1972)
20. Graovac, S., Goma, A.: Detection of Road Image Borders Based on Texture Classification. First Published 1 Jan 2012, vol. 9(6)
21. Belongie, S., Carson, C., Greenspan, H., Malik, J.: Colour- and texture-based image segmentation using EM and its application to content-based image retrieval. In: Proceedings of International Conference on Computer Vision 1998. IEEE, Computer Science Division, University of California at Berkley, Berkeley (1998)

22. Diaz-Escobar, J., Kober, V., Gonzalez-Fraga, J.A.: LUIFT: luminance invariant feature transform. Mathe. Prob. Eng. **2018**(3758102), 17 (2018). https://doi.org/10.1155/2018/3758102

Virtual Police Station System with Chat Bot Using FSM

Bhoomi Gupta, Parth Rustagi, Priyanshu Sinha, and Rohit Sroa

Abstract We can notice that innovation has contacted numerous circles of our lives in India. There is innovation in business, in instruction, in mingling and keeping up human relations, in buying, in horticulture, in banking, correspondence, and pretty much all aspects of our lives. This interruption of innovation has supported the work in every one of these areas, and has demonstrated valuable, and time. The solitary significant piece of our general public that actually remains significantly without this extravagance is the Indian Police System. The authorities have been receiving the essential thing techniques for completing the procedures with the customary "pen and paper" strategy being exceptionally common. In the present India, when the insidious components of the public are in a blast and such countless cases being enlisted each day, it has become a monotonous assignment to deal with the case and all its connected records, physically. The conventional strategy for visiting a police headquarters for enlisting a police officer grumbling and persuading refreshes should be supplanted with an online cycle. The means of this paper is to develop an E-police reporting and a framework with the help of a chat bot which is based on finite state machine (FSM), and NodeJS development which is effortlessly available to people in general and the police division. In this paper, we concentrate on the establishment of an e-police framework and moreover its means, difficulties of usage and its need.

Keywords Virtual police · Finite state machine · Chatbot · Homomorphic encryption · Cryptography · FIR · Artificial intelligence · Automation

1 Introduction

The idea is to provide an easy-to-use environment where a user can register his complaint, but at the same time the system should be secure to keep the complainants data secure and also the details of the complaints from falling into the wrong hands. The following aspects have been kept in mind, while defining the idea of the project.

B. Gupta (✉) · P. Rustagi · P. Sinha · R. Sroa
Maharaja Agrasen Institute of Technology, Delhi, India

© The Author(s), under exclusive license to Springer Nature Singapore Pte Ltd. 2022

A. K. Nagar et al. (eds.), *Intelligent Sustainable Systems*, Lecture Notes in Networks and Systems 333, https://doi.org/10.1007/978-981-16-6309-3_23

1. Easy UI/UX: The system will be used by people of different educational background, familiarity and prior knowledge of computers.
2. Authentication and Verification: While registering a complaint, it is important to verify the person who is registering the complaint. It is a necessity to ensure than the complaint filed is genuine and not redundant or false complaints [1].
3. Database Storage and Connectivity: As catering to the Indian population, it is fair to assume that there will be a large volume of complaints being lodged. A reliable database is important which should also be fast in accessing the required information as when required [2].
4. Chat bot for ease of use: The environment would be a present with a chat bot that would help the user in all it accepts whenever it is stuck at particular page and cannot move ahead [3, 4].
5. Cryptography technique for security of data: Just storing the data will not be essential as the data stored is sensitive and if fallen into the wrong hands can also be a threat to harmony and peace. Thus, an encryption is important.

2 Feasibility

The system saves the person from the requirement to truly go to a station base camp to hold up a protesting. Using our application in a cell phone or computer, one can adequately register their complaint with the authorities. It is consistently seen that people go without going to the police central command. Many think it is monotonous and that they would have to take care of the police to finish the work, while many are basically hesitant to hold up a protesting as a result of social components. Since simply the investigating official can get to the particular FIR id, the information is private and secure. The cycle finishing on the web, in full data on the complainant ensures truthfulness. As the police needs to straightforwardly refresh the plaintiff over the next steps about the procedures of the case, with confirmation, any deferral in the work is immediately seen by the residents and accordingly the extents of bogus guarantees is profoundly diminished.

3 Literature Survey

See Table 1

Table 1 Literature survey

S. no	References taken	Research applied/gap
1	Adesh Jamnik [1]	Use of Aadhar card for authentication and verification
2	Almansor and Hussain [5]	Understanding the fundamentals of interaction between human and chatbot
3	Varma et al. [6]	Implementation of Aadhar card verification and authentication process
4	Devarajan and SudalaiMuthu [2]	Use of cloud storage and implementation of database in our system
5	Patel et al. [3]	Enhancement of emotional relationship understanding between human and a chatbot
6	Prasad and Ranjith [7]	Security and automation aspects of an A.I based chatbot
7	Pratibha Chaudhary [8]	Implementation of homomorphic encryption in our system
8	Yi and Jung [4]	Understanding the architecture and working of a chatbot made with help of finite state machine

4 Methodologies

4.1 Authentication and Verification

The client approval is through two-venture measure utilizing OTP and email address. This step takes place in the registration part where the users create themselves a profile for the registration of a complaint/FIR. Users are required to input their primary details like name, phone number, address, etc., as well as their aadhar card number. The users are required to input the above fields same as their data given in their respective aadhar card. The next step in the user registration is the user will receive an OTP on their mobile number which is registered with the government and linked to their aadhar database. Completion of this step will only allow the user to continue on the online complaint/FIR registration [6].

4.2 Database Storage and Connectivity

Virtual police station's principal thought is to carry out a chat bot for keeping up the legitimate ordinary citizens complaints for various police divisions. By utilizing this application individuals who are apprehensive or need an ideal opportunity to go police headquarters for grumbling about their own lawful issues here they can give their objection through online to enroll any sort of grievance. This proposed framework attempts to take out or diminish troubles dependent upon some degree.

This framework will assist the citizens that want to lodge a complaint or an FIR (First Information Report) with lessening the responsibility and mental struggle. It assists the client with working easy-to-use, and he can undoubtedly take care of his responsibilities without time slacking.

4.3 Cryptographic Techniques for Security of Data

While filing complaints on the virtual police station system, a lot of private and sensitive data is filled. This type of encryption allows you to perform computation or analysis of data without actually decrypting it. The authorities can share the data in its encrypted format and perform their necessary analysis and provide the results accordingly. This makes sharing of data and working in collaboration much more trustworthy and secure. However, homomorphic encryption is an expensive and slow process. It takes much more time and resources to do computations. But with increasing technological advances and much cheaper and faster hardware, it is soon to overcome those difficulties as well [8].

4.4 Chat Bot Using FSM (Finite State Machine)

Our conversational framework is a blended activity framework that oversees discourse utilizing limited state technique and a recovery model. Blended activity implies that discoursed can be constrained by a framework as well as a client. The framework can step up in discourse to get the necessary data; however, the client can likewise step up to the plate and pose inquiries whenever they need. Discourse the board based on limited state machine implies that a client's inquiry will travel the present status of the framework and the framework will react dependent on its state and the progress [4, 7] (Fig. 1).

5 Architecture

Figure 2 represents the architecture of our proposed software. The initial states the user will be asked to login/register in our website to use the chat bot or register his complaints. Details of his authentication will be sent and saved into the database. After login, the user will then have options to whether he wants to register a new complaint or view/check his previous complaints. If the user chooses to register a new complaint, then after giving the desired input all the data given by the user will be encrypted through the process explained in the methodology section.

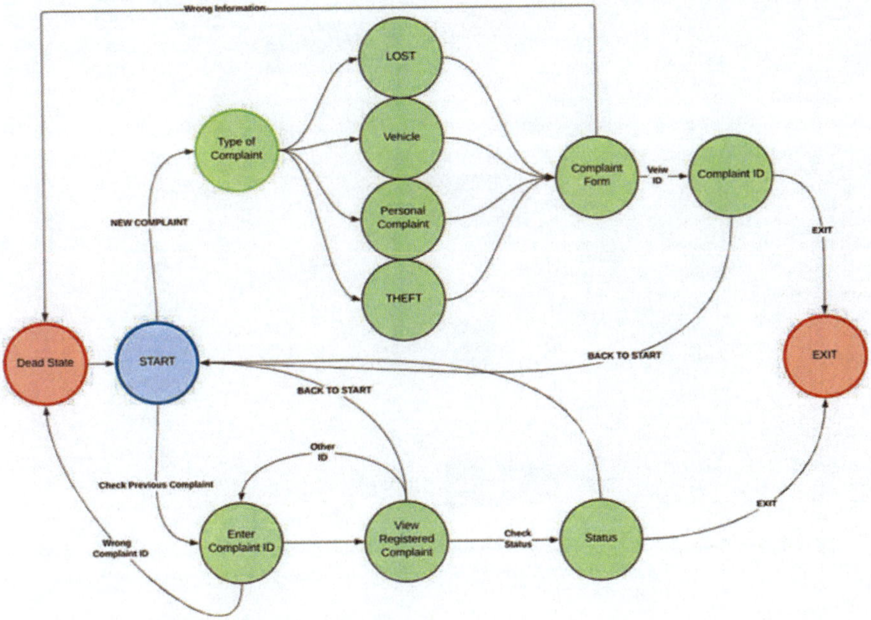

Fig. 1 Finite state diagram of virtual police station system

6 Model Developed

This is the home/main screen of the proposed system, which will open after the user successfully logs in or register with our website (Fig. 3).

After clicking on the respective inputs in the home message, the user is shown the next messages and inputs (Fig. 4).

The user is sent back to the main screen or home screen by the end of the finite state machine. This happens when either user is sent to the dead state by the means of wrong information or when the user completes his transaction with the system (Fig. 5).

7 Result

As you can see the screenshots of our proposed interface in the above section. We have developed a chatbot using the functionality of a finite state machine. In Fig. 2, it is the starting state of the FSM. In Fig. 3, it shows the second and third states of the FSM with their respective input buttons and display messages. In Fig. 4, it demonstrates another second state with its respective input and output options.

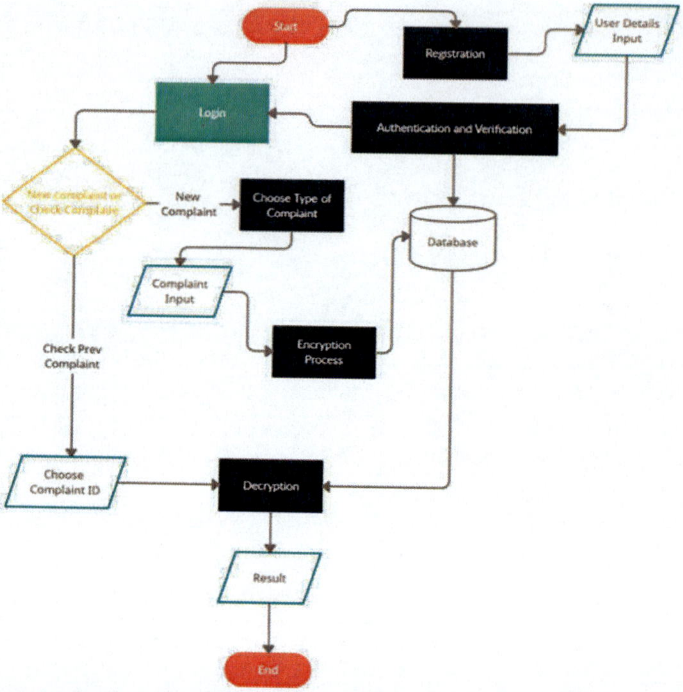

Fig. 2 Architecture of virtual police station system

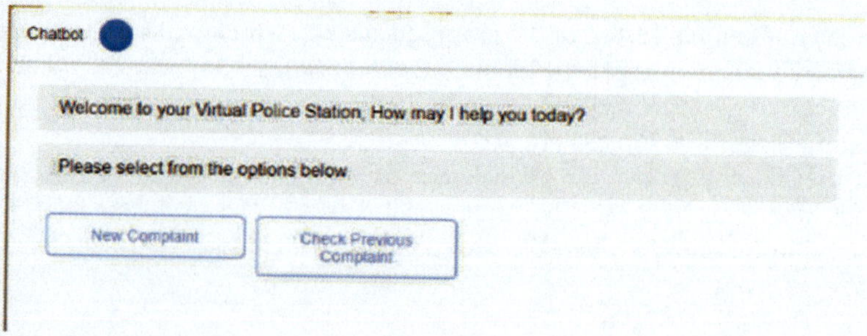

Fig. 3 Screenshot of the developed interface–chat bot

8 Future Scope

The developed system would serve as a foundation for all prospects and systems relating to the Virtual Policing. The system is built with keeping scalability in mind and with the increasing domain of internet services the system can be extended to be used in rural areas. With developments in cryptography data storage will be much

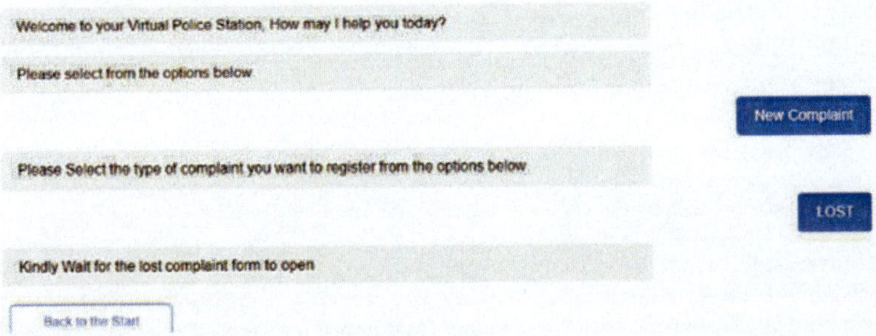

Fig. 4 Screenshot of interface in the next state of FSM

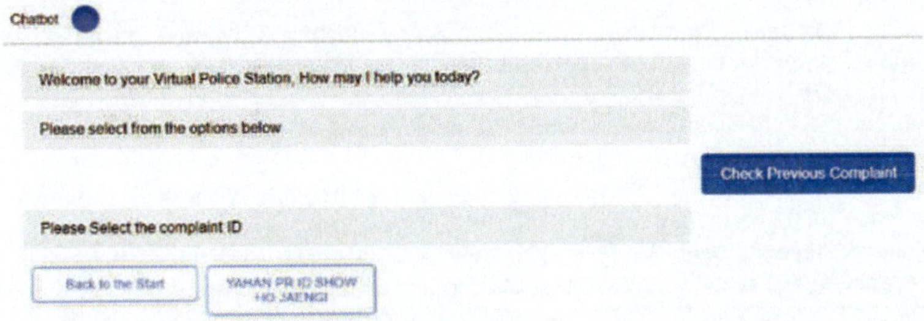

Fig. 5 Screenshot of interface in another state of FSM

secure with faster accessibility at the same time. It would also make collaboration on cases, without disclosing confidential information, much easier. Faster policing would also help with faster judicial procedures on cases. The system with some modification and added functionality can also be used to acquire police documents such as, police verification, permits, permissions, inspection requests, etc. Cloud storage facility for the betterment of database connectivity and one central storage space for easy accessibility of data.

9 Conclusion

The Indian Policing system is of those areas where technology has yet to put its footprint. Our research is followed by analyzing existing papers and identifying their pros and cons. Based on our analysis, we have proposed and developed a suitable model that aims to deliver efficiency with keeping ease of use in mind. The security aspects have been kept in mind and discussed thoroughly by implementing and proposing various cryptographic techniques. If the chat bot crashes, then there exists no alternative to our application. Also, for every action, the chat bot needs to be access from the beginning, which can be tiresome for some users. However, we believe that this paper can serve as a benchmark for future systems and discussions to building a much efficient policing system.

References

1. Adesh Jamnik, M.S.: Digital ticket booking and checking using aadhaar card or fingerprint and android application. In: 3rd International Conference on Recent Developments in Control, Automation & Power Engineering (RDCAPE), 2019
2. Devarajan, A.A., SudalaiMuthu, T.: Cloud storage monitoring system analyzing through file access pattern. In: International Conference on Computational Intelligence in Data Science (ICCIDS), 2017
3. Patel, F., Thakore, R., Nandwani, I., Bharti, S.K.: Combating depression in students using an intelligent chatbot: a cognitive behavioral therapy. In: IEEE 16th India Council International Conference (INDICON), 2019
4. Yi, S., Jung, K.: A chatbot by combining finite state machine, information retrieval and bot-initiative strategy. Proc. Alexa Prize (2017)
5. Almansor, E., Hussain, F.K.: Survey on intelligent chatbots: state-of-the-art and future research directions (2020). https://doi.org/10.1007/978-3-030-22354-0_47
6. Rahul, D.S., Jose, J., Samhitha, B.K., Mana, S.C.: Aadhar card verification base online polling. In: Proceedings of the 4th International Conference on Trends in Electronics and Informatics, ICOEI 2020, pp. 479–483, 2020. https://doi.org/10.1109/ICOEI48184.2020.9142965
7. Prasad, V.A., Ranjith, R.: Intelligent chatbot for lab security and automation. In: 11th International Conference on Computing, Communication and Networking Technologies (ICCCNT), 2020
8. Pratibha Chaudhary, R.G.: Analysis and comparison of various fully homomorphic encryption techniques. In: International Conference on Computing, Power and Communication Technologies (GUCON), 2019

Khon Gesture Detection with Deep Learning Algorithms

Sathit Prasomphan, Pannawat Niyomrat, Burapa Tainsakukdej, Titiwat Kuarkamphun, and Noppawee Deejungwiphat

Abstract This research aims to develop a Khon gesture detection with deep learning algorithms. Thai traditional art culture of "Khon" was used. The main contribution of this research was to develop an algorithm for retrieving information from image for telling story inside that image. The interesting information inside the image will be retrieved to present to the interested people who is interested in its contents. The development consists of telling stories from image. The appearance of the shape inside image can be used to distinct characteristics of image such as type or characteristic of image. The type was created including the story of Thai traditional art culture of "Khon" through the learning of machine learning and image processing. The classification result for classifying gesture of Khon was conducted. We tested the purposed algorithms by using deep learning algorithms compared to other classification methods. The experimental results were evaluated with images from Khon's performance in Thailand. The accuracy of purposed method gives 89.84%. From the results, we found that the purposed method gives more accuracy compare to the other methods.

Keywords Content-based image retrieval · Deep learning · Gesture detection

1 Introduction

The target industries that have potential to the growth of the economy in countries and be able to increase the ability of competitiveness includes 10 target industries. They have the gain to the interested investors from all over the world. These 10 industries can be divided into 2 groups as follows: The first group is the extension of existing five industries which consisting of modern automotive industry, smart electronics industry, tourism industry and health tourism, agriculture and biotechnology, and food industry. The second group is the adding five future industries

S. Prasomphan (✉) · P. Niyomrat · B. Tainsakukdej · T. Kuarkamphun · N. Deejungwiphat
Department of Computer and Information Science, Faculty of Applied Science, King Mongkut's University of Technology North Bangkok, 1518 Pracharat 1 Rd. Wongsawang, Bangsue, Bangkok 10800, Thailand
e-mail: sathit.p@sci.kmutnb.ac.th

© The Author(s), under exclusive license to Springer Nature Singapore Pte Ltd. 2022

A. K. Nagar et al. (eds.), *Intelligent Sustainable Systems*, Lecture Notes in Networks and Systems 333, https://doi.org/10.1007/978-981-16-6309-3_24

that comprise a new industry in which Thailand has competitive potential. There consisting of: industrial robots for industry, aviation and logistics industry, biofuels and bio-chemical industries, digital industry, and comprehensive medical industry. In the target industries mentioned above, the tourism industry is another industry that will affect the income of a country.

At present, KHON shows are becoming less popular in Thai society. The reason may be that the new generation is raised with modern drama, communication by speaking and natural gestures. Therefore, it is easier to understand modern drama than KHON show, which was more popular in the past. For this reason, we present the idea to apply current technology into KHON performance, which is an art performance that has a long time in Thailand. The Thai traditional art culture of "Khon" is slowly fading from the world due to it is difficult to understand. Therefore, in an attempt to help retain this traditional art culture, we decided to develop a project to make "Khon" more accessible to those who interest in the art and provide a mini knowledge base, which is easily accessible. Retain "Khon" traditional art culture for the next generation. We used a program to detect pictures using "machine learning" with gesture pictures called a "dataset" to train a module. This was achieved by applying a "you only look once" (YOLO) algorithm to our module, in order to detect the gestures with accuracy. We developed a program based on the module that we trained and put it on a website. The website includes news, third party links connecting to the reservation of Khon tickets, and background knowledge related to Khon art culture. Our method of combining machine learning with Khon will enable art culture enthusiasts and visitors have an easier way to understand and access Khon traditional art.

We organize the rest of section as follows. Theory of convolution neural network is introduced in Sect. 2. The proposed method is detailed in Sect. 3. In Sect. 4, experimental setup is described. Results and discussions are explained in Sect. 5. After that we concludes the paper in Sect. 6.

2 Related Theory

2.1 *Deep Learning [1, 2]*

One challenge in developing artificial intelligence is to develop machines to learn and predict or create knowledge. Several technique can be used to teach machine to learn and to create knowledge such as deep learning, artificial neural network, support vector machine, etc. The concept of artificial neural network is simulated from neural network system in the nervous system of human. The architecture of artificial neural network composed of three layers, which are input layer, output layer, and the hidden layer. The benefit of artificial neural network is to make decision in the difficult problem that human cannot solve with that problem. One of the most popular technique in the present is deep learning [1, 2]. Deep learning is the extended version of artificial neural network that are created for machine learning. The hidden layer of deep learning and artificial neuron networks is different in which deep learning has many hidden layer than artificial neuron networks [1, 2]. Deep learning can be applied in the face recognition problem. In this technique, a large dataset can be learned from different face in dataset. It can be used for identifying individual faces, tagging a picture on Facebook or to extract a non-human object. In addition, this technique can be used as part of the unmanned vehicle system, etc. [1].

Convolution neural networks, commonly known as deep learning, are a related theory in this study. Deep learning [1–3] is a type of machine learning that replicates the human brain's neuronal system. As a result, its future ability may be supernatural. It can boost processing power, allowing it to replicate nerves and neurons in an artificial neural network. Deep learning is a type of artificial neural network that has a large number of hidden layers. The capacity to process more information than usual. Numerous hidden layers make it quite difficult.

2.2 *Content-Based Image Retrieval [4–7]*

Previously, image retrieval approaches relied on retrieving image file names or using words that describe images (image caption), which is a non-real characteristic of the image. As a result, it is incompatible with the existing huge database. Because each user's description of a picture varies and may not always match, defining each image based on a few words becomes increasingly challenging as the database grows. Furthermore, certain picture databases may contain a large number of image data from the same group, such as a geographic image database, which may have many ground photographs; various tree images, and so on. It must also rely on humans to characterize the image, which takes a long period, while working with large databases. The features of an image-processing algorithm are known as image characteristics. Color, shape, and texture are the three primary elements of the image. Image indexing is the process of introducing image attributes, such as the histogram of colors, into a database to construct and store a distinct vector within each image. Each image's features will be expressed as numeric values, n numbers (depending on the conditions of each method), or n-dimensional vector, which is the image's particular vector. These unique vectors will be utilized as a picture index, with points in n-dimensional space representing them. The foundation of content-based image retrieval is feature extraction [3]. Text-based and visual characteristics are both examples of features in a broad sense (color, texture, shape, faces). We will limit ourselves to visual feature extraction techniques because there is already a large literature on text-based feature extraction in the DBMS and information retrieval research fields. The features can be further divided as general features and domain features within the visual feature scope.

2.3 Research on Automatic Sentence Retrieval from Pictures

We will go over the related research on automatic sentence generation from photographs in this section. Karpathy and Fei-Fei [4] described how to create natural language descriptions of image regions by applying algorithms. Farhadi et al. [8] implement a different system for determining a score that correlates a picture to a text. To attach a descriptive sentence to a given image, they assessed the score relating an image. They developed the concept to simulate the link between image, meaning, and sentence space. Kulkarni et al. [9] proposed a platform for their research that would allow them to utilize monument data from the Indian Digital Heritage Space (IHDS). They used transfer-learning techniques to introduce picture classification and query-based retrieval of Image labels. The recurrent neural network-based techniques for generating automatic sentences from pictures were proposed by the authors in [6, 7, 10]. Socher et al. [5] propose the DT-RNN model, which is a recursive neural network-based method for determining image-sentence relationships. The issue arises when the previously mentioned algorithms are applied. The link between picture spaces, meaning spaces, and sentence spaces can be confusing. Because of these issues, the goal of this study is to build algorithms for creating descriptions of Khon images. A deep neural network was used to create this description.

3 Proposed Method

3.1 Detection Algorithm

We have divided the model into 2 models; the first model detects the character in the image, and the second model is then used to predict the gesture, based on the greatest probability in the model. The YOLO that will be used in this project is the YOLOv3 tiny, as it has the best processing time. The structure of CNN composed of 13 convolutional layers which composed of 1 stride, 1 padding, 6 max pooling layers which 2 stride, 2 route and 1 up sample layers, 2 output layers. The learning rate is 0.001. Number of batches is 64 batches. Activation function is ReLU. Once the users upload an image to the website, it will be sent to the processing in model 1. The outputs are images, which have many bounding boxes of predicted characters that have a probability score, which depends on image. The outputs from model 1 are not returned to the user, rather they are sent for processing in model 2, to detect the gestures in each bounding box.

Users can input any size and resolution images as the program will convert the image size to 416 × 416 pixels (using YOLO format). Our algorithm was divided into 2 models: object detection, image classification. Details of these two models are described as follows: Model 1 was for object detection. In this step, the Khon characters were detected. The cropped images were then put into a bounding box size and sent to model two. Model 2 was for image classification. In this step, one

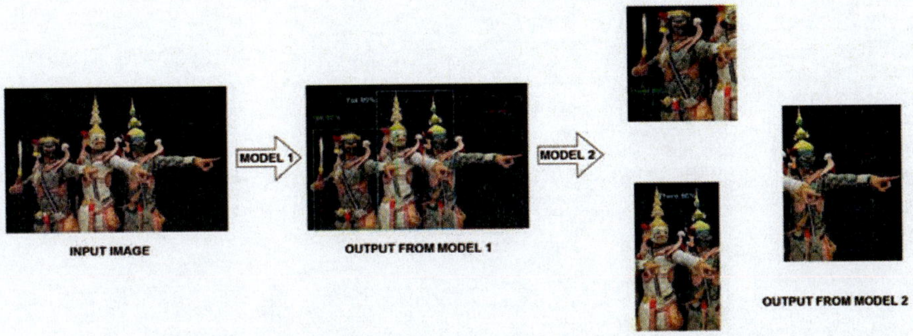

Fig. 1 Illustration of the Khon detection workflow

of the Khon gestures was detected using the greatest gesture probability score. All outputs (predicted images, character type, character score, gestures type, gestures score) were then returned to user. The workflow detection is shown in Figs. 1 and 2. After the datasets were collected, each image was labeled using YOLO annotation tools. The pattern of the annotation format was: <object-class> <x> <y> <width> <height>.

3.2 Image Augmentation

We have also used image augmentation methods for reducing the model's overfitting, using the following steps: At first, we collect the sample image in Fig. 3.

Next, Gaussian Noise was added to the sample images. After that CLAHE (contrast-limited adaptive histogram equalization) was added. In the next step, a black box in image was added. Finally, each image from the previous step was mixed together. We also flipped the images to provide more datasets as shown in Fig. 4.

4 Experimental Setup

4.1 Datasets

In this research, datasets were collected from Google image, Pinterest, YouTube. The requirements are as follows. Detection of 10 basic gestures, including: Angry, Upset, Resent, Offend, Regret, Crying, Lost, Dead, Decline, No, I, Me, Love, Great, Mighty, Shy, Smile, Happy, There, Here, Pointing. To verify the experimental results of gesture detection, the classification results by using the proposed algorithms to classify the type of gesture in Khon was performed.

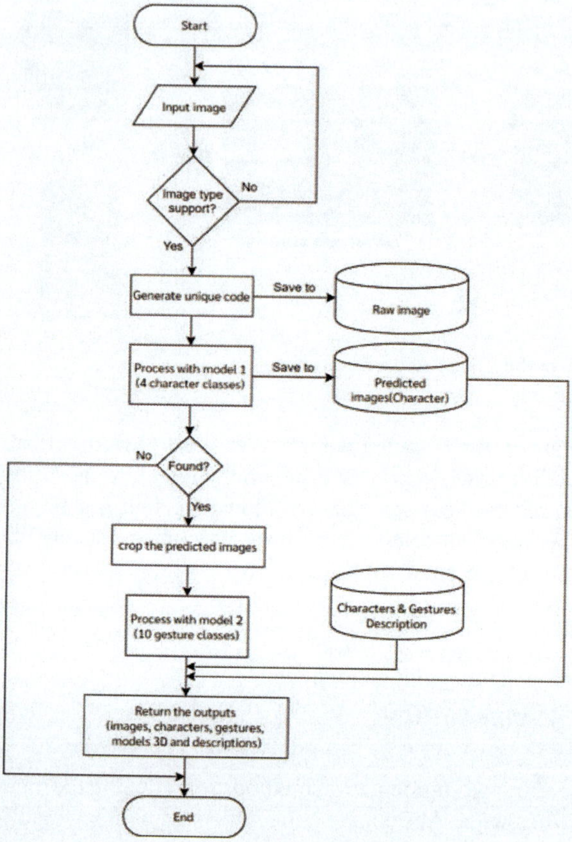

Fig. 2 Image detection flowchart

Fig. 3 Sample images from the dataset

4.2 Training

To train the model, we needed to divide the datasets into 2 sections to train and validate the correction of the model. The best ratio to split the dataset was 80:20 (train:test) and was stored in text file. For each model, we followed the following

Fig. 4 Augmented flipped images

data: "*number of epochs* = *number of class* * *2000*." Each model will take time to train, up to 24 h.

Model 1, we have 4 classes so, number of epoch is 8000 epochs, Model 2 we have 10 classes so, number of epoch is 20,000 epochs, and learning rate is 0.0001. Usually, 2000 iterations are sufficient for each class with no less than a total of 4000 iterations for a more precise definition of when to stop training.

5 Result and Discussion

We used Darknet to evaluate the model. The outputs after running this command are as following: TP (True Positive) for each class, FP (False Positive) for each class, FN (False Negative) for each class. In this research, we also used the precision value, which is the value that measures the number of correct predicting answers by dividing by the total number of images in the dataset. Next recall is calculated. Recall is the value that measures the number of correct predicting answers for each class and is divided by the total number of that class which is the ground truth. The average precision is 0.896 (89.6%); however, this is only for finding 1 class. The program will evaluate all classes repeatedly and find the mean Average Precision (mAP) of the model. The classification results from applying the proposed algorithms to categorize type or characteristics of that Thai traditional art culture of "Khon" image have been used to demonstrate the experimental findings for Thai traditional art culture of "Khon" image content generator. As a result, the classification accuracy ensures the efficiency of our image description generator. The average precision of the existing approach, which is the convolution neural network, is 0.896 (89.6%). The precision of Model 1, which can classify four classes of Khon, with precisions of 89.53, 90.31, 90.59, and 89.28 in the Ling, Yak, Pra, and Nang classes, respectively, using the proposed techniques. The Model 2 precision, which can classify 10 classes of Khon gesture, with precisions of 89.35, 88.30, 92.39, 80.11, 98.07, 77.48, 93.57, 98.41, 84.74, 97.17 in the Angry, Crying, Dead, Decline, I, Love, Mighty, Shy, Smile, and there class in ordering using the provided algorithms. The experimental results indicate that adopting a convolutional neural network for training to acquire the type of Thai traditional art culture of the "Khon" image can be successfully used for generating the image description. The proposed algorithms might well be utilized to classify the type of Thai traditional art culture represented by the image "Khon," with an accuracy rate of 89.84%.

6 Conclusion

This research aims to develop a Khon gesture detection with deep learning algorithms. Thai traditional art culture of "Khon" was used. The main contribution of this research was to develop an algorithm for retrieving information from image for telling story inside that image. The interesting information inside the image will be retrieved to present to the interested people who is interested in its contents. The development consists of telling stories from image. The appearance of the shape inside image can be used to distinct characteristics of image for example the type or character of image. The type was created including the story of Thai traditional art culture of "Khon" through the learning of machine learning and image processing. Using the classification results of the proposed methods to categorize type and attributes of the tested image, the testing results for a Thai traditional art culture of "Khon" with convolution neural network were analyzed. Images from the well-known Khon in Thailand were conducted to analyze the efficiency of the purposed algorithms. The accuracy of the proposed algorithms is 89.84 depending on overall average precision. When compared to the traditional approach, the results indicate that the proposed methodology can efficiently find the correct descriptions.

References

1. Deng, L., Yu, D.: Deep learning: methods and applications foundations and trends. Signal Process. **7**(3–4), 197–387 (2013)
2. Bengio, Y., Courville, A., Vincent, P.: Representation learning: a review and new perspectives. IEEE Trans. Pattern Anal. Mach. Intell. **35**(8), 1798–1828 (2013)
3. Prasomphan, S., Jung, J.E.: Mobile application for archaeological site image content retrieval and automated generating image descriptions with neural network. Mob. Netw. Appl. (2017)
4. Karpathy, A., Fei-Fei, L.: Deep visual-semantic alignments for generating image descriptions. IEEE Trans. Pattern Anal. Mach. Intell. **39**(4), 664–676 (2017)
5. Socher, R., Karpathy, A., Le, Q.V., Manning, C.D., Ng, A.Y.: Grounded compositional semantics for finding and describing images with sentences. Trans. Assoc. Comput. Linguist. **2**(2014), 207–218 (2014)
6. Zaremba, W., Sutskever, I., Vinyals, O.: Recurrent Neural Network Regularization. Preprint arXiv:1409.2329 (2014)
7. Young, P., Lai, A., Hodosh, M., Hockenmaier, J.: From image descriptions to Visual denotations: New similarity metrics for semantic inference over event descriptions. Trans. Assoc. Comput. Linguist. 2 (2014)
8. Farhadi, A., Hejrati, M., Sadeghi, M.A., Young, P., Rashtchian, C., Hockenmaier, J., Forsyth, D.: Every picture tells a story: generating sentences from images. In: Computer Vision—ECCV 2010, Lecture Notes in Computer Science, vol. 6314, pp. 15–29 (2010)

9. Kulkarni, U., Meena, S.M., Gurlahosur, S.V., Mudengudi, U.: Classification of cultural heritage sites using transfer learning. In: 2019 IEEE Fifth International Conference on Multimedia Big Data (BigMM), 11–13 Sept 2019, pp. 391–397
10. Hodosh, M., Young, P., Hockenmaier, J.: Framing image description as a ranking task: data, models and evaluation metrics. J. Artif. Intell. Res. (2013)

Contribution of Robotics in Medical Applications A Literary Survey

Abdul Hadi Himel, Farjana Abedin Boby, Sabrina Saba, Tajim Md. Niamat Ullah Akhund, and K. M. Akkas Ali

Abstract This paper provides an overview of robotics in medical science. After discussing the basic medical robot's idea, the future of medical robotics and the types of medical application are also discussed. One of the ambitions of this work is to deeply discuss the challenges and difficulties of medical robots. This paper presents year-wise discussion of medical robots which provides a throughout idea about what is happening in medical robots from the last 30 years to the next 30 years.

Keywords Robotics · IoT · Robotic surgery · Automated medical system

1 Introduction

In this day and age, the medical framework is very best in class, where medical robots are notable for their jobs in a medical procedure, explicitly the utilization of robots, PCs, and programming to precisely control careful instruments through at least one little entry points for different surgeries. Late many years have seen a perceptible improvement in data and correspondence innovation (ICT) [1]. This advancement has prompted coming of different sorts of robots in dominant part of enterprises, to be specific assembling, military, and medical services, amusement, and family. In the medical area, assistive medical robots and gadgets assume generous part in senior residents' lives [2]. The feeling of inundation that the robot gives implies that the specialist is more engaged, bringing about improved dynamic and patient

A. H. Himel (✉) · F. A. Boby · S. Saba · T. Md. N. U. Akhund
Daffodil International University, Dhaka, Bangladesh
e-mail: hemal15-2080@diu.edu.bd

F. A. Boby
e-mail: farjana15-2470@diu.edu.bd

S. Saba
e-mail: sabrina15-2342@diu.edu.bd

K. M. A. Ali
Institute of Information Technology, Jahangirnagar University, Dhaka, Bangladesh
e-mail: akkas@juniv.edu

© The Author(s), under exclusive license to Springer Nature Singapore Pte Ltd. 2022
A. K. Nagar et al. (eds.), *Intelligent Sustainable Systems*, Lecture Notes in Networks and Systems 333, https://doi.org/10.1007/978-981-16-6309-3_25

results. The ergonomics of the robot imply that the specialist is less focused and tired, bringing about better dynamic and diminished change to open a medical procedure. This would help effectiveness an extraordinary arrangement by killing tedious home visits [3]. The most inescapable careful robot, Intuitive careful da Vinci framework, has been examined in more than 4,000 friend investigated distributions, was cleared by the United States' Food and Drug Administration (FDA) for numerous classes of tasks, and was utilized in 80% of extremist prostatectomies acted in the USA for 2008, only nine years after the framework went available [4]. New uses for medical robots are made routinely, as in the underlying phases of any innovation driven transformation [5]. Medical robots have advanced from self-sufficient frameworks to tele-worked stages and precisely grounded, helpfully controlled robots [6].

2 Background Study

An assortment of characterizations for various sorts of robots helps to depict these heterogeneous gadgets. Mechanical innovation is discovering its way into different surgeries, both reconsidering the manner in which current systems are executed and empowering new methodology [7]. Robots can be described as computerized arms, cell phones, plants, or tele-mechanical gadgets. Moreover, they can be dynamic, semi-dynamic, or aloof. Dynamic gadgets are absolutely programmable and complete errands autonomously. One can envision a doctor entering three-dimensional (3D) registered tomography information into a PC and afterward programming the PC to guide a plant to eliminate specific regions of bone. The majority of the negligibly intrusive assignments by and large and thoracic medical procedure show similar fundamental qualities, which incorporate restricted admittance to the working volume inside the body, frequently moved toward utilizing little entry points through the stomach or thoracic dividers. These cuts go about as rotate focuses for careful devices, yet additionally confine the developments of devices inside the body. This segment will outline the best in class in automated frameworks intended to beat these constraints. The creators express that this improved model expanded turning capacity and gave higher foothold during headway, close to the initial phase in planning a remote portable robot for heart mediations [8], including, yet not restricted to, laparoscopic medical procedure in instinctive medical procedure, genecology, MI heart medical procedure, and urologic medical procedure, for example, laparoscopic prostatectomy, revolutionary prostatectomy, and extremist cystectomy [9]. IoT based robotic systems are making change in the field of EVM [10], EVS [11], neuropatient management [12], COVID-19 affected people management [13], hotel waiter systems [14], and virus-affected patient management [15]. Robotics can also make change in farming [16], nursing in hospitals [17], and remote sensing. An overview of modern surgical robots applying the SEBMA acronym is shown in Fig. 1.

Robots are wherever from sci-fi to your neighborhood emergency clinic, where they are evolving medical care. Generally, these robots take after R2D2 from Star Wars more than they do a humanoid, yet they are having a major effect on the

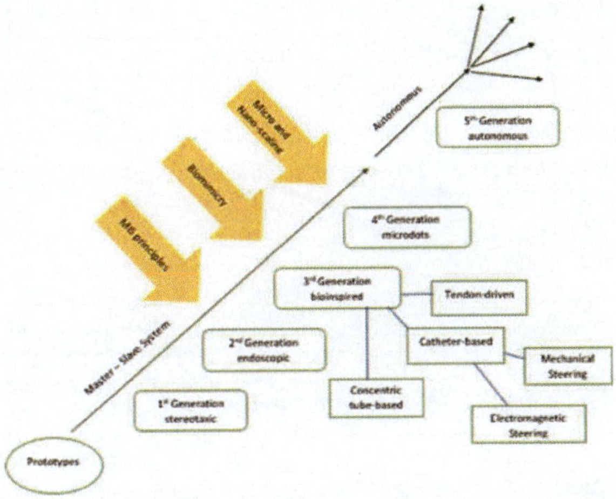

Fig. 1 Evolution of five generations of surgical robotics

field of medication [18]. The strategy is known as "container endoscopy" got FDA endorsement year's prior, yet current innovation may at long last convey on the entirety of its guarantees.

3 Robotics in Medical Science: Past, Present, and Future

Robots for clinical software has been at first gotten from modern robots Job based classifications can be more helpful in light of the fact that they are sweeping and address innovation engineers just as end-clients [18]. There are some role like:

1. Passive role: The job of the robot is restricted in extension or its contribution is generally okay.
2. Restricted role: The robot is answerable for more obtrusive assignments with higher danger, and however, is as yet confined from fundamental bits of the technique.
3. Active role: The robot is personally engaged with the technique and conveys high obligation and danger.

The initially recorded mechanical surgery—a CT-guided mind biopsy—occurred on April 11, 1985, at the Memorial Medical Center, Long Beach, CA, USA [19]. Figure 2 shows the surgical and oncology overview.

Mechanical innovation plans to improve the urologist's careful results by adjusting his/her own human specialized deficiencies; the incorporation of present-day PCs and computerized imaging will just take these advances further [20]. Contemporary robots are right now being utilized to do the positions which were risky, filthy, or exhausting. All that we have imagined about robots 10 years prior has gradually

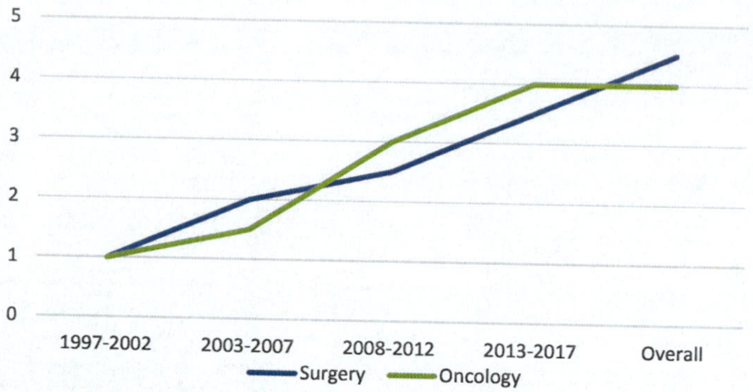

Fig. 2 Surgical and oncology overview

Table 1 Past, present, and future of robotics in medical science

Year	Description
1985	First robotic surgery performed
1988	Transurethral resection performed by PROBOT
1992	ROBODOC performed total hip arthroplasty
1993	First performed AESOP
1995	Intuitive surgical is founded
1997	First laparoscopy performed using a robotic system
2000	Da Vinci and ZEUS surgical systems approved by FDA
2001	First transatlantic robot-assisted procedure
2003	Computer motion and intuitive surgical merge into one company ZEUS is disconnected
2008	Sensei & Spine assist received FDA approval for intravascular and spinal procedures respectively
2012	23,000 hip and knee procedures conducted with MAKO globally
2013	Over 500,000 surgical procedures performed worldwide with the Vinci
2016	Over half of prostatectomies are performed robotically in the USA
2019	Over 5,000 da Vinci surgical systems installed worldwide 6 million surgeries globally

become a reality [21]. The past, present, and future of robotics in medical science is discussed in Table 1.

4 Robotics Application

Robots can be arranged utilizing a few measures like the application territory, control methods, actuators, mathematical setup, shrewd robots, and others [22].

4.1 Medical Transportation Robots

Supplies, drugs, and dinners are conveyed to patients and staff by these robots, subsequently upgrading correspondence between specialists, medical clinic staff individuals, and patients [23]. There is, in any case, a requirement for profoundly progressed and practical indoor route frameworks dependent on sensor combination area innovation to make the navigational capacities of transportation robots heartier [24].

4.2 Sanitation and Disinfection Robots

With the expansion in anti-microbial safe microorganisms and episodes of destructive contaminations like Ebola, more medical services offices are utilizing robots to clean and sanitize surfaces. "Presently, the essential techniques utilized for sanitization are UV light and hydrogen peroxide fumes," says Sahi. These robots can clean a room of any microorganisms and infections in no time [25].

4.3 Robotic Prescription Dispensing Systems

The greatest favorable circumstances of robots are speed and exactness, two highlights that are vital to drug stores. "Robotized administering frameworks have progressed to where robots would now be able to deal with powder, fluids, and exceptionally gooey materials, with a lot higher speed and exactness than previously," says Sahi [26].

4.4 Cardiac Devices

When a patient is determined to have cardiovascular breakdown (HF), the heart cannot siphon adequate blood to keep up tissue perfusion [27].

4.5 Soft Grippers

Another concentration in the advancement of delicate careful applications, particularly for negligibly obtrusive medical procedure, is delicate careful graspers. Sensitive withdrawal and control are regularly basic goals in effective surgeries; however, metallic forceps have not exactly ideal pressing factor conveyance [6].

4.6 *Wearable Soft Robots*

Dynamic versatility can be accomplished either by utilizing consistence and impedance control on robots that depend on unbending connections or by utilizing delicate robots made of materials that have inborn versatile qualities [28]. Biomimetic approaches, for example, muscle-like dynamic advancements, including electro-active polymers (EAPs), SMAs, and FFAs, have been utilized with the point of abusing their natural variable solidness.

5 Future of Robotics in Medical Science

Early usage of medical mechanical technology was troublesome in light of the fact that specialists expected to have an exact specification of the errand [29]. In this manner, engineers needed to visit the working room and construe the estimations of actual boundaries they thought suitable to a technique [30]. Medical advanced mechanics moderately effectively in the research facility by methods for modern frameworks, medical application is a lot of really requesting [31]. The robot was planned with a unique reason structure that compelled slices to the ideal district [32]. In spite of the fact that specialists had believed that this element was attractive, their disquiet with being eyewitnesses of a system that was generally in the control of the robot developer before long became apparent [33]. The monetary favorable circumstances, expanded accuracy, and improved nature of item relentlessly exhibited by mechanical robots invigorated the utilization of robots in the wellbeing area [34]. The joining of automated innovation in the neurosurgical working room is by all accounts inevitable, as careful advantages have been evaluated for ongoing robot-helped intercessions [35]. The recently depicted AI developments in medical care are mechanically progressive. This insurgency will advance in gradual advances [36]. For instance, a calculation will group their tumor at a granularity far better than us as of now unrefined TNM organizing framework to make an individualized treatment plan [37]. It will probably begin with basic direction, for instance, to improve laparoscopic port position or help connect preoperative imaging [38]. It will likewise offer a "telephone a-companion" usefulness to associate with specialists for help. With proceeded with advancement, this innovation will eventually form into an intraoperative "GPS," managing a specialist through an activity bit by bit [39].

The current market share of robotics is illustrated in Fig. 3. Since these innovations will join information from across emergency clinics and even nations [40]. Other little assignments may incorporate fascial conclusion or anastomoses. Effectively, the smart tissue [41]. In the more inaccessible future, these steady self-sufficient advances will join until maybe completely self-governing a medical procedure is figured it out. The expression "Working Room of the Future" (ORF) first showed up in PubMed in a 1992 by Jolesz and Shtern [42].

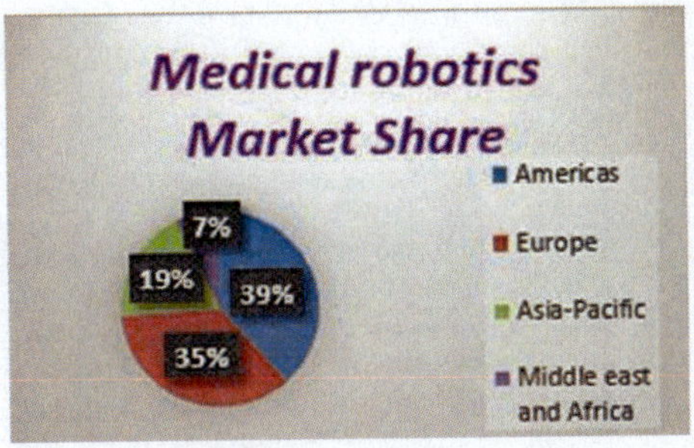

Fig. 3 Medical robotics market share

6 Conclusion

This work represented the history of robotics in medical sciences. This study focused on robotics applications in medical fields, IoT based robotic systems in medical applications, robotic surgery, Automated medical system, and so on. This work presented a literary survey of robotics applications related with medical sciences in a structured way from past to future world. In some extent, this work is limited because lack of data was found. But in future, this work will lead to new inventions in IoT based robots in medical field.

References

1. Goher, K.M., Mansouri, N., Fadlallah, S.O.: Assessment of personal care and medical robots from older adults' perspective. Robot. Biomimetics **4**(1), 1–7 (2017)
2. Beasley, R.A.: Medical robots: current systems and research directions. J. Robot (2012)
3. Payne, C.J., Yang, G.Z.: Hand-held medical robots. Ann. Biomed. Eng. **42**(8), 1594–1605 (2014)
4. Okamura, A.M., Matarić, M.J., Christensen, H.I.: Medical and health-care robotics. IEEE Robot. Autom. Mag. **17**(3), 26–37 (2010)
5. Taylor, R.H., Menciassi, A., Fichtinger, G., Fiorini, P., Dario, P.: Medical Robotics and Computer-Integrated Surgery. Springer Handbook of Robotics, pp.1657–1684 (2016)
6. Horvath, M.A., Wamala, I., Rytkin, E., et al.: An intracardiac soft robotic device for augmentation of blood ejection from the failing right ventricle. Ann. Biomed. Eng. **45**(9), 2222–2233 (2017)
7. Villoslada, A., Flores, A., Copaci, D., Blanco, D., Moreno, L.: High-displacement flexible shape memory alloy actuator for soft wearable robots. Rob. Auton. Syst. **73**, 91–101 (2015)
8. Copaci, D., Cano, E., Moreno, L., Blanco, D.: New design of a soft robotics wearable elbow exoskeleton based on shape memory alloy wire actuators. Appl. Bion. Biomech. **2017**, 1605101 (2017)
9. Galiana, I., Hammond, F.L., Howe, R.D., Popovic, M.B.: 2012 IEEE/RSJ International Conference on Intelligent Robots and Systems, Vilamoura, Portugal, pp. 317–322 (2012)
10. Akhund, T.M.N.U., Mahi, M.J.N., Tanvir, A.H., Mahmud, M., Kaiser, M.S.: ADEPTNESS: Alzheimer's disease patient management system using pervasive sensors-early prototype and preliminary results. In International Conference on Brain Informatics, December 2018, pp. 413–422. Springer, Cham (2018)

11. Akhund, T.M.N.U., Jyoty, W.B., Siddik, M.A.B., Newaz, N.T., Al Wahid, S.A., Sarker, M.M.: IoT based low-cost robotic agent design for disabled and Covid-19 virus affected people. In: 2020 Fourth World Conference on Smart Trends in Systems, Security and Sustainability (WorldS4), July 2020, pp. 23–26. IEEE (2020)
12. Akhund, T.M.N.U., Siddik, M.A.B., Hossain, M.R., Rahman, M.M., Newaz, N.T., Saifuzzaman, M.: IoT waiter bot: a low Cost IoT based multi functioned robot for restaurants. In: 2020 8th International Conference on Reliability, Infocom Technologies and Optimization (Trends and Future Directions) (ICRITO), June 2020, pp. 1174–1178. IEEE (2020)
13. Akhund, T.M.N.U., Newaz, N.T., Hossain, M.R.: Low-Cost Remote Sensing IoT based Smartphone Controlled Robot for Virus Affected People (2020)
14. Akhund, T.M.N.U., Snigdha, S.R., Reza, M.S., Newaz, N.T., Saifuzzaman, M., Rashel, M.R.: Self-powered IoT-based design for multi-purpose smart poultry farm. In: International Conference on Information and Communication Technology for Intelligent Systems, May 2020, pp. 43–51. Springer, Singapore (2020)
15. Akhund, T.M.N.U.: Study and Implementation of Multi-Purpose IoT Nurse-BoT (2011)
16. Akhund, T.M.N.U., Sagar, I.A., Sarker, M.M.: Remote Temperature Sensing Line Following Robot with Bluetooth Data Sending Capability (2011)
17. Akhund, T.M.N.U.: Remote sensing IoT based android controlled robot. Methodology **9**(11) (2018)
18. Badaan, S.R., Stoianovici, D.: Robotic systems: past, present, and future. In Robotics in Genitourinary Surgery, pp. 655–665. Springer, London (2011)
19. Gomes, P.: Surgical robotics: reviewing the past, analysing the present, imagining the future. Robot. Comput.-Integr. Manuf. **27**(2), 261–266 (2011)
20. Zemmar, A., Lozano, A.M., Nelson, B.J.: The rise of robots in surgical environments during COVID-19. Nat. Mach. Intell. **2**(10), 566–572 (2020)
21. Bai, L., Yang, J., Chen, X., Sun, Y., Li, X.: Medical robotics in bone fracture reduction surgery: a review. Sensors **19**(16), 3593 (2019)
22. Hauser, K., Shaw, R.: (2020) How medical robots will help treat patients in future outbreaks. IEEE Spectrum
23. Almurib, H.A.F., Al-Qrimli, H.F., Kumar, N.: A review of application industrial robotic design. In: Ninth International Conference on ICT and Knowledge Engineering, pp.105–112 (2011)
24. Olanrewaju, O.A., Faieza, A.A., Syakirah, K.: Current trend of robotics application in medical. In: IOP Conference Series: Materials Science and Engineering, vol. 46(1), p. 012041. IOP Publishing (2013)
25. Veerabhadram, P.: Applications of robotics in medicine. Int. J. Sci. Eng. Res. **2**(8) (2011)
26. Boone, A.C., Gregory, S.D., Wu, E.L., et al.: Evaluation of an intraventricular balloon pump for short term support of patients with heart failure. Artif. Organs. 10 (2019). https://doi.org/10.1111/aor.13454
27. Wamala, I., Roche, E.T., Pigula, F.A.: The use of soft robotics in cardiovascular therapy. Expert Rev Cardiovasc Ther. **15**(10), 767–774 (2017)
28. Gafford, J., Ranzani, T., Russo, S., et al.: Snap-on robotic wrist module for enhanced dexterity in endoscopic surgery. In: 2016 IEEE International Conference on Robotics and Automation (ICRA). IEEE (2016)
29. Ranzani, T., Cianchetti, M., Gerboni, G., et al.: A soft modular manipulator for minimally invasive surgery: design and characterization of a single module. IEEE Trans. Robot. **32**(1), 187–200 (2016)
30. Wang, H., Zhang, R., Chen, W., et al.: A cable-driven soft robot surgical system for cardiothoracic endoscopic surgery: preclinical tests in animals. Surg. Endosc. **31**(8), 3152–3158 (2017)
31. Gafford, J., Ding, Y., Harris, A., et al.: Shape deposition manufacturing of a soft, atraumatic, and deployable surgical grasper. J. Mech. Robot. **7**(2), 1–11 (2015)
32. Shintake, J., Sonar, H., Piskarev, E., et al. Soft Pneumatic Gelatin Actuator for Edible Robotics. arXiv preprint arXiv:170301423 (2017)

33. Ohta, P., Valle, L., King, J., et al.: Design of a lightweight soft robotic arm using pneumatic artificial muscles and inflatable sleeves. Soft Rob. **5**(2), 204–215 (2018)
34. Yang, D., Mosadegh, B., Ainla, A., et al.: Buckling of elastomeric beams enables actuation of soft machines. Adv. Mater. **27**(41), 6323–6327 (2015). (Advanced Robotics 13)
35. Lin, H.-T., Leisk, G.G., Trimmer, B.: GoQBot: a caterpillar inspired soft-bodied rolling robot. Bioinspir. Biomim. **6**(2), 1–14 (2011)
36. Anderson, I.A., Gisby, T.A., McKay, T.G., et al.: Multifunctional dielectric elastomer artificial muscles for soft and smart machines. J. Appl. Phys. **112**(4), 1–20 (2012)
37. Ji, X., El Haitami, A., Sorba, F., et al.: Stretchable composite monolayer electrodes for low voltage dielectric elastomer actuators. Sens. Actuators B. **261**, 135–143 (2018)
38. Krishnan, G., Bishop-Moser, J., Kim, C., et al.: Kinematics of a generalized class of pneumatic artificial muscles. J. Mech. Robot. **7**(4), 1–9 (2015)
39. Bishop-Moser, J., Kota, S.: Design and modeling of generalized fiber-reinforced pneumatic soft actuators. IEEE Trans. Robot. **31**(3), 536–545 (2015)
40. Miriyev, A., Stack, K., Lipson, H.: Soft material for soft actuators
41. Shepherd, R.F., Stokes, A.A., Freake, J., et al.: Using explosions to power a soft robot. Angew. Chem. **125**(10), 2964–2968 (2013)
42. Sarker, M.M., Akhund, T.M.N.U.: The roadmap to the electronic voting system development: a literature review. Int. J. Adv. Eng. Manage. Sci. **2**(5), 239465 (2016)

Development of Virtual Reality Training System Based on EEG Biofeedback

Pei-Jung Lin and Man-Lai Lam

Abstract This study integrates virtual reality (VR) and electroencephalography technology in the design of a virtual space for multi-user rehabilitation. It allows therapists to conduct remote care and provides easy-to-use functions for stroke patients during their rehabilitation process. To analyse patient rehabilitation data, Azure ML Studio, a cloud service for big data prediction and analysis was employed to evaluate and validate the model. The aim of this study is to provide appropriate rehabilitation procedures and plans through a personalised numerical analysis and improve the immediacy of the clinical evaluation of physical therapy. This study covers rehabilitation mechanisms ranging from individual home-based rehabilitation to remote multi-user virtual network rehabilitation. A brain-computer interface is integrated into the design of the VR rehabilitation function and introduces a personalised music recommendation mechanism, allowing patients to adjust their rehabilitation plans according to their physical movements and specific changes in brainwaves during the rehabilitation process. The system can be used to provide healthcare services for seniors in their communities. With the global spread of COVID-19, many medical institutions in Taiwan and other countries have established VR medical simulation centres in the post-pandemic era to provide educational training for medical staff. By investing artificial intelligence (AI) and VR technology, the remote care of smart hospitals can result in innovative processes and opportunities.

Keywords Virtual reality · Stroke · Rehabilitation · Mirror therapy · Music therapy · EEG biofeedback

1 Introduction

With the vigorous development of 5G applications, the market of extended reality (XR) in the health management and medical field will increase substantially. Major

---

P.-J. Lin (✉) · M.-L. Lam
Department of Computer Science and Information Engineering, Hungkuang University, Taichung, Taiwan
e-mail: pjlin@hk.edu.tw

© The Author(s), under exclusive license to Springer Nature Singapore Pte Ltd. 2022

A. K. Nagar et al. (eds.), *Intelligent Sustainable Systems*, Lecture Notes in Networks and Systems 333, https://doi.org/10.1007/978-981-16-6309-3_26

global manufacturers are making every effort to seize business opportunities. Considering that social health care and smart medical services have been emerging trends in recent years, many countries have invested many human and material resources into hardware and software research and development for health promotion and care. Globally, XR technology has been gradually applied within the digital medical field to provide visual-motor learning, induce specific action responses in a real environment, and provide focussed interaction and gesture-based interaction [1]. The market predicts that, by 2023, the compound annual growth rate of VR applications in the field of health care will reach 54.5%. VR is widely used in the medical field. It can transfer severe pain in patients and functions similar to hospital-prescribed opioid analgesics, which are used to affect the pain transmission and block pain reception in the brain. Scholars have studied how to rewrite the relationship between the patient's brain and motor nervous system through virtual images created in VR [2]. In addition, there are also studies analysing how to use feedback training to improve the possibility of improving the physical or cognitive performance. The so-called neurofeedback or electroencephalogram (EEG) feedback [3] refers to the recording of an EEG while providing instant auditory and visual feedback, or both, to better understand specific cortical arousal and induce active state control. The brain-computer interface (BCI) has currently been extended and applied in many in areas of research, such as signal analysis, instrument manipulation and games. Scholars are also interested in integrating EEG and music therapy to assess the emotional response of patients with advanced cancer. Experiments have found that such an integration has a positive emotional effect on patients with advanced cancer, reducing their fatigue and anxiety, enhancing their well-being, and helping to improve their emotion [4]. VR training integrating EEG feedback can provide operators with feedback when completing a task action, helping to improve the training effectiveness. According to research findings, adding brainwave neurofeedback to VR devices can help stroke patients adapt to their actual situation through a motion simulation and improve the effectiveness of the rehabilitation [5]. In recent years, with the rapid and diversified development of VR technology, medical equipment using VR has begun to appear for rehabilitation [6]. This study uses Photon Unity Networking (PUN) to construct a dedicated multi-user rehabilitation platform, allowing users to perform remote multi-user rehabilitation with doctors through the Internet. Unity 3D was used to provide connection functions and environments to create VR equipment tracking the effectiveness of the rehabilitation.

2 Related Work

Many chronic and stroke patients need to face a long rehabilitation process. Therefore, a well-developed home rehabilitation system can save patients from travelling to hospitals for training and reduce the potential risk of going to medical institutions during the pandemic. VR technology creates the possibility of a common experience for users in different locations. By detecting voice, hands and head positions, the users' eyes, body and emotion are transmitted to create a more natural community interaction, allowing users to feel the real presence of others. VR has a stronger sense of social presence, and virtual objects and environments facilitate more interactions. In recent years, VR has applied computer and multimedia technology to design a realistic rehabilitation sports training environment, providing stroke patients with a gradual and personalised treatment for their upper limbs [7]. An increasing number of researchers are developing a rehabilitation gaming system (RGS) to understand the effects of acute and chronic strokes on patient rehabilitation and daily management. It allows users to receive immediate feedback on their motor performance in an environment with multiple sensory stimulations, which significantly increases the usage motivation and opportunities for practice [8]. Research teams from North Carolina State University and the University of Chicago have jointly created an online VR clinic platform called virtual environment for rehabilitative gaming exercises (VERGE), allowing patients to interact with other patients and therapists on the Internet at home for physical and occupational therapy. Research has confirmed that VR combined with multi-user mode can indeed encourage patients to receive treatment and achieve good results [9].

Under the current influence of the continuously rising medical costs, the perspectives of rehabilitation, care and prevention have shown different changes. With the advancement of information and communication technology, the health industries have extended their services from mainly disease diagnosis and treatment in the past to mobilised smart home health management. The treatment applies science, theory and an activity analysis to understand the factors affecting the patients' performance and provide training for their physical, psychological and social functions. Environmental reconstruction, posture training, balance training and moving training were used to help patients achieve rehabilitation tasks and resume meaningful daily activities to ensure their quality of life. In addition, EEG biofeedback technology was added to the system to understand the users' brainwave activities. An EMOTIV Insight brainwave sensor was used in the system design to monitor different behaviours and facial expressions of the users [10]. To understand the effects of video and sound stimuli on EEG signals, some scholars have evaluated the emotional dimensions of the participants (valence-arousal-dominance) and successfully predicted the emotional adjustment methods with high accuracy [11]. A personalised music recommendation system has even been developed that analyses the brainwaves and sentiment of users (for example, the Spotify online music streaming service platform) [12], demonstrating the application of brainwave monitoring in many different fields. Because the brain innervates the human nerves throughout the body, EEG is currently widely used in the medical community to diagnose certain neurological disorders. Clinically, it can be used to diagnose lesions such as epilepsy, infectious diseases, tumours, trauma and hematomas. An EEG was used to record the electric field generated by nerve cell activities to investigate the recovery mechanism of treatment after the acute stroke phase [13]. An EMOTIV Insight is a brainwave sensing device based on BCI technology that can measure the pattern and frequency of bioelectric signals and collect brain analogue electrical signals [14]. Through integration with brainwave technology, it is used to monitor the changes in activity in the left and right hemispheres of the brain of the participant during the rehabilitation phase to guide the user whenever appropriate. Through a dynamic analysis, the device provides customised training procedures of rehabilitation tasks and integrates and displays such information in the VR task screen in real-time. It also allows telemedical staff to adjust the rehabilitation movements and frequency in a timely manner, thereby enhancing the effectiveness of the rehabilitation [15].

3 Introduction to System Operation Method and Function

The system supports single and multiple connection rehabilitation modes and provides rehabilitation guidance and explanations during the training. Patients can also perform remote multi-user connection with the therapist and follow the physiatrist's guidance on their postures and movements to improve the accuracy. Multi-user training includes the reaction capability, memory and sense of equilibrium. After completing the rehabilitation training, the results will be recorded in the cloud to allow medical institutions and patients to track the training progress.

3.1 Brainwave Signal Detection

To analyse the brainwave status and changes to the patients during the rehabilitation process, this study employed an EMOTIV Insight to detect the brainwaves of users. The positions of the extracranial electrode provided by the device are shown in Fig. 1a, which include AF3, AF4, T7, T8 and Pz. The device has two reference sensors that cover the critical areas of the cerebral cortex. Figure 1b shows the international 10–20 EEG system. The red electrode positions indicate the five sensing positions of the electrodes, which can read EEG data. AF3 and AF4 are electrodes for the frontal lobe, which are mainly responsible for planning, organising, problem-solving, selective attention, personality and behaviours and emotions related to high-level cognitive functions. Therefore, when users think about moving their upper

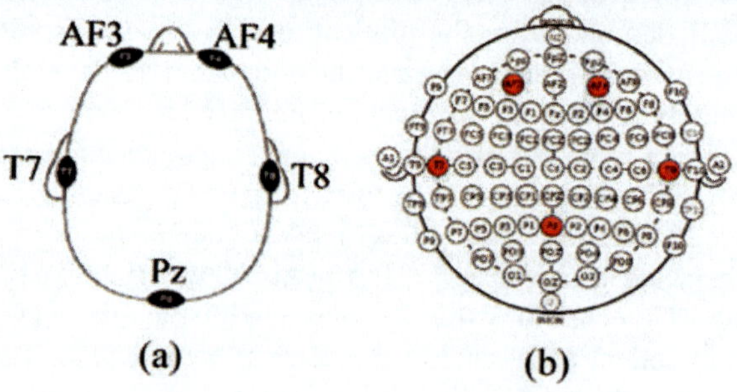

Fig. 1 **a** EMOTIV insight electrode placement. **b** International 10–20 electrode placement

limbs during rehabilitation, there will be brainwave signals in this region. T7 and T8 are responsible for the electrodes of the temporal lobe. Located near the ears, they provide hearing, smell and the ability to arrange new information. To understand the brainwave status of stroke patients when working on rehabilitation programmes, the VR rehabilitation system was integrated to analyse whether the user understands the operations of spatial orientation, sense of balance and rehabilitation treatment in the virtual environment of the constructed images and to confirm whether there are any obvious changes in the brainwave signal.

3.2 VR Rehabilitation Integrated with BCI Design

Figure 2 illustrates the integration of VR and an EEG. The figure shows VR mirror therapy. A canvas object was embedded in the upper left of the screen, and the render mode was set as an overlay to display an overview of the EEG signal on the screen in real-time. The sampling rate of the EEG sensor in this study was 64 Hz, with 128 sampling points applied each time, in 1-s units. A fast Fourier transform (FFT) was used to transform the collected brain raw signals from the time domain into the frequency domain. The amplitude of the corresponding frequency band of each brainwave band was then squared and added as reference for recommending and adjusting the personalised rehabilitation treatment.

After the data were transformed, a waveform calibration with a Hanning Window Function was required to reduce the spectrum leakage produced by a periodic signal truncation. After a waveform calibration, brainwaves such as theta, alpha, low beta, high beta and gamma were extracted using band-pass filter according to their frequencies. The wave data were then transferred to the system. The arousal–valence module is then used to complete the valence and arousal coordinates, as shown in Fig. 3 Based on the affective computing theory, the arousal–valence module can analyse the positive/negative affect of the patients as well as their arousal level.

Based on AF3 and AF4 in the frontal lobe, and the βH (high range) and βL (low range) beta waves divided by electrodes, the values of the arousal and valence

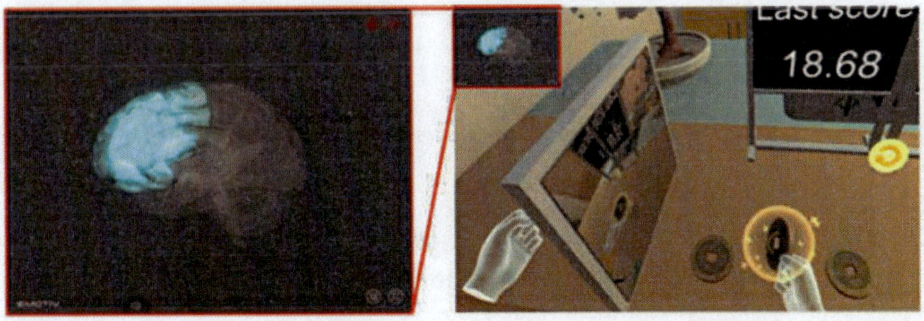

Fig. 2 Integration of VR and EEG

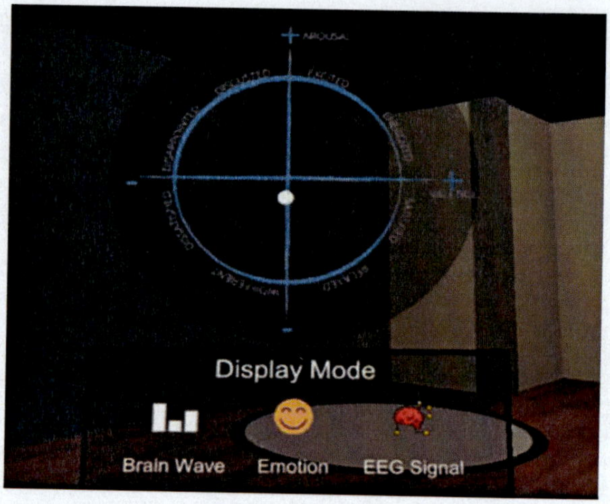

Fig. 3 Two-dimensional valence–arousal space

were calculated. The arousal can be calculated using Eq. (1), and the valence can be calculated through Eq. (2). The valence and arousal levels were calculated using the raw data obtained from the patients' EEG headset and then applied Cortex API to output data such as αF4 and αF3.

$$\frac{(\beta F3 + \beta F4 + \beta AF3 + \beta AF4)}{(\alpha\ F3 + \alpha F4 + \alpha AF3 + \alpha AF4)} \tag{1}$$

$$\alpha F4 - \alpha F3 \tag{2}$$

Next, the inverse trigonometric function (Arctan) was applied to convert the angle and regard each 45° area in the 2D arousal–valence module as an emotion space to show a patient's emotional trend. Then Pearson's product-moment correlation and standard deviation were then used to set the 2D boundary between −1 and 1. The brainwave data of the patients were analysed to see whether there was a significant difference in the brainwave data and determine the difference in the degree of change of the brainwave of the patients during the test, training, rehabilitation and resting stages.

3.3 *Music Elements to Assess Brainwave Activities*

Based on the brainwave feedback, the brainwave activities were measured to judge the effects of music on the brainwave activation of the users. The system established a customised music list for users to help them create a better rehabilitation scenario in a music-enhanced environment. The system architecture mainly consists of three parts: brainwave parameter extraction, a personalised brainwave assisted measurement system, and a database. The descriptions for each module of the music recommendation system architecture are as follows: (a) A brainwave parameter extraction is applied in which the brainwave signals are transferred to the system through a receiver to establish a default music list and then recommend a more suitable music list using the reasoning module. (b) A personalised assistive measurement system includes is used, which features a transformation module and a detection module, of which the detection module comprises a reasoning module and music list recommendation module. After extracting the brainwave signals, the feature transformation module filters out the noise of the brainwave signals and transforms the time domain signals into the frequency domain signal through a Fourier transform, and characteristic frequency bands (α and β waves) of the brainwaves are finally extracted. After transforming the brainwave signals, the detection and reminding module analyse their activation grades. The reasoning module obtains the initial and activated values to determine whether music affects the users' rehabilitation. The music list will adjust its music weight according to the values calculated by the reasoning module. When the value of the α wave increases, the weight increases, and when the β wave increases, the weight decreases. When the weight value of the music is below the threshold, the music will be removed from the list. The database includes a training module and a music list storage module. The training module processes and transforms the brainwave signals into characteristic frequency bands and then transforms them into monitorable values according to the energy characteristics of different bands. The music list storage module stores the music list established by the system, the music weight calculated by the personalised assisted measurement module and the users' brainwave information.

4 Conclusion

This study used VR to implement rehabilitation therapy for patients. A stroke may lead to physical impairment of the patients and affect their behaviours or cognitive abilities to a certain extent. This system mainly guides patients to apply the rehabilitation movements of occupational therapy and assist with their training. Under the current severe conditions of the COVID pandemic, many hospitals are facing a capacity shortage. Chronic disease patients need remote care resources to avoid the potential risks caused by travelling to the hospital for rehabilitation. The results of this study will allow patients to rehabilitate in communities or at home. Patients can conduct their rehabilitation according to the instructions of the system or a remote therapist. Through cross-disciplinary cooperation, we understand the dilemma faced by physical therapists in assisting stroke patients under actual conditions. In the case of fewer medical resources or social distancing during the pandemic, the system can help patients learn rehabilitation autonomously and cooperate with professional physical therapists. The system also allows patients to make remote connections and support their communications with physiatrists with greater convenience. In the future, we aim to develop more accurately movement trajectories of limbs and skeletal. In order to achieve the goal, using mixed reality technology can improve the timeliness of clinical evaluation of physical therapy. The result of this research is expected to introduce into the medical care industry for commercialization and provide value-added health care services.

Acknowledgements The authors would like to thank the Ministry of Science and Technology (MOST) of Taiwan for financially supporting this research project under grant numbers MOST 109-2221-E-241-006. The authors would also like to thank Prof. Han Yu Chen, Mr. Shing Fai Steven Lam, Yu-Ming Huang, and Cho-Jung Chang, for their valuable comments, which helped improve the quality of this work.

References

1. Zweifach, S.M., Triola, M.M.: Extended reality in medical education: driving adoption through provider-centered design. Digit. Biomarkers **3**(1), 14–21 (2019)
2. Seyed Esfahani, M., Heydari Khajehpour, S., Manku, T.: Virtual reality in healthcare, what stops hospitals and patients adopting the technology? Case study of National Health Service in Dorset. In: International Society of Professional Innovation Management Conference (2019)
3. Pinter, D., Kober, S.E., Fruhwirth, V., Berger, L., Damulina, A., Khalil, M., Enzinger, C.: MRI correlates of cognitive improvement after home-based EEG neurofeedback training in patients with multiple sclerosis: a pilot study. J. Neurol. 1–9 (2021)
4. Hunt, A.M., Fachner, J., Clark-Vetri, R., Raffa, R.B., Rupnow-Kidd, C., Maidhof, C., Dileo, C.: Neuronal effects of listening to entrainment music versus preferred music in patients with chronic cancer pain as measured via EEG and LORETA Imaging. Front. Psychol. **12**, 390 (2021)
5. Chang, W.C., Ko, L.W., Yu, K.H., Ho, Y.C., Chen, C.H., Jong, Y.J., Huang, Y.P.: EEG analysis of mixed-reality music rehabilitation system for post-stroke lower limb therapy. J. Soc. Inform. Display **27**(6), 372–380 (2019)
6. Li, J., Jin, Y., Lu, S., Wu, W., Wang, P.: Building environment information and human perceptual feedback collected through a combined virtual reality (VR) and electroencephalogram (EEG) method. Energy Build. **224**, 110259 (2020)
7. Camargo, A., Carmo, J.F., Rosa-Castro, R.M., Rodrigues, C., Mazzei, L.G., Scalha, T.B., Andrade, A.C.N.B.: The influence of virtual reality on rehabilitation of upper limbs and gait after stroke. J. Innov. Healthcare Manage. **2**, 1–16 (2019)
8. Kern, F., Winter, C., Gall, D., Käthner, I., Pauli, P., Latoschik, M.E.: Immersive virtual reality and gamification within procedurally generated environments to increase motivation during gait rehabilitation. In: 26th IEEE Conference on Virtual Reality and 3D User Interfaces on Proceedings, Osaka, Japan, pp. 500–509 (2019)
9. Pereira, F., i Badia, S.B., Jorge, C., Cameirão, M.S.: The use of game modes to promote engagement and social involvement in multi-user serious games: a within-person randomized trial with stroke survivors. J. NeuroEng. Rehabil. **18**(1), 1–15 (2021)
10. López-Hernández, J.L., González-Carrasco, I., López-Cuadrado, J.L., Ruiz-Mezcua, B.: Framework for the classification of emotions in people with visual disabilities through brain signals. Front. Neuroinf. 15 (2021)
11. Caldas, O.I., Aviles, O.F., Rodriguez-Guerrero, C.: Effects of presence and challenge variations on emotional engagement in immersive virtual environments. IEEE Trans. Neural Syst. Rehabil. Eng. **28**(5), 1109–1116 (2020)

12. Marimpis, A.D., Dimitriadis, S.I., Goebel, R.: A multiplex connectivity map of valence-arousal emotional model. IEEE Access **8**, 170928–170938 (2020)
13. Elmali, A.D., Ekizoglu, E., Ciftci, I., Yesilot, N., Coban, O., Baykan, B.: Periodic electroclinical seizures following an ischemic stroke revealed by continuous-EEG. Epilepsy & Behav. Rep. **15**, 100428 (2021)
14. Zabcikova, M.: Visual and auditory stimuli response, measured by Emotiv insight headset. In: MATEC Web of Conferences, vol. 292, pp. 01024. EDP Sciences (2019)
15. Kumar, J.L.M., Rashid, M., Musa, R.M., Razman, M.A.M., Sulaiman, N., Jailani, R., Majeed, A.P.A.: The classification of EEG-based wink signals: a CWT-transfer learning pipeline. ICT Express, 1–5 (2021)

Adapting Automatic Speech Recognition to the Radiology Domain for a Less-Resourced Language: The Case of Latvian

Normunds Gruzitis, Roberts Dargis, Viesturs Julijs Lasmanis, Ginta Garkaje, and Didzis Gosko

Abstract Automatic speech recognition (ASR) is becoming available to more and more languages. Training a quality ASR system typically requires significant amount of language resources—representative speech and text corpora. Recent advances in transfer learning and fine-tuning of transformer models show that state-of-the-art ASR models can be derived for low-resourced languages from neural ASR models pre-trained on high-resource language data. While it is relatively straightforward to implement such a scenario for a general-domain language, the language of the medical domain is very complex, which requires extra preprocessing steps to be taken before training or fine-tuning a language model. We particularly focus on radiology dictations. We make an indicative evaluation of our approach by using the standard metric of word error rate (WER); however, the actual evaluation in practice is most relevant in terms of time efficiency, which makes a crucial role for a tailored and convenient user interface for post-editing the automatic transcriptions.

Keywords Speech recognition · Language modeling · Medical transcription · Post-editing · Speech corpus · Text corpus · Grammar · Latvian

1 Introduction

We present an ongoing collaborative research and innovation project between a research group in language technology at the University of Latvia and a multi-field medical treatment institution in Latvia. Although state-of-the-art medical equipment is widely used in Latvia, especially in the field of radiology and diagnostic imaging, examination reports are still produced completely manually. The largest medical

N. Gruzitis (✉) · R. Dargis · V. J. Lasmanis · G. Garkaje
Institute of Mathematics and Computer Science, University of Latvia, Riga, Latvia
e-mail: normunds.gruzitis@lumii.lv

R. Dargis
e-mail: roberts.dargis@lumii.lv

D. Gosko
Riga East University Hospital, Riga, Latvia

© The Author(s), under exclusive license to Springer Nature Singapore Pte Ltd. 2022
A. K. Nagar et al. (eds.), *Intelligent Sustainable Systems*, Lecture Notes in Networks and Systems 333, https://doi.org/10.1007/978-981-16-6309-3_27

treatment institutions in Latvia outsource production of medical reports to transcriptionist centers, or maintain their own transcriptionist centers. However, there is a constant growth in the number of diagnostic examinations—clinicians and patients have to wait for the reports up to several days if they are not considered urgent. Services of transcriptionist centers are also expensive, and regional medical treatment institutions cannot afford it. A more efficient and sustainable approach and technical infrastructure are clearly needed.

The goal of this industry-driven research project is to create essential Latvian language resources and technology components for automatic speech recognition (ASR) in the medical domain, radiology in particular, and to showcase the use of these resources and technology by developing and validating an automated platform for dictation and transcription of radiology examination reports. The platform is being customized for two user scenarios. In the first scenario, we expect that for certain radiology subfields (e.g., X-ray and ultrasonography) that produce relatively simpler and more fluent dictations which result in more accurate ASR transcriptions (in comparison to computed tomography (CT) and magnetic resonance imaging (MRI) dictations), radiologists will use the dictation platform in a self-service manner, i.e., they will prefer to do quick post-editing of the automatic transcriptions rather than wait for a transcriptionist center to produce such reports. In the second scenario, we expect that services of a transcriptionist center would still be preferred by many radiologists in the case of CT, MRI and other more complicated examinations and therefore less fluent dictations which typically require more post-editing work. Transcriptionist centers, however, will become more productive as part of their workload will be transferred to the self-service user scenario, and draft examination transcripts (generated by the ASR platform) will be made available for post-editing instead of transcribing whole reports from scratch.

Like for many other languages, with the creation of a sufficiently large general-purpose speech corpus for Latvian [3], several general-domain ASR systems have been developed for Latvian [5, 9]. Such ASR systems, however, cannot be used in the the medical domain and radiology in particular since the medical language is completely different from the general language. Adapting ASR language models to the medical domain requires creation of certain language resources, and it also involves extra preprocessing steps to be taken before training or fine-tuning a language model, and extra post-processing steps after an automatic transcription have been generated from an input dictation. The required language-specific resources are:

- a text corpus of anonymized historical medical reports for language modeling—excerpts from a large archive of historical reports, primarily radiology examination reports, but we also include epicrisis reports;
- a pronunciation lexicon of terminology, abbreviations and acronyms, named entities (e.g. drug names) and other terms used in the radiology domain for their automatic recognition and consistent transcription;
- an orthographically transcribed and anonymized speech corpus for evaluating the adapted ASR system and for developing and testing grammar rules for the automatic pre-editing and post-editing;

- pre-editing and post-editing grammar rules (finite-state transducers) to automatically rewrite the text corpus of historical reports into a fully verbalized form (as if dictated and exactly transcribed without using numbers, abbreviations, etc.) and to automatically rewrite the ASR-generated transcriptions to the expected written form (using numbers, abbreviations, etc.).

Although adaption of the acoustic model to the radiology domain (by training it on the speech corpus of dictated radiology examinations) would slightly impact the accuracy of ASR transcripts [2], our hypothesis is that a domain-adapted language model and a domain-adapted pronunciation lexicon (both derived from the text corpus of historical radiology examination and epicrisis reports) will give the most significant improvement in the ASR accuracy.

Section 2 briefly describes our work on creation of the required text and speech corpora, pronunciation lexicon, and rewriting grammar rules. Section 3 describes experiments that we have conducted to test our hypothesis and the initial results that we have achieved. Section 4 highlights our work on an open-source editor for efficient post-editing of the automatic transcripts. Section 5 concludes the paper discussing current results and options for the next tasks.

2 Domain-Specific Language Resources

Creation of language-specific resources (datasets) usually takes the biggest portion of man-hours required to develop a natural language processing (NLP) system for a new language. There are two general approaches to dataset creation for machine learning: creating from scratch or bootstrapping from existing data. Creating from scratch allows to acquire cleaner and better structured data. Bootstrapping from existing data is usually quicker and gives more diversified data but also rises more challenges in data preparation. We focus on creation of language-specific datasets to develop a medical speech recognition system for Latvian using the bootstrapping approach. Required datasets are bootstrapped from an archive of radiology examination reports of Riga East University Hospital (REUH), which is the largest hospital in Latvia.

The current speech and text data flow at REUH is as follows. Audio files dictated by radiologists, clinicians, and their assistants are submitted to the transcriptionist center through multiple channels, including telephone call-ins, thus the quality of audio recordings vary a lot. The audio files are manually transcribed, creating Microsoft Word documents based on predefined document templates. The final documents are sent back to the doctors for their approval. Once approved, documents are added to the archive.

At the beginning, documents were stored in the DOC format. In 2010, switching to the new Microsoft Word Open XML document format (DOCX) was made at REUH. We use only documents in the open DOCX format to create a text corpus for language modeling, which still gives us a decade of historical reports.

Audio recordings older than 6 months are moved to a cold storage as there is no use for them in the everyday hospital operations. The old audio files are retrieved only in rare occasions if there is a dispute about some transcriptions. Due to the high volume of audio files being processed at the transcriptionist center every day, the audio data available in the hot storage was sufficient to create the required speech corpus.

Creation of the domain-specific datasets is briefly described in Sects. 2.1, 2.2, and 2.3; see [1] for more detail (incl. our data anonymization approach). Section 2.4 describes an additional language resource for ASR development, which is particularly important in the radiology domain.

2.1 *Text Corpus*

All archived reports in the DOCX format were used to produce the text corpus. In the last decade, more than 100k reports have been produced by REUH's transcriptionist center. The number of documents has recently reached 15k per year.

It is important to preserve the text segmentation in the text extraction process because language modeling is done at the sentence level. A custom tool was developed to extract plain text from the internal XML structure of Word documents.

When text segments are extracted, they have to be anonymized to avoid any sensitive personal data (forename, surname, identification number, address, etc.) leakage in the text corpus. Data anonymization is done at the whole-segment level because personal data privacy is more important than text preservation. Segments that possibly contain any potential personal data are automatically excluded from the corpus. The final version of the anonymized text corpus contains more than 1GB of text (more than 135M tokens).

2.2 *Speech Corpus*

A 30-hour in-domain speech corpus of audio dictations aligned with manually provided exact transcriptions was created in the project. The selected set of dictations was designed to be a representative 10% subset (w.r.t. the various acoustic conditions and speaker variation) of the approx. 300-hour collection of recordings available in the above-mentioned hot storage.

It should be noted that audio recordings were filtered by quality since there is no benefit to include recordings of low quality (highly compressed, low bitrate, or sample rate). Low-quality data can be generated from high-quality recordings. In the result, the corpus represents about 70 unique speakers.

To avoid overfitting of the acoustic model and to get better generalisation, approximately, equal number of audio recordings were selected for each speaker. Recording and acoustic conditions often differ from recording to recording for the same speaker; therefore, up to one recording per day was selected per speaker, and the time intervals between the selected recordings were spread out evenly.

Audio file anonymization is a significantly more challenging task than text anonymization. Using inaccurate ASR transcripts to detect segments with personal data would lead to incomplete anonymization. Since personal data should occur only at the beginning of a recording, a simpler but reliable approach was used instead. ASR result for each recording was automatically aligned to the corresponding report in the anonymized text corpus to find the part of the audio which corresponds to the main body of the report and the rest of the audio file was automatically trimmed.

2.3 Pronunciation Lexicon

The lexicon contains a list of words and word forms with pronunciation that the ASR system should recognize. We have derived the in-domain lexicon from the text corpus.

The text corpus was tokenized to find a list of unique words (including different word forms). Word pronunciation is generated using a rule-based system. In some cases, expanded forms or pronunciation hints are added manually beforehand. Words can be divided into four categories regarding the pronunciation generation:

- words which in Latvian are pronounced exactly as written (majority of Latvian words);
- words with typos, which should be excluded from the lexicon;
- words that first have to be expanded, such as abbreviations and symbols;
- words in foreign languages, like Latin terms and names of drugs, for which pronunciation hints have to be provided, using Latvian phonemes.

The text corpus contains 1.8M unique tokens of which 1.1M tokens contain at least one letter and, therefore, can be considered as words for the lexicon. Categorization of these words into the four groups manually would be impossible in terms of man-hours; therefore, multiple existing Latvian language resources were used to categorize most of the words automatically. Words that were not classified automatically and occur at least 1000 times in the text corpus (almost 14k words) were selected for manual categorization. This threshold was set empirically based on the word frequency distribution in the text corpus.

2.4 Rewriting Grammar

The role of rewriting rules is two-fold as illustrated in Fig. 1. First, they are applied to the text corpus of original reports to acquire a derived text corpus of expanded reports for language modeling. This is a text preprocessing task which is performed only during the development phase while trying to acquire the best language model for the medical domain. Second, the same rules are reversely applied to the automatic transcriptions to acquire concise reports and minimize the post-editing work. This is

1	CT krūškurvja	orgāniem	,
2	CT krūšu kurvja	orgāniem	komats
3	CT chest	organs	comma
4	CT krūšu kurvja	orgāniem	,

1	pēc	Ultravist	300	100	ml	i/v	ievades
2	pēc	Ultravist	trīssimt	simt	mililitru	intravenozas	ievades
3	after	Ultravist	three-hundred	hundred	mililiter	intravenous	injection
4	pēc	Ultravist	300	100	ml	i/v	ievades

1	nieres	konkrements	1,89				cm	/	ø
2	nieres	konkrements	viens	komats	astoņdesmit	deviņi	centimetri		diametrā
3	renal	concretion	one	point	eighty	nine	centimetres		diameter
4	nieres	konkrements	1,89				cm	/	ø

Fig. 1 Sample fragments from examination reports illustrating the role of a rewriting grammar. Lines: 1—an excerpt from the text corpus (original); 2—the corresponding expanded for language modeling (preprocessing); 3—a literal translation to English; 4—the expected output of an adapted ASR system (post-processing)

a text post-processing task which is performed by the production platform for each new ASR-generated transcription.

The rewriting grammar is implemented as a cascade of finite-state transducers using the OpenFST-based Thrax framework [8]. It consists of a few hundred pattern-matching rules, some of which are context-dependent. In the case of Latvian which is a highly inflected language, context-matching is required not only to resolve expansion of some ambiguous abbreviations but also to ensure that the expanded word forms are in syntactic agreement. The text preprocessing is therefore done in two sequential applications of finite-state transducers. In the first step, the content of the target token or text span is matched with a set of grammar patterns, and non-expanded tokens are rewritten into their full base forms. Context-dependent grammatical case, number and gender markers are attached to the expanded tokens, to be resolved in the next step. In the second step, a separate set of transducers conjugates all the expanded tokens (base forms with the attached grammatical markers), generating the necessary word forms and removing the grammatical markers.

It is important to note that the grammar rules (transducers) are defined in a corpus-driven manner—based on the alignment between the manual transcriptions in the speech corpus and the corresponding text segments of the original reports in the text corpus.

3 Domain-Adapted ASR System

We have trained the medical ASR system using the Kaldi framework [4]. The acoustic model has been trained on a general-domain Latvian speech corpus containing 100 h of broadcast recordings [3] augmented with various noisy recordings and musical recordings from the MUSAN corpus [6]. The TDNN+LSTM neural network is trained on 40-dimension FBANK vectors. Language models are trained using the SRILM toolkit [7]. Trigram language models pruned to 1e-8 are used in all experiments. A rule-based system is used to generate the pronunciation lexicon based on 52 phonemes.

3.1 Experiments

The impact of the different language models was measured both directly in terms of perplexity (PPL) over the test set transcriptions and indirectly as part of the modified ASR system in terms of word error rate (WER) over the test set.

The test dataset contains audio recordings and their transcriptions from the domain-specific speech corpus, which were not included in the language model training dataset. To evaluate the impact of the created domain-specific language resources on the overall performance, various experiments were conducted on the test dataset:

- First, a general language model trained on a subset of the CommonCrawl[1] archive was used as a baseline (CC).
- Second, an in-domain language model was trained on the created text corpus of medical reports without any preprocessing of the corpus (MED).
- Third, the first two language models were interpolated, calculating the best mix on a development dataset (MED+CC). The best mix was achieved at weights of 0.2 for CC and 0.8 for MED.
- Fourth, the rewriting grammar was used to preprocess the in-domain text corpus by contextually expanding abbreviations and regular expressions and by verbalizing numbers (MED_EXP).
- Fifth, the language model trained on the expanded text corpus was once again interpolated with the generic model (MED_EXP+CC). The best mix was achieved at weights of 0.19 for CC and 0.81 for MED_EXP, showing that the preprocessing step made the text corpus more similar to the desired result but not as much as expected.

3.2 Results

The results given in Table 1 show that an in-domain text corpus gives a major improvement compared to a generic language model. Although the in-domain text corpus was large, the interpolation with generic language model still gave a significant improvement. Despite the fact that text preprocessing had significant improvement on perplexity, it had almost no effect on word error rate (WER). This, however, is

[1] CommonCrawl—https://commoncrawl.org/.

Table 1 The impact of different language models on the overall ASR performance

Language model	Perplexity	WER (%)
CC	6797	62.18
MED	246	44.04
MED + CC	244	38.03
MED_EXP	159	38.28
MED_EXP + CC	142	38.24

also related to the noisy test dataset and inconsistencies in the pronunciation lexion which have to be improved.

Although the WER score is rather high for the overall test set, the WER using the latter version of the language model (MED_EXP+CC) varies greatly for individual reports—from 20 to 60%. The biggest contributor to the high WER score are epicrisis reports which contain dictation of complex blood analysis tables—information that should not be dictated in the first place and should be copied directly from source. Therefore, a cleaner test dataset will be created to measure the WER score more objectively.

4 User Interface for Post-Editing

At REUH, 10 radiologists, on average, use the current transcriptionist center services every day to produce around 10 reports each. The average length of a dictation submitted to the transcriptionist center is 5 min. Consequently, 10,000 min (166.7 h) are dictated and transcribed monthly. There are 30 operators employed at REUH's transcriptionist center. Each operator transcribes 320 min (5.3 h) or 160,000 characters per month on average: 4,800,000 characters or 685,714 words are transcribed each month in total (assuming that a word consists of seven characters on average).

Currently, the transcriptionist center uses a general-purpose text processor (such as Microsoft Word) and a separate (non-integrated) audio player to produce the reports. A more convenient and integrated post-editing platform would both make the work of the transcriptionist center more efficient and encourage radiologists and clinicians to post-edit comparatively simple dictations by themselves.

We are developing such an integrated dictation platform to improve the user experience and efficiency. The new workflow is design in a way to disrupt the old workflow as little as possible, making the transition process easier. In the new workflow, the user (i.e., the transcriptionist) receives recordings in the same way as previously, but instead of using an audio player and a text processor separately to make a report from scratch, the user imports recordings into the platform which generates a draft report automatically. The platform provides automatic alignment between text and audio segments which facilitates the post-editing process. After the user has finished editing a report, he or she can download it in the DOCX format or copy the content

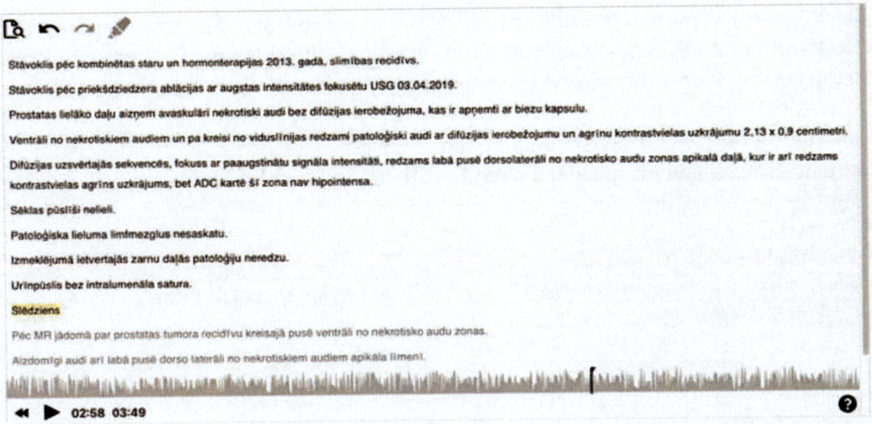

Fig. 2 Illustration of the integrated post-editing interface

to a document template, which is compliant to the current workflow. The new platform can be further integrated into the workflow by supporting the management of dictation records and transcripts.

The integrated editor (illustrated in Fig. 2) provides synchronization between audio and text, allowing to navigate to a specific place in a recording by clicking anywhere in the text and vice versa. The editor provides all essential editing and text formatting features, such as bold, italic, underline, superscript, subscript, paragraphs, and lists, to eliminate the need for post-editing in another text processor outside the platform.

5 Conclusion

Our experiments have shown the impact of in-domain data for ASR performance in the radiology domain. We have demonstrated how a general-purpose ASR system can be adapted to a specific domain by creating and exploiting the required language resources for a less-resourced language. Apart from speech and text corpora for acoustic and language modeling, a rewriting grammar is an important component in the post-processing of ASR transcripts, reducing the amount of manual editing. Regarding language modeling for improved ASR accuracy, text preprocessing with the same bidirectional rewriting grammar gave significant reduction in perplexity; however, it gave the same improvement as interpolation of an in-domain language model with a generic language model—an approach which requires significantly less man-hours to implement.

To improve the evaluation of the various development options of the ASR system, the test dataset has to be significantly revised to make it more consistent. To improve the integrated editor, a comprehensive user study has to be conducted. This has unfortunately been postponed due to COVID-19 restrictions. The time required by a transcriptionist to process one minute of dictation with the new platform compared to the current workflow will be the most important key performance indicator.

A future task is to conduct a comparative study by using transfer learning and fine-tuning of transformer models to derive a Latvian ASR model for the radiology domain from a general-purpose neural ASR model which has been pre-trained on high-resource language data.

Acknowledgements This work has received financial support from European Regional Development Fund under grant agreement No. 1.1.1.1/18/A/153 (*Latvian Speech Recognition and Synthesis for Medical Applications*).

References

1. Dargis, R., Gruzitis, N., Auzina, I., Stepanovs, K.: Creation of language resources for the development of a medical speech recognition system for Latvian. In: Human Language Technologies - The Baltic Perspective, Frontiers in Artificial Intelligence and Applications, vol. 328, pp. 135–141. IOS Press (2020)
2. Paats, A., Alumäe, T., Meister, E., Fridolin, I.: Retrospective analysis of clinical performance of an Estonian speech recognition system for radiology: effects of different acoustic and language models. J. Digit. Imaging **31**(5), 615–621 (2018)
3. Pinnis, M., Auzina, I., Goba, K.: Designing the Latvian speech recognition corpus. In: Proceedings of the 9th International Conference on Language Resources and Evaluation (LREC), pp. 1547–1553. Reykjavik, Iceland (2014)
4. Povey, D., Ghoshal, A., Boulianne, G., Burget, L., Glembek, O., Goel, N., Hannemann, M., Motlicek, P., Qian, Y., Schwarz, P., et al.: The Kaldi speech recognition toolkit. In: IEEE 2011 Workshop on Automatic Speech Recognition and Understanding. IEEE Signal Processing Society (2011)
5. Salimbajevs, A., Strigins, J.: Latvian speech-to-text transcription service. In: Proceedings of the 16th Annual Conference of the International Speech Communication Association (INTERSPEECH), pp. 722–723. Dresden, Germany (2015)
6. Snyder, D., Chen, G., Povey, D.: Musan: A music, speech, and noise corpus. arXiv preprint arXiv:1510.08484 (2015)
7. Stolcke, A.: SRILM—an extensible language modeling toolkit. In: Seventh International Conference on Spoken Language Processing (2002)
8. Tai, T., Skut, W., Sproat, R.: Thrax: An open source grammar compiler built on OpenFst. In: IEEE Automatic Speech Recognition and Understanding Workshop (2011)
9. Znotins, A., Polis, K., Dargis, R.: Media monitoring system for Latvian radio and TV broadcasts. In: Proceedings of the 16th Annual Conference of the International Speech Communication Association (INTERSPEECH), pp. 732–733. Dresden, Germany (2015)

Contextualization of Personal Data Discovery and Anonymization Tools

Hassane Tahir and Patrick Brézillon

Abstract Today for most of the companies, the volume of data continues to increase and in an incredible way. Many of them do not automatically ask customers for permission to use their data, which is illegal. In an increasingly connected and digitized world, it is necessary to establish a precise and strict legal framework so that each piece of data is used wisely and protected. The legal framework is called GDPR (General Data Protection Regulation). The objective is to reassure and protect the user when using internet particularly social networks and ecommerce applications. Anonymization is a technique that irreversibly prevents the identification of an individual's personal data. Our work is about how to contextualize the personal data discovery tools to improve the anonymization process adopted by a company to protect its corporate applications and those of its international subsidiaries. The anonymization is reinforced by different contextualization of connected data groups through user practices added to well-established procedures. For instance, some contextual elements about personal information (Customer Bank Account, zip codes, name of streets and cities, etc.) can be used to change the anonymization rules to mask data differently from one subsidiary to the other. We use a contextual graphs formalism to improve sensitive personal data discovery and anonymization tool.

Keywords Anonymization · Contextual graphs · GDPR · Personal data · Practice · Procedures · Sensitive data · Discovery tool

1 Introduction

Protecting sensitive and personal data is a major challenge that companies must face, and which is gaining ground since the publication of the European Data Protection

H. Tahir (✉)
Links Consultants, Paris, France

P. Brézillon
LIP6, Sorbonne University, 4 Place Jussieu, 75005 Paris, France
e-mail: Patrick.Brezillon@lip6.fr

© The Author(s), under exclusive license to Springer Nature Singapore Pte Ltd. 2022

A. K. Nagar et al. (eds.), *Intelligent Sustainable Systems*, Lecture Notes in Networks and Systems 333, https://doi.org/10.1007/978-981-16-6309-3_28

Regulation. On May 25, 2018, the European Union adopted the General Data Protection Regulation (GDPR) to avoid the abuse of personal data when using the Internet, on commercial sites and on social networks. European countries and the United States have also strengthened personal data protection by developing comprehensive data protection procedures that give people more control over their own data. Personal data is defined by the CNIL (French regulatory institution for personal data) as: "All information related to a physically identifiable person. Is considered to be a "physically identifiable person" any physical person who can be identified, directly or indirectly, in particular by reference to an ID, such as a name, an identification number, location data, online ID, or one or several specific elements linked to his physical, physiological genetic, psychic, economic, cultural or social identity". The GDPR calls for security measures that must be adapted according to the risks weighing on the data subjects, in the event of non-consented use of personal data.

Most business production environments offer sufficient measures to protect against data theft, but current uses in the IT world generate much larger and less secure data display surfaces than environments of production. Among these exhibition areas, some are processed by cryptography, such as the encryption of communications with Internet browsers or inter–site communications (the famous HTTPS). Still others come from test environments, which in many cases are copies of production without the proper security. In this case, even incomplete anonymization (thus pseudonymization) drastically reduces the surface area of the data, which improves the overall state of security and legal compliance. Anonymization also provides solutions that improve the level of compliance with the GDPR and lays the foundations for an ideal solution. Most anonymization solutions are constantly evolving and improve the situation by successive optimizations of the state of anonymity. Our work is about how to contextualize a sensitive personal data discovery tool used at the beginning of the anonymization process of applications of a company and the those of its international subsidiaries. For instance, some contextual elements (zip codes, name of streets and cities) should be used to change the masking rules, and some data can be anonymized differently from one subsidiary to the other (a female first name can be replaced by male in an application of a subsidiary while in applications of other subsidiaries a first name must respect the person's gender). We will use a contextual graph formalism as an assisted tool to manage all the contexts involved in the process of data discovery of personal data. This will help in adding in an incremental way practices (performed by business analysts) to the already set procedures in order to build an experience database useful for the potential involved actors (security managers and anonymization developers) and thus assisting them in sharing context and resolving conflicts. This paper starts by presenting sensitive personal data discovery and the anonymization tool, with an introduction to the anonymization process, and then presenting related work about some existing data discovery and anonymization tools found in research and industry. After this we present how to contextualize the process of personal data discovery in our tool that automatically detects and specify personal and sensitive data in production source databases. We will use the contextual graph platform as an assisted tool to improve

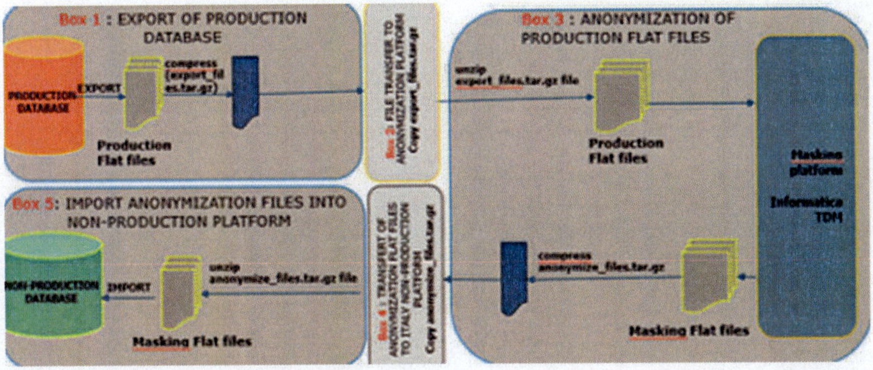

Fig. 1 Typical data anonymization process

sensitive personal data discovery. Finally, we conclude our paper by an evaluation and future works.

2 Personal Data Discovery and Anonymization Tool

2.1 Description of an Anonymization Process

For companies handling sensitive data, developers should not use and manipulate production data about personal customer information in their development and test environments to avoid data exposure, vulnerability, and other risks of illegal sales and data theft. The best solution to protect data is to anonymize sensitive data in environments other than production to ensure that data privacy policies are not violated. The process of data anonymization includes identifying personal data, selecting the associate data masking technique, launching the masking process, and finally checking integrity. Figure 1 shows an example of the stages (or boxes) of the anonymization process of data stored in flat files exported from a production database. These anonymized data are then imported into a non-production database.

2.2 Sensitive Personal Data Discovery Tool

We have used a tool based on Python that can connect to database sources like Oracle, IBM DB2, and SQL Server. The objective is to analyze data contained in the different tables of a database to determine the fields containing personal data using regular expressions and dictionaries of names and first names of people, city names, etc. Figure 2 shows the platform of data discovery tool where the anonymization

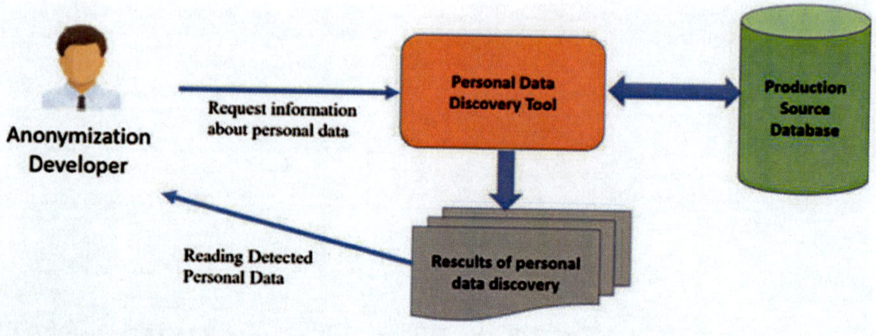

Fig. 2 Platform of the sensitive personal data discovery tool

developer requests the personal data read from the production source database and displayed in suitable forms to users. Table 1 shows an example of a result sheet with a classification of personal data as well as the rate of how many times they are found in the database.

The discovery tool cannot always perform an accurate classification of the personal data because the context about the meaning of the data stored in each field. For example, a field named "TEXT" may store names classified by a tool as first names (i.e., personal data) but may also store names of cars (i.e., Zoé, Mégane) which should not be considered as a personal data. The following section presents related work about personal data discovery and anonymization tools.

Table 1 Example of results of personal data detected by a discovery tool

Column name	Location	Number of rows in table	Number of rows analyzed	Percentage of personal data	Data found
First name	['TEXT']	534	534	4.S5	13
Firstname	['USER NAME']	56,366	1000	35.4	354
First rrame	['cm']	1124	1000	2.0	20
Town	['cm']	1124	1000	1.9	19
Email	['EMAIL']	1124	1000	100	23
Firstname	['NAME']	1124	1000	8.4	84
Lastname	['NAME']	1124	1000	6.7	67
Email	['COMMENT']	1124	1000	1.2	12

3 Related Work

Tools for the automatic detection and discovery of sensitive and personal data are of great use and support to business analysts in companies and organizations in order to comply with common industry regulatory standards for data protection and privacy including the General Data Protection Regulation (GDPR) and other important regulations. The following are some related research works. Securiti [13] is an artificial intelligence-based tool that helps organizations in discovering sensitive data stored in different platforms (Cloud, SaaS, and on-premises environments), protecting that data and automate many of the privacy functions. Nightfall [9] is a machine learning-based tool that supports a business company to manage and protect critical information such as personal data about customers, through data infrastructure, SaaS, and APIs. IBM Security Guardium [5] is tool to analyze, discover, classify, and control access to sensitive data. Other related research work discussions can be found in [1, 6, 7, 14, 19]. Finally, it is important to cite some anonymization tools and approaches to anonymize data such as SMD Anonymizer [12] developed to generate anonymized files using the k-anonymity algorithm to improve the data privacy. In [11], an anonymization tool named ARX can efficiently implement various privacy techniques and thus be able to create an intuitive multi-platform graphical interface. Other related research discussions can be found in [2, 4, 8, 10, 15, 18].

4 Contextual Graphs for Improving Personal Data Discovery

4.1 Introducing Contextual Graphs

The basic components of a contextual graph (abbreviation CxG) are actions and contextual elements. Activities and temporal branching are two other elements of a CxG. An action can be represented on one or more paths, but it will be in different contexts. A contextual graph has these properties: (1) is a directed acyclic graph, (2) has one input, (3) has one output and (4) has spindle structure. It can model the different ways to solve a problem. A user practice can be represented in each path of a CxG, to show how an action is performed in each context. References [3, 16, 17] present more details about the CxG formalism as well as the way it is implemented.

Figure 3 shows a contextual graph to model how a business analyst checks personal data detected and classified by the data discovery tool to adjust and contextualize the automatic detection of the data. The activity is also a CxG and can appear on different paths (i.e., different contexts) of contextual graphs (as in the case of actions).

For example, Fig. 4 represents the main activities used to perform a data anonymization using an adapted masking tool. The first activity is about specifying personal data to be anonymized. It corresponds to the phase of identifying sensitive personal data by business analysts. Automatic data discovery tools can be used to

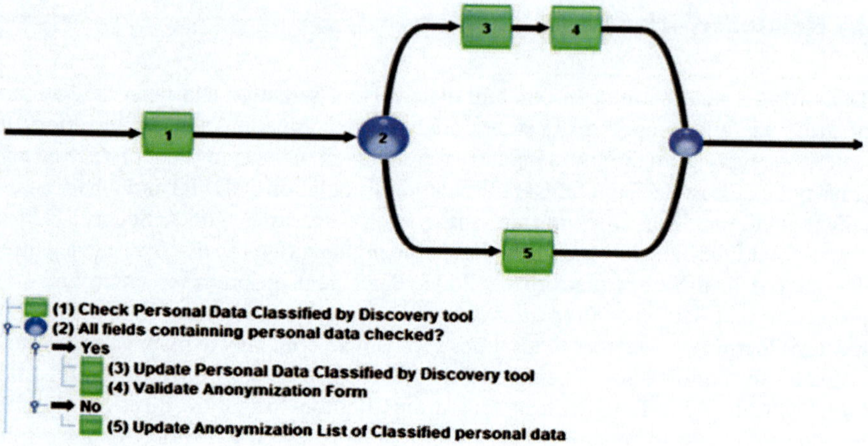

Fig. 3 Contextual graph for checking personal data

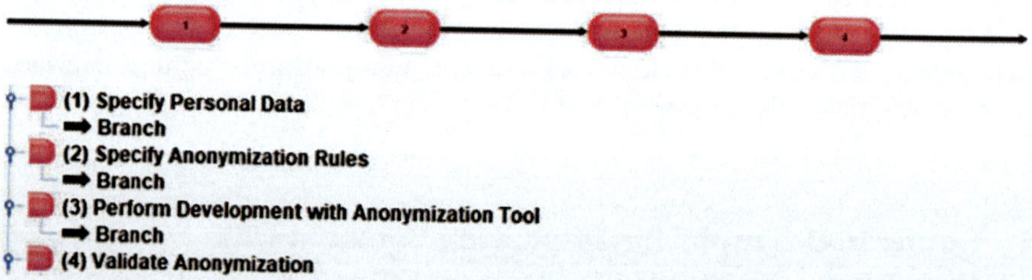

Fig. 4 Contextual graph to perform a data anonymization using activities

support the analysts in this activity. Sometimes these tools do not produce accurate classification of personal data, and they should be contextualized by analyst to confirm or invalidate the data specified as sensitive by these tools. Our illustration focusses on this activity. The remaining three activities (2–4) in Fig. 4 are out of the scope of this paper.

4.2 *Contextualization of Personal Data Discovery*

In this section, we present an example of contextualization of personal data discovery used with anonymization tools. In this example, the discovery tool classifies fields in a database as personal data such as: Last name, First name, Email, Town (or City). But when a business analyst examines the result of personal data classification by the discovery tool (Fig. 5), he must verify each field or column to determine whether

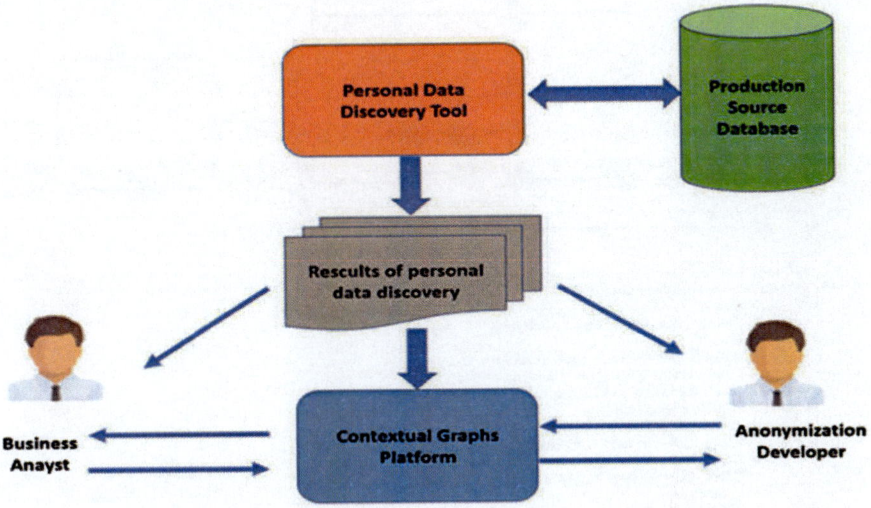

Fig. 5 Contextual graphs platform connected to the sensitive personal data discovery tool

data stored in that field is sensitive personal data or not and if it will be therefore anonymized or not in order to be GDPR compliant.

Contextual graphs software can be used as a decision-making tool to assist the business analyst in personal data classification when working on databases with too many tables as shown in Table 2.

Figure 6 presents a contextual graph used by a business analyst to adapt the result classification of personal data (by the discovery tool) to the context about the meaning of the data stored in each field of the database table. For instance, a contextual element (Node 7) is used to determine if the field named "TEXT" is a First Name or No. If this field contains most of times First Names (70% in scanned rows), the analyst confirms a "TEXT" field as a First Name which is a personal data. But if this field contains names of cars (for example ZOE which is a name of Renault Car), this field is not considered by the analyst as personal data. Contextual elements must be distinguished from their instantiations. It would not be enough to give the instantiations for a person/company. Two contextual elements are considered relevant together based on their instantiations. The next section concludes our work.

Table 2 Example of personal data to verify by business analysts

Column	Data Type	Data Lenght	Column ID	Num_rows	To be masked
EMAIL_TO	VARCHAR2	2000	11	3243580	To verify
EMAIL_CC	VARCHAR2	2000	12	3243580	-
EMAIL_SUBJECT	VARCHAR2	512	13	3243580	Yes
TXT_DMD	VARCHAR2	1750	5	156	No
POC_LIB	VARCHAR2	50	4	41	
PAR_DEST_NAME	VARCHAR2	40	8	41	To verify
ADRESSE_LIB	VARCHAR2	1000	10	41	To verify
ORI_PAR_DEST_NAME	VARCHAR2	40	9	46	To verify

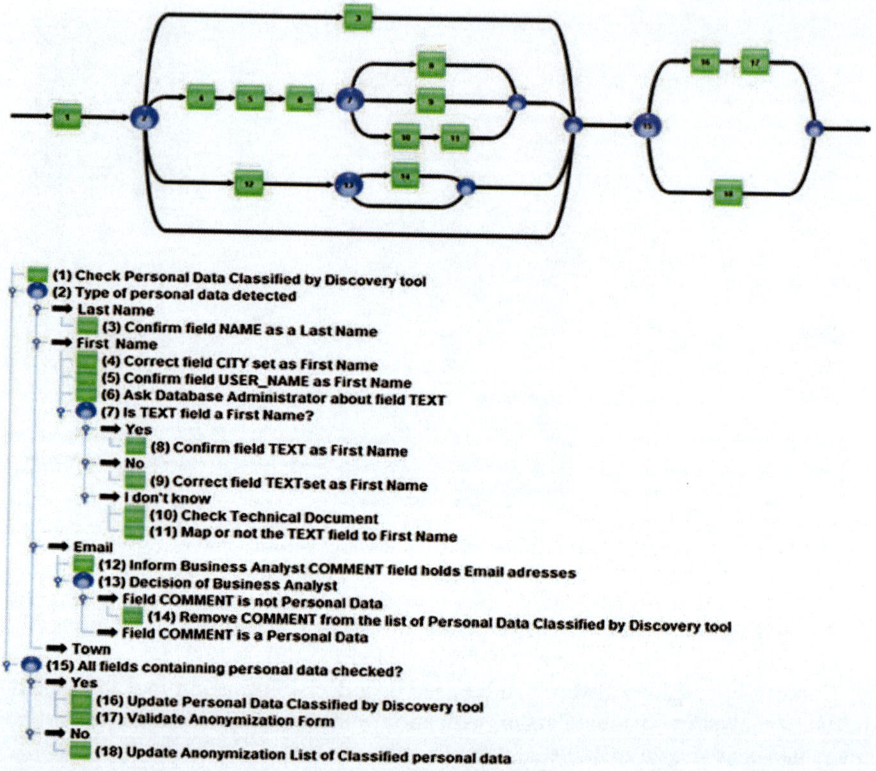

Fig. 6 Contextual graph to assist users in specifying personal data

5 Conclusion

This paper has presented how to contextualize personal data discovery and anonymization. Business analysts' practices can be simply represented by the different actions in the paths of a contextual graph. These practices correspond to checking personal data by the analyst based on the context about the meaning of the contents of each field in a database table. Practices are added in contextual graphs to contextualize official automated procedures for personal data discovery tools. All the data stored in each field and classified by the automatic data discovery tool as personal will then be contextualized by analysts using the CxG formalism representation to produce the final set of fields confirmed as containing personal data. This set of fields will then be used by developers to establish the masking rules required in the implementation (or development) phase of the anonymization process. Our work can be extended and applied to other area particularly in computing, industry, mobile security, and medicine.

References

1. Abdalla, A.: Personal GIS—a tool to process our personal data, KN. J. Cartography Geogr. Inf. **63**(2), 86–88 (2013)
2. Bezzi, M., Montagnon, G., Salzgeber, V., Trabelsi, S.: Sharing Data for Public Security, HAL Id: hal-01061070. https://hal.inria.fr/hal-01061070, Submitted on 5 Sep 2014
3. Brézillon, P.: Task-realization models in contextual graphs. modeling and using context. In: Dey, A., Kokinov, B., Leake, D., Turner, R. (eds.) Springe, LNAI 3554, pp. 55–68 (2005)
4. Elliot, M.J., Mackey, E., O'Hara, K.: The Anonymisation Decision-Making Framework, 2nd edn. European Practitioners' Guide (2020)
5. IBM Security Guardium: https://www.ibm.com/security/data-security/guardium. Last accessed 10 May 2021
6. Kudim, K.A., Proskudina, G.Yu.: Methods and tools for extracting personal data from theses abstracts. https://doi.org/10.15407/pp2019.02.038 (2019)
7. Microsoft, Azure Information Protection: https://azure.microsoft.com/. Last accessed 15 Apr 2021
8. Nassar, M., Orabi, A.A., Doha, M., Al Bouna, B.: An SQL-like query tool for data anonymization and outsourcing. In: 2015 International Conference on Cyber Situational Awareness, Data Analytics and Assessment (CyberSA), June 2015
9. Nightfall: https://nightfall.ai/. Last accessed 10 Apr 2021
10. Pawar, A., Ahirrao, S., Churi, P.P.: Anonymization Techniques for Protecting Privacy: A Survey. IEEE, Pune, India, 30 Nov–2 Dec 2018
11. Prasser, F., Kohlmayer, F., Lautenschlaeger, R., Kuhn, K.A.: ARX-A comprehensive tool for anonymizing biomedical data. In: AMIA Symposium 2014, November 2014, pp. 984–993
12. Santos, I., Coutinho, E., Moreira, L.: SMDAnonymizer: a web tool for data anonymization. In: Conference: 6th International Workshop on ADVANCEs in ICT Infrastructures and Services (ADVANCE'2018), Santiago, 2018
13. Securiti: AI Powered Data Security & Privacy across Multi-Cloud, SaaS and On-Premise. https://securiti.ai/. Last accessed 10 Apr 2021
14. Siteimprove.com: Making a Personal Data Inventory and Data Map in the GDPR Era (2020). GDPR, Last updated: 8 Mar 2020
15. SoftwareTestingHelp. 10 Best Data Masking Tools and Software in 2021. https://www.softwaretestinghelp.com/data-masking-tools/. Last accessed 10 Apr 2021
16. Tahir, H., Brézillon, P.:Contextual graphs platform as a basis for designing a context-based intelligent assistant system. In: Brézillon, P., Blackburn, P., Dapoigny, R. (eds.) CONTEXT 2013, LNAI 8175, pp. 259–273 (2013)
17. Tahir, H., Brézillon, P.: Improvement of database administration by procedure contextualization. In: Conference: CEUR Proceedings 4th Workshop HCP Human Centered Processes, Genoa, Italy, 10–11 February 2011
18. Vardalachakis, M., Kondylakis, H., Koumakis, L., Kouroubali, A.: ShinyAnonymizer: a tool for anonymizing health data. In: International Conference on Information and Communication Technologies for Ageing Well and e-Health (ICT4AWE), March 2019
19. Vianna, D., Yong, A., Xia, C., Marian, A., Nguyen, T.: A tool for personal data extraction. In: 30th International Conference on Data Engineering Workshops, IEEE, Chicago, USA, 2014

Early Lungs Tuberculosis Detection Using Deep Learning

Shiv Pranathi Kale, Jigisha Patil, Akshada Kshirsagar, and Varsha Bendre

Abstract Tuberculosis (TB) is a bacterial infection in lungs that causes the greater number of deaths throughout the world than other infectious diseases. TB is an airborne disease caused by the bacteria *Mycobacterium tuberculosis* or tubercle Bacillus. Severe infection of this disease in lungs leads to coughing, fever, and chest pains. The existing research in the last decade is providing a guideline into TB transmission, diagnosis, and treatment. However, early detection of TB will help to efficiently decrease the severity of disease upon identification leading to reducing the mortality rate. This will in turn decrease the strain on the public health and eventually exterminate the TB. Early detection is a screening pathway resulting in identification of susceptible persons. This paper discusses the recent breakthroughs made in diagnostics and treatment of existing TB detection by studying the X-ray images for identifying dominant features for TB in early stages using predefined algorithms, software's and image processing. In this paper, a capable method for early detection of TB using Deep Learning is proposed.

Keywords Convolutional neural network · Image preprocessing · Image classification · Feature extraction · Tuberculosis

1 Introduction

Tuberculosis (TB) is a curable bacterial infection, if detected early followed by effective treatment. There are various radiological procedures which are the primary tools to check the lung infection of TB, procedures such as Chest Radiography, Magnetic Resonance Imaging (MRI), and Computed Tomography (CT) [1]. In addition to this, the suspicious person has to undergo the sputum test for determining the presence of TB [2]. Out of the processes which are used for detection of tuberculosis, screening and chest radiographs, i.e., CXR are the methods widely used [3]. These methods are

S. P. Kale (✉) · J. Patil · A. Kshirsagar · V. Bendre
Pimpri Chinchwad College of Engineering, Pune, India

V. Bendre
e-mail: varsha.bendre@pccoepune.org

© The Author(s), under exclusive license to Springer Nature Singapore Pte Ltd. 2022

A. K. Nagar et al. (eds.), *Intelligent Sustainable Systems*, Lecture Notes in Networks and Systems 333, https://doi.org/10.1007/978-981-16-6309-3_29

low cost, and their deployment is also easy [4]. Robert Koch discovered the cause of TB in 1882. It has been more than 125 years since then, and this disease, to this date, has a global presence. It affects everyone getting in contact with the bacterium through any means irrespective of all ages, classes, and societies [5].

Detection of especially latent TB can be challenging. Treatment of a late diagnosed complicated disease is complex. Hence, 13 Sustainable Development goals were set by the United Nations under which goal 3 talked about health and wellbeing. This goal predominantly is about ending 3 endemics by 2030, namely AIDS, TB, and Malaria.

In many of the infected people, there is a lack of active symptoms and hence would not be detected [6]. Due to this, chances of spreading and contamination among the normal person increases. Thus, the late detection of TB increases the transmission rate and mortality rate. Progress in controlling TB and mitigating its penalties can be accelerated through early diagnosis and treatment [7].

This paper aims at accurate and early detection of TB using Chest X-ray. Section 2 provides the review and comparison of existing techniques used for TB detection. Deep learning approach for early diagnosis of TB using Convolutional Neural Network (CNN) is explained in Sect. 3. Experimental results and observations are mentioned in Sect. 4, and Sect. 5 provides the conclusion of the paper.

2 Background

Shah et al. [8] have presented a paper on lung tuberculosis detection from sputum smear images. In their research as mentioned in their paper, the chest X-ray (CXR) is taken as the input data. Noise in CXR is removed using a median filter followed by image segmentation and feature extraction methods. Segmentation methods like watershed segmentation and gray level segmentation are applied to filtered CXR images. Later, efficiency is calculated and compared using different classifiers based on their accuracy. The disadvantage of watershed segmentation method is that it is highly sensitive to local minima, since at each minima a watershed is created. It is sensitive to noise, and over segmentation is also there due to Region Merging Algorithms [9].

In [10], Gabriella et al. have explained a Computer Aided Design (CAD) system used for tuberculosis detection using CXR images. CNN is used in this paper which can assist the doctors and radiologists to make better analysis of images for accurate diagnosis. This system may increase the specificity of an early TB diagnosis. The disadvantage of CAD systems is vulnerability to hacking and high production or purchasing cost for new systems, and it is a time-taking process to understand how to operate and run the software.

Ekata et al. [11] have proposed a fuzzy inference system to detect pulmonary TB. A system which makes decisions based on a set of rules is developed for this research. The disadvantage of the fuzzy inference system is that it does not stand at par with the pattern recognition presented by machine learning and neural networks.

Table 1 Comparison of methods for diagnosis of TB

Methods	Disadvantages
Watershed method [8]	Highly sensitive to minima, sensitive to noise
Automated CAD system [10]	Vulnerability to hacking, high production, or purchasing cost for new systems
Fuzzy inference system [11]	It does not stand at par with the pattern recognition presented by machine learning and neural networks
SVM classification system [12]	Not suitable for large volumes of dataset

Fuzzy logic does not always give accurate results, because the results are perceived mostly based on assumptions and are not always accepted by many.

Junaedi et al. [12] have proposed a method for TB detection in CXR images using optimized gray level co-occurrence matrix features. TB detection using X-ray images has been developed to reduce the subjectivity factor. Support Vector Machine (SVM) classification system with gray level co-occurrence matrix input optimized by Principal Component Analysis (PCA) is better than SVM without optimization. PCA is a method better than gray level co-occurrence matrix feature in SVM classification system. The disadvantage of SVM classification is that it is not suitable for large volumes of dataset. SVM does not perform very well if the dataset has more noise. The comparison of various methods used for diagnosis of TB is given in Table 1.

Considering various disadvantages of techniques reviewed in this section, a deep learning approach using CNN is used in the proposed work. CNN has an advantage of higher accuracy among all the algorithms which predict images and also has computational efficiency.

3 Material and Methods

3.1 Deep Learning Using Convolutional Neural Network (CNN)

Methodology followed in this paper is presented in Fig. 1

Following are the steps that describe the block diagram:

Step 1: Input Image Acquisition
Step 2: Image Preprocessing
Step 3: Classification using CNN
Step 4: Detection of Tuberculosis Positive or Negative

In this project, the dataset is acquired from "Kaggle Database" which consists of 1000 X-rays images for TB detection. Out of these 1000 images, 800 images are taken for training purpose and 200 for testing purpose. In these 200 images, 100 images

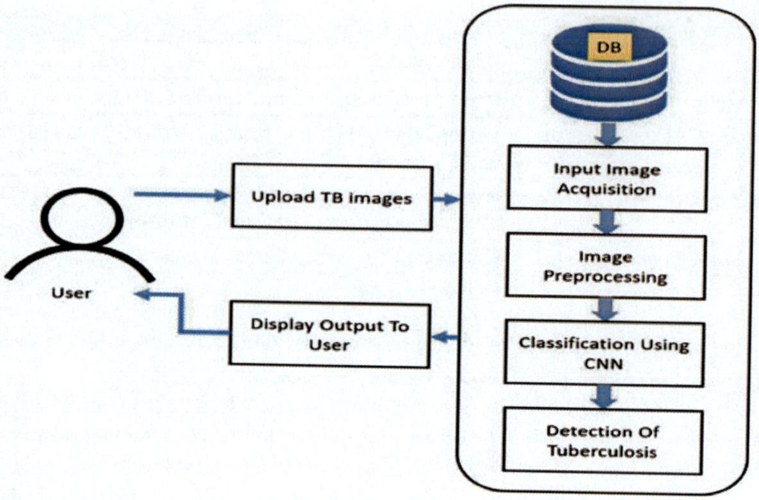

Fig. 1 Block diagram

are of normal patients and 100 images are of TB infected patients. The distribution of the dataset diagrammatically is shown in Fig. 2.

The input image to be tested is accepted from the user. This is done using an interactive interface. After image acquisition, preprocessing is done on the input image.

Preprocessing involves RGB to gray-level conversion and gray level to binary-level conversion. Once this conversion is done, the threshold values are generated. The threshold values generated from the input image are partially 0 matrix and 255 matrix values. Wherever there is a 0 value in the matrix, it represents the strong intensity area of the X-ray and value of 255 indicates weak intensity area of the input image. The strong intensity or 0 value represents the X-ray of an infected patient,

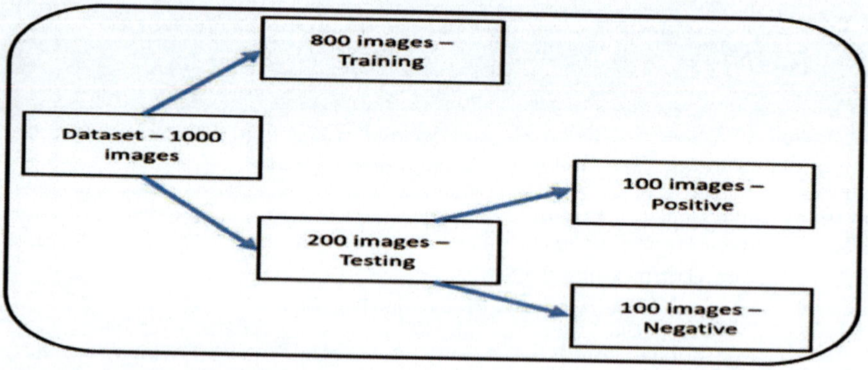

Fig. 2 Distribution of dataset

whereas the weak intensity or 255 value represents an X-ray of a normal patient. After preprocessing required features are extracted from that image automatically by using CNN.

In this project, CNN algorithm is used as a classifier. Once the algorithm is designed, it is fed with the above 800 training images. Now this classifier itself decides the required features and gives appropriate results. This is used as it gives a higher percentage of accuracy to classify whether the given input image is of a TB infected patient or the normal patient.

CNN is a deep learning algorithm consisting of multiple layers for image analysis and its classification. Most generally, CNN is an artificial neural network (ANN) that has the ability of learning a large number of datasets under some constraints and then predicting the pattern from it. This pattern recognition ability of CNN makes it more fascinating for image analysis [13].

A CNN has many layers, broadly classified as convolutional and non-convolutional layers. The layers of CNN used in this project are: Convolutional Layer, MaxPooling Layer, Flattening Layer, Dense Layer, Dropout Layer.

A convolutional layer is what detects patterns. These layers receive input from the preceding layer, transform the input, and transfer them to the next layer. This transformation is called convolving. This convolution layer is a hidden layer and uses a filter. A filter is like a relatively small matrix with some number of rows and columns, and this is the basis of convolving. The deeper the network goes, the more sophisticated these filters become. This project uses three convolution layers which makes it three layers deep.

Each convolution layer is followed by a MaxPooling layer. The function of this layer is to reduce the image size. This is done by taking the maximum value of the elements in the window and mapping the image of a given window into a single result.

In the flattened layer, the pooled featured map is flattened. Flattening layer involves transforming the feature map matrix into one dimensional matrix then it is given to the final classification model, which is called a fully connected layer.

Dense layer is a deeply connected layer of convolution neural network. It means each neuron present in the dense layer receives the input from all neurons which are present in the previous layer. An "m" dimensional vector is generated by the dense layer. The term "dropout" refers to dropping out units in a neural network. Dropouts are the regularization technique that is used to prevent overfitting in the model.

The code snippets used for each layer are shown below in Fig. 3.

The libraries and packages that have been used in this project are predominantly Keras, OpenCV, NumPy.

In this project, the OpenCV library is used for resizing the images and creating feature vectors out of it, which can be achieved by converting the image data to NumPy arrays.

CNN is an abstract concept of building an algorithm equivalent to human vision. This is implemented in this project by importing various CNN layers packages from Keras library. Keras is built on TensorFlow and is written in Python.

```
#Convolution Layer
classifier.add(Convolution2D(32, 1, 1, input_shape = (64,64,3), activation = 'relu'))

#MaxPooling Layer
classifier.add(MaxPooling2D(pool_size =(2,2)))

#Flattening Layer
classifier.add(Flatten())

#Full Connection
classifier.add(Dense(256, activation ='relu'))
classifier.add(Dropout(0.5))
classifier.add(Dense(2, activation ='softmax'))
```

Fig. 3 Snippet of code

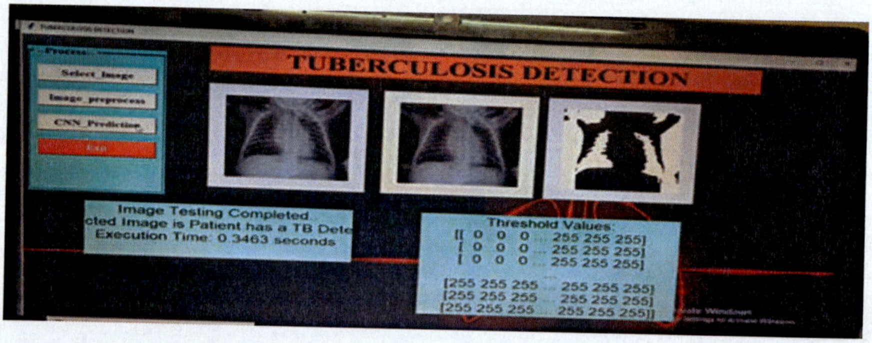

Fig. 4 TB positive output

The final TB detection results obtained from training and testing are discussed as follows.

Here a total dataset of 1000 images of X-rays for TB detection are used. Out of which 800 images are taken for training purposes and 200 for testing purposes. In these 200 images, 100 are of normal patients and 100 of TB positive patients. CNN algorithm has been used for this project and using this algorithm results in accurate results for normal and TB positive patients. Figure 4 shows the output of TB prone patients, and Fig. 5 shows the output of normal patients.

4 Results

Following are the output results of the TB positive and TB negative patients from the project.

From the result, it can be observed that the image that consists of threshold values partially 0 and partially 255 are the images of tuberculosis infected patients. As 0 value indicates strong intensity area which means, the image consists of more

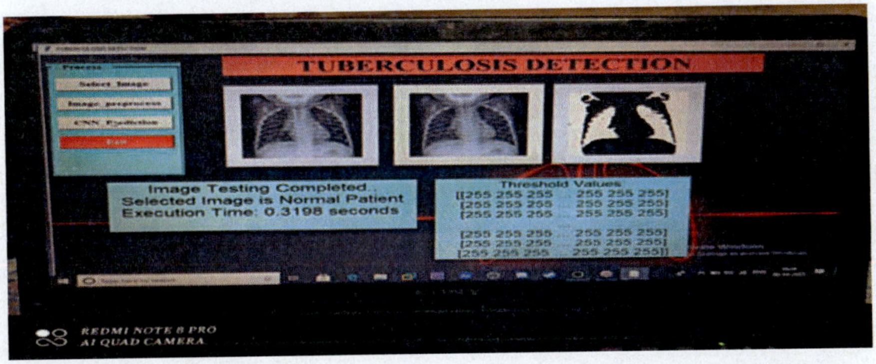

Fig. 5 TB negative output

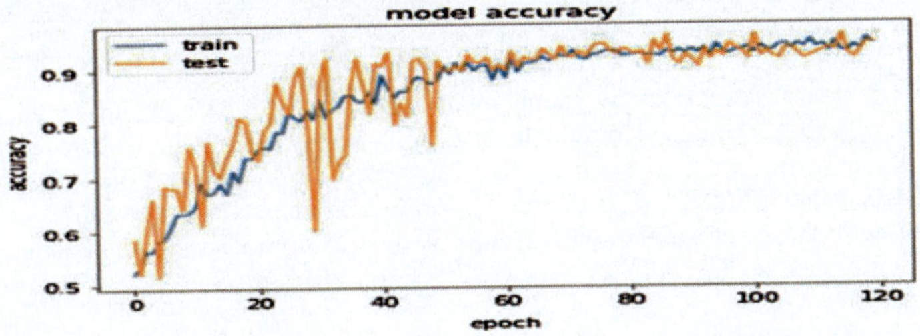

Fig. 6 Accuracy versus epoch graph

white patches or white shades that indicate the tuberculosis strains in the X-ray. The images with all the threshold values as 255, then that image is of normal patients as mentioned in the above (Fig. 6).

5 Conclusion

The method to detect early lung tuberculosis using chest X-ray is presented in this paper. Here the early lung tuberculosis has been detected with training accuracy of 84.00% and testing accuracy of 82.75%. So, by implementing CNN, it is possible to detect tuberculosis at early stages so that further devastated conditions can be avoided. As an extension of this work, fully automated computer-aided system can be obtained to detect the chances of TB from the analysis of chest X-rays.

References

1. Bomanji, J.B., Gupta, N., Gulati, P., Das, C.J.: Imaging in tuberculosis. Cold Spring Harb Perspect. Med. **5**(6), a017814. Jan 20; (2015). https://doi.org/10.1101/cshperspect.a017814. PMID: 25605754; PMCID: PMC4448708
2. Fogel, N.: Tuberculosis: a disease without boundaries. Tuberculosis. **367** (2015). https://doi.org/10.1016/j.tube.2015.05.017
3. Nyein Naing W.Y.Z., Htike, Z.: Advances in automatic tuberculosis detection in chest x-ray images. Signal Image Process An. Int. J. **5**(6), 41–53 (2014)
4. Necho, M., Tsehay, M., Seid, M., Zenebe, Y., Belete, A., Gelaye, H., Muche, A.: Prevalence and associated factors for Alcohol Use Disorder among tuberculosis patients: a systematic review and meta-analysis study. Research Square (2020)
5. Hooda, R., Sofat, S., Kaur, S., Mittal, A., Meriaudeau, F.: Deep-learning: A potential method for tuberculosis detection using chest radiography. In: 2017 IEEE International Conference on Signal and Image Processing Applications (ICSIPA), 2017, pp. 497–502
6. Monsod, T.M.C., Gochoco-Bautista, M.S.: Rethinking “economic fundamentals” in an era of global physical shocks: Insights from the philippines experience with covid-19. Asian Economic Papers **20**, 110–140 (2021)
7. Ho, L.-J., Yang, H.-Y., Chung, C.-H., Chang, W.-C., Yang, S.-S., Sun, C.-A., et al.: Increased risk of secondary lung cancer in patients with tuberculosis: a nationwide, population-based cohort study. PLoS ONE **16**(5), e0250531 (2021). https://doi.org/10.1371/journal.pone.0250531
8. Shah, M.I., Mishra, S., Sarkar, M., Rout, C.: Automatic detection and classification of tuberculosis Bacilli from ZN-stained sputum smear images using watershed segmentation. International Conference on Signal Processing (ICSP 2016), 2016, pp. 1–4, https://doi.org/10.1049/cp.2016.1459
9. Kim, L.E.: Automated detection of early lung cancer and tuberculosis based on x-ray image analysis. Proceedings of the 6th WSEAS International Conference on Signal, Speech and Image Processing, Lisbon, Portugal, September 22–24, (2006)
10. Gabriella, I., Kamarga, S.A., Setiawan, A.W.: Early detection of tuberculosis using chest x-ray (CXR) with computer-aided diagnosis. In: 2018 2nd International Conference on Biomedical Engineering (IBIOMED), 2018, pp. 76–79, https://doi.org/10.1109/IBIOMED.2018.8534784
11. Ekata, Tyagi, P.K., Gupta, N.K., Gupta, S.: Diagnosis of pulmonary tuberculosis using fuzzy inference system. IEEE Second International Innovative Applications of Computational Intelligence on Power, Energy and Controls with their Impact on Humanity (CIPECH) (2016, November), pp. 3–7, https://doi.org/10.1109/CIPECH.2016.7918726
12. Junaedi, I., Yudaningtyas, E., Rahmadwati, R.: Tuberculosis detection in chest x-ray images using optimized gray level co-occurrence matrix features. In: 2019 International Conference on Information and Communications Technology (ICOIACT), 2019, pp. 95–99, https://doi.org/10.1109/ICOIACT46704.2019.8938584
13. Rahman, T., et al.: Reliable tuberculosis detection using chest x-ray with deep learning, segmentation and visualization. In: IEEE Access. vol. 8, pp. 191586–191601, (2020). https://doi.org/10.1109/ACCESS.2020.3031384

Localising the Privacy Requirement to Improve the Scalability Issue of Blockchain

Ajay Singh and Preetida Vinayakray Jani

Abstract Blockchain is one of the most ice breaking technology of this century. To make fast integration of this technology in various domains some prominent issues like scalability, high throughput, processing delay need to be resolved. Blockchain architecture is decentralized with distributed system approach, where maintaining the privacy of every block comes at the cost of scalability of the network. Scalability is the problem associated with transaction rate (throughput) of the system. Therefore to make a blockchain technology embedded within the mainstream it is necessary for it to be secure and scalable at the same time. To deal with the issue of scalability proposed framework attempts to localize the privacy requirement on a block level along with a simple but efficient consensus algorithm. Therefore this paper shows a novel framework that is built around a blockchain network that uses distributed *nano node* architecture. These *nano node* assist to verify blocks coupled with an efficient consensus algorithm and maintain the privacy of the blocks.

Keywords Blockchain · Distributed network · Nano node · NodeMCU · Scalability · Peer-to-peer network · Peer-to-peer storage

1 Introduction

In conventional information technology, transactions or data processed in a centralized manner. However with this centralized approach there are various risk associated with it such as manipulation, security, hidden cost, single point failure etc. These issues can be resolved through decentralized distributed architecture like Blockchain. Blockchain provides a peer to peer connection and stores the data into a decentralized manner. Data decentralization, transparency, security, immutability, and privacy are the salient principles of Blockchain Technology. Blockchain is a data structure by

A. Singh (✉) · P. V. Jani
Sardar Patel Institute of Technology, Mumbai, India
e-mail: ajaykumar.singh@spit.ac.in

P. V. Jani
e-mail: preeti.vinayakray@spit.ac.in

© The Author(s), under exclusive license to Springer Nature Singapore Pte Ltd. 2022
A. K. Nagar et al. (eds.), *Intelligent Sustainable Systems*, Lecture Notes in Networks and Systems 333, https://doi.org/10.1007/978-981-16-6309-3_30

which digital ledger of transaction can be created, shared, or stored among a distributed network of computers. Block chain protocol controls access so that it does not require trust from third party. Along with salient features of Blockchain it is having the major issue of scalability which can be categorized into throughput, cost, capacity, privacy and networking. Improvement in scalability affects the application of blockchain in business which may compromise throughput and privacy of blocks. Scalability is also limited due to some other factors like block interval time and block size which may also challenge the security of the block. The System may become vulnerable to different attacks if we blindly focus on the scalability.

1.1 Limitations Due to Lack of Scalability

To make the blockchain technology the spine of all the major systems that are currently based on centralized approach is not a very easy task since the complexity of these system can range between low to very high with various dependency such as server capacity, limited computational power etc. To accommodate or migrate such a system like emerging blockchain network will create a bottleneck due the below mention limitations in the existing blockchain networks.

- Slow transaction rate—Currently the transaction rate for blockchain lag way behind its centralized counterpart. Statistically this is 80–85% slower and this limits the blockchain to be used up to its full potential.
- Inefficient Consensus algorithm—The excessive Overhead involved to reach to the consensus and this factor overall affects the efficiency of the blockchain network.
- High energy consumption—A high computational needs for adding a new block in the blockchain network the power consumption and heat output is also very high.

1.2 Prime Objectives

The prime goal of this paper to make a blockchain system that is scalable while maintaining privacy of block nodes. The proposed framework is light weight, flexible and computationally efficient as energy consumption is low with low computational power requirement. To make scalable blockchain system with localizing privacy through *nano nodes*, the proposed framework attempts:

- To reduce the computation needed for a blockchain network and accommodate it on *nano nodes*.
- To make use of efficient consensus algorithm to improve the Scalability.

2 Literature Survey

In the paper presented by Kohad et al. [15], the author analyzed and compared different ways to overcome the scalability of the blockchain system. The basis of this comparison was the usage of different consensus algorithm and its efficiency and performance. The author observed that to increase scalability the throughput plays an essential role and increasing the block size has its own downsides. The author proposed factors such as transaction rate, block size and total chain size is what scalability of blockchain rely on.

In the paper presented by Qin et al. [26], the author discussed various type of blockchain architecture and concluded that the Scalability of permissionless blockchains requires novel concepts, and this is due to its availability in the public domain security and privacy are as major concern. So it id very important to treat the security and trust of the blockchain with priority while dealing with the scalability issue.

In the paper presented by Hazari et al. [11], the author outlines the traditional transactional system that revolves around a third party that verifies every single transaction at a very rapid pace. He also pointed that there are many flaws in this approach due to its decentralized behavior. Then the author highlighted how to over come the issue that are faced by the usage of a centralized application by a decentralized solution that is blockchain. Author shed light on types of blockchain network with its merit and demerits. And also their security and privacy concerns.

In the paper presented by Hafid et al. [10], the author classifies blockchain scalability into two layered categories

- First layer solution—It deals with structural changes in the blockchain ex. block size.
- Second layer solution—It deals the methodology used outside blockchain.

It deals the methodology used outside blockchain. The author also showcased an analytical comparison between throughput, block size, latency and the merits and demerits of the present solution to the scalability issue.

In the paper presented by Nikitinc et al. [23], the author gave an overview of the existing and popular blockchain technology such as bitcoin and Ethereum. It mentioned the main advantage to using advantage and performance improvement when using a blockchain network instead of centralized application. The author also provide some solution namely SegWit and Lightning to deal with the scalability issue of the blockchain.

In the paper presented by Eyal et al. [31], the authors review is based on the comparative analysis of the existing consensus algorithm it is based on the efficiency and performance under different work load and conditions. The author proposes an in-depth review of each and every algorithm that was picked for this particular study.

In the paper presented by Novo et al. [24], the author contributed to an architectural design of decentralized system based on IoT coupled with blockchain system. The approach they followed with new and very different from some of the existing approaches which was to exclude inclusion of blockchain technology within IoT

Table 1 Comparative analysis of different blockchain infrastructure [3, 15, 25]

Properties	Public	Consortium	Private
Consensus determination	All miners	Selected set of nodes	One organization
Read permission	Public	public or restricted	public or restricted
Immutability	No tampering	Can be tampered	can be tampered
Efficiency	Low	High	High
Centralized	No	Partial	Yes
Consensus process	Permissionless	Permissioned	Permissioned

Table 2 Comparative analysis of different blockchain network

Properties	Bitcoin	Dogecoin	Litecoin	XPP
Release year	2008	2015	2011	2015
Block generation time (min)	9.7	5	2.5	1
Hash rate	28.6*106 TH/s	570.14 TH/s	213.6 TH/s	167.6
Cryptographic algorithm	ECDSA	ECDSA	Scrypt	Scrypt
Power consumption	Very high	Medium	Moderate	low

devices. It was based on smart contracts and thus proposes a light weighted operation on the blockchain network with minimum communicational overhead.

In the paper presented by Sun et al. [28], the author proposed the use of IoT to deal with the current scalability issue and suggest the use of IoT devices as a node that verifies and authenticate transactional block before adding it to the blockchain network.

In the paper presented by Zhang et al. [23], focuses on providing a scalable solution to the blockchain for health records. The basis of this study was to outline the limitation of the blockchain system in terms of security, privacy and scalability of healthcare records. And they put forth a demonstration of a decentralized application showing the same. And they provide with insight of where and how the improvement can be induced in the current standards (Tables 1 and 2).

3 Statement of Problem

As per the aforementioned limitations of the Blockchain system's acceptance as a standard across various field is stranded. Blockchain dilemma which consists of security, scalability, and decentralization pushes the system down from its optimal performance. The mining of a single block in a Blockchain requires a certain set of hardware with high computational power hence high energy consumption. Since a specialized hardware is required the participants in the process of mining becomes low due to the cost involved to invest in the hardware and maintaining it. Eventually this affects the scalability of Blockchain. A consensus algorithm is a procedure through which all the peers of the Blockchain network reach blockchain-based ledger. Many consensus algorithms have meaningless energy costs. A large number of nodes make calculations, but in reality, only the first one to do so receives, a reward. The consensus process already consumes a lot of computational power. Meanwhile, resource accumulation can lead to being more network centralized rather than decentralized and distributed. To lower the energy consumption and to speed up the computational process of consensus with low power consumption the proposed framework makes use of *nano nodes*. Such networked *nano nodes* are low-powered IoT devices with low computational power.

4 System Design

4.1 Architecture

The proposed architecture shows four major components as an extended architecture for basic Blockchain system. Such components include Blockchain network, client, storage and *nano node* known as nodemcu. Here we elaborate on the participation of various components in the proposed solution and their functionality.

1. Blockchain network
2. Client device
3. Storage
4. Nano node (nodemcu)

The following description explains the functionality of all the architectural components

- Blockchain network—It is the distributed peer-to-peer network of blocks that are spread across multiple devices in the network. Since it is distributed there is no centralized control of the network and this lays the foundation stone for any Decentralized applications (dApps).
- Client device—It is the initiator of the transaction and acts as the entry point to its blockchain network. It may also be used to hold the local copy of its blockchain that resides on the client's machine.
  A transaction here is termed as the exchange of information, events, or monetary value between two-party. The client is the initiator of the transaction and the receiver is at the other end of the transaction process.
- Nano node—It the most important part of the proposed solution. All the major tasks such as fetching the blocks from the network, verifying and validating each block, computing a new block, and adding it to the blockchain network to mark the completion of the transaction are done on the *nano node*.

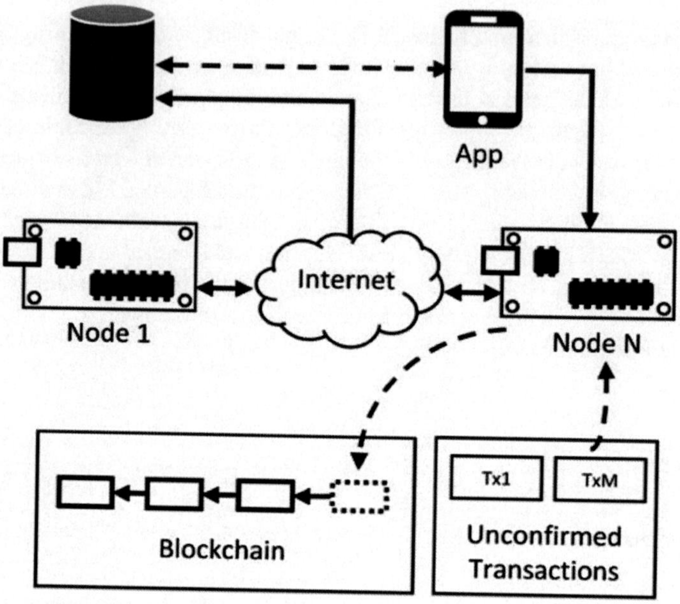

Fig. 1 Architecture of the proposed solution

As shown in Fig. 1 As soon as the new transaction takes place it is moved to an unconfirmed state until the transaction is verified and added to the blockchain network. This triggers an event in the neon node i.e. to fetch the entire block of the user that initiated the transaction and verify them. Once the entire response block is verified now it's time to add the new transaction to the block and then to the blockchain network. Now, this process is done parallel across multiple nodes all nodes are interconnected via the Internet so that they can communicate with one another despite not using public IP addresses. Once all the node reaches to a consensus on which block is to be added. This newly created block is added to the blockchain network and alongside the peer-to-peer network, storage is also updated with the latest transaction record. The transaction is now on the blockchain network and this marks the end of the transaction.

4.2 *Logical View*

A new Blockchain system based on nano node was proposed to deal with the blockchain's scalability issue. The following steps were used to implement the proposed system that is mentioned in Fig. 2 are as follows

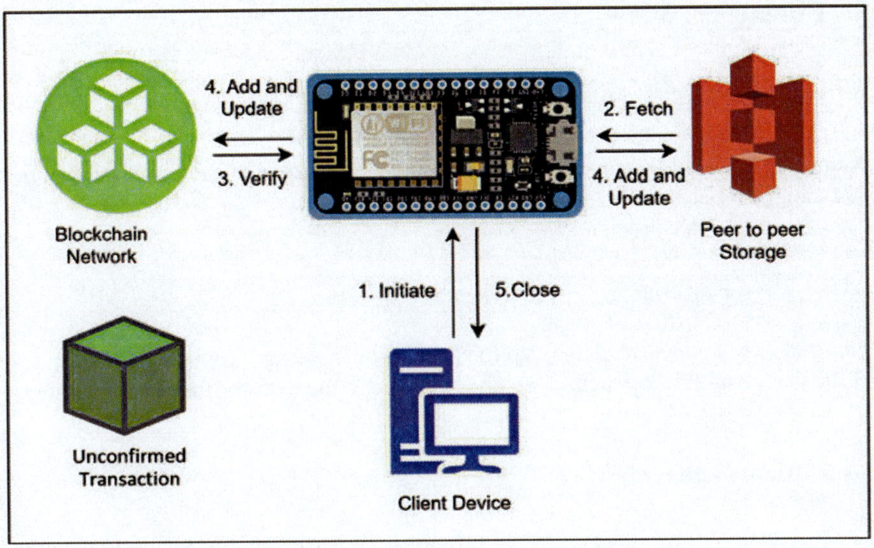

Fig. 2 Logical view of the proposed solution

- The client initiates the transaction and the request sent to the *nano node*.
- The *nano node* fetches the block of the client using its unique identifier from the peer-to-peer storage.
- The *nano node* verifies the received blocks and starts the consensus.
- After the consensus is reached the newly generated block is added to the client's blockchain and update in the storage.
- Once the blockchain is updated with the new transaction. It is marked as completed.

4.3 Data Acquisition and Processing

So the data of the blocks are stored in a distributed peer-to-peer storage in the form of encrypted *JSON* object which is easier to process and takes very little storage. The reason to opt with *JSON* object was due to its lightweight and the way data is structurally stored in it.

So the received data is processed and used to check the integrity of the entire blockchain. It is calculated based on the chain that connects two different blocks on a blockchain i.e. the hashes. The hashes of every block are check and matched and if the data is non-tampered the blockchain is verified.

5 Functional View

5.1 Process

In this paper we follow set of steps to complete the process of the proposed solution that are given below.

5.1.1 Data Fetching

This step is to collect the data from the peer-to-peer storage. By using the key i.e. id of the user. And this key acts as an endpoint to get the data for further processing.

5.1.2 Information Flow

The information flow of the proposed solution is shown in the Fig. 3 it has four subject i.e. User, *Nano node*, Blockchain network, Storage. The flow of information is as follows:

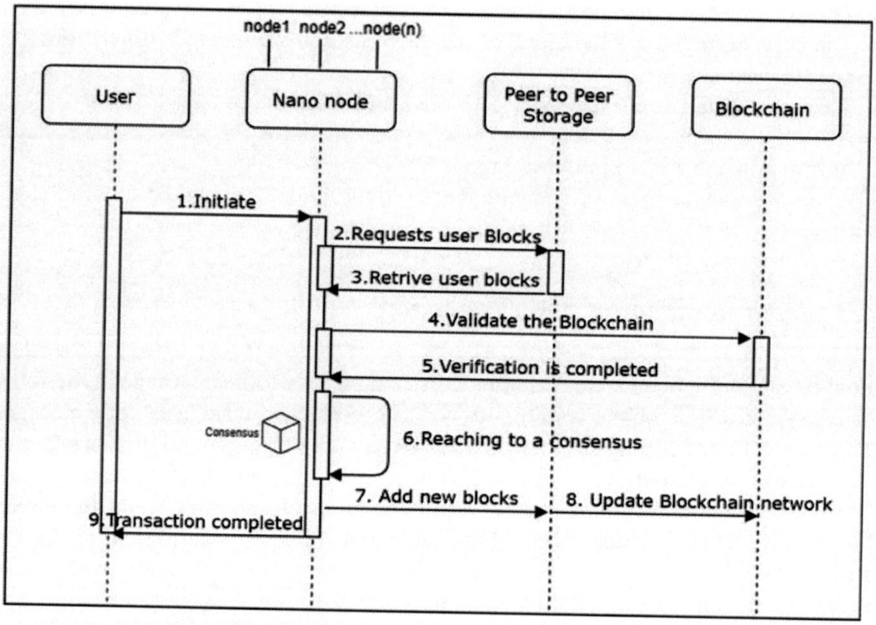

Fig. 3 Information flow of the proposed solution

1. Initiate—The user/client initiates the transaction i.e. an event/action between the user (sender) and the receiver.
2. Requests user block—The node then requests the blocks using the id provided by the Initiator.
3. Retrieve user block—The nodes get a response object for the request is made.
4. Validate the Blockchain—The blockchain is validated for the integrity of the data. This validating process makes use of the primitives of the blockchain that is the chain of hashes, thus validating this link of hashes assures integrity.
5. Verification is Completed—On successful validation of the hashes, the blockchain is verified.
   Upto this step the process that was followed is as per standard Blockchain processing.
6. Consensus—To create the new block Consensus mechanisms is triggered which is explained in the later section.
7. Add new block—After the new block is generated both the blockchain
8. Update Blockchain network—the peer-to-peer storage is updated with the latest block data.
9. Complete—After updating the Blockchain network the transaction is marked as completed.

6 Consensus Mechanisms

The consensus algorithm plays an integral part in any blockchain network. So choosing a good consensus to ensure the goals that a blockchain wanted to achieve becomes essential for both performance and stability.

The flow step of the consensus algorithm for the proposed solution is shown in Fig. 4. The process of this consensus algorithm consists of selecting a leader among a set of nodes, generating hash for the block, and verifying the blockchain before updating the blockchain network with the generated block.

So for the proposed solution, we developed a consensus algorithm that will that is optimized and fast and requires very less computing. The steps involved in the consensus algorithm are stated below

Step I

The Unconfirmed (unsigned) transaction is broadcasted to the available node. As shown in Fig. 5a there are 4 available nodes during the transaction processing.

Step II

The available nodes at randoms pick a leader to create the hash for the new block to be generated. As seen in Fig. 5b node (2) is the chosen leader. If the selected leader is down then another leader is chosen.

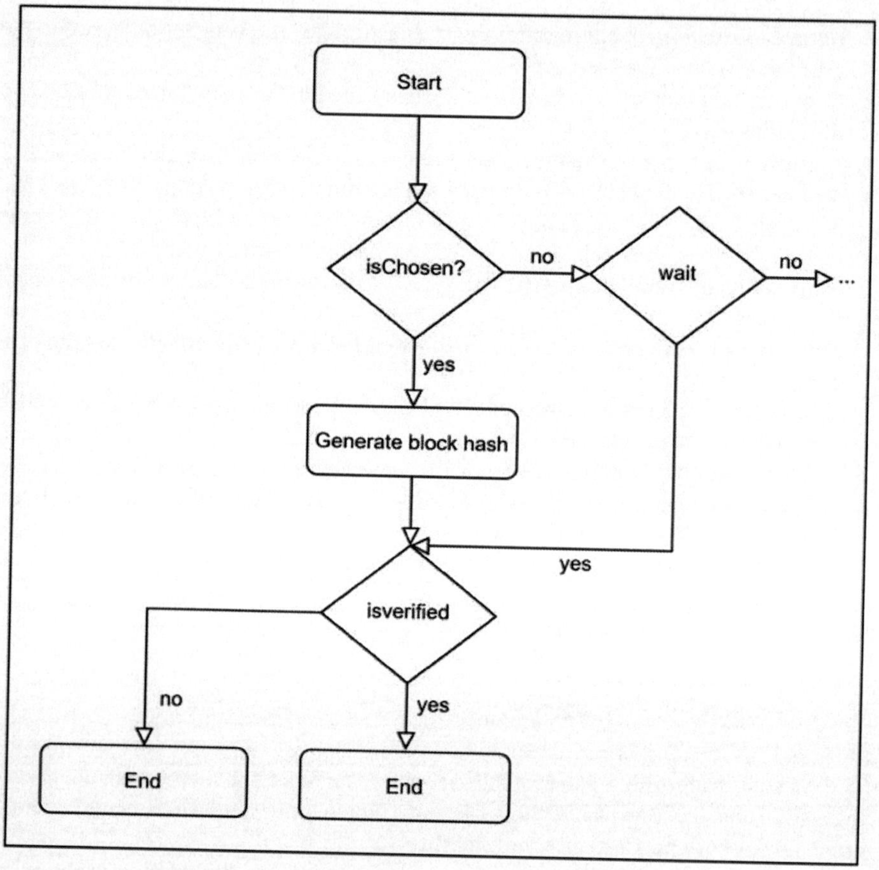

Fig. 4 The flow step of the consensus algorithm

Step III
The leader node signs the block and broadcasts it to all the available nodes in the transaction. As shown in Fig. 5c node (2) sends the singed block to all the active nodes.
Step IV
As shown in Fig. 5d All the nodes add this node to their copy of the blockchain and then reverify the updated blockchain network. In which if the verification is successful the created block is ready to be added to the blockchain network and storage.

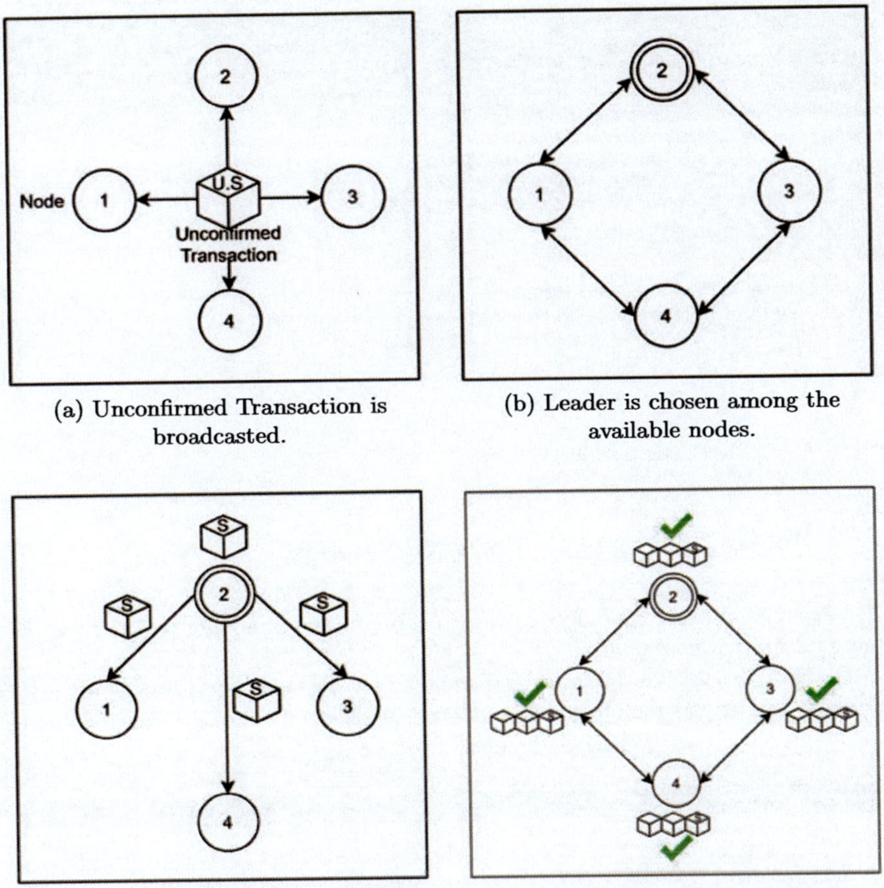

(a) Unconfirmed Transaction is broadcasted.

(b) Leader is chosen among the available nodes.

(c) Leader generates the block and broadcasts it.

(d) All the blocks are verified

Fig. 5 Consensus algorithm of the proposed solution

7 Experimental Results

7.1 Data Set

In the proposed solution for a demonstration, I have created multiple users and assigned them a blockchain with transactional details and hashes. This Data is distributed and is in *JSON* format.

In Fig. 6 the mentioned user data set is of the user "Ajax" with the 50 blocks in it and the Id assigned to each block is auto-generated.

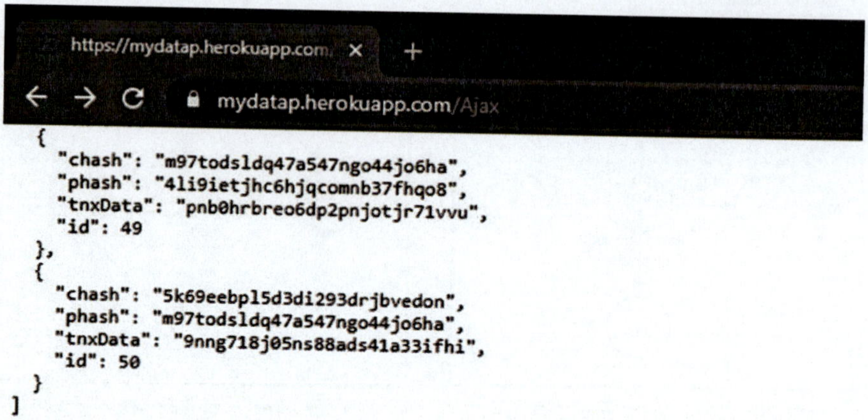

Fig. 6 User's BlockChain data (blocksize=50)

7.2 *Working Example for Proposed Solution*

The demonstration of the proposed system is on the serial monitor which is the I/O interface for our *nano nodes*.

In Fig. 7, I enter the Id as "Ajax" and along with transaction data that I need to verify and add to the blockchain network. So as soon I hit send the *nano node*

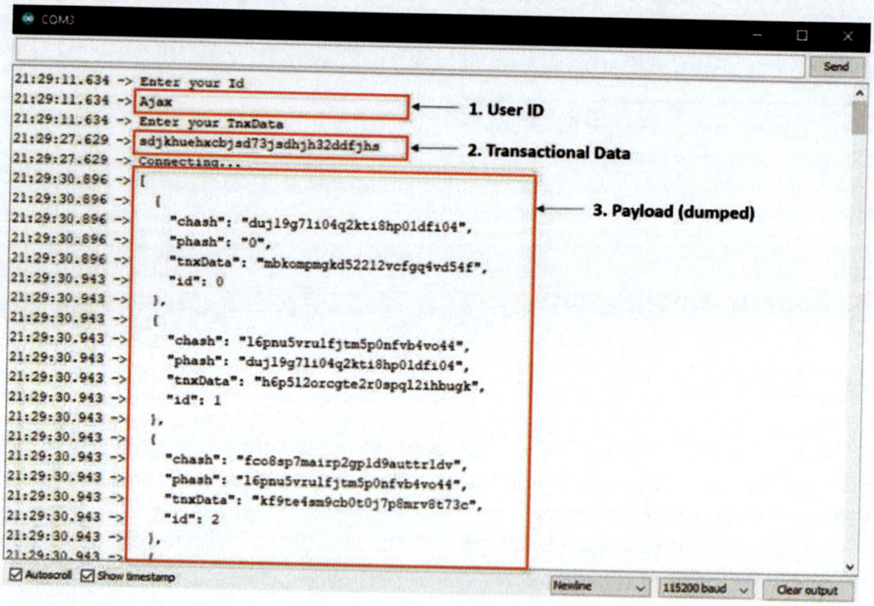

Fig. 7 Passing the data to the nodes and getting the payload

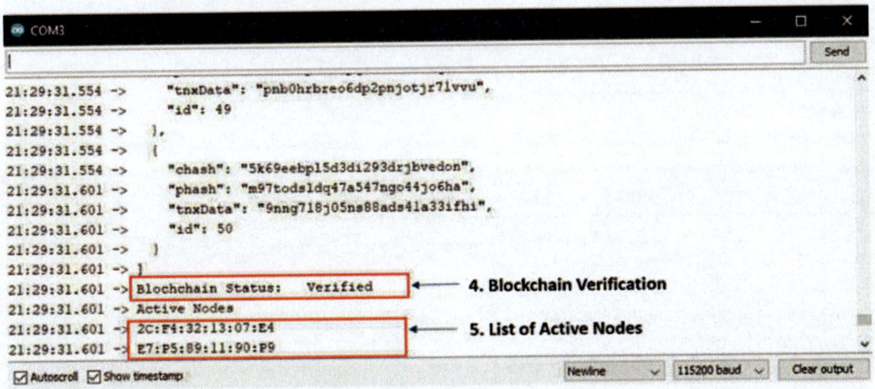

Fig. 8 Verification of blocks and list of available node

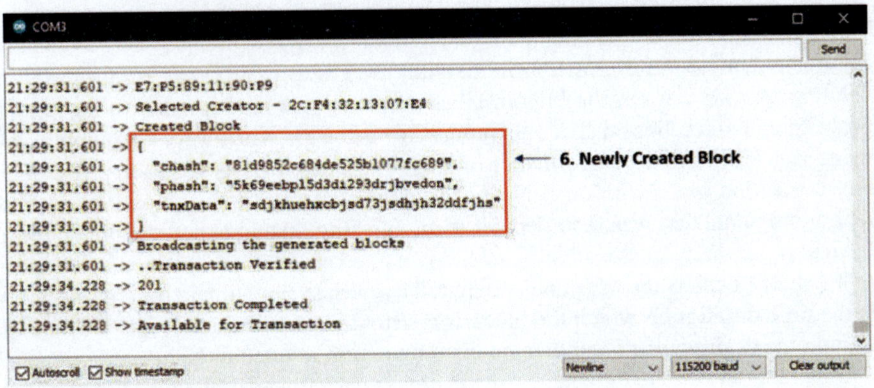

Fig. 9 Newly generated block

connects to the network to get the blockchain details for the ID “Ajax”. The response is called a payload which is dumped on the serial monitor.

In Fig. 8, we can see the result of verifying each block for integrity and after successful verification, the node searches for the other active node to be a part of this transaction and lists it out as shown.

In Fig. 9 Show the newly generated which is further verified by all the node and on successful verification the block is added to the blockchain with a postcode of 201. After which the nodes mark themselves as available for other transaction processing.

In Fig. 10 we can see the newly generated block added to the blockchain system making the length of the blockchain 51.

```
  {
    "chash": "5k69eebpl5d3di293drjbvedon",
    "phash": "m97todsldq47a547ngo44jo6ha",
    "tnxData": "9nng718j05ns88ads41a33ifhi",
    "id": 50
  },
  {
    "chash": "81d9852c684de525b1077fc689",
    "phash": "5k69eebpl5d3di293drjbvedon",
    "tnxData": "sdjkhuehxcbjsd73jsdhjh32ddfjhs",
    "id": 51
  }
]
```


Fig. 10 Updated user's BlockChain data

7.3 Experiment Results

Figure 11 depicts the transaction processing time of Bitcoin, Ethereum, proposed solution. As we can see the bitcoin takes the most time and it's due to the heavy processing that it relies on. The Ethereum is the same too with a very slight improvement over BTC. But the performance of the proposed is was sky high and processing the transaction under a minute. Note: All the processing time is supported by the real mining data that was recorded while mining Bitcoin and Ethereum on the local device.

Figure 12 depicts the time that is required by each system to reach a consensus. It is the time duration in which the block is distributed to the miner by the ledger and mining the desired block and then confirming it with the network.

Figure 13 shows the time taken by the *nano nodes* to verify the range of transactions. As we can observe when we increase the transaction count the time duration

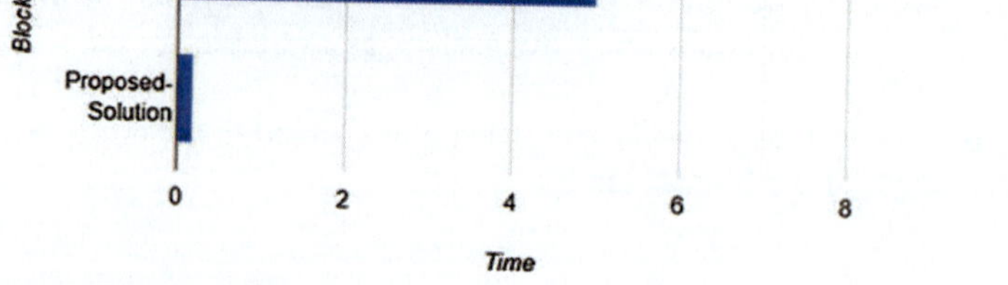

Fig. 11 Comparing transaction processing time

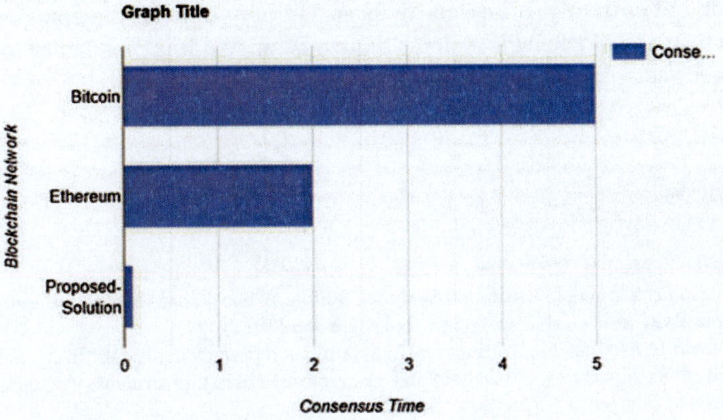

Fig. 12 Comparing consensus processing time

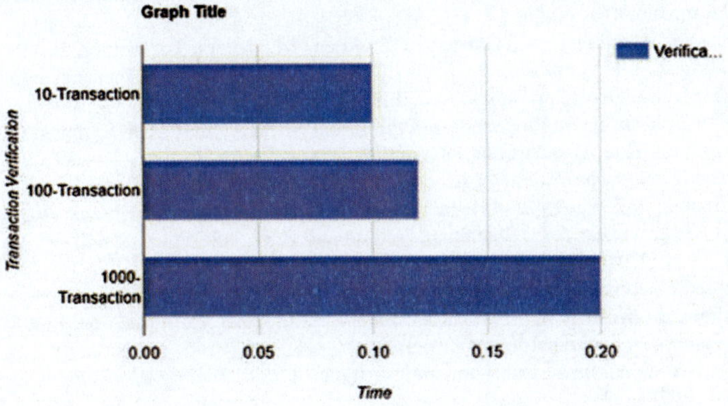

Fig. 13 Block verification time

has also increased the margin of this is very nominal compared to the other network. Even when we increase the number of transactions by a large amount the system accommodates it rather than choking it increases throughput.

8 Conclusion

The backbone of the entire blockchain network is based on its utilization of the peer-to-peer network. So when having a light *nano node* connected to a network of *nano nodes* where each of the nodes can validate the blocks with communication with the other block it gives high availability and thus less time to authenticate and add a new transaction to the blockchain network and for blockchain technology to thrive it is necessary to increase the transaction rate to accommodate more and more transaction but currently, it limited by its scalability while also keeping security in mind. So to make blockchain systems more sustainable it is mandatory to come up with a more scalable and secure solution.

References

1. Abraham, I., Malkhi, D., Nayak, K., Ren, L., Spiegelman, A.: In the 21st International Conference on Principles of Distributed Systems. Solida: A Blockchain Protocol Based on Reconfigurable Byzantine Consensus, pages 1–17. Lisbon (2017)
2. Zhumabekuly Aitzhan, N., Svetinovic, D.: Security and privacy in decentralized energy trading through multi-signatures, blockchain and anonymous messaging streams. ieee trans. Dependable Secure Comput. **15**(5):840–852, 2016
3. Kaushik, A.A.: In: Chaudhary "Blockchain Literature Survey "2017 IEEE Conference on Recent Trends in Electronics Information & Communication Technology (2017)
4. Ali, S., Wang, G., White, B., Cottrell, R.L.: A blockchain-based decentralized data storage and access framework for pinger **17**, 1303–1308 (2018)
5. Ben-Sasson, E., Chiesa, A., Garman, C., Green, M., Miers, I., Tromer, E.: Virza, M. In: Proceedings–IEEE Symposium on Security and Privacy. Zerocash: Decentralized Anonymous Payments from Bitcoin, pp. 459–474. IEEE, San Jose (2014)
6. Canetti, R., Rabin, T.: In: Proceedings of the Twenty-fifth Annual ACM Symposium on Theory of Computing. Fast Asynchronous Byzantine
7. Christidis, K., Devetsikiotis, M.: Blockchains and smart contracts for the Internet of Things. IEEE Access. 4, 2292–2303. 4. Principles and Practices. Chap. In: Knirsch, F., Unterweger, A., Eibl, G., Engel, D. (eds) Sustainable Cloud and Energy Services (2016)
8. Eyal, I., et al.: Bitcoin-ng: A scalable blockchain protocol. In: 13th USENIX Symposium on Networked Systems Design and Implementation (NSDI 16) (2016)
9. Greenspan, G.: Multichain private blockchain—white paper. Coin Sciences Ltd. MultiChain-White-Paper.pdf, Technical report (2015)
10. Hafid, A.: Abdelhakim senhaji hafid and mustapha samih. Scaling Blockchains Compr. Surv. **10**, 1109 (2020)
11. Hazari, S.S., Mahmoud, Q.H.: A parallel proof of work to improve transaction speed and scalability in blockchain systems, 916–0921 (2019)
12. Hearn, M., Corda.: A distributed ledger. Version 0.5. Technical report, Technical report. Corda. Accessed September 2018 (2016)
13. Katz, J., Lindell, Y.: Introduction to Modern Cryptography, 1st edn., pages 1–498. Chapman and Hall/CRC. New York (2007)
14. Knirsch, F., Unterweger, A., Engel, D.: Privacy-preserving blockchain-based electric vehicle charging with dynamic tariff decisions. In: J, pages 71–79 (2018) [32] Munsing, E., Mather, J., Moura, S. (eds) In: 2017 IEEE Conference on Control Technology and Applications (CCTA). Blockchains for Decentralized Optimization of Energy Resources in Microgrid Networks (IEEE, Mauna Lani pp. 2164–2171, 2017. Comput. Sci. Res. Dev. (CSRD) 33(1)
15. Kohad, H., Kumar, S., Ambhaikar, A.:. Scalability issues of blockchain technology. Int. J. Eng. Adv. Technol. (IJEAT) **9**. ISSN: 2249–8958 (2020)
16. Krampe, L., W"unsch, M.: Schlussbericht Mieterstrom—Rechtliche Einordnung. Organisationsformen, Potenziale und Wirtschaftlichkeit von Mieterstrommodellen (MSM)
17. Lamport, L., Shostak, R., Pease, M.: The byzantine generals problem. ACM Trans. Program. Lang. **4**(3), 382–401 (1982)
18. Lesser, J.A., Su, X.: Design of an economically efficient feed-in tariff structure for renewable energy development. Energy Pol **36**(3), 981–990 (2008)

19. Mengelkamp, E., G"arttner, J., Rock, K., Kessler, S., Orsini, L., Weinhardt, C.: Designing microgrid energy markets: a case study: the Brooklyn microgrid. Appl. Energy **210**, 870–880 (2018)
20. Mengelkamp, E., Notheisen, B., Beer, C., Dauer, D., Weinhardt, C.: A blockchain-based smart grid: towards sustainable local energy markets. Comput. Sci **33**(1), 207–214 (2018)
21. Mihaylov, M., Jurado, S., Avellana, N., Van Moffaert, K., De Abril, I.M., Nowé, A.: Virtual currency for trading of renewable energy in smart grids. In: 11th International Conference on the European Energy Market, EEM. IEEE, Krakow, NRGcoin (2014)
22. Mihaylov, M., Jurado, S., Moffaert, K.V., Now, A.: In Proceedings of the 3rd International Conference on Smart Grids and Green it Systems. NRG-X-Change—A Novel Mechanism for Trading of Renewable Energy in Smart Grids, pages 101–106. SciTePress, Barcelona (2014)
23. Nikitin, K., Kokoris-Kogias, E., Jovanovic, P., Gasser, L., Gailly, N., Khoffi, I., Cappos, J., Ford, B.: 26th usenix security symposium (usenix security 17). In: CHAINIAC: Proactive Software-Update Transparency via Collectively Signed Skipchains and Verified Builds, pages 1271–1287. USENIX Association, Vancouver (2017)
24. Novo, O.: Blockchain meets iot: an architecture for scalable access management in Iot. J. Internet Things Class Files **14**, 8 (2018)
25. Panda, S.S., Mohanta, B.K., Satapathy, U., Jena, D., Gountia, D., Patra, T.K.: Study of blockchain based decentralized consensus algorithms. TENCON **908–913**, 2019 (2019)
26. Qin, K., Gervais, A.: An Overview of Blockchain Scalability. Interoperability and Sustainability. Imperial College, London (2019)
27. Sikorski, J.J., Haughton, J., Kraft, M.: Blockchain technology in the chemical industry: machine-to-machine electricity market. Appl. Energy **195**, 234–246 (2017)
28. Sun, X., Ansari, N.: Edgeiot: Mobile edge computing for the internet of things. IEEE Commun. Mag. **54**(12), 22–29 (2016)
29. Unterweger, A., Knirsch, F., Leixnering, C., Engel, D.: In: 2018 9th IFIP International Conference on New Technologies, Mobility and Security (NTMS). Lessons Learned from Implementing a Privacy-Preserving Smart Contract in Ethereum. IEEE, Paris (2018)
30. Vukolic, M.: In 'open problems in network security. In: The Quest for Scalable Blockchain Fabric: Proof-of-Work vs. BFT Replication, pages 112–125. Cham (2016)
31. Wood, G., Ethereum, A.: Secure decentralised generalised transaction ledger. Technical report, Technical report, Ethereum ethereum.github.io/yellowpaper/paper.pdf [35] Croman, K., Decker, C., Eyal, I., Gencer, A.E., Juels, A., Kosba, A., Miller, A., Saxena, P., Shi, E., G"un Sirer, E., Song, D., Wattenhofer, R. in International Conference on Financial Cryptography and Data Security (2017)
32. Xie, J., Richard Yu, F., Huang, T., Xie, R., Liu, T., Liu, Y.: A survey on the scalability of blockchain systems "0890-8044/19 2019 IEEE

Performance Analysis of Wi-Fi and LTE-LAA/LTE-U

Prajwal S. Telkar, Deekshith Nayak, D. N. Rahul Raj, and V. Kiran

Abstract Extensive proliferation of Internet and data services over the past decade has turned the interest of researchers toward unlicensed spectrum in radio space. The unlicensed band is especially appealing as a result of enormous accessibility of uninhibitedly available range. LTE portable organizations has seen a staggering ascent in the course of recent years and to serve the immensely expanding information clients and to give upgraded insight to versatile broadband customers, utilizing unlicensed range which has increased an inescapable interest. But the impact of such a utilization on the long existing unlicensed users such as Wi-Fi is significant and accountable, and hence, design of such network topology is challenging. In the context of LTE, LTE-U and LAA-LBT have been emerged as the technology of choice in order for it to coexist with other wireless technologies which has been long existing in unlicensed band. With introduction to coexistence of LTE and Wi-Fi and deliberating coexistence scenarios, the performance of various network deployments such as LTE-U vs Wi-Fi and LTE-LAA vs Wi-Fi has been evaluated. Deployment scenarios of LTE and Wi-Fi in 5 GHz band with duty cycle-based mechanism and LBT based mechanism in the 5 GHz band are evaluated in terms of various performance parameters. The coexistence network topology was simulated in the basic scenario using NS3, and the improvement in performance was demonstrated and evaluated in terms of throughput, latency and jitter. Performance of transport layer protocols were compared using a coexistence model in an indoor scenario.

Keywords Long-term evolution (LTE) · LTE-unlicensed (LTE-U) · Licensed-assisted access (LAA) · Listen-before-talk (LBT) · Throughput · Contention window

---

P. S. Telkar (✉) · D. Nayak · D. N. Rahul Raj · V. Kiran
R V College of Engineering, Bengaluru, India

V. Kiran
e-mail: kiranv@rvce.edu.in

© The Author(s), under exclusive license to Springer Nature Singapore Pte Ltd. 2022

A. K. Nagar et al. (eds.), *Intelligent Sustainable Systems*, Lecture Notes in Networks and Systems 333, https://doi.org/10.1007/978-981-16-6309-3_31

1 Introduction

The idea of deploying LTE in unlicensed band has pulled researchers toward exploration of various mechanisms which ensures fair or friendly coexistence of LTE and Wi-Fi. Wi-Fi follows a decentralized algorithm by using CSMA protocol for channel access, in which all the end nodes analyses the channel for any possible contention before the transmission. But LTE has a centralized architecture, where a base station will take care of scheduling of channel access. In LTE, base station acts identify applicable funding agencies here. If none, delete this as a central entity that governs the communication in the spectrum. LTE-U employs duty cycle-based technique for friendly coexistence of LTE with Wi-Fi, in which it has a specific active duration for Wi-Fi, and the remaining time slot will be allotted for LTE. LTE without any protocol modifications can transmit without worrying about the conditions of surrounding Wi-Fi users. On the other hand, every time Wi-Fi node sensing the transmission in the channel may cause it starve for longer period. This is certainly a disadvantage to Wi-Fi users and can experience severe deterioration in their performance. Hence, it is a need to deploy suitable protocol changes to LTE in order to ensure fair and amicable coexistence with Wi-Fi users. The fundamental idea of using unlicensed band is to increase spectral efficiency and not to dismount the Wi-Fi users from that band. Therefore, several variants of LTE have been proposed as the solution to the said problem. LTE-LAA employs a fairness -based mechanism for coexistence, which allows the LTE to transmit by sensing the channel using LBT. In LTE, no specific channel sensing mechanism exists, while in Wi-Fi, the Media Access Control (MAC) layer is based on carrier sense multiple access with collision avoidance (CSMA/CA) in which each station or the Wi-Fi access point senses the channel for idleness. If the idle channel is found, then it will be allotted for LTE to transmit. The main objective of this paper is to analyze LTE-U and LTE-LAA mechanisms in unlicensed band in terms of performance parameters like throughput, mean delay, and mean jitter. Performance analysis of transport layer protocols has been done in indoor coexistence scenario by considering the cumulative distributive function of the traffic generated through FTP.

2 Methodology

The paper implements LTE-U [1] and LTE-LAA [2] as mechanisms for coexistence of LTE and Wi-Fi. LTE-U is a duty cycle-based approach. Hence, it has specific active duration in time division fashion for LTE to transmit. However, LAA-LBT follows the LBT algorithm, which is based on CCA technique. CCA stands for clear channel assessment, which analyzes the spectrum to search for a clear channel. Energy in the channel will be taken as a constraint to define it as clear or not. The detected energy in the channel will be compared against the predefined threshold value and will be marked clear only if it falls less than the threshold. Both the designs will be implemented in NS3 software, [3] using namespace in C++. Design flow of NS3 will be followed to integrate various layers of LTE and Wi-Fi architecture. Starting from creating nodes and then installing protocol stacks, NS3 will ensure the normal functionality of LTE and Wi-Fi along with coexistence. A basic scenario will be designed to analyze the performance parameters, and the values will be noted from monitors in NS3. Tabulated values of those parameters will be plotted for visualization purposes. In order to analyze the variation in performance between TCP and UDP, an indoor scenario will be built in a building with predefined specification, such that all the cells are centrally spaced within the building.

3 Design

Basic coexistence scenario

The model is designed to implement a basic scenario for coexistence of LTE with Wi-Fi in unlicensed band [4, 5]. The executable program is written in C++ language in the NS3 coding environment. The scenario consists of two cells, which are placed such that their radio environment overlaps. Notionally it represents two operators such that their transmission will affect different mechanisms required for coexistence such as CCA, coding, and adaptive modulation. Both the cells represented notionally in this scenario consists of one UE and one base station per cell. When modeling a cell as LTE cell, BS will be modeled as eNB and UE as UE. While modeling a cell as Wi-Fi, BS will be modeled as an AP and UE as station (STA). Along with the two cells in the model, both BS are connected to a client node, which is a backhaul node. The connected client node creates data required for transmission in the downlink direction from BS to UE. The downlink path from client's to user equipment (UE) was modeled as low latency paths. In the model, d1 and d2 symbols represent intracell and intercell distances in meters as shown in Fig. 1. The runtime execution of the

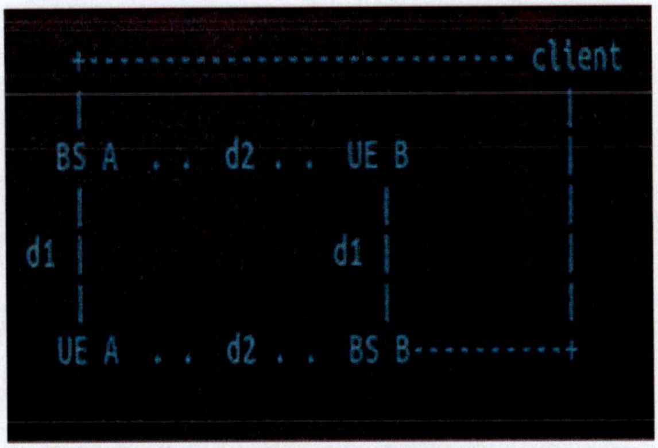

Fig. 1 Design of basic scenario

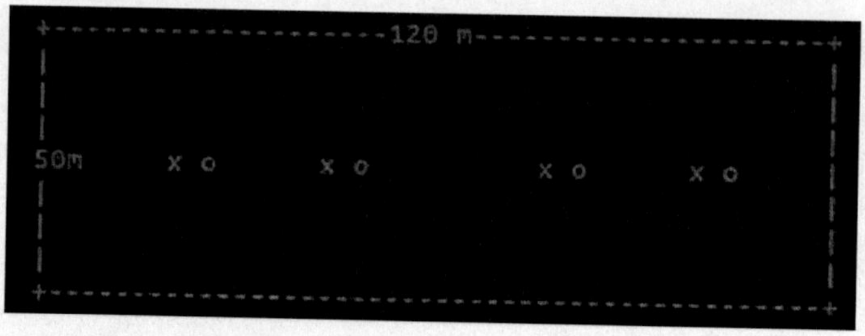

Fig. 2 Design of indoor scenario

program can be done by passing necessary command-line arguments. The output can be obtained by passing the command "./waf –run laa-wifi-basic."

Coexistence in indoor scenario

Indoor scenario is an extended model of the code designed in basic scenario [6]. The program is written in C++ language in the NS3 coding environment. The design contains two operators with four cells each deployed in a single floor building. All the cells are equally spaced and centered as per the shorter dimension of the building, one cell from each operator is added together to make a set and these sets are centered as per the longer dimension of the building. In the design, 'x' and 'o' denote the small cell center points of each operator as shown in Fig. 2. In Wi-Fi, the 'x' and 'o' correspond to access points. The UEs (STAs) move around within the bounding box at a speed of 3 km/h. In general, the program can be configured at runtime by passing command-line arguments. In addition, some other variables may be modified at compile time. For simplification, an IEEE 802.11n configuration with the MinstrelWifi (with the 802.11a basic rates) is used. The runtime execution of the program can be done by passing necessary command-line arguments. The output can be obtained by passing the command "./waf –run laa-wifi indoor." Analysis of various parameters like user perceived throughput, throughput, TxOffered, delay, jitter, and latency of the network in the scenario was done using monitors in NS3.

4 Implementation

LTE and Wi-Fi models were designed in NS3 namespace. In C++, namespace is a declarative region that resolves scopes for variables defined inside it. Access manager used in the design to govern the spectrum management was added as a class named ChannelAccessManager. The class ChannelAccessManager was provided with subclasses such as LteEnbRrc, LteEnbMac, and LteEnbPhy to have control over actual layers of LTE. The class ChannelAccessManager was integrated with the LTE module in the NS3 namespace, such that it will not affect the original functionality of LTE. An LteEnbRrc class instance was added in each LTE eNodeB, so that each eNodeB will have its own access manager, which can be set by the attribute ChannelAccessAttribute. ChannelAccessManager was added as global value as well as command-line parsing argument, so that the access method can be specified at runtime, which can take three different values as follows duty cycle, LBT, and default. DutyCycleAccessManager is a subclass in the main class ChannelAccessManager, such that it was configured with three attributes, which are OnDuration, OnStartTime, and DutyCyclePeriod. Access manager for LBT is implemented as one more subclass named LbtAccessManager, which handles exponential back off and listen before talk as per 3GPP standards. WifiPhy and MacLow are the two listeners for LbtAccessManger class. Coexistence was modeled using the LaaWifiCoexistenceHelper method.

5 Results and Discussion

Various performance parameters like throughput, mean delay, and mean jitter are analyzed. Tabulating the results for LTE-U duty cycled mechanism, and optimizing this technology with LAA mechanism, graphs have been plotted. For LTE-U and Wi-Fi deployment scenario, the coexistence environment has been created in NS3 simulator consisting of two cells, and the performance parameters like throughput, mean delay, and mean jitter have been analyzed for duty cycling mechanism. These parameters have been analyzed with duty cycle approach with TCP protocol and non-voice constant. (1) Throughput: The throughput of the duty cycle approach was plotted for TCP protocol with non-voice, and the energy threshold was maintained at –72.0 db with Cell A as Wi-Fi and Cell B as LTE. From Fig. 3, it can be inferred that

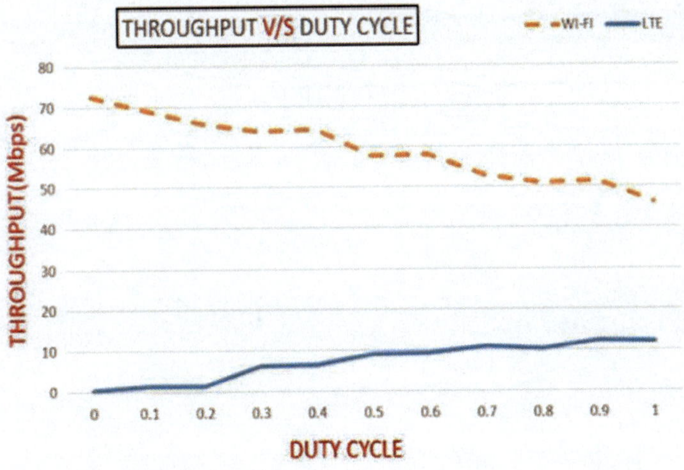

Fig. 3 Throughput versus Duty cycle LTE-U

as the duty cycle time increases, there is drop in throughput of Wi-Fi but increase for LTE since this makes LTE access the channel more than Wi-Fi and thus the transmission of Wi-Fi is affected in unlicensed band. (2) Mean delay: Mean delay has been tabulated and plotted for duty cycle approach with Cell A as Wi-Fi and Cell B as LTE with optimum energy threshold of −72.0 db for TCP protocol with no voice. As observed in Fig. 4, initially mean delay is high for LTE since the duty cycle is less since the channel transmission for LTE is less but as the duty cycle increases, delay decreases due to more availability of bandwidth for LTE and rise in delay for Wi-Fi is observed. (3) Mean jitter: Mean jitter has been tabulated and plotted for duty cycle approach with Cell A as Wi-Fi and Cell B as LTE with optimum energy threshold −72.0 db for TCP protocol non-voice. From Fig. 5, it is observed that initially mean

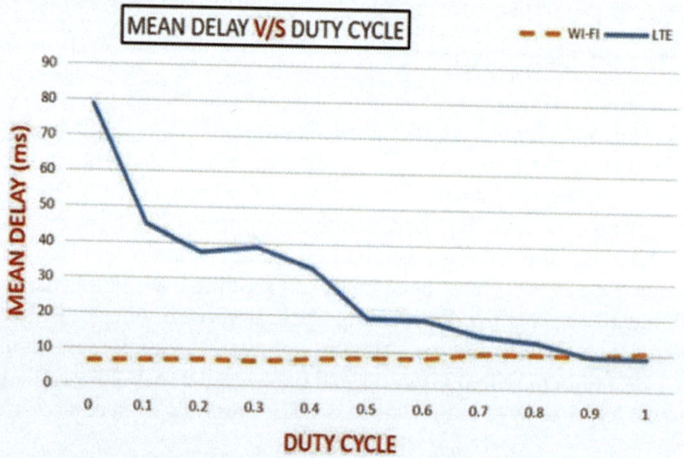

Fig. 4 Mean delay versus Duty cycle LTE-U

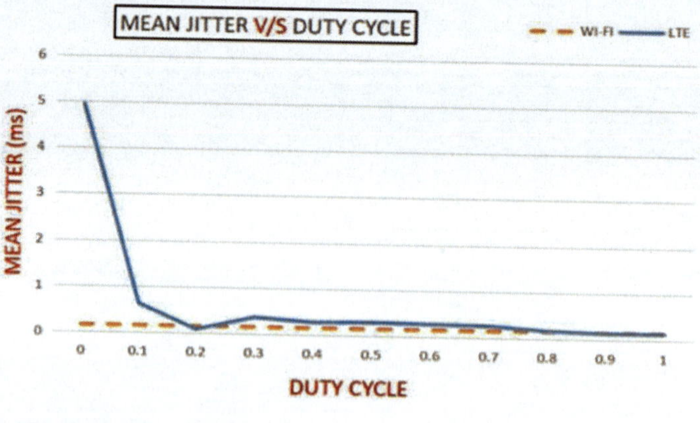

Fig. 5 Mean jitter versus Duty cycle LTE-U

jitter is high for LTE due to small availability of transmission space in the channel, and this becomes nearly constant with low value as the duty cycle increases.

LTE-LAA and Wi-Fi

The coexistence between LTE-LAA and Wi-Fi has been observed by plotting the data by simulating the coexistence environment in NS3 and has been analyzed. In this approach, it can be seen that LBT mechanism has been used and the performance parameters like throughput, mean delay, and mean jitter. The performance parameters are plotted against TXOP. (1) Throughput: The throughput for LTE-LAA and Wi-Fi has been plotted such that Cell A is Wi-Fi and Cell B is LTE-LAA. This has been analyzed with TCP protocol with non-voice and optimum threshold energy –72.0 db. From Fig. 6, it is observed that as the transmission opportunity increases for LTE, the throughput increases since the availability of bandwidth in the channel for LTE to access is more since there is minimal transmission of Wi-Fi in the channel. Thus, the transmission opportunity indicates the increased throughput for LTE when TXOP increases. (2). Mean delay: Mean delay has been tabulated and plotted for LAA-LBT approach with Cell A as Wi-Fi and Cell B as LTE-LAA with optimum energy threshold of –72.0 db for TCP protocol with no voice. From Fig. 7, a downward trend exists such that mean delay decreases for LTE when the TXOP duration increases since the available bandwidth for LTE is more, and thus, the mean delay decreases. (3) Mean jitter: Mean jitter has been plotted for LAA-LBT approach with Cell A as Wi-Fi and Cell B as LTE-LAA with optimum energy threshold –72.0 db for TCP protocol non voice. Mean delay: Mean delay has been tabulated and plotted for LAA-LBT approach with Cell A as Wi-Fi and Cell B as LTE-LAA with optimum energy threshold of –72.0 db for TCP protocol with no voice. From Fig. 8, it can be seen that mean jitter decreases as the transmission duration increases since the available bandwidth in unlicensed channels is more for LTE and thus the jitter value decreases with increase in TXOP duration.

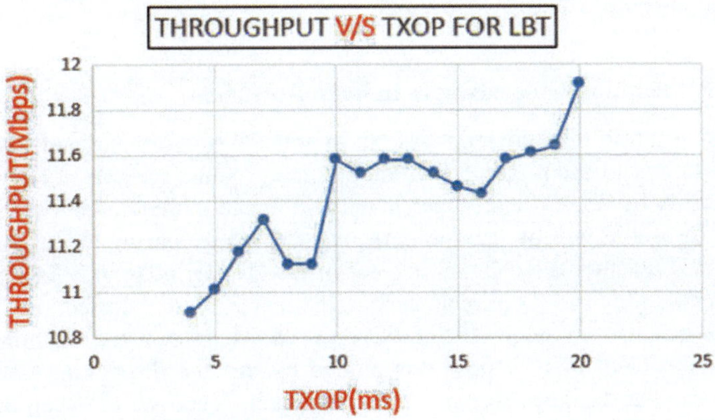

Fig. 6 Throughput versus TXOP for LTE-LAA

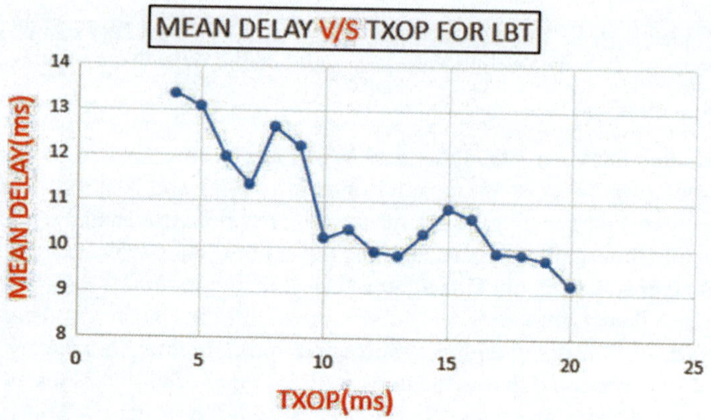

Fig. 7 Mean delay versus TXOP LTE-LAA

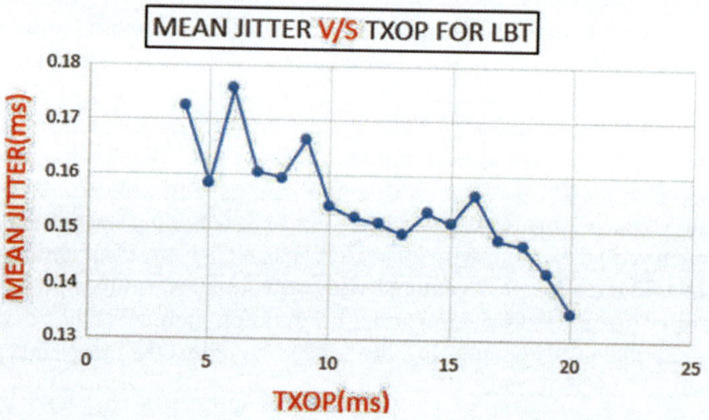

Fig. 8 Mean jitter versus TXOP for LTE-LAA

Results of simulation of coexistence in indoor scenario

The indoor scenario implemented was put in an NS3 environment and CDF of the FTP flows tabulated and plotted using gnuplot tool. Lamba variable in FTP controls traffic intensity by default; it has been given a value 0.5 in the model. Energy detection (ED) threshold will be used in detecting the energy during the CCA process and to decide whether the channel is clear or not. Using TCP: TCP is a transport layer protocol, which gives reliable service and ensures error correction. TCP is a connection-oriented protocol which relies on acknowledgment for reliability.

Throughput CDF of FTP flow was plotted against the throughput obtained in both the cells. The designed scenario had LAA in cell A and Wi-Fi in cell B. Energy threshold was kept at −72.0, and traffic intensity was kept at 0.5. FTP Model 1 of NS3 was used for traffic generation. (1) Throughput: CDF of FTP flow was plotted against

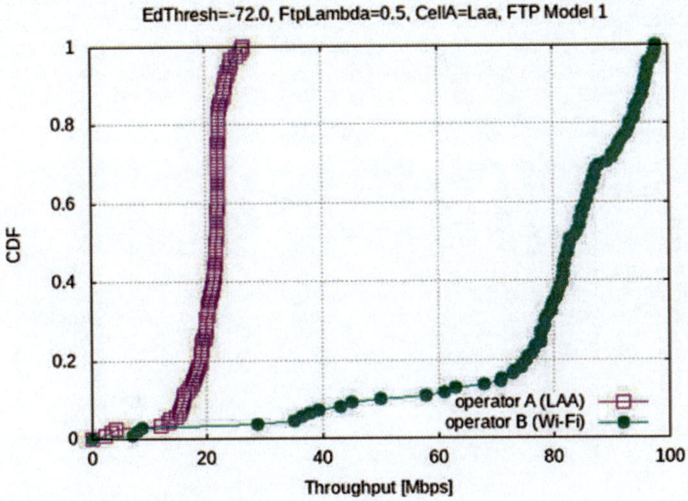

Fig. 9 CDF versus Throughput using TCP

the throughput obtained in both the cells. The designed scenario had LAA in cell A and Wi-Fi in cell B. Energy threshold was kept at –72.0, and traffic intensity was kept at 0.5. FTP Model 1 of NS3 was used for traffic generation. As observed in Fig. 9, as the traffic in the spectrum increases, throughput of Wi-Fi increases drastically because LAA senses the spectrum channels energy above the threshold level and declares the channel as unclear. Hence, every channel will be available for LTE to transmit. (2) Latency: CDF of FTP flow was plotted against the latency obtained in both the cells. The designed scenario had LAA in cell A and Wi-Fi in cell B. Energy threshold was kept at –72.0, and traffic intensity was kept at 0.5. FTP Model 1 of NS3 was used for traffic generation. As observed in Fig. 10 plot, as the traffic in the spectrum increases, latency of both the cells increases slowly because of increase in packet numbers. However, latency for Wi-Fi cells will be less compared to LAA cells because LAA has to follow LBT algorithm to gain access to spectrum. Using UDP: UDP is a communication protocol at the transport layer. It is a connectionless protocol, and it cannot ensure in-order delivery of data; hence, it provides unreliable service. However, it is a very useful protocol in real-time applications like video streaming. (1) Throughput: CDF of FTP flow was plotted against the throughput obtained in both the cells. The designed scenario had LAA in cell A and Wi-Fi in cell B. Energy threshold was kept at –72.0, and traffic intensity was kept at 0.5. FTP Model 1 of NS3 was used for traffic generation. As observed in Fig. 11, as the traffic in the spectrum increases, latency of both the cells increases slowly because of increase in packet numbers. However, latency for Wi-Fi cell will be less compared to LAA cell because LAA has to follow LBT algorithm to gain access to spectrum. However, the throughput in the TCP case was lower than the UDP case because it expects acknowledgment packets for reliable service. (2) Latency: CDF of FTP flow

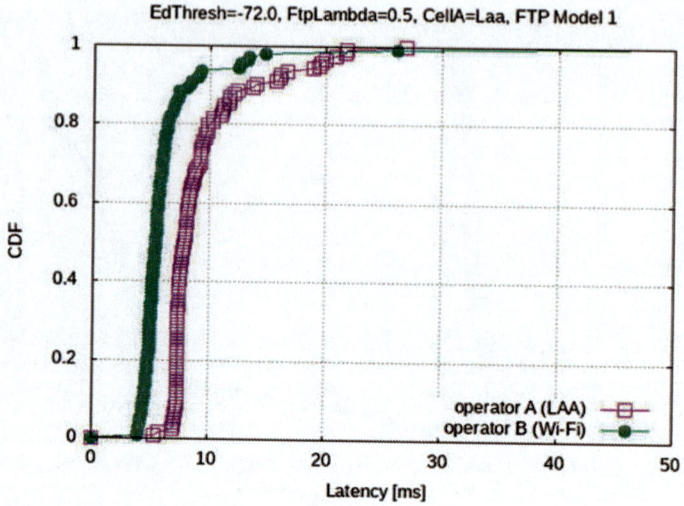

Fig. 10 CDF versus Latency using TCP

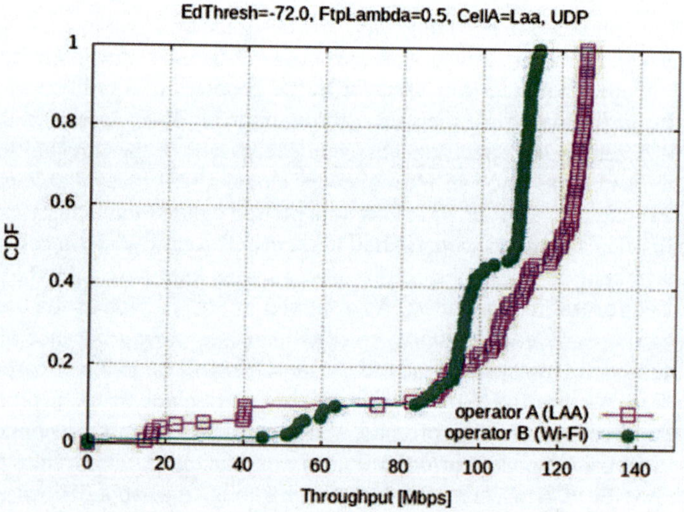

Fig. 11 CDF versus Throughput using UDP

was plotted against the latency obtained in both the cells. The designed scenario had LAA in cell A and Wi-Fi in cell B. Energy threshold was kept at –72.0, and traffic intensity was kept at 0.5. FTP Model 1 of NS3 was used for traffic generation. As observed in Fig. 12, as the traffic in the spectrum increases, latency of both the cells increases slowly because of increase in packet numbers. However, latency for Wi-Fi cell will be less compared to LAA cell because LAA has to follow LBT algorithm to

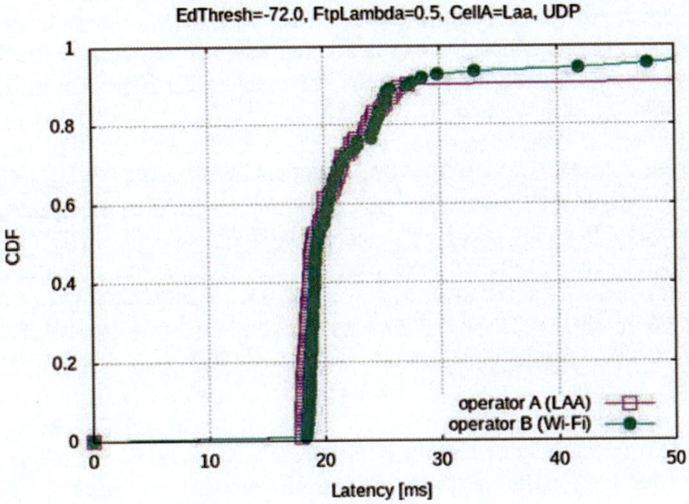

Fig. 12 CDF versus Latency using UDP

gain access to spectrum. However, the latency in the TCP case was higher than the UDP case because the acknowledgment packets increase the traffic in the spectrum.

6 Conclusion

The work has been dedicated to the field of communication to enhance the user experience, since a huge traffic is experienced in the licensed spectrum. It was diffi cult to meet the demands of exponentially growing users and thus decreasing the bandwidth to be allocated in the licensed spectrum was the only option. The codes were designed to meet the specification of our coexistence technology [7] and were successfully implemented and compared with existing technology. Firstly, the duty cycled approach mechanism for coexistence technology was implemented in NS3 which consisted of two cells with Cell A as Wi-Fi and Cell B as LTE and measured the parameters like mean delay, throughput, and mean jitter. The same procedure was repeated on the novel coexistence technology which involved LAA-LBT [8], and these results were compared and observed. Throughput was good for LAA-LBT, since LTE is accessing the channel depending on the traffic coming into Wi-Fi. Energy detection techniques implemented in LBT make sure that LTE can access channels in the shared spectrum only if the energy detected is less than the threshold. Hence, a significant improvement was shown in throughput by LAA without compromising with Wi-Fi throughput, as what happens in case of LTE-U. Mean delay and mean jitter follow the same trend of decreasing with increase in the transmission opportunity [9] for LBT. These variables will not get affected in Wi-Fi cell because sufficient spectrum for carrying the Wi-Fi data will always be made available to Wi-Fi cell. In case of LTE-U, mean delay and mean jitter decrease for LTE with increase in duty cycle, because higher the duty cycle gives more bandwidth to LTE to transmit but lesser to Wi-Fi cell hence as duty cycle for LBT increases, mean delay and mean jitter for Wi-Fi decreases. All the performance parameters were in their improved value in the LAA-LBT mechanism in contrast to duty cycle approach, by considering that Wi-Fi should always be given higher priority for transmission in unlicensed band. For analysis of performance parameters between UDP and TCP, an indoor scenario was designed by considering FTP as the application protocol with default traffic intensity. CDF was taken as a measure of FTP traffic in the design. As the FTP CDF increases, Throughput of Wi-Fi cell increases drastically because the underlying LBT algorithm for LAA cell will govern the channel access. Latency in case of UDP increases slowly with respect to CDF, because CDF directly represents traffic load. Higher throughput in case of UDP in comparison with TCP was observed because TCP expects acknowledgment packets to give reliable service, which in turn account for increase in the traffic and increase in latency.

References

1. Qualcomm.: Lte in unlicensed spectrum: Harmonious coexistence with Wi-Fi. Qualcomm Technologies, Inc.5775, Morehouse Drive, San Diego, CA 92121, U.S.A (2014)
2. Pei, E., Jiang, J.: Performance analysis of licensed-assisted access to unlicensed spectrum in lte release 13. IEEE Trans. Veh. Technol. **68**(2), 1446–1458 (2019). https://doi.org/10.1109/TVT.2018.2886332
3. Giupponi, L., Henderson, T., Bojovic, B., Miozzo, M.: Simulating lte and wi-fi coexistence in unlicensed spectrum with ns-3 (2016). arXiv: 1604.06826 [cs.NI]
4. Cano, C., Leith, D.J.: Coexistence of wifi and lte in unlicensed bands: a proportional fair allocation scheme. In: 2015 IEEE International Conference on Communication Workshop (ICCW), pp. 2288–2293 (2015). https://doi.org/10.1109/ICCW.2015.7247522
5. Mukherjee, A., Cheng, J., Falahati, S., Falconetti, L., Furusk¨ar, A., Godana, B., Kang, D.H., Koorapaty, H., Larsson, D., Yang, Y.: System architecture and coexistence evaluation of licensed-assisted access lte with ieee 802.11. In: 2015 IEEE International Conference on Communication Workshop (ICCW), pp. 2350–2355 (2015). https://doi.org/10.1109/ICCW.2015.7247532
6. Dalili Shoaei, A., Derakhshani, M., Le-Ngoc, T.: Efficient lte/wi-fi coexistence in unlicensed spectrum using virtual network entity: optimization and performance analysis. IEEE Trans. Commun. **66**(6), 2617–2629 (2018). https://doi.org/10.1109/TCOMM.2018.2801789
7. Yuan, X., Qin, X., Tian, F., Hou, Y.T., Lou, W., Midkiff, S.F., Reed, J.H.: Coexistence between wi-fi and lte on unlicensed spectrum: a human-centric approach. IEEE J. Sel. Areas Commun. **35**(4), 964–977 (2017). https://doi.org/10.1109/JSAC.2017.2680819
8. Li, Y., Baccelli, F., Andrews, J.G., Novlan, T.D., Zhang, J.C.: Modeling and analyzing the coexistence of wi-fi and lte in unlicensed spectrum. IEEE Trans. Wireless Commun. **15**(9), 6310–6326 (2016). https://doi.org/10.1109/TWC.2016.2582866
9. Xu, S., Li, Y., Gao, Y., Liu, Y., Gaˇcanin, H.: Opportunistic coexistence of lte and wifi for future 5g system: experimental performance evaluation and analysis. IEEE Access **6**, 8725–8741 (2018). https://doi.org/10.1109/ACCESS.2017.2787783

Blockchain-Based Decentralized Electronic Voting System

Mahima Pawar, Mansi Dabriwal, Raj Rathi, Shrey Kumar Sinha, and Sachin Suresh Pande

Abstract India is a constitutional democracy, and it holds regular, free and fair elections. According to the current voting system of India, every person casts a vote using an electronic voting machine, in a voting booth. Once the voting time has elapsed, the votes are counted. However, the transparency is not maintained. There is also a possibility of system rigging and delay in results. Blockchain being a distributed ledger and a transparent system is hence a better alternative to the current voting system. This ensures a secure, reliable and decentralized voting environment. E-voting system using blockchain works as a step towards creating a secure and transparent environment for elections where the users will be able to cast their votes only once and will be able to view the total votes cast in real time without having the permission to edit the same after elections get over. The working of blockchains will ensure the votes are maintained and the systems are not rigged by any third party. The secure electronic voting system uses blockchain which is a decentralized, peer-to-peer transaction ledger. The system will create permissioned blocks from the authorized nodes where the casted votes will be considered as individual transactions. These blocks are counted, and the results are then announced.

Keywords Blockchain · E-voting · Electronic voting machine · Encryption · Decentralized · Secret key · Hash · Authentication · Voter · Admin

1 Introduction

The primary goal of the election is to channel supreme authority as a representative. Every voter taking part in the election goes to the polling station with his voter card to cast his/her vote [1]. After verification by the election committee and supervisors, the voter gets a chance to vote. And to protect this right of the citizens, conducting

M. Pawar (✉) · M. Dabriwal · R. Rathi · S. K. Sinha · S. S. Pande
Department of Information Technology, Pune Institute of Computer Technology, Pune, India

S. S. Pande
e-mail: sspande@pict.edu

© The Author(s), under exclusive license to Springer Nature Singapore Pte Ltd. 2022

A. K. Nagar et al. (eds.), *Intelligent Sustainable Systems*, Lecture Notes in Networks and Systems 333, https://doi.org/10.1007/978-981-16-6309-3_32

fair elections is the fundamental requirement for any country. But in the traditional system, there are many challenges faced which are discussed in the next section.

In a democratic system, every vote counts. But in the coming generation, voters are reluctant while casting their vote. This could be due to many reasons like proximity, long queues, their unavailability or the belief that their vote does not count [1]. Situations like COVID-19 also result in further complications as mass gatherings need to be pragmatically controlled. For attracting voters, E-voting can be a viable alternative that can address these concerns. Challenges faced by online voting technology are large-scale manipulation, risk of hacking activities, coerced and influenced voting [2]. Including functional and security requirements like transparency, secrecy, privacy, data integrity and distribution of authority can make the e-voting system more robust [3].

The usage of blockchain technology can help to mitigate voting system difficulties. A continuously growing distributed decentralized ledger which maintains blocks of data that cannot be manipulated, tampered or changed is called blockchain [2]. Blockchain is a peer-to-peer network where any user is able to join the network of nodes and has roles like verifying the new transactions and adding blocks to the chain. Every block can be identified by its unique cryptographic hash, and this hash is calculated by combining the hash of the previous block with the current block's data and its corresponding hash. Thus, by changing information in any block, the following blocks will become invalidated. A hash of previous blocks is stored, effectively creating a chain of blocks and making a blockchain secure. This is paired with proof of work, making it extremely difficult to tamper with the blocks [4]. The development of blockchain technology has piqued the interest of many people. Soon, others realized the various use cases of the technology like in the medical field storing patients' records, creating a digital currency or maintaining untamperable electricity bills using smart grids. Blockchain can help implement an e-voting system that is distributed, immutable and transparent and cannot be tampered to change the results. Blockchain is an effective means to conduct fair elections.

Our research will look at key issues like record immutability, voter confidentiality, voter anonymity and end-to-end verification. Using the proposed approach, we are able to overcome the challenges of the traditional e-voting system while keeping the integrity intact. In this paper, we discuss the implementation and challenges of the e-voting system in departmental elections.

2 Traditional Voting Systems

2.1 *Process*

The vote counting process may take five to nine working days in traditional elections, depending on how fast the data is sent to the higher level [2]. The issues of data manipulation, transparency and security are some of the most common issues faced in elections.

Electoral systems in which general elections are managed by only one organization are termed as centralized systems. Some of the issues that can occur in traditional electoral systems is that an organization may have full control over the database and system, and it is possible to tamper with this database and system.

2.2 Security Issues

Electronic voting machines can be tampered, primarily based on physical security concerns. A study was conducted regarding the same which showcased some ways by which the EVM could be tampered. Following are some of the ways to do the same.

Substituting look alike. Attackers might also try to substitute original CPUs with the look alike CPUs which might contain the software for dishonest vote counting [5].

Tampering with software before CPU manufacture. If the software in the microcontroller chip is altered or if a different version of the compiled program is loaded on the chip, then it could lead to unfair elections [5].

Clip on memory manipulator. A small device was constructed which could directly be clipped to the EEPROM memory chip which recorded votes inside the EVM. It was so small it could fit discreetly in a shirt's pocket [5].

Absentee voting. When a vote is cast by someone who is unable or unwilling to attend the normal process by voting, in the assigned voting station to which the voter is normally allocated is called absentee voting. Many methods have been contrived to facilitate this. Providing the availability to absentee ballots is one of the many ways to improve voter turnout. Though the voter is allowed to participate in absentee ballot only if there is a valid reason like disability or travel [6].

At present, an absentee ballot system is not available for all citizens in India. It is only available for people on polling duty and serving in army, and head of state like president is allowed to vote in absentia through postal means according to the 'The Representation of the People Act-1950 (RPA) section20(8)'[6].

3 Proposed System

Certain properties must be held by the system in order for it to be deemed secure for E-voting.

Fairness. Results of the election should not be made public before the deadline of the election is over [4].

Authentication. This property ensures that only those voters will be allowed to vote who are eligible and gives them the right to vote only once [4].

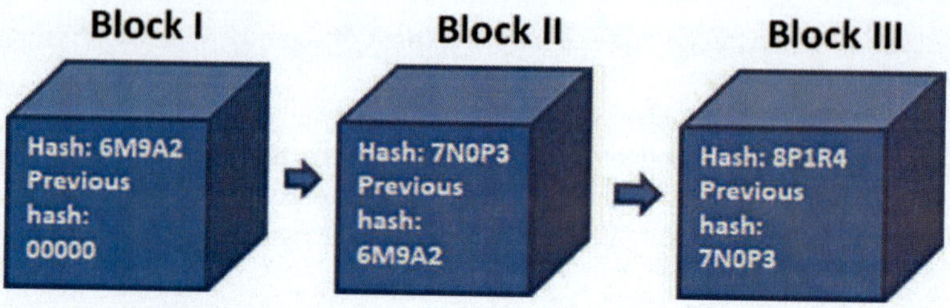

Fig. 1 Illustration of blocks in blockchain

Privacy. This property ensures that the identity of the individual voter is not revealed to anyone [4].

3.1 Blockchain

A blockchain is a time-stamped block of data managed by a cluster of computers not owned by any single entity.

Each group of a transaction is a unique block in a blockchain. As you can see in Fig. 1, all three blocks have different cryptographic hash making them unique. The hash of the preceding block helps to compute the value of the cryptographic hash of the block. In Fig. 1, there is a chain of three blocks in which block 3 with cryptographic hash 8P1R4 also contains a cryptographic hash(7N0P3) of the preceding block 2. The first block is genesis block which does not have the hash of the previous block. Tampering any data in a block will cause a significant change in its cryptographic hash. The hash value in the following block will become invalid which again requires to be changed. Recalculating all the cryptographic hash will make the blockchain valid that needs lots of computing power. With the advancement of devices, it is possible to recalculate cryptographic hash in reduced time to make blockchain consistent. To avoid the above scenarios, the concept of proof of work came into the picture. It helps to achieve consensus in the blockchain network. For adding a block in the blockchain, miners have to provide nonce that requires lots of computational power. Other miners in a peer-to-peer network validate the block hash produced by passing nonce to a hash function and matching it to the given requirements. After validation, the block gets broadcasted to the network with block hash denoting the work done by a particular miner.

3.2 Working of the System

Login and validation of electors. Electors need to use their registered email ID and password for logging into the system, and after the successful authentication of the credentials, they will receive 16-character secret key in their same registered email id, then it will be verified, and the elector will be redirected to their main dashboard.

Cast a vote. The elector will get a list of candidates in their dashboard from which they can choose anyone. For each elector, a Boolean flag is associated which is initially set as false, and once the vote is casted it is changed to true [3]. To avoid re-voting, the Boolean flag is validated, and if it is false, then only they can cast their vote.

Encrypting votes. Once the elector has successfully cast their vote, the input will be created by the system which will hold the elector's unique ID along with the elector's information and the hash value of the preceding vote [3]. This method will confirm the uniqueness of both input and output.

Appending the vote to the blockchain. After the block is generated, it will be broadcasted, and then it will be validated by the nodes in the network and thus appended to the blockchain.

Declaration of result. After the end of the election period, admin will make the results available to all the electors in their respective dashboard.

4 Design and Architecture of System

4.1 Detailed Description

See Fig. 2.

User interaction and authentication. It provides the user interface to the voter as well as to the administrator to interact with the blockchain. It validates the voter so that only approved users can access the system.

Security and access control management. This layer acts as a foundation for all individual layers. Validation is done using a 16-character secret token received in the user's registered email ID. It is responsible for defining roles, access rights and E-voting transaction services. E-voting transaction services help the functioning provided by layer III, whereas access and roles rights support layer I.

E-voting transaction management. Every vote is mapped as a blockchain transaction to be mined. The transaction holds the details of the voter as well as the ID of the candidate to which the vote is casted. The transaction is mined by virtual nodes called miners to get the transaction within the chain [7].

Ledger synchronization. In this layer, the ledger is perpetuated. Electors can follow their casted vote once their vote is mined and attached to the blockchain [7].

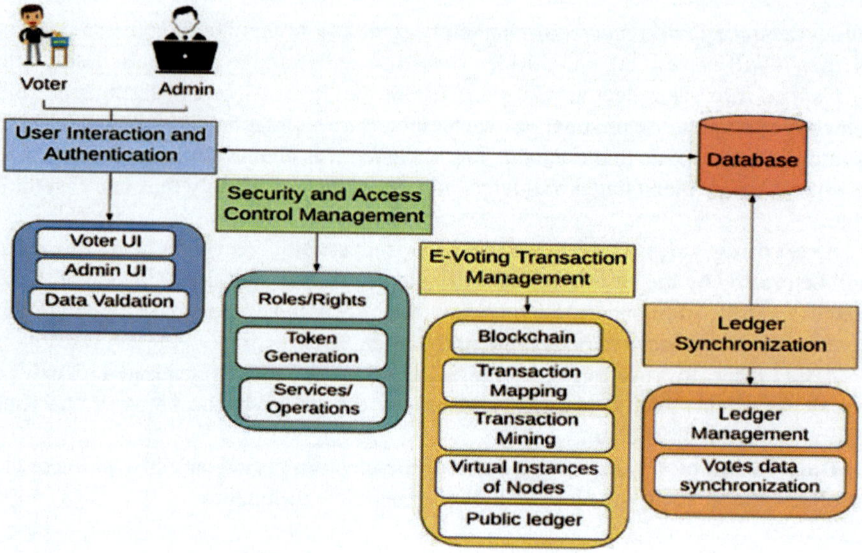

Fig. 2 Architecture of proposed e-voting system

4.2 Voting Process

The elector will login into the system using the 16-character secret token which the user receives in their registered email ID. If all the credentials along with the secret token are approved, then the elector is redirected to the dashboard where all the candidates participating in the election are displayed. If the entered credentials are not approved, then further access will be denied.

The elector can cast their vote, and each vote is mapped into a transaction. The transaction is mined by virtual nodes and attached to the chain. Thus, each vote is attached as a fresh block in the chain.

5 Implementation and Evaluation

5.1 Smart Contract

A smart contract is a set of rules that are managed by a P2P connection of nodes. Smart contracts are agreements between network participants which provide a coordination and enforcement framework, without the need for traditional legal contracts [8].

A structure representing every candidate is initially created which will have attributes like the name of the voter along with a unique ID and the number of votes received by him/her.

```
struct ElectionCandidate {
    uint cId;    //Election Candidate Id
    string cName; //Election Candidate Name
    uint cVotes; //votes received by the candidate
   }
```

Assuring single vote per voter. Same as normal voting, every voter will get a chance exactly one time for voting. After successfully voting by the voter, the voter will not be allowed to vote again. This can be established by mapping the address of every voter with a Boolean value.

Initially, the value of the Boolean flag of all the voters will be false, indicating that no vote has been casted yet. This value will be made true after successful voting of the voter.

For successful voting to take place, the first and foremost condition which is checked is if the Boolean value is false. If yes, then and only then the voter will be able to cast the vote. After the successful voting is done from that particular voter, the Boolean value which is associated with that voter is changed to true, indicating that the voter has successfully cast the vote.

Thus, that voter is no longer allowed to cast the vote after casting the vote successfully once.

```
//Each Voter is assigned a boolean flag mapping (Voter's
address => bool) public voter_profile;

Function castVote(){

   //Verify that voter hasn't voted till now
     require(!voterProfile[msg.voter]);
      ...
   //Function to cast a vote
      ...
   //Make the flag to true
     voterProfile[msg.voter] = true;

}
```

Casting vote phase. This function is directly called from the front end after the voter casts his/her vote to their desired candidate. A number of validations are performed over here before the casting of the vote.

The first condition which is checked is if the voter has already voted in the past. This is checked by referring to the Boolean value associated with the address of each voter. If this value is false, then it represents that the voter has not voted, and if it is true, then it represents that the vote has already been casted by the voter. If the voter has not voted, then and only then will he be allowed to cast his vote.

The next condition which is checked is if the candidate which is chosen by the voter is a valid candidate. This is done by referring to the candidate ID of the candidates. If this candidate ID lies between zero and the number of total candidates, then that candidate will be considered as a valid candidate, if not then it reverts an error.

After the successful checking of both of these validations, the voteCount attribute of the chosen candidate is incremented, indicating successful casting of the vote. After the vote has been casted, the Boolean value associated with the address of the candidate is updated to true, which will not allow the voter to vote again in the future.

```
function castVote (uint _candidate) public {

   //Verify that voter hasn't voted till now
    require(!voterProfile[msg.voter]);

//Check for a valid candidate
     require(_candidate > 0 && _candidate <= maxCandi-
date);

   //Make the flag to true.
     voterProfile[msg.voter] = true;

//update candidate vote count
    candidates[_candidate].cVotes++;
    emit votedSuccess(_candidate);
}
```

5.2 *Evaluation*

The aim of the evaluation was to verify the working of the system with respect to the existing voting system.

The evaluation is done on the basis of verification of voters, voting only once per voter, vote converting into blockchain transactions, adding candidates standing for the elections and displaying results after the election is over.

6 Conclusion

The centralized nature and the security for E-voting are some of the most fundamental issues looked at by governments and associations. Blockchain brings decentralized features to the E-voting application with the least trust in third parties. Voting systems might be inclined to altering, but utilization of blockchain diminishes the danger of altering and mistakes.

The utilization of encryption algorithms enables us to encrypt the final vote details which gives it resistance against tampering. In this manner, each individual can cast their vote through a portable and reliable application.

A secret token of 16 characters which is valid for the next 5 min is used to authenticate any voter. With the help of a Boolean flag, we can decide if the voter has voted once, thus providing a platform which is completely secure and reliable to conduct voting for elections.

Additionally, the system can be made more robust by doing authentication of voters through biometric as well as web cams.

So, our system can ensure that voters can cast their votes from distant places in a completely secure manner, thus providing a better alternative of conducting the elections in which any situation which requires gathering of people at any given place will be avoided, which will prove to be the most useful in the near future given the current scenarios.

References

1. Freya Sheer Hardwick, F., Apostolos Gioulis, S., et al.: E-voting with blockchain: an E-voting protocol with decentralisation and voter privacy. In: 2018 IEEE International Conference on Internet of Things (iThings) and IEEE Green Computing and Communications (GreenCom) and IEEE Cyber, Physical and Social Computing (CPSCom) and IEEE Smart Data (SmartData), pp. 1561–1567. IEEE, Canada (2018). https://doi.org/10.1109/Cybermatics_2018.2018.00262
2. Rifa Hanifatunnisa, F., Budi Rahardjo, S.: Blockchain Based E-Voting Recording System Design. In: 2017 11th International Conference on Telecommunication Systems Services and Applications (TSSA), pp. 1–6. IEEE, Lombok, Indonesia (2017). https://doi.org/10.1109/TSSA.2017.8272896
3. Harsha, V., Patil, F., Kanchan, G., Rathi, S., Malati, V., Tribhuwan, T.: A Study on decentralized E-voting system using blockchain technology. Int. Res. J. Eng. Technol. IRJET **5**(11) (2018)
4. Ahmed Ben Ayed, F.: A conceptual secure blockchain-based electronic voting system. Int. J. Netw. Secur. Appl. IJNSA. https://doi.org/10.5121/ijnsa.2017.9301
5. Hari, K., Prasad, F., Alex, J., Halderman, S., Rop Gonggrijp, T., et al.: Security analysis of India's electronic voting machines. In: Proc.17th ACM Conference on Computer and Communications Security (CCS'10), pp. 1–14. ACM (2010). https://doi.org/10.1145/1866307.1866309
6. Absentee Voting. https://en.wikipedia.org/wiki/Absentee_ballot
7. Kashif Mehboob Khan, F., Junaid Arshad, S., Muhammad Mubashir Khan, T.: Secure digital voting system based on blockchain technology. In: International Journal of Electronic Government Research January 2018, pp. 53–62. ACM (2018). https://doi.org/10.4018/IJEGR.2018010103
8. Venkata Naga Rani, F., Akshay, S., Arun kumar, T., Ishwar Kumar, F.: Decentralized E-voting system. Int. Res. J. Eng. Technol. IRJET **6**(3) (2019)

Human–Robot Intermittency and Invariants

Bernhard Heiden, Bianca Tonino-Heiden, and Volodymyr Alieksieiev

Abstract Isac Asimov's famous "I, Robot" was an inspiration for many generations, and we are still before increasing daily applications of robots in our lives. From generation to generation, the idea of a robot becomes more realistic and less dystopic, as robots are becoming increasingly human and animal-like, hence, similar to our environment. Therefore, it becomes more and more a legitimate question: what is the difference between humans and robots? This paper investigates fundamental principles of some invariants between humans and robots. It gives a two-dimensional order map that relates robot and similarly human properties, where it seems that the human–robot gap becomes increasingly smaller. The interrelatedness of humans and robots increases robotic applications in Industry 4.0 and leads to a new definition of culture and humanity. Since Cartesian times, the notation of rationality is ceasing increasingly chaotic irrational behaviour. We systematically get an order increase by focussing on the paradoxically irrational and creative human part, his creative potential. This creative potential is what Roger Penrose relates to as non-algorithmic. Moreover, we will argue with six axioms why this is of human–robot existential importance. We then give a short logistic example as application and a conclusion and outlook.

Keywords Robots · Human-o triangle · Robot-o · Human-o · Human-robot-o · Robot-human-o · Humans · Chaos theory · Ethics · Artificial intelligence · Computational intelligence

---

B. Heiden (✉)
University of Applied Sciences, 9524 Villach, Austria
e-mail: b.heiden@cuas.at

B. Heiden · B. Tonino-Heiden
University of Graz, 8010 Graz, Austria

V. Alieksieiev
Leibniz University Hannover, 30823 Garbsen, Germany

© The Author(s), under exclusive license to Springer Nature Singapore Pte Ltd. 2022

A. K. Nagar et al. (eds.), *Intelligent Sustainable Systems*, Lecture Notes in Networks and Systems 333, https://doi.org/10.1007/978-981-16-6309-3_33

1 Introduction

When Wittgenstein was in the war, he wrote his famous book Tractatus Logico Philosophicus (TLP) [1]. In his diary, he says that he was soldier on a shipping boat and using the searchlights. As can be followed from many indicators in the artificial intelligence (AI) research, the searching process is itself an intrinsic artificial and natural intelligence tool, beginning with primitive life forms over rape animals up to humans. It is essential throughout life for all species and hence an intrinsic property of intelligence and life. This is quite important in terms of economics because, in the recent decades, search companies like Google have had enormously global growth and application rates, now earning most of the money with all sorts of AI techniques. Only last year, when the founder divided the company into market segments that originally should have been equally distributed with relation to internal economic distribution, it turned out that the search-machine application segment outperformed all other segments. So we can see from this modern example that the searching process is like a motor in life, society, industry and economics—and it is an AI-core process.

When Wittgenstein has searched in the war the progressing field of the ship's passage, he was a soldier, a human-robot, executing commands of a people. He did not "work" at all. But when he was in his privacy, he has written down the astonishing TLP, which is now, as it also builds on foundations of Freges "Begriffsschrift" [2], the basis of modern AI, and will also be of that AI-robot applications. And he made this ground-laying, long before computers had been invented.

In this sense, Wittgenstein was a human, opposed to a robot, as he was creative to foresee future applications by the thinking process.

This open-to-the-future-life attitude is hence a main differentiating feature of humans. In a lot of applications, this applies. For example, in the political field, Arendt [3] said approximately that political acting is only legitimate if the outcome is open. Similarly, von Förster [4] has argued for the examination process in education that it is only legitimate when the answer is not known or open. This strong link of human properties to an open system deeply requires humans to be creative existentially.

Today creativity is one of several eminent parameters that are increasingly entangled in human-socio-cultural evolution [5]. It is becoming more important and one of the mysterious properties that humans are just beginning to understand. Daniel Kahnemann, who got the Nobel Prize, explains this in his book [6] with a particular focus on fallacies, or human imperfections, that can be overcome. Astonishingly, those imperfections are related to the creative process in the brain, and it is just opposed to the perfection mechanism (see also [7]). It seems to be a complex interplay of both, according to his theory, that remarkably resembles results of number and chaos theory [8], and hence is a bridging between quite different fields of research.

Content In this work, we will give, in the following paragraph, the goal and the research methods of this work. In Sect. 2, we discuss human applications for beauty and what this is meaning in terms of the human capability of being uniquely human, also in contrast with currently available robots. In Sect. 3, we first give the analytical reasoning for six axioms or invariants which apply partly general to humans and robots to finally depict the path of human–robot intermittency. Then we show how this seamlessly applies to modern and future logistics. In Sect. 4, we conclude the results and give a short outlook.

Goal and Research Methods This work aims to provide, using systemic analysis, an axiom set that can be used as a tool to describe the robot–human interaction qualitatively and, in some regards, quantitatively. For this, hints are given how to possibly quantify mainly the status quo and the potential of the human–robot connectivity. The logistics application then demonstrates the given method.

2 Beauty as an Approximate Measure of Life

All successful companies have this in common: successful advertising works. Because what and how they advertise it in real is appealing, authentic and of high quality. This is beauty, authenticity and brilliance. But in the bad companies, there is nothing adequate behind it. Those people mostly pulling corruptly the strings in the background have this corruption, the deviousness, written all over their faces. Therefore, at a glance, it is enough to judge companies by their beauty. These are happy people on the outside and the inside. These are good companies through and through: the appearance, the language and the tone. You can recognise the good by all of these. And that is why, all this is used in advertising. Only the clever customer checks whether the advertising delivers what it promises. You can tell by observing, by feeling, through the effect.

The advertising, the effect at the beginning at the entrance of the process is on the surfaces. The product is the effect inside when you interact with the company day after day. Only a living company can hence be beautiful.

The most irrational number is the golden mean [8]. This is because it contains the highest potential for order, which is contained in it, the indeed greatest promise in terms of a living future. It best guarantees sustainability and the future of life. Therefore, beauty is a, in summary, sustainable guide, rightly towards sustainable life. Beauty is when one can discover the order of different natures in it. Beauty can therefore be recognised precisely by the fact that the perfect approach is infinitely distant.

The deceptive can be recognised by the fact that, in comparison to the beautiful, it becomes finitely distant and thus runs empty. That means, however, that after a certain point, it no longer provides any new information—it repeats itself in the same way. In chaos theory, this corresponds to a period with a rational number.

Beauty is a measure of how one determines, measures, different orders and their simultaneity. To measure beauty means to measure infinity, and that is impossible in the end. But it is possible to come closest to it in relative terms. But infinity is the potential of life, the chaotic potential, the strange attractor that leads out into the infinity of life (see also Axiom 1).

The beautiful child is the child who holds the future of life the most.

The deformed child is the child that has been corrupted. Abused, violated, cheated. It has been cheated of the future.

Therefore, the resilient child must unfold its beauty from within itself. And every child does this more readily than any adult.

In a sense, the beauty is found in the infinite.

The Greeks made myths around beauty. The beautiful Helene is a common word. And her name, the word, stands for the Greek people, and it stands therefore also for the future of life, the power of the beauty. Hence, the eternity of the Hellenic philosophy of Socrates lies in it.

The far-reaching, relative to human life's eternity, shows in its almost 3000 years' existence.

Beauty, truth, dialogic and justice are core elements of it.

Ugliness, on the other hand, means determinacy, finiteness. Death is ugly because life ends. The gaze is beautiful because it is the expression of the best life. Therefore, we should strive for the flowering of life, strive until the last moment, as Socrates did. This is how we achieve immortal beauty in the best approach.

Life is functional imperfection. That is, it functions and it is infinitely imperfect. But it is precisely through this that it exists. So it has infinite "escape" regions. A robot that does not function is dysfunctional. A robot that functions incorrectly or imperfectly is also dysfunctional. To make a robot human is to make it more imperfect but at the same time, more functional. Therefore, the human operates at the thin line between functionality and imperfection.

Crime is therefore bad because it is a social aberration. The path does not go any further on it. Therefore, justice is increasingly important, but the intrinsic culturalisation goes towards freely self-implemented equilibration conditions. Thus, humanity, especially in the human–robot interaction, will lead to minor criminality and higher transparency concerning aberration paths. By increasing robotic use, dangerous, fast, complex and diverse combinatorial applications become feasible in the future societal control of human peace standards.

3 Robots and Humans

In this section, we first investigate some invariants of humans and robots. Then we give an application in logistics in broad outlines. But before we begin, we formulate our first axiom 1:

Axiom 1 When the system is chaotic or has a strange attractor, it has the greatest potential to build dynamic order.

This could be regarded as a rough and not final definition of life and chaos. As we will see later on, chaos and life are strongly interwoven, if not even a mystery. With regard to the discoveries in non-linear dynamics in the last decades, it seems that the number theory is not only a game, like in game theory but is the foundation for critical future applications, and in our focus today, for the dynamics of the future robots and humans. Not only the pure mechanical facts that arise from dynamics, but also the life property even seem to depend on it. And this is an interesting question when we ask ourselves what is the essential gap between being a robot and a human.

3.1 *Invariants of Robots and Humans*

When we approach the robots from the side of the humans, that would be called human-os. When we approach the humans from the side of the robots, this could be called robot-os. With this notation, we adhere to the orgiton grammar, according to orgiton theory [9]. The notation "-o" refers to an orgiton, a cybernetic unit of mass, energy and information, which is evolutionary evolving in order steps, which are further denoted by an index in forward (+) and backward (−) direction (cf. also, e.g. Human-o_i, Robot-o_j (i[−1,+3], j[−4,0]) in Fig. 1).

In the future, this might lead to quite an intermixture of elements and creations.

So this might be the first invariant robot-os and human-os that merge in a robot–human-os or human–robot-os, like fashion.

Axiom 2 Humans or human-os and robots or robot-os are getting an increasing similarity or human–robot equity level resulting in uprising robot–human-os with merged and intermittently enhanced properties.

The to Axiom 2 related intermittency is depicted in Fig. 1.

The important thing today is, as the words robot and human indicate, that they are entirely separated. And there is another notion of robots that they are tools or better extended tools. In this notation, robots ever since were external extensions of the humans like all the human tools, which applies even to some animals that are using tools quite effectively (cf. also [10, 11]).

What is new here in humanity is that the richness in those tools' variations has led to a fantastic panoply of produced goods. And among those goods are also and in our context above all, tools (cf. Fig. 1).

Those tools are so important as they are the limiting factors of growth.

So we can postulate the second invariant as Axiom 3:

Axiom 3 As was the computing ability, the limiting factor of growth will be that of the human-os or robot-os in the future.

The reason for this axiom is two-fold. First (1), while humans have used computers, they were used for computations, and computations in principle create a virtual space. Concerning automation and humans, they are an open-loop system when applying these systems to real applications. These systems are enormously powerful, and we have seen many applications in the last century, which were possible only due to computing devices since approximately the foundation of the UN amid

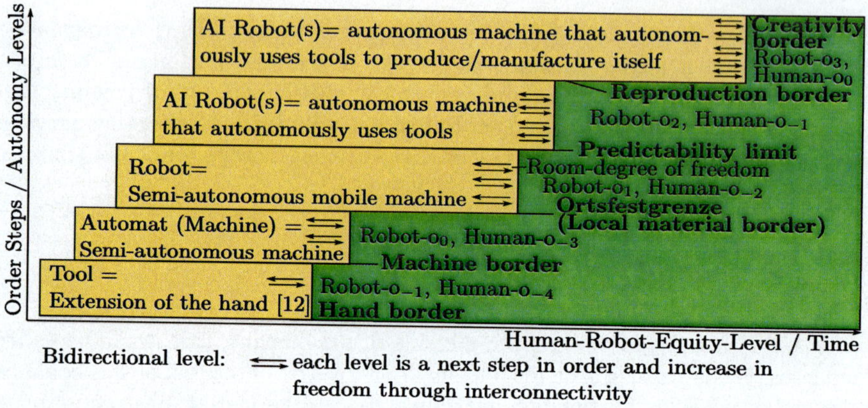

Fig. 1 Human-os and robot-os—how they merge in steps of order and human–robot-intermittency

the previous century. Everybody knows the von Neumann architecture, although it is breaking down now into more chaotical substructures, due to diverse limitations, like computability borders, atomic borders for transistors and global borders, to mention only a few that lead to a crisis that is to begin now. So we have great hopes for quantum computers [12], but, even in this case, the solution room stays computational (versus robotical), which means that nothing changes towards humans from the systemic point of view. It even means that the system is unidirectional. Towards that, the computational tools are in use (control) of humans, and humans are staying the critical factors in all operations and above all in the existence of humanity, which is essential now and in the future, as humans tend seriously to dominate all other species. And among the humans, for some, there is non-equilibrium in multiple dimensions like power, intellectual capabilities and many others. Let us call this existentiality or survivability condition.

Regarding the cybernetic cycle, those machines, that close the cycle and that act like humans do, are robots. In this sense, the robots and the robot-os are more human-like and of a higher order or bidirectional, which goes well with orgiton theory [9, 13].

With relation to the previous section, we can formulate the Axiom 4 towards higher human–robot-os functions across the creativity border in Fig. 1:

Axiom 4 Beauty and form are related in relation to their species.

Let us formulate another axiom for the order increase with regard to robot–human intermittency as Axiom 5:

Axiom 5 Information versus truth content: robot versus human. The truth content W changes with the level of the category K: W=f(K).

With relation to Axiom 5 we may state that the categorial difference can be expressed as self-organisational levels of a higher order like indicated in Fig. 1. To get a higher level of truth result, new categories of increasing universality are needed, implying higher order.

The category of robot versus computer is a higher order category as robots are bidirectional, and computers are unidirectional regarding a material application, like serving a human. So intelligence processes are always subordinated to categorial material processes. Hence, we could say that the information retrieving process is a second-level material process.

While today humans are focussed on each other emotionally, the industry of "social platforms" uses the psychological knowledge widely and exploits human weaknesses to motivate to work for it. This subtracts humans' energy from each other, making them even more splendid isolated. This is hence a potential dangerous robotisation of society with enormous political impact.

"Work" is something that only a human being can do, i.e. a robot or a human being that is required to be like a robot is understood usually today does not work. This means as Axiom 6:

Axiom 6 A person "who is a robot" potentially harms himself and others (e.g. a perfectionist). Hence, the critical (existential) potential, or sustainable optimal operation of human–robot-os (humans, robots and human–robot hybrid forms), is to use the creative potential across the creativity border in Fig. 1.

Essentially, the damage arises here because the human being does not make full use of his human potential, with regard to Fig. 1 the order steps decrease towards a robotic action in the lower human-o steps. It follows directly from this that if a human exploits this potential and is supported by a robotic application in robotic or periodic activities, he or she can potentially achieve the most excellent performance in terms of humanity.

3.2 *Towards Quantification of Higher Human–robot-os Properties*

Regarding quantification, you can use steps in Fig. 1 qualitatively and assign the tasks, the time, the energy or similar quantities assigned to humans and robots. By this, the potential of robotic work can be quantified with regard to applications like economics, technical applications, human interests and many other categoric assignments related to the human–robot interaction. Moreover, the order steps could be quantified by the number of relevant cybernetic cycles, decision steps, material efforts and energy use for the overall autonomy level; although, this may be difficult.

In the creativity region, which is difficult to assign due to its intrinsic high level of intermittency, measures of "beauty" of products, processes, with relation to the species may play a future key role, for which certain sorts of AI algorithms shall be suitable. And last but not least, the irrationality measurement as a chaoticity measure of non-linear dynamics may support this process by measuring the systematic imperfection, the irrationality as a number and hence computational achievable measure. So it might be foreseen that future robot-os will have a sense of beauty, especially regarding their species (see Axiom 4), which will also be linked to existential questions, as this is the case in humans now, and is this the culmination of the ever since human evolution.

3.3 Application: What Is the Role of Humans in the Future Logistics?

As human life is a dynamic process, the human always stays in front of different logistical problems. With the increasing amount of data, many of these problems cannot already be solved manually but need to use, for example, computer-based systems. Here applies Axiom 3, hence robotic applications are of increasing importance. This is practically the case with Amazon who founded a special company only for the production of the intensively used warehouse robots [14]. Due to the increasing complexity and speed of processes, the number of decision-making processes is increasing so that automated decision-making processes are widespread in logistics. Automated systems took over many human tasks and are becoming smarter within the years. However, there are still many obstacles in the automation of logistic processes. In [15], it is mentioned that despite the automation of many human tasks, humans play and will play an important role in logistic processes because the deterministic character of computer-based support systems cannot consider all human-related factors (e.g. different positions of goods in logistic process in a warehouse, which cannot be detected by a computer algorithm and, therefore, the computer solution fails). Here applies Axiom 5 and 6. This can be crucial for the general system performance. In [16], the human factors of the order-picking process were analysed. According to this research, such an ordinary logistic process is carried out manually in up to 80% of the order-picking in warehouses due to flexibility reasons, especially in the case of "heterogeneous and changing product portfolios" [16]. The human factor increases, therefore, and hence, humans play still an essential role in general system performance [16]. In [17], the human-o triangle as the interactional triangle of human, machine and product is given. In this triangle, the human can occupy different positions (e.g. manager, user and performer). However, its role and significance remains the same, as each knot in the triangle graph is directly connected with all other knots. Hence, the role of humans will remain crucial, regardless of automation degree, role and functions.

In this example, we also see that Axiom 1 always applies, whereas Axioms 2–6 are more related to future scenarios of a higher level of robotisation, development and human–robot interaction.

4 Conclusion and Outlook

In this paper, we have introduced several axioms that relate to the invariants of robots and humans.

Robots have evolved in the midst of the rationalistic age, and they clearly show the side of rationality as a mechanistic paradigm and the perfect machine is impersonating it. In the now increasingly merging behaviour of robots and humans, the difference seems to become smaller, which is quite clearly related to the latest industrial revolution, Industry 4.0.

Among the need to build, construct new robots, humanity's aim is to perfect these machines for the good of humanity. In the meantime, it seems that we may reproduce humans in their function with robots and even replace them, as is an old fear that goes back to the first machines at the end of the middle ages. Although this fear in history, all the industrial revolutions so far have built on machines, and always old professions have ceased new ones, and humanity even has spread more than ever before.

The aim of those machines has been and was successful in lessening the burden of human suffering, and this will even continue.

But for continuing, we have argued that the gap between humans and robots has to become smaller with regard to functionality and larger with regard to technological multiplicity.

This means we will need robots or robot-os, humans or human-os and their intermittency or human–robot-os increasingly necessarily (Axioms 2, 3).

One reviewer asked for answering the question of time complexity. This question seems to be itself quite complex and is still not answered yet here, although it is an essential pressing question. To give at least an idea in which direction the answer could guide, we suggest thinking of time intermittency. A first straightforward interpretation would be to allow chaotic regions in the direction towards the future (see also Axiom 1), like in the bifurcation analysis of Poincaré planes. A second now quite utopic interpretation, including time as an emergent parameter, possibly with regard to Penrose [18] or Heiden and Leitner [19], would allow for time intermittency and (space-)time singularities.

With this new paradigm, which can be foreseen clearly, the equality conditions between species (see also Axiom 4) are drastically increasing, as this process is running towards unity of all living species and hence of life as an expression of a gigantic universal ecosystem.

The reason is quite simple. With robots and their integration into our lives, the material dimension gets perfectionated, and by this, the suffering of life and in life can be further reduced by means of this technology (cf. Axioms 5 and 6).

The material dimension is from the point of view of self-organisational theory, an order increase by means of a long-term trajectory, although material processes through their inertness also tend to decrease or overrule local order.

Concerning an outlook of humanity, the argument from Carter [20] applies that through the human–robot intermittency, the possibilities even increase for both humans and robots as species (Axiom 4) and finally as a living world (Axiom 1). In the end, the survivability of life as a whole is at stake.

References

1. Wittgenstein, L.: Tractatus Logico-Philosophicus Logisch-philosophische Abhandlung Side-By-Side-By-Side Edition, Version 0.25 (2011), containing the original German, alongside both the Ogden/Ramsey, and Pears/McGuinness English translations. (First published by Kegan Paul (London) 1922)
2. Frege, G.: Begriffsschrift, eine der arithmetischen nachgebildete Formelsprache des reinen Denkens. Halle. Accessed 05/02/2021. https://gdz.sub.uni-goettingen.de/id/PPN538957069 (1879)
3. Arendt, H.: Vita activa oder Vom tätigen Leben. Piper, München, Berlin, Zürich (2016)
4. von Förster, H.: Sicht und Einsicht, 252 pp. Vieweg+Teubner Verlag. ISBN 3528084685 (1985)
5. Götschl, J.: Zur Epistemologie der Selbstorganisation: Von Konvergenzen zu Korrelationen zwischen Systemwissenschaften der Natur und Systemwissenschaften vom Menschen. In: Die Welten von Psyche und Soma - Zur Verbindung von Psychoanalyse und Neuropsychiatrie - Gedenksymposion für Universitätsprofessor Dr.in med. Dr.-in phil. In: Fabisch, H., Fabisch, K. Kapfhammer, H.-P. (eds.) Margarete Minauf im Meerschein- schlössl in Graz am 25. November 2017, pp. 85–104. Verlag Dr. Kovac, Hamburg (2020)
6. Kahneman, D.: Schnelles Denken, langsames Denken, 624 pp. Siedler Verlag (2012)
7. Beck, H.: Irren ist nützlich! (To err is useful!, orginal in German), 316 pp. Hanser Verlag (2017)
8. Hilborn, R.C.: Chaos and Nonlinear Dynamics–An Introduction for Scientists and Engineers. Oxford University Press, New York (1994)
9. Heiden, B., Tonino-Heiden, B.: Philosophical Studies—Special Orgiton Theory/Philosophische Untersuchungen - Spezielle Orgitontheorie (English and German Edition) (unpublished) (2021)
10. Heiden, B., Walder,S., Winterling, J., Perez, V., Alieksieiev, V., Tonino-Heiden, B.: Universal language artificial intelligence (ULAI). In: Schulz, F. (ed.) Advances in Artificial Intelligence Research, Nova Science Publishers, New York (2020)
11. Eccles, J. C. Evolution of the Brain Creation of the Self, 282 pp. Routledge, London etc. ISBN: 0-415-02600-8 (1989)
12. IBM Research.: The World's First Integrated Quantum Computing System. Accessed 07/01/2021. https://www.youtube.com/watch?v=LAA0-vjTaNY#
13. Heiden, B., Tonino-Heiden, B., Alieksieiev, V.: Artificial Life—Investiga- tions about a Universal Osmotic Paradigm (UOP). In: Arai, K. (ed.) Intelligent Computing, LNNS Computing Conference 2021, 15.-16. July 2021, 285 (Springer Nature, Virtual/London, 2021), 1–11. https://doi.org/10.1007/978-3-030-80129-842
14. Day One Autor. In: Roboter in Zahlen: Daten und Fakten zur Robotertechnik bei Amazon, Amazon. Accessed 06/20/2021. https://blog.aboutamazon.de/logistikzentren/roboter-in-zahlen-daten-und-fakten-zur-robotertechnik-bei-amazon (2019)
15. Sgarbossa, F., Grosse, E.H., Neumann, W.P., Battini, D., Glock, C.H.: Human factors in production and logistics systems of the future. Ann. Rev. Control **49**, 295–305 (2020)
16. Grosse, E.H., Glock, C.H., Neumann, W.P.: Human factors in order picking: a content analysis of the literature. Int. J. Prod. Res. **55**, 1260–1276 (2016)

17. Heiden, B., Alieksieiev, V., Tonino-Heiden, B.: Communication in human-machine-product triangle-universal properties of the automation chain: witness simulation example. In: Advances in Intelligent Systems and Computing, pp. 987–998. Springer International Publishing (2021). https://doi.org/10.1007/978-3-030-73103-8_71
18. Penrose, R.: Gravitational collapse and space-time singularities. Phys. Rev. Lett. **14**, 57–59 (1965)
19. Heiden, B., Leitner, U.: Additive manufacturing—a system theoretic approach. In: Drstvenšek, I. (ed.) ICAT 2018, Maribor 10.-11. Oct. Interesansa - zavod, Ljubljana, 136–139 (2018). ISBN: 978-961-288-789-6
20. Carter, R.J.: Cognitive advantage—how artificial intelligence is changing the rules, 261 pp. In: Carter, Richard J. (ed.) Cognitive Advantage: How Artificial Intelligence Is Changing The Rules For Winning In Business And Government (S.4). Mayhill Publishing. Kindle-Version. Mayhill Publishing, United Kingdom. isbn: 978-1-9168723-0-1 (2021)

Breast Cancer Detection Using Machine Learning Algorithms

Kapil Tajane, Siddhesh Sheth, Rushikesh Satale, Tanmay Tumbare, and Omkar Panchal

Abstract Cancer detection at an early stage is important so that the patient can be ready to deal with it. There are many machine learning architectures that help in predicting cancer. Gene expressions are the most prominent thing in classification of breast cancer (Fan et al in Recursive SVM biomarker selection for early detection of breast cancer in peripheral blood [1]). The gene expression dataset currently available includes fewer records and a huge number of features (very high dimensionality). We are going to implement three different approaches on different datasets to find out the best performing approach. In the first approach, we have used a recursive feature elimination (RFE) and random forest classifier (RBF) on a gene expression dataset. In the second approach, we have used the ResNet-34 model over histopathology image dataset. And in the third approach, we are proposing a CNN architecture which has a small number of convolutional layers for the classification of cancerous genes from healthy ones using gene expression data available at Pan-Cancer Atlas. Due to high dimensionality of gene expression data, that too with small count of records and high noise, it brings more challenges (Elbashir et al in IEEE Access J Mag 7, 2019 [9]).

Keywords Gene expression · RFE · CNN · Histopathology · ResNet-34

1 Introduction

RNA-Seq converts RNA into a sequence of DNA parts called c-fragments that gives gene expression profile [1]. There were challenges in early detection as obtaining tissue biopsies with traditional factors was not easy, mammography also does not detect small tumors and as this disease has numerous states and stages [2].

K. Tajane (✉) · S. Sheth · R. Satale · T. Tumbare · O. Panchal
Department of Computer Engineering, Pimpri Chinchwad College of Engineering, Nigdi, Maharashtra, India

© The Author(s), under exclusive license to Springer Nature Singapore Pte Ltd. 2022

A. K. Nagar et al. (eds.), *Intelligent Sustainable Systems*, Lecture Notes in Networks and Systems 333, https://doi.org/10.1007/978-981-16-6309-3_34

Here presenting the system uses gene expression profiles (genomic data) to predict breast cancer at an early stage, prediction using histopathological images as well using feature selection technique. Pathologists usually use hematoxylin and eosin (H&E)-stained tissue images to identify invasive ductal carcinoma (IDC) [3]. IDC detection takes a lot of time because in this process, a pathologist scans large area of benign parts to lastly recognize the parts of malignancy. In the process of grading, tumor aggressiveness and classifying tumor correct delineation of IDC are important.

Generally, Bloom–Richardson and Nottingham grading schemes are used to grade tumors. Yinan Kong [4] used local clustering to do classification on breast histopathology dataset. CNN model is commonly used for image-type classification. It decreases the image variables making it easy to process without data loss [1]. CNN is made up of different layers (i. convolutional, ii. pooling, iii. fully connected). It takes an image matrix as input having x rows, y columns and color RGB value [5].

After downloading the gene expression data, it is sent to preprocessing. After preprocessing, the gene expression data is converted into 2D images [5]. At last, we determined the best values for hyper-parameters using search-type grid, and filtering is applied. We got great results with such light architecture.

2 Datasets

The breast cancer gene expression data available at Pan-Cancer Atlas from "Illumina HiSeq" is acquired using TCGAbiolinks library [6]. Dataset has two sample types: a) solid primary tumor and b) solid normal tissue. This data has **1208 records** with a huge feature set of **19,948** types of **genes**. *These high dimensions make data noisy.* Initially, we removed the outliers from data [7] and then used normalization techniques to make sure that biases are avoided in the data [1]. At last, filtering is based on threshold 0.25. For each gene, mean is calculated, and if it is less than threshold, then that gene is removed from the data frame. Now we are left with 1208 records and 14,477 genes. This data is converted to 2D images [5] of size 127 × 114 (Fig. 1).

In another approach, we have used the breast histopathology dataset available on kaggle. This dataset consists of 162 WSI images of breast cancer cases scanned at 40x. In that, 277,524 image patches of dimension 50 × 50 were stored (198,738 IDC benign and 78,786 IDC malignant) (Figs. 2 and 3).

We have split above dataset into training, testing and validation dataset. Out of 277,524 patches, each 38,000 patches is used for testing and validation. Remaining images are used for training the model.

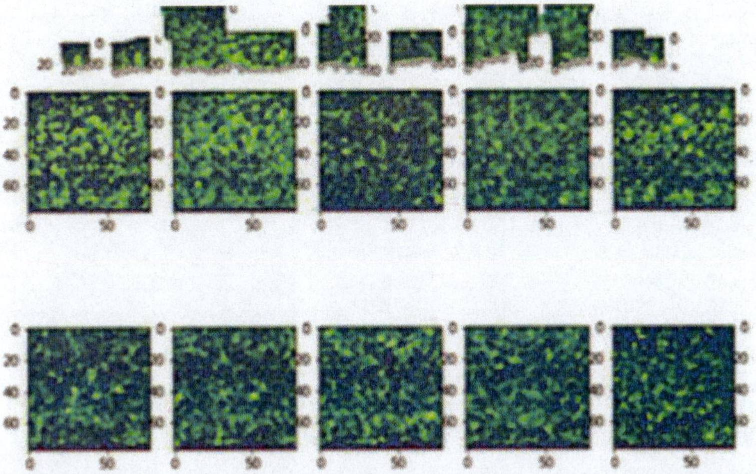

Fig. 1 2D images of genes constructed from gene expression dataset. First row shows healthy genes, and the second row shows infected genes

Fig. 2 Statistics of benign and malignant samples present in histopathology image dataset

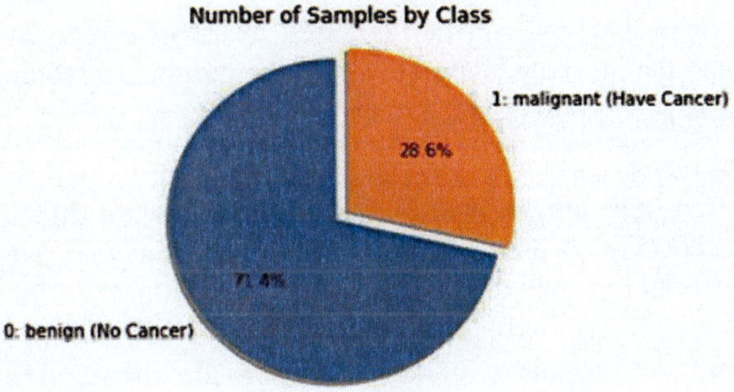

3 Algorithmic Approach

We have implemented three separate methodologies on two distinct datasets which are listed below.

1. (Random Forest + RFE).

Using random forest with RFE where RFE led to faster outputs with increase in accuracy levels too, here we apply random forest classifiers along with RFE and select the particular genes that have a major impact in output, and then we apply cross-validation and hence obtain the desired values.

First we process the data to remove outliers and noisy variables if any using label encoder, then as the dataset is fit for use, we apply random forest where we need to state the number of random states which are taken into consideration, and then we apply RFE to extract best features [6]. Here RFE has a number of steps as an

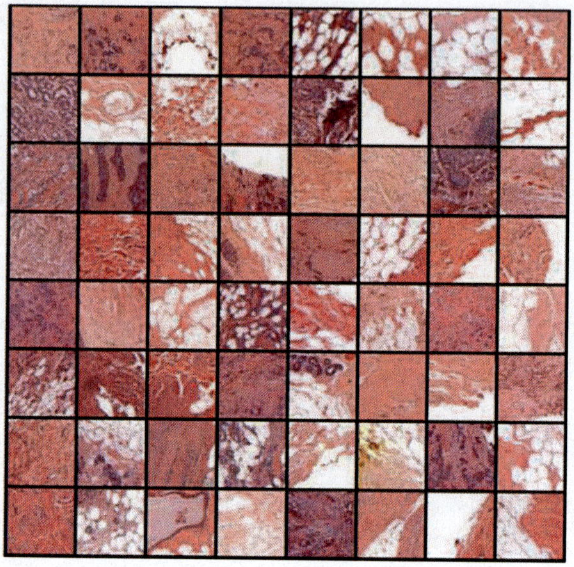

Fig. 3 Sample images from histopathology image dataset

attribute which can be increased to traverse recursively to get better results, and another attribute is used to select the number of features we need further which play a key role to decide the output.

What is RFE?

A selection approach works in a backward direction and helps to select genes according to their status or how influential they were on random forest. First it checks through all the mentioned states and then traverses and finds the optimum states. Finally to reassure that the features selected are optimum, it repeats the process in order to achieve higher accuracy in classification [8].

Steps of Random Forest + RFE

1. Train random forest on the training set;
2. Check the random states taken;
3. Remove features which get ranked least criterion; and
4. Repeat the entire process.

Small perturbations of the training set can also be detected effectively using recursive feature elimination (RFE).

2. ResNet-34.

This model is built using different machine learning algorithms like a highly versatile ResNet-34, CNN, random forest classifier and recursive feature elimination to predict breast cancer. In this approach, they have done grid sampling of image patches, and a CNN is trained from available patches to show the probability of a patch belonging to IDC and a probability map built over the WSI highlighting the

Fig. 4 Skip connection of ResNet-34

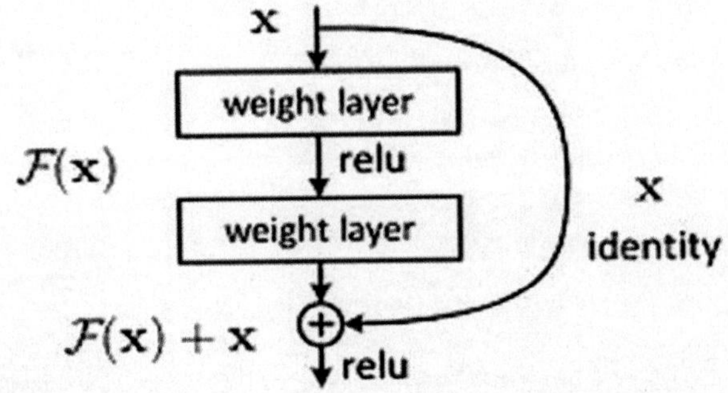

predicted IDC regions. In this approach, maximum accuracy was 90.20%. Anant Madabhushi presented a method for automatic detection of IDC in WSI using CNN [9].

In the residual network, we use skip connections technique. It skips training from a few layers and connects to the next layer as an input (Fig. 4).

The advantage of skip connection is if any layer hurts performance of architecture, then it will be skipped by regularization. This network uses 34 layer network architecture inspired by VGG-19.

4 CNN

In this approach, we are proposing a CNN architecture with only two convolutional layers for classification of breast cancer using gene expression dataset. The input given to it passes through all layers sequentially. When input goes through convolution operation, a feature map gets created. Values of the filter (kernel) are multiplied with pixel values of the input image as shown in the given equation (Fig. 5).

$$O(i, j) = \sum_{k=1}^{f} \left(\sum_{l=1}^{f} \text{input}(i + k - 1, j + l - 1)\text{kernel}(k, l) \right) \quad (1)$$

The CNN model has two convolution layers. In which, one layer finds out aspects like curves and contour (low features), and another layer builds features on the basis of features identified by the first layer. Rectified linear unit (ReLU) activation function which is very much known because of its simplicity and effectiveness is used. Equation of ReLU function is given below:

$$F(x) = \max(0, x) \quad (2)$$

Another one is sigmoid which is nonlinear. Its value varies a lot even when a small change is done in input variable x. It is preferably used to drive its value to the

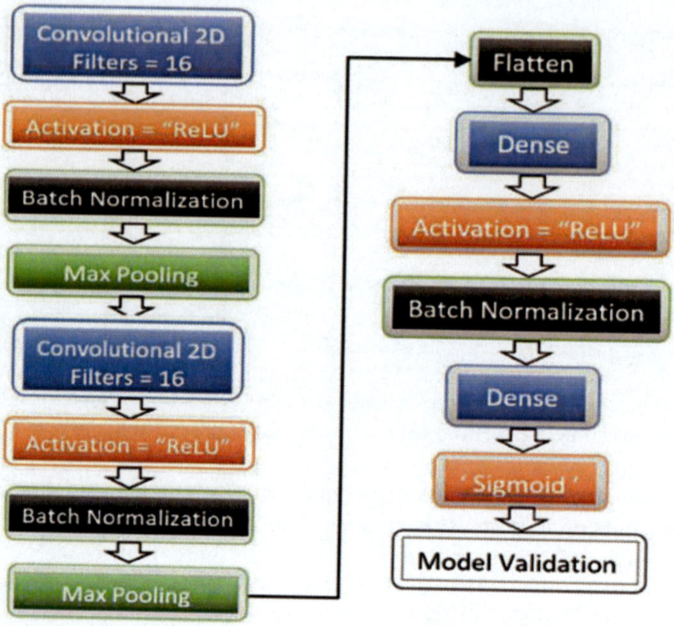

Fig. 5 Proposed CNN architecture

extremity of the graph to make classification easy. It is used at last stages just before carrying out classification output.

$$F(x) = 1/(1 + e^{-x}) \tag{3}$$

Architecture includes convolution layer followed by activation function which is followed by max pooling layer. The convolution layer filters the input image features and forms a feature matrix. Feature matrix includes the specific spot of the feature in the image. Max pooling layer recapitulates the features identified by convolution layer. After that fully connected layer is constructed. It maps each individual input to individual output.

We used kernel size as 3 and 16 filters because these values gave us best performance. We used fivefold cross-validation, in which the dataset is divided into five parts. Out of these five parts, four parts are used as training data, one part is used as testing data, and this process is repeated five times.

5 Performance

We evaluated the proposed models based on standards like accuracy, F-score, precision, specificity and recall. These parameters are calculated using true positive (TP)—output is actually positive and predicted correctly as positive, true negative (TN)—output is actually negative and predicted correctly as negative, false negative (FN)—output is actually positive but predicted wrongly as negative and false positive (FP)—output is actually negative but wrongly predicted as positive.

$$\text{Sensitivity} = \frac{\text{TP}}{\text{TP} + \text{FN}}$$

$$\text{Specifivity} = \frac{\text{TN}}{\text{TP} + \text{FP}}$$

$$\text{Precision} = \frac{\text{TP}}{\text{TP} + \text{FP}}$$

$$\text{Accuracy} = \frac{\text{TP} + \text{TN}}{\text{TP} + \text{TN} + \text{FP} + \text{FN}}$$

6 Results and Discussion

After implementing RFE along with random forest, we get a stellar accuracy of around 95.8% which is much better than the previous error prone techniques.

The ResNet-34 model over histopathology images dataset led us to an accuracy of 91%.

CNN model with kernel size = 3, filter count = 16, dense units = 1024 performed best among all. This model obtained an accuracy of 98.87%, F-score of 98.79%, sensitivity (Recall) of 100%, specificity of 97.64% and precision of 99.98% (Fig. 6; Table 1).

Result Table:

7 Conclusion

In this paper, we have studied and implemented three different approaches to breast cancer classification. So we can conclude from our observations which match our expected results as accuracy is increased compared to the previous technique. Hence, recursive feature elimination technique is quite efficient. Another is based on a ResNet-34 model over histopathology images that gave us accuracy of 91%. Next approach was based on RBF + RFE based on gene expression data which led us to accuracy of 96%.

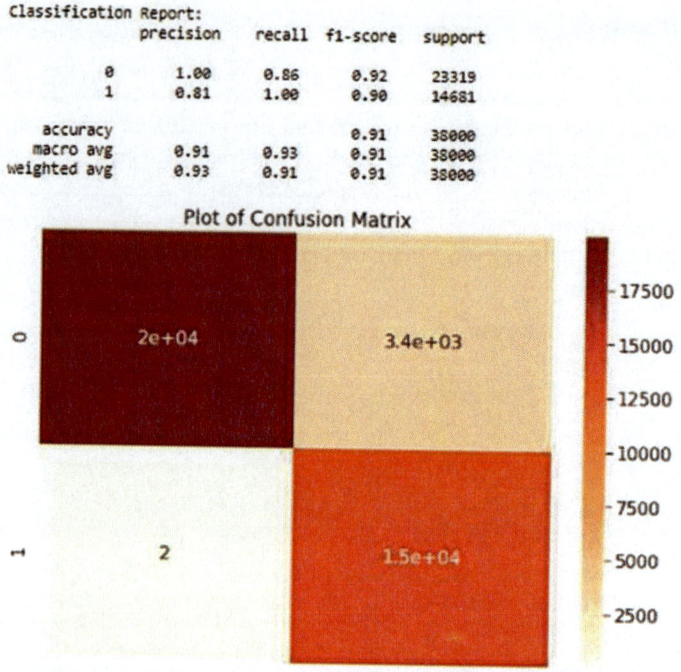

Fig. 6 Performance metrics of ResNet-34

Table 1 Comparison of all techniques

Approaches	Techniques used	Dataset	Accuracy
1	RFE + Random forest	Gene expression (TCGA)	95.8
2	ResNet-34	Histopathology image dataset	91
3	CNN	TCGA-BRCA gene dataset	98.87

Third approach was a two-layered CNN model over gene expression data available at Pan-Cancer Atlas which is a RNA-Seq data that led us to the accuracy of 98.08%. So, from above results, we can infer that we got great outcomes from two approaches based on gene expression data. But, the only problem is that the creation of gene expression data is a little tedious and time consuming because this kind of data has a very huge feature set with comparatively less number of samples.

Acknowledgements We pay our sincere gratitude to our principal sir for being the backbone during our study. We also thank our guide Prof. Kapil Tajane sir for helping us in tough times and guiding us throughout. We thank all the teaching and non-teaching staff and even our fellow mates for inspiring us throughout.

References

1. Elbashir, M.K., Ezz, M., Mohammed, M., Saloum, S.S.: Lightweight Convolutional Neural network for breast cancer classification using RNA-Seq gene expression data. IEEE Access J. Mag. **7** (2019)
2. Zhang, F., Chen, J.Y.: Discovery of pathway biomarkers from coupled proteomics and systems biology methods. BMC Genomics, **11**(Suppl 2), S12 (2010)
3. Fischer, A.H., Jacobson, K.A., Rose, J., Zeller, R.: Hematoxylin and eosin staining of tissue and cell sections. CSH Protoc. 1 May 2008. pdb.prot4986. https://doi.org/10.1101/pdb.prot4986. PMID: 21356829
4. Nahid, A.A., Mehrabi, M.A., Kong, Y.: Histopathological breast cancer image classification by deep neural network techniques guided by local clustering. Biomed Res Int. **7**(2018), 2362108 (2018). https://doi.org/10.1155/2018/2362108.PMID:29707566;PMCID:PMC5863327
5. Lyu, B., Haque, A.: Deep learning based tumor type classification using gene expression data. In: Proc. ACM Int. Conf. Bioinf., Comput. Biol., Health Inform. (BCB) (2018)
6. Zhang, F., Kaufman, H.L., Deng, Y., Drabier, R.: Recursive SVM biomarker selection for early detection of breast cancer in peripheral blood: Affiliations expand PMID: 23369435 PMCID: PMC3552693 (2013). https://doi.org/10.1186/1755-8794-6-S1-S4
7. Sharma, P., Sahni, N.S., Tibshirani, R., Skaane, P., Urdal, P., Berghagen, H., Jensen, M., Kristiansen, L., Moen, C., Zaka, A., et al.: Early detection of breast cancer based on gene-expression patterns in peripheral blood cells. Breast Cancer Res.: BCR **7**(5), R634-644 (2005)
8. Aaroe, J., Lindahl, T., Dumeaux, V., Saebo, S., Tobin, D., Hagen, N., Skaane, P., Lonneberga, S., Borresen-Dale, P.A.L.: Gene expression profiling of peripheral blood cells for early detection of breast cancer. Breast Cancer Res. BCR **12**(1), R7 (2010)
9. Cruz-Roa, A., Basavanhally, A., González, F.A., Gilmore, H., Feldman, M., Ganesan, S., Shih, N., Tomaszeweski, J., Madabhushi, A.: Automatic detection of invasive ductal carcinoma in whole slide images with convolutional neural networks. Medical Imaging (2014)

A Machine Learning-Based Congenital Categorization Approach in Early Recognition of Breast Carcinoma

Sandeep Singh, Zunaid Aalam, Amritpal Kaur, and Prabhjot Singh Manocha

Abstract Nowadays, breast carcinoma is one of the maximum leading reasons for cancer dying inside with the less evolved international locations in the world. This is partially due to a change in the lifestyle that has resulted in a rise in the number of cases. The cancer of the breast develops from the internal part of milk ducts and may be invasive or non-invasive. Mammography, especially when incorporating with computer-aided detection, can now generate a figure that is accurate adequate for diagnosis, and virtual mammography allows for three-dimensional transmission. The proposed system's motive is the creation of a novel computer-assisted diagnostic method for detecting destructive and dangerous (cancer generate) mammograms. This research has focused to create a figure processing algorithm that can reliably identify and classify breast lesions automatically. The radiologist was able to improve his assurance accuracy with the aid of a CAD method. To classify the breast malignant sarcoma as cordially, a supported vector-based machine (SVM) morpheme is used and included the best features of the discovery that play a key role in attaining the ideal turnout, which is then appointed and linked to artificial neural network (ANN) to train and identify.

Keywords Breast cancer · Entropy classification · Artificial neural network (ANN) · Image processing · Bayesian classification

1 Introduction

The breast cancer should be detected at early ages and treated effectively, which reduces mortality [1]. A bigger tumour has been discovered after sometime and the

S. Singh (✉) · Z. Aalam · A. Kaur
SGT University, Gurugram, Haryana, India
e-mail: sandeep_feat@sgtuniversity.org

A. Kaur
e-mail: amritpal_fet@sgtuniversity.org

P. S. Manocha
Chandigarh University, Chandigarh, India

© The Author(s), under exclusive license to Springer Nature Singapore Pte Ltd. 2022
A. K. Nagar et al. (eds.), *Intelligent Sustainable Systems*, Lecture Notes in Networks and Systems 333, https://doi.org/10.1007/978-981-16-6309-3_35

higher chances of long-term survival and permanent cure [2]. If the breast cancer is removed until it spreads to lymph nodes and other organs which include the brain, liver, lungs as well as bones, so the chances of recovery are better [3]. The accuracy of breast cancer screening and diagnosis has increased with the help of imaging technology. Each examine modality is advantageous based on human characteristics such as breast density, obesity, HER to status (human epidermis growth receptor factor), hormone receptors, age risk factor (as you get older, breast cancer risk increases), family history and hereditary caused by abnormal genes in genetic passed from parent to child [4]. For detecting breast cancer, a new medical imaging technique breast thermography is used for the female persons in early ages. The long infrared range radiations based on their electromagnetic spectrum thermographic cameras are used for usual detection and producing images of that spectrum for the breast malignant sarcoma. Various distinct methods are used for revealing breast malignant sarcoma development with the aid of ANN, support vector-based machine (SVM) and other techniques [5]. The mammogram is a reliable and widely used method for detecting breast cancer [6]. It detects even the tiniest variations of breast cancer when irregular mammograms are detected. A biopsy is a well-known surgical technique for identifying breast cancer. It is an expensive, time-consuming and troublesome procedure. At this point, where there is a chance of incorrect assurance, a radiologist's advice is critical, and the patient does not undergo an unfavourable biopsy [5]. For an auxiliary client through a programmed test, the radiologist is urged to improve diagnostic accuracy for identifying breast cancer. The CAD scheme categorizes mammograms into two categories: hazardous (to other human cells that expand and constructing the cancer patients departing this life) and less dangerous (unrightfully to human beings and does not extend to others human bodies).

2 Related Work

Berbar [7] suggests using a ray-tracing supporting contour shaping form method to segregate mammogram order highlights. Before isolating highlight widened segregate limits, strechlim is used for initialization. Comparison, inverse refinement moment energy, homogeneity, absolute sum and collapse were included among the highlights that were extracted in GLCM [8].

ST-GLCM integrates seven measurable highlights (e.g. kurtosis, uncertain, standard deviation, collapse, imbalance, autocorrelation, precision and contingency) with GLCM textual features. To label mammograms, reinforce SVL is used as regular or affected.

To evaluate the structure, Mini MIAS compiled descriptive [9] is used. The network architecture is assessed in terms of affectability, identity and consistency. The structure's affectability and exactness are 97% and 97.89%, respectively.

Preetha [10] identified a system for masterminding mammograms that consists of four stages, the first of which is a pre-processing stage used in the centre channel to enhance the image concept and eliminate disturbance in the picture. Histon-based fuzzy c classifier was used to assign the hazard locale of the enhanced picture. After partitioning, such as texture, a few elements, strength and form were removed from the divided picture. As a valid class for chronicle photograph, ANN categorized were used for investigating the multiple mammograms standards. Accuracy, affectability and contrast are both guaranteed at 93.6%, 72.72% and 88.66%, respectively.

Kumar et al. [11] developed a method for pre-processing that makes use of the wavelet-based adaptable sigmoid capability. Pre-processing is divided into three stages: multiple images were created using different increment balanced sigmoid limits and wavelet disintegration, and the final advanced image was created using a versatile histogram. The region of interest for an image (ROI) once altered the preliminary processing of malignant tumour data which was completed using 13 linked GLCM with the textual features in the proposed approach. To prepare the lesson, the feed forwarded escalation strategy is used. The mini MIAS data collection achieves plan precision of approximately 95%.

Xie et al. [12] developed a method to disintegrate the progression of breast cancer using machine learning. Pre-processing consisted of two stages of establishment, and removal of the pectoral is muscle. To fill ROI, the Hough shift approach is used. In a mammogram, all 32 grey measurements and textual features are limited. Precision is claimed to be 96.02% when using a mini MIAS informative array.

3 Proposed Work

Figure 1 depicts the analysis methodology used in this study. The overall method consists of four stages:

1. Image acquisition,
2. Isolating features from the mammograms,
3. Selecting consistently ideal features,
4. Training the classifier to recognize the correct kind of mammograms.

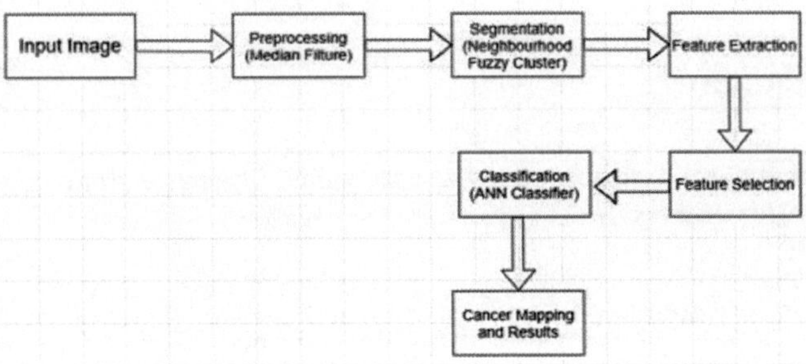

Fig. 1 Suggested system block diagram

In this paper, there is no correlation between the pixels of a picture class of tissues; segmentation is needed to enhance image analysis. The features of each image that is educated are derived from the divided image. The best few features from the bring out features were chosen and fed into the classifier. The main goal of image pre-processing is to remove unnecessary contortion from the original image. It is frequently used to improve the transparency of certain image features when detecting anomalies. Nonlinear digital filtering is also known as a median filter that is used as the smoothing technique. To keep the image's edges intact, it is used as a median filter in order to eliminate salt and pepper noise along with the speckle noise. The edges of the images since it are not a linear filter, with retaining the edges of the images by using the median filtering technique.

4 Features for Classifying Pattern Based on GLCM

In classifying pattern, feature extraction plays a fundamental role for entire mammograms by choosing grey level co-occurrence matrix (GLCM) feature in 0°. For recovering and developing, the optical content of pictures capturing feature extraction is a very crucial and important process. The low-level ancient picture features includes the extraction of colour, shape texture or domain specific in an image. Using the feature selection after the completion of the segmentation process, the needed features are available to be chosen. During the selection feature, it consists of a lot of multiple redundant features of image data inside it. The GLCM features are stored in the database after being extracted.

4.1 *Contrast*

$$\text{Contrast} = \sum_{m=0}^{i=1-1} \sum_{n=0}^{j-1} [m - K]^2 Q(m, k)$$

4.2 *Correlation*

Connection delineates the direct needy of grey value. The correlation value is vast if the mammogram contains direct shape for extensive sum.

$$\text{Corelation} = \sum_{m=0}^{i-1} \sum_{j=0}^{i-1} \frac{(m - u_x)(j - u)_q(m, j)}{\partial_x \partial_y}$$

4.3 Sum Entropy

$$\text{SumofEntropy} = -\sum_{J=0}^{2m-2} (K)Q_{x+y}\log_2 Q_{m+n}(k)$$

4.4 Difference in Entropy

$$\text{DiffrenceofEntropy} = -\sum_{J=0}^{2n-1} Q_{x-y(K)}\log_2 Q_{m-n}(k)$$

4.5 Entropy

$$\text{Entropy} = \sum_{i=0}^{m-1}\sum_{J=0}^{2n-1} Q(i, j)\log_2 Q_{m-n}(i, j)$$

4.6 Feature Selection

Feature set guarantee reduces feature area to reduce estimating time. It works by removing noise, excess and insignificant features, i.e. it selects the persuading characteristic to persuade the user to yield.

For the analysis, the rank half methodology selects items as an alternative that contributes most to the realized yield. This constraint completely increases the possibilities as surely as their commitment throughout this approach high Vi located choices were chosen for mending the network. Table 1 shows an outline of the selected feature.

Table 1 Best feature selection by rank method

Optical features	
FK1	Contrast
FK2	Correlation
FK3	Sum entropy
FK4	Difference in entropy
FK5	Entropy variance

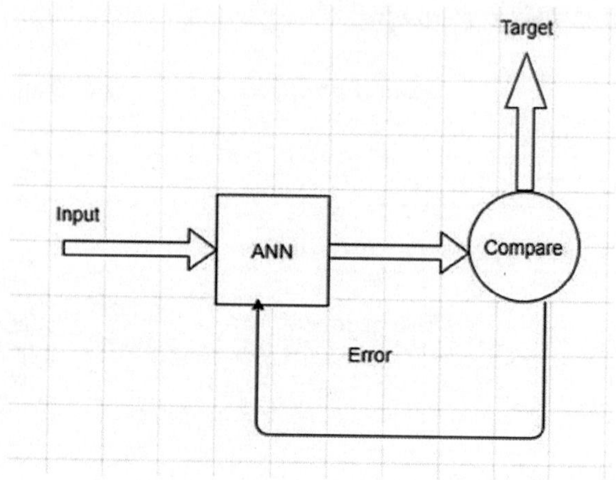

Fig. 2 Functioning in ANN Classifier

4.7 Classification

Neural networks are made up of vital components that are squarely animated by natural vegetative cell function. The neural network is taught to execute a variety of functions through the use of dynamic weighting between components. To get the needed yield, a neural framework was constructed. This state is depicted in Fig. 3. The framework was altered based on the relationship between the yield and searching in the centre until the framework output organized the aim (Figs. 2 and 4).

5 Experimental Results

The retro gradation examination is a proper methodology used between all the relapse graph's elements for computing the relationship yield of the goal set vs planned framework. The immaculate match showed as a noted line in the relapse graph, and the output robust line is indicated by strong line that are luminescent and indistinguishable to dash line which is processed only when the classifier forecasts 100% accurately (Table 2).

Total 70% of the data from the selected information referred is utilized from the preparatory work for preparation, 15% for trial assessment and 15% for affirmation. Neural systems are made up of three layers: the input level, the shrouded layer and the relent layer. The specifications used for the ANN are shown in the preparation window. The framework was prepared using a preparation capacity based on Liebenberg–Marquardt, which produces exceptional outcomes in preparing and characterization. Other preparation capabilities such as conjugate descent with Powell, flexible back spread and so on are used. Using various classification techniques,

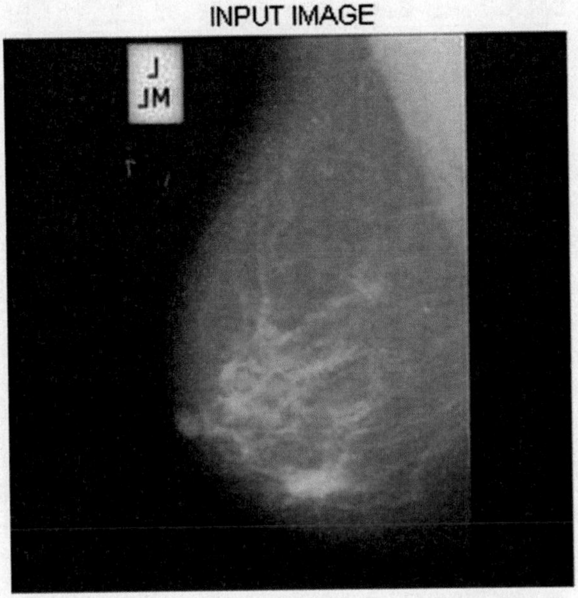

Fig. 3 Input image of proposed method

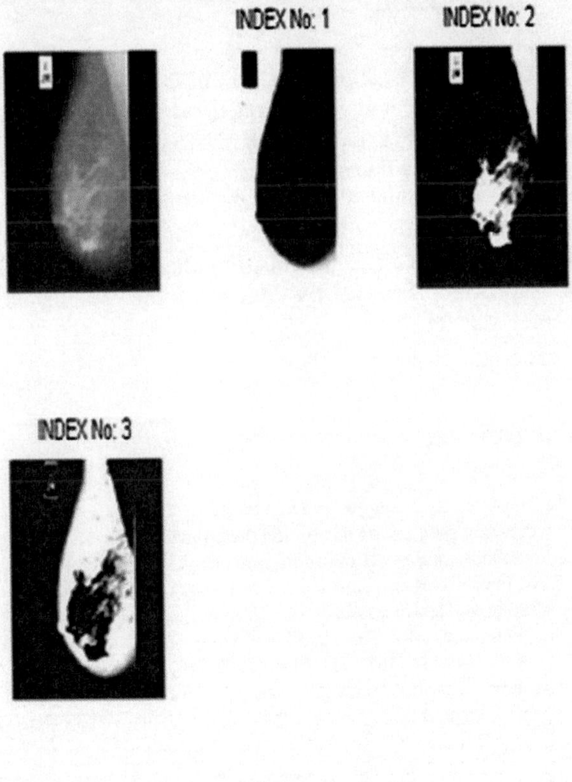

Fig. 4 Index recognition of the input image regression analysis

Table 2 Complete summary of the result

	Data Division	Precision (%)	Susceptivity (%)	Fidelity (%)
ZOverall results by using mini MIAS database	Priming	99.00	98.61	98.10
	Affirmation	99.89	100.00	99.40
	Trial assessment	97.84	97.89	100
	All-inclusive	99.30	100	98.40

Table 3 Extracted GLCM features of Image

Parameterization	Precision	Susceptivity	Fidelity
Decision tree	62.59	77.6	70.2
ANN	60.3	72.9	68.4
SVM	79.92	96.66	93.02

Table 3 displays from the wavelet feature extraction approach the instances numbers categorized such as Naive Bayes, decision tree, support vector machine and ANN.

6 Conclusion

Since the breast malignant carcinoma is a very fundamental illness, so for minimizing the fatality rate, it should be regarded as a planning stage. For the examination, an X-ray of a smaller than foresee MIAS conceptualization is used in the investigative activities. These datasets include 322 mastography screening X-ray, 270 of which were prevalent and 52 of which are carcinogenic. The ten surface highlights are determined on 0° using the rank highlights methodology and possibilities area to boot aperient to six highlights. For checking data and approval, the outcome shows 100% correctness, and by employing the established technique, overall precision achieved is 99.4%.

References

1. Daisy, P.A., Nagappa, A.N., Udupa, N., Mathew, N., Carla, B.: Effectiveness of a planned teaching program on improving the knowledge on warning signs, risk factors and early detection methods. Indian J. Pharm. Pharmacol. **2**, 6–9 (2015)
2. Fujitaa, H., et al.: Computer-aided diagnosis: the emerging of three CAD systems induced by Japanese Health Care Needs. Comput. Methods Programs Biomed. **9**(2), 238–248 (2008)
3. Bhanumathi, R., Suresh, G.R.: Advances in computer-aided detection of breast cancer by mammography. Int. J. IT Eng. IJITE **1**(6), 1–13 (2013)
4. Andreea, G.I., Pegza, R., Lascu, L., Bondari, S., Stoica, Z., Bondari, A.: The role of imaging techniques in diagnosis of breast cancer. Curr. Health Sci. **37**(2), 55–61 (2011)

5. Jalalian, A., Mashohor, S., Mahmud, R., Karasfi, B., Saripan, M.I.B., et al.: Foundation and methodologies in computer-aided diagnosis systems for breast cancer detection. Excli **16** (2017)
6. Singh, A.K., Gupta, B.: A novel approach for breast cancer detection and segmentation in a mammogram. Procedia Comput. Sci. **54**, 676–682 (2015)
7. Berbar, M.A.: Hybrid methods for feature extraction for breast masses classification. Egypt. Inform. J. (2017)
8. Albregtsen, F.: Statistical texture measures computed from grey level co-occurrence matrices. Image Processing Laboratory, Department of Informatics, University of Oslo 5:5 (2008)
9. Rajesh, A., Mohan, E.: Classification of mammogram using wave atom transform and support vector based machine classifier. Int. J. Comput. Sci. Inf. Technol. **7**(2), 467–470 (2016)
10. Preetha, K.: Breast cancer detection and classification using ANN with partical swarm optimization. IJARBEST **2**, 19 (2016)
11. Kumar, S., Chandra, M.: Detection of microcalcification using the wavelet based Adaptive sigmoid function and neural network. J. InforProc. Syst. **13**, 703–715 (2017)
12. Xie, W., Li, Y., Ma, Y.: Breast mass classification in digital mammography based on extreme learning machine. Neuro-computing **173**, 930–941 (2016)

Review of Powering and Actuation in Bio-robotic Walking Orthoses

M. Shankar Lingam and A. Arun Kumar

Abstract This paper presents the powering and actuation requirements for an active bio-robotic walking orthoses (BRWO) for people with a spinal cord injury (SCI). The main functions required include the capabilities to walk on flat ground, sit and stand and negotiate stairs. Identification of these functions has enabled the selection of potential components for experimental testing and characterisation. In this respect, some actuators are described.

Keywords Actuator · Biowalker · Orthoses · Transmission · Power sources comparison

1 Introduction

The INSPIRE foundation conducted a survey among young people with a spinal cord injury (SCI) to gauge the importance of standing and walking and to identify how well existing aids addressed these needs; see Tylor [1]. The abilities to stand and upright mobility have been identified as very important, but respondents felt that current devices failed dismally to achieve this. The aims of this research are to explore the potential of robotic technologies available to reduce the gaps between desired and available functions. The intention is to specify, design and develop active bio-robotic walking orthoses that use present-day industrial automation and robotic technologies to provide the capability to stand, walk, sit and use stairs. One of the most important aspects in this respect is to assess active actuation technology; to do this, different types of actuators have been investigated.

The outline of this paper is as follows. Section 2 gives the requirements for an orthoses. A brief review of current powering and actuation is presented. Section 4 discusses the characteristics of potential technologies. Sections 5 and 6 describe the experimentation of actuator and discussion of the results, respectively.

M. Shankar Lingam (✉)
NIRDPR MoRD, Government of India, Hyderabad, India

A. Arun Kumar
BITS, Warangal, India

© The Author(s), under exclusive license to Springer Nature Singapore Pte Ltd. 2022

A. K. Nagar et al. (eds.), *Intelligent Sustainable Systems*, Lecture Notes in Networks and Systems 333, https://doi.org/10.1007/978-981-16-6309-3_36

2 Requirements

Actuators for orthoses and associated power supplies must meet a range of requirements successfully. A wide range of actuators are available on the market, yet few have been tested in the laboratory for this purpose; some of these results are presented in this paper.

2.1 *Size and Weight*

Actuators and their power supplies should be as small as possible to enable them to be incorporated in the orthosis structure without causing restriction of movement or unsightly bulk. The size, weight and positioning of the actuators are particular concern. They must not compromise good posture and load distribution when sitting down. Nor must they hinder fitting into seating and negotiating narrow openings. The use of transmissions allowing the actuator to be remote from the joint may help avoid some of these complications.

Equally, the weight of the actuation and powering components should not be burden to the user. In addition, the size and weight of supporting hardware, e.g. actuator drivers, recharging/fuelling equipment, should be minimised.

2.2 *Efficiency*

The efficiency of the actuators and power supplies affects their size and weight. The power consumption should be as low as possible to maximise the range and time of use of the orthosis. Hence, specific power and specific energy are important factors.

2.3 *Specific Power and Specific Energy*

Many publications in the literature mentions specific power and energy. The peak powers required at the hip or knee during walking for an average male are 56 W, the mean power is 10 W, the average torque is 72 Nm, and angular speed is 45°/s; see Winter [2]. An average able bodied adult walks for 2.5 h a day; see Goldsmith et al. [3]. If a 1 kg power source is to run four actuators with an efficiency of 60% for one day, then it must have specific energy and power greater than 170 Wh/kg and 190 W/kg, respectively; see Wright et al. [4].

2.4 Actuator Control and Power Supply Management

The actuators will be managed by an embedded control system, which may require additional interfacing hardware (e.g. A/D and D/A cards, terminal blocks and transducers). The system will also need an effective and easy-to-use user interface with one hand left free. In this respect, single chip technology may be well suited.

2.5 Safety

The orthosis must be designed with safety in mind, and there are many aspects to be considered. For example:

(i) If the power source fails, the controller must have a backup power supply to enable a safe shutdown of the system.
(ii) A situation where the actuators are powered, but the controller is not, must not be allowed to occur.
(iii) The power supply must be safe from potential hazards depending on the source, i.e. electric shocks, burns, chemical damage, fire, exhaust, etc. So, an appropriate safety level must be considered.

2.6 Maintenance, Sound and Vibration

An aesthetic design and smooth operation is more important; hence, it should produce minimal noise and vibration. Otherwise ideal orthosis could become unacceptable to users. It should require minimal servicing or maintenance. The majority of users will be unable to carry out all, but the most basic maintenance (e.g. cleaning, recharging/fuelling) can be done.

3 Power and Actuation Review

There are many prototype rehabilitation systems that have been reported recently. A brief review on current powering and actuation methods for both rehabilitation and mobile robots is presented in Wright, et al. [4]. One of these is a spring brake orthosis that produces a natural-looking swing phase built at the University of Sheffield; see Gharooni et al. [5]. Recently, a computer controlled brake orthosis for FES-aided gait is built by Vanderbilt University; see Goldfrab et al. [6]. The drawback in these systems and some electrical power is required to operate the FES and compressed air to operate the brake. The Louisiana State University Reciprocating gait orthosis (LSU-RGO-II) has been used by many patients; see Solomonow et al. [7]. They allow only coupled flexion/extension of the hips and lock the knees. An active orthosis developed at Politecnico di Torino is a device that controls each knee with a pneumatic actuator via a chain and sprocket drive; see Belforte et al. [8]. It is believed that the power was supplied via a compressed air line. A device developed in Switzerland (driven gait orthosis (DGO) or Lokomat) enables paraplegics to walk on a treadmill; see Hocoma-Medical-Engineering [9] and Colombo et al. [10]. All of them have used passive device and have their own merits and demerits.

Among the mobile robots, the Shadow Biped developed by the Shadow Robot Company is a life size robot consisting of legs and torso; see [11]. The joints are actuated by 28 air muscles, controlled by 12 V input 56 solenoid valves and powered by an external compressed air. The RS-01 dog is an impressive walking robot demonstrator produced by a UK company RoboScience, and details are available in its website [12]. The joints are actuated by Epidex gear motors and RCPE power electronics. Honda has developed several humanoid robots; see [13]. The joints are operated by DC servomotors with harmonic speed reduction. Silo 2 is the latest bipedal robot developed by Caballero et al. [14], and the joints are operated by specially developed SMART actuators. Although it is known that the power supply is carried on board, no other details of the battery are available. The aim of this project is to use this mobile robotic technology to develop an orthosis.

4 Characteristics of Potential Technologies

In this section, the potential technologies for powering and actuation, particularly those used by the projects above, are compared to the requirements of the walking orthosis.

4.1 Functional Electrical Stimulation

FES uses external electrical signals to stimulate the muscles. In this process, the power required to stimulate the muscle is low compared to the power required to move the limb. This technique would result in significant downsizing in the power supply and any auxiliary actuator requirements. Stimulation of the muscles can also be beneficial to the user. However, some spinal cord injuries prevent the use of FES and insert of electrode on the body, so a decision was made not to consider FES in this first phase of the project.

4.2 Electrical Power

All of the examples that do not use FES are non-tethered and use electrical power. This is perhaps an indication that this power source is well suited to mobile robotic systems and the control inherent in them.

4.2.1 Electrical Power Sources

Due to advances in battery technology, the specific energy and power available are higher than most other power sources. Some examples are in the following:

(i) Nickel zinc and nickel metal hydride batteries can have specific energies of as much as 70 Wh/kg, specific powers of up to 150 and 175 W/kg, respectively; see Sharkh [15].

(ii) Li-SO2 and Li-SOCl2 batteries have specific energies of 460 W/kg, specific powers of 330 and 660 Wh/kg, respectively. Thus, batteries meet the power requirements as well as power supply management, safety, sound and vibration.

Fuel cells also meet these requirements; however, they require hydrogen or other fuel sources and produce water, sound and heat as by-products, which may have bad effects to the human being. Nevertheless, they have specific energies of 1500 Wh/kg and specific powers of 480 W/kg, so could be much lighter than batteries [4].

4.2.2 Electric Actuators

Electric actuators are most common in robotic technology. There are subdivided into rotary or linear and direct or transmission drives. The largest category is directly driving rotary actuators, i.e. electric motors of which there are numerous kinds. Some of them are discussed here.

Electric Motors

Electric motors usually run at high speed and develop low torques. A reduction gearbox can be added to reduce the speed and increase the torque. For this project, a peak power of 56 W is required. A motor with this output running at 2000 rpm will produce only 0.26 Nm; therefore, a reduction of >200:1 is required to increase the torque to 70 Nm (i.e. average man, stair ascent; see Berchuck et al. [16]). Gearboxes with this reduction and with output torque rated at 70 Nm tend to be too large and heavy (1.5 kg or higher Maxon Motor) [17]. These are not only true of traditional gearboxes, but to a lesser extent (1.1 kg) of harmonic drives too; see Harmonic Drive Technologies [18].

An electric motor driving a joint directly will stick out sideways. This may make it difficult to negotiate narrow spaces and seating. A flat motor may be considered for this purpose.

RoboScience used standard DC brush motors in Robodog but avoided the problems of size and weight by developing their own gear reduction. A larger version of the actuator would meet the requirements of an orthosis prototype, but would extend laterally from the joints.

The Honda [13] humanoid robots (P2, P3, Asimo) use a combination of DC servomotors and harmonic speed drivers. However, these robots are notoriously heavy. Silo2 also uses DC motors, but again as part of a specially developed (SMART) actuator, which includes a linkage that may be inappropriate for this.

Exceptions to the high-speed drawback include stepper motors, servomotors and torque motors. These motors have controls which allow low speed and avoid the need for gearboxes. However, the motor itself is larger to produce the required torque, and so problems of bulk are encountered. Servomotors are therefore over-sized DC motors with additional control circuitry, stepper motors for this duty weigh around 6 kg and torque motors 8 kg, and details are available in Kollmorgam [19] and EAD _motors [20]. Torque motors, however, do have the advantage that their length is short in comparison with their diameter.

Current technology, consisting of brushed DC motors and gearboxes with the latest battery technology, would allow for an acceptable power and weight, provided that the peak torque and power requirement are accommodated without resorting to continuously rated machines.

Gear Motors

The second-largest category of electric motors is gear motors. A right-angle drive can reduce the distance that the motor extends. Another advantage is that they reduce weight compared to suing separate components, but these reductions are small. An appropriate gear motor might weighs 8 kg; see Bodine Electric Company [21].

Linear Motors

A linear motor function in the same way as rotary motors, but as the "rotor" and "stator" are flat, linear motion occurs. They are based on rotary motor types (e.g. AC or DC, brushed or brushless, etc.).

The main advantage of linear motors is that they can be aligned nearly parallel to the leg. However, there is a trade-off between force output (hence size and weight) and the offset at the joint to produce the required torque. Too large an offset causes problems when the joint is flexed. The maximum joint torque required is 70 Nm; therefore, limiting the offset to 50 mm requires an actuator force of 1.5 kN; see EAD_Motors [22]. The limited data availability suggests that linear motors capable of these loads are likely to be considerably heavier than 1 kg.

Linear Actuators

Rotary electric motors can be used to drive transmissions that convert rotary motion to linear motion. Linear actuators involve some form of screw, and its pitch determines the relationship between displacement and rotation. Thus, a form of gear reduction is implicit. Linear actuators can be supplied as a single item or as separate components. A linear actuator capable of generating 1.5 kN weighs 1.4 kg; see EAD Motors [22].

Ball screws are used successfully by the Lokomat [9, 10] driven gait orthosis, but the orthosis does not carry the full user weight so the load on the actuators is reduced. Also the joints are fairly bulky and may not be acceptable to users of a walking orthosis.

4.3 *Pneumatic Power*

Compressed air may be used for the source of power. This could be supplied from a reservoir, charged externally or by an on-board compressor powered by an electric motor or engine. If battery power is used, losses would occur in the energy conversions making the overall system less efficient. Also it is likely that the compressor would cause sound, vibration and heat. An engine used to drive the compressor would also produce exhaust gas.

A reservoir could not store sufficient energy. Many recharges would be needed for 1 day's use unless the pressure is raised above 800 psi. In this case, safety of the orthosis is mainly concerned.

4.3.1 Pneumatic Actuator

Piston cylinder or inflated tube can be used as actuation device and compressed air for powering them. A cylinder capable of generating 1.5 kN will weigh 1.2 kg; for details see SMC Pneumatics [23]. Hence, the specific power and energy produced are normally low. The active orthosis uses a double acting cylinder, the most common form of pneumatic actuator; see Belforte et al. [8, 13].

Another type of pneumatic actuator is the fluidic or air muscle. These actuators shorten as they inflate. They are therefore single acting. The Shadow Biped uses air muscles of Shadow's own design in antagonistic pairs; see Shadow Robot Company [11]. An air muscle of length 30 mm version weighs only 80 g, but can generate upto 0.7 kN; see RoboScience [12].

4.4 Hydraulics

Hydraulic systems must include a pump and the means to run it. Pumps have the same drawbacks as compressors (sound, vibration, heat, weight, inefficiency of conversions, maintenance) and additionally leaking hydraulic oil. The most common form of hydraulic actuator is the hydraulic cylinder. Hydraulic systems operate at higher pressures than pneumatic systems; therefore, the size of actuator is smaller. The reduction in weight is not so pronounced since the cylinder walls are thicker to cope with the higher pressure and loads.

Hydraulic motors use very similar control technology, but drive vanes on a shaft rather than a piston. These can either have a fixed or continuous rotation. Hence, hydraulic actuation appears low priority to use in orthosis.

4.5 Direct Fuel Actuators

Hydrocarbons can be the power source in these systems. Hardware is required for safe storage and delivery of the fuel to the actuator. Many fuels could be used for this purpose, but it should be selected from among those which are easily obtained (e.g. petrol, diesel, bottled gas, etc.) but need storage and safety.

The energy of hydrocarbon fuel is normally converted into mechanical energy by combustion in an engine, although fuel cells could be used. Traditional combustion engines would be inappropriate as they produce heat, exhaust gas, sound and vibration. They also require air preparation, cooling and transmission. A micro-combustion engine may be proposed for this purpose, but more research is needed to make this usable.

4.6 Transmissions

Design of transmission system of force, pneumatic and electricity are most important factors. Researchers have developed many types of linkage to transmit the forces from source to the point of actuation. Some of these are reviewed by Wright et al. [4].

One transmission option not employed in the examples given is flexible shafts; see [13, 24]. In this transmission, a flexible shaft revolves inside a sheath. Shafts of this type could increase flexibility in the actuator placement. Another type is chain and sprocket transmission as used in the active orthosis; see Belforte et al. [8]. This would also allow some flexibility in actuator placement. Furthermore, it could be used to provide some reduction. However, chains can be noisy and dirty and require maintenance.

Further design of transmission line may be simplified to some extent using latest technology like use of silicon wafer, printed cloths and flexible capillary, etc. This will reduce the unnecessary wiring and minimise the weight of the orthosis.

4.7 Emerging Technologies

Development of new technology, use of MEMS and miniaturisation of actuators will drive the future production of all mechanical systems. Shape memory alloys (SMA) are one of the potential developments; see Hesselbach [25]. This has the property that the crystal structure changes (martensitic to austenitic) as they are heated and pass through a transition temperature. This results in a shape change, which if resisted, and develops a significant force. Although temperature is the controlling variable, heat can be induced in the material electrically.

However, the disadvantages are relatively small stroke and velocity, low efficiency and generally high transition temperatures. Successful applications have tended to use small (milli- or micro-metre scale), low-force actuators, which can be quickly heated and cooled due to their low heat capacity.

Although polymers are generally electrically non-conducting, there is a class which do conduct electricity. When the voltage across such polymers is altered, there is a corresponding change in dimension. This phenomenon allows actuators to be constructed from these conducting polymers; see Wurmus [26]. One of the most commonly used polymers for this purpose is polypyrrole (PPy).

Micro-actuators have been constructed by depositing PPy on a gold electrode. Applying a voltage causes bending of the actuator as the polymer changes length relative to the gold layer. A synthetic muscle could be constructed by connecting layers of silicon with numerous micro-actuators. The application of voltage would cause translation of the silicon layers similar to the operation of sacromeres in human muscle.

This technology is promising as a means of actuating rehabilitation devices, but is in its infancy. There are currently no known commercial polymer actuators of the size required for this project.

The piezoelectric effect of ceramics, commonly used in spark igniters and in force sensing, where it converts force to electrical charge, can be used in the opposite direction in actuators, to convert voltage to force or motion [27]. The most commonly used materials are sintered lead–zirconate–titanate (PZT) ceramics.

The main advantage of piezoelectric actuators over more traditional options is their speed of operation; hence, they use in shakers, valve actuation and active vibration damping.

Commercial actuators are generally constructed from a stack of ceramic discs in series. This generates high loads for relatively small actuators (1 kN for 5 mm diameter); however, displacements are low (0.2 mm for an actuator length of 200 mm). An alternative method is to use a two-layer actuator and cause opposing strains to create bending. This method can produce large deflection but relatively small forces (1 mm for 20 mm actuator, less than 1 N).

Emerging technology is not restricted to actuators; advances are also being made in the area of power supply. For instance, micro-combustion generates heat by burning very small amounts of hydrocarbon fuel. The primary advantage of this technique is that the efficiency of the conversion from chemical to thermal energy is very high. The heat is then converted into electricity by direct energy conversion techniques. In this way, micro-combustion batteries could provide the same power and operating time as conventional batteries 10 times their size.

One form of direct energy conversion uses the thermoelectric effect. Free electrons in hot material have higher kinetic energy and so diffuse to cold areas creating negative charge and hence a potential difference.

Practical sources have been made by stacking n and p pairs in series to generate higher voltages and used in space, military, remote, sub-sea applications, powered by radioisotopes or fossil fuels. These sources have no moving parts and can generate approximately 1 kW; however, efficiency is low (5%), and they are expensive.

Another form of direct energy conversion uses the thermionic effect. Free electrons are "boiled off" from a heated cathode (2000 K) and flow to a cool anode (<1000 K) in a vacuum or plasma. The electrode materials are dissimilar so that free electrons in the cathode have more energy (higher Fermi level) than in the anode. So after the transfer between electrodes, energy is available to do work.

Thermionic power sources have been used in space and military applications, powered by solar or nuclear power or fossil fuels. They have many of the same advantages and disadvantages as thermoelectric sources, but slightly higher efficiency (up to 25%).

The commercial actuators and power sources in these emerging areas do not appear to meet the requirements of this project at present, but successful ongoing research suggests that small highly efficient power sources will become commercially available in the next few years.

5 Experimental Results

In order to find a feasible solution in the process of selecting an actuator, a test rig is developed using a mass attached to a bar and the actuator is a forcing member. During actuation, the mass moves forward; the maximum displacement and time are measured. The required parameters like torque, power and specific power are calculated to identify the potential use of the actuator.

A flat motor of 24 V and 90 W of weight 1.03 kg has been tested. The maximum output power of 17 W and torque 19 Nm for a test weight of 20 kg have only been achieved while maximum voltage and current are applied to the DC motor. It also appears that a test load of 20 kg requires 23 W power and 27 Nm torque which can be achieved over a motor size of 36 V. This reduces the specific power and increases risk for power shock. However, an orthosis needs at least 72 Nm torque and 56 W

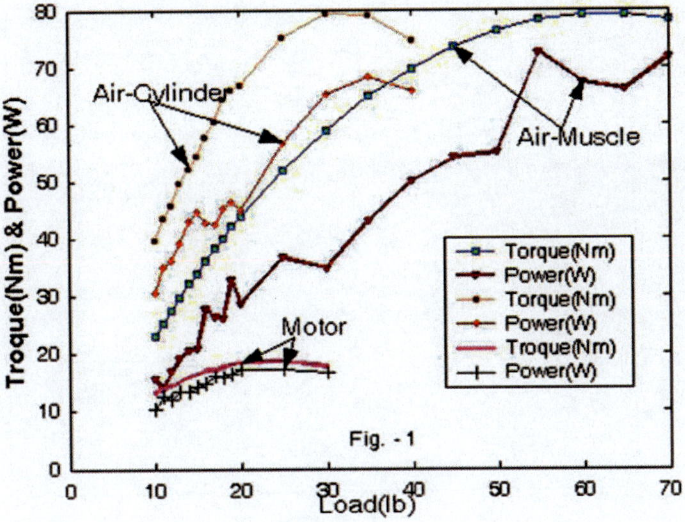

Fig. 1 Torque and power generated by actuators

power at knee and hip joint, which is not possible to achieve using this class of DC motors. The output torque and power developed in this system are shown in Fig. 1.

A cylinder of bore diameter 50 mm and stroke length 100 mm has been tested; it has given a maximum output torque 76 Nm and power 65 W which are above the average requirements for the test load of 20 kg. In another test with constant load, it has been shown that a cylinder of this type has capabilities of developing a power of 55 W and a torque 63 Nm which are much closer to the requirements. Hence, this actuator may be useful with certain precautions and recharging facilities.

An air muscle actuator is tested at 2 bar pressure and can achieve power up to 56 W and the torque up to72 Nm for a test load of 20 kg. The test results are presented in Figs. 1 and 2. The specific power and energy are higher than requirements. Hence, if suitable powering devices and compressed air are available, then these actuators may be used for the development of bio-robotic walking for orthosis.

6 Discussion

The most common and standard actuators available in the market have been tested. A 24 V DC servomotor has shown to be incapable of developing sufficient power and torque. The test shows that a motor over 50 V is necessary to achieve the desired power and torque but needs a high-voltage battery. This will increase size and weight as well as risk of electric shock and decrease the specific power and energy. In the history of robot technology, the use of electric motors is well known, but in this occasion these are not suitable for the development of active orthosis independently. However, it is

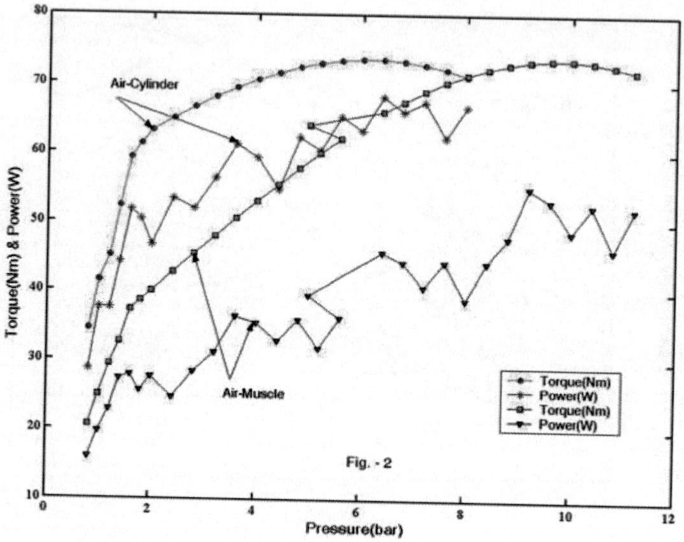

Fig. 2 Torque and power of pneumatic actuators

possible to use electric motor of appropriate size and weight in combination with some passive devices.

The test result of pneumatic cylinder has shown significant power and torque. This may be useful for development of an active orthosis. However, the size and weight of the reservoir, its charging facilities, frequency and limitation of pressure are great concern which may decrease the specific power and energy. However, new technology and development of high strength materials may improve possibility.

The pneumatic air muscle could be the best choice for this purpose with the same limitation for reservoir and recharging facilities. However, due to its low weight, it increases the specific power and energy.

The angular speed of the joint is another factor to be considered. The pneumatic cylinder and air muscle show the maximum angular speed of 50°/s and 45°/s at 2 bar pressure, respectively, while a human gait has an average angular speed of 45°/s. Hence, this investigation shows some possibilities of development of self-powered orthosis using these actuators.

The use of FES and active actuators or combined passive and active actuators or combination of all these three may be another possibility of developing the active orthosis. This will reduce the size and weight of the actuators and risk in many ways as in the case of carrying of heavy battery or big air reservoir.

7 Conclusions

Rehabilitation systems developed to date have significant differences from the walking orthosis envisaged by this project. Some use FES and have no actuators and hence low-power consumption. Others are not autonomous and do not carry their power source. It follows that there is little experience of meeting the criteria for power sources and actuators listed above.

These key requirements for the walking orthosis are however well covered by the examples cited of mobile robots. In particular, size, weight and power which are an issue with all mobile robots have been addressed by the specific hardware employed. The main functional requirements are discussed.

The rig test results described in this paper showed some hints and process of selection of suitable actuators for the development of bio-robotic walking for the orthoses. Further rig test on able body person will be carried out with these actuators to find the suitability and final development of the orthosis.

References

1. Tylor, R.I.: Bio-Robotic Walking Orthoses—Phase 1, Appendix C, EPSRC Research Proposal SI3/06 (2000)
2. Winter, D.A.: The Biomechanics and Motor Control of Human Gait: Normal, Elderly and Pathological, 2nd edn. Waterloo University Press (1991)
3. Goldsmith, A.A., Dowson, D., et al.: Comparative study of the activity of total hip arthroplasty patients and normal subjects. J. Arthroplasty **16**(5), 613–619 (2001)
4. Wright, P.J., et al.: Powering and Actuation Technology for a Bio-robotic Walking Orthosis, ISCMR (2003)
5. Gharooni, S., Heller, B., Tokhi, M.O.: A new hybrid spring brake orthosis for controlling hip and knee flexion in the swing phase. IEEE Trans. Neural Syst. Rehabil. Eng. **9**(1), 106–107 (2001)
6. Goldfrab, M., et al.: Preliminary evaluation of a controlled-break orthosis for FES-aided gait. IEEE Trans. Neural Syst. Rehabil. Eng. (2003)
7. Solomonow, M., Baratta, R., D'Ambrosia, R.: Standing and walking after spinal cord injury: experience with the reciprocating gait orthosis powered by electrical muscle stimulation. Topics Spinal Cord Injury Rehabil **5**(4), 29–53 (2000)
8. Belforte, G., et al.: Active orthosis—a rehabilitation device for disable people. In: Proceedings of the 3rd International Conference on Climbing and Walking Robots. 2000: Professional Engineering Publishing
9. Hocoma-Medical-Engineering, Lokomat driven gait orthosis (DGO) (2001)
10. Colombo, G., Wirz, M., Dietz, V.: Driven gait orthosis for improvement of locomotor training in paraplegic patients. Spinal Cord **39**(5), 252–255 (2001)
11. Shadow-Robot-Company.: Shadow biped (2003). http://www.shadow.org.uk/projects/biped.shtml
12. RoboScience.: RS-01 Robodog (2003). http://www.roboscience.com/
13. Honda.: Humanoid Robot (2003). http://world.honda.com/robot/
14. Caballero, R., et al.: On the modelling of SMART non-linear actuator for walking robots. In: Fourth International Conference on Climbing and walking Robots. CLAWAR 2001. 2001. Professional Engineering Publishing Ltd., Karlsruhe, Germany

15. Sharkh, S.A.: Power sources for untethered unmanned underwater vehicles. 1994, University of Southampton, Department of Electrical Engineering: Southampton
16. Berchuck, M., Andriacchi, T.P., et al.: Gait adaptations by patients who have a deficient anterior cruciate ligament. J. Bone Joint Surg. Am. **72**(6), 871–877 (1990)
17. Maxon_Motor.: Planetary Gearhead GP62. http://www.maxonmotor.co.uk/
18. Harmonic_Drive_Technologies.: PSA 14 Precision Gearhead. http://www.harmonic-drive.com/indexproducts.html
19. Kollmorgan.: Direct Drive DC Torque Motor QT-11301. http://www.motionvillage.com/products/product_literature/product_brochure/pdfs/KOL1154.pdf
20. EAD_Motors.: Stepping motor LA42BLK-*. www.eadmotors.com
21. Bodine_Electric_Company.: 42A-GB Series DC Right Angle Gearmotor Model N4866. http://www.bodinemtr.com/Products
22. EAD_Motors.: Linear actuator LA34AGK. http://www.eadmotors.com/pdf/LinearActSize34ConvTable.pdf
23. SMC_Pneumatics.: Cleanline Cylinder CDG1BN50. http://www.rscomponents
24. SS_White_Industrial.: Third Avenue, Bletchley, Milton Keynes MK1 1EJ
25. Hesselbach, J.: Shape memory actuators. In: Janocha, H. (ed.) Adaptronics and Smart Structures, pp. 143–160. Springer, Berlin (1999)
26. Wurmus, H., Kallenbach, M.: Chemomechanical actuators. In: Janocha, H. (ed.) Adaptronics and Smart Structures, pp. 207–217. Springer, Berlin (1999)
27. Clephas, B.: Piezoelectric actuators. In: Janocha, H. (ed.) Adaptronics and Smart Structures, pp. 106–123. Springer, Berlin (1999)

Study of Device-to-Device Communication in the Next Generation of Wireless Communication

Sapana Dhanvijay and Rajneesh Karn

Abstract The next generation of wireless communication set the goal of high-speed data rate without delay and low energy. The next generation of wireless communication's primary challenge is the full utilization of a dedicated spectrum and minimizing resource waste. The journey of all communication decade, such as 1G to 5G, works on the mode of high speed and low latency. The device-to-device communication model fulfills all criteria of next-generation communication with specific limitation. The mode of device-to-device communication is direct and short range. The direct communication link increases data speed, and short range reduces the possibility of interference in a cellular network. The process of device-to-device communication works in all mode of the spectrum, such as licensed and unlicensed of the spectrum. This paper presents the extensive review of device-to-device communication in various heads, such as communication mode, resource utilization, interference management, power control, and mode of selection. The limitation of cellular network scale with the model of device-to-device communication. The primary issues and challenges faced are in terms of low bandwidth and low energy. The maximization of data throughput required the process of cooperative communication model with the cellular network. This paper also focusses on the limitations and challenges of device-to-device communication in the area of 5G network.

Keywords Wireless communication · D2D · Cellular network · Mode selection · Interference

1 Introduction

The complexity of cellular network cannot support the requirement of high-speed data and utilization of spectrum. The various upcoming communication models scale the capacity of cellular network communication, and among these one is called device-to-device communication [1]. The device-to-device communication works without the infrastructure of cellular network. The process of communication of device to

S. Dhanvijay (✉) · R. Karn
Madhyanchal Professional University, Bhopal, Madhya Pradesh, India

© The Author(s), under exclusive license to Springer Nature Singapore Pte Ltd. 2022
A. K. Nagar et al. (eds.), *Intelligent Sustainable Systems*, Lecture Notes in Networks and Systems 333, https://doi.org/10.1007/978-981-16-6309-3_37

device follows the principle of direct communication with short range. The very famous communication mode is Wi-Fi and Bluetooth applied in direct communication mode to operate in device-to-device communication [1–3]. The characteristics of device-to-device communication are dynamic, and it does not require any infrastructure and access points [4]. Hence, the operator has new opportunity to design new business models for device-to-device communication without paying extra spectrum as it provides commercial, proximity service, and emergency. The growth of traffic in 4G and 5G of almost 30 times has been detected, in 2019 more than 11 extra bytes of traffic growth per month overload in cellular network [5–8]. The dynamic cooperative transmission methods allow the switching technique between different groups of device-to-device communication. The integration of device-to-device communication is with third-generation partnership project (3GPP). They cannot support the need of application such as data rates and distance between nodes of device-to-device communication and device discovery process. The maximum data rate is a major issue in concerned network and devices such that Bluetooth 5 supports a maximum data rate of 50Mbps and range close to 240 m. Despite these processes, Wi-Fi direct provides up to 250 Mbps rate and 200 m range. Instead of that, LTE model provides 13.5 Mbps and range 500 m [9–11]. The process of device-to-device communication increases the spectral efficiency and reduces the factor of energy and cost per bit. Despite a conventional cellular system, it required all that communication process through the base station, and it does not permit device to communicate directly with one another. The objective of device-to-device communication increases the signal strength of mobile device in a sparse environment. The most challenging factor is designing of D2D communication for sharing of resource in terms of energy and spectrum between cellular and D2D. One of the best benefits of device-to-device communication is privacy and anonymity. The sharing data of device-to-device communication is not central responsibility to handle storage and sorting [12–14, 19]. The functioning difficulties in processing of device-to-device communication are mainly complex resource management, device discovery approach, process of node selection methods, security protocol, and mobile management.

The various authors' survey reported the efficiency of spectrum utilization, and management of interference is successfully achieved.

The authors [10] describe the utilization of resource in different condition in local communication group. The local communication group depends on the process of power control and many other factors such as interference management and optimization of resource (Fig. 1).

The rest of the paper is described as follows: Section 2, describes the related work in area of device-to-device communication, Sect. 3 describes the comparative study of issue and methods, and finally, Sect. 4 describes conclusion and future work.

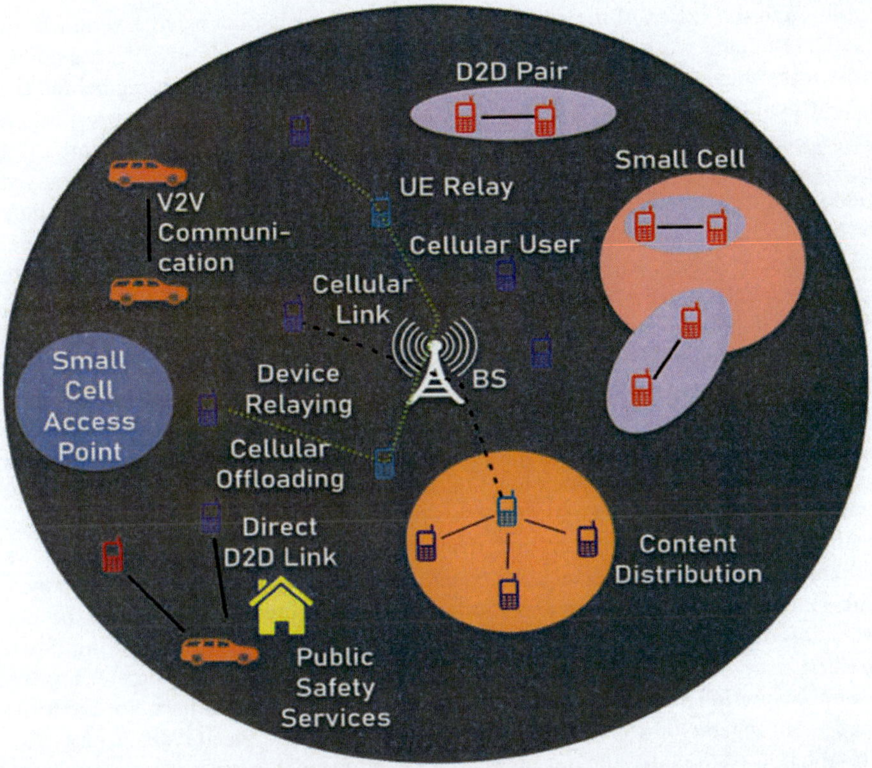

Fig. 1 Process block diagram of device-to-device communication mode. The model shows the direct connection with device to device and also shows device-to-cellular base station

2 Related Work

The challenge of next generation of communication device and algorithms is resolved in each and every day, but new challenges and issues are generated. The process of improvements in side of algorithms and device increases the possibility of success. In successful utilization of device-to-device communication process, various authors contribute. The contribution of authors are described here.

Loussaief et al. [1] The V2X interchanges, as empowered by cell D2D joins, has as of late acquired more prominent consideration attributable to the striking limit it shows in improving rush hour gridlock well-being, viability, and solace. Here setting up the current paper can be set, whereby another innovation is progressed, identifying with the 5G-related D2D depended on vehicular correspondence, as of late affirmed by the 3GPP normalization local area. As a result, V2X correspondence seems to raise various basic issues that are intently between related, principally, those connected to severe idleness and unwavering quality prerequisites. A thorough examination of the cutting edge is advanced, and D2D-based vehicular correspondence is nitty gritty, all through the extent of the current work. In any case, an intensive assessment of the D2D-related works, treating such an investigation region from alternate points of view, is shown. In a subsequent spot, the significant explanation lying behind the use of the D2D correspondence for vehicular help aims, including the prospects it gives, alongside the normal imminent possibilities, it could offer, are explored [2]. A whole throughput amplification issue for a CR helped D2D network is displayed calculating the RF-EH component. A joint improvement issue is detailed for whole rate amplification of cell and D2D clients by calculating power designation, channel task, client blending, and transmission time proportion portion. At that point, the issue is changed into a standard curved streamlining issue subject to control requirements at singular hubs, impedance limitation, and the separate rate. Moreover, the mystery limit necessities of cell, and D2D hubs are additionally thought of. The duality hypothesis is utilized to deteriorate the issue into various sub-issues, and Karush–Kuhn–Tucker conditions are misused to give the arrangement of the sub-issues. The recreation results accommodated the approval of their talked about plans [3]. SCMA innovation is applied to a D2D empowered cell organization, directed at using the over-burdening highlight of the SCMA plan to help an enormous gadget availability while upgrading the general organization execution. The SCMA conspire is executed to mutually enhance the codebook and force designation in the downlink D2D empowered cell organization, with the expectation to boost the framework information rate. This joint advancement issue is tackled by breaking down the first issue into two sub-issues: codebook portion and force assignment. For the codebook portion, the RACBS-D2D is examined utilizing strife chart. For the force portion arrangement, a GWF strategy is used to examine the IGWFPA plot. The presentation of the talked about plans is assessed through recreations that uncover the advantages of the examined arrangements under various situations [4]. Authors first present an orderly review of the socially uninformed specialized difficulties and the planned applications to solidify a foundation on D2D interchanges. Then, all the social highlights used in handling and improving the specialized issues and the applications are informed, individually. At that point, authors give an exhaustive review on socially mindful D2D interchanges, where different highlights of social space are utilized, for example, ties, local area, trust, and self-centeredness to handle the D2D specialized issues and improve the presentation of planned D2D applications. Creators arrange best in class socially mindful D2D correspondences into two principal segments. The principal segment thinks about specialized issues, while the subsequent segment relates to their applications to give smart data and propel more top to bottom examinations here. For each previously mentioned area, authors outline specialized perspectives and order the center commitments dependent on methodologies [5]. Authors center around strategies for dealing with these difficulties concerning mode choice, power control, and asset assignment. As contrasted and other contemporary chips away at this subject, authors talk about these issues in accordance with probably the latest exploration patterns. Furthermore, authors assemble appropriate plan contemplations of D2D talked about in writing while at the same time extricating new examples to acclimate per users with applications, models, techniques, and measurements concentrated recently.

Do et al. [6] Helpful NOMA network is examined to help D2D correspondence. In such D2D NOMA, AF and DF modes are conveyed to flag preparing at transfer who means to serve NOMA far client. To guarantee the QoS for the NOMA gadgets, a base rate prerequisite is pre-characterized for every gadget. Two primary plans are considered keenly: The principal situation is that the main gadget expects to speak with the second gadget through the help of AF/DF-helped BS, where it prepares singular force; and the subsequent situation is that the remote fueled gadget can speak with non-energy collecting gadget. Due to the QoS prerequisite, it is first important to decide execution hole in quite a while: hand-off and direct mode; NWPT situation and WPT situation, by contrasting the blackout execution and the necessary force for fulfilling the QoS of the gadgets. To additional assessment, authors expand their investigation on non-straight energy gathering strategy and various NOMA far clients which are conveyed in such D2D transmission. On the off chance that is attainable, a shut structure arrangement is accommodated framework execution assessment. Mathematical outcomes are introduced to approve the adequacy of the talked about D2D transmission methodologies [7]. Device discovery works and techniques are introduced and evaluated utilizing various methodologies. Furthermore, a gadget disclosure method is talked about for single cell and a multi-cell which is skilled to furnish precise and quick revelation with energy effective and improved directed way. D2D revelation procedures are assessed, its principle challenges are investigated, and potential arrangements are likewise recommended [8]. Creators present a social trust conspire that improves the security of MSNs. While considering the trust-based MSNs with MEC, reserving, and D2D, creators apply a novel profound support learning way to deal with naturally settle on a choice for ideally allotting the organization assets. Google TensorFlow is utilized to carry out the talked about profound support learning approach. Reenactment results with various organization boundaries are introduced to show the adequacy of the talked about conspire [9]. D2D correspondence is required to assume a huge part in impending cell networks as it guarantees super low inertness for correspondence among clients. This new mode may work in authorized or unlicensed range. It is anything but a novel expansion to the customary cell correspondence worldview. Its advantages are, in any case, joined by numerous specialized and business given that should be settled prior to coordinating it into the cell environment. This paper talks about the primary attributes of D2D correspondence including its use situations, engineering, specialized highlights, and spaces of dynamic exploration [10]. Authors consider the joint advancement of force distribution and channel task of the D2D joins and the CUs with the plan to give security to the CUs and improve the ghostly efficiency of the D2D connects at the same time. Creators center around the difficult downlink asset sharing issue and research both single-channel and multi-channel D2D interchanges. The subsequent asset the board plan issues end up being difficult nonlinear blended whole number issues. By the by, by abusing the characteristic properties of the formed advancement issues, authors can scientifically describe the ideal force portion of the CUs and D2D connects and create efficient strategies for joint improvement of their channel tasks. Reenactment results show that the examined asset and the executives strategies beat a few standard plans for sure.

Shafique et al. [11] A thorough audit for these key-empowering advances furthermore examines the new arising use instances of 5G-IoT driven by the advances in artificial insight, machine and profound picking up, continuous 5G drives, QoS necessities in 5G, and its normalization issues. At the long last, the paper examines difficulties in the execution of 5G-IoT because of high information rates requiring both cloud-based stages and IoT gadgets-based edge processing [12]. 5G, the cutting-edge media communications will be striking the business sectors in not-so-distant future. D2D correspondence would be a piece of 5G to serve correspondence needs for billions of associated gadgets to help high information rate ultrareliable low-dormancy interchanges. Indoor 5G will depend on appropriated little cell arrangements and D2D alongside machine-to-machine associations. AI is perhaps the most encouraging instrument for giving the best arrangement of answers to get familiar with the persuasive situations and certain boundaries of the correspondence organizations. This examination talked about RL-LCDC calculation and its Q-learning approach in an indoor D2D correspondence network for solid 5G availability with least idleness. The examined approach, RL-LCDC, productively finds the neighbors, chooses the D2D interface, and adaptively controls the correspondence range for most extreme organization network. The outcomes show that RL-LCDC upgrades the availability with lower start to finish delay and better energy productivity with effective assembly time when contrasted and other regular plans [13]. Authors examined a hereditary calculation-based technique to limit the impedance and boost the otherworldly efficiency. One of the upsides of hereditary calculation is that it escapes from nearby maximums and advances toward worldwide greatest via looking through changed pieces of search space all the while. Since D2D underlay cell network debases the SINR, a base SINR is considered for cell clients. Mathematical assessments show the prevalent presentation of the examined method as far as ghastly efficiency and impedance alleviation [14]. The leftover equipment debilitations starting from handset equipment, for example, stage commotion, speaker nonlinearity, and in stage and quadrature lopsided characteristics make twisting clamors producing a jumble between a planned and a got signal. This letter explores the blackout likelihood execution of a gadget-to-gadget correspondence which helped mmWave network by considering obstruction and reasonable hardware distortion noises.

Li et al. [15] A semi-brought together agreeable control technique is talked about for the cell uplink transmissions, where the UE transfers are haphazardly chosen by a specific thickness chosen by the eNB. Two explicit agreeable plans dependent on D2D correspondences are examined, which are the irregular UE transfer plot and the one further applying network coding. The D2D impedance is thought of and is demonstrated dependent on the stochastic calculation. The examined plans are investigated dependent on two particular traffic models, i.e., the MTC traffic with little information included and the full-cradle traffic. Broad Monte Carlo reproductions have been led for the little information traffic, and the shut structure hypothetical outcomes have been determined for the full-cushion traffic. Execution gains are recognized in different situations, and the correlations between two helpful plans are made also. The outcomes give a significant rule to the eNB to decide the ideal thickness of the UE transfers [2]. A whole throughput expansion issue for a CR helped

D2D network is displayed considering the RF-EH instrument. A joint advancement issue is planned for total rate boost of cell and D2D clients by considering power allotment, channel task, client blending, and transmission time proportion assignment. At that point, the issue is changed into a standard arched streamlining issue subject to control imperatives at singular hubs, impedance limitation, and the individual rate. The duality hypothesis is utilized to deteriorate the issue into numerous sub-issues, and Karush–Kuhn–Tucker conditions are misused to give the arrangement of the sub-issues. The recreation results accommodated the approval of their talked about plans [16]. An incompletely associated design is where the choice ought to be made on the bundle blends as well as on the arrangement of sending gadgets. To start with, the various occasions happening at every gadget are recognized to determine an articulation for the likelihood conveyance of the translating delay. The joint advancement issue over the arrangement of sending gadgets and the bundle blends of each is, at that point, detailed. The ideal arrangement of the joint enhancement issue is determined utilizing a chart hypothesis approach by presenting the collaboration diagram and reformulating the issue as a greatest weight coterie issue in which the heaviness of every vertex is the commitment of the gadget recognized by the vertex.

Poulakis et al. [17] To examine the ramifications of impedance, densification and range partaking in D2D execution acquire. The in-band D2D approach, where heritage clients exist together with potential D2D sets, is considered in a multi-cell framework. Overlay and underlay range sharing methodologies are utilized all together for the potential D2D sets to get to the range. Given that two of the most basic issues in the D2D idea are mode choice and client planning, authors mutually address them, targeting boosting the complete framework uplink throughput. Consequently, authors present a radio asset and the executive's component for intra-cell and cross-cell overlay/underlay D2D correspondences empowered in a multi-cell framework. Framework level re-enactments are executed to assess the framework execution and analyze the patterns of D2D correspondence acquiring for the distinctive range sharing methodologies and different densification situations. At the long last, genuine SDR-based examinations are performed to test and survey D2D interchanges for overlay and underlay range sharing [18]. Driven by arising big data methods, authors examined to plan a major information stage, named D2D big data, to empower the remote D2D correspondences among clients viably, to advance substance for suppliers precisely, and to complete offloading insight for administrators effectively. Creators convey a major information stage and further use an enormous scope dataset from a well-known D2D sharing application, which contains 866 million D2D sharing exercises on 4.5 million records scattered by means of almost 850 million clients in 13 weeks. By abstracting and examining multi-dimensional highlights, including the web practices, content properties, area relations, underlying qualities, meeting elements, social arborescence, protection conservation strategies, etc., authors confirm and assess the D2D big data stage with respect to prescient substance proliferating inclusion [19]. In versatile crowdsensing, clients have a focal job giving detected information utilizing their convenient gadgets. Versatile crowdsensing applications have become very well known these days. Creators address the collaboration issue with regard to D2D correspondence to upgrade versatile crowdsensing stages. Creators initially talk about how D2D correspondence can improve portable crowdsensing. Then, authors talked about and assessed an overall system joining portable crowdsensing and D2D correspondence [20]. To full the prerequisites of different arising applications, the future 6G versatile organization is required to be a naturally smart, profoundly unique, super thick heterogeneous organization that interconnects everything with very low idleness and fast information transmission. It is accepted that AI will be the most imaginative strategy that can accomplish astute mechanized organization activities, the board, and upkeep in future complex 6G organizations.

Huynh et al. [21] Mobile interpersonal organizations and D2D correspondences have arisen as promising strategies to help better neighborhood progressed administrations in 5G organizations [22]. To examine dynamic asset dividing among network substances in a downlink transmission plan to amplify EE of the CUs served by either (μW) macrocells or mmWave little cells while keeping a base QoS for the D2D clients. To address this issue, first, a self-versatile force control component for the D2D sets is detailed, subject to an impedance edge for the CUs while fulfilling their base QoS level. Consequently, an EE advancement issue, which is pointed toward amplifying the EE for the two CUs and D2D sets, has been settled. Reproduction results exhibit the viability of their examined calculation, which considers the inborn trade-offs between framework EE, framework total rate, and blackout likelihood for different QoS levels and changing densities of D2D sets and CUs [23]. A convergence with four legs and creators encoded the relating set of sign lights of every leg utilizing a hereditary calculation. To assess the efficiency of stage timing plan in this paper, creators have directed broad recreations, and the outcomes show that their methodology can react to the considered traffic flow inside one second. In contrasted and other traffic signal control frameworks, the presentation is improved nearly by 67% concerning the line length holding up at the crossing points during traffic signal light cycle [24]. Future age organizations will oblige a lot of information traffic and lower dormancy. To satisfy these needs, it is fundamental for investigating current ghastly utilizing or presenting new recurrence groups. Presentation of new recurrence groups requires an incomplete or complete difference in, as of now, sent framework, which will have high activity consumption and capital use. It is more helpful to find different arrangements by focusing on gadget-related arrangements. One of the answers for accomplishing higher spectra efficiency is through D2D correspondence. Here it looks at late phantom efficiency procedures in 5G through D2D correspondence. The principle center is around the usage of various procedures to improve range efficiency.

Hayat et al. [25] Security and protection issues in DD are featured. It is extensive and demonstrates that in-band is obviously superior to out-band with pragmatic and innovative reasons. To upgrade the extent of the exploration, network level, and framework level, S&P issues in the conveyed and incorporated frameworks climate with or without focal administration are reviewed [26]. D2D correspondence is quickly developing into a feasible technique for data trade in a cell organization.

It has an exceptionally low start to finish inertness and can increment ghostly proficiency of a cell organization. The most recent arrivals of 3GPP detail have focused on normalizing this method of correspondence and incorporating it in the biological system of LTE progressed. This will give more stimulus to the advancement of D2D innovations and their reception by versatile administrators. This paper presents a conversation and basic investigation of the primary highlights of D2D correspondence as characterized in Release 12 and resulting arrivals of 3GPP details [27]. D2D correspondence has arisen as a promising innovation for streamlining ghastly proficiency in future cell organizations. D2D exploits the nearness of conveying gadgets for proficient usage of accessible assets, improving information rates, diminishing inertness, and expanding framework limit. The exploration local area is effectively researching the D2D worldview to understand its maximum capacity and empower its smooth reconciliation into the future cell framework design. Existing overviews on this worldview generally center around impedance and asset the executives. Creators survey, as of late, examined arrangements in over investigated and under investigated regions in D2D. These arrangements incorporate conventions, calculations, and structures in D2D.

Shah et al. [28] Creators examine the utilization case situations of D2D correspondence by grouping its applications into two sorts: business and public well-being administrations. This is trailed by an inside and out conversation on the cutting-edge arrangements examined in different exploration considered resolving various issues related to every order. While examining an enormous number of past works, creators feature a portion of the open exploration issues and difficulties in D2D correspondences [29]. To clarify the advantages and difficulties for shrewd D2D correspondence will be needed to achieve the prerequisites for IoT. Numerous businesses and normalization bodies have shown gigantic interest toward the execution of D2D approach in remote organizations. D2D approach works to work without the fundamental control of the unified oversight which makes remote organizations more range and energy proficient with traffic offloading. Between systems administration, gadgets offer different applications toward the advancement of IoT. Actual requirements of the gadgets and asset imperatives of the organizations bring about a few issues which should be settled [30]. Authors considered a situation where D2D clients convey within the sight of cell clients in an overlay network arrangement. To examine the income of specialist organizations in money-related terms, the paper gives accurate articulations of administrator benefit for both D2D and cell clients. All the more explicitly, authors consider diverse organization boundaries including client thickness, communicate force, and channel varieties to comprehend their effect on the absolute income of the administrator. At the long last, authors determine the adjusting estimation of recurrence dividing factor and give important conversation on the scientific articulation. Their discoveries show that D2D interchanges beat the regular cell correspondences as far as income age capacity. Their outcomes have been checked by performing broad re-enactments.

Bonjorn et al. [31] Authors present advances of SL D2D interchanges as a key-empowering innovation for 5G upgraded eV2X correspondences. Creators give an outline about the asset portion and planning of various SL D2D modes under in-inclusion and out-of-inclusion application situations. Recreations were done to assess the presentation as far as impact likelihood expecting various upsides of the key boundaries, for example, RRI and asset choice window. Creators at the long last examine about the open specialized difficulties for super dependable and low-dormancy interchanges as refined from the 5G V2X use cases presented in 3GPP Rel.15 Waqas et al. [32] Creators researched PLS for SKGR in D2D interchanges dependent on helpful trusted and non-confided in transfers. By utilizing social ties, authors misuse three social wonders for secure interchanges, i.e., confided in situation, non-confided in situation, and part of the way confided in situation. The alliance game hypothesis is additionally used to choose the ideal hand-off sets for improving SKGR. Based on friendly ties, creators foster a calculation for SKGR that shields the keys mysterious from both snoop and non-confided in chosen transfers. Creators join secure transfers determination and framework wide security for D2D correspondences. The security and combination of the examined calculation are additionally demonstrated in their work. Both mathematically and scientifically results confirm viability and consistency of their talked about plot, which guarantees better SKGR execution in D2D interchanges [33]. Authors apply stochastic math hypothesis to determine an insightful articulation of downloading execution for the D2D storing network. Specifically, a social relationship model concerning the actual distance is embraced in their examination to get the normal download postpone execution utilizing arbitrary and deterministic reserving systems. Second, to accomplish a superior presentation in more reasonable and specific situations, authors build up a socially mindful appropriated storing procedure dependent on a decentralized learning robot, to streamline the reserve arrangement activity in D2D organizations. Not the same as the current reserving plans, the talked about calculation not just considers the file demand likelihood and the closeness of gadgets as estimated by their actual distance yet additionally considers the social connection between D2D clients. Their recreation results show that the examined calculation can join rapidly and beat the arbitrary and deterministic reserving techniques. With these outcomes, their work sheds experiences on the plan of D2D storing in the pragmatic sending of 5G organizations.

Debroy et al. [34] SpEED-IoT likewise utilizes a transformative game hypothetical course distribution model to support equal D2D correspondence. SpEED-IoT guarantees authorized occupant security, IoT gadget energy protection, powerful start to finish information rate enhancement, and quick union and reasonable course task among meddling D2D interchanges. Through reproduction-driven GENI-based IoT testbed, authors assess SpEED-IoT's presentation as far as guaranteeing availability and reachability among the IoT gadgets under changing range use conditions, information rate advancement of the relegated courses and the generally speaking IoT organization, viability in authorized officeholder insurance, and level of reasonableness while appointing courses to numerous meddling gadgets [35]. Authors have contemplated the different methods of asset assignment, mode determination for underlay interchanges regarding gadget to gadget and helpful correspondence procedures as far as LTE and LTEA stage. Better usage of the range is likewise portraying which is done based on investigation if appropriate asset assignment whether it is force, recurrence or time and mode choice is done in the customized way which would

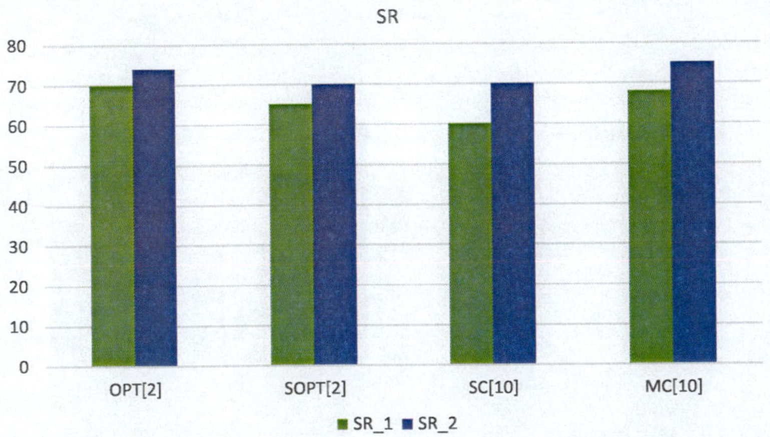

Fig. 2 Comparative analysis of SR using different techniques OPT[2], SOPT[2], RACBS[3, SC[10], and MC[10].

bring about the decrease of obstruction, and it will likewise prompt the protected framework.

3 Comparative Study

This section analyzes the results of some author in terms of SR, number of user, power (PQ), data rate, subcarrier degree, no. of users, SNR, outage probability, and iteration index [2, 3, 6, 10, 12] (Figs. 2, 3, 4, 5, 6, 7, 8, and 9; Tables 1 and 2).

4 Conclusion and Future Work

Speculating the device-to-device communication, they integrate with next-generation wireless communication and fulfill all requirements such as data rates, utilization of spectrum, and management of the resource. The user and vendor's concern benefited both traffic offloading, throughput, and low-power consumption—the flexibility of device-to-device communication to increase the reliability and availability of the network. The paper study focuses on all respect for device-to-device communication. The primary concern of this study is a resource management and interference management. The management of these points ensures the efficiency of device-to-device communication. Various authors focus on discovering the device process for the minimization of energy in terms of signal connection. The spectral efficiency of device-to-device communication benefited vendors for creating a new business plan

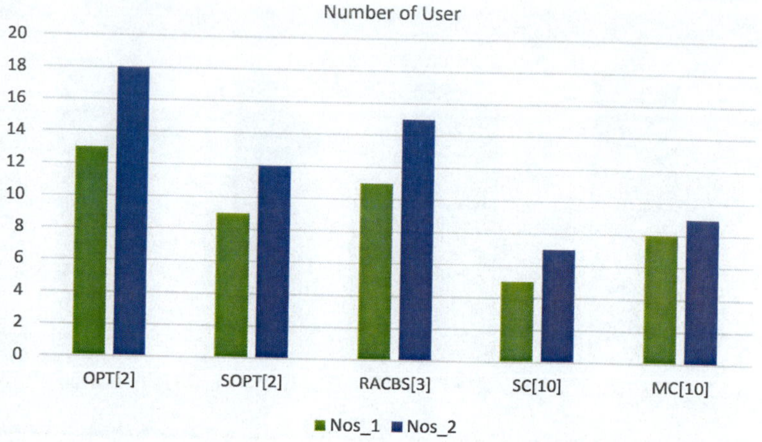

Fig. 3 Comparative analysis of number of user using different techniques OPT[2], SOPT[2], RACBS[3], SC[10], and MC[10]

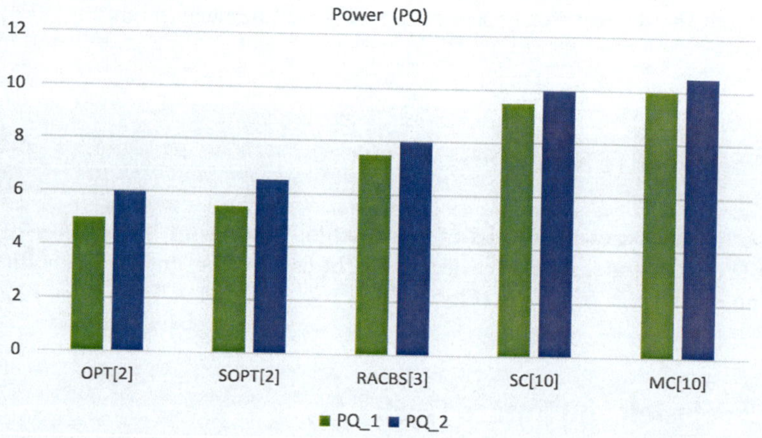

Fig. 4 Comparative analysis of power (PQ) using different techniques OPT[2], SOPT[2], RACBS[3], SC[10], and MC[10]

without allocating extra spectrum. The comparative table of authors work in the direction of results shows several users and their efficiency. The table focuses the work in the area of interference and synchronization are updated and need modification. The analysis process also focuses on the relation of the area of signal strength of user equipment. Most of the authors analyzed the results on the biases of numerical simulation. The numerical solution is the only way to theoretically validate the process of communication. It also indicates the implementation and real-time simulation of the device-to-device communication model.

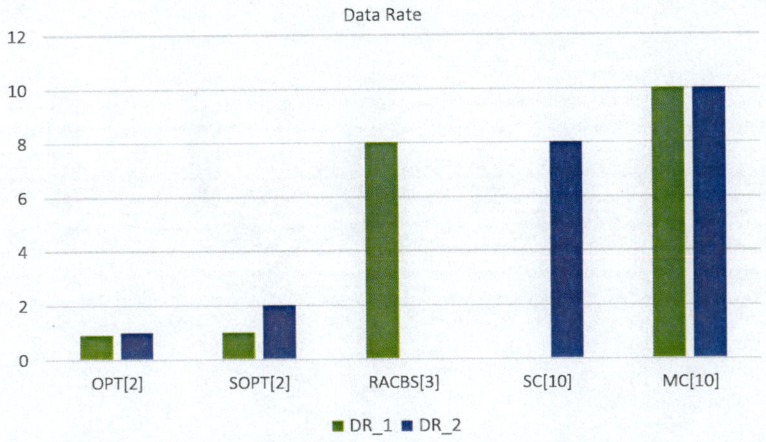

Fig. 5 Comparative analysis of data rate using different techniques OPT[2], SOPT[2], RACBS[3], SC[10], and MC[10]

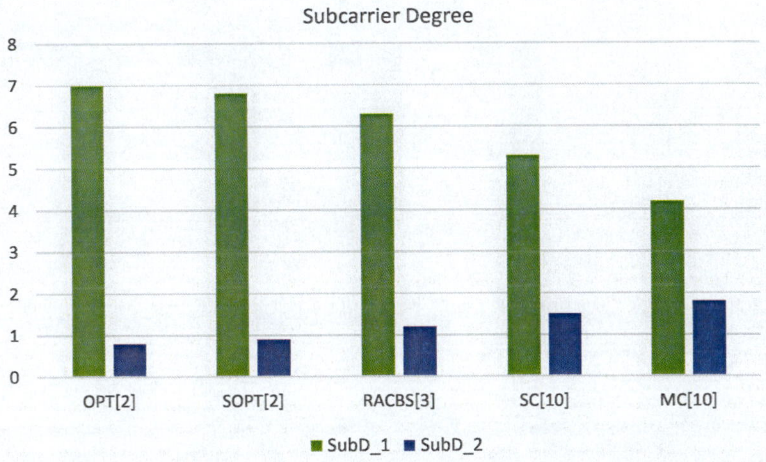

Fig. 6 Comparative analysis of subcarrier degree using different techniques OPT[2], SOPT[2], RACBS[3], SC[10], and MC[10]

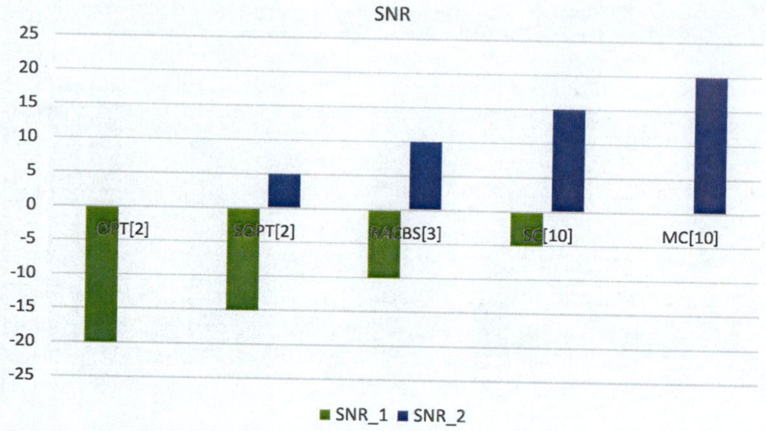

Fig. 7 Comparative analysis of SNR using different techniques OPT[2], SOPT[2], RACBS[3], SC[10], and MC[10]

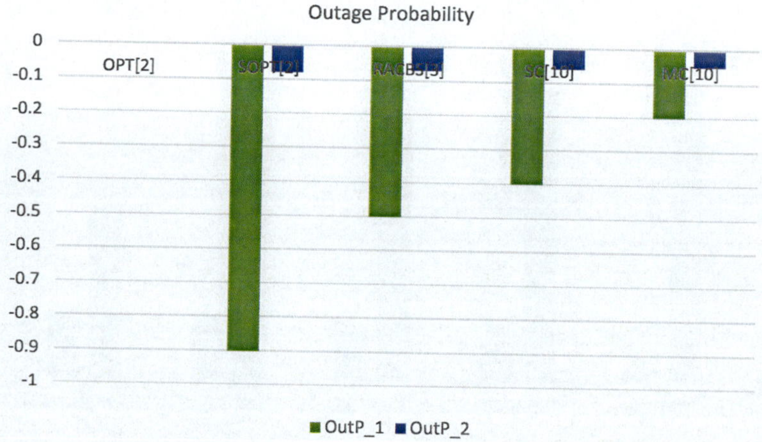

Fig. 8 Comparative analysis of outage probability using different techniques OPT[2], SOPT[2], RACBS[3], SC[10], and MC[10]

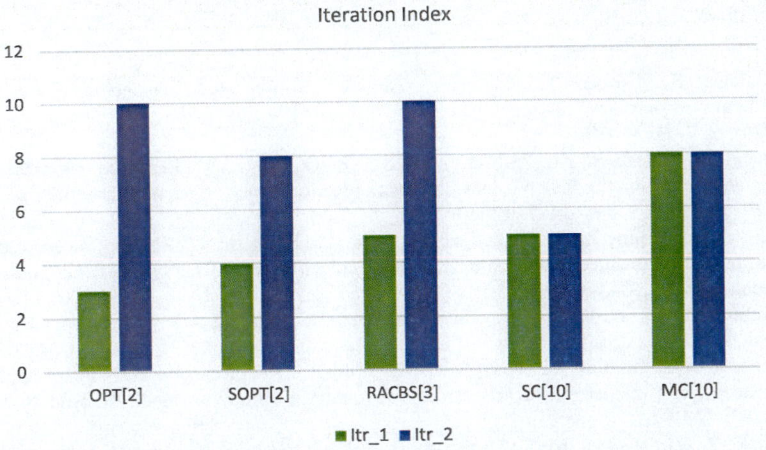

Fig. 9 Comparative analysis of iteration index using different techniques OPT[2], SOPT[2], RACBS[3], SC[10], and MC[10]

Table 1 Comparative analysis of different techniques OPT[2], SOPT[2], RACBS[3], SC[10], and MC[10]

	OPT[2]	SOPT[2]	RACBS[3]	SC[10]	MC[10]
SR_1	70	65	–	60	68
Number of User_1	13	9	11	5	8
Power (PQ) _1	5	5.5	7.5	9.5	10
Data rate_1	0.9	1	8	–	10
Subcarrier Degree_1	7	6.8	6.3	5.3	4.2
SNR_1	–20	–15	–10	–5	0
Outage Probability_1	0	–0.9	–0.5	–0.4	-0.2
Iteration Index_1	3	4	5	5	8

Table 2 Comparative analysis of different techniques OPT[2], SOPT[2], RACBS[3], SC[10], and MC[10]

	OPT[2]	SOPT[2]	RACBS[3]	SC[10]	MC[10]
SR_2	74	70	–	70	75
Number of User_2	18	12	15	7	9
Power (PQ) _2	6	6.5	8	10	10.5
Data rate_2	1	2	–	8	10
Subcarrier Degree_2	0.8	0.9	1.2	1.5	1.8
SNR_2	0	5	10	15	20
Outage Probability_2	0	–0.08	–0.07	–0.06	-0.05
Iteration Index_2	10	8	10	5	8

References

1. Loussaief, F., Marouane, H., Koubaa, H., Zarai, F.: Radio resource management for vehicular communication via cellular device to device links: review and challenges. Telecommun. Syst. **73**(4), 607–635 (2020)
2. Waqas, M., Aslam, S., Ali, Z., Sidhu, G.A.S., Xin, Q., Jang, J.W.: Resource optimization for cognitive radio-based device to device communication under an energy harvesting scenario. IEEE Access **8**, 24862–24872 (2020)
3. Sultana, A., Woungang, I., Anpalagan, A., Zhao, L., Ferdouse, L.: Efficient resource allocation in SCMA-enabled device-to-device communication for 5G networks. IEEE Trans. Veh. Technol. **69**(5), 5343–5354 (2020)
4. Ahmed, M., Li, Y., Waqas, M., Sheraz, M., Jin, D., Han, Z.: A survey on socially aware device-to-device communications. IEEE Commun. Surv. Tutor. **20**(3), 2169–2197 (2018)
5. Amodu, O.A., Othman, M., Noordin, N.K., Ahmad, I.: A primer on design aspects, recent advances, and challenges in cellular device-to-device communication. Ad Hoc Netw. **94**, 101938 (2019)
6. Do, D.-T., Van Nguyen, M.-S.: Device-to-device transmission modes in NOMA network with and without Wireless Power Transfer. Comput. Commun. **139**, 67–77 (2019)
7. Hayat, O., Ngah, R., Zahedi, Y.: In-band device to device (D2D) communication and device discovery: a survey. Wirel. Pers. Commun. **106**(2), 451–472 (2019)
8. He, Y., Yu, F.R., Zhao, N., Yin, H.: Secure social networks in 5G systems with mobile edge computing, caching, and device-to-device communications. IEEE Wirel. Commun. **25**(3), 103–109 (2018)
9. Kar, U.N., Sanyal, D.K.: An overview of device-to-device communication in cellular networks. ICT Express **4**(4), 203–208 (2018)
10. Wang, J., Huang, Y., Jin, S., Schober, R., You, X., Zhao, C.: Resource management for device-to-device communication: a physical layer security perspective. IEEE J. Sel. Areas Commun. **36**(4), 946–960 (2018)
11. Shafique, K., Khawaja, B.A., Sabir, F., Qazi, S., Mustaqim, M.: Internet of things (IoT) for next-generation smart systems: a review of current challenges, future trends and prospects for emerging 5G-IoT scenarios. Ieee Access **8**, 23022–23040 (2020)
12. Sreedevi, A.G., Rama Rao, T.: Reinforcement learning algorithm for 5G indoor device-to-device communications. Trans. Emerging Telecommun. Technol. **30**(9), e3670 (2019)
13. Takshi, H., Doğan, G., Arslan, H.: Joint optimization of device to device resource and power allocation based on genetic algorithm. IEEE Access **6**, 21173–21183 (2018)
14. Tlebaldiyeva, L., Maham, B., Tsiftsis, T.A.: Device-to-device mmwave communication in the presence of interference and hardware distortion noises. IEEE Commun. Lett. **23**(9), 1607–1610 (2019)
15. Li, Y., Cai, L.: Cooperative device-to-device communication for uplink transmission in cellular system. IEEE Trans. Wireless Commun. **17**(6), 3903–3917 (2018)
16. Douik, A., Sorour, S., Al-Naffouri, T.Y., Yang, H.-C., Alouini, M.-S.: Delay reduction in multi-hop device-to-device communication using network coding. IEEE Trans. Wireless Commun. **17**(10), 7040–7053 (2018)
17. Poulakis, M.I., Gotsis, A.G., Alexiou, A.: Multicell device-to-device communication: a spectrum-sharing and densification study. IEEE Veh. Technol. Mag. **13**(1), 85–96 (2018)
18. Wang, X., Zhang, Y., Leung, V.C.M., Guizani, N., Jiang, T.: D2D big data: content deliveries over wireless device-to-device sharing in large-scale mobile networks. IEEE Wirel. Commun. **25**(1), 32–38 (2018)
19. Mota, V., Thiago, F.S., Silva, H., Macedo, D.F., Ghamri-Doudane, Y., Nogueira, J.M.S.: Towards scalable mobile crowdsensing through device-to-device communication. J. Netw. Comput. Appl. **122**, 99–106 (2018)
20. Zhang, S., Liu, J., Guo, H., Qi, M., Kato, N.: Envisioning device-to-device communications in 6G. IEEE Network **34**(3), 86–91 (2020)

21. Huynh, D.-T., Wang, X., Duong, T.Q., Vo, N.-S., Chen, M.: Social-aware energy efficiency optimization for device-to-device communications in 5G networks. Comput. Commun. **120**, 102–111 (2018)
22. Naqvi, S.A.R., Pervaiz, H., Hassan, S.A., Musavian, L., Ni, Q., Imran, M.A., Ge, X., Tafazolli, R.: Energy-aware radio resource management in D2D-enabled multi-tier HetNets. IEEE Access **6**, 16610–16622 (2018)
23. Tang, C., Wei, X., Hao, M., Zhu, C., Wang, R., Chen, W.: Traffic signal phase scheduling based on device-to-device communication. IEEE Access **6**, 47636–47645 (2018)
24. Iqbal, J., Iqbal, M.A., Ahmad, A., Khan, M., Qamar, A., Han, K.: Comparison of spectral efficiency techniques in device-to-device communication for 5G. IEEE Access **7**, 57440–57449 (2019)
25. Hayat, O., Ngah, R., Kaleem, Z., Hashim, S.Z.M., Rodrigues, J.J.: A survey on security and privacy challenges in device discovery for next-generation systems. IEEE Access **8**, 84584–84603 (2020)
26. Kar, U.N., Sanyal, D.K.: A critical review of 3GPP standardization of device-to-device communication in cellular networks. SN Comput. Sci. **1**(1), 1–18 (2020)
27. Jameel, F., Hamid, Z., Jabeen, F., Zeadally, S., Javed, M.A.: A survey of device-to-device communications: research issues and challenges. IEEE Commun. Surv. Tutor. **20**(3), 2133–2168 (2018)
28. Shah, S.T., Hasan, S.F., Seet, B.C., Chong, P.H.J., Chung, MY.: Device-to-device communications: a contemporary survey. Wirel. Personal Commun. **98**(1), 1247–1284 (2018)
29. Pawar, P., Trivedi, A.: Device-to-device communication based IoT system: benefits and challenges. IETE Tech. Rev. **36**(4), 362–374 (2019)
30. Jameel, F., Kumar, S., Chang, Z., Hamalainan, T., Ristaniemi, T.: Operator revenue analysis for device-to-device communications overlaying cellular network. In: 2018 IEEE Conference on Standards for Communications and Networking (CSCN), pp. 1–6. IEEE (2018)
31. Bonjorn, N., Foukalas, F., Pop, P.: Enhanced 5G V2X services using sidelink device-to-device communications. In: 2018 17th Annual Mediterranean Ad Hoc Networking Workshop (Med-Hoc-Net), pp. 1–7. IEEE (2018)
32. Waqas, M., Ahmed, M., Li, Y., Jin, D., Chen, S.: Social-aware secret key generation for secure device-to-device communication via trusted and non-trusted relays. IEEE Trans. Wireless Commun. **17**(6), 3918–3930 (2018)
33. Ma, C., Ding, M., Chen, H., Lin, Z., Mao, G., Liang, Y.-C., Vucetic, B.: Socially aware caching strategy in device-to-device communication networks. IEEE Trans. Veh. Technol. **67**(5), 4615–4629 (2018)
34. Debroy, S., Samanta, P., Bashir, A., Chatterjee, M.: SpEED-IoT: spectrum aware energy efficient routing for device-to-device IoT communication. Futur. Gener. Comput. Syst. **93**, 833–848 (2019)
35. Malik, P.K., Wadhwa, D.S., Khinda, J.S.: A survey of device to device and cooperative communication for the future cellular networks. Int. J. Wirel. Inf. Netw. 1–22 (2020)

Atherosclerosis Disease Prediction Based on Feature Optimization and Ensemble Classifier

Brajesh Kumar and Harsh Mathur

Abstract The early prediction of cardiovascular disease (CVD) saves millions of lives worldwide. The early symptoms of cardiovascular disease are usual and cannot be predicted. Atherosclerosis is a significant contributor to cardiovascular disease. Atherosclerosis hardens the coronary artery, reduces blood flow and increases severe heart attack and stroke. The automated detection of cardiovascular diseases plays an essential role in heart disease. The machine learning-based classification algorithm plays a vital role in accurate classification and detection. This paper proposed ensemble-based classifier methods prediction of atherosclerosis disease. The proposed ensemble classifier built on the principle of boosting methods uses two classifiers: support vector machine and KNN. The support vector methods work as a base classifier and KNN as a variable classifier. The ensemble classifier works with feature optimization methods. The glowworm optimization (GSO) algorithm is applied for the optimization of features. The optimized features increase the rate of classification accuracy. The proposed algorithm is simulated in MATLAB software and tested with three reputed datasets of atherosclerosis. The proposed algorithm compares with the existing three algorithms for atherosclerosis disease detection: support vector machine, KNN and decision tree (DT). The difference of results suggests that the proposed algorithm is efficient instead of the existing algorithms.

Keywords CVD · Heart disease · Atherosclerosis · SVM · KNN · DT ensemble classifier · GSO

1 Introduction

The mortality and morbidity rates increase due to cardiovascular diseases (CVD). The World Health Organization reported annually 30% of deaths due to cardiovascular diseases worldwide [1]. The main reason for cardiovascular disease is atherosclerosis, and it also affected future cardiovascular disease [2, 3]. The conventional assessment process of cardiovascular diseases cannot estimate and control the impact of

B. Kumar (✉) · H. Mathur
Madhyanchal Professional University, Bhopal, Madhya Pradesh, India

© The Author(s), under exclusive license to Springer Nature Singapore Pte Ltd. 2022

A. K. Nagar et al. (eds.), *Intelligent Sustainable Systems*, Lecture Notes in Networks and Systems 333, https://doi.org/10.1007/978-981-16-6309-3_38

atherosclerosis. Atherosclerosis causes harding in coronary arteries and built plaques that thicken large and medium-sized arterial walls, narrowing the lumen, obstructing blood flow [4–6]. The affected flow rate of blood causes severe life-threatening diseases such as cardio arrest, heart stroke and many others disease related to cardiovascular [7–10]. However, several parameters could not be monitored such as age, family history, gender and behaviors of patients. Patients neglect the initial symptoms of cardiovascular disease, and that negligence is converted into a high risk for cardio arrest and stroke [11, 12]. Caring for cardiovascular disease is a very costly and long-term treatment, so most patients left the treatment suggested by doctors. The early-stage prediction of cardiovascular disease saves many lives around world. The computer-aided diagnosis (CAD) predicted the symptoms and nature of heart disease [2, 13]. The collection of data related to computer-aided diagnosis is a challenging task. The variation of data changes the accuracy of heart disease detection [14]. The various authors and research scholars reported the approach of machine learning to improve the capacity of the computer-aided diagnosis process. The process of machine learning provides a different algorithm for the classification and detection of heart diseases [15–17]. The method of heart disease detection involves oral data such as high blood pressure, diabetes, age and many others factors. In other ways, applied imagery diagnosis process such as X-ray of heart, ECG, RMT and some different imagery process are to measure the risk level of cardiovascular disease [18–20]. Most authors recommended the oral and statistical parameters for the diagnosis of heart disease [21]. Machine learning has provided a practical approach for various healthcare applications such as medical image segmentation, computer-aided diagnosis (CAD), tools for liver cancer diagnosis and many more critical disease detections [22, 23]. Real-time medical data is very complex and diverse, so processing data is an important phase for diagnosis and prediction. The processing of data is an imbalance the error occurred; this error affected the accuracy of detection. The recent trends of machine learning algorithms applied include decision tree (DT), Naïve Bayes (NB), KNN, support vector machine and neural network-based algorithms [24, 25]. The reported literature also focuses on clustering algorithms for the processing of medical data for the prediction process [26, 27]. Despite several classifications and predictions, models face the problem of conducting accurate predictive models for medical data diagnosis. The massive imbalance of dataset distribution is another problem of heart disease diagnosis [28]. The feature optimization and selection processes improve the issue of data imbalance in heart disease prediction. The methods of feature selection and optimization apply swarm intelligence algorithms such as ant colony optimization, particle swarm optimization and many more inspired algorithms. This paper spreads glowworm optimization (GSO) for data imbalance and optimization of features [27]. The GSO algorithm works as a global optimization algorithm. This paper proposed an ensemble-based classifier for predicting heart and atherosclerosis symptoms—the process of ensemble classifier based on the principle of boosting methods. The support vector machine works as a primary classifier, and the KNN algorithm applies as a variable classifier [24, 27]. The proposed ensemble classifier enhances the prediction capacity of atherosclerosis prediction. The rest of paper is organized as follows: Section 2 describes the background of classification algorithm. Section 3 describes the proposed methodology. Section 4 describes the experimental analysis and finally conclusion in Sect. 5.

2 Background of Atherosclerosis and Classification

The accuracy of atherosclerosis disease prediction plays a vital role in cardiovascular diseases (CVD). The hardening of coronary artery invites various critical abnormal functions of bloods in artery system. The low flow of bloods also influences the presence of oxygen in blood vessel [12, 14]. The various authors contribute various algorithms for the classification and detection of atherosclerosis diseases. The machine learning-based algorithm influences the process of detection and classification of heart disease. The machine learning provides various algorithms such as supervised, unsupervised and reinforced learning algorithm [15]. The major dominated contribution we got is through a supervised learning-based classification algorithm [16, 17]. The reported survey of machine learning algorithms such as support vector machine, KNN, Naïve Bayes, decision tree and other methods of ensemble classifier also draws the attention for the utility of classification. The most of classifier depends on the process of feature selection cum optimization process [18]. The large number of features of atherosclerosis disease declines the accuracy of classifier algorithms. Many authors suggest that the feature optimization is a better option to reduce unwanted features of atherosclerosis disease. This section describes the algorithm of machine learning applied for the classification of atherosclerosis disease [18, 19].

2.1 Support Vector Machine

Support vector machine is regressive as well binary classifier. The process of classification depends on the function of kernel and separable plan. The support vector machine uses linear as well as nonlinear kernel function for the classification of data. The nature of support vector machine is linear, nonlinear and sigmoid. The nonlinear support vector machine maps the feature data with respect to one plane to another plan [2, 3, 15]. The separation of data plan is nonlinear, and decision factor correlates with margin function of support vector. The hyperplane of equation is derived as

$$\begin{aligned} WD.xi + b &\geq 1 if yi = 1 \\ WD.xi + b &\leq -1 if yi = -1 \end{aligned} \tag{1}$$

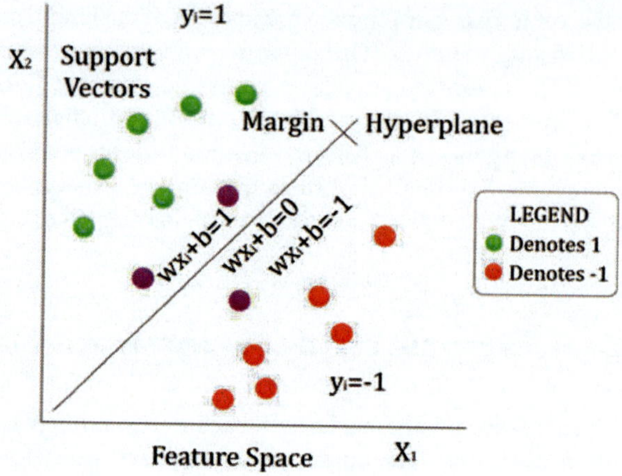

Fig. 1 Process block diagram of support vector machine

Here W is weight vector, x is input vector yi label o class, and b is bias (Fig. 1).

The minimization formulation of support vector

$$\begin{aligned} &\text{Minimize}\frac{1}{2}||w||2 + C\sum_{i=1}^{n}\varepsilon i, \quad i = 1, 2, \ldots, n \\ &\text{subject to } y_i\left(w^{\mathrm{T}}D.x1 + b\right) \geq 1 - \varepsilon 1 \\ &\varepsilon_i \geq 0i = 1, 2, \ldots, n. \end{aligned} \tag{2}$$

Here C is constant, n is number of observation, and $\varepsilon 1$ is slack variable.

The rule of decision function is

$$f(x) = \sum_{i=1}^{n} yi\alpha i\, K(xi, xj) + b. \tag{3}$$

2.2 *Decision Tree*

A decision tree is a predictive model of data classification. The processing of decision tree algorithm follows the tree structure of feature set. The partition of branch applied the concept of maximum entropy. The feature attribute has maximum entropy treat as leaf node class. The learning process of decision tree algorithm is very simple and suitable for the prediction of medical data analysis [11]. The decision tree rule formation according to the behaviors of feature attribute is narrated in terms of maximum occurrence of probability. The core algorithm of decision tree is C4.5. The processing of C4.5 algorithm on design of tree to provide output in manner of top to down as class label. The entropy and gain are two functions and manage the processing of decision tree algorithm [11].

2.3 K Nearest Neighbor Classifier

The KNN classifier is simple algorithm of machine learning, and it also knows as lazy classifier. The classification accuracy of KNN classifier varies in range of 70–80%. The major utility of KNN classifier is in case of pattern recognition [24]. The KNN classification algorithm is applied on the case of continuous nature of attribute. The processing of KNN algorithm is described here.

1. Estimate K training attribute which belongs to unknown attribute
2. Chose the common occurring classification of K

For the estimation of similarity in class of K instance applied different distance equation, the very famous distance equation is Euclidean distance equation.

2.4 Glowworm Optimization Algorithm

The process of feature selection and feature optimization used various algorithms such as genetic algorithm and particle swarm optimization. For the process of feature selection using glowworm optimization algorithm, the glowworm optimization algorithm is new swarm intelligence algorithm without the processing of memory [26]. The process of algorithm used the concept of local sharing information and updated the feature set of parameters.

Algorithm steps

Luciferin update: The value of luciferin update depends on fitness value and pervious value of luciferin [14, 15]

$l_i(t+1) = (1-\rho)l_i(t) + \gamma \text{fitness}(xi(t+1)).$

here $l_i(t)$ denotes the luciferin value of glowworm i at time t.

Neighborhood selection

The selection of neighborhood $N_i(t)$ as

$$N_i(t) = \{J : d_{ij}(t) < r_d^i(t); l_i(t) < l_j(t)\}$$

here $d_{ij}(t)$ euclidean distance between glowworm i and j at time t.

Compute probability

Glowworm applied function of probability to measure the movements of glowworm.

$P_{ij}(t) = \frac{l_j(t) - l_i(t)}{\sum_{k \in Ni(t)} l_k(t) - l_i(t)}$.

Movement process

$$x_i(t+1) = x_i(t) + s\left(\frac{x_j(t) - x_i(t)}{\left|x_j(t) - x_i(t)\right|}\right)$$

Decision rule update

$$r_d^i(t+1) = \min\left\{r_{s,max}\{\ 0, r_d^i(t) + \beta(n_t - |N_i(t)|.)\right\}\}$$

Here β is constant and $r_{s,}$ shows radius of glowworm.

3 Proposed Methodology

The proposed ensemble classifier is based on boosting methods. The support vector machine works as a base classifier, and KNN classifier works as variable classifier and collects the similar features of atherosclerosis data and process to support vector machine. The processing of features of atherosclerosis dataset is by glowworm optimization algorithm. The glowworm optimization algorithm circulates the value of luciferin and collects relevant features to mapping of class. The process of ensemble classifier used K features and S classifier. The data search space of classifier is K × S. Process of boosting ensemble classifier is explained as under showing flow of operation in detail.

The process of boosting is explained as below:

Consider $K = \{k1, k2, k3,\ldots, kn\}$ be the atherosclerosis dataset.

$S = \{\}$ ensemble classifier set.

$C = \{c1, c2, c3,\ldots, cn\}$ classifier set

$$X = \text{trainingset},\ X \in K$$
$$Y = \text{testset},\ Y \in k$$

$M = n(K)$

Consider init = 1

$L(\text{init})$ = random subset of X; $L(\text{init}) \in X$

$P(0) = \{\}$

For $i = 1$ to P do

if $i > 1$

$f(i)$ = irrelevant features of data $P(I - -1) + M(i)$

$P(i)$ = training process of classifier $C(i)$ and $S(i)$
$S = S\ C(i)$
end if
next i
for $i = 1$ to M
$R(i) = Y$
next i
estimate = max($R(i)$: i = 1,2, ..., n)
Rule of ensemble classifier

1. The base class features = $\{k_1, k_2, \ldots, kn\}$., where each k_j = $\{z_{j1}, z_{j2}, \ldots, z_{js}\}$. is collection of features of *Em* class $\{C_1, C_2, \ldots, C_k\}$., and the classes of features are non-overlapping the process of ensemble describe as follows:

$$Em(C_k) = \{\{z_j, z_k, \ldots, z_m\} \subseteq \{zf_1, zf_2, \ldots, z_s\}\} \quad (4)$$

2. Let the kernel mapping of support vector machine classifier class

$$M_{i+1}[EM(C_g) = Em(C_g) \cup z_r] > S_i[Em(C_K)] \quad (5)$$

3. Prediction of ensemble classifier is $Em(C_k)$ (Fig. 2).

$$\mathrm{Pre}_{Em(C_K)} = \mathrm{pre}(Em(C_K)) \leq S \quad (6)$$

4 Experimental Analysis

To evaluate the performance of proposed ensemble classification algorithm with MALTAB software, the version of software is R2014a, and the configuration of system is I7 processor, 16 GB RAM and windows10 operating system. The MATLAB provides the basic support library file of support vector machine and other classification algorithm. But the other function of classifier is defined and programmed with function file and compiled with library file. For the process of detection applied UCI machine learning dataset, the process of sample of data applied ten cross folds for the processing of prediction and measurement of parameters such as accuracy, sensitivity and specificity.

DESCRIPTION OF DATASET

To evaluate the performance of machine learning algorithms, use various datasets of heart disease prediction. The resource of dataset is UCI machine learning repository. All dataset is freely available for study purpose. The description of dataset is mentioned below.

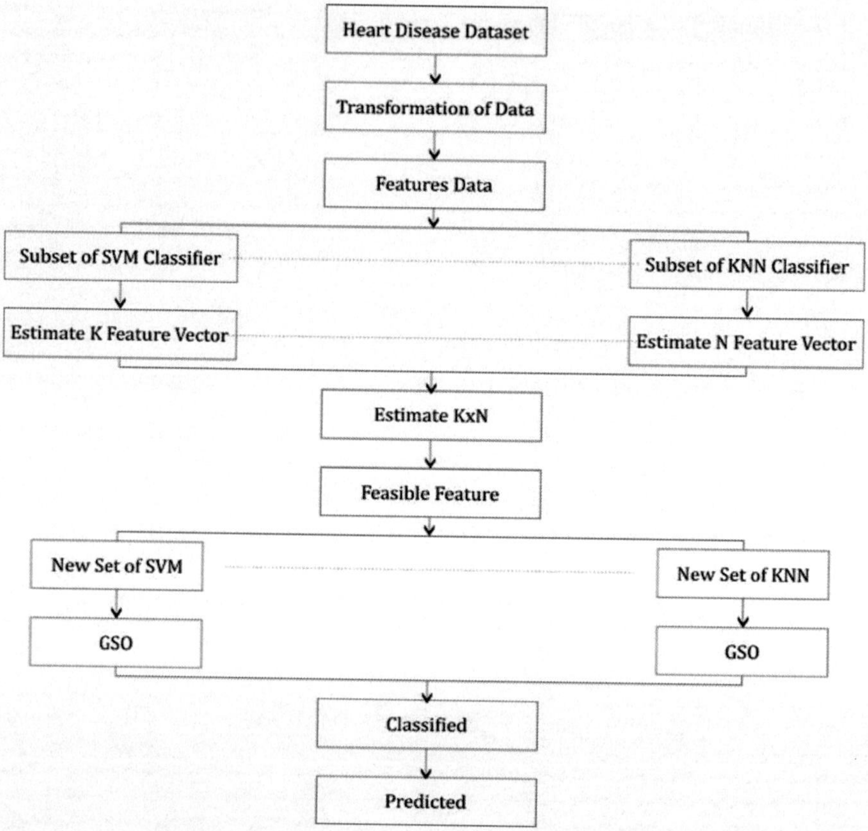

Fig. 2 Process model of proposed work

- Hungarian: The Hungarian dataset has been collected at the Hungarian Institute of Cardiology, Budapest, by Andras Janosi. This database contains ten features. Through the 294 dataset samples, 34 samples were discarded because of missing values, and 262 records were commonly used and segregated in 62.21% healthy subjects and 37.78% with heart disease.
- Cleveland: This database contains 76 attributes, but considers only of 14 of them. The total number of instances is 303.
- Z-Alizadeh Sani: This dataset contains 270 instance and 13 attributes. Each patient could be in two possible categories: CAD or normal. A patient is categorized as CAD, if his/her diameter narrowing is greater than or equal to 50%, and otherwise as normal.
- Statlog: This dataset has been 270 instances and 13 attributes. The missing attribute of dataset is null (Fig. 3, 4 and 5; Tables 1, 2, 3 and 4).

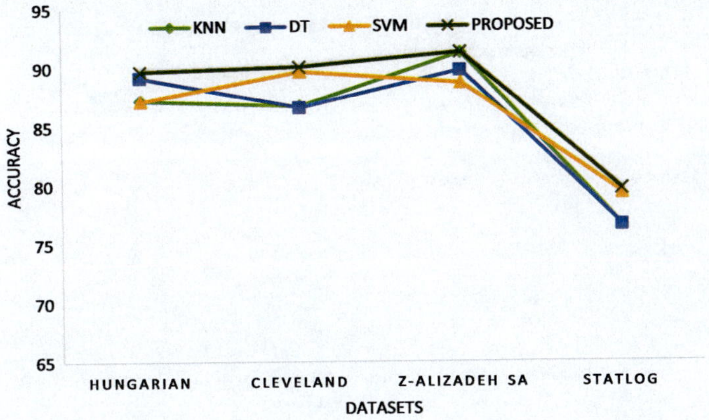

Fig. 3 Comparative performance of KNN, DT, SVM and proposed using accuracy with Hungarian, Cleveland, Z-Alizadeh Sani and Statlog datasets. Here we observe the comparative better performance of accuracy: in case of Hungarian dataset for proposed technique, in case of Cleveland dataset for proposed technique, in case of Z-Alizadeh Sani dataset for proposed technique and in case of Statlog dataset for proposed technique

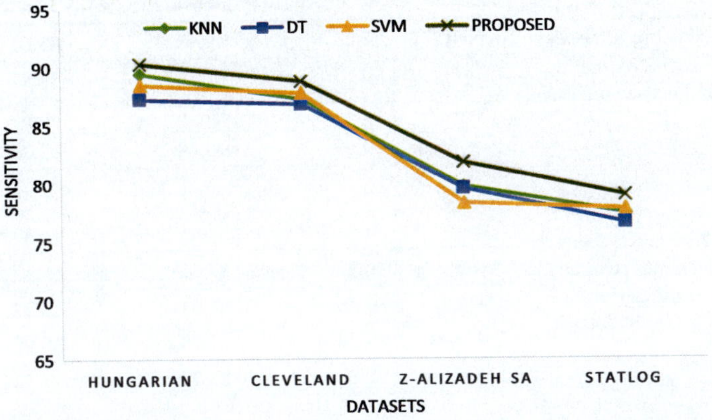

Fig. 4 Comparative performance of KNN, DT, SVM and proposed using sensitivity with Hungarian, Cleveland, Z-Alizadeh Sani and Statlog datasets. Here we observe the comparative better performance of sensitivity: in case of Hungarian dataset for proposed technique, in case of Cleveland dataset for proposed technique, in case of Z-Alizadeh Sani dataset for proposed technique and in case of Statlog dataset for proposed technique

5 Conclusion and Future Work

A large number of features and the size of the dataset decline the performance of classification algorithms. The feature selection cum optimization algorithm glowworm optimization reduces the numbers of features and increases the classification ratio of

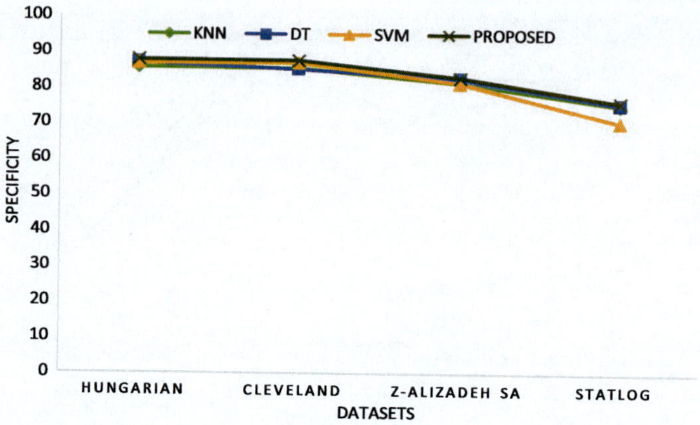

Fig. 5 Comparative performance of KNN, DT, SVM and proposed using specificity with Hungarian, Cleveland, Z-Alizadeh Sani and Statlog datasets. Here we observe the comparative better performance of specificity: in case of Hungarian dataset for proposed technique, in case of Cleveland dataset for proposed technique, in case of Z-Alizadeh Sani dataset for proposed technique and in case of Statlog dataset for proposed technique

Table 1 Comparative result analysis of KNN, DT, SVM and proposed using accuracy, sensitivity, specificity and AUC with Hungarian dataset

Method	Accuracy (%)	Sensitivity (%)	Specificity (%)
KNN[24]	87.23	89.42	85.61
DT[11]	89.15	87.24	86.92
SVM[30]	87.17	88.46	86.83
Proposed	89.69	90.28	87.74

Table 2 Comparative result analysis of KNN, DT, SVM and proposed using accuracy, sensitivity, specificity and AUC with Cleveland dataset

Method	Accuracy (%)	Sensitivity (%)	Specificity (%)
KNN[24]	86.73	87.25	85.37
DT[11]	86.59	86.81	85.22
SVM[30]	89.63	87.74	86.95
Proposed	90.06	88.77	87.58

Table 3 Comparative result analysis of KNN, DT, SVM and proposed using accuracy, sensitivity, specificity and AUC with Z-Alizadeh Sani dataset

Method	Accuracy (%)	Sensitivity (%)	Specificity (%)
KNN[24]	91.21	79.72	81.41
DT[11]	89.73	79.54	82.62
SVM[30]	88.65	78.26	81.33
Proposed	91.27	81.78	82.95

Table 4 Comparative result analysis of KNN, DT, SVM and proposed using accuracy, sensitivity, specificity and AUC with Statlog dataset

Method	Accuracy (%)	Sensitivity (%)	Specificity (%)
KNN[24]	76.53	77.38	75.29
DT[11]	76.56	76.56	75.58
SVM[30]	79.27	77.74	70.47
Proposed	79.65	78.92	76.16

atherosclerosis disease detection. The glowworm optimization reduces the unwanted features of datasets. The proposed ensemble classifier overcomes the limitation of the binary classification algorithm. The ensemble classifier used two classifiers: support vector machine and KNN. The process of boosting increases the mapping of features and reduces the noise and outlier data of atherosclerosis patients. The proposed algorithm is assessed on different datasets and tried with a combination of datasets. The classification accuracy improves 5% instead of support vector machine and decision tree. The difference in classification rate in KNN is 8%. The efficiency of the proposed algorithm is suitable for a small segment of the dataset. The cross fold fixed in all cases is 10. The proposed algorithm is very significant for clinical data analysis and prediction of atherosclerosis diseases detection. Delineate that our ensemble-based classifier scores maximum accuracy such as 89% instead of another classification algorithm. The GSO algorithm selects the most relevant features and increases the possibility of classification. The proposed ensemble classifier also applies other disease detection such as cancer and skin disease.

References

1. Terrada, O., Cherradi, B., Raihani, A., Bouattane, O.: A novel medical diagnosis support system for predicting patients with atherosclerosis diseases. Inform. Med. Unlocked **21**, 100483 (2020)
2. Terrada, O., Cherradi, B., Raihani, A., Bouattane, O.: Classification and Prediction of atherosclerosis diseases using machine learning algorithms. In: 2019 5th International Conference on Optimization and Applications (ICOA), pp. 1–5. IEEE (2019)
3. Yilmaz, N., Inan, O., Uzer. M.S.: A new data preparation method based on clustering algorithms for diagnosis systems of heart and diabetes diseases. J. Med. Syst. **38**, 48–59 (2014)
4. Bhatla, N., Jyoti. K.: A novel approach for heart disease diagnosis using data mining and fuzzy logic. Int. J. Comput. Appl. **54**(17) (2012)
5. Munger, E., Hickey, J.W., Dey, A.K., Jafri, M.S., Kinser, J.M., Mehta, N.N.: Application of machine learning in understanding atherosclerosis: Emerging insights. APL Bioeng. **5**(1), 011505 (2021)
6. Parameswari, C., Siva Ranjani, S.: Prediction of atherosclerosis pathology in retinal fundal images with machine learning approaches. J. Ambient Intell. Humanized Comput. 1–11 (2020)
7. Kolossváry, M., De Cecco, C.N., Feuchtner, G., Maurovich-Horvat, P.: Advanced atherosclerosis imaging by CT: radiomics, machine learning and deep learning. J. Cardiovasc. Comput. Tomogr. **13**(5), 274–280 (2019)
8. Singh, N., Firozpur, P., Jindal, S.: Heart disease prediction system using hybrid technique of data mining algorithms. Int. J. Adv. Res. Ideas Innov. Technol. **4**(2), 982–987 (2018)
9. Rao, V. Hari, S., Naresh Kumar, M.: Novel approaches for predicting risk factors of atherosclerosis. IEEE J. Biomed. Health Inform. **17**(1), 183–189 (2012)

10. Qawqzeh, Y.K., Otoom, M.M., Al-Fayez, F., Almarashdeh, I., Alsmadi, M., Jaradat, G.: A proposed decision tree classifier for atherosclerosis prediction and classification. IJCSNS **19**(12), 197 (2019)
11. AnimeshHazra, S.K., Mandal, A.G., Mukherjee, A., Mukherjee, A.: Heart disease diagnosis and prediction using machine learning and data mining techniques: a review. Adv. Comput. Sci. Technol. **10**(7), 2137–2159 (2017)
12. Shouman, M., Turner, T., Stocker, R.: Integrating Naive Bayes and K-means clustering with different initial centroid selection methods in the diagnosis of heart disease patients. CS & IT-CSCP, 125–137 (2012)
13. Han, D., Kolli, K.K., Al'Aref, S.J., Baskaran,L., van Rosendael, A.R., Gransar, H., Andreini, D., et al.: Machine learning framework to identify individuals at risk of rapid progression of coronary atherosclerosis: from the PARADIGM registry. J. Am. Heart Assoc. **9**(5), e013958 (2020)
14. Nikan, S., Gwadry-Sridhar, F., Bauer, M.: Machine learning application to predict the risk of coronary artery atherosclerosis. In: 2016 International conference on computational science and computational intelligence (CSCI), pp. 34–39. IEEE (2016)
15. Serrano, J.I., Tomeckova, M., Zvárová, J.: Machine learning methods for knowledge discovery in medical data on atherosclerosis. Eur. J. Biomed. Inform. **2**(1), 6–33 (2006)
16. Georga, E.I., Tachos, N.S., Sakellarios, A.I., Kigka, V.I., Exarchos, T.P., Pelosi, G., Parodi, O., Michalis, L.K., Fotiadis,D.I.: Artificial intelligence and data mining methods for cardiovascular risk prediction. In: Cardiovascular Computing—Methodologies and Clinical Applications, pp. 279–301. Springer, Singapore (2019)
17. Kumar, A., Kumar, P., Srivastava, A., Ambeth Kumar, V.D., Vengatesan, K., Singhal, A.: Comparative Analysis of Data Mining Techniques to Predict Heart Disease for Diabetic Patients. In: International Conference on Advances in Computing and Data Sciences, pp. 507–518. Springer, Singapore (2020)
18. Magesh, G., Swarnalatha, P.: Optimal feature selection through a cluster-based DT learning (CDTL) in heart disease prediction. Evol. Intell. 1–11 (2020)
19. Rémy, N.M., Martial, T.T., Clémentin, T.D.: The prediction of good physicians for prospective diagnosis using data mining. Inform. Med. Uunlocked **12**, 120–127 (2018)
20. Repaka, A.N., Ravikanti, S.D., Franklin, R.G.: Design and implementing heart disease prediction using naives Bayesian In: 2019 3rd International Conference on Trends in Electronics and Informatics (ICOEI), pp. 292–297. IEEE (2019)
21. Kazerouni, F., Bayani, A., Asadi, F., Saeidi, L., Parvizi, N., Mansoori, Z.: Type2 diabetes mellitus prediction using data mining algorithms based on the long-noncoding RNAs expression: a comparison of four data mining approaches. BMC Bioinformatics **21**(1), 1–13 (2020)
22. Shaji, S.P.: Prediction and diagnosis of heart disease patients using data mining technique. In: 2019 International Conference on Communication and Signal Processing (ICCSP), pp. 0848–0852. IEEE (2019)
23. Atallah, R., Al-Mousa, A.: Heart disease detection using machine learning majority voting ensemble method. In: 2019 2nd International Conference on new Trends in Computing Sciences (ICTCS), pp. 1–6. IEEE (2019)
24. Bashir, S., Khan, Z.S., Khan, F.H., Anjum, A., Bashir, K.: Improving heart disease prediction using feature selection approaches. In: 2019 16th International Bhurban Conference on Applied Sciences and Ttechnology (IBCAST), pp. 619–623. IEEE (2019)
25. Amin, M.S, Chiam, Y.K., Varathan, K.D.: Identification of significant features and data mining techniques in predicting heart disease. Telematics Inform. **36**, 82–93 (2019)
26. Zhu, X., Ni, Z., Xia, P., Ni, L.: Hybrid ensemble pruning using coevolution binary glowworm swarm optimization and reduce-error. Complexity 2020 (2020). Dhar, S., Roy, K., Dey, T., Datta, P., Biswas, A.: A hybrid machine learning approach for prediction of heart diseases. In: 2018 4th International Conference on Computing Communication and Automation (ICCCA), pp. 1–6. IEEE (2018)

27. Dwivedi, A.K.: Performance evaluation of different machine learning techniques for prediction of heart disease. Neural Comput. Appl. **29**(10), 685–693 (2018)
28. Rathnayakc, B.S.S., Ganegoda, G.U.: Heart diseases prediction with data mining and neural network techniques. In: 2018 3rd International Conference for Convergence in Technology (I2CT), pp. 1–6. IEEE (2018)

Damage Estimation of Reed Fibers During Recycling

Ming Li, Kaitang Hu, and Suying Shao

Abstract Recycled fibers retrieved from waste paper products have been widely utilized in the pulp and paper industry. However, the properties of fibers deteriorate after successive recycling processes, and the final products made of recycled fibers may not have the required quality. Thus, it is important to evaluate the properties of recycled fibers, especially tensile strength, during recycling. This paper aims to estimate and predict the tensile strength of reed fibers using machine learning methods. Reed fibers, as well as other non-wood fibers, have not attracted enough research interests. To evaluate properties of reed fibers during recycling, handsheets were formed, defibrated, and formed again to simulate recycling processes. After each recycling, density, water retention value, and tensile strength were measured. Gaussian process regression was utilized to estimate and predict the tensile strength values through supervised learning. It was chosen because of limited training data. The models with squared exponential kernel and matern 5/2 kernel were trained, and the validation root mean square errors were 1.7188 and 2.1616, respectively. However, the performance of the models with a testing dataset was not as expected, and the root mean square error was up to 4.67. One reason might be the testing data were obtained from filtered reed pulp.

Keywords Damage estimation · Recycled fibers · Tensile strength · Gaussian process regression

M. Li (✉)
University of Michigan-Flint, Flint, MI 48502, USA
e-mail: minglilm@umich.edu

K. Hu
Yuncheng Polytechnic College, Yuncheng, China

S. Shao
Tianjin University of Science & Technology, Tianjin, China

© The Author(s), under exclusive license to Springer Nature Singapore Pte Ltd. 2022

A. K. Nagar et al. (eds.), *Intelligent Sustainable Systems*, Lecture Notes in Networks and Systems 333, https://doi.org/10.1007/978-981-16-6309-3_39

1 Introduction

The pulp and paper industry has to satisfy human requirement as the demand for paper products continues to rise. It faces a massive challenge because of the rise of the awareness of environmental pollution and recourses shortage [1]. Using recycled fibers retrieved from waste paper products is a key solution. Recycled fibers can replace virgin fibers extracted from wood as an important fiber source, which can reduce overexploitation of existing forests. The papermaking processes using recycled fibers consume less energy and water, and at the same time produce less by-products to pollute air and water. Thus, the pulp and paper industry can benefit greatly from recycled fibers in resolving resource shortages, saving energy, and protecting the environment.

Recycled fibers have long been widely utilized in manufacturing of paper products, and the usage will continue to grow as technologies develop rapidly. However, using recycled fibers causes some issues. Because recycled fibers have to undergo mechanical and chemical treatment during the paper making and recycling processes, the properties of recycled fibers deteriorate to such an extent that the final product made of recycled fibers may not have the required quality. Recycled fibers have to endure defibrating, grinding, drying, deinking, etc., when their mechanical and chemical properties have been altered, including polymerization degree of cellulose, structure of fibers, inter-fiber bonds, and so on [2].

Recycling has been thought to incur loss of fiber swelling, change of fiber morphology, reduction of bonding surface area, breakage of fibers, and so on [3, 4]. The loss of fibers' capability of swelling with water after drying is ascribed to hornification, though some researchers have reservations [4]. Hornification is induced by partially irreversible closure of pores in fiber walls, which cannot reopen to contain water while rewetting. Thus, the flexibility and conformability of fibers decrease, and in turn, the fibers become stiffer and more brittle. Hornification is considered as a main cause of the reduction of mechanical strength and product quality of recycled fibers.

Researchers have been exploring various mechanical and chemical treatments to improve the strength of recycled fibers, so that the paper product can satisfy the quality requirement. For example, blending virgin fibers with recycled fibers, using ultrasound and enzymes in deinking processes, adding numerous types of cellulose nanofibers, etc., have been proved to be efficient in strengthening recycled fibers and improving properties of paper products [5].

However, recycled fibers retrieved from waste paper may comprise fibers that have already been recycled various times. Even though virgin wood fibers can be recycled five to seven times, the properties of recycled fibers deteriorate after each recycling [6]. The recycled fibers may be too weak or too short to compose paper products with acceptable quality. Thus, it is necessary to evaluate the conditions of recycled fibers in order to determine if they are suitable for paper products of certain grades.

Wood fibers are the dominant material source in the paper and pulp industry, while non-wood fibers only account for a small portion as the raw material. The reasons behinds this are regional and seasonal availability, various physical and chemical properties, transportation and storage difficulties, and so on. Thus, the research on recycled fibers is mainly based on wood fibers, and non-wood fibers seldom can attract interests. However, in some developing countries, especially in China and India, non-wood plants account for up to 70% of the raw materials used in the paper and pulp industry [7]. Reed is one of the most important non-wood materials in China, and it makes up about 14% of all the pulp manufactured [8, 9]. As the demand for paper products continues to grow, it will play a more important role to address the shortage of wood fibers.

In this paper, the deterioration of reed fibers during recycling was examined. The properties of recycled reed fibers, especially tensile strength, were evaluated. Handsheets made from reed fibers were defibrated and formed again into handsheets for four times. The properties such as tensile strength, density, and water retention value (WRV) were obtained from test sheets after each simulated recycling process.

Gaussian process regression (GPR) is a nonparametric Bayesian approach to solve nonlinear regression and classification problems. Instead of focusing on individual data samples, GPR considers the probability distribution over functions to determine the estimation and prediction of uncertainty [10]. In addition, GPR can take advantage of prior knowledge to optimize the base kernels to capture the underlying properties in the data samples. Thus, GPR can fit well on fewer data samples and be applied in various scenarios for the estimation and prediction of uncertainties in real life [11]. In this paper, GPR was utilized to estimate and predict the tensile strength of handsheets made from recycled reed fibers, and the performance of different kernels was evaluated for comparison.

2 Experiment

The reed pulp was obtained from a pulp and paper mill in NE China. Handsheets were formed following the standard procedure to prepare for tests of reed pulp properties. Tensile strength, density, and water retention value were measured to determine the quality of handsheets. Then, the handsheets were defibrated to simulate recycling processes, and following the previous procedure, new handsheets were formed and tested. The simulated recycling processes were repeated four to five times.

3 Results and Discussion

To evaluate the property deterioration during recycling, tensile strength, density, and water retention value of the test handsheets were measured after each recycling. Figures 1 and 2 show the change of the tensile strength, density, and water retention

Fig. 1 Tensile strength during recycling

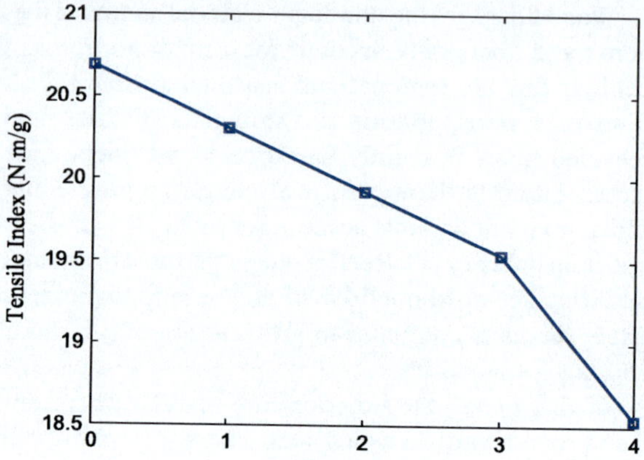

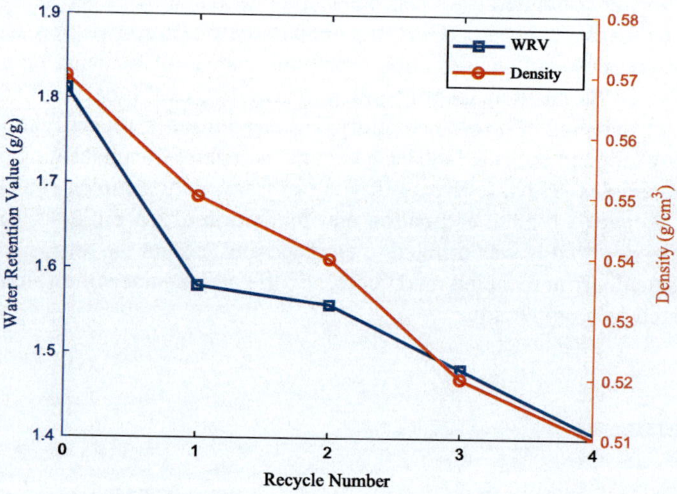

Fig. 2 Density and WRV during recycling

value during successive recycles of one experiment. From the figures, it is obvious to see the effect of recycling on these properties. The properties of reed fibers deteriorated after each recycling, and in most cases, the properties deteriorated the most after the first recycling.

Tensile strength is an important property of fibers, which usually is an indicator if the quality of paper products can satisfy their purpose or not. Tensile strength is a big concern for recycled fibers, since it deteriorates during recycling processes as well as other properties. Thus, it is important to estimate and predict the tensile strength of fibers during recycling in order to determine the usage of recycled fibers for products with acceptable quality. Density and water retention value are two other properties indicating the condition of fibers, which are also important to the quality of paper products. They are considered highly correlated with tensile strength, so they were chosen for the estimation and prediction of tensile strength.

GPR is a nonparametric supervised machine learning model based on probability distributions, which can be used to solve nonlinear regression and classification problems. GPR can make an estimation and prediction with prior knowledge using kernel functions. It aims to identify the underlying relationship in the given training dataset, and predict the response values of new input testing dataset assuming it has the similar relationship.

Equation 1 shows the joint distribution of the function values $\boldsymbol{y}$ and $\boldsymbol{f}_*$, according to the training dataset and the testing dataset, respectively.

$$\begin{pmatrix} \boldsymbol{y} \\ \boldsymbol{f}_* \end{pmatrix} \sim \mathcal{N}\left(0, \begin{bmatrix} \boldsymbol{K} + \sigma_n^2 \boldsymbol{I} & \boldsymbol{K}_* \\ \boldsymbol{K}_*^T & \boldsymbol{K}_{**} \end{bmatrix}\right) \tag{1}$$

$\boldsymbol{K}$, $\boldsymbol{K}_*$, and $\boldsymbol{K}_{**}$ are the covariances for the training set, training-test set, and test set, respectively. The predictive distribution shown in Eq. 2 is the conditional distribution based on training input $\boldsymbol{X}$, training output $\boldsymbol{y}$, and testing input $\boldsymbol{X}_*$.

$$\overline{\boldsymbol{f}_*} | \boldsymbol{X}, \boldsymbol{y}, \boldsymbol{X}_* \sim \mathcal{N}(\overline{\boldsymbol{f}_*}, cov(\boldsymbol{f}_*)) \tag{2}$$

The covariance functions can be defined by various kernel functions which reflect prior knowledge. Equations 3 and 4 are two frequently used kernel functions, squared exponential kernel and matern 5/2 kernel, respectively.

$$k(\boldsymbol{x}i, \boldsymbol{x}j) = \sigma_f^2 \exp\left[-\frac{1}{2}\frac{(\boldsymbol{x}_i - \boldsymbol{x}_j)^T(\boldsymbol{x}_i - \boldsymbol{x}_j)}{\sigma_l^2}\right] \tag{3}$$

$$k(\boldsymbol{x}i, \boldsymbol{x}j) = \sigma_f^2\left(1 + \frac{\sqrt{5}r}{\sigma_l} + \frac{5r^2}{3\sigma_l^2}\right) exp\left(-\frac{\sqrt{5}r}{\sigma_l}\right) \tag{4}$$

where $r = \sqrt{(\boldsymbol{x}_i - \boldsymbol{x}_j)^T(\boldsymbol{x}_i - \boldsymbol{x}_j)}$. σ_f and σ_l are hyperparameters, σ_f is the standard deviation of the signal span, and σ_l is the characteristic length scale [12].

In this paper, squared exponential kernel and matern 5/2 kernel were used to estimate and predict the tensile strength values. Four experiment results during the recycling of reed fibers were input as the training data. The indicators included recycle number, density, and water retention value, and the response was the tensile strength after each recycling. Because of the limited training data, the three cross-validation folds was chosen to protect against overfitting. The training results are shown in Fig. 3. Figure 3 a shows the validation predicted value vs the actual value based on the squared exponential kernel and b is based on the matern 5/2 kernel. From the figures, the performance of the matern 5/2 kernel was better than that of the squared exponential kernel, though they were similar. The validation root mean square error

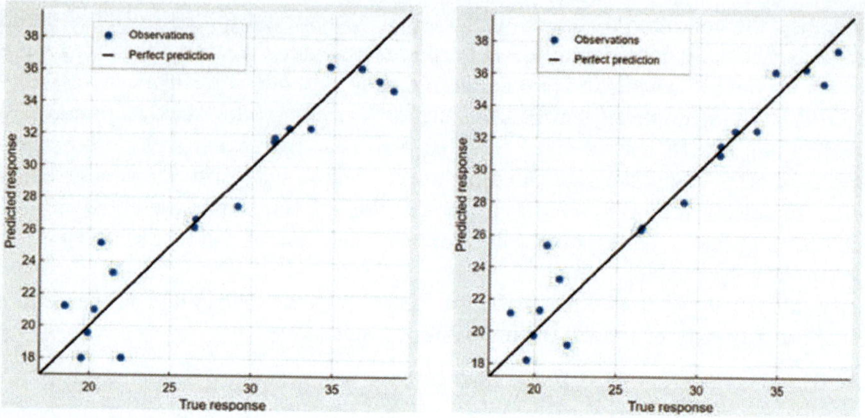

a) Squared exponential kernel b) Matern 5/2 kernel

Fig. 3 Performance of training

(RMSE) and the R-squared value for the matern 5/2 kernel were 1.7188 and 0.94, respectively, and they were 2.1616 and 0.91 for the squared exponential kernel.

Another set of experimental data was utilized as testing data, and it was input to the trained models with the squared exponential kernel and the matern 5/2 kernel for estimation and prediction. The testing results are shown in Fig. 4. It is obvious that the testing results for both the squared exponential kernel and the matern 5/2 kernel had no significant difference. However, the performance of the prediction was not as expected, and the RMSE for the model with matern 5/2 kernel was 4.5256 and 4.6662 for the squared exponential kernel. The reasons for the difference between the prediction and the actual values might be:

Fig. 4 Testing results of GPR

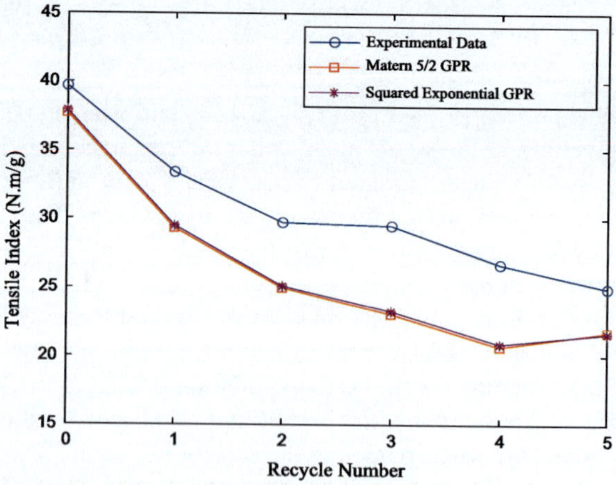

- Limited training data could not express the underlying relationship
- During some recycles, the properties did not deteriorate that much
- Testing data was from filtered reed pulp from which fine components were removed by a 200 mesh filter

4 Conclusion

The pulp and paper industry has long been utilizing recycled fibers retrieved from waste paper products in order to address issues such as shortage of raw materials, limitation of energy, pollution of environment, and so on. However, fibers retrieved from waste may already have been recycled. After successive recycling processes, damage developed in fibers causes the properties of fibers to deteriorate, which cannot guarantee the final products made of recycled fibers have acceptable quality. Tensile strength is a key factor to determine the quality of paper products. Thus, it is important to evaluate the deterioration of tensile strength during recycling.

This paper explored using machine learning methods to evaluate the tensile strength of recycled reed fibers. Reed fibers and other non-wood fibers will play a more important role as demands for paper products increase. Handsheets made of recycled reed fibers were formed, defibrated, and formed again to examine properties of recycled fibers. Tensile strength, density, and water retention value were measured during the successive recycling processes.

Gaussian process regression is a nonparametric kernel-based approach for supervised learning. It examines probability distribution over functions to solve nonlinear regression and classification problems. Thus, it is suitable for limited training datasets. GPR models with squared exponential kernel and matern 5/2 kernel were trained to estimate and predict tensile strength based on density and WRV. Four experimental results composed the training dataset and the training performance indicated by the validation root mean square errors was 1.7188 for squared exponential kernel and 2.1616 for matern 5/2 kernel. However, the models did not work well with a testing dataset based on reed fibers whose fine particles were removed. Though the training results of both models bore a close resemblance, the RMSE could be as high as 4.67. The possible reasons might be limited training datasets and different treatments of the filtered reed fibers. In the future, more experiments will be carried out to test the proposed approach, and other kernels will be explored to improve the representation of data.

References

1. Adu, C., Jolly, M., Thakur, V.K.: Exploring new horizons for paper recycling: a review of biomaterials and biorefinery feedstocks derived from wastepaper. Curr. Opin. Green Sustain. Chem. **13**, 21–26 (2018)

2. Čabalová, I., Kačík, F., Geffert, A., Kačíková, D.: The effects of paper recycling and its environmental impact. In: Broniewicz, E. (eds.) Environmental Management in Practice, vol. 17, pp. 329–350. InTech, Rijeka (2011)
3. Nazhad, M.M.: Recycled fiber quality-a review. J. Ind. Eng. Chem. **11**(3), 314–329 (2005)
4. Hubbe, M.A., Venditti, R.A., Rojas, O.J.: What happens to cellulosic fibers during papermaking and recycling? A review. BioResources **2**(4), 739–788 (2007)
5. Bajpai, P.K.: Solving the problems of recycled fiber processing with enzymes. BioResources **5**(2), 1311–1325 (2010)
6. Misman, M., Alwi, S.W., Manan, Z.A.: State-of-the-art for paper recycling. In: International Conference on Science and Technology (ICSTIE), vol. 12(13), pp. 1–5. Malaysia (2008)
7. Liu, Z., Wang, H., Hui, L.: Pulping and papermaking of non-wood fibers. In: Kazi, S.N. (eds.) Pulp and Paper Processing, vol. 1, pp. 4–31. Intech (2018)
8. Peng, Y.X., Lin, H., Liu, D.Y., Ma, W.J.: Effects of recycling on physical and chemical properties for reed pulp. J. cellul. Sci. Technol. **4**(4), 44–50 (1996)
9. Jiao, R., Peng, Y.X.: Study on properties and improvement of recycled reed fibers. In: 8th Proceedings of Annual Conference of China Technical Association of Paper Industry, pp. 43–52. China (1997)
10. Rasmussen, C.E.: Gaussian processes in machine learning. In: Summer School on Machine Learning, pp. 63–71. Springer, Heidelberg (2003)
11. Zhang, N., Xiong, J., Zhong, J., Leatham, K.: Gaussian process regression method for classification for high-dimensional data with limited samples. In: Eighth International Conference on Information Science and Technology (ICIST), pp. 358–363. IEEE (2018)
12. Zhang, Y., Xu, X.: Relative cooling power modeling of lanthanum manganites using Gaussian process regression. RSC Adv. **10**(35), 20646–20653 (2020)

Sentiment Analysis in Teachers Performance Rating Using Naïve Bayes Algorithm

Irish C. Juanatas, Ma. Corazon G. Fernando, Ace C. Lagman, and John Benedict C. Legaspi

Abstract Sentiment analysis has been extensively researched for the purpose of analyzing qualitative data using a computation technique. However, there are only a few research papers that focus on analyzing sentiments in terms of teachers' evaluation. Analyzing comments on teacher evaluations can lead to understanding more of what faculty development programs can be provided to improve teachers' academic performance. Thus, this study presents an opinion-mining system utilizing the Bayesian technique of the Naïve Bayes algorithm. The descriptive research method was used in this study, with a questionnaire serving as the instrument for testing the acceptability of the application. One hundred (100) evaluators were surveyed for the teachers' evaluations. The application performance attributes are defined using the functionality, usability, reliability, performance and security (FURPS) model. The mean formula was used to analyze the data. The usability and security were evaluated as perfectly acceptable, with a weighted mean of 4.64, and 4.56, respectively. Furthermore, the functionality, reliability, and performance were evaluated at acceptable evaluation ratings with a weighted mean of 4.29, 4.02, and 4.11. The overall quality of the system was given an acceptable rating with a weighted mean of 4.32, indicating that the application provided and managed ratings and comments on individual teachers' performance.

Keywords Sentiment analysis · Naïve Bayes algorithm · Teacher evaluation

1 Introduction

The interest in sentiment analysis has been increasing rapidly in recent years because of its great number of applications. Sentiment analysis deals with analyzing qualitative data using set computation algorithm that can determine whether a certain

I. C. Juanatas (✉)
FEU Diliman, 1101 Quezon City, Philippines
e-mail: icjuanatas@feudiliman.edu.ph

Ma. C. G. Fernando · A. C. Lagman · J. B. C. Legaspi
FEU Institute of Technology, 1015 Manila, Philippines

© The Author(s), under exclusive license to Springer Nature Singapore Pte Ltd. 2022
A. K. Nagar et al. (eds.), *Intelligent Sustainable Systems*, Lecture Notes in Networks and Systems 333, https://doi.org/10.1007/978-981-16-6309-3_40

comment be tagged as positive, neutral, and negative. [1]. With the integration of sentiment analysis, it helps to identify and classify opinions expressed in text, to determine the polarity namely positive, negative, or neutral [2]. The flexible framework and reasonable performance of Naïve Bayes algorithm is normally utilized for text classification [3]. Importantly, the objective to know what levels of the attributes are suggestive of individuals who perform well.

Many Higher Education Institutions (HEIs) are gradually becoming more involved in quality measurements that offer evidence for decision-making, policy direction, and funding requirements. As the most significant resources in an education system, teachers are critical to raise quality standards. Teaching quality is determined not only by a teacher's qualifications and knowledge, but also by their dedication and engagement in the classroom [4]. The trainings and workshops attended by teachers serve as their steps to improve their professional growth contribute to improve their teaching pedagogy in delivering high quality of education. [5]. The evaluation of teachers has been crucial in the field of education to provide comprehensive development programs. It is considered an important role in teachers' professional development [6], and it is necessity to have a close look at the teacher evaluation [7]. Different teacher evaluation systems have been developed and conducted to guide HEIs to review and rate teachers' performance and effectiveness in the classroom.

Evaluation of teachers is one of the core foundations in improving the quality of education. Assessing teachers' performance can identify relevant strength and weakness, which can be used in developing faculty development program suited and personalized to teachers [8].

Assessing teachers' performance in academic institution is a needed mechanism to improve the quality of teaching. This academic activity is very important to create necessary policy in terms of faculty development program and teaching pedagogy activities [9]. A procedure and process of faculty evaluation is in-placed to provide a direct link of communication to multiple stakeholders [10]. However, most faculty evaluation systems are focused on quantitative or numerical ratings provided by the students. There are few papers that used the qualitative or comments made by the students. With this, the study aims to extract and analyze the comments made by the students using sentiment analysis.

Also, given the culture in the teacher evaluation, most teachers received "outstanding" on their evaluation regardless of the bad feedback of students on the actual quality of their teaching. As a result, HEI administrators only see the values and its corresponding interpretation from the Likert scale leaving the bad feedback and deficiencies behind.

With this, providing a platform that can provide to analyze both qualitative and quantitative ratings and comments from evaluation will pave way in comprehensive analysis in helping what needed trainings are necessary for teachers. This study introduces an application with combined analysis of the quantitative and qualitative responses using opinion-mining techniques to provide comprehensive understanding of the teacher's performance.

2 Methodology

The system architecture uses a corpora which holds all necessary datasets in terms of comments made by the students extracted in an evaluation system (see Fig. 1). The system allows the student to provide quantitative ratings and feedback in the form of comments or suggestions. Learning-based techniques are used to create a model by training the classifier with labeled examples. Then the opinions extracted are analyzed to find out the polarity using Naïve Bayes Algorithm. Naïve Bayes has been one of the most effective opining mining algorithms because of its probabilistic approach. With the algorithm, the sentiment can be classified whether it is positive, neutral, or negative. The algorithm used the training examples to test the probability of a certain statement belongs to a certain class. The algorithm converts the documents into features sets, where the attributes are the possible words, and the values are the number of times a word occurs in the given document. 1 and 0, which represent if the word exists under vector of words, represents feature sets. It computes the conditional probability of negative and positive and compare to determine the sentiment classification.

To determine the probability of a certain class, their combined probability is obtained by multiplying the probabilities of each evidence where:

$$\Pr[\text{yes}|E] = \frac{\Pr[E_1|\text{yes}] \times \Pr[E_2|\text{yes}] \times \Pr[E_3|\text{yes}] \times \Pr[E_4|\text{yes}] \times \Pr[\text{yes}]}{\Pr[E]} \quad (1)$$

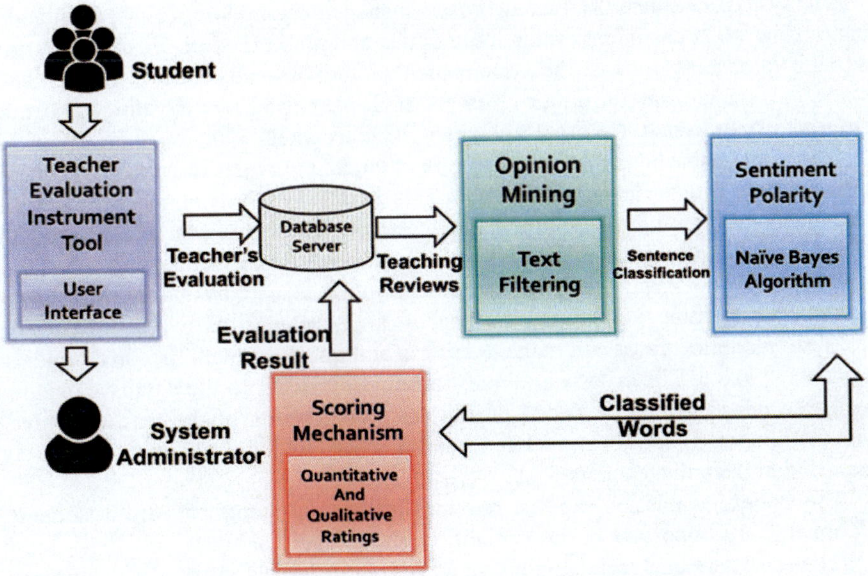

Fig. 1 System architecture of the study

Table 1 Numerical rating and descriptive interpretation of the mean

Numerical scale	Mean numerical scale	Interpretation
5	4.51–5.00	Perfectly acceptable
4	3.51–4.50	Acceptable
3	2.51–3.50	Neutral
2	1.51–2.50	Slightly unacceptable
1	1.00–1.50	Unacceptable

A total of one hundred (100) respondent participated in the study. Purposive sampling was used to select respondents until the desired number of participants was reached. This used the mean performance to interpret the results of based on the survey questionnaire (Table 1).

The developed system was created using latest web development frameworks. It is an application to be used on the Internet or localhost and was written using HP suited for web development and embedded into HTML. MySQL server is used for database operations, including the storage and retrieval of login information. The application includes four (4) major features of the study which include users management module, teacher evaluation instrument module, students evaluation module, and report module.

Login page is provided to restrict access to parts of the application for authorized users of the system, namely the system administrator, human resource (HR) head, directors, teachers, and students. The system administrator can log in and access the backend of the application and granted with full administrative privileges.

The system administrator manages users management module, teacher evaluation instrument module, student evaluation module, and report module. First, the system administrator adds users of the system such as the HR head, director, faculty, and student accounts using the single entry and multiple entries by uploading the list of students in.csv format. No other file type will be accepted.

The manage teacher evaluation instrument module provides the system administrator in managing the teacher evaluation. This module is only available in the system administrator. It includes component of add, edit, delete school year and department to show the frequency of evaluation and the different departments involved. Also, assignment of courses to teachers is included in the teacher evaluation instrument module to determine who among the teachers are covered in the evaluation. Lastly, the system administrator can manage criteria and questions with its corresponding percentage value. The criteria and questions are inserted into the database dynamically, so new sets can be added during the configuration of the teacher evaluation instrument. These are considered standards used to define and measure quality expectation from the teachers.

The student evaluation module provides ratings and comments from students. Comments are important in the generation of sentiment analysis (see Figs. 2 and 3). The sentiment analysis using Naïve Bayes algorithm classifies these comments to positive, negative, or neutral category. An overall positive, negative, and neutral evaluation of a teacher are given out from students' perspective.

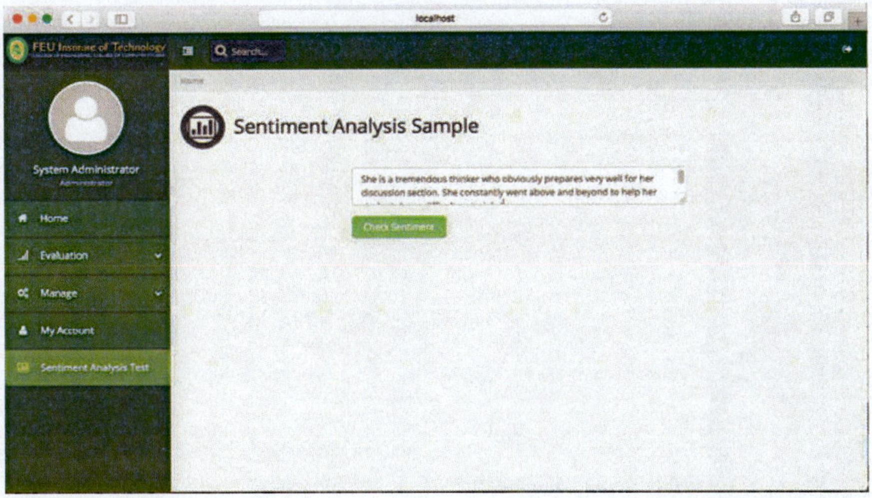

Fig. 2 Sentiment analysis module

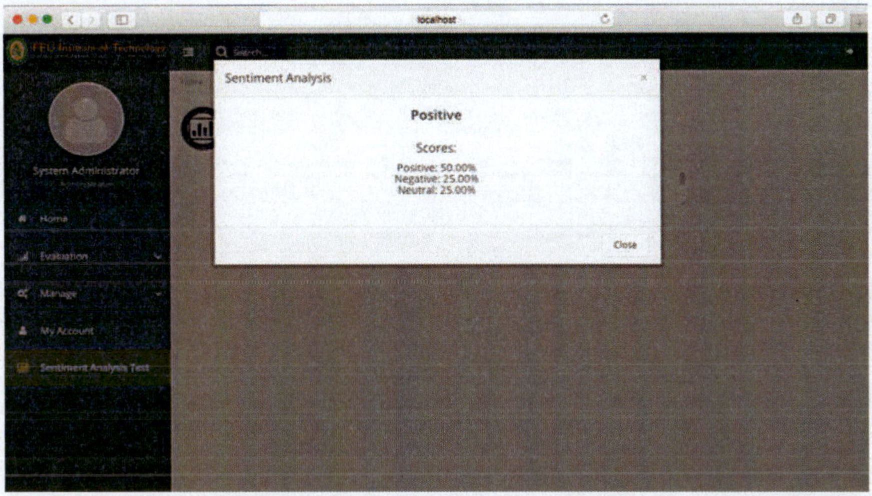

Fig. 3 Sentiment analysis result

The report module will create the final report of the teacher evaluation (see Fig. 4). The purpose of this report communicates the final rating which has been generated from the rating and comments of students. The result of sentiment analysis is combined in the final rating of the teacher (see Fig. 5).

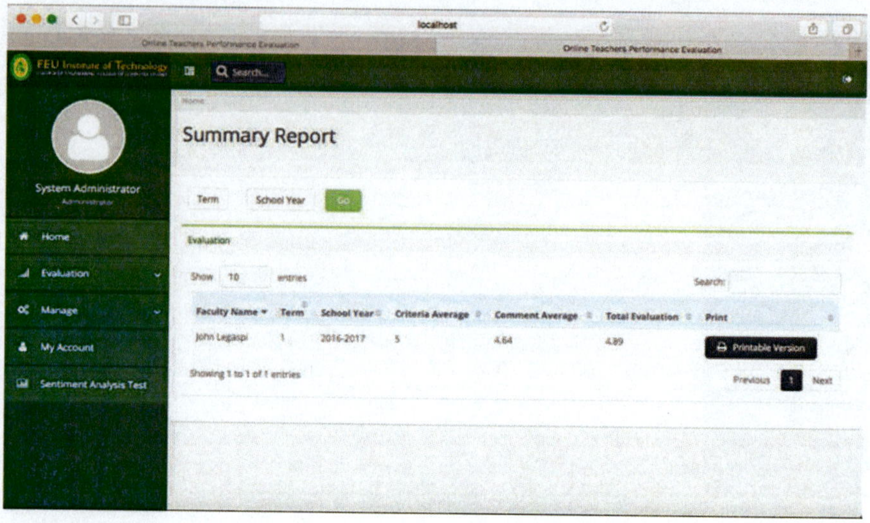

Fig. 4 Report module

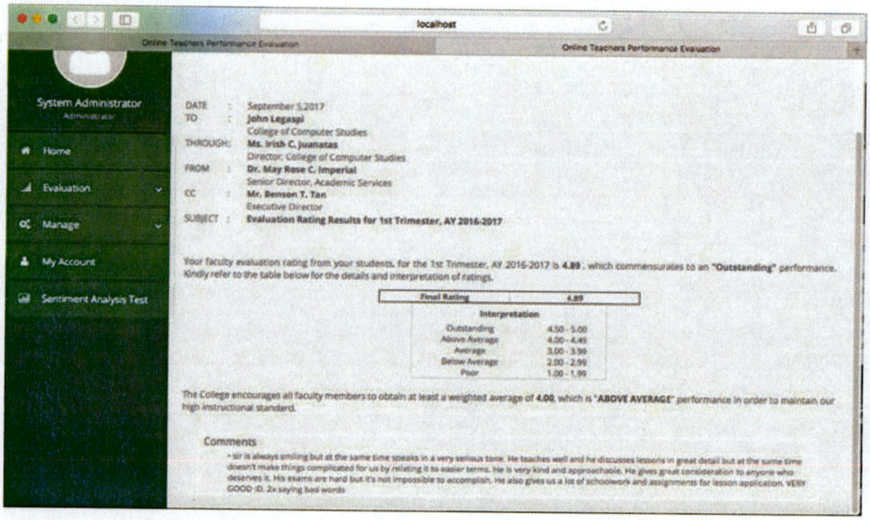

Fig. 5 Details of the report

3 Results and Discussion

Table 2 presents the summary of the weighted mean of FURPS for the sentiment analysis in teachers performance rating using Naïve Bayes algorithm based on the

Table 2 Weighted mean and description on the security indicator of sentiment analysis in teachers performance rating using Naïve Bayes algorithm

Criteria	Weighted mean	Experts' response description
Functionality	4.29	Acceptable
Usability	4.64	Perfectly acceptable
Reliability	4.02	Acceptable
Performance	4.11	Acceptable
Security	4.56	Perfectly acceptable
Overall weighted mean	4.32	Acceptable

evaluators response. The overall acceptability of the system based on experts' evaluation is 4.32 with an interpretation of acceptable. This means that the system is ready for deployment and implementation.

4 Conclusion

This evaluation was able to provide and handle ratings and comments in the performance of individual teachers to identify their strengths, potentials, and weaknesses. The overall indicator of the application was rated acceptable (4.32) which indicates that the application performed well and ready for operation. It was acceptable (4.29) in terms of functionality as all the features and functionalities of the application were being developed. It was perfectly acceptable (4.64) in terms of usability which shows that the application is easy to use. It was acceptable (4.02) in terms of reliability as the application provides the correct results. It was acceptable (4.11) in terms of performance which shows the ability of application to provide results immediately and application's speed to classify quantitative and qualitative result. Lastly, it was perfectly acceptable (4.56) in terms of security which shows that the application is protected by security features and level of system access privileges. The researcher aims to use other natural language processing algorithms in creating sentiment analysis as future works.

Researchers can add a grammar and spelling checker to validate the right and valid words for giving feedback in the future to further develop and improve the study. It also requires more data in the dataset dictionary as well as new corpora to improve classification analysis. HEIs encourage the facilitation of this type of teacher evaluation system to validate the acceptability of the study and to provide feedback for future system enhancements. Furthermore, additional algorithms such as Multinomial Nave Bayes may be integrated into the system to increase the percentage of accuracy in evaluating the system's performance when multiple occurrences of a word are present in text classification problems.

References

1. Petz, G., Karpowicz, M., Fürschuß, H., Auinger, A., Stříteský, V., Holzinger, A.: Computational approaches for mining user's opinions on the Web 2.0. Inf. Process. Manage. **51**, 510–519 (2016).
2. Pandey, S., Astya, P.: Sentiment analysis: approaches and open issues. Comput. Commun. Autom. (ICCCA), (2019) https://doi.org/10.1109/ccaa.2017.8229791
3. Ko, Y.: How to use negative class information for Naive Bayes classification. Inf. Process. Manage. **53**, 1255–1268 (2017)
4. Sang, J., Naeem, H.: A Generic Feedback System for Better Evaluation of Teacher Performance (2015)
5. Ko, Y.: Desirable characteristics defining to describe an effective teacher. Procedia-Soc. Behav. Sci. **53**, 1223–1229 (2015)
6. Delvaux, E., Vanhoof, J., Tuytens, M., Vekeman, E., Devos, G., Petegem, P.: How may teacher evaluation have an impact on professional development? A multilevel analysis. Teach. Teacher Educ. **36**, 1–11 (2013)
7. Sun, H., Liu, Y., Zheng, H., Wang, Y.: A research on the indicators for evaluating teachers in Hong Kong. Procedia Soc. Behav. Sci. **116**, 1459–1463 (2014)
8. Marsh, H.: Students' evaluation of university teaching: research findings, methodological issues, and directions for future research. Int. J. Educ. Res. 253–388 (1987)
9. Noor, S.: Students' Perception on Good Lecturer Based on Lecturer Performance Assessment (2013)
10. Hammond, L., Beardsley, A., Haertel, E., Rothstein J., Kappan P.: https://www.edweek.org/ew/articles/2012/03/01/kappan_hammond.htm. Last accessed 20 June 2019

The Lack of Focus Given to Ethical Training in the IT Industry

Pradnya Chitrao, Pravin Kumar Bhoyar, Rajiv Divekar, and Janhvi Pant

Abstract Through this study, the authors wish to understand the importance given to and the level of ethical training in the Information Technology (IT) industry in India. This is important as specific training for ethical belief and action is necessary at all levels of employees in all industries so that they may make the right decisions when in a dilemma and to benefit all involved persons not just one or a few. The authors recognize the heterogeneous nature of the industry studying the same in detail before going on to understanding ethical training as per the employees and then as per the companies. This study is taken up to understand how much focus is given to the same. This study understands these questions through interviews with employees at various levels in the various branches of the IT industry. As the study is qualitative and not quantitative focus was laid on subjective understanding of the respondents and analysis of the same. The result from this analysis is being discussed in detail in the study.

Keywords Executives · IT industry · Ethical training

1 Introduction

Through reports and industry knowledge, it is a known fact that the IT sector in all its models lays a lot of focus on learning and development. The reason for this today has also to do with the rapid change in technology, the need of the customer and the market requires skills to be acquired and learnt at a fast pace. Technical skills in the IT sector thus become the mode and means of staying relevant in the market. This is as true for people as it is for companies.

P. Chitrao (✉) · P. K. Bhoyar · R. Divekar · J. Pant
Symbiosis Institute of Management Studies (SIMS), A Constituent of Symbiosis International Deemed University, Pune, India
e-mail: pradnyac@sims.edu

P. Chitrao
Symbiosis International Deemed University, Pune, India

© The Author(s), under exclusive license to Springer Nature Singapore Pte Ltd. 2022
A. K. Nagar et al. (eds.), *Intelligent Sustainable Systems*, Lecture Notes in Networks and Systems 333, https://doi.org/10.1007/978-981-16-6309-3_41

Learning and development are an aspect that is relevant at all levels of employees in any company. The noteworthy aspect here is that when a person moves from the role of an executive to that of a leader, the need and want of the kind of training undergoes a transition at each step of the way.

Being a strategist does not come in one go or in one set of trainings as one reaches middle and senior management. It is rather a set of skills and a way of thinking that is imbibed over time. This is equally true for soft skills as well. One does not learn these over the course of a single training plan [3].

As the maturity and place of the employee increases in this sector, it is seen that the training rightfully takes a shift from technical to soft skills. These skills include communication skills, presentation skills, emotional intelligence, leadership management, and many more. The reason this becomes relevant lies in the understanding that as one becomes a more senior employee in the company, it is not their technical know—how that is of prime importance but rather their skills as the representative of the company, their skills as a leader, their skills as a communicator and most importantly their skills as the face of the project/team/company.

It is in this aspect that soft skills training differs and must differ from technical training of employees, executives, and management [4].

Based on my understanding of the IT sector and knowledge of the people there; and by going through research on the issue, it came to light for me that there is a possible mismatch in the training given to the management and employees on these very soft skills that are this crucial.

It was particularly noticed that the concept of ethical working and leadership is something that is neglected in the name of value systems training necessarily given to employees by legal or company mandate.

It becomes important here to note the difference between abiding by the values or a company and being an ethical person at work. The values of a company denote the core concepts that are believed to make up the composition of the company senior management. These are those beliefs that the founders of the company thought must be exhibited by the company and its employees. Mostly employees are given standard run off the mill trainings with respect to these values.

Ethics is not only a much broader, but also a much deeper concept. An employee is ethical not only when they abide by the values of the company but by the values of society in that time. Understanding what business or corporate ethics are is a presupposition to the analysis carried on as it is the presence or absence of this particular training that is the focus of the study here.

Corporate or business ethics are a form of applied ethics. These check the moral or ethical principles that get shined on or the problems that arise in any business setting. They are often the written or unwritten values and principles and most often get defined by the culture of the organization. They govern the actions and decisions that are taken by the people in an organization. In a nutshell, business ethics boil down to the knowledge between right and wrong and to making the right decision.

Like other soft skills, this is often believed to be an aspect inbuilt in a person and no more than a cursory focus is given to these trainings. Lack of focus in no way means a lack of necessity and this is then where the problem begins. Be it employees or senior management, the ethical decisions of the company need to be taken with care. This needs to be imbibed in them through proper simulations and trainings.

It becomes relevant to note here as well that this may not be uniformly true in all aspects of the IT sector. The industry is far from homogeneous and that is likely to have an effect on the kind of training imparted to the employees. It is keeping that in mind that the study we undertake has been on the basis of the three major distinctive types of work that we noticed in the sector. These distinctions are drawn on the basis of the service or product provided by the companies as well as their team structure. It is of importance to note that given the differences in the mode of functioning and the tasks of these three, the level of ethical training imparted in each may differ. This, thus, became another point of study here.

Through the research, this study has undertaken, the wish is to verify, validate, and if needed question the IT sector with respect to the reasons behind the lack of ethical training. The intent here also lies in seeing this from the broader scope of soft skills trainings and the real gravity of the mismatch if there is one.

2 Past Literature on the Issue

Having gone through the limited research papers on the learning and development patterns in the IT industry and on the ethical components in leadership roles, along with papers on what personality traits are needed in good employees and leaders, a conclusion is drawn that there is a gap in the learning and development process for ethical understanding in most roles, and these aspects of the business are not given due importance in the IT industry especially in India.

The studied papers pointed at the behavioural traits that were necessary to make a good and charismatic leader.[1] An important trait for a good leader and therefore good employee is strong ethical values. The analysis of the learning and development systems in the IT industry revealed, that this was a quality seemingly ignored in the later stages of L&D for senior positions while focus was laid on strategic planning and soft skills development and even then it was not as high as that laid on technical skills development.[2]

While there are very limited papers on the topic and questions at hand, it has become clear that there is immense need of a strong ethical value system for all employees working today. This takes greater relevance as a person heads to more senior roles. It has also become irrevocably clear that while learning and development focusses on technical skills, the makeup of the fixed schedules hardly includes ethical trainings in the IT industry. Value-based trainings are imparted yet may not be the same as ethical training. This seeming gap has become visible from the lack of study on the topic as well as the visible absence of the same from the resources available.

[1] Den Hartog Deanne [1].

[2] Doh Jonathan [2].

It is crucial to note here that the literature is found grossly missing in this aspect of trainings in India in IT sector. The literature available is also foreign in origin and not Indian. The presence of literature for industries other than IT or extensively for technical trainings even more clearly portrays the need for this study.

A clear need can thus be established for the study of this issue of ethics and ethical training in the IT industry in India.

3 Objectives

Having understood the likelihood of a discrepancy in the content of the learning and development modules in the various forms of IT companies in India, this study looks to uncover the same if it truly exists.

The main question that has led to the study here being carried out is this: Are ethical trainings given less than needed importance across the IT sector?

This question, while appearing straightforward, has multiple layers to it. Understanding the depth of this led to the conclusion that there can be many facets that need to be explored here.

Studying the level of ethical training could not be done in silos. It needed to be understood with respect to the position of the employee in the company, the level of training otherwise given to the employees, the amount of time and resources dedicated to these trainings, the importance and weight given to other soft skills trainings, among other things.

Based on this, the aims of the study can be listed in the following way:

- Understanding the differences in the three kinds of companies in the IT industry based on their work
- Understanding the definition of ethics as seen by employees in comparison to the definition they believe their organization holds
- Seeking to find the focus laid on soft skills training vis-á-vis technical training given to IT employees.

The research would help in clear analysis of the current situation and how the same needs to be changed or improved.

4 Methodology Used in the Study

Having read through the research carried out till around this topic, it became clear that the ideal method would be to interview those working in various IT companies from all three kinds in all roles present there. This data collection is in the form of a subjective questionnaire sent to employees across the IT industry. Their responses were collected maintaining their anonymity. This allowed them to answer without fear or favor of discussing their companies.

The hypothesis finally formed and evaluated is:

Ethics Training is not given due importance in L&D in the IT sector in India.

It is my belief that in-depth interviews from employees from 3 different companies, one in each kind of IT company as explored seemed the right way to get the industry trends in detail.[3]

The data collection was by means of subjective interviews with the employees.

The analysis of the data in this case was qualitative and not quantitative and would thus did not need any specific technological tools.

5 Scope and Limitations

It is important to state here that the study below is carried out in the form of interviews of select persons as is a qualitative and not quantitative analysis of the issues at hand. It is also necessary to note that the analysis here is on things as they are for the questions asked and not on the reasons behind the same or possible solutions. It is also important to note that the study focuses only on the IT industry in India, and no analysis or study is done of any other industry or sector.

6 Understanding the IT Industry

When any industry is looked into, it is easy to assume that most jobs in any given role would be nearly homogeneous. An assumption is also made often that the team working and structuring would be likely similar with certain differences between companies based on their model of working. As this study discovered while interviewing people in the industry, the same is very far from the truth.

The IT industry is not as homogeneous as we believed. On the contrary, having spoken to executives and management in IT companies, it became clear that the industry can be divided into three models/kinds of companies at a broad level that function there. These are divided on the basis of the work they do, the kind of clients they have, the team formation and commitments among other factors. It also became increasingly clear that this distinction is important to understand the base issue of this study, the level of ethical training given to the employees. This distinction in the provided training stems from multiple factors which will be explored further in the study.

Before we dwell on the training and the details of the same, it thus becomes extremely important to understand these three models of business.

The first model of work in the IT sector is the consulting model. This mostly includes either technical or otherwise consulting to clients of the business. The

[3] https://www.researchgate.net/publication/249629660_How_Many_Interviews_Are_Enough.

employees in such companies are often experts in their understanding of core software concepts and programs. These companies' help their clients find answers to specific issues being faced with respect either directly to their IT needs or indirectly stemming from the same. This model often has expertise based teams consisting of software experts, legal experts, and trouble shooters. The structure in these companies is mostly of permanent full time employees, while they also take into advisement the skills of experts from outside the company at times. This makes up nearly 20% of the industry in India.

The second work model that is widespread is the technical model. These are companies where the customers are provided with IT-based solutions about softwares and other services for their needs. The employees in these companies are mostly coders, developers, and testers along with certain persons being in ideation. These employees look to give to the customers the IT products in terms of softwares that they need. These companies often function on skill-based teams working from project to project for clients. The employees here at the entry level are often contractual or are employed on project to project basis. The concept of benching employees is very common in this part of the IT industry. This makes up 19% or so of the IT industry in India.

The final model that this study has found and taken in consideration is one that has gained momentum in the past few years but today makes up a major chunk of the IT industry. These companies offer to their customer web and application-based solutions. Mobile application development, website development, etc., are the forte' of these companies. Often these companies also focus on digital marketing and content creation. The employees in such companies are often software or application experts who can develop websites, pages or applications on demand. Another set of employees in such companies are digital marketing experts based on the platform of choice or the content. These companies also have people focused on content creation for all these services that can be provided to clients. These companies too are often made up of skill-based teams working from client to client or project to project. However, unlike the earlier model, the employees here are not benchers but rather freelancers or remote working persons. This model often has the highest percentage of freelancers employed with them. Today, this makes up more than 50% of the IT industry in India.

7 Training in Ethics versus Training in Company Values

Many of the respondents when asked the question assumed that ethical training was the same as the compulsory trainings at their work which dealt with the values of the company. This, more than anything else, also became proof of how little importance is given to ethical training for the most part if the agenda includes value training in the same.

However with this coming out in the study, it becomes important to clarify the difference between the two before the rest of the results of the study are looked into.

Company values, also known as corporate values or core values, are the fundamental beliefs upon which a business and its behavior are based. They are the guiding principles that any business uses to manage its internal affairs as well as its relationship with customers. These are often defined at the outset when the company is being formed or when the company grows substantially for the first time and are often unchanged during the lifetime of the company. While these are based on ethical standards of the founders, these may or may not in its entirety define the ethics of the company or more importantly of the employees of the company.

The corporate value trainings mostly focus on imparting specific guidelines to employees through theory videos, simulations or games, where the focus is to have the employees recognize and act as per the values.

On the other hand, ethnical training is much more extensive. Generally ethical training includes in its ambit corporate value training but the two are in no manner synonymous. If we look beyond the IT industry, we find that ethical training often includes programs devised to cover a plethora of aspects, such as:

- Keeping the workplace respectful and friendly
- Protecting the interests and resources of the company, the clients and the other employees
- Keeping financial data and activities honest and thorough
- Being fair in all dealings and avoiding all unethical practices
- Abstaining from political or lobbying activities between employees or clients
- Avoiding conflict of interest
- and many more.

This understanding, therefore, brought to light even more the need of this study and validated to some extent how a mismatch exists between what is ethical training in general and what is followed in the IT industry.

8 Definitions in Context

Through the interviews held with all respondents in this study, one of the focus areas in terms of the questions were how they thought ethics was defined and how they believed their organization viewed the same.

Some of the common definitions with respect to their own way of looking at ethics can be seen in line with the three types of companies the respondents belonged to. This becomes relevant also with respect to the actual training imparted which shall be explored later in the study.

Some common definitions from employees in the consulting aspect of IT are as follows:

- "doing what is right; and NOT what is convenient or easier"
- "Doing the right thing, staying within the confines of the rule of the land as much as possible"

- "Ethics is having not only characteristics as honesty and accountability but also how one does his responsibilities. it involves attitude, behavior, respect, communication, and interaction; how one gets along with others. Ethics demonstrate many things about how a person is"
- "Sticking to your words, commitment, sincerity, honesty, respect for individual perspective"
- "Ethics according to me is a reflection of one's behavior. It includes things like being honest about the leaves (and their reason) one takes to no. of hours required on a request."

From the software development part of the IT industry came some of the following definitions:

- "Principles that define the moral and working conduct."
- "Being honest, morally sound, and not playing politics."
- "A clear understanding and adherence to the right things to do, based on implicit and explicit norms set for being in certain role."
- "This helps me decide the path to take whenever I have dilemma on what is the right way."
- "Dedicated and commitment to work. The way one treats the juniors in the team and works in collaboration with the seniors."

And finally, the digital aspect of the IT industry had the following views on personal definitions of ethics:

- "values to live by, something that differentiates humans from rest of the species."
- "Personal and professional values that drive your day to day decisions."
- "They are the morals, values we bring which govern how we behave in different professional and personal environments."
- "Ethics are core to any genuine transaction and behavior. When anyone is conscious about the consequences of actions or inaction and undue impact on individuals they display core ethics."
- "Ethics are discipline that deals with moral values and obligation. Relates to principle of fairness in behavior."

As can be seen from the above statements, which form a randomly chosen 25% of responses from each set of respondents, it becomes clear that when their own way of looking at ethics is concerned, there is negligible, if any, difference in their thought process. All employees across the industry view the term and it is implied meaning in a similar fashion.

Looking now at the responses when asked how they believed their organization defined the term should help shed some light on how the two compare. To disallow any error to be noted by taking separate responses, these replies below belong to the same randomly chosen respondents as above in the same order.

From the consulting aspect of IT industry:

- "Company defines Ethics very correctly about what's right and what's not"
- "Doing the right thing"

- "Respect for individual"
- "Respect for others"
- "Respect for the individuals around us."

The responses from the people in the software development companies are:

- "How you manage yourself & conduct while in office both personally & professionally"
- "Doing the right thing"
- "Not only do we do 100% right by our client, but we also fulfill our obligations everyone around us, giving due respect to the weight carried by all of our roles, no matter how small or big."
- "Knowing and following Company values, best practices and knowing well about Company clients work culture."
- "NA."

Those respondents in the digital sector of the IT industry believed:

- "integrity above all is practiced well"
- "Company defines ethics as both professional and personal conduct in office boundaries (not limited to physical space)"
- "Not sure"
- "It covers most of the areas it applies in Company. and examples used in training cover real life scenarios"
- "This company does not know ethics."

It was surprising to find that people who had such categoric and well placed ideas on what they believed to be ethics often seemed to struggle to find an answer to the same question from the viewpoint of their companies. Answers showed uncertainty or a vague definition which pointed toward a lack of clarity on this by their respective organizations.

This while by itself may not seem like a huge finding, in the context of this study it becomes extremely relevant as it allows us to see how little attention is paid to the topic of ethics. This painted a grim picture for the analysis of ethics training that is next detailed.

9 Level of Training

This study has the primary goal of looking into the level of training given to all levels of employees across the three major distinctive types of companies in the IT industry. Keeping that in mind, certain questions in the interview directly sought out details of how much non-technical training was provided in the companies to these employees and what was the focus on ethical trainings in particular. The responses can again be segregated on the basis of the type of company.

To the questions " At your position in the company, what is the level of training given to you by Company in non- technical skills?" and "Is training on ethics included in the set of trainings given to you over the past financial year?" respectively, the role of the employees answering and their responses are as under.

In consulting:

- Senior Analyst—"Good level of training"—"No"
- Analyst—"Low. Quite low infact. Just catharsis"— "None received so far"
- Senior Analyst—"It's good. There are 2–3 training like these around the calendar"—"No"
- Senior Analyst—"Few training's are organized"—"Yes"
- Analyst - "Seminars and meetings" - "No".

In software development:

- Team lead—"There is more than sufficient amount of training in non-technical skills given at my level"—"No"
- Project lead—"Did not attend any training this year except for POSH"—"Company values were covered on multiple occasions"
- BDE—"Regular meet-ups and discussions"—"No.. But I have learnt it in my past companies"
- BDE—"Training by inviting all employeess' opinions on various strategies."—"Yes"
- Team lead—"Until few years back we used to get these trainings, but in the last 2 years no such trainings being happening"—"Nope. There has been no training given."

In digital IT services:

- Manager—"1 per year"—"POSH was there"
- Expert—"I have not been part of any training for past two years, except Agile which was applicable to all Company *name omitted"* leads."—"No. Was there any training like that?"
- Manager—"Only 1 training in a year focused on leadership development is given."—"No"
- Content Marketer—"Soft skill trainings are organized."—"Yes"
- Content Marketer—"I think being a content writer, I get special training to improve my skills."—"No".

It is imperative that it be noted that these responses have been randomly selected keeping in mind any extreme answers needed to be included.

The picture painted by the responses tells a clear answer to the original question. The differences in the three kinds of companies become clear at this point and need to be understood.

In the consulting services, it is evident that enough emphasis is laid on soft skills trainings to be given to its employees. However, it is also categorically evident that ethics training is not of focus in these companies. This is surprising given the amount of integrity and compliance that is at the focus of this business model. As mentioned earlier, it seems that the companies believe these aspects are ingrained in their employees and there is no need of training apart from the legal compliance rules that are informed to them.

In the software development companies, there is lesser focus on soft skills as the jobs need high technical work. In the same vein, ethical training is of low importance in this aspect as well. Employees often confuse value training to ethics training as has been mentioned before in this study.

Finally, in the digital sector of the IT industry, the responses are not as homogeneous as the other two sectors but what can be understood is that a moderate level of focus is laid on soft skills trainings. With regards to ethics trainings, the same pattern continues wherein employees receive no such training or believe legally mandated trainings such as POSH (Prevention of Sexual Harassment) to be ethics training.

10 Conclusion

The results of the study showed that as believed, the IT industry in India is not homogenous but rather is made up of various kinds of companies, three kinds being the most prominent. The consulting, software development, and digital services segments of the IT industry thus make up nearly 90% of the industry with 20%, 19%, and 50%, respectively.

The study also concludes that the companies often neglect defining ethics leading to a gap in definition between the ideas of the employees and what they believe the company thinks. This is specifically true in terms of ethics. Many employees also feel that ethics and company values are synonymous terms but the same is not true actually.

Finally the study also brought to light the alarming conclusion that there is little to no focus on the aspect of ethics when it comes to organizing trainings for the employees in the IT industry in general. This is a fact across the industry with little to no variation among various companies or sectors. Many employees feel that compulsory trainings such as POSH are the only aspects of ethical training needed. That this is a mandated training for all organizations throws light on the severity of the issue.

11 Significance

Understanding the gravity of this is important to see why leadership at senior and junior level suffers from ethical dilemmas and why unethical decisions are found to be taken in this industry. A lack of clarity by companies on what defines ethics and a lack of training for the same for employees can be seen as the root for the same.

As past literature on related issues has taught us, ethics is an intrinsic part of having good employees and leadership which makes the results found here all the more grave.

12 Recommendations

Based on the findings of the paper and the severity of these findings, further research must be carried out into the reasons behind the existing gaps in ethical training, the mismatch of understanding between the employees and the company and the heterogeneous nature of the industry. In depth quantitative research can also be carried out to secure knowledge of the variations between specific companies or between these three kinds of companies in the industry. Another aspect of research that may be taken up is to have a comparative analysis of the IT industry with other industries in India to understand the length and breadth of the issue at hand here. Finally, an analysis may also be done on the possible changes that need to be brought about in the IT industry with regards to ethics trainings.

Appendix

Questionnaire

1. Your age?
2. Your experience in the industry in years?
3. Present/ last held position in the industry?
4. How do you define ethics?
5. How much importance do you feel should be given to ethical/value training at each position in your industry?
6. Do you believe there exists a gap between the ethical expectations and ethical training given in Company?
7. How do you believe Company defines ethics?
8. At your position in the company, what is the level of training given to you by Company in non-technical skills?
9. How much importance is given to personal and professional ethics at Company?
10. Is training on ethics included in the set of trainings given to you over the past financial year?

References

1. Den Hartog Deanne, N., Hoogh Annabel, H.B., Koopman Paul, L.: Linking the big five—factors of personality to charismatic and transactional leadership, perceived dynamic work environment as a moderator. J. Organ. Behav. **26**(7), 839–865
2. Doh Jonathan, P.: Can leadership be taught? Perspectives from management educators. Acad. Manage. Learn. Educ. **2**(1), 54–67
3. Karthik, K.S., Rahul, K.: Learning and Development at Workplace: Changing Paradigms, Emerging Trends; Grant Thornton Thinking, pp. 14–18 (2015)
4. Sandler, S.E.: Learning and Development for Managers, Perspectives from Research and Practice, pp. 429–432 (2006)

Heuristic Approach to Sustainable Agile Development with Secure Indicators

G. Priyalakshmi, R. Nithin, and N. M. Tharanitaran

Abstract Sustainable agile development is the act of cohesively adding sustainable elements to a dynamic agile development cycle. While agile development in general talks about an iterative approach to software delivery that builds software incrementally by taking modular steps, sustainability adds in few characteristics to the cycle making it environment friendly. This is also integrated with a few security indicators to ensure that the cycle is not prone to major attacks in a heuristic approach.

Keywords Sustainable · Agile · Heuristic · Security

1 Introduction

Agile is a repetitive method to software development that constructs software from the start of the project in a fixed time period one step at a time. It splits the project into small segments of user functionality termed as user stories and ranks them based on importance, and continuously delivers them in short two week cycles called iterations. Working software is the key metric of progression in agile software development. Agile processes control change toward the customer's competing edge.

Sustainability of software is a type of software characteristic which directs the development process to sustain for a prolonged tenure by employing the rules of the sustainability model [24]. Earlier in 2008, a series of projects were undertaken by Pyxis Technologies, which were disseminated between Canada and France. It involved contributors from different companies and skills. The issues faced by them varied from team structuring, collaboration with Scrum and product evolution. The Scrum project team has solved issues faced toward the project termination by concentrating on problems to maintain sustainability and readiness to change. In this case, study Scrum which is an agile framework has been predominantly used for the sustainable development of the product.

G. Priyalakshmi (✉) · R. Nithin · N. M. Tharanitaran
PSG College of Technology, Coimbatore, Tamil Nadu 641004, India
e-mail: gpl.amcs@psgtech.ac.in

© The Author(s), under exclusive license to Springer Nature Singapore Pte Ltd. 2022
A. K. Nagar et al. (eds.), *Intelligent Sustainable Systems*, Lecture Notes in Networks and Systems 333, https://doi.org/10.1007/978-981-16-6309-3_42

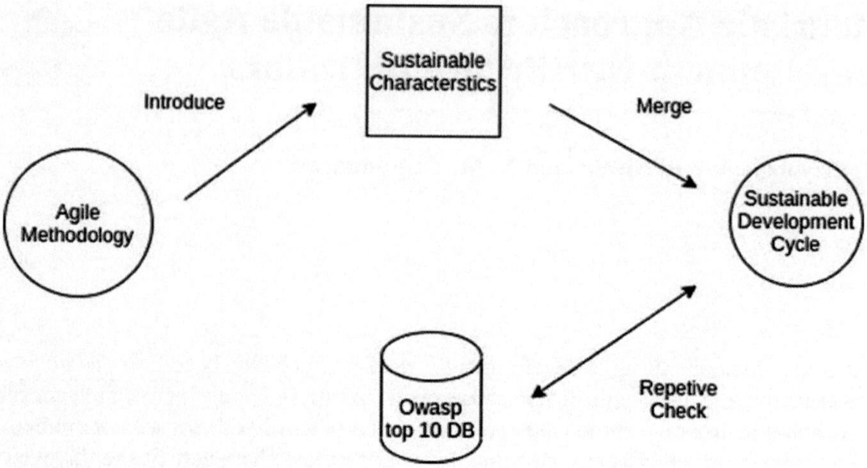

Fig. 1 Workflow

Secure software development is performing a list of security engineering activities along with the traditional software development practices. Presumably this is carried out by abiding to a security model. However, in agile software development, the execution of the process development cannot be determined prior. Also, they are not even foreseeable at the start of a project.

2 Workflow

The idea is to first create an agile environment without adding sustainable characteristics to it. This will be a sample environment which will be later used as a comparison metric with the final agile cycle. A replica of the same is taken, and it is introduced to the three major sustainable features, which a software life cycle needs: human resources, economic resources, and energy resources. Managing them efficiently is the end goal. This sustainable environment is later given repetitive heuristic checks with the OWASP top 10 attacks database [1] and is pinged to and fro often as shown in Fig. 1. Once this is done, the end model is taken and compared with the initial model and a relative study is done.

3 Case Study

The conventional security methodologies are usually sequential rather than iterative. This often leads to a "big design up front" in order to assess the security of a system which is considered as an anti-pattern in the Agile Manifesto [2]. Furthermore, classification of vulnerabilities in terms of proprietary and open-source software is being done in a heuristic manner [1].

3.1 Agile Security Testing for Web Based Applications

The usage of agile development methodologies is a rising tendency for Web application development [3]. This has further led to the idea of agile security engineering, which follows the same philosophy as agile software engineering in order to reduce security risks in software. It is a phenomenally iterative process for delivering the security solution and converting security objectives (requirements) into automated security test cases [4]. Furthermore, it develops the notion of designing security test cases prior to the existence of the system, i.e., Test Driven Development (TDD).

Agile Security Testing. The agile security testing methodology consists of three main steps. Step 1, the modeling of security requirements, is executed by creating harmful stories and/or misuse cases in order to bring forth security requirements. These are then used as reference sites when testing for security in order to prove or disprove a given security requirement [5]. Step 2, an exceptionally testable architecture, is attained by adding a test layer on top of each of the three layers that Web applications commonly comprises, i.e., presentation layer, business service layer, and data service layer. The ensuing architecture is skillfully suited to agile development methodologies because of its numerous test layers. Additionally, it is beneficial for security testing because the architecture makes it feasible to utilize many security testing techniques within any number of the test layers. Step 3, executing automated security tests, which is mandatory in order to get the best reward from agile security testing [6].

Penetration Testing. Penetration testing is the most commonly applied security testing methodology. There are five essential steps to be carried out here. Step 1 is to create a threat model in order to get a detailed, written description of the risks that threatens the application. The goal is to get an overview of the various conditions (vulnerabilities) that have to be present in order to realize a given threat [7]. Step 2 is to build a test plan. The test plan functions as a work plan for the total security testing effort. It is designed to acquire a high-level outline of the security test cases, a run-through of how exploratory testing (i.e., simultaneous learning, test design, and test execution) will be administered, and to get a view of the components that will be tested [8]. Step 3 is to execute the security tests. These are divided into four main groups; dependency testing, user interface testing, design testing, and implementation testing. Step 4 is to create a report of the findings from the security testing process. The report must at least cover reproduction steps, severity and exploit scenarios. Step 5 is to execute a postmortem evaluation [9].

This is later integrated with the OWASP testing framework which has 5 essential phases of work. They are: Phase 1 (before development begins): Review policies and standards, and develop measurement and metrics criteria (ensure traceability). Phase 2 (during definition and design): Analyze security requirements, assess design and architecture, create and review UML models, and design and evaluate threat models. Phase 3 (during development): Code walkthroughs and code reviews. Phase 4 (during deployment): Application penetration testing and configuration management testing. Phase 5 (maintenance and operations): Conduct operational reviews, conduct periodic checks, and ensure change verification [1].

3.2 Agile Security Testing for Web-Based Applications

This study considers the software sustainability testing practices listed in the below subsections.

Maintainability. Maintainability is the ability of the system to reflect the changes with a degree of ease. The occurrence of the changes can impact the components, services, features, and interfaces. Once the responses to fix the errors are done, the changeability of the application plays a role, which must be minimized. Maintainability is the main characteristic that needs to be highlighted in improving the systems function, in increasing the availability, and in reducing the effects of run time detection. In the context of sustainability, the sub-characteristic for maintainability are broken down into evolvability, analyzability, changeability, and testability [3, 4].

Potential Security Impacts: In general, with the advent of dynamism and a broader scope, there will be a possibility for threats in corners that have been added recently. It has to be assured that the new versioning follows the same security protocols which the previous state in versioning follows.

Functionality. The second proposed characteristic is functionality in which the ability of the system to do the task that has been intended. This characteristic is the system task which is the element involved in working in a coordinated manner in completing the job cooperatively [3, 4]. The element involved in the system will be assigned into the correct responsibilities for coordinating with other elements in the system [4]. In the context of sustainability, the functionality characteristic is needed to decompose into some modules in order to make it understandable and to support varieties of purpose in the system. This characteristic is divided into sub-characteristics such as suitability, accuracy, and interoperability.

Potential Security Impacts: One major vulnerability that accounts to interoperability is exploiting privilege access. The potential of software to interchange information which is of further use is called interoperability. Since there are access restrictions and permissions to be followed, we have to make sure that the permission settings for each user or entity is properly defined and also governed. Especially in situations where there is an admin-employee situation, we have to keep in mind that proper authorization and access control mechanisms are implemented else they can be leveraged for a greater loss.

Portability. Portability is the system capability to execute under various environments, for example, one application can be used, understood in different languages of the source code, and different platforms or protocols. Typically, the environment categories pertain to the merging of system hardware and system software. The sub-characteristics for portability are installability, reusability, adaptability, and replaceability. In order to be sustainable, this characteristic is very important to be highlighted due to the level of monitoring, coordinating and communication for the standards and procedures in handling the asset. For example, in disaster management such as land and acquisition, re-housing, compensation development for flood victims, all the sustainable benefits must be involved in the management.

Potential Security Impacts: Portability is a major concern when it is integrated with security principles. There are various vulnerabilities such as open port access, remote control execution leveraged from open ports, zone transfer, etc. All these have to be taken into account when designing the portability mechanics, and they have to be installed more carefully. Several protocol transmission vulnerabilities have been identified in the past. This ranges from attacks such as host leveragement using open redirect to the biggest of names such as Wannacry that caused severe damage. The working of Open Redirect is depicted in Fig. 2. As fundamental support, security across all network ports must incorporate defense-in-depth. Ports that are not employed should be closed, and organizations should install a firewall on each single host and also scan and refine port traffic.

Efficiency. It is the ability of the system function to complete any task given in the required time. It is related to the utilization of processor capacity, disk space, and memory efficiency that can be used in real time applications [4]. Efficiency is a very important characteristic to be highlighted in order to achieve sustainability [2,

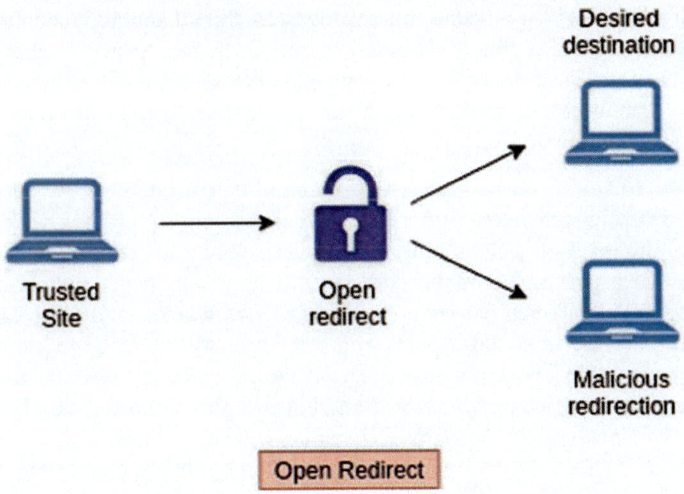

Fig. 2 Open redirect vulnerability

10]. The researchers have studied that the achievement of efficiency characteristic of a software system is higher performance in achieving sustainability of software systems. This is because the efficiency characteristic enables the completeness of system functions in the required task rapidly. It is associated with the temporal system behavior, the consumption of hardware and software of the system, and level of speed of the system running overtime and gives the best accurate results to the users. This leads to the sub-characteristics divided into time behavior, resource utilization, and efficiency compliance.

Potential Security Impacts: Efficiency is a characteristic that would not possibly have any deep security concerns as it only measures time behavior, resource utilization, etc. Hence, this part of sustainable development can be overlooked and can be made as a quality trait irrespective of the others with security impacts.

Usability. It is categorized into the user qualities experience which has involved the application interfaces in the software system. The selection of the interfaces in the software system is based on the user and consumer requirements [3, 4, 7, 8]. It provides access for disabled users and provides a good overall user experience throughout the application. This function is extremely important to produce the software system with the understandability function, which is, considering the incorporating workflows where appropriate to simplify multi-step operation. Through this function, the software system can find out whether any other changes have occurred and can give feedback to the user, especially for errors and exceptions. This is the way to improve the software development toward a long living system by assessing the level of usability of the software system in strengthening the regional project planning capabilities. Consequently, this sub-characteristic is divided into understandability, learnability, and operability.

Potential Security Impacts: In certain cases, a user-friendly design may lead to data leakage. If your program launches other programs, then it should launch them with the minimum privileges that are needed to run it. If your program shows a user interface that calls for the user to resolve whether to execute a potentially dangerous action, the default choice should be the safe alternative. If there is no safe option, there must be no default.

User Conformity. In order to achieve sustainability, user conformity decomposes into user perception and user requirement as their sub-characteristics, which is used to measure the level of sustainability in environmental and social aspects [1]. The environmental aspect involves the consensus feature which the stakeholders will involve and act as affected parties in handling the software systems to related decision making. While in social aspects, this characteristic involves in changing the perception of people about the effect caused by the software system, for instance in disposing of the software equipment and encouraging people in building awareness [9]. User perception measures the demand, level of achievement, confidentiality, regulatory act, advocacy, environment, and flexibility, while user requirement measures the user acceptance and satisfaction.

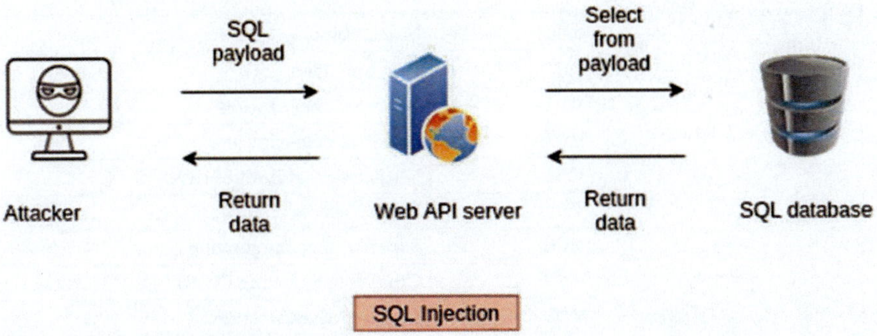

Fig. 3 SQL injection

Potential Security Impacts: If user conformity is achieved in a software program, then it means more number of users will be prone to cyber-attacks. Therefore, it is critical to use additional layers of encryptions and system checks to find out if there is any suspicious user behavior. It includes monitoring failed login attempts, investigation of network traffic spikes, monitoring the network for any bots, etc.

Integrity. Integrity is invalid system data detection and management. In other words, it is the ability to protect the software system from unauthorized access to system functions, prevent information loss, and ensure software is protected from virus infections. Integrity characteristics are focused to achieve sustainability in software systems due to their best practices in assessing security in software engineering. It measures application architecture that has been used in the system such as Software-as-a-Service (SaaS), Enterprise knowledge sharing (know-how), Enterprise Application Integration (EAI), and Service Oriented Architecture (SOA).

Potential Security Impacts: Data integrity is confirming entry of authentic and precise data, and not data protection. Since this involves protection and usage of data, this generally has a huge backend database. The security constraints associated with these will be related to database modification using the likes of SQL injection, leakage of credentials, etc. The working of SQL injection is depicted in Fig. 3. Therefore, in order to achieve security, additional data protection methods have to be employed such as backup and recovery, data integrity constraints, and validation checks.

3.3 Case Study Evaluation

The sustainable characteristics are taken into account, and they are classified based on whether these characteristics will be accountable in a proprietary surrounding and open-source surrounding. A comparison is being made between those two, and a distinction is being done in Table 1. The characteristics are majorly tested using the OWASP top 10 vulnerabilities such as:

Table 1 OWASP top 10 vulnerabilities

Index	Vulnerability
V.01	Injection flaws
V.02	Broken authentication
V.03	Sensitive data exposure
V.04	XML External Entities (XXE)
V.05	Broken access control
V.06	Security misconfiguration
V.07	Cross-Site Scripting (XSS)
V.08	Insecure deserialization
V.09	Known vulnerabilities components
V.10	Lack of logging and monitoring

Table 2 Comparison of sustainable characteristics

Characteristic	Proprietary	Open-source
Maintainability	Yes	Yes
Functionality	Yes	Yes
Portability	Yes	No
Efficiency	No	No
Usability	Yes	Yes
User conformity	Yes	Yes
Integrity	No	Yes

A comparison on whether the characteristic will be vulnerable to proprietary software or if it would be vulnerable to free software is shown in Table 2.

This metric allows us to select which characteristic can be used depending on the project chosen. This classification is being done in heuristic terms and depending upon the history and functionality of the above discussed vulnerabilities.

4 Conclusion

In this paper, an analysis is performed on how various phases of the agile development cycle are prone to vulnerabilities when integrated with sustainable characteristics. Sustainability is the future of computing, but it must be made sure that this does not make it easy for somebody out in the wild to exploit the industry that supports sustainable architecture or sustainable cycle. In this paper, a heuristic approach is taken and sustainable characteristics that are prone to different vulnerabilities are discussed. With this information, a classification on characteristics that might affect open-source software or proprietary software is done. This is done by leveraging the history of these vulnerabilities and the industries they have hit previously.

References

1. OWASP: Top 10 vulnerabilities. https://owasp.org/www-project-top-ten/. Last accessed 01 Mar 2021
2. Agile Manifesto: https://agilemanifesto.org/. Last accessed 01 Mar 2021
3. Jazayeri, M.: Some trends in web application development. In: International Conference on Software Engineering, pp. 199–213. IEEE Computer Society, Washington (2007)
4. Kongsli, V.: Towards agile security in web applications. In: Companion to the 21st ACM SIGPLAN Symposium on Object-Oriented Programming Systems, Languages, and Applications (2006)
5. Ge, X., Paige, R.F., Polack, F.A.C., Chivers, H., Brooke, P.J.: Agile development of secure web applications. In: Proceedings of the 6th International Conference on Web Engineering. ACM, New York (2006)
6. Siponen, M., Baskerville, R., Kuivalainen, T.: Integrating security into agile development methods. In: Proceedings of the 38th Annual Hawaii International Conference on System Sciences, vol. 7, p. 185a (2005)
7. Sindre, G., Opdahl, A.L.: Eliciting security requirements with misuse cases. Requirements Eng. **10**, 34–44 (2005)
8. Røstad, L.: An extended misuse case notation: including vulnerabilities and the insider threat. In: The Twelfth Working Conference on Requirements Engineering: Foundation for Software Quality (2006)
9. Arkin B., Stender, S., McGraw, G.: Software penetration testing. IEEE Secur. Priv. **3**, 84–87 (2005)
10. McGraw, G.: Software Security: Building Security. Addison-Wesley, Reading (2006)

Using Machine Learning Based CNN Architectural Models for Breast Ductal Carcinoma Recognition

Prashant Vats, Reenu Batra, Faraz Doja, Manu Phogat, Piyush Kumar Gupta, and Siddhartha Sankar Biswas

Abstract The ductal carcinoma has been represented as a diverse illness with several subgroups. Early recognition of a ductal carcinoma kind is critical in determining the patient's course of medical therapy. In the bioinformatics and biomedical, several research projects are being focused on the identification of the ductal carcinomatous cells as malignant benign tumors. In recent years for ductal carcinoma diagnosis, various machines learning (ML) approaches have been employed by the various researchers for the prediction of recurrence and survival of the ductal carcinomatous patients. Furthermore, by identifying essential highlights, ML tools may be used for the determination of the ductal carcinomatous causing malignant cells and tumors based on their relevance complicated data set classification in the mastography. Classification of masses in mammograms remains a significant problem, yet it is critical in assisting radiologists in making correct diagnoses. In this study, we present a deep learning methodology called Convolution Neural Networks (CNN) based categorization. CNN architectural models like as Mobile Net and Inception are used to classify mammography pictures as normal or abnormal.

Keywords Naive Bayes · Logistic regression · Feature selection · Neural networks (NN) · Decision tree · Machines learning (ML) · Convolution neural networks (CNN) · Breast ductal carcinoma

---

P. Vats (✉)
Department of Information Technology, Fairfield Institute of Management and Technology, GGSIP University, New Delhi, India

R. Batra · M. Phogat
Department of Computer Science and Engineering, SGT University, Gurugram, India
e-mail: manu_feat@sgtuniversity.org

F. Doja · P. K. Gupta · S. S. Biswas
Department of Computer Science and Engineering, SEST, Jamia Hamdard, New Delhi, India
e-mail: farazdoja@jamiahamdard.ac.in

S. S. Biswas
e-mail: ssbiswas@jamiahamdard.ac.in

© The Author(s), under exclusive license to Springer Nature Singapore Pte Ltd. 2022

A. K. Nagar et al. (eds.), *Intelligent Sustainable Systems*, Lecture Notes in Networks and Systems 333, https://doi.org/10.1007/978-981-16-6309-3_43

1 Introduction

Many people have tried over the years to get to the bottom of how to pronounce and medicament the breast ductal carcinoma, which is think of to be the detrimental disease that is to be considered as incurable disease. In the field of medicine with the introduction of new technology, using methods such as ML, we obtain fresh ideas for healing diseases. Malignant tumors or ductal carcinomatous cells may be roughly categorized into three groups. First and foremost, the challenge is predicting that at a specific stage of ductal carcinoma whether a person has a probability of survival or not. Second, it is difficult to anticipate a person who has already faced the condition and been treated is that whether he would experience it again in future or not. Third, the arena in which we are operating involves the primeval recognition of ductal carcinoma. We focus on various ML algorithms that may be worn to diagnose the presence of the ductal carcinoma at a genesis, because prevention is always better than treatment. We all have a fundamental understanding of what ductal carcinoma is. Ductal carcinoma is the malignant development of tissues caused by accelerated cellular division. We have a data file consisting of data sets here that specifies the numerous parameters for the patients to determine that the ductal carcinoma tumor is benign or malignant. As we are working in the field of ML, and this investigation targets on various types of conventional algorithms during our literature review, such as Naive Bayes, logistic regression, NNs, Decision Trees and so on, for predicting the presence of breast ductal carcinoma in a given data set. By using the data set's, symptomatic analysis as the context for the variable that is or should be the output. For more than 500 patients, we have analyzed with a portion of the data set that's used as the basis of training data set and the remainder data serving as the testing data sets. In its plot, this study we have presented the results and contrast differentiation for each type of learning method.

2 Related Work

Yao et al. [1] proposed his work for developing a NN for breast ductal carcinoma recognition. The pessimistic correlation instruction approach was used to dissect and solve the ductal carcinoma recognition ductal carcinoma recognition automatically. The author has described two ways in this article: the evolutionary method and the ensemble technique, both of which may be used to automatically create small NNs. The ensemble strategy was intended to tackle enormous challenges, although it was still in the works.

Afzan Adam et al. [2] created a computerized breast ductal carcinoma diagnosis by integrating a back propagation NN with a genetic algorithm, which was designed as a quicker classifier model to minimize diagnostic time, while enhancing accuracy in identifying breast mass as benign or malignant.

Tuba et al. [3] proposed his work for showing that how statistical NNs may be used for the successful detection and diagnosis of the ductal carcinoma. On the WBCD database, the researcher compared statistical NNs to Multi-Layer Perceptron (MLP). For classification, General Relapse NN (GRNN), Probabilistic NN (PNN) and Radical Base Function (RBF) were utilized, and their overall performance was 98.8 percent for GRNN 96.18% for RBF, 98.4% for PNN, and 95.74% for MLP. As a result, it has been shown that to diagnose the ductal carcinoma these statistical NN topologies may be used.

Sasikala et al. [4] conducted a survey on for gene selection categorization using the data mining approaches. They have discussed about the most commonly used data mining strategies for ductal carcinoma classification using cancerous gene selection, with a focus on four major growing topics. They are ML algorithms, NN-based algorithms, cluster-based algorithms, and genetic algorithms.

Valerie et al. [5] submitted the study in which they compared artificial NNs to logistic regression. To compare the Standard Logistic Regression (SLR) with the multilayer perceptron NNs (NNs), the author has the used Area Under Receiver-Operating Characteristics (AUROC) for identifying the key covariates that influences the impermanence from the ductal carcinoma causes, disease Recurrence and the Disease-Free Survival (DFS) in breast ductal carcinoma patients.

Punitha et al. [6] described the adaptive resonance theory NN and a genetic algorithm for the ductal carcinoma recognition utilizing Wisconsin Breast Ductal carcinoma Data (WBCD). They lined up around 650 samples that were acquired from a diagnostic process for examining lumps or masses in which a hollow needle is introduced into the mass to collect cells, which are then stained and studied under a microscope called FNA (Fine Needle Aspirates) with 16 missing data, and 683 samples with ductal carcinoma tumors were used in this work, with 35% proving to be malignant tumor and 65% proving to be benign tumor. They have compared the results of Radial Basis Function (RBF), Probabilistic NN (PNN), Multi-Layer Perceptron (MLP) with Adaptive Resonance Theory (ART) by finding that the unite method not only decreased the time required to train the network but also strengthen accuracy for the early detection of the ductal carcinoma.

Using two breast ductal carcinoma data sets Chih-Lin Chi et al. [7] provided his work on breast ductal carcinoma survival analysis. The survival analysis problem is tackled using Artificial NNs (ANNs) in this article. ANNs provide more flexible survival time prediction than previous approaches because they can readily develop a nonlinear prediction model by including variable interactions. This study by employing nuclear morphometric characteristics compares the ANN results from two distinct breast ductal carcinoma data sets. The findings demonstrate that ANNs may correctly estimate recurrence likelihood and distinguish between patients with a favorable and unfavorable prognosis.

By the use of the SVM classifier methodology Iraneus et al. [8] published their work on early diagnosis of ductal carcinoma. They have explained that how to identify tumors in mammograms. The authors of this study outlined an algorithm for tumor recognition and provided a technique that comprises mammography images based on matrix dimensions such as rows and columns and the standard deviation that was filtered with a Gaussian filter.

Ilias et al. [9] have proposed an intelligent system utilizing based on SVM-based classifiers with Bayesian classifiers and ANN for automated ductal carcinoma diagnosis and prognosis. To train an SVM model, the Wisconsin diagnostic breast ductal carcinoma records were utilized to differentiate between benign breast lumps and the malignant tumors, and they were having files that consist of estimations that were collected by utilizing the Fine Needle Aspirates (FNA) [10].

3 Proposed Approach

The convolutional NNs are the most often used deep learning architecture for image categorization. This paper's purpose is to determine which class the supplied image belongs to. The process of creating a convolutional NN always consists of four basic parts as shown in Fig. 1.

- Convolution
- Pooling
- Flattening
- Fully connected layer.

The proposed model is model for the Ductal Carcinoma is shown in Fig. 2 using convolutional NNs.

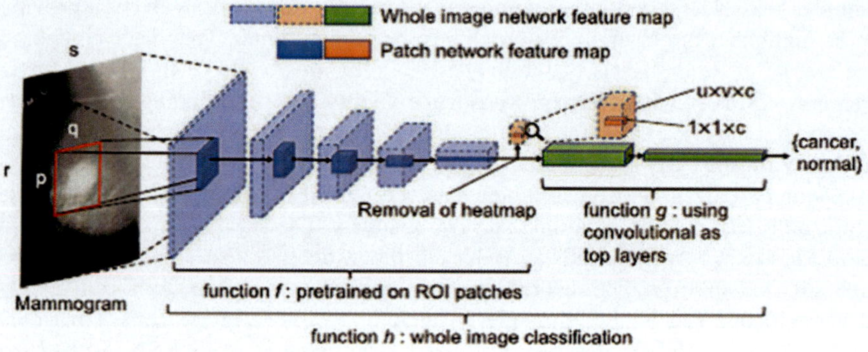

Fig. 1 Convolutional NNs

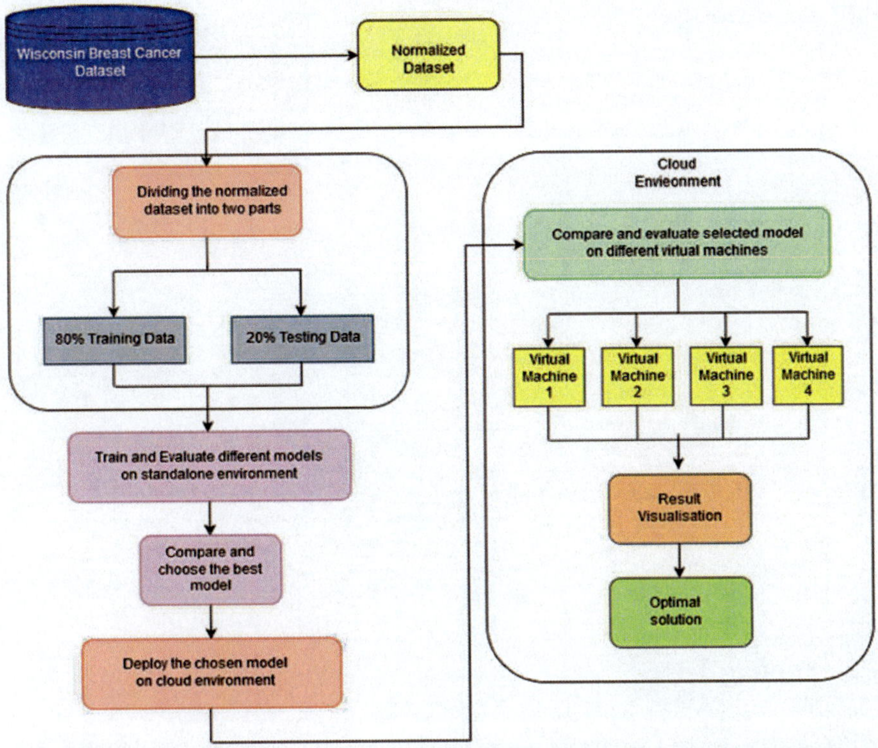

Fig. 2 Proposed model is model for the Ductal Carcinoma using Convolutional NNs

4 Types of Layers

All neurons in one layer do comparable mathematical calculations, which is how the layer gets its name.

4.1 Convolution Layer

Convolution is a mathematical procedure used in image processing to filter a signal, recognize a pattern in a signal, and so on. Convolution is performed on inputs by all neurons in this layer. The filter size is the most critical parameter in a convolutional neuron. To compute this output over the picture, we will slide the convolution filter across the whole input picture, and here we will slide our window by one pixel at a time; this value is called Stride. In most cases, more than one filter is used in a single convolution layer as shown in Fig. 3.

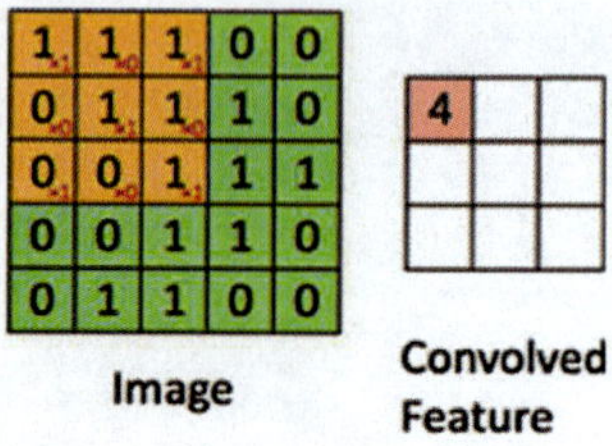

Fig. 3 Convolution layer

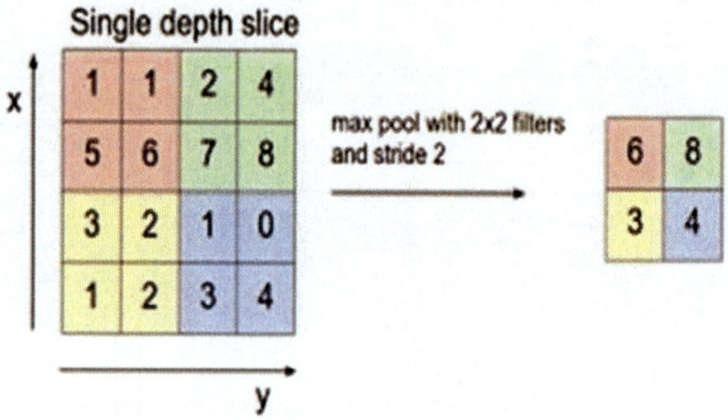

Fig. 4 Pooling layer

4.2 Pooling Layer

In Fig. 4 to lower the spatial size, the pooling layer is typically utilized directly after the convolutional layer (only height and breadth, not depth). This lowers down the parameters, and therefore, the computation. Furthermore, using fewer parameters prevents over fitting. The most popular type of pooling is Max pooling, in which we take a 3×3 filter and apply the maximum operation to the 3×3 sized portion of the picture.

4.3 Flattening Layer

In Fig. 5 Flattening is converting the data into a 1-dimensional array for inputting it to the next layer. We flatten the output of the convolutional layers to create a single long feature vector. And it is connected to the final classification model, which is called a fully connected layer.

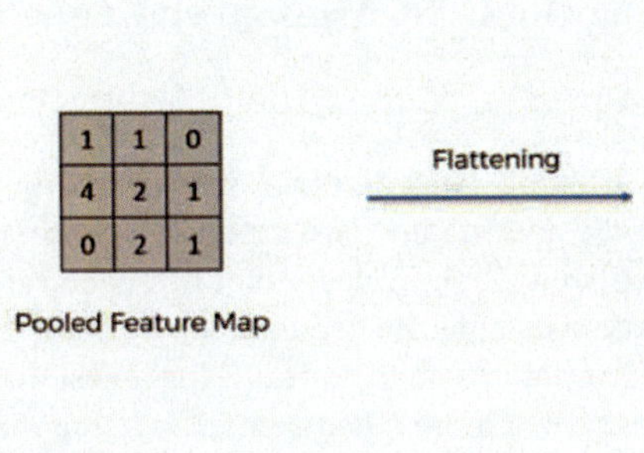

Fig. 5 Flattening layer

4.4 Fully Connected Layer

In Fig. 6 the preceding layer from every neuron if every neuron in a layer receives input, the layer is said to be completely linked. This layer's output is calculated by matrix multiplication, followed by bias offset.

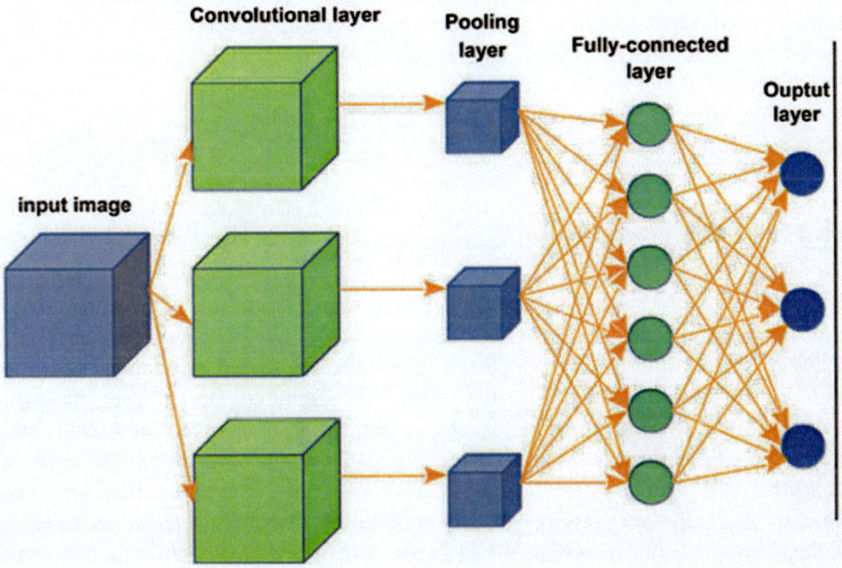

Fig. 6 Fully connected layer

5 Testing and Validating the CNN Maneuver Real-World Data

The actual real time data was 1024 by 1024 pixels in size as in Fig. 7A and B. We began training by categorizing the data set into two categories: normal and aberrant. Filters of various sizes were utilized (2, 3, 5). Data that has been randomly trained produces better outcomes than data that has not been randomly trained. The proposed technique is effective, and it incorporates deep learning for breast ductal carcinoma screening. It is a continuing study, and more advancement is being made by refining the CNN architecture and deploying pre-trained networks, which will most likely result in improved accuracy. The data set was pre-processed, and the photos were scaled to 224 by 224 from 1024-by-1024. The data set was then subdivided into seven sub-classes, six of which had various forms of abnormalities and one of which had only normal photos. Architectural distortion, circumscribed masses, calcification, speculated masses, asymmetry, and various anomalies were used to designate the classes (images which were neither identified as benign or malignant ductal carcinoma). We examined and evaluated the post processing data set with the same three filter sizes (2, 3, 5), as well as with and without randomization. In Fig. 8, the image shows the categorization by noticeable mapping for benign and malignant tumor detection in ductal carcinoma. The contours show the radiologist-annotated samples within a data set identified for a particular purpose that have been confirmed during the biopsy process of Breast mammary gland to have either ductal carcinoma in blue region or benign tissue in green region. The gradients of the input picture with respect to the ductal carcinoma class result are shown by the red dots.

6 Experimental Results

The suggested CNN-based breast ductal carcinoma recognition system produced satisfactory results. The data set was classified into 7 classes in total, with the aberrant classes subsequently subdivided into 6 further classes. Two approaches were used for training and testing. The first approach separated the data set into two classes, normal and aberrant as in Fig. 9. The second technique featured additional sub-division of the abnormal groups as in Fig. 10, which comprised six categories of abnormalities present in Breast mammary gland, such as calcification deposits, asymmetrical malignant tumor growths, architectural distortion, confined masse, speculated masses, and miscellaneous factors. Various miscellaneous photos were ones in which it was unclear whether they were benign or malignant. Pre-processing is used to improve NN's performance and accelerate learning. Accuracy of raw pictures acquired in CNNs with varied filter sizes. When the model parameter that are basically configuration variables that are internal to the model and whose value can be estimated from the given data are learnt and fixed, then no additional learning occurs and the accuracy is calculated for the detection of the ductal carcinoma.

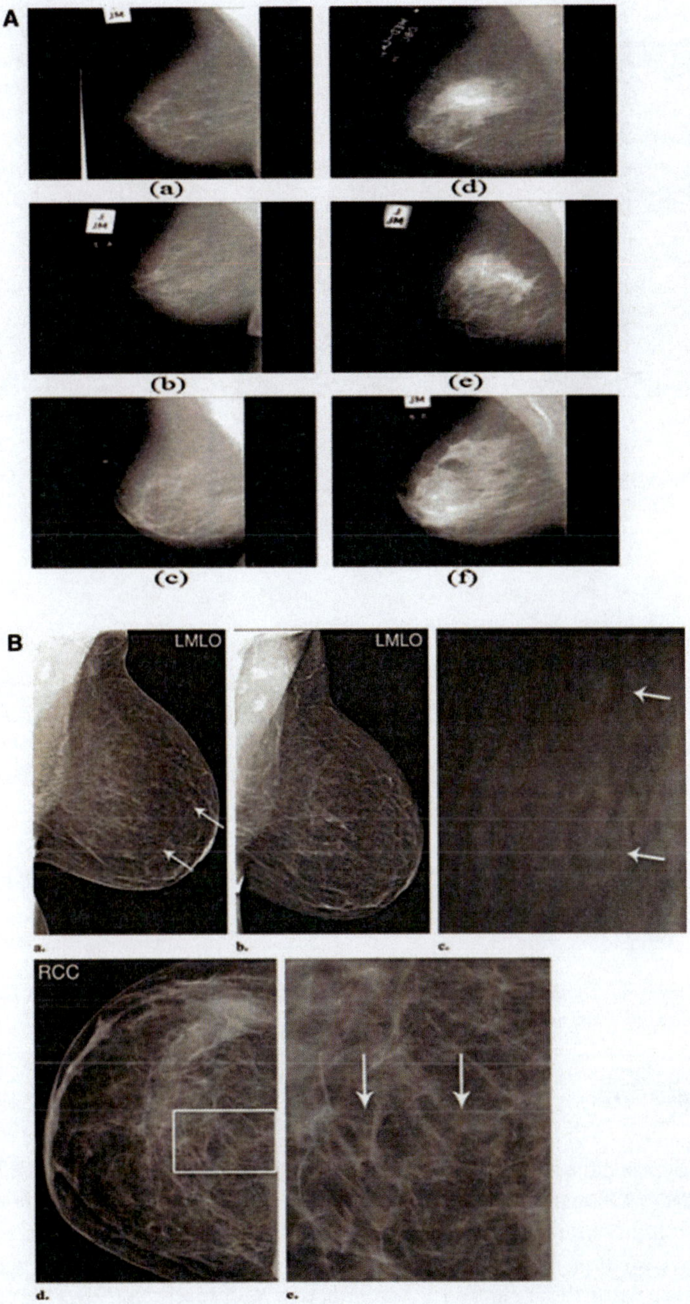

Fig. 7 **A** Images from the MIA's mammography data set. Normal pictures (**a–c**), asymmetry (**d**), calcification (**e**), and speculated bulk (**f**). **B** Breast ductal carcinoma screening CNN architecture 1024 by 1024 scale

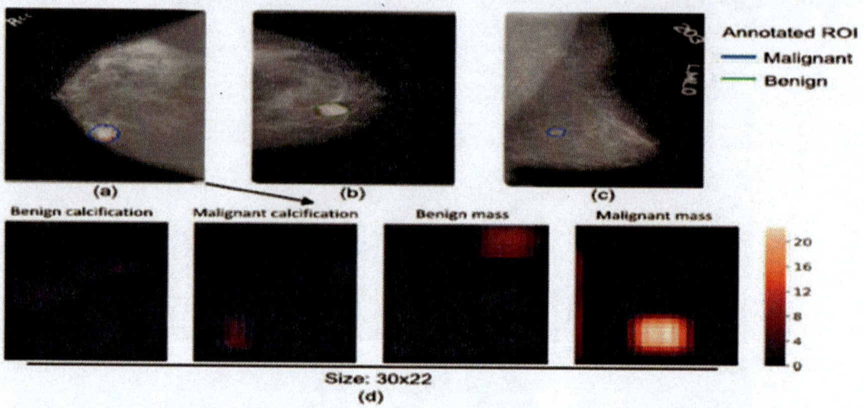

Fig. 8 Image categorization noticeable mapping for benign and malignant tumor detection in Ductal Carcinoma

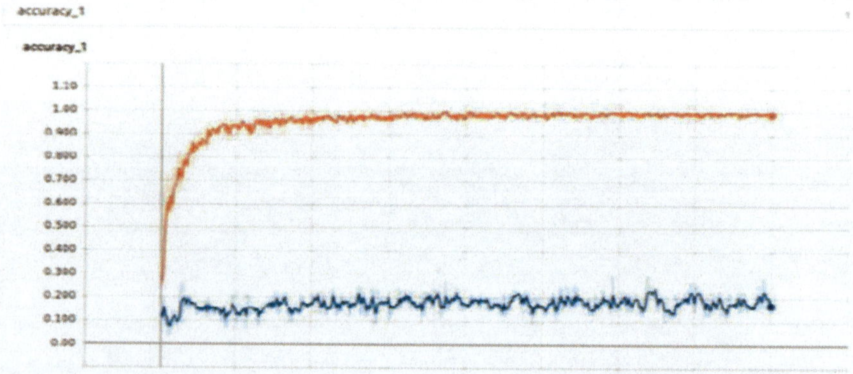

Fig. 9 Training and testing in first method where data set were differentiated into two classes named abnormal and normal

7 Conclusions

In this study, we have suggested the convolution NN as a computer-aided diagnostic (CAD) system. On mammograms, convolution NN's is used to recognize normal and abnormal mammograms. It is usually ideal to diagnose ductal carcinoma early and counsel the patient on how to treat it. We identify breast ductal carcinoma with 83 per cent accuracy using the Inception V3 architecture. Our research shows that convolution NN models trained from end to end of may be very much precise and certainly easily conveyable across various Mammographic scanning platforms. Convolution

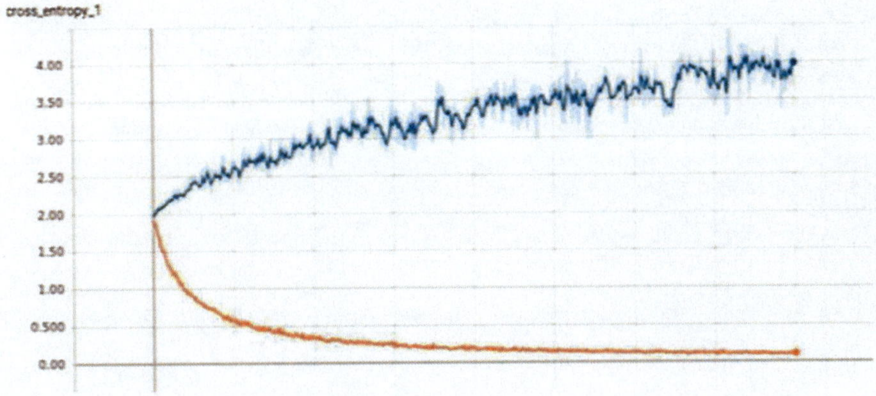

Fig. 10 Second procedure involves further sub-classification of the abnormal classes

NN's approaches have huge potential to enhance the correctness of the ductal carcinoma diagnosis on mammographic scanning as computational resources and the training data sets become more available. Our method might aid in future expansion of improved CAD systems that can be used in prioritizing the most vulnerable cases to be scrutinized by a computed tomography specialist radiographers after an initial independent interpretation.

References

1. Yao, X., Liu, Y.: Neural networks for breast cancer diagnosis. In: Proceedings of the 1999 IEEE Congress on Evolutionary Computation-CEC99 (1999)
2. Khalid, A., Noureldien, A.: Determining the efficient structure of feed forward neural network to classify breast cancer dataset. (IJACSA) Int. J. Adv. Comput. Sci. Appl. (2013)
3. Kiyan, T., Yildirim, T.: Breast cancer diagnosis using statistical neural networks. J. Electr. Electron. Eng. (2003)
4. Santhosh Baboo, Sasikala, S.: A survey on data mining techniques in gene selection and cancer classification. Int. J. Comput. Sci. Inf. Technol. (2010)
5. Bourdes, V., Stephane, B.: Comparison of artificial neural network with logistic regression as classification models for variable selection for prediction of breast cancer patient outcomes. Adv. Artif. Neural Syst. (2010)
6. Punitha, A., Sumathi, C.P., Santhanam, T.: A combination of genetic algorithm and ART neural network for breast cancer diagnosis. Asian J. Inf. Technol. **6**, 112–117 (2007)
7. Chi, C.-L., et al.: Application of artificial neural network-based survival analysis on two breast cancer datasets. In: AMIA Annual Symposium Proceedings, AMIA Symposium, vol. 2007, pp. 130–134 (2007)
8. Ireaneus, Y., Rejani, A., Thamarai Selvi, S.: Early detection of breast cancer using SVM classifier technique. IJCSE **1**(3), 127–130 (2009)

9. Maglogiannis, I., Zafiropoulos, E., Anagnostopoulos, I.: An intelligent system for automated breast cancer diagnosis and prognosis using SVM based classifiers. Appl. Intell. **30**, 24–36 (2009)
10. Zhang, Q., Wang, S., Guo, Q.: A novel SVM and its application to breast cancer diagnosis. In: 1st IEEE International Conference on Bioinformatics and Biomedical Engineering (2007)

Improving Collaborative Filtering Based Recommender System with Season and Style Features

Raghav Modi and Ronakkumar Patel

Abstract People have gained greater shopping interest as their living standards have risen, owing to an increase in demand for fashion products. Fashion e-commerce has risen at a rate of 12.2% in the last year as a result of this demand, users often have trouble identifying desired items. The fashion product depends not only rating but also on other features. The season and style are two important features while recommending the fashion products. This research aims to provide recommendations to a wide range of users through collaborative filtering based system by incorporating season and style features of fashion products. The deep learning based technique, convolutional neural network, is used to extract season and style features of fashion product. The dataset used to validate the proposed approach is the amazon product dataset for clothing, shoes, and jewelry. The results are compared using standard evaluation measures like MAE, MSE, RMSE.

Keywords Fashion product recommendation · Convolutional neural network · Collaborative filtering

1 Introduction

In the global economy, the fashion industry plays a significant role and holds a significant position. By the end of 2021, it is estimated that 2.14 billion people will have purchased merchandise online, with e-commerce transactions reaching $4.5 trillion. The apparel industry accounts for a large portion of this growth in e-commerce. Thus, we can claim that the fashion industry plays an integral role in the global economy.

R. Modi (✉) · R. Patel
U & P U. Patel Department of Computer Engineering, Chandubhai S. Patel Institute of Technology (CSPIT), Charotar University of Science and Technology (CHARUSAT), Anand, India
e-mail: 18ce056@charusat.edu.in

R. Patel
e-mail: ronakpatel.ce@charusat.ac.in

© The Author(s), under exclusive license to Springer Nature Singapore Pte Ltd. 2022
A. K. Nagar et al. (eds.), *Intelligent Sustainable Systems*, Lecture Notes in Networks and Systems 333, https://doi.org/10.1007/978-981-16-6309-3_44

Besides, the increment in number of user, there is acceleration in the number of products available online. As the number of products online increased, users spend a large amount of time searching for their desired products. In the vast majority of instances, people are unable to find the fashion products of their choice. E-commerce store owners, on the other hand, have to use a variety of tactics to boost sales and assist users in finding their desired product.

Both the seller and the customer will profit from a recommender system that helps the consumer to find similar products. Recommender system is algorithm which suggest related products to the user based on the data. Recommender system is an essential aspect of today's society.

The collaborative filtering, content-based, and hybrid methods are the main classification of recommender system. The collaborative filtering approach is based on similarity between users/items based on past experience. It solely focuses on previous experiences with the user and objects to generate new suggestions. In contrast to collaborative filtering, which is solely focused on user-item interaction, the content-based approach is more dependent on the data provided for the user or item. As a consequence, the core principle is to build a model based on the user or object's functionality. The hybrid recommendation system can be described as a combination of collaborative filtering and content-based methods combined into a single roof to avoid the drawbacks in each case, such as the cold start problem and the sparsity problem.

The ratings are essential part of collaborative filtering based recommender system but there are other features which play a vital role in the fashion products recommendation. These features are season, style, gender, geographic location, etc. This paper examines the use of season and style features in the fashion product recommendation with the help of Convolutional Neural Network (CNN).

The remainder of the paper is laid out as follows: Sect. 2 enlightens the literature in the domain. The proposed approach is discussed in Sect. 3, which includes feature extraction, similarity measurement calculation, and a method for predicting outcomes. Experiments and result analyses are presented in Sect. 4, and the Sect. 5 concludes the paper.

2 Literature Review

There has been a lot of work in the past few years on the topic of recommendation systems, with a focus on the fashion industry. Customer ratings of products are used to create collaborative filtering systems [1], and skip the product other information which is available. Item and user-based collaborative filtering are two forms of collaborative filtering. The two methods are based on calculating the similarity between the product and the consumer.

Item-based collaborative filtering compares each of the user's purchased products to similar ones, then adds them to the recommendation list [2]. This strategy has the advantage of being able to precompute product similarities and show them as soon as a consumer clicks or buys a product. In user-based collaborative filtering, it aims to recommend items to users that are close to the target user [3]. Users-based and item-based collaborative filtering can be combined using multiple linear regression and support vector regression, as suggested in Ref. [4].

For specific domains such as books, movies, and music, a recommendation system has been created. There has also been a significant increase in fashion recommendation systems, with numerous studies conducted in this area. Chen et al. [5] used neural networks to understand the input image as one of the product categories and recommend products to the user using Jaccard similarity. Liu et al. [6] suggested recommending products to the user by learning the style features of the items and the preference of the user.

Li et al. [7] proposed a method by calculating the similarity between the item ratings and used to recommend products. Xu and Chen [8] proposed a collaborative filtering recommendation system that uses a user-item linked list to overcome the problem of large time complexity. Stefani et al. [9] suggested a trend-driven recommendation system that calculates trend scores and recommends a product based on it. He and Jin [10] suggests recommending products using user-item rating and user-attribute rating similarity matrix.

3 Proposed Approach

We have proposed an improved collaborative filtering method with the aid of season and style for this research work. As you can see the proposed approach in Fig. 1 which consists of three stages: (1) Using CNN, extract features from product images

Fig. 1 Flowchart of proposed approach

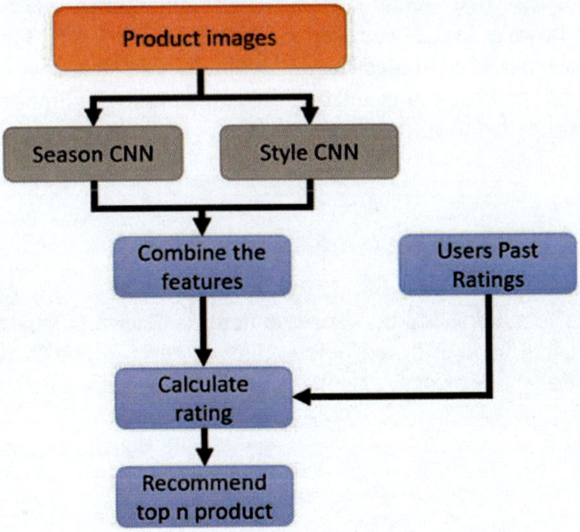

Fig. 2 Model architecture for season CNN

based on season and style of the products (2) These features are combined in this stage (3) The combined features and past ratings are used to calculate the rating of products and then top n items are recommended to the user.

3.1 Feature Extraction Using Convolution Neural Network

Convolutional Neural Networks can be used to decide on a category or extract features from an image. Two different convolutional neural networks have been trained to extract season and style features from an image. The features are extracted from the CNN's dense layers; the CNN model was trained over product images of size 224 × 224. To extract features, two different models were developed. The first model, as shown in Fig. 2, was used to train the season CNN. The style CNN was trained using a separate model as shown in Fig. 3. These models are trained on fashion product dataset [11] and tested on Amazon dataset [5]. Fine-tuning of the hyper-parameters is needed to achieve maximum accuracy and extract relevant features.

3.2 Combine the Features

The relationship between two items is determined using ratings provided by various users in item-based collaborative filtering. The Pearson correlation, provided in Eq. (1), is used to calculate the similarity between the products.

$$\text{sim}(i, j) = \frac{\sum_u^U r(u, i) \times r(u, j)}{\sqrt{\sum_u^U r_{(u,i)}^2}\sqrt{\sum_u^U r_{(u,j)}^2}} \tag{1}$$

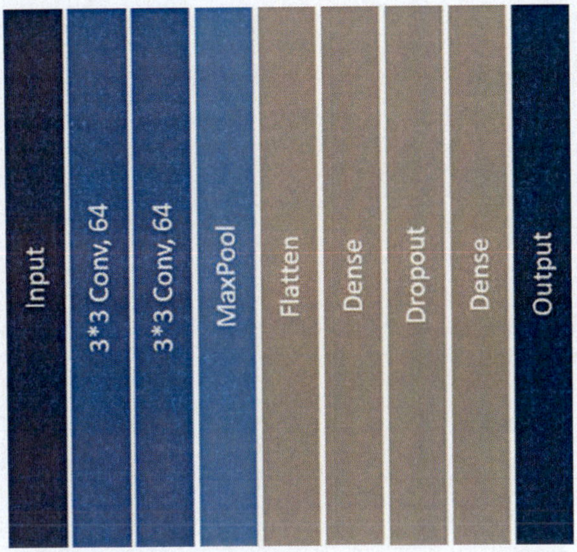

Fig. 3 Model architecture for style CNN

Here, sim (*i*, *j*) stands for the similarity between product *i* and *j*. Here, *U* is a set of users who have rated both items *i* and *j* and *r* (*u*, *i*) and *r* (*u*, *j*) represent the rating given by user u to item i and j, respectively.

The disadvantage is that both the consumer and item rating similarity matrices are very sparse, resulting in a situation in which accurate similarity is not measured. To address this flaw, this study uses Eq. (1) to calculate the Pearson correlation Similarity between products using the features extracted from season and style CNN. These features extracted from the dense layer of all CNN's can be used in one of two ways to determine the final similarity between product images: (1) the concatenation of features and (2) Finding the mean of features. The product rating can be calculated using one of these two final similarity matrices.

3.2.1 Concatenation of Features

Once the features are extracted from two CNN, all of the images would have two sets of features, season and style. Moving forward, all of the season and style features images will be concatenated into a single matrix, and the Pearson correlation formula will be used to determine the similarity between different products. The size of feature vector will increase as you add new feature in it.

3.2.2 Mean of Features

Another way to combine the features are element wise mean of the features. The advantage here is though you increase the number of different feature the length of feature vector will remain same as we are taking mean of all the features.

3.3 Rating Calculation

Using the similarity matrix between all the items the rating for the user is predicted using Eq. (2)

$$\hat{r}(u,i) = \frac{\sum_j^J sim(i, j)*r(u, j)}{\sum_j^J |sim(i, j)|} \tag{2}$$

In the above equation, *u* denotes user and *i* and *j* represent the product. Here, we are trying to predict a rating for user *u* and product *i* which is represented by $\hat{r}(i, j)$. *J* denotes a set of products rated by user *u* and sim (*i*, *j*) is the similarity between product *i* and *j*. The similarity is calculated with the help of features extracted using after feature concatenation or mean of features. After predicting the rating for all the unseen products, we will recommend top-n products with maximum rating.

4 Experiment Setup and Result Analysis

4.1 Dataset

The method's efficiency is measured using Amazon product data for clothes, shoes, and jewelry [5]. This dataset contains reviews for all products, which include user reviews and ratings, as well as metadata, which contains the product summary. Here, the review data is used to gather ratings by different users for products and metadata to gather season, style, gender, and type of clothing category for each product from the description of the given product. Summer, winter, spring, and fall are the four seasons. Casual, party, ethnic, formal, sports, and travel are some of the styles available. Men, women, girls, boys, and unisex are all gender groups. Shirts, jeans, watches, athletic shoes, casual shoes, dresses, coats, and several other types of clothing products are only a few examples. The dataset is divided into two parts, train, and test. The proportion of the training dataset is 0.8% and the test dataset is 0.2%. To evaluate the approach, we have used the Mean Absolute Error (MAE), Mean Square Error (MSE), and Root Mean Square Error (RMSE).

Table 1 Comparison of three methods with MAE

Fold	M-1	M-2	M-3
1	0.77	0.22	0.24
2	0.75	0.22	0.24
3	0.77	0.23	0.24
4	0.77	0.23	0.24
5	0.77	0.23	0.24
Average	0.77	0.22	0.24

Table 2 Comparison of three methods with MSE

Fold	M-1	M-2	M-3
1	2.68	0.38	0.41
2	2.60	0.37	0.42
3	2.66	0.38	0.41
4	2.65	0.38	0.41
5	2.66	0.37	0.42
Average	2.65	0.37	0.41

4.2 Result Analysis

We have compared three approaches using these evaluation metrics. The tradition item-based collaborative filtering (M-1), the similarity based on concatenation of season and style features (M-2), and the similarity based on mean of season and style features (M-3). We have used five-fold cross validation to evaluate proposed model in different distribution.

The result from Tables 1, 2 and 3 clearly indicates the improvement in proposed methods (M-2, M-3). The result for M-2 and M-3 are nearby and comes with trade-off between number of features and error. The M-2 has very low error but feature vector size is large which will impact similarity calculation time, while M-3 has small feature vector size but error is increased compare to M-2.

Table 3 Comparison of three methods with RMSE

Fold	M-1	M-2	M-3
1	1.64	0.61	0.64
2	1.61	0.61	0.65
3	1.63	0.61	0.64
4	1.63	0.61	0.64
5	1.63	0.61	0.65
Average	1.63	0.61	0.64

5 Conclusion and Future Work

The season and style are the essential features of the fashion product recommendation. This research concludes that the use of season and style features and proper combination of it, will definitely improve the tradition collaborative based recommender systems. The CNN can effectively extract the season and style feature with the use of transfer learning as product images are available. In future, other features of fashion products like gender, location, material can also be considered. Also, other deep learning techniques like sequential model, Auto Encoder, GAN, etc., can be explored depending on the features available in fashion products.

References

1. Herlocker, J.L., Konstan, J.A., Riedl, J.: Explaining collaborative filtering recommendations. In: CSCW '00: Proceedings of the 2000 ACM Conference on Computer Supported Cooperative Work (2000)
2. Linden, G., Smith, B., York, J.: Amazon.com recommendations: item-to-item collaborative filtering. IEEE Internet Comput. (2003)
3. Ying, Z., Zhou, Z., Han, F., Zhu, G.: Research on personalized web page recommendation algorithm based on user context and collaborative filtering. In: 2013 IEEE 4th International Conference on Software Engineering and Service Science (2013)
4. Thakkar, P., Varma, K., Ukani, V., Mankad, S., Tanwar, S.: Combining User-Based and Item-Based Collaborative Filtering Using Machine Learning, Information and Communication Technology for Intelligent Systems (2019)
5. Chen, L., Yang, F., Yang, H.: Image-based product recommendation system with convolutional neural networks. In: International Conference on Machine Learning, Optimization, and Data Science Machine Learning, Optimization, and Data Science 4th International Conference (2018)
6. Liu, Q., Wu, S., Wang, L.: DeepStyle: learning user preferences for visual recommendation. In: 40th International ACM SIGIR conference on research and development in information retrieval (2017)
7. Li, Z., Huang, M., Zhang, Y.: A collaborative filtering algorithm of calculating similarity based on item rating and attributes. In: 14th Web Information Systems and Applications Conference (2017)
8. Xu, L., Chen, Y.: Clothing recommendation system based on advanced user-based collaborative filtering algorithm. Signal Inf. Process. Netw. Comput. (2018)
9. Stefani, M.A., Stefanis, V., Garofalakis, J.: CFRS: a trends-driven collaborative fashion recommendation system. In: 2019 10th International Conference on Information, Intelligence, Systems and Applications (2019)
10. He, X., Jin, X.: Collaborative filtering recommendation algorithm considering users' preferences for item attributes. In: 2019 International Conference on Big Data and Computational Intelligence (2019)
11. Fashion Product Images (Small) https://www.kaggle.com/paramaggarwal/fashion-product-images-small. 26 Apr 2019

Predicting Sentiment and Emotion on ASD Using Machine Learning and Deep Neural Networks Based on Twitter Data

T. Lakshmi Praveena and N. V. Muthu Lakshmi

Abstract Autism spectrum disorder (ASD) is by birth neurological disorder. The research on ASD summarizes causes as prenatal conditions, postnatal conditions, and genetic. This paper analyzes the sentiment and emotion of tweets posted in twitter related to ASD causes and treatments. These analytical results provide awareness about treatments and causes of ASD. Natural language processing methods are efficient to perform sentiment and emotion analysis on tweets, but it requires a lot of repeated text processing. So, to reduce the processing time of text, machine learning models are applied on tweets to predict sentiment and emotion. The recent research proved that deep neural networks models are efficient in sentiment analysis and emotion analysis. For efficient results, deep neural networks models are applied on tweets to predict sentiment and emotion. Comparative analysis is performed between ML models results and DNN models results. It is proved that DNN is efficient than ML.

Keywords Autism spectrum disorder · Sentiment analysis · Emotion analysis · Machine learning · Deep neural networks · Natural language processing · Preprocessing methods

1 Introduction

ASD is by birth neurological disorder, and children affected to ASD are approximately 2.5% of the total world population [1]. ASD symptoms given in Refs. [2, 3] are hyperactivity, impulsiveness, carelessness, less social communication, and repeated behavior. ASD is generally affected at birth, but symptoms are often observed after 2 years of age [4]. The underlying cause of this disorder is widely thought to be due to prenatal, postnatal, and genetic [5] and environmental influences [6]. The manual and standard ASD diagnosis tool is DSM-5 [7] and is used to diagnose ASD symptoms. ASD treatments or therapies suggested by physicians are different for different ASD individuals. Finding best treatment for ASD individuals is difficult

T. Lakshmi Praveena (✉) · N. V. Muthu Lakshmi
Sri Padmavati Mahila Visvavidyalayam, Tirupati, India

© The Author(s), under exclusive license to Springer Nature Singapore Pte Ltd. 2022

A. K. Nagar et al. (eds.), *Intelligent Sustainable Systems*, Lecture Notes in Networks and Systems 333, https://doi.org/10.1007/978-981-16-6309-3_45

due to the variety of symptoms observed in ASD individuals. Physicians, parents, and care givers play an important part in finding the right treatment program for ASD individuals [8]. Analysis of advantages and disadvantages of treatments helps to take right decision. This paper presents sentiment analysis and emotion analysis of ASD causes and treatments performed on data collected from familiar social media Web site twitter.

Twitter has 284 million active users per month and 500 million messages are exchanged daily in twitter, which places twitter at top position in social media Web sites and micro-blogs [9]. Twitter data became the main source for research in natural language processing and sentiment analysis. ASD individuals, parents, care givers, special educators, doctors, therapists, and pharmacists share information regarding ASD in twitter.

Sentiment analysis and emotion analysis are performed on tweets collected from twitter related to ASD causes, treatments, or therapies to know the opinion of different individuals. This analysis helps to know the advantages and disadvantages of treatments of ASD to choose right treatment. Machine learning and deep neural networks models reduce the time involved in this analysis by using trained model with pre-analyzed tweets. Pre-analyzed tweets are undergone with machine learning and deep neural networks models. Machine learning models like support vector machine, k-nearest neighbor, Naïve Bayes, decision trees, and random forest are used to predict sentiment from pre-analyzed tweets to predict sentiment and emotion accurately. The recent research proved that deep neural networks models are efficiently working in sentiment analysis and emotion analysis. For efficient results, deep neural networks models like recursive neural networks (RNN), deep neural networks (DNN), and deep convolutional neural network (DCNN) models are applied on tweets to predict sentiment and emotion. Deep learning models are efficient and reliable compared to machine learning models.

Further sections of the paper are organized as follows: Sect. 2 briefs about related work performed in sentiment analysis, emotion analysis, ASD, machine learning for sentiment analysis, and deep learning for sentiment analysis. Section 3 is about proposed methodology and flow diagrams. Section 4 presents analysis results and comparison of different machine learning models and deep learning models.

2 Related Work

2.1 Machine Learning

Machine learning algorithms are efficient, reliable, and accurate compared to manual methods. The work related to machine learning in sentiment analysis is given in this section. This paper [10] proposed a hybrid approach that involves machine learning as a sentiment analyzer. This paper compares sentiment analysis approaches in the analysis of political opinions using supervised machine learning algorithms like nave

Bayes and support vector machines (SVM). In paper [11] proposed an intelligent clustering-based classification and topic extraction model named TClustVID that analyses COVID-19-related series of posts to extract relevant sentiments with high accuracy using the Latent Drichlet Allocation (LDA) algorithm for topic modeling.

2.2 *Deep Learning and Deep Neural Networks*

Deep learning is considered as application of artificial neural networks (ANN). ANN is generally constructed with few layers, and it solves simple problems of classification and prediction. Some works related to deep learning and NLP are given in this section. Paper [12] presented a study of different deep neural networks for sentiment analysis. The performance of different deep neural networks is checked by changing hyperparameters of networks. In paper [13], the main goal is to analyze whether the deep learning improves the performance of standard sentiment analysis.

2.3 *Autism Spectrum Disorder*

Autism spectrum disorder (ASD) is a neurological disorder, effected mostly at birth. ASD individuals generally has social communication deficiency, language deficiency, repetitive behavior, and so on. Some of the works done in this area of sentiment analysis on ASD and neurological disorders are given in this section. In paper [14], sentiment analysis is performed on multiple neurodevelopmental disorders. By showing two videos to people who have different disorders and sentiments are predicted based on the opinions and emotions shown by people. In paper[15], the author presented results on sentiment analysis based on lexical affinity-based emotions in posts on ASD. This analysis helped to China government to know the spreading of ASD in China. The author in paper [16] is aimed to find a set of linguistic and semantic properties of messages posted by people who are related to ASD research and also aimed to motivate ASD research by providing future enhancements in ASD research.

2.4 *Emotion Analysis*

Emotions are used to know the feelings of people on anything. The general emotions are happy, sad, surprise, fear, and angry. The emotion and sentiment are similar, but emotion specifies detailed feeling of person, whereas sentiment species brief opinion of person. Emotion analysis is performed in paper [9, 12] to know the impact of tweets posted related to ASD causes and treatments. Machine learning and deep neural networks are also applied for emotion analysis for better results. Some of the works related to emotion analysis are given in this section. Paper [17] proposed a network model based on bilingual attention network for code-switched emotion classification. The model used in this paper is LSTM for monolingual and bilingual contexts. Paper [18] proposed a model for emotion-based chatting system for large conversations based on GRU. Paper [19] proposed an emotion detection method and large emotion detection dataset. Paper [20] proposed a question-answering methodology based on deep memory network, to find cause of emotion.

2.5 *Sentiment Analysis*

Sentiment analysis is the study of opinions, emotions, and sentiments expressed by people. Sentiment analysis is applied on reviews or comments posted in e-commerce Web sites on products they have purchased or services they utilized. Some works performed in sentiment analysis and deep neural networks are given. Paper [21] compared SVM and ANN for sentiment analysis at document level and stated that ANN gave better results than SVM in the number of cases. The paper [22] proposed a CNN model for bag of word conversion in CNN and also created a new model as seq-CNN for sequential words concatenation.

3 Proposed Methodology for Sentiment and Emotion Prediction

The proposed sentiment prediction and emotion prediction methodology is divided into different processes. These processes are explained with the help of algorithms.

3.1 *Preprocessing Tweets*

Twitter provides vast information in different forms with multiple characters like emojis, hash tags, and URLs. The preprocessing step is very important step in the analysis of tweets, and well-preprocessed tweets can produce analytical and prediction results accurately.

```
Algorithm Tweets Preprocessing(tweets dataset - DT )
do
for each document, di in DT
        di = eliminate duplicates in d
        di = substitute special characters, hashtags, URLs, numbers in di
        di = remove punctuation symbols in di
        di = remove stop words in di
        di = tokenize the di using unigram, bigram and 1-3 grams
        di = stemming of words in di using snow-ball stemmer method
        add di into preprocessed tweets dataset DP.
until all documents in DT are preprocessed.
return DP.
```

3.2 Sentiment Analysis

Sentiment analysis is the process of analyzing sentences to extract sentiment of sentence. This process includes calculating polarity score based on the number of positive words, negative words, and neutral words and then assigning sentiment.

```
Algorithm Sentiment Analysis(preprocessed tweets dataset – DP)
do
for each entity ei in DP
        calculate polarity scores ps_neg – negative, ps_pos – positive, ps_neu – neutral of ei
        compare polarity scores and assign max score to ei
        if ps_neg>ps_pos then
                ei.sentiment="Negative"
                append ei to negative list Dneg
                increase count of negative sentiment tweets c_neg =c_neg+1
        else if ps_pos>ps_neg then
                ei.sentiment="Positive"
                append ei to positive list Dpos
                increase count of negative sentiment tweets c_pos =c_pos+1
        else
                ei.sentiment="Neutral"
                append ei to positive list Dneu
                increase count of negative sentiment tweets c_neu =c_neu+1
        Add sentiments into Ds dataset.
until all entities in DP are assigned with sentiment.
calculate overall sentiment percentages.
calculate sentiment wise word count and mean value.
create a data frame DST with DT -actual tweets , DP preprocessed tweets, Ds - sentiments and
sentiment scores.
return DSA
```

3.3 Emotion Analysis

Emotion analysis is the process of analyzing the sentences to extract words related emotions of sentence. This process includes calculating emotion score based on the number of respective emotion words of *happy, sad, fear, surprise, and angry.* Respective emotion assigned is based on the max emotion score.

```
Algorithm Emotion_Analysis Algorithm(preprocessed tweets dataset – D_P)
do
for each entity e_i in D_P
        calculate emotion scores es_happy, es_sad , es_fear, es_angry, es_surprise of e_i
        compare emotion scores and assign max score to e_i
        e_i.emotion = emotion[max( es_happy, es_sad , es_fear, es_angry, es_surprise)]
        append e_i to max emotion list like D_H | D_S | D_A | D_SR | D_F.
        increase count of emotion count of max emotion tweets.
        Add emotion into D_E dataset.
until all entities in D_P are assigned with emotion.
calculate overall emotion percentages.
calculate emotion wise word count and mean value.
create a data frame D_EM with D_T -actual tweets , D_P -preprocessed tweets, D_E - emotions and emotion scores.
return D_EM
```

3.4 Machine Learning Algorithms for Sentiment and Emotion Prediction

Machine learning algorithms are used to automate the prediction process and predict sentiment and emotion using the previous data which improves accuracy of prediction. Proposed prediction method accuracy is compared with accuracy of different machine learning algorithms.

```
Algorithm ML_Sentiment & Emotion_Prediction(sentiment data D_SA ,emotion data D_EM)
do
        Load dataset D_SA or D_EM
        Feature selection based on correlation coefficient calculated.
        Divide features into categorical and numerical set.
        Apply label encoder to convert categorical features to numerical features.
        Combine numerical features with converted numerical features.
        Select class label / independent attribute.
        Split dataset into trainset and test set.
        Call Apply_ML(trainset,testset)
Algorithm Apply_ML(trainset,testset)
do
        DT_model=Decision_Tree(trainset)
        DT_prediction=DT_model.predict(testset)
        NB_model=Naïve_Bayes(trainset)
        NB_prediction=NB_model.predict(testset)
        KNN_model=KNearestNeighbor(trainset)
        KNN_prediction=KNN_model.predict(testset)
        LSVM_model=Linear_SVM(trainset))
        LSVM_prediction=Linear_SVM.predict(test set))
        NLSVM_model=NonLinear_SVM(trainset))
        NLSVM_prediction=NonLinear_SVM.predict(test set))
        RF_model=Random_forest(trainset)
        RF_prediction=RF_model.predict(testset)
        LR_model=Logistic Regression(trainset)
        LR_prediction=LR_model.predict(testset)
```

3.5 *Deep Learning Models for Sentiment and Emotion Prediction*

Sentiment and emotion prediction of ASD tweets is implemented using different deep neural networks models. The models are base model, reduced model, drop model, deep model, RNN, DCNN, and DNN.

```
Algorithm DL_Sentiment_Emotion_Prediction (sentiment dataset DSA ,emotion dataset DEM)
do
    Load dataset DSA or dataset DEM .
    Select tweet text and related features.
    Divide dataset into train and test sets.
    Tokenize tweet text and apply filters.
    Generate word embeddings for tokenized words.
    Generate one hot sequences for word embeddings.
    Create validation set from one hot sequences.
    Apply deep learning models on train dataset.
    Create sentiment prediction model.
    Predict sentiment by testing model with test set.
```

4 Analysis Results

Actual work of this paper finds opinion and emotion based on tweets posted by parents, care takers, and special educators. The Deep learning models used in this method are given in Table 1. Different causes of ASD are prenatal, genetic, prenatal, vaccines, and environmental conditions. Compared to all other causes, genetic cause for ASD has higher negative sentiment than others, prenatal conditions have highest sentiment which is negative, vaccination can cause ASD which has highest negative sentiment, and environmental conditions and medication also have highest negative sentiment. This analysis shows that frequent causes of ASD are listed in Table 2 but only less than 40% people believing that ASD caused by genetic, prenatal, vaccination, environmental conditions, and medicines used.

Emotion analysis of ASD causes is given in Table 3. The emotion analysis gives percentage of five emotions like happy, sad, fear, angry, and surprise. People expressed surprise about genetic cause, happy emotion has highest value for prenatal conditions cause, and vaccination cause shows surprise, emotion, and medication give happy emotion. Emotion analysis shows people reactions on causes of ASD. Sentiment analysis of treatments shows the opinion of different members on different therapies used for ASD people. All treatments or therapies have highest positive sentiment. Based on this sentiment analysis, people can choose therapy required (Table 4).

Emotion analysis of treatments is given in Table 5. The highest emotion of therapies includes happy, sad, and surprise. Emotion analysis gives in detail opinion of people on ASD treatments which helps in choosing therapies by parents for their kids. Table 6 gives accuracy of different machine learning algorithms on twitter data

Table 1 Deep learning models used in work and the model details

Model	N_Layers	Layer size	Total Params	Trainable Params
Base model	Dense:2	(None,64)	644,419	644,419
	Output: 1	(None,3)		
Deep model (optimizer: rmsprop, loss: categorical cross enropy) added	Dense:2	(None,64)	644,419	644,419
	Output: 1	(None,3)		
Reduced model	Dense:1	(None,32)	320,131	320,131
	Output: 1	(None,3)		
Regulariser model (l2 regularizer: 0.001 LR)	Dense:2	(None,64)	644,419	644,419
	Output: 1	(None,3)		
Drop model	Dense:2	(None,64)	644,419	644,419
	Output: 1	(None,3)		
	drop:2	0.5		
RNN	Embedding:1	(None,50,100)	644,419	644,419
	Masking:1	(None,50,100)		
	Lstm:1	(None,64)		
	Dropout:1	(None,64)		
	Dense:1	(None,64)		
	Output: 1	(None,3)		
DCNN	Conv:3	256,128,64	644,419	644,419
	Pooling:3			
	FC:1			
	Dropout:1			
	Output: 1	(None,3)		
DNN	Dense:3	256,128,64	644,419	644,419
	Dropout:3	256,128,64		
	Output: 1	(None,3)		

for sentiment and emotion prediction. Machine learning algorithms helps to predict sentiment and emotion in less time without repeating the sentiment analysis and emotion analysis processes. Naïve Bayes, decision trees, and random forest algorithms gives highest accuracy compared to other algorithms. This comparison helps to researchers to choose best algorithm for sentiment and emotion prediction from twitter data.

Table 7 gives the accuracy of neural networks models and deep learning models for sentiment and emotion prediction of preprocessed ASD data provided by sentiment analysis and emotion analysis modules. The neural networks models give similar accuracy for all models except for drop model. Deep learning models, RNN, DCNN, and DNN, give highest accuracy compared to neural networks models (Figs. 1 and 2).

Table 2 Sentiment analysis of ASD causes

ASD causes	Positive (%)	Negative (%)	Neutral (%)
Genetic	23	49	28
Prenatal	26	45	30
Vaccines	30	40	30
Environmental conditions	32	43	25
Medication	40	42	18

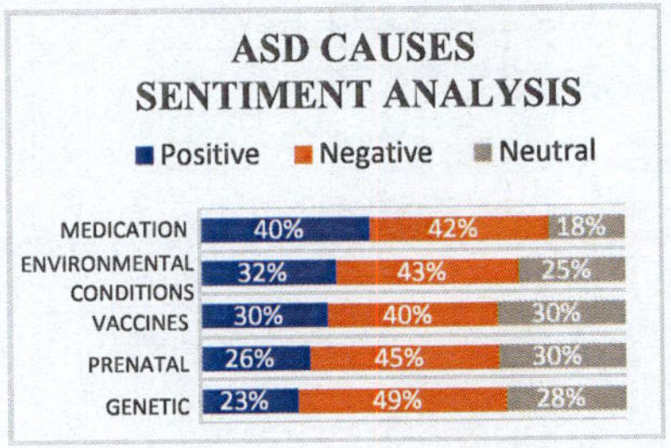

Table 3 Emotion analysis of ASD causes

ASD causes	Happy (%)	Sad (%)	Fear (%)	Angry (%)	Surprise (%)
Genetic	26	14	23	4	30
Prenatal	42	33	17	0	8
Vaccines	0	0	25	0	75
Environmental conditions	23	15	25	5	18
Medication	30	20	25	10	15

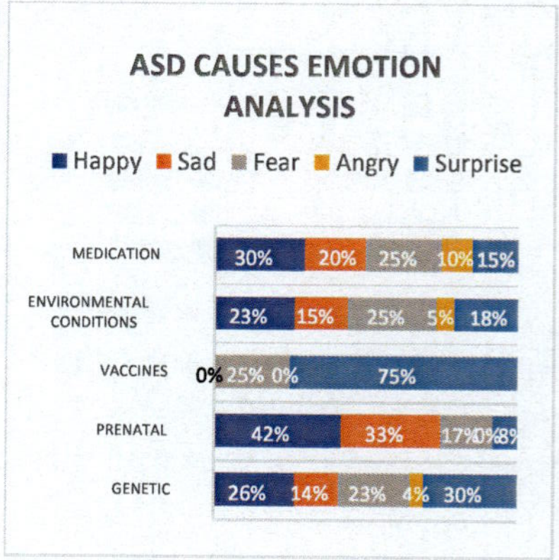

Table 4 Sentiment analysis of ASD treatments

ASD treatments	Positive (%)	Negative (%)	Neutral (%)
Behavior therapy	55	9	36
Occupational therapy	56	0	44
Speech therapy	66	14	26
Physical therapy	71	0	29
Sensory therapy	46	0	54

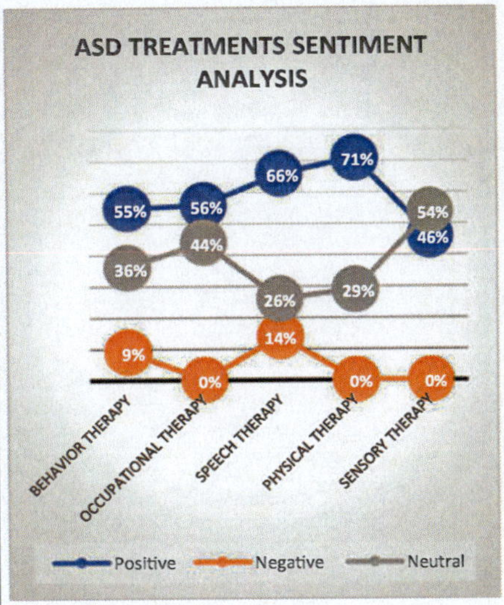

5 Conclusion and Future Scope

In this work, new methodology is proposed for sentiment and emotion prediction for ASD causes and treatments for ASD using twitter data. Compared to the standard methods of emotion prediction and sentiment prediction, machine learning models are predicted accurately with the less computation time. Deep learning models are used to build more efficient and accurate model than machine learning-based prediction model for ASD data. The proposed methodology results are used to create awareness on ASD causes and ASD treatments by predicting the sentiment and emotion on twitter data. The proposed methodology can also be applied to create prediction models for different diseases, and to help people to understand feelings and emotions of others on specified disease.

Table 5 Emotion analysis of ASD treatments

ASD treatments	Happy (%)	Sad (%)	Fear (%)	Angry (%)	Surprise (%)
Behavior therapy	25	25	25	0	25
Occupational therapy	38	0	25	0	38
Speech therapy	21	37	21	5	16
Physical therapy	50	0	50	0	0
Sensory therapy	31	15	31	0	23

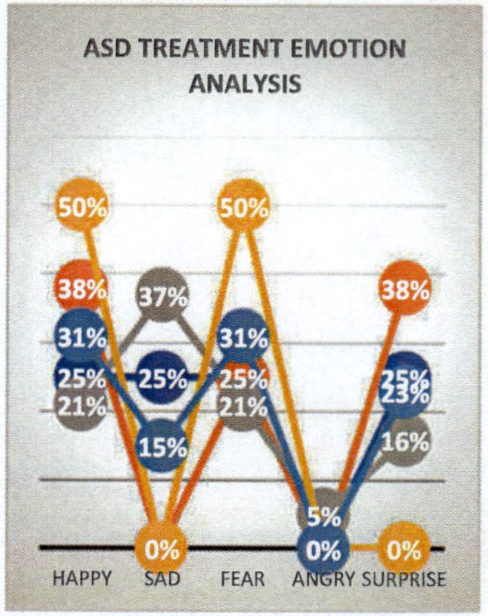

Table 6 ML and DL algorithms accuracy for sentiment and emotion prediction of ASD

MLAlg	ML_Sentiment Prediction (%)	ML_Emotion Prediction (%)
Logistic Regression	51	50
KNN	42	45
Naïve Bayes	80	81
Linear SVM	37	40
Non-Linear SVM	51	55
Decision Trees	100	100
Random Forest	100	100

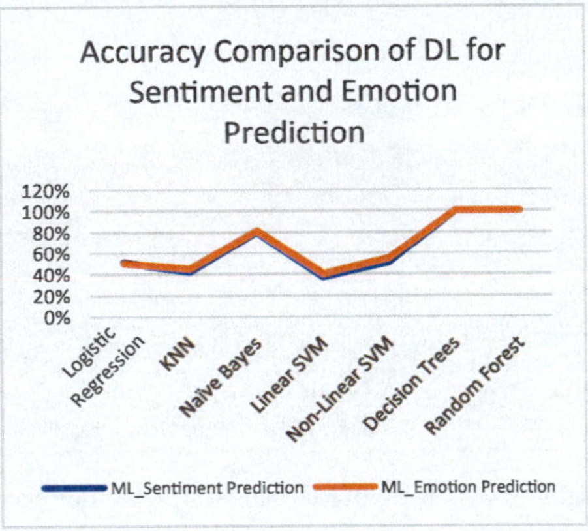

Table 7 ML and DL algorithms accuracy for sentiment and emotion prediction of ASD

Model	Sentiment prediction (%)	Emotion prediction (%)
NN Base Model	66.60	63.60
NN Deep Model	66.60	63.60
NN Reduced Model	66.60	63.60
NN Regularized Model	66.60	63.60
NN Drop Model	55.50	54.50
RNN Model	86	82
Deep-CNN model	84.50	82.50
DNN model	85	83

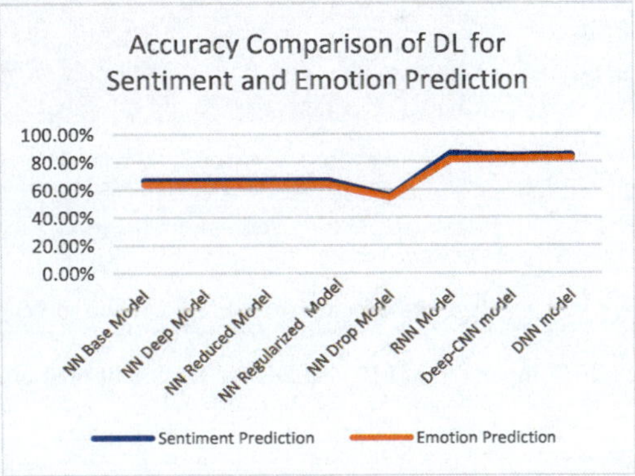

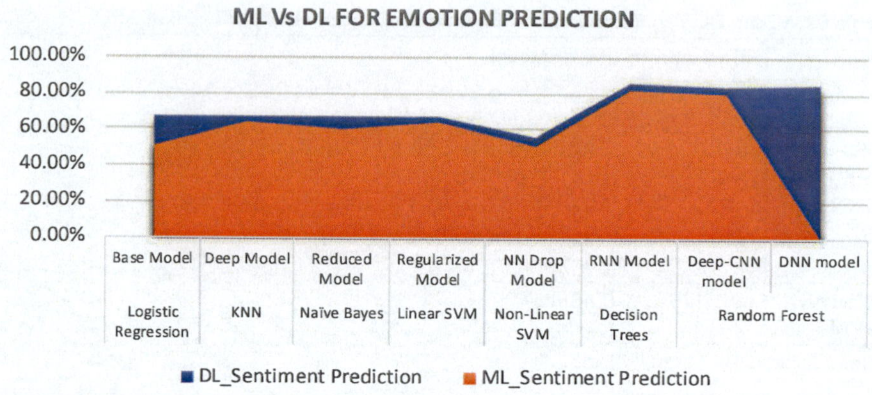

Fig. 1 Comparison of ML and DL for emotion prediction on ASD twitter data

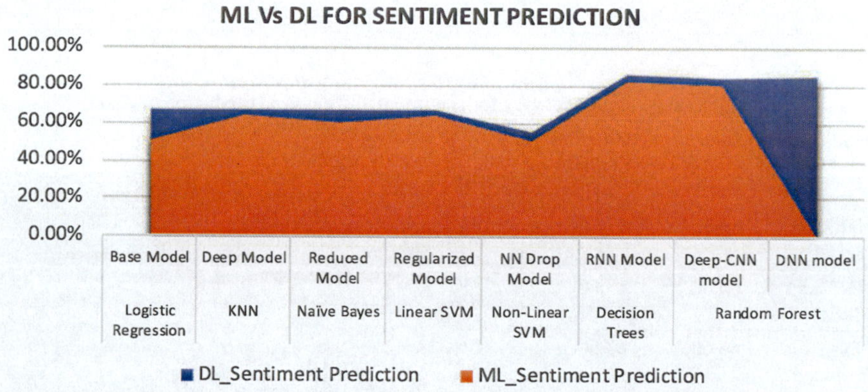

Fig. 2 Comparison of ML and DL for sentiment prediction on ASD twitter data

References

1. Simon, V., Czobor, P., Bálint, S., Mészáros, Á., Bitter, I.: Prevalence and correlates of adult attention-deficit hyperactivity disorder: meta-analysis. Br. J. Psychiatry **194**(3), 204–211 (2009)
2. Barkley, R.A.: Behavioral inhibition, sustained attention, and executive functions: constructing a unifying theory of adhd. Psychol. Bull. **121**(1), 65 (1997)
3. Spencer, T.J., Biederman, J., Mick, E.: Attentiondeficit/hyperactivity disorder: diagnosis, lifespan, comorbidities, and neurobiology. J. Pediatr. Psychol. **32**(6), 631–642 (2007)
4. Weiss, G., Hechtman, L., Milroy, T., Perlman, T.: Psychiatric status of hyperactives as adults: a controlled prospective 15-year follow-up of 63 hyperactive children. J. Am. Acad. Child Psychiatry **24**(2), 211–220 (1985)
5. Stevenson, J., Asherson, P., Hay, D., Levy, F., Swanson, J., Thapar, A., Willcutt, E.: Characterizing the adhd phenotype for genetic studies. Dev. Sci. **8**(2), 115–121 (2005)

6. Genetic and environmental contributions to stability and change of adhd symptoms between 8 and 13 years of age: A longitudinal twin study. J. Am. Acad. Child Adolesc. Psychiatry **43**(10), 1267–1275 (2004)
7. Dsm. diagnostic and statistical manual of mental disorders (dsm-5). http://www.dsm5.org/
8. Treatments and Therapies for ASD. https://www.nimh.nih.gov/health/topics/autism-spectrum-disorders-asd/index.shtml#part_145441
9. Stojanovski, D., et al.: Twitter sentiment analysis using deep convolutional neural network. Conference: HAIS 2015 at Bilbao, Spain (2015). https://doi.org/10.1007/978-3-319-19644-2_60
10. Hasan, A., Moin, S.: Machine learning-based sentiment analysis for Twitter accounts. Math. Comput. Appl. **23**, 11 (2018)
11. Satua, M.S., Khanb, M.I., TClustVID: A Novel Machine Learning Classification Model to Investigate Topics and Sentiment in COVID-19 Tweets. https://doi.org/10.1101/2020.08.04.20167973
12. Wadawadagi, R., Pagi, V.: Sentiment analysis with deep neural networks: comparative study and performance assessment. Artif. Intell. Rev. **53**, 6155–6195 (2020). https://doi.org/10.1007/s10462-020-09845-2
13. Colneric, N., Demsar, J.: Emotion recognition on Twitter: comparative study and training a unison model, February 2018. In: IEEE Transactions on Affective Computing vol. 99, p. 1, (2018). https://doi.org/10.1109/TAFFC.2018.2807817
14. Vaucheret Paz, E., Martino, M., Hyland, M., Corletto, M., Puga, C., Peralta, M., Deltetto, N., Kuhlmann, T., Cavalié, D., Leist, M., Duarte, B., Lascombes, I.: Sentiment analysis in children with neurodevelopmental disorders in an ingroup/outgroup setting. J. Autism Dev. Disord. **50**(1), 162–170 (2020). https://doi.org/10.1007/s10803-019-04242-3. PMID: 31571066
15. Tang, T.Y., Wang, R.Y., Chen, C.G.: "The sum of all our feelings!": sentimental analysis on Chinese autism sites. In: Antona, M., Stephanidis, C. (eds.) Universal Access in Human–Computer Interaction. Human and Technological Environments. UAHCI 2017. Lecture Notes in Computer Science, vol 10279. Springer, Cham (2017). https://doi.org/10.1007/978-3-319-58700-4_10
16. Beykikhoshk, A., Arandjelovic, O., Phung, D., Venkatesh, S., Caelli, T.: Data-mining twitter and the autism spectrum disorder: a pilot study. In: 2014 IEEE/ACM International Conference on Advances in Social Networks Analysis and Mining (ASONAM), China, pp. 349–356 (2014). https://doi.org/10.1109/ASONAM.2014.6921609
17. Wang, Z., Zhang, Y., Lee, S., Li, S., Zhou, G.: A bilingual attention network for code-switched emotion prediction. In: Proceedings of the International Conference on Computational Linguistics (COLING 2016) (2016)
18. Zhou, H., Huang, M., Zhang, T., Zhu, X., Liu, B.: Emotional chatting machine: emotional conversation generation with internal and external memory. arXiv preprint. arXiv:1704.01074 (2017)
19. Abdul-Mageed, M., Ungar, L.: EmoNet: fine-grained emotion detection with gated recurrent neural networks. In: Proceedings of the Annual Meeting of the Association for Computational Linguistics (ACL 2017) (2017)
20. Felbo, B., Mislove, A., Søgaard, A., Rahwan, I., Lehmann, S.: Using millions of emoji occurrences to learn any-domain representations for detecting sentiment, emotion and sarcasm. In: Proceedings of the Conference on Empirical Methods on Natural Language Processing (EMNLP 2017) (2017)
21. Moraes, R., Valiati, J.F., Neto, W.P.: Document-level sentiment classification: an empirical comparison between SVM and ANN. Expert Syst. Appl. (2013)
22. Johnson, R., Zhang, T.: Effective use of word order for text categorization with convolutional neural networks. In: Proceedings of the Conference of the North American Chapter of the Association for Computational Linguistics: Human Language Technologies (NAACL-HLT 2015) (2015)

Incidence of Dialectal Accents in the Students' Listening Understanding

Wilmer Patricio Collaguazo Vega, Gina Alexandra Albán Rocha, Paulina Maricela Jácome Santana, and Luis Gonzalo Maiza Vayas

Abstract The didactics of foreign languages have been changing according to the needs of the contexts, presenting new models and resources, as well as learning objectives aimed at communicative competence. The exposition to different accents and dialects of the English language can improve the audible range of its apprentices due to the possibility of locking on different intonations and pronunciations that allow the development of listening comprehension for the development of oral expression. This article addresses the importance of integrating the practice of pronunciation and understanding of different dialectal accents in the teaching–learning process. A non-experimental design of the research model was used, including in the field research the application of techniques whose collected information allowed evaluating the level of development of the communicative competence and the level of use of the dialectal accents in the auditory comprehension. The analysis and statistical processing were performed for the interpretation of the results.

Keywords English language teaching · Listening comprehension · Hearing process · Dialectal accents · Auditory resources

W. P. C. Vega (✉)
Universidad Técnica de Cotopaxi, Latacunga, Ecuador
e-mail: wilmer.collaguazo@utc.edu.ec

G. A. A. Rocha
Departamento de Seguridad Y Defensa, Universidad de Las Fuerzas Armadas ESPE, Av. General Rumiñahui Y Ambato, 170501 Sangolquí, Ecuador
e-mail: gaalban2@espe.edu.ec

P. M. J. Santana
Escuela de Educación Básica General Córdova, Ambato, Ecuador
e-mail: paulina.jacome@educacion.gob.ec

L. G. M. Vayas
Universidad Técnica de Ambato, Ambato, Ecuador
e-mail: lg.maiza@uta.edu.ec

© The Author(s), under exclusive license to Springer Nature Singapore Pte Ltd. 2022

A. K. Nagar et al. (eds.), *Intelligent Sustainable Systems*, Lecture Notes in Networks and Systems 333, https://doi.org/10.1007/978-981-16-6309-3_46

1 Introduction

The teaching of foreign languages has been evolving and, at present, adjusted to the needs of the contexts; it presents new models, resources, as well as learning objectives aimed at communicative competence when considering the importance of oral skills.

The exposition to different accents and dialects of the English language can improve the listening range of their apprentices due to the possibility of catching different intonations and pronunciations by speakers from various areas of the world compressed in a recording accessible to students that allow them to develop listening comprehension for the development of oral expression. From this perspective, the need to learn the English language and, in particular, the development of listening skills imply the knowledge of dialectal accents.

This research was developed at the Technical University of Cotopaxi to determine the incidence of oral expression development based on the knowledge of dialectal accents in the students' listening comprehension in the academic period March–August 2020.

The researchers aimed to provide a solution to the flawed communicative methodology of the English language and its relationship with speaking and listening skills, besides the need to improve the teaching–learning process and, therefore, the development of students' listening understanding.

2 State of the Art

The communicative competence presents important dimensions such as listening and speaking, while its fundamental indicators are the following: comprehension; listen carefully; understand the message; make sense of the message; utilize gestures, movements, expressions, and fluency at speaking; respect toward others' opinions; and show persuasive qualities.

Listening is conceived as an invisible mental process, which makes it difficult to describe. The listeners must discriminate among the different sounds, understand the vocabulary and grammatical structures, interpret the emphasis and intention, and retain and interpret all these within the immediate context [1]. In this definition, although it is true, the reference is made to the socio-cultural aspect, albeit the current importance is given to the phonological, lexical, and structural aspects of language.

Reference [2] Some researchers affirm that listening comprehension is not a skill but a series of skills marked by the fact of involving the auditory perception of oral signs [also] … it is not passive. A person can hear something but not listening to it …

Cognitive strategies have been defined as specific listening tasks that directly operate in the text. To make inferences, the listener uses all the linguistic and non-linguistic information at his or her disposal: to guess the meaning of what he or she has heard, to predict what will happen, or to complete what the person has not heard. The probing strategy to find the main ideas (skimming) includes selecting information to relate it to other concepts already understood [3]. Then, the purpose of probing to find details (scanning) is to identify or select individual words and short phrases. The fourth and last cognitive strategy of the model, to make summaries, is defined as producing abbreviated versions of the text, or parts of the text, in the listener's words.

The different types of listening that students should attend in a conversation could be synthesized: (1) Basic listening that refers to the occasions in which students want to know the general idea of what is being talked about; (2) listen to specific information that relates to the circumstances when students do not have to understand everything, only a very specific part; (3) listen in detail that refers to the hearing students do when they need to find errors or determine the differences between one and another listening didactic resource; and (4) inferential listening that refers to the hearing students do when they want to know how the speaker feels.

Therefore, it can be affirmed that the local or foreign accent is not only the product of a characteristic or inadequate pronunciation of the sounds but a characteristic of international integration of all the phonic matter, whether or not it is pronounced correctly [4]. Although the English heard in the classroom is British in the different audios, teachers should consider presenting all the possibilities of the language to their students. It means more opportunities exist for the students to grow, mature, and enrich into a second language when they study an extensive range of English forms.

It is essential to offer a range of language variants, and students from their freedom should choose which variant they feel most comfortable expressing themselves and communicate in another language [5].

To distinguish, more clearly, one dialect from another, it consists of a "combination, which often avoids the analysis of segmental and supra-segmental phonetic features." Although the term is often linked to the social origin or the condition of being native or foreign of the speaker [6], it is also essential to circumscribe to the geographical area.

The dialect or dialectal, diatopic, geographical, geolectal, geolinguistic, regional variety and its equivalent denominations do not carry any pejorative connotation; it is preferably "a variety of a language circumscribed to a specific geographical area" [7] or "manifestation of a language in a given territory" [8]. The language is not made or updated in its entirety but precisely through these modalities [9].

To understand without difficulty extensive and complex oral texts, articulated at normal or fast speed and live transmitted or by technical means, even in a noisy environment, such as operating instructions, announcements, conferences, lectures, debates, conversations or movies, [10] students can deal with both concrete and abstract topics at the time, even of a technical nature or not belonging to the field of specialization. In addition, they have to appreciate registry changes and recognize idiomatic and colloquial expressions, even in cases in which the accent is not familiar.

Despite the advances and innovations in education, the pronunciation is still unattended, and it does not receive the appropriate study due to teachers do not know how to integrate the teaching of pronunciation in the communicative approach or the most recent task approach because they do not have enough materials and those available do not separate the pronunciation of phonetic correction, and therefore, its integration in the communicative class is impossible and can only be presented as an annex or appendix that bears no relation to the development, content, and focus of the classes.

Reference [11] Accent and intonation are responsible for integrating speech in phonic blocks. The accent plays its role in the words (paradigmatic level): the accent highlights a vowel (the tonic vowel) in front of the other vowels that are unstressed or it can also highlight a tonic vowel (with tonal inflection) in front of the different tonic vowels of the same phonic group, then one can refer to the syntagmatic accent [12]. While the accent acts on the words, the intonation exerts its influence on the phrase (syntagmatic level). Each sound of speech (and, especially, the vowels) has a particular tone, and the succession of these tones is called intonation.

3 Methodology

The general objective of this research is to determine the incidence of the oral expression development based on the knowledge of the dialectal accents in the students' listening comprehension at Learning a Foreign Language Program—English in the Technical University of Cotopaxi during the period March–August 2020.

The preceding shows that research corresponds to a development project because it aims to solve practical problems through a proposal. According to the origin of the study, it is quantitative and qualitative, taking into account that the qualitative methodology allowed making inferences, subjective valuations around the research results, while the quantitative method permitted to establish the percentage analysis and the descriptive statistics. In correspondence with the objectives of the research, it constitutes a descriptive/explanatory study. Among the main techniques used in field research, the questionnaire stands out. Also, it comprised the description, registry, analysis, and interpretation of the problem. Therefore, as indispensable elements of the research design, a bibliographic review, approach, and formulation of the problem, objectives, and scientific questions were considered as a starting point; later, the characterization of the population and selection of the study group, development of instruments, study field, data processing and analysis, and conclusions.

For the structure of the study and fieldwork, a representative sample of 50 students was established. As a measurement instrument, a questionnaire was selected and applied based on the Likert scale with indicators that allowed to measure the teaching of the phonological level in English, based on the development of listening comprehension. The analysis and interpretation of results were carried out in the Statistical Program SPSS, from which descriptive statistics were obtained on the dataset, establishing general assessments and for each of the items.

4 Results

The multiple-choice option for the items is considered, considering the Likert-type scale, a type of measurement instrument or data collection that constitutes a social research tool (Table 1).

Regarding the academic resources used by the teacher in the classroom, the development of auditory competence is still insufficient. In each of the indicators analyzed, the frequency is mostly rarely to almost always, from 2.4 to 2.6 in most cases. The standard whether the teacher emphasizes the importance of differentiating dialects and accents as a strategy for a better understanding of the English language, based on the respondents' answers is 2.4, which shows teachers consider these strategies, but the development of auditory competence is still insufficient (Fig. 1).

Similarly, whether the teacher considers the messages, for the development of auditory competence, the importance and need to recognize different accents and dialects, the average of responses is 2.6, so the teacher must be oriented and generate tools that allow them to improve its development and, therefore, of a better oral expression.

Regarding whether the messages used to promote the listening skill and the auditory competence are structured, there is an average of 2.4; therefore, it is essential to guide teachers for their performance in terms of developing auditory competence.

A favorable aspect, although it must continuously be improved, is the fact that the teachers, with an average of 3.2, in the development of the classes, rely on audiovisual media for their development.

One aspect of improving with an average of 2.6, the respondents acknowledge that the teacher develops mainly analytical and appreciative listening through audiovisual resources to promote communicative competence and listening skills, being necessary a planned and structured work that allows developing these activities more frequently (Fig. 2).

Concerning the educational resources used by the teacher in the classroom for English language learning, the development of auditory competence is still insufficient. In each of the indicators analyzed, the frequency is mostly never to rarely in Block 13–19, with averages of 1.6–2.2 in most cases.

The average whether the teacher applies the top-down model for the development of listening comprehension, based on the responses with an average of 1.6, shows teachers consider these strategies, but auditory competence development is still poor.

Likewise, whether the hearing process is comprised chiefly of auding, the average of responses is 1.6, so the teacher should be oriented and generate tools that allow them to improve the development of auditory competence.

As for whether the teacher for the development of a correct pronunciation develops in students the ability to catch different intonations, dialects, and accents, there is an average of 2.2; therefore, it is essential to guide teachers for improving their performance. On the other hand, it must continuously be improved, a favorable aspect that teachers, with an average of 3.6, propose leveling and recovery activities.

Table 1 Items with Likert scale rating

Item	Rating			
Motivation	1	2	3	4
Does the teacher encourage the collaboration and participation of students inside and outside the classroom in a pleasant and favorable environment?		20	30	
Does the teacher encourage the positive behaviors and strengths of the student in the classroom?		10	40	
During the development of the classes, does the teacher teach teamwork habits, encourage interest and motivation in the students?			40	40
During the development of the classes, does the teacher explain clearly and speak fluently?			30	20
Does the teacher have an adequate pronunciation that favors the listening competence of the students?			30	20
Does the teacher adapt to the needs and personal characteristics of the students?		20	30	
Use of Teaching Resources				
Is the teacher supported by audiovisual means (recorder, TV, DVD, computer) to develop his or her classes?			40	10
Do the teachers develop analytical and appreciative listening mainly through audiovisual resources to promote communicative competence and listening skills?		30	20	
Do the listening resources used in classes contain sequences of time, space, and deduction?		20	30	
Are the messages used to promote the listening skill and the auditory competence structured, previously planned by the teacher?		20	30	
Are the importance and need to recognize different accents and dialects taken into account in the messages to develop auditory competence?		20	30	
Does the teacher emphasize the importance of differentiating dialects and accents as a strategy for a better understanding of the English language?		30	20	
Accents and dialects influence the listening comprehension of the English language			20	30
Does the teacher propose leveling and recovery activities in those objectives and contents not achieved by the students?			20	30
Does the teacher use the hours assigned to practice in the laboratory to strengthen auditory competence in students?			20	30
Does the teacher, for the development of auditory competence, develop students' ability to capture different intonations and pronunciations from different parts of the world?	20		30	
Does the teacher, for the development of correct pronunciation, develop students' ability to capture different intonations, dialects, and accents?	20		30	
Primarily the listening process is structured by auding	20	30		
The teacher applies the top-down model for the development of listening comprehension	20	30		

Source Field research. Elaborated by: The researchers

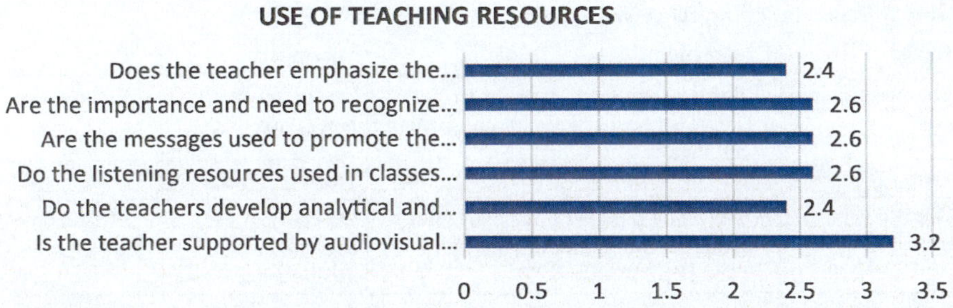

Fig. 1 Arithmetic stockings by item. Assessment. *Source* Field research. Elaborated by: The researchers

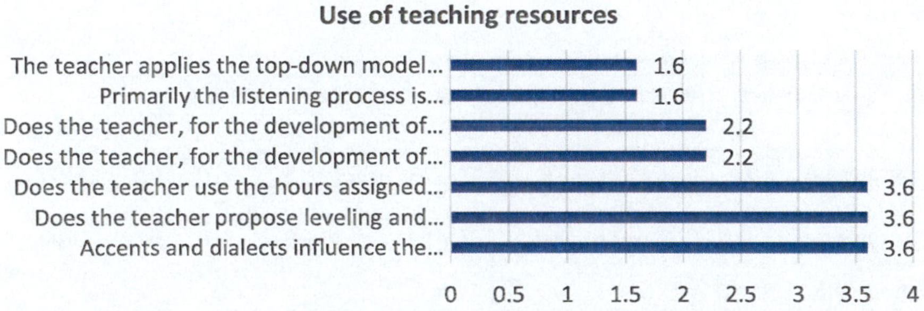

Fig. 2 Use of teaching resources. *Source* Field research. Elaborated by: The researchers

In the same assessment, with an average of 3.6, students recognize that accents and dialects influence listening comprehension, being favorable to the actual perception of the importance of these resources for the development of listening skills.

Finally, an aspect of improving with an average of 2.2, the respondents affirm that teachers develop their students' ability to capture different intonations, dialects, and accents that allow creating these activities with a more significant frequency.

4.1 Hearing Discrimination Test

After the application of a final standard sheet of performance criteria to the sample under study, it was possible to determine the following structure based on the structure designed for the verification of the model (Tables 2 and 3).

Table 2 Topics for the hearing discrimination test

Topics	Explanation	Vocabulary	Level
Countries United	• Shirley (Scotland) explains to Yuri (Italy) the terms: The UK and Britain and their relationship to Scotland	On-the-fence Not my forte To change the subject To go on all-day	Elementary
The Big Mix	• Cle Ann (Trinidad y Tobago) talks about her multicultural country to Todd (USA)	Descent Slavery To abolish	Elementary
Korean Wedding	• Sunny (Korea) talks about marriage traditions in their country to Isthma (USA)	Average age Stigmatized Suit	Elementary

Source Field research. Elaborated by: The researchers

Table 3 Comparative report

Accents and dialects in the listening comprehension of the English language											
Criteria	A	A (%)	B	B (%)	C	C (%)	D	D (%)	E	E (%)	SUM
Listening for gist	0	0	19	38	27	54	4	8	0	0	100
Listening for specific information	0	0	23	46	27	54	0	0	0	0	100
Listening in detail	0	0	15	30	31	62	4	8	0	0	100
Inferential listening	8	16	34	68	4	8	4	8	0	0	100
Understanding	8	16	23	46	15	30	4	8	0	0	100
Vocabulary development	0	0	19	38	31	62	0	0	0	0	100
Connect fine ideas	4	8	27	54	15	30	4	8	0	0	100
Appropriate use of paraphrasing	0	0	23	46	27	54	0	0	0	0	100

Source Field research. Elaborated by: The researchers

5 Conclusions

It can be affirmed that students possess functional knowledge and an adequate understanding of the purposes of the application for this research. Additionally, they have efficient and purposeful use of the spoken language that will allow them to listen correctly. On the other hand, they have functional knowledge and a good understanding of the grammatical construction of the proposed listening resources. They also possess an appropriate and competent use of the elements of spoken language to catch essential information in context, making use of the last phase of the listening process, "auding."

The criterion of inferential listening was taken as a performance reference. While the listening range criterion was listening in detail due to the academic level of the sample object of study where it was denoted that the students have a low listening range, not indicating that their auditory competence has not been adequately developed, but the reduced exposure that students have had to different accents and dialects. Several students have considerable difficulty in performing activities whose purpose is hearing because they cannot determine keywords or clues.

Even though the final sample used to check the model was reduced, the results are decisive and show the importance of the academic listening resources within the teaching–learning process of English as a foreign language. The interaction with audio recordings coming from different areas of the world, facilitated by interlocutors who have English as their mother tongue, their second language or receive a marked influence of it, are ideal for extending the range of hearing of the participants, and therefore of people interested in learning English.

References

1. Hamad Al-khresheh, M.: The impact of cultural background on listening comprehension of Saudi EFL students. Arab World Engl. J. **11**(3.22), 349–371 (2020)
2. Bozorgian, H.: Listening skill requires a further look into Second/Foreign language learning. Int. Sch. Res. Not. **2012**, 10 (2012)
3. Liu, Y.-C., core.ac.uk: Texas A&M University, December 2009. (Online). Available: https://core.ac.uk/download/pdf/4281447.pdf
4. Gilakjani, A., Ahmadi, S., Ahmadi, M.: Why is pronunciation so difficult to learn? Engl. Lang. Teach. 4(3), 74–83 (2011). www.ccsenet.org/elt
5. Council of Europe, Common European Framework of Reference for Language: Learning, teaching, assessment. The English edition is published by Cambridge University Press (2011)
6. Byram, M., Gribkova, B., Hugh, S.: Developing the intercultural dimension. Strasbourg (2002)
7. Robins, R.H., Crystal, D.: Encyclopedia Britannica. https://www.britannica.com/topic/language, 09 Mar 2021. [Online]. Available: https://www.britannica.com/topic/language
8. Ivić, P., Crystal, D.: Encyclopedia Britannica. 22 July 2014. [Online]. Available: https://www.britannica.com/topic/dialect
9. Brohy, C., Ferrando, V.C., Pagett, A.O., Ramallo, F.: The European charter for regional or minority languages. Hung. Yearb. Int. Law Eur. Law **7**(1), 259–269 (2020)
10. Wudthayagorn, J.: Mapping the CU-TEP to the common European framework of. LEARN J.: Lang. Educ. Acquisition Res. Netw. J. **11**(2), 163–180 (2018)
11. Cantero Serena, F.J.: La cuestion delacento en la enseñanza de lenguas. Researchgate (1994)
12. Cantero Serena, F.J.: Conceptos clave en lengua oral. Researchgate, pp. 141–153 (1998)

Classification of Handwritten Indic Scripts Using Deep Learning

Aaditree Jaisswal, Prachi Ghalsasi, Aabha Jain, Misha Batra, and Sandhya Arora

Abstract Old handwritten documents written in regional languages are difficult to interpret. It is also hard to identify the script that the characters are conforming to. Script recognition is an integral part of researching the anthropology of a region and understanding its geography and culture. Current research in handwritten script recognition focuses on character level recognition of at most two or three languages. It is also a lot harder to identify scripts in Indian languages due to the similarity between their characters. The aim of this paper is to be able to create a deep learning model that can classify up to twelve different languages, at document level and display the output based on the result by performing line-level segmentation, followed by word-level segmentation and finally with character recognition of the document.

Keywords Convolution neural network · Handwritten characters · Languages · Indic scripts · Segmentation · Image classification

1 Introduction

Script classification refers to the process of identifying the scripts in which a given document or text corpus is written. In a country like India with a rich chronicle and the presence of multiple languages, it is difficult to preserve the remains of scriptures and other age-old documents by digitization. It is infeasible to develop an OCR for numerous scripts. Thus, script recognition is a prelude to digitizing said documents. Once the script is identified, an OCR can be implemented to machine editable form. In our paper, we focus on character level recognition of handwritten documents in the 12 Indian languages, i.e., Malayalam, Gurumukhi, Gujarati, Bangla, Telugu, Devanagari, Kannada, Urdu, Manipuri, Odia, Tamil and Roman. The main challenge in the process is the variation in the language due to different handwriting. This variation makes it difficult to detect features in the script that may be pertinent to the classification. In the past few years, a lot of work has been done in the field of

A. Jaisswal (✉) · P. Ghalsasi · A. Jain · M. Batra · S. Arora
M.K.S.S.S's Cummins College of Engineering for Women, Pune, India
e-mail: sandhya.arora@cumminscollege.in

© The Author(s), under exclusive license to Springer Nature Singapore Pte Ltd. 2022

A. K. Nagar et al. (eds.), *Intelligent Sustainable Systems*, Lecture Notes in Networks and Systems 333, https://doi.org/10.1007/978-981-16-6309-3_47

Indic handwritten script classification. However, this work has been majorly focused on the recognition of a particular language. Little or no work has been done in the classification of multiple languages. Moreover, the majority of the work has been done on recognition of characters only and not on the script as a whole. The paper has been divided into two parts. In the first part, the papers written by different researchers have been reviewed, who have implemented Handwritten Indic Script Recognition of different Indian languages. In the second part, the data collection process, the preprocessing of the document images and the deep learning algorithms created to build the classification model have been analyzed.

2 Literature Review

Thulasi et al. [7] proposed a model that classified cursive handwritten Malayalam characters using an ANN with Gradient descent and then used an HMM to recognize these characters. Feature extraction was done using median filters and the feature vector used, served as the input to the ANN. The absolute accuracy of the proposed model was 93.4%; however, it focused only on Malayalam.

Dessai et al. [4] proposed a CNN-based OCR for recognition of Devanagari characters. Letter segmentation was performed by identifying the local minima of column-wise pixels. This was a significant challenge in this paper because of the header line characteristic of the Devanagari script. The final accuracy of the proposed model was 89.3% due to the different styles of writing of the same letter.

Saha et al. [10] proposed a model that performed the classification of handwritten Bangla characters and numerals by calibrating a deep convolution neural network that contained two techniques called Divide and Merge Mapping and Optimal PathFinder, and this architecture reduced the required number of parameters learned during training and functional complexity. The final accuracy of the model was 97.21%, but it was not tested with a larger dataset.

Siddique et al. [12] aimed to observe the variation of accuracies of CNN used to classify handwritten digits using various numbers of hidden layers and epochs. The experiment was done using the MNIST dataset. Every neuron in the convolution layer was connected to the local receptive field. The network was trained using stochastic gradient descent and the backpropagation algorithm. The maximum accuracy in the performance was found to be 99.21% for 15 epochs.

Sethi et al. [11] performed the classification of Odia characters by using transfer learning techniques such as ResNet50, VGG16 and compared these with the traditional CNN-based method. The accuracy was 96% for character classification using transfer learning and 70% using CNN. There was, however, limited development in handwritten character recognition due to the lack of adequate literature on the identification of Odia characters.

Maitra et al. [8] implemented a feature extraction strategy using CNN. The model was trained using samples from the basic Bangla characters database, and the feature extraction was done for five different languages, i.e., English, Devanagari, Bangla,

Telugu and Oriya. In this paper, the support vector machine was used as a classifier. The training of the model was terminated as soon as it achieved 85% accuracy in order to avoid the overtraining of the CNN. Since the model was slow and computationally expensive, only one CNN was trained for the basic Bangla characters. Furthermore, there was another drawback of the learning phase of CNN requiring an extensive database for efficient feature extraction.

Acharya et al. [1] proposed classification of Devanagari handwritten characters using a deep convolutional neural network and the use of dropout and dataset increment approach to improving test accuracy. The architecture scored an accuracy of 98.47% on the character dataset. However, the dataset did not include all the vowel characters or the derived cast of consonant characters and special characters in the language.

The paper [3] written by Dr. Sandhya Arora et al. focused on recognizing Devanagri characters, using regular expressions, to classify the distinct characters. As a result, the use of preprocessing procedures is reduced. The proposed paper does not require any training examples; it extracts the feature to recognize the character's shape as a regular expression. These letters are represented by 1s and 0s. Some characters in the script overlap each other; therefore, approaches like tree classifiers and percentage component matching are used to handle the problem. The model's highest accuracy is found to be 89%. However, the paper was only able to classify Devanagari consonant letters.

3 Data Collection

Our model works on character level recognition of scripts for which we needed a dataset of 12 languages, each language containing images of all the characters in the script. Such a consolidated dataset was not available for use so we aggregated various freely available datasets of individual languages from open-source websites, as seen in Fig. 1. For Devanagari [2], Malayalam [6], Manipuri [13], Telugu [9], Gujarati [5] and Urdu, we made use of Kaggle and Github. Due to the unavailability of datasets for languages such as Gurumukhi, we created our own by writing the characters of the language ourselves and also asking a few others to do the same, as seen in Fig. 2.

We also made use of Google's Teachable Machine to create a bigger dataset. This consolidated dataset of all 12 languages contained approximately 3000–4000 images of handwritten characters per language. This yields a complete dataset of around 40,000 images. Out of these, 35,000 were used for training of the model while the remaining were used to test the model. The images were also augmented to increase the dataset size further. All of these images were converted into grayscale and a size of 100×100 before passing them to the model.

Fig. 1 Data collection

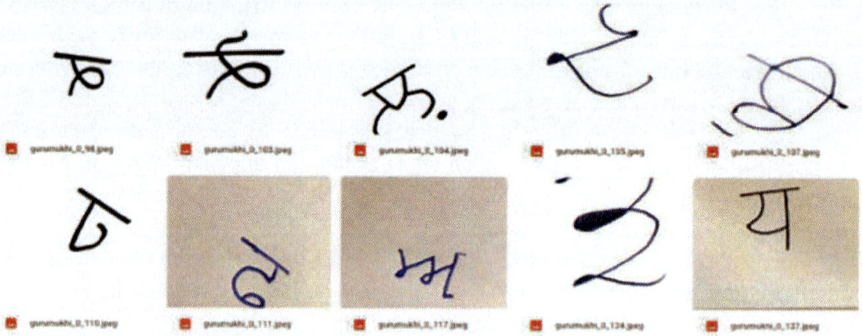

Fig. 2 Gurumukhi dataset

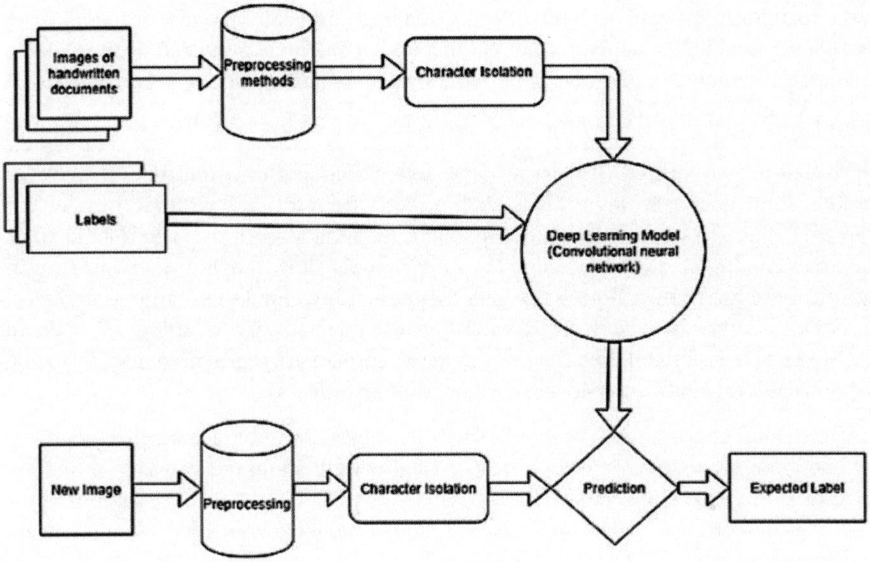

Fig. 3 System architecture

4 System Approach

Our system takes a handwritten document image as the input that is converted into grayscale, and further preprocessing techniques such as skew correction and brightness correction are applied. The preprocessed image is then passed to a deep learning model that performs identification of the script in which the document was written making use of a convolutional neural network. The goal of the recognition-based system is a good classification accuracy of the script language and fast processing of the result. Strategies like continuously improving the rate of accuracy of the document images have also been aimed for. The model makes use of multiple convolutional layers and pooling layers for the feature extraction followed by a dense layer for classification.

At the character level, for the task of handwritten Indian script recognition, a deep learning architecture is presented in Fig. 3. Then, the model is applied explicitly to the script classification issue for the pictures of handwritten documents written in one of the twelve selected scripts.

5 Algorithm Design

Deep learning is a subfield of machine learning concerned with algorithms inspired by the structure and function of the brain called artificial neural networks. In fields such as computer vision, machine vision, speech recognition, natural language processing, audio recognition, social network filtering, machine translation, bioinformatics, drug design, medical image analysis, content inspection and inspection and deep-learning architectures such as deep neural networks, deep belief networks, recurrent neural networks and convolutional neural networks have been implemented.

The word "deep" commonly refers to the number of layers in the neural network that are concealed. By using large sets of labeled data and neural network architectures, deep learning models learn directly from the data without the need for manual extraction of features. Other machine learning algorithms tend to plateau as they become larger; however, deep learning models show a continuous increase in performance as the neural network gets bigger and is trained with larger datasets.

For the characters to be isolated, we followed a three step methodology. We take in the image of a document then the lines in the document are segmented first, followed by word and then character-level segmentation as follows:

1. Horizontal contours are detected. The line is split along the least dense contour.
2. Each line is separately checked for vertical contours and the words in a line are split at the least dense contour.
3. For each word, vertical as well as horizontal contours are detected. Characters are separated by the spaces between them.

   a. For hollow characters, the boundary is checked.
      (a) If the boundary is closed, a split does not take place.
      (b) else, make the split along the least dense contour.

This makes use of OpenCV, a popular python library for image processing. The isolated characters go through the classification model and output is the script that most characters in the document belonged to.

The classification model is designed in the following manner:

1. Four convolutional layers have been implemented with the activation function as "ReLu," i.e., rectified linear unit.
2. Regularization has been implemented to prevent overfitting.
3. Four max pooling layers following each convolutional layer to reduce dimensions of the feature map.
4. Batch normalization has been added to accelerate the training as well as normalize the inputs.
5. In each layer, some nodes are skipped for better generalization. This is done via dropouts.
6. Dense layers are used to reduce the dimensions of feature vector obtained after flattening feature maps.
7. Activation function, Leaky ReLu, is used which solves the problem of "Dying ReLu." Dying ReLu is when the neurons become inactive and only give the output as 0 irrespective of the input.
8. The final output layer is a dense layer of 12 units representing each output class.
9. The model is then compiled using the Nadam optimizer with categorical cross_ entropy as its loss function and accuracy as its metric.

Table 1 Optimizer table

Optimizer	Accuracy(%)	Remarks
Adam	92.40	Adam combines the finest features of the AdaGrad and RMSProp methods to create an optimization technique for noisy issues with sparse gradients
Nadam	97.41	Nadam is Adam with Nesterov momentum
Adamax	89.25	Adamax is sometimes superior to adam, especially in models with embeddings

6 Results and Discussion

In neural networks, optimizers are the methods that are utilized to alter the parameters of the neural network such as weights and learning rate, to minimize losses. The optimizers used most often for image classification are Nadam, Adam and Adamax. We have observed our model with these optimizers and the highest accuracy is observed with the Nadam optimizer.

The accuracy rate of these optimizers are mentioned in Table 1.

Based on these results, we decided to go ahead with Nadam as our optimizer for training the model for image recognition of the scripts. This gave good results for languages such as Telugu and Odia while mismatching languages such as Urdu due to its similarity with some characters in English.

7 Conclusion

Owing to the diversity in Indic scripts, the classification of these has been a challenge. Most research has been conducted on isolated script identification. There is thus a need for a single software capable of identifying all 12 scripts and allow for the digitization of documents in Indian scripts. There are various algorithms that can be used for this, including feature extraction techniques, followed by machine learning classification algorithms. However, deep learning has produced the most reliable results in working with images. Our work focuses on using convolutional neural networks (CNN) to classify the Indic scripts, which are handwritten, at the character level. The classification model that has been implemented has an accuracy of 95.85%. It proved useful in the classification of all 12 scripts. However, the accuracy is found to be dependent on how the character isolation was done.

In the near future, an optical character recognition system can be developed and handwritten documents even in Indic scripts can be digitized.

References

1. Acharya, S., Pant, A.K., Gyawali, P.K.: Deep learning based large scale handwritten devanagari character recognition. In: 2015 9th International Conference on Software, Knowledge, Information Management and Applications (SKIMA), pp. 1–6. IEEE (2015)
2. Deore, S.P., Pravin, A.: Devanagari handwritten character recognition using fine-tuned deep convolutional neural network on trivial dataset. Sādhanā **45**(1), 243 (2020)
3. Deshpande, P.S., Malik, L., Arora, S.: Recognition of hand written Devnagari characters with percentage component regular expression matching and classification tree. In: TENCON 2007—2007 IEEE Region 10 Conference, pp. 1–4 (2007). https://doi.org/10.1109/TENCON.2007.4428775
4. Dessai, B., Patil, A.: A deep learning approach for optical character recognition of handwritten devanagari script. In: 2019 2nd International Conference on Intelligent Computing, Instrumentation and Control Technologies (ICICICT), vol. 1, pp. 1160–1165. https://doi.org/10.1109/ICICICT46008.2019.8993342
5. Gandhi., M.: Gujarati database: handwritten Gujarati database for numerals and characters. https://github.com/MikitaGandhi/Gujarati-Database- Version 1 (2020)
6. James, A.: Malayalam handwritten character dataset. https://www.kaggle.com/ajayjames/malayalam-handwritten-character-dataset/metadata Version 1 (2019)
7. Kishna, N.P.T., Francis, S.: Intelligent tool for Malayalam cursive handwritten character recognition using artificial neural network and hidden markov model. In: 2017 International Conference on Inventive Computing and Informatics (ICICI), pp. 595–598. https://doi.org/10.1109/ICICI.2017.8365201
8. Maitra, D.S., Bhattacharya, U., Parui, S.K.: Cnn based common approach to handwritten character recognition of multiple scripts. In: 2015 13th International Conference on Document Analysis and Recognition (ICDAR). pp. 1021–1025. IEEE (2015)
9. Rachapudi, M.: Telugu character recognition dataset. https://github.com/medini-r/teluguChar Version 2 (2020)
10. Saha, S., Puja, N.S.: A lightning fast approach to classify Bangla handwritten characters and numerals using newly structured deep neural network. Procedia Comput. Sci. **132**, 1760–1770 (2018). https://doi.org/10.1016/j.procs.2018.05.151
11. Sethi, R.K., Mohanty, K.K.: Optical Odia character classification using CNN and transfer learning: a deep learning approach (2008)
12. Siddique, F., Sakib, S., Siddique, M.A.B.: Recognition of handwritten digit using convolutional neural network in python with tensorflow and comparison of performance for various hidden layers. In: 2019 5th International Conference on Advances in Electrical Engineering (ICAEE), pp. 541–546 (2019). https://doi.org/10.1109/ICAEE48663.2019.8975496
13. Singh, P.: A benchmark dataset for Manipuri meetei-mayek handwritten character recognition (2019). https://doi.org/10.21227/fwax-yr43

Classification of Recyclable Waste Generated in Indian Households

Harshith MohanKumar, Mohammed Zayd Jamadar, S. Natarajan, and H. Nithin

Abstract Detecting and classifying recyclable material is one of the hardest computer vision problems. In this study, we present an optimized residual network with the inclusion of an SVM classification layer to categorize single object images containing recyclable waste. The model is trained on three thousand images of which $\frac{1}{3}$rd were of recyclable products commonly found in India. The results of the trained model have shown to contain competitive accuracy rates and reliably fast inference times.

Keywords Recycle · Waste · Detection · Classification · Machine learning · Artificial intelligence

1 Introduction

The production of plastic has seen a constant growth rate for well over 65 years. It has easily dominated every other manufactured material. Each year, billions of metric tons of plastic and various other recyclable material are accumulating across various landfills and aquatic ecosystems. A study conducted in 2017 showed that of the 2500 million metric tons (Mt) of plastic produced, only 9% have been recycled [1]. In the United States of America, the recycling rate has remained stagnant; however, the plastic market has seen a growth of 23% between the years 2008 and 2015 [2]. At this rate, by the year 2050, the amount of plastic waste that would fill up landfills would be around 12,000 Mt [1]. To prevent the world from piling up with products that will never decompose, we need to take immediate action. The two most popular methods of slowing down plastic waste accumulating in nature are to shift to an

H. MohanKumar (✉) · M. Z. Jamadar · S. Natarajan · H. Nithin
PES University, Bengaluru, Karnataka 560085, India
e-mail: harshithmohankumar@pesu.pes.edu

S. Natarajan
e-mail: natarajan@pes.edu

H. Nithin
e-mail: nithinha@pes.edu

© The Author(s), under exclusive license to Springer Nature Singapore Pte Ltd. 2022
A. K. Nagar et al. (eds.), *Intelligent Sustainable Systems*, Lecture Notes in Networks and Systems 333, https://doi.org/10.1007/978-981-16-6309-3_48

alternate material and increasing recycling rates. In this paper, we will improve on the latter by introducing a deep learning model that can improve the efficiency of recycling factories.

A significant reason why the recycling rates are so low is due to a lack of automation. Automation is not popular in recycling factories due to the lack of accuracy at which the robotic arms operate. Unlike humans, mechanical systems often fail when trying to distinguish objects which have been contaminated and deformed during the process of transportation to the factory. Generally, robots do not have an adaptive intuition on recyclable items. In order to improve these robotic systems, we have specifically included images of common Indian household items and utilize the power of deep neural networks to classify materials into five different categories—cardboard, glass, metal, paper, and plastic. During the training process, we augmented the images in several ways to mimic all sorts of color and shape deformations that may occur during the process of collecting these waste items. The results achieved have a competitive accuracy and inference time making applicable to implementation in autonomous recycling factories.

The past few years has shown fruitful development in the field of deep learning. Each year, a new architecture is proposed with extraordinary state of the art results. We decided to utilize the power of one of these architectures known as ResNet-50. Through ensemble learning, we propose to modify the architecture to suit our needs of classifying recyclable waste.

2 Related Work

Over the past decade, there has been a significant improvement in image classification and detection models. Image classification and detection were popularized from the creation of the convolutional neural network by Yann LeCun in 1989. Since then, many large-scale architectures have utilized the power of ConvNets. The ImageNet large-scale visual recognition challenge gave rise to new and outstanding architectures that have only gotten better over time. Architectures like ResNet, Inception, AlexNet, VGG, and YOLO have all shown to have exceptional performance in image detection and classification.

So far, there has not been much progress toward image classification projects that are particular to classifying recyclable materials. Two researchers at Stanford created a dataset and sparked interest in classifying images for recyclability status [3]. Their work was used later on by another group of researchers from the University of South Africa. They improved on the research by introducing ResNet-50 [4].

More board image-based classification models were created and tested on the Flicker Materials Database [5]. These models are more focused on general material classification and detection. Although the scope of their project is similar to ours, the dataset used is not suitable for our requirements.

3 Dataset and Data Collection

Common products are heavily dependent on the location and time period. Materials from 20 years ago may no longer be popular today. Also, materials from another country may not be seen in your country. Keeping this in mind, we decided to take an existing dataset and expand on it by including our images of our own to make the dataset more relevant for our needs.

To start off, we took the work collected from researchers at Stanford who created a dataset consisting of several hundred images of recyclable items for each category [3]. These images were taken on a white background and consisted of popular products found in the United States of America. They were grouped into five different categories: cardboard, glass, metal, paper, and plastic. We decided to expand on their work by introducing more images of our own.

First, we built a small contraption to make our images consistent in size and quality. We used a black background to mimic the background of a conveyor belt in a recycling factory. We collected around 1000 images of recyclable products commonly found in India. Some of these products range from masala boxes to Maggi packets. A few of the images can be observed in Fig. 1. When combined with the dataset found online, we had a total of 3104 images divided into five categories, namely cardboard, glass, metal, paper, and plastic.

Fig. 1 Images from dataset

In order to use recyclable materials collected from Indian households as our end inputs, we needed images from the same source. By including these images, we hope that our work would be more relevant to the location in which we stay.

Since deep neural networks require huge amounts of data to train, we decided to add augmentation during training to tackle the problem of small dataset. Data augmentation allows us to use the same images to make several more by applying several image processing techniques.

4 Architecture

After doing intensive research on various classification algorithms, we decided to start out experimentation using ResNet-50 and a multilayer SVM as its top layer. We use the ResNet-50 architecture as a basis for extracting high-level features and then feeding that to the SVM as input for classification.

4.1 ResNet-50

Research has shown that deep neural networks are more difficult to train due to the vanishing gradient problem. In order to work around this, we use residual networks as they are easier to optimize and can gain accuracy from considerably increased depth [6]. We used ResNet-50 as the feature extractor for our model. The primitive features are extracted with the help of the first few layers. As the training progresses through the layers, more complex and detailed features are extracted. The ResNet-50 was pretrained on the ImageNet dataset. The ImageNet dataset has over 15 million labeled high-resolution images belonging to roughly 22,000 categories.

4.2 SVM

After extracting the features from the ResNet-50, they are fed to a multi-class SVM. Support vector machines are incredibly popular and are considered to be one of the best classification algorithms to date. The data item is plotted as a particular point in n-dimensional space against the feature value of a specific coordinate. The items in SVM are classified based on the separation of hyperplane for each of the multidimensional data. It finds the hyperplane which the minimum distance is greater for the training data.

5 Experiments

For our experiments, we utilized the ResNet-50 model packaged from the Keras library with the pretrained weights from imageNet. We created a Python script to extract the features from the convolutional layers and use them for training the SVM classification model. To build the SVM model, we utilized the scikit-learn library. After setting up our environment to log each iteration of our code, we performed a sequence of various different combinations of tests. For our testing, we used an RTX 3080 optimized using TensorFlow gpu. Our testing was split into two different sections: 5.1 ResNet-50 Optimization and 5.2 SVM Optimization.

5.1 ResNet-50 Optimization

We started off by removing the top layers from the ResNet-50 model and replacing them with a dense layer followed by an activation layer. We decided to switch this up after getting poor results. We then used a combination of a flatten layer and a dense layer. We noticed that this had some incremental improvement; however, it was still under 90% during training time. We then decided to add a pooling layer and more dense layers. This happened to do the trick and we were receiving results of over 95% accuracy. But when taking a look at the validation accuracy, it was clear that the model was overfitting to the data. To counter this, we added dropout layers with the rate parameter set to 0.5 within the convolutional block of the ResNet-50 model.

5.2 SVM Optimization

Initially, when training the SVM layer, we were looking at an average accuracy around 30%. We made adjustments to the way which the features were extracted from the trained ResNet-50 model. We decided to extract the features from earlier convolutional layers to minimize the error in accuracy. After deciding on the right layers, we finalized the model and moved on to implement a grid search algorithm to fine-tune the parameters for the SVM model.

6 Result

While training the ResNet-50, we noticed that the validation accuracy was an average of 30%. The SVM classification model outperformed the CNN layer by over 30%. The accuracy results from our SVM model had a high variance with the low being 60% and the high being 75%. This high difference in accuracy between CNN can

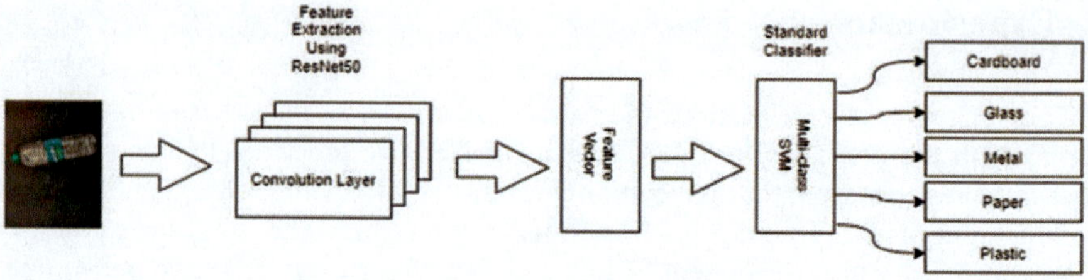

Fig. 2 Training pipeline for custom model

Fig. 3 Confusion matrix from the SVM run

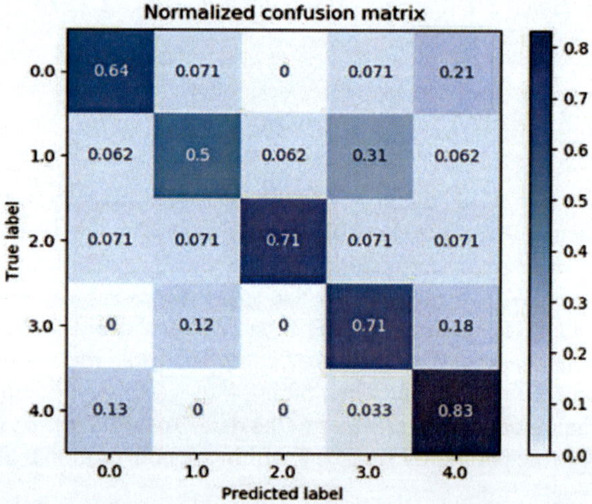

be explained due to the simplicity of the SVM algorithm as compared to CNN. We noticed that the class which had the lowest accuracy rate was glass. On average, this glass would receive an accuracy score of around 50%. We assume that this is due to the translucent property found in many glass materials which are making pattern-finding from the features extracted difficult to differentiate from the background and the glass. The material which performed the highest was plastics. On average, the accuracy of plastics was upwards of 80%. Unlike glass, we had combinations of opaque and semi-translucent plastics, which we believe played a significant role in helping the model learn. We can observe the confusion matrix for the SVM model in Fig. 2. The dark blue spots across the diagonal of the matrix indicate the correct prediction accuracies, whereas the other squares in the matrix represent the confusion.

While training the modified ResNet-50 model, we observed a final accuracy of 94% with a loss of 0.2. This was a perfect run. The results from this run are shown through the tensorboard graphs in Figs. 3 and 4. Therefore, for our final result of our model, we conclude that it has an average accuracy of 68% (Fig. 5).

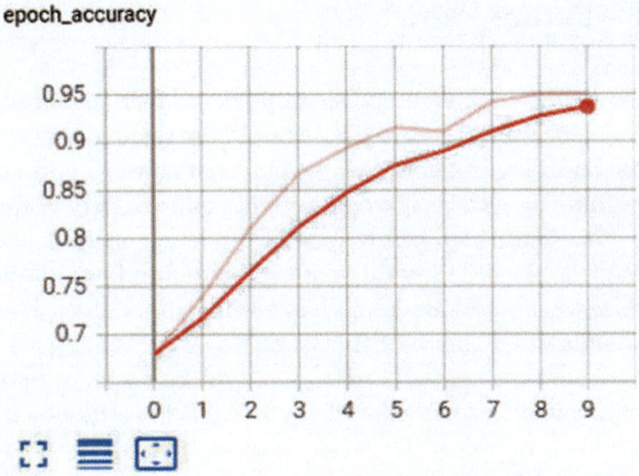

Fig. 4 Epochs versus accuracy for ResNet-50

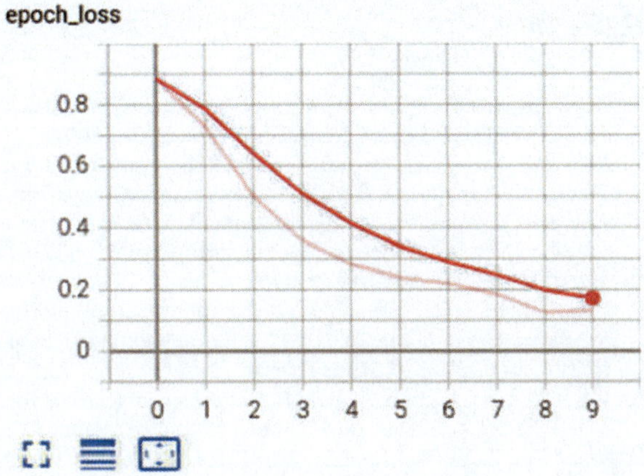

Fig. 5 Epochs versus loss for ResNet-50

7 Conclusion

The experiments performed in this study have shown that the power of current deep learning networks can be utilized for the classification of trash into various recycling categories. These models can predict thousands of pictures in a matter of seconds which provides a viable option as an efficient alternative to humans. With little more tweaking of the model to achieve higher accuracy, this model can be implemented in autonomous robotic arms to sort out the trash.

A significant challenge to combat in the future will be procuring a dataset of reasonable size to train larger neural networks. Inevitably, the trade-off between accuracy and speed of the model would become more sensitive. These obstacles must be systematically dealt by analyzing the catalog of multiple state-of-the-art neural network architectures or coming improving on the our proposed custom architecture [7].

8 Future Work

Going forward, we hope to implement other advanced algorithms like xgboost and test what the difference would be as compared to a more traditional SVM. To increase the accuracy of the SVM, we believe that having more images would also help the training process as it would provide more variety in the materials.

We would also like to move from image classification to image detection. Using models like YOLOv4 to achieve bounding boxes around multiple images [8]. For this, we would have to collect an incredibly large dataset. We would utilize crowd sourcing to tackle such an issue.

Another important addition could be including specific algorithms that detect materialistic properties to help increase the accuracy of sorting materials.

References

1. Geyer, R., Jambeck, J., Law, K.: Production, use, and fate of all plastics ever made. Sci. Adv. **3**, e1700782 (2017). https://doi.org/10.1126/sciadv.1700782
2. BIO Intelligence Service (2013) Final Report prepared for the European Commission, D
3. Thung, G., Yang, M.: Classification of Trash for Recyclability Status (2016)
4. Adedeji, O., Wang, Z.: Intelligent waste classification system using deep learning convolutional neural network. In: Procedia Manufacturing, vol. 35, 607–612. Elsevier B.V. (2019). https://doi.org/10.1016/j.promfg.2019.05.086.
5. Liu, C., Sharan, L., Adelson, E.H., Rosenholtz, R.: Exploring features in a Bayesian framework for material recognition. In: 2010 IEEE Conference on Computer Vision and Pattern Recognition (CVPR), pp. 239–246. IEEE (2010)
6. He, K., Zhang, X., Ren, S., Sun, J.: Deep residual learning for image recognition. In: Proceedings of the IEEE Conference on Computer Vision and Pattern Recognition (CVPR) (2016)
7. Wang, Y., Zhao, W.J., Xu, J., Hong, R.: Recyclable Waste Identification Using CNN Image Recognition and Gaussian Clustering. CoRR abs/2011.01353 (2020)
8. Aishwarya, A., et al.: A waste management technique to detect and separate non-biodegradable waste using machine learning and YOLO algorithm. In: 2021 11th International Conference on Cloud Computing, Data Science and Engineering (Confluence) (2021). https://doi.org/10.1109/confluence51648.2021.9377163

A Hybrid Model on Deep Learning for the Diagnosis of Diabetic Retinopathy Using Image Cropping

A. Aruna Kumari and Santosh Kumar Henge

Abstract The Diabetic Retinopathy is one of the eye condition that may cause vision loss, blindness to the eyes quickly if not treated at the earliest in the patients having Diabetes Milletus type-1 and Type-2. This can be treated when diagnosed at the early stages of the disease with screen monitoring and treatment like laser etc. If not diagnosed in early stages can cause the impaired vision and may also lead to blindness. To avoid blindness, vision loss, the "Early Treatment Diabetic Retinopathy Study (ETDRS)" helps to detect eye disease at the earliest and prevent the consequences of the disease. As a progressive disease, The Diabetic retinopathy divided into two stages one is "non proliferative diabetic retinopathy" (NPDR) and "proliferative diabetic retinopathy" (PDR). Non-proliferative diabetic retinopathy has four stages, those are normal, mild, moderate, severe which has the threshold as 0, 1, 2, 3 respectively for No Diabetic Retinopathy, mild, moderate, severe, but if the threshold value is 4 it is a proliferative diabetic retinopathy is last stage of the DR. Unlike the available methods, the proposed work aims at diagnosis of all the stages of Diabetic Retinopathy where the retinal image preprocessing followed by the analysis using Convolutional Neural Network Classifier which classifies into different stages of the disease based on the threshold. RESTNET is used for the image tuning and the Hyper parameter tuning is done for detailed analysis. The Kaggle public dataset is used to build the model and would give the higher performance when compared to the existing ones.

Keywords Diabetic retinopathy · Eye disease · Convolutional neural networks · Deep learning · Medical retinal images · Fund images

A. Aruna Kumari (✉) · S. K. Henge
Department of Computer Applications and Computer Science and engineering, Lovely Professional University, Jalandhar, Punjab, India
e-mail: santosh.24372@lpu.co.in

© The Author(s), under exclusive license to Springer Nature Singapore Pte Ltd. 2022
A. K. Nagar et al. (eds.), *Intelligent Sustainable Systems*, Lecture Notes in Networks and Systems 333, https://doi.org/10.1007/978-981-16-6309-3_49

1 Introduction

1.1 Diabetic Retinopathy Treatment

It can be treated with injections where the medicines like anti-VEGF drugs and corticosteroids that slows down the damage, using Laser treatment where the shrinkage of blood vessels is done by stopping the leakage using eye surgery called vitrectomy. When diabetic retinopathy is diagnosed, shrinkage of adaptable danger issues like age, blood glucose, blood pressure can slow down the disease evolution. When sight-threatening retinopathy is diagnosed, laser supervision and dealing with vascular endothelial evolution issue inhibitors decreases the risk of vision loss. The DR (Diabetic Retinopathy) has binary major Stages: The Non-Proliferative Diabetic Retinopathy (NPDR) and Proliferative Diabetic Retinopathy (PDR) [1]. The DR in the primary stages is called NPDR, NPDR divided into No Diabetic Retinopathy, Mild, Moderate, and Severe stages. A minor circular red dot at the end of blood vessels, Where the mild stage has one micro-aneurysm (MA). Large-scale global screening programs have been implemented by hospitals and local establishments [2, 3] with excessive accomplishment, to provide primary disease diagnosis and appropriate eye care. In order to attain eye color images with a retinal fundus camera, In the condition of such programs, patients are called to clinical settings. Some images are very dark and difficult to visualize, Images arise with many different lighting conditions. The image to gray scale we can try to convert for better visualize. These images are then submitted to experts, perform diagnosis based on them and who look for visual signs of the occurrence of scratches. Deep learning is the important machine learning paradigm in computer vision and image processing domains, particularly as admirations convolutional neural networks (CNNs) [4]. In such conditions, experiences of doctors and examination accuracy greatly relies on the abilities. The association of retinal or other medical images, Considering the patients suffering

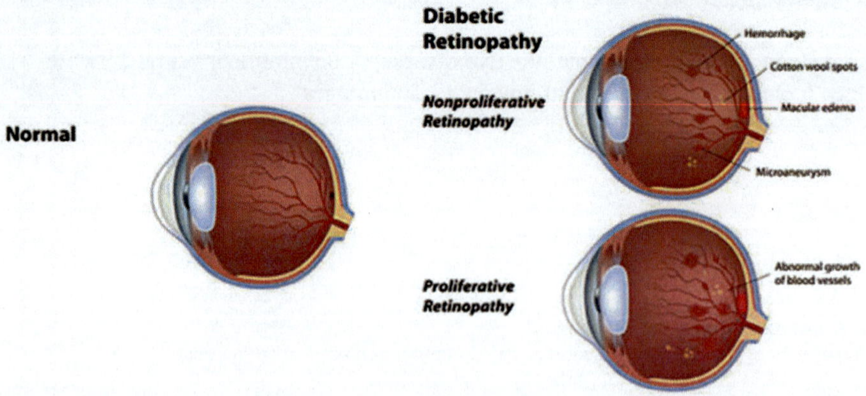

Fig. 1 Differences between the types of diabetic retinopathy

from DR are a very huge population, it is meaningful to explore the automatic classification of retinopathy images for the diagnosis of DR. Figure 1 shows the differences between the types of diabetic retinopathy.

1.2 Label Class Distribution

The dataset is used to train and test predictions using deep learning models [5]. The 'Asia Pacific Tele Ophthalmology Society'-APTOS. The dataset have size 3662 color fundus images Fig. 2. The images Threshold value from 0 to 4, After Threshold classification dataset is imbalanced. Figure 2 shows the class distribution is not uniform. Threshold value Legend: 0 represents No Diabetic Retinopathy 1 represents Mild 2 represents moderate 3 represents Severe 4 represents Proliferative Diabetic Retinopathy Fig. 2. The data set instances of all the predicts of binary classification occurrences are either positive or negative representation of prediction outcomes are true positive, true negative, false positive, false negative. The prediction outcomes represent the True positive [TP] represents the correct positive prediction. The True negative [TN] represents the correct negative prediction. False positive [FP] represents incorrect positive prediction. The False negative [FN] incorrect negative prediction [6]. Accuracy. Accuracy (ACC) equation (1, 2, 3) is calculated as the number of all correct predictions divided by the total number of the data set. The perfect accuracy is 1.0, whereas the worst accuracy is 0.0

$$\text{accuracy} = (\text{TN} + \text{TP})/(\text{TP} + \text{FP} + \text{TN} + \text{FN}) \tag{1}$$

$$\text{Precision} = \text{TP}/(\text{TP} + \text{FP}) \tag{2}$$

$$\text{Recall} = \text{TP}/(\text{TP} + \text{FN}) \tag{3}$$

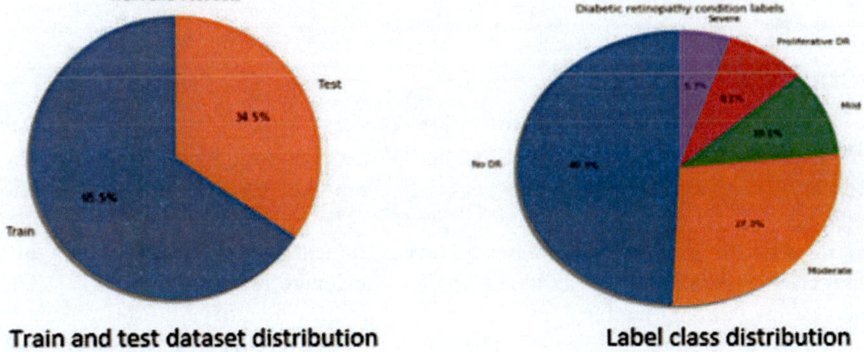

Fig. 2 Class label distribution

2 Related Work

Qummar et al. [1], is proposed method is ensemble method means weak entity to make a stronger entity, is a combination of different algorithms and it gives better accuracy but can training the model in different methods is the draw of ensemble technique. After applying all the methods, it got accuracy 90%.

De et al. [7], the proposed method involves input image is consider as neuron, every neuron called as object. And apply these two stages for every object one is segment and second one is tracing and separating the filamentary. Here every time the graph can perform joint and disjoin the neuron, it is very difficult. The disadvantage of this model is separating.

AL-saedi et al. [2], the proposed technique is the image segmentation, after segmentation it can use the optic disc, this proposed method applies the datamining techniques using random forest it gives the low accuracy value. The disadvantage of is using old datamining techniques.

Yabe et al. [3], the method proposes the dietary instructions, but the disadvantage is the age eligibility (40). If age <40 is not mentioning the dietary instructions.

Patil et al. [8], The proposed algorithm is segmentation is texture based, the drawback of this method hemorrhages is variable size, the author only calculates the F1-score, drawback is image classification.

Kamil et al. [5], proposed method is RGB scale images, the disadvantage of this method is optic disc.

Pal et al. [9], the method is mainly focused on biometric data, it stores the personal identification, the disadvantage is digital image into multiple segments.

Gambhir et al. [10], the author mainly focuses on segmentation extraction, the draw of this proposed method is only considering the minimum instances of blood vessels.

Kamil et al. [4], the proposed method works on two stage process, but on for text classification and second method is image classification, the disadvantage is only the image gives normal, abnormal information. The proposed only type 2 diabetic, the draw of method is lack of characteristic of type2 DR.

3 Existing Method

RESENT 50: Resnet50 stands for Residual Network that is 50 layers. Refers the Resnet is quick name for Residual Network that supports Residual Learning. The 50 indicates the quantity of layers that are present. The ResNet-50 is a convolutional neural network that contains 50 layers deeper. The rest net have syntax as weights, ImageNet . Initially the Resents applied for the image recognition task [2]. The Deep Convolutional neural networks are very good at identifying low, mid ,high level features from the images [6]. The training of deeper neural networks are very difficult. Now, upload a pretrained type of the network trained almost a million images from the ImageNet database.

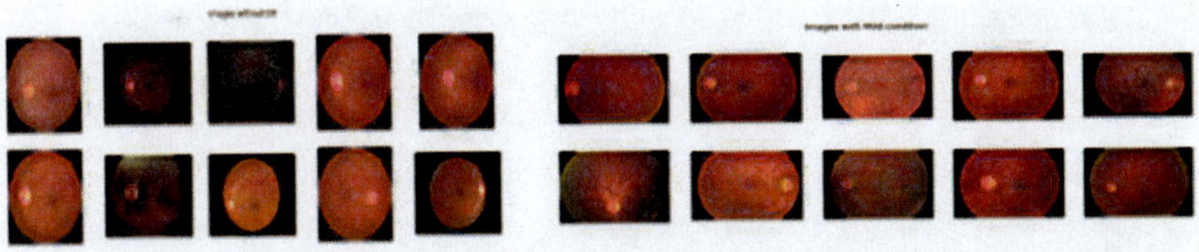

Fig. 3 The transfer learning on label images with no DR, mild

Input Layer: The input layer is start of the road map process for the artificial neural network. The network of Layers is a input layer of a neural network is collected of a artificial input neurons, The input layer takes the initial layer information to the structure for the next processing by sub sequent layers of artificial neural network (1).

Convolution padding: Padding is very simple process of adding of zeros to input images. To avoid this problem [2], apply the shrinking as, if p is the number of layers of zero added to the border of the image. Pad-dings are two types they are valid padding, same padding. Valid padding is implying no padding to all.

Activation Function: The activation function is a serious part of the design the neural network, when choice of the activation function in the hidden layer controller how the network model learns the training of the data set [9]. The option of activation function in the output layer can define the type of predictions of the model. Layers refers the Activation function plays an important role in neural network design, then afterwards the modern defaulting activation function for hidden layers are the ReLU function [11], When the activation function for output layer depends on the classes of prediction problem.

Max pooling The max pooling is also named as sample-based discretization process. The main aim for max pooling is downcast sample of participation of image , hidden layer, output matrix, dropping its dimensionality [4]. Basically, the max pooling is used to reduce the image size because larger number of pixels are contributing or participate to more parameters it can involve the large chunks of the data in Layers. The Features of max pooling adds a bit of slight, shift variance, rotational invariance, scale invariance.

Batch normalization: The batch normalization technique is used to make artificial networks faster, extra unchanging complete normalization of the layers input by recentering and rescaling [6]. The deep neural networks are stimulating to train and the transfer learning Fig. 3 shows the transfer learning on label images with no DR, Mild.

4 Proposed Method

The proposed algorithm is circle cropping image Fig. 4, the image gives the better quality because the proposed method can be spotting the all the corners of the image i.e., Fig. 4 Left, right, top, bottom, corners, edges. The preprocessing of the Ben

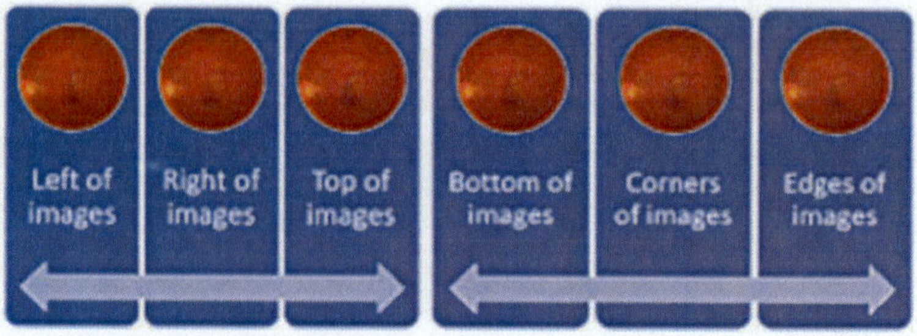

Fig. 4 Circle cropping algorithm architecture

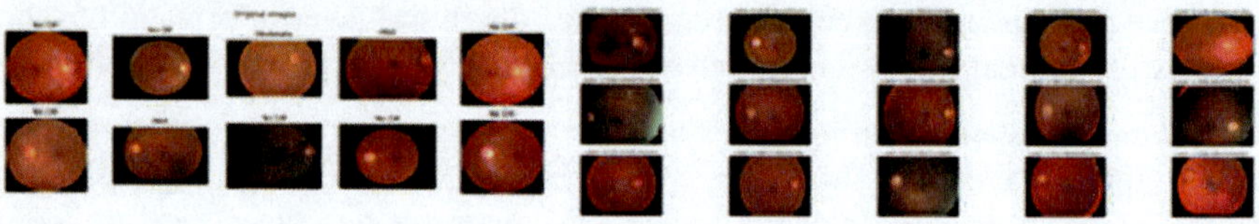

Fig. 5 Actual dataset images

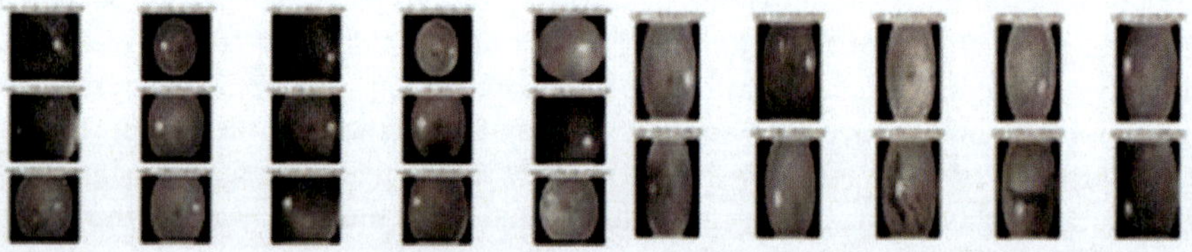

Fig. 6 The original images convert into gray scale images

Graham's approach is to adopt the input fundus. Ben can evaluate both image and scaling and add the circular crop to input fundus image.Rescale the images, If using rescale on images they have the same radius, that is 500 pixels or 700 pixels, Almost 50% grey is mapped to the local average. The average of local color is subtracted; When you are apply clipping the size of image 90% to eliminate the boundary effects.

Image cropping: Actual image of the eye in the data set have a black section image around, so black section impacts the output of the method it consists no data, then cut black portion out of picture Fig. 5. Resizing: The dataset image may vary in size, so we can use radius of image exactly equal to 500 pixels to render image images of the size Fig. 6. Let we have look at the actual inputs first. In the Fig. 6 Every row can display the severity level. Here we have two problems which it makes the difficult to spot on the severity. For the First some images are very dark. Examples in fig 0th row 2nd image and 4th row 4th image, Different colors confusing sometimes. Example 3rd row 3rd image. And Second, we can get dark areas for some pictures in uninformative. Example in the fig 0th row 1st image and 3rd image. When we reduce the size of picture this is important, as informative areas are too small. So, to crop the uninformative areas in the case second.

```
Ben Graham's Algorithm
To generate image aspect ration (path,old_res,new_res = None):
pass
(old_x -X)  / (old_y- y)   = new_x  / new_y
Old_x * old_y = resolution of the actual image
New_x *new_y = complete resolution
X = the group  of pixels or points that should be cropped off
the y axis to achieve the complete resolution
Y = the group of pixels or points that should be cropped off
the y axis to achieve the complete resolution
```

Fig. 7 screen shot refers the Ben Graham's pre processing algorithm

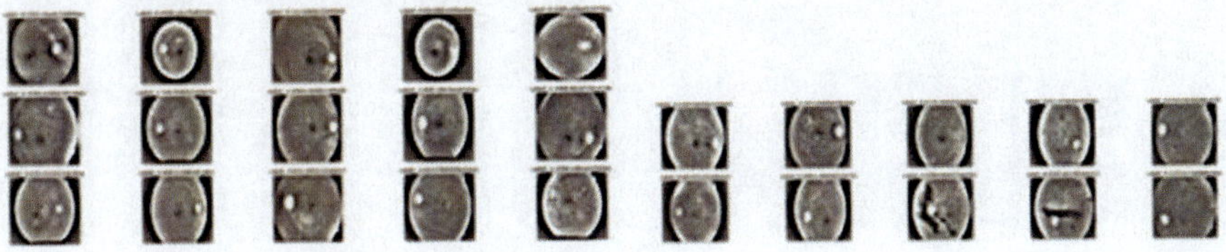

Fig. 8 The images as Ben Graham's images

Step 1: Try to apply gray scale and feel to better understanding for some images, as a color distraction is not there. Figures 7 and 8 example 4th row 4th image we can see more blood clearer in the upper part, the image has severity level 4, Here two examples are difficult to spot on 4th row 1st image and 4th row 4th image. As can we try Increasing the image size after you can observe actual image size, then you can find some abnormal conditions are cotton wool spot, hard exudates in eyes clearer are observe the lower right of the eye part. The image dimension, size of the value is important.

Ben Graham's Algorithm:
The Ben's further improve by auto cropping, an important update is color version of cropping step (1). Before Ben's algorithm we try gray scale for better image quality but ben's color version cropping gives the better image quality. In Figs. 9 and 10 show the quality comparison between these two methods, we can use ben's color version cropping because it gives quality image.

Gaussian Blur Algorithm:
Step 2: Gaussian Blur Algorithm Step (2) is calculate every point can multiplies its weight value, after add the nine values up, then get the Gaussian Blur value of the point center and repeat the process for all the points of images , then we will get graph after Gaussian blur. Figures 11 and 12 shows that image is cropped of the version color apply the argument of sigmax = 50 of GaussianBlur,when Ben actually use sigmax = 20 which may have better performance.This sigmax = 20 or sigmax = 50 make beautiful, sometimes bloody yellow moon images.

```
Gussian Blur Algorithm
step 1:The blur can be understood as taking a pixel as the average value of its surrounding pixels
step 2:Weight of normal distribution
step 3:To calculate the Gaussian function
step 4: To evaluate Weight matrix
step 5:To Calculate Gaussian Blur,grayscale value(0-255)
step 6:The process for the points at borders
```

Fig. 9 Screen shot refers the Gaussian Blur algorithm

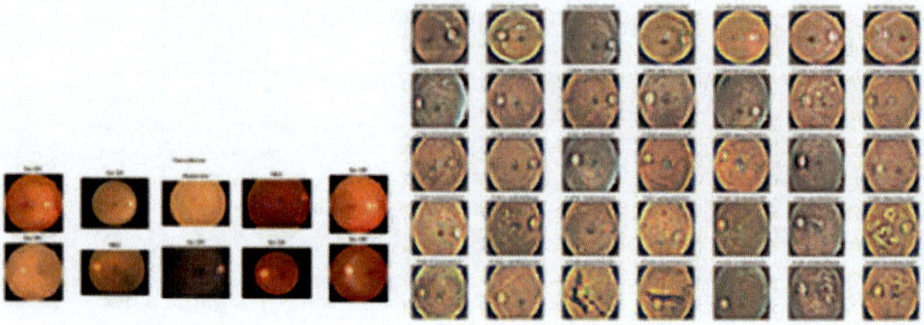

Fig. 10 Actual images with Gaussian noise, the Gaussian blur images

```
Circle Cropping Algorithm
When apply the image autocrop<-image, threshold=0:

"""Crops any edges below or equal to threshold
Crops blank image to 1x1.
Returns cropped image.
"""
if len<-image.shape == 3:
flatImage = np.max<-image, 2
else:
flatImage = image
assert len<-flatImage.shape == 2
rows = np.where<-np.max(flatImage, 0) <- threshold)[0]
if rows.size:
cols = np.where<-np.max(flatImage, 1) <- threshold)[0]
image = image[cols[0]: cols[-1] + 1, rows[0]: rows[-1] + 1]
else:
image = image[:1, :1]
end if
end if
return image
```

Fig. 11 Screen shot refers the circle cropping algorithm

Circle Cropping Algorithm:

Step 3:

Circle Cropping Algorithm: An Algorithm can be converting the image has some resolution: The old resolution of image is

$$L * M \tag{4}$$

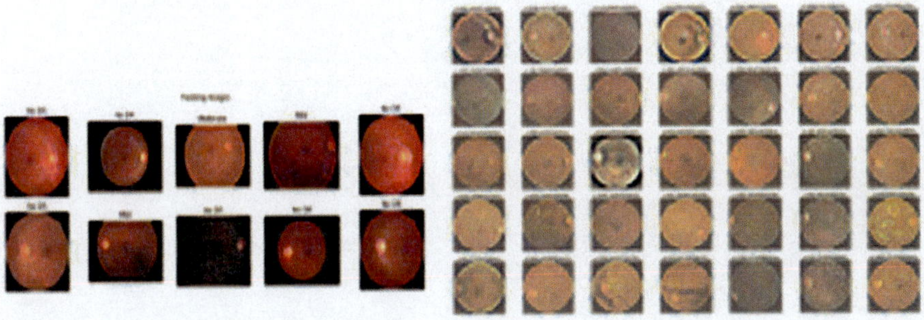

Fig. 12 The padding images, circle cropping images

Let the image resolution mainly focus on cropping of the image or resize of the image this resolution is old. Now, apply the new resolution of image is

$$P * R \tag{5}$$

The step (3) refers the new resolution of image covers the all the corners of the image or maximum number of points or pixels from the actual image. It can apply a generic way to all the images in the website. The resolution or optimization of the image is re-scaling or reduce the size of the actual image's applications are like Facebook profile image. Figure 12 shows the new resolution of the crop image. To take actual image it covers the maximum size of image, then cropping of the image follows the ratio of the pixel from the actual image. Then apply the resolution to the image to get full ratio. This scenario is very close to the actual image with new resolution (Fig. 13).

5 Result Analysis

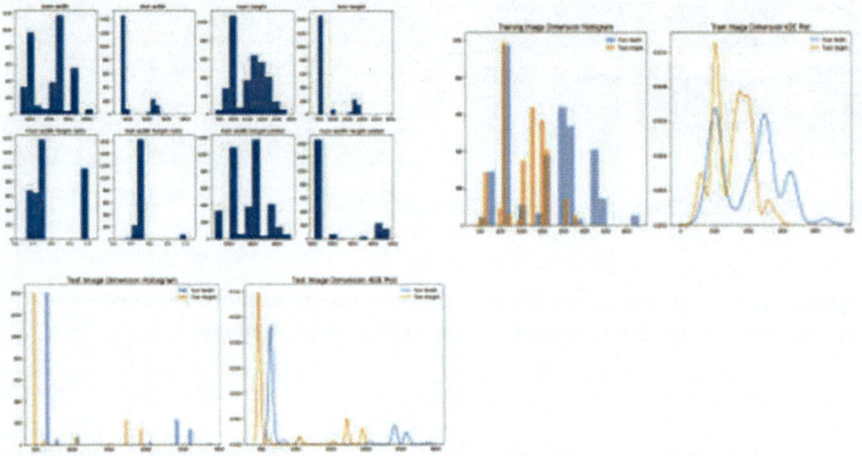

Fig. 13 Histogram refers the train-test data set image's width and height, result analysis on train images, test images

6 Conclusion and Future Work

The image quality is here three step process they are, (i) Ben Graham's (ii) blur image (iii) cropping for perfect spot. The Ben's further improve by auto cropping, an important update is color version of cropping and Ben's Algorithm Then after apply Ben model gives the excellent quality in gray scale images and Gaussian blur algorithm can calculate every point multiplies its weight value, finally color image classification and for good resolution. After that apply the Circle Cropping algorithm. This cropping algorithm gives the better quality on blood vessels and cotton wool spot and it provides excellent vision quality for the image. And it provides all the angles of the images like left-side, right-side ,top-side, bottom-side, edges of the image and corner of the image, Where the wool spot it clearly mentions in the form of the circle cropping. The circle cropping can easily recognize the which stage in Diabetic Retinopathy images. In the future, the application of classification models improves the proposed algorithm.

References

1. Qummar, S., Khan, F.G., Shah, S., Khan, A., Shamshirband, S., Rehman, Z.U., Khan, I.A., Jadoon, W.: A deep learning ensemble approach for diabetic retinopathy detection. IEEE Access. **15**(7), 150530–9 (2019)
2. AL-Saedi, K. H., Jelinek, H. F. : A data mining framework for the classification of retinopathy images based on a new multistage prediction algorithm. Int. J. Eng. Technol. **7**(4), 4201–4206 (2018)
3. Yabe, D., Kuwata, H., Fujiwara, Y., Sakaguchi, M., Moyama, S., Makabe, N., ...and Seino, Y.: Dietary instructions focusing on meal-sequence and nutritional balance for prediabetes subjects: an exploratory, cluster-randomized, prospective, open-label, clinical trial. J. Diabetes Complications **33**(12), 107450 (2019)
4. Kamil, R., Al-Saedi, K., Al-Azawi, R.: An accurate system to measure the diabetic retinopathy using SVM classifier. Ciência e Técnica Vitivinícola **33**, 135–139 (2018)
5. Kamil, R.A., Al-Saedi, K.H., Al-Azawi, R.J.: A novel approach for optic disc detection in RGB retinal fundus images. Int. J. Sci. Res. **6**(8), 1263–1268 (2017)
6. Qiao, L., Zhu, Y., Zhou, H.: Diabetic retinopathy detection using prognosis of microaneurysm and early diagnosis system for non-proliferative diabetic retinopathy based on deep learning algorithms. IEEE Access **8**, 104292–104302 (2020)
7. De, J., Cheng, L., Zhang, X., Lin, F., Li, H., Ong, K.H., Yu, W., Yu, Y., Ahmed, S.: A graph-theoretical approach for tracing filamentary structures in neuronal and retinal images. IEEE Trans. Med. Imaging **35**(1), 257–72 (2015)
8. Patil, V.B., Rai, M.V., Barve, A.: Blood vessel segmentation from color retinal images using unsupervised texture classification. Int. J. Sci. Res. (IJSR) **4**(1), 2319–7064 (2016)
9. Pal, M., Dey, B.: A Study on Retina: A Unique Identification Using Stereoscopic Retinal Image
10. Gambhir, A.: Advanced practices on detection and classification of diabetic retinopathy from fundus images. Int. J. Innov. Res. Comput. Sci. Technol. (IJIRCST) **8**(3) (2020)
11. Zong, Y., Chen, J., Yang, L., Tao, S., Aoma, C., Zhao, J., Wang, S.: U-net based method for automatic hard exudates segmentation in fundus images using inception module and residual connection. IEEE Access **8**, 167225–167235 (2020)

Automation of Translating Unit-Level Verification Scenarios for Test Vector Generation of SoC

Rahul Anilkumar, B. K. S. V. L. Varaprasad, and K. Padmapriya

Abstract System-on-chips (SoCs) incorporate a multitude of analog and digital circuit blocks with complex functionalities into a single chip. Verification of SoC designs is a humongous task executed by a large team over the course of a few months' time frame. Prior to integration of each block to the SoC fabric, which is usually controlled by a processor, the block needs to be thoroughly verified for its functionality. Toward this, methodologies like Universal Verification Methodology (UVM) have enabled verification engineers to verify such complex designs with relative ease. While detailed verification of the individual modules is possible at the unit level, verification of all functionalities at the top level still remains a challenge. The limited number of tests run at SoC level is dwarfed by the large number of tests performed at unit levels. This is primarily due to the processor being driven by a software stimulus, while the unit tests with constrained-random verification are written as System Verilog UVM sequences. The real deficit comes when testing the manufactured device against functional specifications due to lack of sufficient tests for individual blocks. Achieving complete functional testing of SoC device is impossible without covering the verification space exercised by the stand-alone tests at unit level. In this paper, we propose a method to utilize the stand-alone functional test scenarios verified at unit level, by converting these scenarios into top-level functional tests that can be used to derive functional test vectors for testing the SoC device. Including these unit-level test scenarios into the top-level functional vectors increases the confidence on the device for its total functionality.

Keywords System-on-chip (SoC) · UVM · Stand-alone verification · Verification scenarios · Test vectors · Device testing

R. Anilkumar (✉) · B. K. S. V. L. Varaprasad · K. Padmapriya
U R Rao Satellite Centre, Indian Space Research Organization, Bangalore, India
e-mail: rahulak@ursc.gov.in

© The Author(s), under exclusive license to Springer Nature Singapore Pte Ltd. 2022

A. K. Nagar et al. (eds.), *Intelligent Sustainable Systems*, Lecture Notes in Networks and Systems 333, https://doi.org/10.1007/978-981-16-6309-3_50

1 Introduction

The complexity of SoC designs has grown exponentially over the last decade to keep up with the increasing demand placed on the computing power of portable platforms. Incorporating everything from accelerometers to digital signal processors, todays SoCs are expected to handle billions of operations per second. The cost and time taken for developing an SoC are huge, and hence, it is necessary to ensure the first-time success of fabricating such a device by ensuring it meets its total functionality.

Verification of an SoC design is a mammoth task given the complexity and communication between the various components, and it is done with a divide-and-conquer approach [1]. Multiple engineers with different views of the design need to work in parallel to contribute to the overall verification goals. Toward this, the verification is tackled at multiple levels, split on the basis of hierarchy. Subsystem or IP-level verification treats each individual block as a stand-alone unit and verifies it thoroughly. Verification engineers spend a lot of time to verify individual units in the SoC design at RTL level using techniques like UVM to identify functional bugs. Constrained random verification along with UVM enables the thorough verification of individual blocks in a stand-alone environment. Top-level verification is focused more on the interconnection and interaction of the various subsystems. Given that top-level verification eventually includes the subsystems whose stand-alone verification has been completed, there needs to be a way in which the verification effort put toward stand-alone verification become useful during the top-level verification time. While the UVM-based monitors and checkers used for unit-level testing can still be used at the top-level verification, they miss out on the wide variety of the stimulus generated by the drivers in such stand-alone environments. This is because the top-level test cases are usually driven by software running on a processor, as opposed to the UVM agents in a stand-alone test case.

Studies in [2–4] detail the verification of SoC at the top level, but the extent of verification is limited to overall functionality and not on the detailed functionality of each block. Individual blocks are verified in detail in a stand-alone environment [5–14], but focus is not given on reusing the unit-level test cases at top level. To ensure correctness of the manufactured device functionality, there is a need to utilize the verification space covered by the stand-alone test cases. Formal methods as explored in [15–18] are static methods suitable for design verification and cannot be used to derive test vectors for device testing. Hence currently, there are no techniques with which the large number of verification scenarios at unit level can be automatically taken to a top-level testbench to generate automatic test equipment (ATE) vectors for device testing.

Considering a generic SoC with a processor and associated peripherals, the verification is carried out at top level in a manner where the processor is driven by software stimulus, while stand-alone verification of the internal modules is done using System Verilog UVM sequences. Since there is no technique to use the stand-alone scenarios to validate the device by porting to a top-level test scenario, due to the variation in verification stimulus, it leads to limited functional verification of the overall SoC.

In this paper, we propose a method to utilize stand-alone test cases to validate the manufactured device, in an automated manner with minimal effort, by converting unit-level tests written as SV UVM sequences to top-level software tests, from which functional test vectors can be derived.

The motivation behind developing the proposed system is the limited functional testing of the manufactured device possible with Automatic Test Pattern Generation (ATPG) and few top-level functional test vectors. In advanced technology nodes, there is a higher probability for faults during manufacturing of device [19], and only thorough testing of functionality along with ATPG vectors after fabrication can bring out these issues. Therefore, by providing an automated method to utilize the stand-alone functional verification scenarios for functional testing of the device reduces the burden on the verification engineers to manually add more functional tests, as well as improve the confidence on the device when using it for high-reliability applications like aerospace.

The rest of this paper is organized as follows: Sect. 2 gives a background of the current UVM-based SoC verification methodology. Section 3 details the proposed technique by taking I2C master protocol controller integrated with a SPARC-v8-based processor as a case study. Section 4 presents the strengths and weakness of the proposed technique for reusing the stand-alone test cases at top level of SoC. Conclusions are drawn in Sect. 5.

2 Background of UVM-Based SoC Verification

Current SoC verification methodologies rely on techniques like UVM to enable quick and scalable verification. UVM is a library of System Verilog (SV) base classes that dictate a clear structure and separation of work in the verification architecture. By providing a standard and scalable framework that can suit highly complex designs, that separates testbench structure from testbench stimulus [20], it opens up opportunities for using SV language features like constrained random verification to enable faster functional verification and coverage closure. Figure 1 outlines the structure of a general UVM testbench [20], showing the different layers and components typically used in a UVM-based verification environment.

UVM is widely used in the verification of SoCs. In [2], Jiayi Wang et al. detail the verification of a RISC-V SoC using UVM. Verification is partitioned as module level and system level, and the authors acknowledge the versatility of UVM that enabled the reuse of components horizontally across modules and vertically toward the higher hierarchies. While the monitors and checkers from the stand-alone environment proved useful at top level, the constrained random stimulus that was generated by the UVM components were not reusable since there was no way to scale them to the stimulus supported by the processor.

In [3], Khaled Salah explores the features and components of UVM that makes it suitable for verification of complex digital designs and demonstrates the use of UVM to verify a WISHBONE SoC. Using UVM to verify the SoC design was listed to be

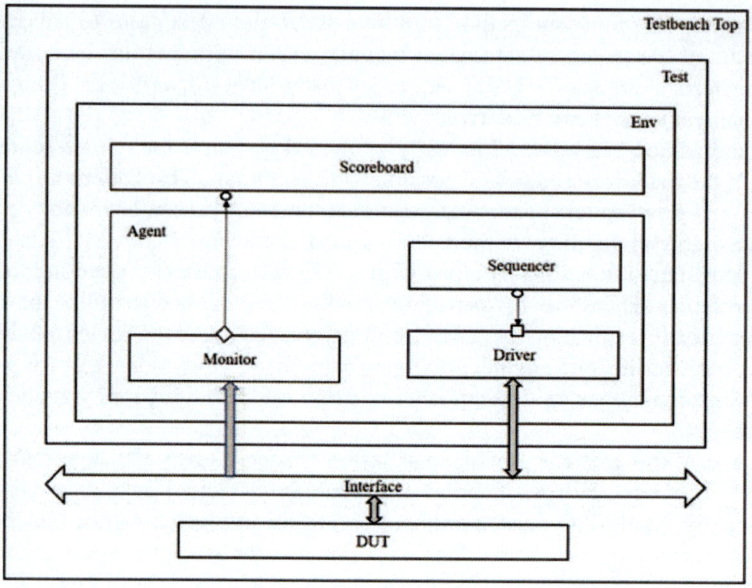

Fig. 1 General UVM testbench architecture

12 times faster than using other directed verification techniques to achieve similar verification completeness. However, the study brings attention to the drawback of using UVM for SoC verification, since there was no way to derive the software test cases running on the embedded processors from the UVM testbench sequences in an automated manner.

On the stand-alone verification of IPs for SoCs, detailed verification scenarios involving the verification of network-on-chip (NoC), cache controller, SPI, I2C, CAN, SWP, JTAG, and FIFO modules are elaborated in [5–14]. The authors detail the verification environment and techniques like constrained random excitation used to verify the functionality of the design using UVM and also explain including the monitors and checkers used in stand-alone verification to the top-level SoC testbench. However, the potential of using the stimulus generators present in the stand-alone test cases is still missed out. The SoC top-level verification of blocks is limited to data connectivity and integrity checks of interconnected modules, but not on the overall functionality of the individual module.

Authors in [15] use an alternate approach to verify a RISC-V-based SoC design. Instead of relying on software test cases to drive the processor, UVM sequences are hardcoded to provide stimulus in an instruction format that simulates the behavior expected from a software sequence. While this technique does away with stand-alone and top-level verification, the specific requirements placed on the stimulus generation would limit the verification completeness achievable in a reasonable time. Further, development of such instruction generators is time consuming and tailored to the exact processor instruction set which it is trying to verify.

In Sect. 3, our proposed method for converting the UVM stand-alone tests to top-level software tests for generating device test vectors is presented.

3 Automation of Software Test Case Generation from UVM Sequence

In the proposed technique, the UVM sequence test cases written in SV are converted to software test cases in C language, which would then be used to drive the processor, to simulate the stand-alone verification scenario at SoC top level. The technique is demonstrated through the verification of an I2C master protocol controller IP integrated as part of a SPARC-V8-based SoC. A typical configuration of the SoC, representative of the configuration used in our study, is shown in Fig. 2. Details of other peripherals and blocks are excluded from this block diagram.

In our study, the device-under-test (DUT) is an I2C controller with an APB wrapper. A typical write transfer on the I2C bus for this master is shown in Fig. 3. The stand-alone verification environment for the I2C master is built using the UVM framework. The stimulus is provided from a UVM agent which drives the APB side, that would be driven by the processor in the integrated SoC, and an I2C slave which would sit outside the SoC. The testbench structure of the stand-alone verification environment is shown in Fig. 4. The unit tests are written as UVM sequences.

Fig. 2 Configuration of SPARC V8-based SoC

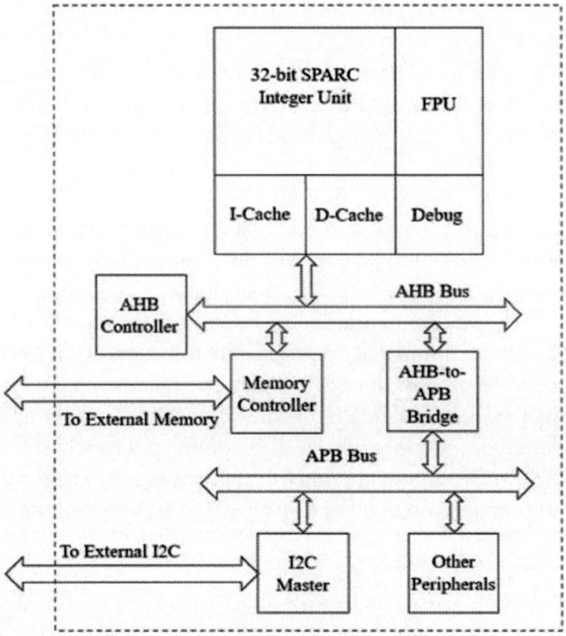

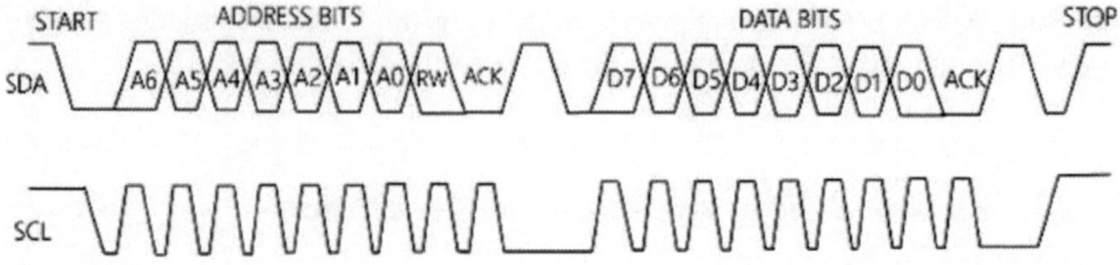

Fig. 3 Transfer on the I2C bus

Fig. 4 Stand-alone verification environment for I2C master

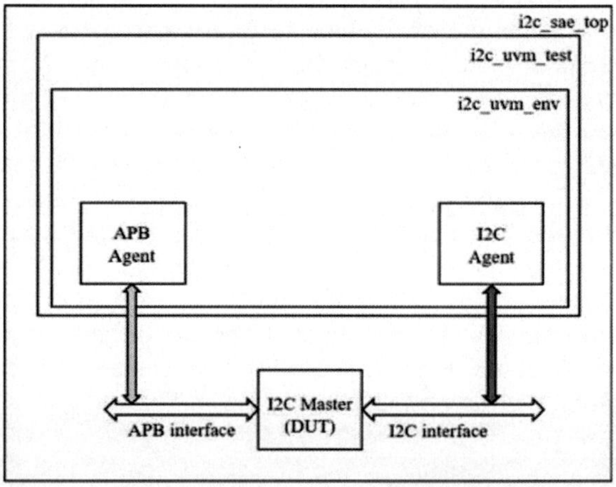

Multiple UVM tests were written to drive each sequence into the DUT to test each function of the I2C block. The next step is to enable the use of these sequences to generate top-level test cases that can contribute as functional test vectors during device testing.

Toward this goal, the UVM sequence is modified to capture the data and control information going into the DUT from the UVM agent (the side which would be driven by the processor core once integrated), into a log file while the sequence is being run. Figure 5 shows an example code snippet.

The data and control information from the UVM agent captured in the log file is extracted and fed to an automated tool. Along with extracted information, the address space of the SoC is used to generate an equivalent C test case that can drive the processor to create the stimulus created by the UVM agent. Figure 6 shows the output of the tool with the code converted to a C test case.

Test cases generated in this manner are then run at the top-level testbench to cover all functional scenarios exercised by the stand-alone UVM sequences.

```
class i2c_basic_read_test extends uvm_test;
  `uvm_component_utils(i2c_basic_read_test)

  ....

  // Configure I2C register for read operation

  i2c_slave_addr = // constrained random address
  i2c_operation  = `I2C_READ;
  i2c_command    = {24'h000000, i2c_slave_addr, i2c_operation};

  apb_write(`I2C_CTRL_REG,i2c_command);
  `uvm_info("SAE2TOP", $sformatf("REG_WRITE :: ADDR : %h :: DATA : %h",\
      `I2C_CTRL_REG, i2c_command), UVM_LOW)

  ....

  // Wait for transaction to finish
  while(apb_read(`I2C_STATUS_REG) != 'h00000001);

  `uvm_info("SAE2TOP", $sformatf("WAIT_FOR_REG :: ADDR : %h :: DATA : %h", \
      `I2C_STATUS_REG, 'h00000001), UVM_LOW)

  ......

  // Read the readdata
  readdata = apb_read(`I2C_RECEIVE_REG);
  `uvm_info("SAE2TOP", $sformatf("REG_READ :: ADDR : %h : VAR : %h", \
      `I2C_RECEIVE_REG, "readdata"), UVM_LOW)
```

Fig. 5 Code snippet for adding UVM sequence information to log file

Fig. 6 C program generated from the UVM simulation logs

```
// Registers
volatile int * I2C_CONTROL_REGISTER = (int *) 0xA0000000;
volatile int * I2C_STATUS_REGISTER  = (int *) 0xA0000004;
volatile int * I2C_TRASMIT_REGISTER = (int *) 0xA0000008;
volatile int * I2C_RECEIVE_REGISTER = (int *) 0xA000000C;

...

void main() {
    ...

    // REG_WRITE
    *(0xA0000000) = 0x00000057;
    ...

    // WAIT_FOR_REG
    while(*(0xA0000004) != 0x00000001) {
    }
    ...

    // REG_READ
    readdata = *(0xA000000C);
    ...
}
```

These C test case runs at SoC level are further used to generate functional test vectors that in turn are used to test the manufactured device. Using these additional test vectors allows to check the functionality of each module inside the SoC at the device level before declaring as a good device. These additional test cases enhance the confidence on the device, especially when it is slated for use in high-reliability critical applications.

4 Strengths and Weakness of the Proposed Method

The proposed method enables reuse of the huge verification effort spent at module level by translating the verification scenarios to SoC top level. The increased number of functional test vectors generated by this technique helps in testing the fabricated device against functionality of individual SoC blocks in full-fledged manner.

The method requires minimal modification to the UVM sequences to capture the data and control information, and the rest of the process for generating SoC software test cases is automated.

However, the current technique cannot cover stimulus generated by different blocks other than the processor, for the Unit Under Verification (UUV). Toward this, the tool needs to be enhanced to capture data from unit-level tests written for different blocks of SoC interfaced to UUV. Hence, though the technique cannot completely cover stimulus outside the processor's context, by automating the conversion of stimulus on the processor interface to software code, it reduces the burden on the engineer who would otherwise have to manually write the entire software to address the scenario.

5 Conclusions

In this article, a novel technique to utilize the verification space covered by module level test cases to test the manufactured SoC device is presented. The stand-alone test scenarios generated with constrained random stimulus and UVM are converted to top-level software test cases. Finally, these top-level tests are used to generate additional functional test vectors to test the fabricated device. The additional test cases generated by this technique greatly enhance the confidence of the manufactured SoC device against functional requirements and help device utilization for high reliability applications like aerospace.

In an SoC, some blocks are driven by multiple interfaces, and capturing the stimulus from interfaces other than processor is challenging and is not addressed in this article. In the future, this work can be enhanced by adding the automatic conversion of stimulus generated by these other interfaces, which would help to cover even more scenarios.

References

1. Yun, Y.-N., et al.: Beyond UVM for practical SoC verification. In: International SoC Design Conference, pp. 158–162. Jeju, Korea (South) (2011)
2. Wang, J., et al.: A UVM verification platform for RISC-V SoC from module to system level. In: IEEE 5th International Conference on Integrated Circuits and Microsystems, pp. 242–246. Nanjing, China (2020)
3. Salah, K.: A UVM-based smart functional verification platform: concepts, pros, cons, and opportunities. In: 9th International Design and Test Symposium, pp. 94–99. Algeries, Algeria (2014)
4. Subramanyan, P., et al.: Template-based parameterized synthesis of uniform instruction-level abstractions for SoC verification. In: IEEE Transactions on Computer-Aided Design of Integrated Circuits and Systems, vol. 37(8), pp. 1692–1705 (2018)
5. El-Ashry, S., et al.: On error injection for NoC platforms: a UVM-based generic verification environment. IEEE Trans. Comput.-Aided Des. Integr. Circ. Syst. **39**(5), 1137–1150 (2019)
6. Guo, Y., et al.: A SPI interface module verification method based on UVM. In: IEEE International Conference on Information Technology, Big Data and Artificial Intelligence, pp. 1219–1223. Chongqing, China (2020)
7. Jaideep Varier, E.V., et al.: Design of generic verification procedure for IIC protocol in UVM. In: Proceedings of the Third International Conference on Electronics Communication and Aerospace Technology, pp. 1146–1150. Coimbatore, India (2019)
8. Pavithran, T.M., et al.: UVM based testbench architecture for logic sub-system verification. In: IEEE International Conference on Technological Advancements in Power and Energy, Kollam, India (2017)
9. Chinchole, M., et al.: Functional verification of ten gigabytes media independent interface (XGMII) using universal verification methodology. In: 4th International Conference on Computing Communication Control and Automation, Pune, India (2018)
10. Hao, Y., et al.: A SWP interface module verification method based on UVM. In: IEEE 2nd International Conference on Electronics and Communication Engineering, pp. 5–9. Xi'an, China (2019)
11. Goel, A., et al.: UVM based controller area network verification IP (VIP). In: International Conference on Smart Electronics and Communication, pp. 645–652. Trichy, India (2020)
12. Vanaraj, A.T., et al.: Functional verification closure using optimal test scenarios for digital designs. In: Third International Conference on Smart Systems and Inventive Technology, pp. 535–538. Tirunelveli, India (2020)
13. Biswal, B.P., et al.: Cache coherency controller verification IP using SystemVerilog assertions and universal verification methodologies. In: International Conference on Intelligent Systems and Control, pp. 21–24. Coimbatore, India (2017)
14. Elakkiya, C., et al.: Functional coverage—driven UVM based JTAG verification. In: IEEE International Conference on Computational Intelligence and Computing Research, Coimbatore, India (2017)
15. Munir, A., et al.: Fast reliable verification methodology for RISC-V without a reference model. In: 19th International Workshop on Microprocessor and SOC Test Security and Verification (MTV), pp. 12–17. Austin USA (2018)
16. Liao, W.-S., et al.: FVP: A formal verification platform for SoC. In: IEEE International SOC Conference, pp. 21–24. Portland, USA (2003)
17. Roy, S.K.: Top level SOC interconnectivity verification using formal techniques. In: International Workshop on Microprocessor Test and Verification, pp. 63–70. Austin, USA (2007)
18. Saafan, H., et al.: SoC connectivity specification extraction using incomplete RTL design: an approach for formal connectivity verification. In: 11th International Design & Test Symposium (IDT), pp. 110–114. Hammamet, Tunisia (2016)

19. Zorian, Y.: Addressing test challenges in advanced technology nodes. In: IEEE 21st Asian Test Symposium, Niigata, Japan (2012)
20. Salemi, R.: The UVM Primer: An Introduction to the Universal Verification Methodology, 1st edn. Boston Light Press (2013)

Analysis of Firmware Security Mechanisms

Sandeep Gogineni Ravindrababu, Varsha Venugopal, and Jim Alves-Foss

Abstract Attacks, such as Mirai Botnet and Stuxnet, which both target low-level devices, have shown how vulnerable firmware on today's devices is. As a result of these attacks, security mechanisms such as address space layout randomization (ASLR), stack canaries, non-executable stack, data execution prevention, and others have been implemented in firmware in an attempt to mitigate the exploits. Some of these techniques, however, have limits and are vulnerable to particular attacks. In this paper, the standard security mechanisms currently in use are initially described. Following that, we analyze their shortcomings. Finally, we present the findings of the experiments to see if these security techniques are reliable.

Keywords Firmware · Security · ASLR · Data execution prevention · Stack canaries · Non-executable stack · RELRO · PIE

1 Introduction

Firmware, in a broad sense, refers to the software that is installed in computing devices to control how they run. This software is required by the system or device and is usually beyond the user's control. Firmware is also updated much less frequently than most software, and it comes in many different forms whether it is used in an embedded system like a Smart TV or being the basis of how the operating system runs on personal computers or smart phones. With all these different sorts of firmware, determining whether security safeguards should be employed and to what extent defenses should be deployed while maintaining a suitable overhead cost can be tricky.

S. G. Ravindrababu (✉) · V. Venugopal · J. Alves-Foss
Center for Secure and Dependable Systems, University of Idaho, 875 Perimeter Dr. MS 1008, Moscow, ID, 83844-1008, Moscow, Idaho, USA
e-mail: sandeepg@uidaho.edu

V. Venugopal
e-mail: vvenugopal@uidaho.edu

J. Alves-Foss
e-mail: jimaf@uidaho.edu

© The Author(s), under exclusive license to Springer Nature Singapore Pte Ltd. 2022

A. K. Nagar et al. (eds.), *Intelligent Sustainable Systems*, Lecture Notes in Networks and Systems 333, https://doi.org/10.1007/978-981-16-6309-3_51

In addition, for the majority of devices, the exact same firmware version is usually installed across each instance of the same device. For example, in a power station with hundreds of the same type of relay, each is running the exact same firmware image. This is an issue because an exploit that is developed to work on the specific firmware image can be leveraged against every device running that same firmware image. As a result, every device is vulnerable to the same exploit, and the adversary can use the exploit on any device they want, eventually exploiting them all.

Furthermore, firmware developers are typically untrained in security, and there are memory and time constraints in the devices that make security features difficult to implement. Hence, firmware has a limited set of security features, and if the devices get older, security flaws in their firmware or related libraries discovered must be patched to minimize the vulnerability. Modern security mechanisms like ASLR, stack canaries, non-executable stack, data execution prevention, and others are currently used to secure firmware devices. Address space layout randomization (ASLR) implemented within firmware is used to mitigate buffer overflow attacks. ASLR allows programs to be placed at different locations in memory, hopefully thwarting attacks that rely on knowledge of actual memory addresses. Unfortunately, ASLR mechanisms are limited and can be prone to attacks [6, 12, 13, 18]. This brings us to conduct experiments on different security mechanisms and analyze there limitations.

The remainder of the paper is organized as follows: Sect. 2 outlines on modern security mechanisms implemented in firmware. The limitations and experimental results on the current techniques is described in Sect. 3. The conclusion and future work are discussed in Sect. 4.

2 Firmware Security Methods

In this section, the mitigation strategies like address space layout randomization (ASLR), stack canaries, non-executable stack, data execution prevention, relocations read-only, and position independent executable (PIE) are discussed.

2.1 Address Space Layout Randomization

Address space layout randomization (ASLR) is a mechanism in which the base addresses of an executable are randomly assigned in memory during the loading and linking processes. This includes the stack, heap, and code addresses. The ASLR technique ensures randomness of memory address locations for each portion of the executable each time it runs, including any associated libraries. The implementation of ASLR technique is described in detail by Shacham et al. [13]. The main purpose of ASLR is to protect the addresses of the functions on the stack within an executable. Since attackers use the locations of specific functions to create attacks, this technique aids in reducing the attacker's ability to find those addresses. Because each of the sites is randomly generated, an adversary cannot be assured of the exact placements of the functions and hence cannot carry out the exploits. The most common exploit that ASLR protects against is a buffer overflow.

2.2 *Stack Canaries*

A stack canary is a value that is placed on the stack between the current function's return address and the local variables. This value can be anything as long as the system is aware of it in order to check it. While the program is running usually during the return from a function call, this value is checked, and if it is found to have been updated or modified, the program will terminate. The key benefit of this technique is that it can tell if a buffer has been written outside of its assigned space and onto the stack.

2.3 *Non-executable Stack*

The term non-executable stack refers to hardware prevention of treating stack data as instructions. A non-executable stack is used as a defense against attacks that can place malicious code on the stack to be executed, with buffer overflow being the most prevalent method. By removing the ability for code to be executable on the stack, even if code were to be injected, it would fail to execute and accomplish the adversary's purpose. Microsoft's implementation of the non-executable stack, which includes certain other safeguards, is known as data execution prevention. The non-executable stack was included in the PaX package for Linux in 2001, along with ASLR [8].

2.4 *Data Execution Prevention (DEP)*

Data execution prevention (DEP) is a generalization of non-executable stack that allows you to mark any memory pages as non-executable. This approach is successful against buffer overflow attacks because the overwritten bytes are placed on non-executable memory, which prevents them from executing. DEP uses an exception handling status code called STATUS_ACCESS_VIOLATION where, if it is not handled, DEP terminates the process [11]. By default, applications cannot execute code on the stack nor the heap, while DEP is in place and space allocated by malloc() is automatically set as non-executable.

2.5 *Relocation Read-Only (RELRO)*

Relocation read-only (RELRO) is a technique to make control flow sections of the binary read-only. It overcomes the global offset table (GOT) attacks by setting the data sections used by the dynamic linker to read-only when loading the program. There are two types of RELRO: Partial and full. Partial RELRO eliminates the risk of buffer overflows on a global variable overwriting GOT entries. It is considered to be fast as it enables lazy loading mechanism. Full RELRO makes the entire GOT table read-only. It is considered as slow at it disable the lazy loading mechanism.

2.6 *Position-Independent Executables (PIE)*

Position-independent executables (PIE) is a security method that makes code reuse attacks more difficult by randomly placing executable memory locations [5]. Compilation of executables under PIE is position independent. This means that the resulting PIE can be stored and used anywhere in memory. PIEs are similar to shared libraries where the dynamic linker is responsible for relocating them to the virtual memory location. The key benefit of this method is that it makes it more difficult for an adversary to pinpoint specific locations in code because the position of jumps and addresses changes each time the linked executable runs.

3 Analysis on Firmware Security Methods

The security methods discussed in the previous section have their own limitations. With randomized addressing in ASLR technique, there are issues in the form of tracing and debugging an executable. This is because, it is difficult to follow where exactly the problem occurs as it varies each time the executable is loaded into memory, whereas if ASLR was not enabled, it would be easy to observe that an error only occurs in the same exact place. Additionally, it has been shown by researchers that certain attacks can bypass ASLR [3, 7, 8]. Marco-Gisbert et al. [7] discuss how the low entropy of ASLR on 32-bit systems is enough to make brute forcing an address of a library much easier than expected.

Stack canaries are simple to implement, and combining them with ASLR provides additional security, but still it has some downsides. The first and foremost is that if the adversary is able to read the value of the canary and input the value in the buffer that is being overflowed in the location where the result of the buffer overflow places the same value in the exact location, the stack canary check will erroneously succeed and the buffer overflow will not be detected. Another strategy that is effective as shown by [10] is overwriting the global offset table (GOT) entry for different functions; in their example using the __exit() function address and gaining control over the execution flow of the program. When stack canary fails, it allows an adversary to modify execution flow of a program to their own addresses and shellcode.

Non-executable stack is better than ASLR, and the combination of both will make it difficult for the adversary to find valid addresses and does not allow execution of the injected code on stack. Since this technique effects only the stack, the heap is left wide open for attacks. Cowan et al. [2] have demonstrated how an attacker can inject shellcode into the heap buffer and then utilize the stack buffer to overwrite the return address to the heap region where the shellcode is placed. As a result of this consequence, a non-executable stack will be unable to avoid all variations of the attack it is designed to defend against. This problem is overcome by the DEP approach, which effectively covers both the stack and heap by marking data pages as non-executable.

Much like the non-executable stack, DEP is also vulnerable to several different types of attacks [1, 19]. These attacks include Return-into-Libc, Just in Time spraying (JIT spraying) [1], and others. When an executable uses the stack or heap for some of its functionality, DEP may cause problems. Because the memory pages are tagged as non-executable, executables that execute with DEP enabled may have limited to no functionality. This is primarily a problem with older applications that were created before to the implementation of DEP. Another side effect of DEP interfering with execution is that a user can turn off the protection to restore functionality rather than creating an exception case for that specific program.

The effectiveness of RELRO is discussed in detail by [5]. When Full RELRO is implemented, the effect of lazy binding increases as the amount and complexity of the code grows, which has an impact on programs with long startup times. Full RELRO is also ineffective because many functions are not called at runtime, resulting in performance problems.

PIE offers the same security benefit as ASLR, but the disadvantage is not using absolute addressing. Therefore, the execution time is slow since there is no static address for which repeated lookups can occur efficiently [4, 9]. Considering PIE must be taken care for the programs that are sensitive to runtime performance overhead [5].

3.1 Experimental Results

We conducted an experiment for this paper in which numerous attacks were examined in terms of security mechanisms. For each experiment we developed some code with sample vulnerabilities. We then compiled the code with the specific security features enabled. We used attack techniques specified by [14–17].

Tables 1 and 2 summarize the outcomes of this experiment. ASLR, stack canary, and non-execution memory pages were among the security techniques investigated (NX). Each of these attacks assumes some base knowledge, some base set of security controls, and possibly other configurations. Table 2 summarizes these enabling fea-

Table 1 Attacks against common protection mechanisms

Protection mechanisms	Attacks			
	Bypass ASLR, NX, stack canary [16]	Overwrite GOT table [17]	Retum2libc to bypass NX bit [18]	Bypass NX bit using Mnrotect function/Make stack executable [19]
Address space layout randomization (ASLR) is a memory-protection process for OS that guards against buffer overflow attacks)	✓	✓		
Stack canary (secret value placed on stack used to detect a stack buffer overflow before execution of malicious code)	✓			
NX bit (prevents buffer overflow attacks by blocking code execution from memory that is marked as non-executable)	✓	✓	✓	✓
Relocations read-only (RELRO)-(security measure which makes some binary sections read-only. Partial—eliminates the risk of buffer overflows on a global variable overwriting GOT entries. Full—makes the entire GOT read-only)	Full	Partial	Full	Full
Fortify (provides runtime checks of buffer lengths and memory regions) PIE—position-independent executable (it loads binary into random locations within virtual memory each time the application is executed)	✓	✓	✓	✓
PIE-position-independent executable (it loads binary into random locations within virtual memory each time the application is executed)	✓	✓	✓	✓

Table 2 Enabling features

Enabling features	Attacks			
	Bypass ASLR.NX, stack canary [16]	Overwrite GOT table [17]	Retum21ibc to bypass NX bit [18]	Bypass NX bit using Mprotect function/make stack executable [19]
Offset211bc (offset of libc base found by dumping contents of stack using string format specifiers/using address of libc functions such as write/puts/read)	✓	✓	✓	✓
One_gadget offset (offset of one gadget from libc base address to call exective)	✓	✓	✓	
Offset to setuid (setuid function is found in libc which is set to 0, to get root privileges while executing remote shell)	✓		✓	✓
Pop Rdi value (It is a Rop gadget used to execute 'sh'. Address of 'sh' should be in rdi register)	✓		✓	✓
Address of start function (this address is used to restart the program without letting the program to die)		✓		
Address of exit GOT (this is usd to overwrite the GOT entry of exitf) to start of the program whenever exit is called to prevent program from being killed)		✓		
Offset to /bin/sh (offset of /bin/sh in libc to execute shell)			✓	
Offset to Mprotect function (Mprotect junction offset in libc is used to change the permissions of stack to make stack executable)				✓
Offset to memcpy (the offset to memcpy function in libc is used to copy the shellcode from stack to different area in memory)				✓
Pop rdx, pop rsi ROP chain (it is a ROP chain used to execute 'sh')				✓

tures with the goal of determining how much additional mitigation would be needed to prevent successful attacks.

Bypass ASLR, NX, and stack canary are some attacks that are able to bypass those protection mechanisms [14]. Overwriting the GOT table means being able to change the values of the addresses within the GOT, leading to a change in execution flow based on what addresses were changed [15]. Return2libc allows to bypass the non-executable bit which involves utilizing the current library, libc [17]. Finally, using Mprotect allows the bypassing of the non-executable bit and can make the stack executable. It can do this because Mprotect allows the changing of the permissions of an area in memory, effectively allowing execution permissions in memory that is already marked as non-executable [16].

Table 1 shows common protection mechanisms and their effectiveness against certain attacks. A check mark within Table 1 represents that the attack is successful against that defense. For example, the overwrite GOT table attack is successful at exploiting code that has ASLR implemented and the non-executable bit set [15].

For each attack to be successful, certain conditions must be true for the binaries. These are given in Table 2.

4 Conclusion

In this paper, we presented different security strategies that are currently used to secure firmware, as well as their flaws, by assessing each methodology with four different attacks and summarized the results. Based on the findings, we believe that new techniques for mitigating attacks on firmware devices are required. We are now working on a technique called shuffler, which shuffles object files and generates unique firmware versions using the same code base while ensuring that the functionality is preserved. In terms of future work, we intend to put the shuffler to the test against all four attacks, compare its performance to that of current security mechanisms, and assess how reliable the object file shuffling technique is for securing firmware devices.

References

1. Chen, P., Fang, Y., Bing, M., Xie, L.: JITDefender: a defense against JIT spraying attacks. In: Camenisch, J., Fischer-Hübner, S., Murayama, Y., Portman, A., Rieder, C. (eds.) Future Challenges in Security and Privacy for Academia and Industry, pp. 142–153. Springer, Berlin, Heidelberg (2011)
2. Cowan, C., Beattie, S., Day, R.F., Pu, C., Wagle, P., Walthinsen, E.: Protecting systems from stack smashing attacks with StackGuard. In: Linux Expo (1999)
3. Evtyushkin, D., Ponomarev, D., Abu-Ghazaleh, N.: Jump over ASLR: attacking branch predictors to bypass ASLR. In: 2016 49th Annual IEEE/ACM International Symposium on Microarchitecture (MICRO), pp. 1–13. IEEE (2016)

4. Federico, A.D., Cama, A., Shoshitaishvili, Y., Kruegel, C., Vigna, G.: How the ELF ruined christmas. In: 24th USENIX Security Symposium (USENIX Security 15), pp. 643–658. USENIX Association (2015)
5. Jeong, S., Hwang, J., Kwon, H., Shin, D.: A CFI countermeasure against got overwrite attacks. IEEE Access **8**, 36267–36280 (2020)
6. Liu, L., Han, J., Gao, D., Jing, J., Zha, D.: Launching return-oriented programming attacks against randomized relocatable executables. In: 2011 IEEE 10th International Conference on Trust, Security and Privacy in Computing and Communications, pp. 37–44 (2011)
7. Marco-Gisbert, H., Ripoll, I.: On the effectiveness of full-ASLR on 64-bit linux. In: Proceedings of the In-Depth Security Conference (2014)
8. Marco-Gisbert, H., Ripoll, I.: Exploiting linux and PaX ASLR's weaknesses on 32-and 64-bit systems. BlackHat Asia (2016)
9. Payer, M.: Too much PIE is bad for performance. Tech. Rep. **766** (2012)
10. Richarte, G.: Four different tricks to bypass stackshield and stackguard protection. World Wide Web **1** (2002)
11. Satran, M.: Data Execution Prevention—Win32 apps. Win32 apps. Microsoft Docs (2018)
12. Serna, F.J.: The info leak era on software exploitation. Black Hat USA (2012)
13. Shacham, H., Page, M., Pfaff, B., Goh, E.J., Modadugu, N., Boneh, D.: On the effectiveness of address-space randomization. In: Proceedings of the 11th ACM Conference on Computer and Communications Security, pp. 298–307 (2004)
14. Shrirao, S.: Format String Exploits: Defeating Stack Canary, NX and ASLR Remotely on 64 bit (2018)
15. Shrirao, S.: Format Strings: GOT Table Overwrite to Change Control Flow Remotely on ASLR (2018)
16. Shrirao, S.: Make Stack Executable Again (2018)
17. Shrirao, S.: Return to libc on Modern 32 bit and 64 bit Linux (2018)
18. Sotirov, A., Dowd, M.: Bypassing browser memory protections in Windows Vista. Blackhat USA (2008)
19. Stojanovski, N., Gusev, M., Gligoroski, D., Knapskog, S.J.: Bypassing Data Execution Prevention on Microsoft Windows XP SP2. In: The Second International Conference on Availability, Reliability and Security (ARES'07), pp. 1222–1226 (2007)

Mutual Funds Categorization Using Ensemble Clustering

Anureet Bansal, Tushar Sharma, and Gaganmeet Kaur Awal

Abstract Mutual funds are a type of investment with each fund having some investment objective. Funds with the same investment objective should behave in a similar manner and have similar characteristics with respect to risk and returns. Our paper proposes a novel approach to categorize mutual funds using ensemble clustering which is then compared with their stated objectives. Firstly, we applied PCA to identify important latent variables which is then used to classify funds using ensemble clustering. Afterward, discovered clusters are compared with stated objectives which are capital preservation funds, wealth builder funds, aggressive growth funds, steady income funds, and tax saver funds to determine the misclassification. The experiment result demonstrates that capital preservation funds behave similar to its stated objectives, whereas few wealth builder funds, aggressive growth funds, steady income funds, and tax saver funds behave differently from their stated objectives, hence the misclassification.

Keywords Mutual funds · Investment objectives · Ensemble clustering · PCA

1 Introduction

Mutual fund is a type of investment where money from different investors is collected and invested in different securities. In this way, a mutual fund is well diversified which is managed professionally. It is of many types like debt mutual funds, equity mutual funds, and hybrid mutual funds. There is an exponential growth in the industry of mutual funds in India with more than 4.5 times increase in its value in the last ten

A. Bansal (✉) · T. Sharma · G. K. Awal
USME, Delhi Technological University, Delhi, India
e-mail: anureetb28@gmail.com

T. Sharma
e-mail: tushu.network@gmail.com

G. K. Awal
e-mail: gaganmeetkaur.awal@dtu.ac.in

© The Author(s), under exclusive license to Springer Nature Singapore Pte Ltd. 2022

A. K. Nagar et al. (eds.), *Intelligent Sustainable Systems*, Lecture Notes in Networks and Systems 333, https://doi.org/10.1007/978-981-16-6309-3_52

years as stated in [1]. There are over 2500 mutual funds schemes with 44 registered fund houses in India [2].

As we know mutual funds can be classified into different categories based on various criteria, investment objective is one of the criteria used for classification of mutual funds. Investors use these investment objectives to select the mutual funds to meet their financial goals [3]. According to Marathe in [4], mutual funds are misclassified by the fund managers in order to get higher ranking and show poor performing high-risk related funds under low-risk high return generating category.

Though various researches have been conducted on different types of funds using clustering techniques like K-means in [5–7], hierarchical clustering in [8], SOM in [9] which can be utilized to improve performance, however, research on mutual funds misclassification using ensemble clustering is yet to be done. Since every technique has some limitations associated with it, ensemble model can be used to incorporate strengths of various algorithms. Also, most of the researches has been conducted on different types of funds. Therefore, research on Indian mutual funds is in the nascent stage.

To address these research gaps, we propose a novel approach to classify mutual funds using ensemble clustering in Indian context. For this, we identify hidden latent variables using PCA, and then we deploy ensemble clustering in which we use majority voting and k-modes technique and classify funds. Finally, these classified funds are compared with their stated objectives and presented. No work has been done so far for misclassification of mutual funds using ensemble clustering in Indian context for time period of 2016–2020 after extracting important latent variables to the best of our knowledge.

The paper is organized as follows. Section 2 discusses review of a few past studies. Section 3 outlines the underlying methodology. Section 4 describes data preprocessing, data source, and our approach. Section 5 examines the classified funds and their stated investment objectives. Section 6 concludes our discussion on the same.

2 Literature Survey

This section discusses the background of mutual funds categorization, classification, and misclassification.

Kim et al. [10] proposed classification of funds on the basis of their attributes using discriminant analysis and checked for the percentage of misclassification of these funds. They found 50% funds whose stated objectives differ from their attribute-based objectives and 33% funds which show deviation from their stated objectives.

Chen et al. [11] presented an approach to cluster Taiwan equity mutual funds using K-means based on rates of return, standard deviation, turnover rate, and Treynor index and then used a fuzzy optimization model for portfolio optimization which maximized future return and minimized future risk. The authors find four clusters where aggressive funds and good performance funds clusters were considered in the portfolio.

Moreno et al. [9] introduced nonlinear techniques like SOM, KNN, and K-means algorithm for classifying Spanish mutual funds. They have used two clustering methods, KNN and a two-level method in which SOM is applied followed by K-means. The authors concluded that 40% funds were misclassified using first clustering technique and lesser funds were misclassified using two level technique.

Sidana et al. [12] proposed classification of Indian mutual funds taken from May 2002–May 2006. The authors performed K-means clustering on funds from six different sectors which were classified into six categories based on investment style and capitalization using ten attributes. They observed small inconsistencies in style classification and actual performance of funds.

Marathe et al. [4] presented a technique to classify mutual funds using K-means on 28 Morningstar's variables. They used PCA for dimensionality reduction and had 904 different funds with seven investment objectives. The authors observed that 43% of the mutual funds are misclassified.

Sakakibara et al. [13] introduced a new method to classify 551 Japanese mutual funds by forming a network of top ten stocks of these funds and then partitioned the network using K-means and spectral clustering.

Lisi et al. [8] proposed a three-step statistical process to determine the misclassification of European mutual funds. They used hierarchical clustering with Euclidean distance for classification of funds on the basis of risk calculated using threshold GARCH model.

Pattarin et al. [14] presented an approach to classify mutual funds using GAME clustering algorithm after dimensionality reduction. They have then identified the style type of funds using constrained regression model given by William Sharpe in [15].

3 Proposed Model

Here, we explain the method proposed in this study as shown in Fig. 1. First we extract data and preprocess it. Since, we have a large pool of 29 variables to work on, we apply principal component analysis for dimensionality reduction. After we get the significant components, we apply ensemble clustering using these components [16]. The aim is to solve the limitations faced during selection of clustering techniques as no technique is optimal for classification.

3.1 Principal Component Analysis

Principal component analysis (PCA) is a technique from statistics which helps in reducing the dimensions of data. The original data is projected to fewer dimensions with the help of linear combinations of variables which are known as principal components. These principal components are not correlated with each other and are

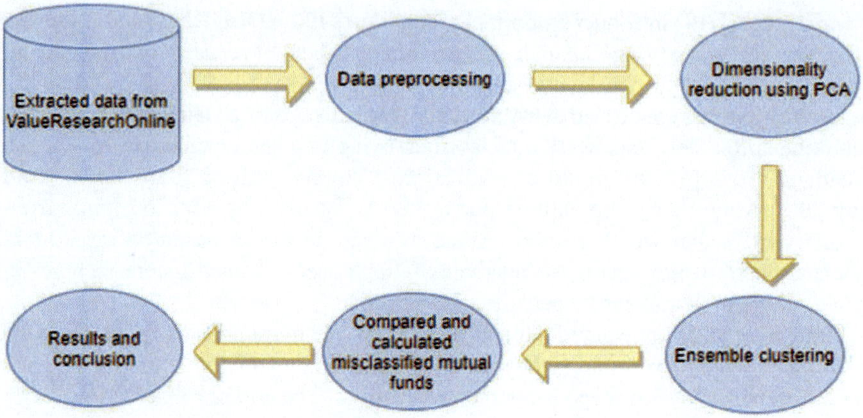

Fig. 1 Proposed framework for the research

sequenced in a way so that the first few components captures most of the variation present in the original variables. Choosing those first few components instead of all the attributes can help solve the problem of multicollinearity. In this paper, we took the 29 variables and reduced it to ten most important components.

3.2 *Ensemble Clustering*

Ensemble clustering helps to produce better results by combining multiple clustering models than that obtained from applying a single clustering model. We took the given models, non-negative matrix factorization (NMF) using Kullback–Leibler divergence or Euclidean distance, hierarchical clustering (HC), divisive analysis clustering (DIANA), K-means clustering (KM), partition around medoid (PAM), and combined them together in ensemble clustering. We then use majority voting and k-modes methods to select the best cluster classification for every fund.

HC and DIANA are hierarchical clustering models which connect objects using distance metric. We have used an average agglomerative hierarchical clustering method which is a bottom up approach which evaluates cluster quality based on similarities between all objects rather than the similarity of a pair of objects. DIANA is a top-down approach where all the points combined are taken as one cluster, and then they are recursively split until distinct clusters are formed. It is a form of hierarchical clustering. KM and PAM are centroid-based models. They use initial cluster centers and find objects near to these points. NMF transforms input data into a non-negative matrix which is used to group together samples.

Majority voting technique selects the cluster for an object based on maximum voting system. The funds are allocated to clusters which receive the highest votes.

K-modes is another technique used as a consensus function [17] which determines a cluster for a fund by clustering categorical variables.

4 Experimental Setup

For this study, we used ValueResearchOnline [18] as a data source. The data has 177 mutual funds of different types with five investment objectives: capital preservation funds, wealth builder funds, aggressive growth funds, steady income funds, and tax saver funds. This dataset consists of 29 attributes which are fund risk grade, fund return grade, investment objective, standard deviation, Sharpe ratio, Sortino ratio, beta, alpha, R-squared, 1-week return (%), 1-month return (%), 3-month return (%), 6-month return (%), 1-year return (%), 3-year return (%), 5-year return (%), 2015 annual return, 2016 annual return, 2017 annual return, 2018 annual return, 2019 annual return, latest NAV, previous NAV, 52-week high NAV, 52-week low NAV, minimum investment, expense ratio (%), 3-year SIP return (%), and 5-year SIP return (%). Fund risk grade is the quality rating of a mutual fund based on the risks associated with it. Fund return grade is a rating of a fund's risk adjusted return in relation to other funds in that category. Standard deviation is the risk of a fund, and alpha is risk-adjusted return performance of a fund. Beta is the systematic risk of a fund. Alpha is a risk-adjusted return in relation to benchmark to gauge the performance. Sharpe ratio is a risk-adjusted return that tells us how a mutual fund archives its returns. Sortino ratio is similar to Sharpe ratio, but it is calculated on negative asset returns. R-squared tells us the percentage of a mutual fund's performance using a benchmark. Net asset value is a fund's per share market value. Expense ratio the total percentage of a fund's assets used for its expenses.

In order to select fewer attributes, either one should have enough knowledge about which attribute is more important or one should apply one of the attribute reduction techniques. For this study, we have used principal component analysis for dimensionality reduction.

4.1 Data Preprocessing

Before applying PCA on the attributes, preprocessing of data is performed as follows: All the missing values were replaced with zeros as prior to imputation there were no zeros in these data variable values. Also, some funds were new so their 5-year missing returns were substituted with zeros as they did not exist then. After dealing with the missing values, we created dummy variables for categorical attributes which were fund risk grade and fund return grade. We used MinMax scaling for normalization of the dataset [19]. It is used to rescale the range of features between 0 and 1. Its formula is as given below.

$$\text{Xscaled} = \frac{x - \min(x)}{\max(x) - \min(x)} \tag{1}$$

4.2 *Model Selection and Parameter Selection*

In this study, we have selected five types of clustering with two variants of NMF clustering method one is nmf_lee and another one is nmf_brunet. We have selected these clustering techniques as per our dataset requirements. While clustering we have also checked for the optimal number of clusters which came out to be four. For distance measure, we have selected Euclidean distance which is one of the most used distances.

5 Results

In this section, we present our observations and their corresponding interpretations. We observe that four clusters are identified giving distinct characteristics instead of five clusters according to the stated objectives (Table 1).

We have attempted to analyze the misclassification of the funds in the investment objectives. Table 2 shows us the alpha, beta, standard deviation, 1-year return, 3-year return, and 5-year return according to the original investment objectives. Aggressive growth funds and wealth builder funds have the highest 5-year return. We can also notice that the steady income funds and capital preservation funds have almost steady returns for 1 year, 3 years, and 5 year. The risk for aggressive growth is the highest as it is supposed to be, followed by wealth builder funds and tax saver funds.

Table 1 Cumulative explained variance

Principal component	Cumulative explained variance (%)
PC 1	30.7
PC 2	45
PC 3	53.3
PC 4	59.4
PC 5	65.2
PC 6	70.2
PC 7	74
PC 8	77.6
PC 9	81.1
PC 10	84.2

Table 2 Cluster of funds as per fund managers

	Alpha	Beta	Standard deviation (Risk)	1-year return	3-year return	5-year return
Aggressive growth funds	5.04	0.838	23.9	12.3	4.71	10.7
Capital preservation funds	2.32	0.760	0.47	5.58	6.05	5.99
Steady income funds	0.134	0.894	6.64	7.85	6.65	7.14
Tax saver funds	1.20	0.940	21.29	6.74	5.73	9.43
Wealth builder funds	1.44	0.954	21.33	7.67	6.16	10.4

Table 3 Cluster of funds as per ensemble clustering

	Alpha	Beta	Standard deviation (Risk)	1-year return	3-year return	5-year return
Cluster 1	2.32	0.751	0.557	5.69	6.10	6.04
Cluster 2	0.38	0.826	6.15	8.31	6.87	7.34
Cluster 3	2.13	0.926	21.9	8.41	5.74	10.20
Cluster 4	-1.90	1.45	10.4	4.22	4.86	5.63

Table 3 provides us risk and returns of the clusters formed after clustering. We can notice that Cluster 1 has the highest alpha, followed by Cluster 3. Standard deviation and 5-year return of Cluster 3 is highest and second highest beta. Cluster 4 has the lowest alpha, 1-year return, 3-year return, and 5-year return but has the highest beta. Cluster 2 has almost steady returns over the years but has low beta and alpha.

On comparing Tables 2 and 3, we notice that the risk of steady income funds as per investment objective is 6.64 and average risk associated with cluster 2 is 6.15 which is close to that of the former and Cluster 2 consists of only 33 steady mutual funds. We can say that funds under Cluster 2 can be categorized as steady income funds. Further risk of capital preservation funds and Cluster 1 is similar and Cluster 1 consists of only 43 capital preservation funds which means we can categorize cluster 1 as capital preservation funds. Since Cluster 3 has highest risk and return, like aggressive income funds, we can categorize Cluster 3 as aggressive funds. Cluster 4 has a risk of 10.4 and lowest returns which does not match with any investment type, so we can categorize these funds under a new objective where there is medium risk and low returns.

Also, we can see that Cluster 3 which behaves like aggressive growth funds comprises of funds whose stated objective was tax saver and wealth builder funds,

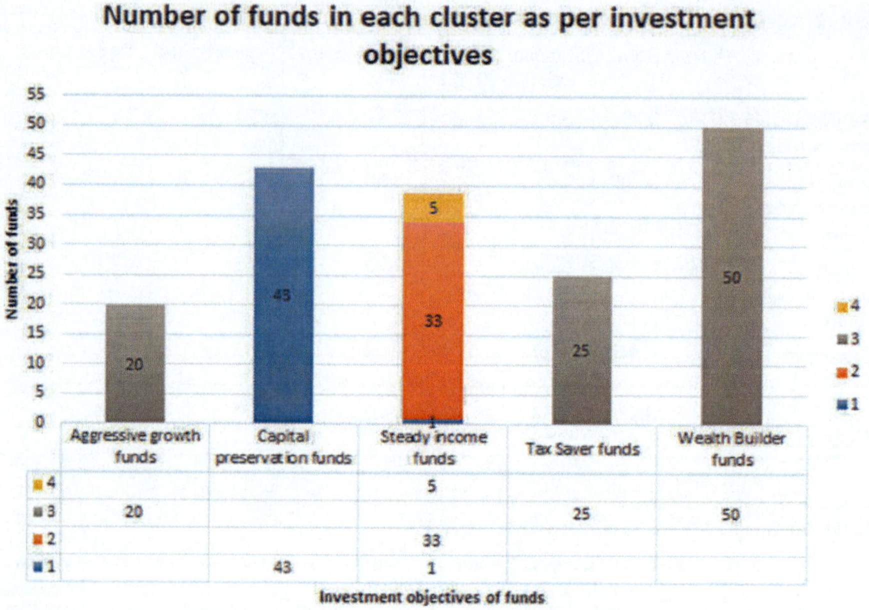

Fig. 2 Number of funds in each cluster as per investment objectives

Cluster 2 which behaves like steady income funds has 33 funds whose stated objectives was steady income only, Cluster 1 which behaves like capital preservation funds have 1 fund whose stated objective was steady income, and Cluster 4 which represents a new and different category from stated objectives has 5 funds from steady income. This shows that funds are misclassified into different categories.

Here, Fig. 2 shows us the number of funds which are present in attribute-based clusters formed by ensemble clustering and shows the misclassification of these funds. It clearly shows that Cluster 1 consists of 43 capital preservation funds and 1 steady funds, and in alignment with our above comparison, we can say here one fund has been misclassified as a steady income fund by the fund manager. In Cluster 2, there are just 33 funds categorized as steady income funds; in Cluster 3, there are three types of funds clubbed together but their risk is high. So, these funds can be categorized under one category instead of three. In Cluster 4, we have 5 funds which have been categorized as steady income by fund manager but actually they behave differently from the other steady income funds.

As we can see from Table 4, there are 75 funds which have been classified into different categories in stated objectives whereas they belong to aggressive growth funds according to clustering. Similarly, out of 6 funds from steady income funds in stated objectives, 1 fund belongs to capital preservation funds and 5 belongs to a new category. Therefore, 81 funds are misclassified according to clustering out of 177 mutual funds that is 45.76% funds are misclassified.

Table 4 Number of funds according to stated objectives and clustering based on funds attributes

Type of funds according to stated objectives	Number of funds in each category	Type of funds according to clustering based on funds attributes	Number of funds in each category
Aggressive growth funds	20	Aggressive growth funds (Cluster 3)	95
Tax saver funds	25	Steady income funds (Cluster 2)	33
Capital preservation funds	43	Capital preservation funds (Cluster 1)	44
Steady income funds	39	New category (Cluster 4)	5
Wealth builder funds	50		
Total funds	177	Total funds	177

6 Conclusion

In this paper, we adopted a holistic approach to capture the misclassification of mutual funds in terms of their stated objectives and attributes. We conducted experiments on publicly available data. We observed that when we cluster funds based on their attributes these funds form 4 clusters with distinct characteristics. However, fund managers have clubbed funds with different characteristics together which should not be the case. Every category should have distinct characteristics, and this shows that fund managers tend to misclassify mutual funds which is not good for an investor.

We also observe how our cluster formed is similar to the stated objectives. Some of the clusters have the ability for a wealth builder funds, and some have the quality of steady income funds. The overall performance of clustering funds is also improved by applying ensemble clustering as compared to that observed in the previous studies of detecting misclassification of funds. In this study, we observed that 45.76% mutual funds are misclassified.

As future work, we will utilize this study for the portfolio selection and optimization [11] and will try to incorporate time series aspect of funds for analysis [14].

References

1. Viswanadham, B.K., Yadav, S.: The growing popularity of mutual funds in India. J. Compos. Theory 0731–6755 (2019)
2. https://www.paisabazaar.com/mutual-funds/mutual-funds-in-india/. Last accessed on 02 Apr 2021
3. Chen, H., Cohen, L., Gurun, U., Lou, D., Malloy, C.J.: Don't take their word for it: the misclassification of bond mutual funds. J. Finance **138**, 118–137 (2017)

4. Marathe, A., Shawky, H.A.: Categorizing mutual funds using clusters. Adv. Quant. Anal. Finance Account. **7**(1), 199–204 (1999)
5. Acharya, D., Sidana, G.: Classifying mutual funds in India: Some results from clustering. Indian J. Econ. Bus. **6**(1), 71–79 (2007)
6. Lytkin, N.I., Kulikowski, C.A., Muchnik, I.B.: Variance-based criteria for clustering and their application to the analysis of management styles of mutual funds based on time series of daily returns. DIMACS Technical Report 2008–01
7. Narabin, S., Boongasame, L.: A cluster analysis of mutual funds data. In: International Conference on Big Data and Artificial Intelligence (BDAI), pp. 1–5. IEEE, China (2018)
8. Lisi, F., Otranto, E.: Clustering mutual funds by return and risk levels. In: Corazza, M., Pizzi, C. (eds.) Mathematical and Statistical Methods for Actuarial Sciences and Finance, pp. 183–191. Springer, Milano (2010)
9. Moreno, D., Marco, P., Olmeda, I.: Self-organizing maps could improve the classification of Spanish mutual funds. Eur. J. Oper. Res. **174**(2), 1039–1054 (2006)
10. Kim, M., Shukla, R., Tomas, M.: Mutual fund objective misclassification. J. Econ. Bus. **52**(4), 309–323 (2000)
11. Chen, L.H., Huang, L.: Portfolio optimization of equity mutual funds with fuzzy return rates and risks. Expert Syst. Appl. **36**(2), 3720–3727 (2009)
12. Sidana, G., Acharya, D.: Classifying mutual funds in India: some results from clustering. Indian J. Econ. Bus. **6**(1), 71 (2007)
13. Sakakibara, T., Matsui, T., Mutoh, A., Inuzuka, N.: Clustering mutual funds based on investment similarity. Proc. Comput. Sci. **60**, 881–890 (2015)
14. Pattarin, F., Paterlini, S., Minerva, T.: Clustering financial time series: an application to mutual funds style analysis. Comput. Stat. Data Anal. **47**(2), 353–372 (2004)
15. Sharpe, W.F.: Asset allocation: management style and performance measurement. J. Portf. Manag. **18**(2), 7–19 (1992)
16. Chiu, D.S., Talhouk, A.: diceR: an R package for class discovery using an ensemble driven approach. BMC Bioinformatics **19**(1), 1–4 (2018)
17. Monti, S., Tamayo, P., Mesirov, J., Golub, T.: Consensus clustering: a resampling-based method for class discovery and visualization of gene expression microarray data. Mach. Learn. **52**(1–2), 91–118 (2003)
18. https://www.valueresearchonline.com/funds/. Last accessed on 25 Oct 2020
19. Pedregosa, F., Varoquaux, G., Gramfort, A., Michel, V., Thirion, B., Grisel, O., Duchesnay, E.: Scikit-learn: machine learning in Python. J. Mach. Learn. Res. **12**, 2825–2830 (2011)

Load Balancing Synchronized Brownout Oriented Approach for Safeguarding Cloud Services Sustainability

Saumitra Vatsal and Shalini Agarwal

Abstract Efficient global networking demands proficient data centres, to cater cloud computing of par excellence. The attainment of this goal is infringed by unexpected events like flash crowds and hardware failures. It degrades the users experience since applications exhibit an unpredictable unresponsive behaviour. This challenge can be negotiated by brownout-enabled cloud applications to control response times and by incorporating replicas which are mirror image facsimile of applications with same functionalities to address redundancy factor. The induction of a load balancer to address direct incoming traffic revamps the objective of curbing this aforesaid challenging problem further. But the load balancers are not reckoned as a good indicator to decipher the replicas-related performance since their functionalities are response time-dependent. Response times themselves are subjugated to self-adaptive features of applications hence unable to express replica-related performance with full authenticity. Therefore, load balancing strategies are needed to be synchronous with brownout applications to address user experience degradation with prime importance with respect to response time. It can be achieved by deactivating application optional components in a dynamic and selective manner to contain response time as close as possible to a set point under a scenario of stress during phase of high resource utilization.

Keywords Brownout · Load balancing · Replicas · Cloud computing

1 Introduction

Cloud computing has embarked to be a global entity and as a paradigm with a difference to provide ubiquitous, convenient, on-demand, shared pool of configurable computing resources related with IT industry [1]. Its importance can be deciphered by the fact that it is reckoned among five technologies which would drive global

S. Vatsal (✉) · S. Agarwal
Shri Ramswaroop Memorial University, Barabanki, Uttar Pradesh, India

S. Agarwal
e-mail: shalini.cs@srmu.ac.in

© The Author(s), under exclusive license to Springer Nature Singapore Pte Ltd. 2022

A. K. Nagar et al. (eds.), *Intelligent Sustainable Systems*, Lecture Notes in Networks and Systems 333, https://doi.org/10.1007/978-981-16-6309-3_53

economy of future [2]. To address this cloud-related necessity, a congregation of recently innovated software applications has been designed and developed [3]. Since the user number is fluctuating and loads also exhibit labile nature the corresponding cloud application-related resource capacity requirements also exhibit themselves of dynamic nature [4]. The unexpected gush of user demand designated as flash crowds poses an additional challenge by expanding a request volume up to five times [5]. An outage of Sydney-based Amazon data centres was registered knocking services pertaining to many companies offline [6]. The aforesaid challenges of unexpected events can be duly addressed by application robustness to face the challenges of sudden flash crowd-related load hike. The challenge is further enhanced due to increased data centre failure incidences quite often which otherwise would be exceptional to happen [7, 8]. Unexpected performance degradation may culminate due to workload consolidation along with resultant interference within co-located applications [9–11]. The events due to sudden onset of unexpected surge of flash crowds are high in magnitude but short in duration hence make it unfeasible to ensure sumptuous spare capacity to properly address them. Hence, such unexpected incidences invoke an array of events which are expressed as infrastructure overload translating into unresponsive services resulting in dissatisfied end-users and revenue loss ultimately. Therefore, cloud services-related self-adaptation techniques such as brownout are of great benefit [12–14].

By virtue of minimizing the amount of computations required for serving a request, the brownout service empowers itself to adapt for maintaining response time close to a given set-point. As a consequence, certain computations are specified as mandatory like display of product information pertaining with an e-commerce website while certain computations are reckoned as optional like products with similar qualitative value being recommended. In accordance with available capacity and response time of previous iteration, the service can preferentially choose whether to execute or not to execute the optional code in response to an end-user request. It is noteworthy that providing an optional code-enabled service directly translates into a satisfied end-user and thus incurring more revenue for service provider [13]. The brownout service faces a limiting challenge if service is enabled by a single replica which permits a single copy of application to be executed within a single virtual machine. This challenge can be suitably addressed by providing independent multiple copies pertaining to same application known as multiple replicas which cater same data to the user and each replica is imbibed in a separate virtual machine and further each virtual machine can be hosted by different physical machines. This results in enhancement of brownout services for two parameters—firstly brownout application-related ability to negotiate more users by inducting more computing resources is designated as scalability exhibits improvement since the limitation of using a single physical machine-related resources no longer exists for an application. Secondly resilience power is enhanced which ensures that for any reason if a physical machine is unable of accepting a replica then other replicas whose virtual machines have their placement on different physical machines can execute a seamless takeover of the related request.

The load balancer is the component that determines which replica should address any specific end-user request. Although in-depth studies have been made regarding load balancing techniques, it has been found that state-of-the-art load balancers pass on requests on the basis of certain metrics. These metrics cannot differentiate between a replica which is exhibiting non-execution of optional code for avoidance of overload and for a replica which is not subjected to any overloading stress. Hence, prime objective is to acquire a load balancing technique that is brownout compliant along with awareness regarding each individual replica-related self-adaptation mechanism.

The paper is organized is as follows:

- Sect. 2 describes the related work.
- Sect. 3 elucidates the problem statement.
- Sect. 4 puts forth the system model.
- Sect. 5 considers the evaluation aspect followed by Conclusion in Sect. 6.

2 Related Work

Load balancers are reckoned as Internet scale services-related standard components [15] which make the applications scalability-enabled and resilience-enabled [16–18]. Numerous load balancing policies have been propounded targeting different optimizations ranging from processor load equalization [19] to memory pool [20, 21] management and further specific optimization to address iterative algorithms [22]. The load balancing policies are usually focussed on web server systems [23, 24] and the most pertinent result is generated to contain the maximum response time within limits for the sake of upholding clients-related interests [25]. Load balancing strategies are subjugated to many factors which may include geographical [26, 27], related to electricity price for minimization of data centre-related operational cost [28] or specific strategies for addressing cloud applications [16, 29, 30]. Load balancing solutions are of two types—static and dynamic. Static load balancing is a non-adaptive, fixed approach for selecting a replica in order to direct traffic [31, 32]. In round-robin technique which is commonly used, each replica is selected turn by turn which could be either probabilistic—selecting a replica at random or deterministic—by virtue of focusing to store the replica last selected. This strategy being static in nature does not exhibit satisfactory performance with respect to brownout compliant applications for the reason of their inability to address inherent fluctuations of cloud environment.

The dynamic load balancing focuses for addressing of system-related current state. Here, a replica having lowest response time is taken into consideration. The algorithm is designated as Fastest-Replica-First (FRF) if the basis of choice is each replica-related last measured response time. The algorithm is designated as FRF-EWMA if choice is made on the basis of past response time of each replica-related exponentially weighted moving average. This algorithm with a variation is referred as Two-Random-Choices (2RC) [33] as it selects randomly two replicas and request is assigned to fastest one having the lowest maximum response time.

Any decision based on response times alone cannot justify the efficiency of algorithm. This is because the brownout application cannot be addressed on response times alone. Thus, FRF-EWMA and 2RC are rendered unsuitable to address brownout applications since they selectively depend on response times at a given set-point and hence reducing the ratio of optional contents as a consequence. It is expressed as inability for discriminating between a replica which is unresponsive for the optional core for avoidance of overloading and a replica which is not overloaded while executing all optional codes although both have secured the required response times.

By another approach, pending requests are tracked by load balancer, and the completion of task is executed by those replicas which have minimal number of requests in waiting. This strategy is designated as shortest-queue-first, and it addresses the architectures which possess replicas of similar capacities of homogeneous requests. To address the non-homogeneity factor, the remaining capacity of replicas is utilized as a load balancing solution for management of next request [34]. The capacity is evaluated by considering remaining available CPU and memory, current pending request count and network-related transmission status together. CPU utilization is indirectly controlled by brownout applications by means of optional content adjustment in order to remain prepared for addressing the challenge of possible request bursts. By combining response times and number of routed requests in a linear profile leads to secure merger of fastest replica with pending request count for providing a dynamic solution [35]. It facilitates a predictive load balancer to ascertain replicas' status on the basis of metrics-related performance difference.

3 Elucidation of Problem Statement

The load balancer is required to adjust the prevailing situation by virtue of distributing the load to adaptive parameters. It incorporates a brownout compliant application which is executed in synergism with multiple replicas as in Fig. 1 [16]. The load balancer offers an access to the cloud application architecture by involving clients to serve a request then to either wait for a reply otherwise to repeat the request

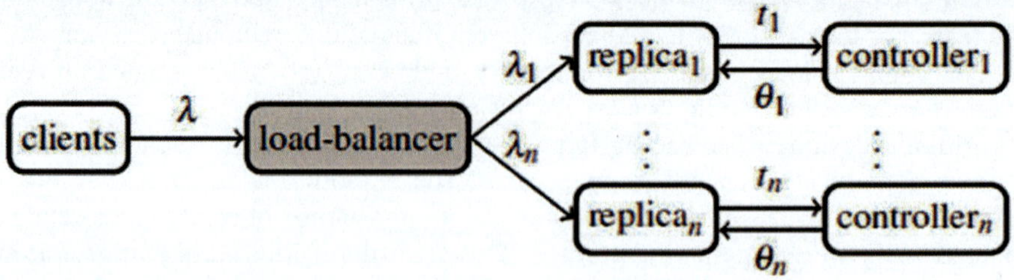

Fig. 1 Multiple replicas-related brownout compliant cloud application architecture

at a distributed time interval. This approach is instrumental while interacting with websites which demand predefined number of requisitions for completing the task like buying a consumer product or a flight booking [36]. The load balancer receives every client request and pass on to one out of 'n' replicas. A replica so chosen in this process responds by reflecting it again back to load balancer which then is passed on to original client. The time required within the replica for execution of request is designated as response time. As the routing of the responses is ricocheted back to load balancer, the information is attached along with, for securing balancing decisions. Since a fraction of incoming traffic is received by each replica, it represents the stand-alone version of application. Each replica is enabled by a controller within it which helps to monitor response time measurement periodically with respect to served requests and adjusts optional component-related percentile regarding control period-related response times. The response time for replica is expressed [13] as follows:

$$t_i^{k+1} = \alpha_i^k \cdot \theta_i^k$$

For establishing synergy with brownout paradigm, a replica response for requests is either partial which include only mandatory contents for response or it may be full when optional and mandatory both contents are included together. Such decision is independent for each request. The rate of partial response generation is more than rate for full response generation hence making partial replies to be computed faster than full ones. It can be expressed as $\mu_i \geq M_i$ where μ_i represents partial response and Mi represents full response. With the assumption for replica not being fully saturated, the requests are served fully at the rate of $\lambda_i \theta_i$ and partially at a rate of $\lambda_i(1 - \theta_i)$ where λ_i stands for measurable traffic rate and θ_i stands for percentage of requests.

For addressing the challenge of synchronicity between load balancer and brownout compliant applications, the load balancer must function in coherence with brownout applications while receiving information about percentage of requests from replicas. This solution culminates in estimation of probability-related execution of the optional components with respect to load balancer. Since a single value is reported on per replica basis, the magnitude of overhead is contained in small limits. For the enhancement of load balancing decisions, the percentage of requests pertaining to replicas is piled up for extending its support to load balancers in curtailing the overhead. During the process, the load balancer is unaware of each replica-related controller adjustments as it only registers the reported value. This facilitates to attain the complete separation of activity of a load balancer from one of the self-adaptive application. The load balancer-related ratio of requests has to be maximized in order to formulate an appropriate load balancer policy. It demands enhancement of percentage of requests pertaining to involved quantities. The load balancer-related values of weights are computed by following expression:

$$\sum_{k=0}^{\infty} \sum_{i} w_i(k)\theta_i(k) \tag{1}$$

where k stands discrete time and w_i stands for value of weights.

4 System Model

4.1 Variation Principle-Based Heuristic (VPBH)

It involves a predictive solution for assessment of variation pertaining with involved quantities. It addresses response time variation with each replica-related request count which is present in pending mode. It is based on changes associated with control variables θ_i whose rising values decipher a less loaded state for more traffic flux to be passed on. Further a decrease in optional content deciphers as less traffic towards replica resulting in load minimization and θ_i value enhancement. The values of θ_i pertaining to replicas assist in periodic update of load balancer-related weight values. It targets procurement of an integral and proportional load balancing action. For assessment of replicas response for a new incoming request, only the positive weight values have to be reckoned. Any increase in integral gain demands a strong action pertaining with load balancing side confirming that load balancer decisions are influenced by instant values of θ_i. As a result it permits utilization of trade-off between robustness and performance. It can be rescaled according to following equation:

$$w_i(k) = \frac{\max(\widetilde{w}_i(k), \varepsilon)}{\sum_i \max(\widetilde{w}_i(k), \varepsilon)} \tag{2}$$

4.2 Equality Principle Based Heuristic (EPBH)

This is a heuristic-based strategy where near-optimal situation is deciphered on the basis of all replicas which should provide identical percentage optional content. It is possible when control variables θ_i are closest to each other. If θ_i value point to a single value, it is inferred that each replica is enabled for serving same percentage of optional content due to routed traffic. It means that more traffic is received by powerful replica as compared to a less powerful one. Hence, the selection of weight should be made judiciously in such a way that convergence of control variable θ_i helps for achieving a mean $\frac{1}{n} \sum_j \theta_j$ and the potential weight is computed by

$$\widetilde{w}_i(k+1) = w_i(k) + \gamma_e \left(\theta_i(k) - \frac{1}{n} \sum_j \theta_j(k) \right) \quad (3)$$

The weight-related positive values indicate that each replica secures input of requests along with incoming traffic.

4.3 *Convex Optimization-Based Load Balancing (COBLB)*

It is an optimization-based approach for updating replica weights with an objective to secure that each replica is destined to follow a queuing system in accordance with processor sharing discipline. To address brownout compliant applications, the requests are provisioned with or without optional content with varying rates. It provides mixture of two exponential distributions with a changing probability function in accordance with following formula:

$$f_{S_i}(t) = (1 - \theta_i) \cdot \mu_i \cdot e^{-\mu_i \cdot t} + \theta_i \cdot M_i \cdot e^{-M_i \cdot t} \quad (4)$$

where τ represents continuous time and θ_i represents probability of activating the optional components. It is the optimization-based load balancer which computes the control variables θ_i. Using the inequalities as constraint, the optimization problem is perceived as concave although the constraints are linear. The optimization-based solution depends on information about service times of requests regardless whether the optional contents are available or not. Hence, it ensures solutions for single time instant problem [23].

5 Evaluation

5.1 *Performance Indicator*

To achieve comparative performance analysis of different algorithms, different parameters are considered as indicators. The first one is percentage $\%_{oc}$ which is regarding total number of optional content-enabled requests in accordance with pertinent metric where users exercise certain number of clicks in order to use the application. The load balancing techniques implemented are further titrated by certain other performance metrics too. The metric user-perceived-stability σ_u evaluates user-related variation of performance which is measured as response time-related standard deviation which is deciphered as replicas ability to respond in time against clients requests (Table 1).

Table 1 Infrastructure resources-related performance variance

Algorithm	$\%_{oc}$ (%)	μ_u	σ_u
COBLB	91.5	0.77	0.98
EPBH	88.7	1.08	1.93
VPBH	88.7	1.01	1.92
SQF	84.5	0.57	0.39

A better user-perceived-stability is secured by stabilizing the response times by brownout framework with load balancer functioning as an adjuvant. The perceived response times are affected by load balancing algorithms. Hence, the algorithms developed more recently with novelty perform better than classical ones for addressing stability factor for achieving response time with enhanced stability and higher values. The average response time is utilized for distinguishing those algorithms which can achieve low response times with high fluctuations.

5.2 *Simulator*

For evaluating load balancing strategies, the brownout compliant applications are processed by CloudSim simulator. The concepts related with replica, request, client, and load balancer are considered. An open-loop client model which upholds Markovian assumption negotiates a new client who initiates a certain number of unrelated requests or it is dealt by a closed-loop client model [37, 38] where client invokes a request after receiving the response to continue for sending request one after another to the application. This makes closed-loop model more realistic while the open-loop model can simulate large number of clients to justify its usefulness. Both models are implemented in the simulator for evaluating applications of diverse nature requiring certain numbers of clicks to be executed by the users.

Load balancer receives the requests for diverting them towards different replicas. The basis of load balancer working rests on either per request or weights-related parameters. The per request basis is used to address simulation policies for tackling random, round-robin and shortest-to-first which are independent of weights concept. The replicas-related functioning is assessed by computational simulation for serving the request and to choose whether execution will be exercised with or without optional components activation. In case optional content is provided then according to Gaussian distribution, service time is taken as a random number having mean value as ϕ_i and a variance of 0.01. In case optional content is not provided, the mean is ψ_i and variance value is taken as 0.001. During the phase of replicas generation, parameters ϕ_i and ψ_i are specified and may exhibit changes during execution. An internal control loop is also executed by replicas in order to select control variables θ_i [13]. For processing the requests in a queued profile, replicas use processor sharing which means that each of the 'n' active requests will be enabled with processing capability of $\frac{1}{n}$ value related with the replica.

The simulator input called scenario supports insertion of new client and erases the existing one. It also monitors execution-related turning on and off the replicas at specific time during the execution phase along with changing the service times for each replica taking both optional and mandatory components into consideration. It exhibits as change in resources back-up for machine hosting the replica. It manages unpredictable changes at architecture level for monitoring the computation capability [39]. For evaluation of changes pertaining to load balancing and replica execution, scenarios offer a complete overview of the changes by virtue of simulations.

5.3 Reacting to Client behaviour

The performance evaluation regarding different algorithms is adjudged with respect to new clients' arrival and disconnecting magnitude of existing clients. For experimental purpose, the first considered replica is taken to be the fastest one with average execution time for mandatory and optional components both. Second replica considered has to slightly slower than the previous one. Third and fourth replicas are the slowest ones. Clients with closed-loop model are considered. Figure 2 depicts each replica-related control variable θ_i in right column and left column depicts effective weights wi which are the weights offered by load balancing strategies. By executing the decreasing percentage of served optional content by means of algorithmic approach, EPBH procures best percentage overall, followed by VPBH, then COBLB, and finally SQF.

The brownout-aware strategies yield better results related with serving the optional contents in terms of percentage, and the SQF algorithm exclusively delivers the same performance related with optional content as such. The advantage of brownout-aware strategy manifests as its ability to constantly underutilize replicas. By the users perspective, an application which exhibits quick response is preferred more than an application which delivers more features but with sluggish response. Therefore, the application-related revenue return to the owner may be enhanced through lower response times, and this trade-off between optional contents and response time-bounds seems to be reasonably addressed by proposed heuristics. It is obvious that optimization-based strategy enhances application-related capability for reacting fast towards changes and achieving high predictive values (Fig. 3).

Load balancing strategies extend their support to enhance functionalities of brownout compliant load applications. It is clearly evident that by considering the application adaptation capability for planning load balancing strategies results in predictable behaviour along with performance maximization. The SQF load balancing strategy encompasses granularity of actuation on a per-request basis. Since current infrastructure status-related information with respect weight-based solutions is updated on priority basis, it facilitates SQF to serve requests faster. It is also evident that EPBH addresses optional contents with highest percentage on the expense of response time-bound.

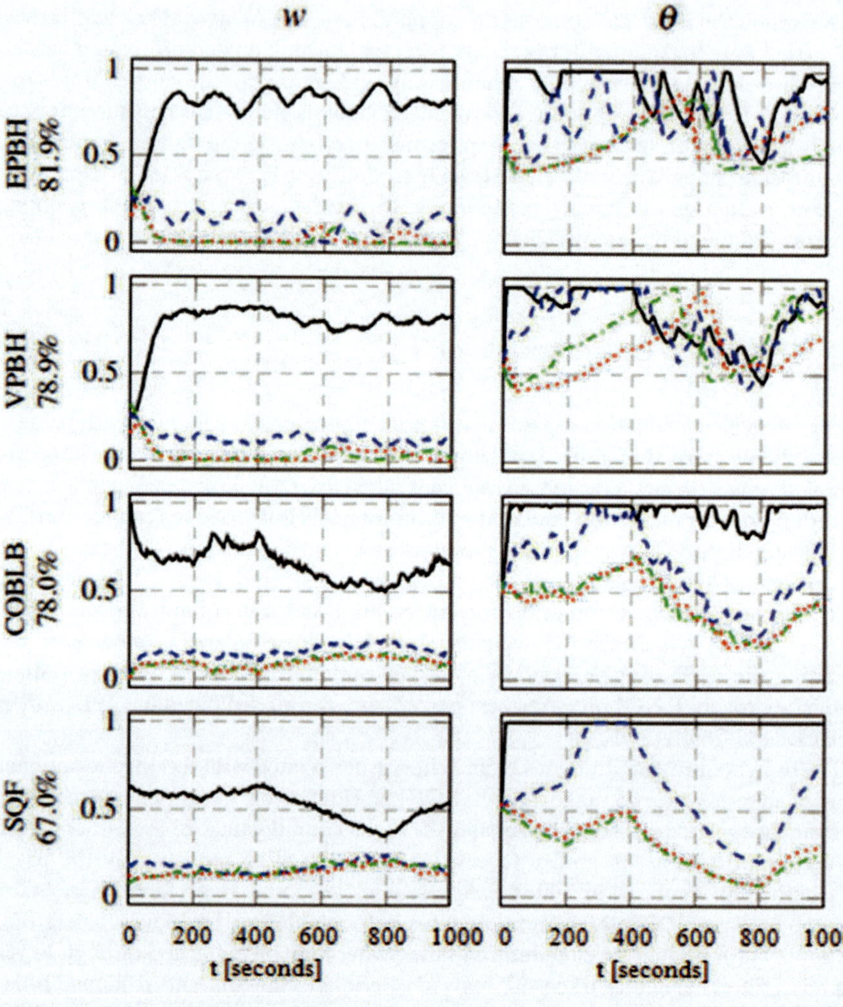

Fig. 2 System-related entry and exit depiction of four replicas with respect to effective weights in right column and control variables in left column at different time instants

6 Conclusion

The ever-expanding scope of cloud computing is inspirational to fulfil the dire need of cloud applications to withstand and negotiate the unexpected challenging events in the form of flash crowds and hardware failures or instant overloads. To alleviate these issues, it is indispensable to establish a synergism between load balancing solutions and self-adaptive applications of brownout. A harmony between these two parameters can suitably address the issue of aforesaid problem by a timely pre-emptive safeguarding manoeuvring. However, in pursuit of achieving the coveted objective

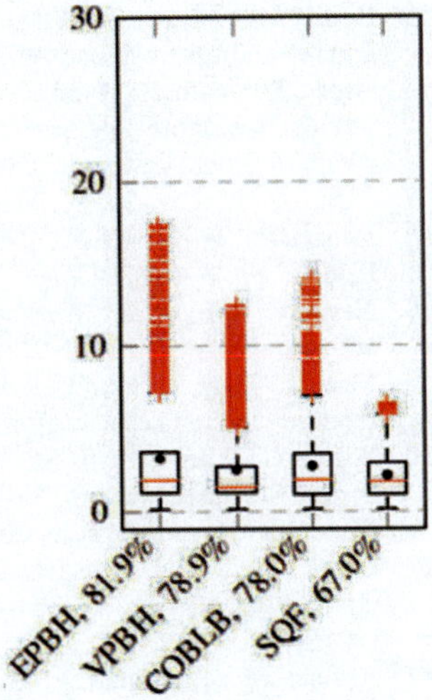

Fig. 3 Replicas-related response time

of efficient networking simply combining load balancing and brownout together without synchronization could lead to poor performance. These two entities, load balancing and brownout compliant applications, are required to exhibit a fusion in such a manner that it could yield a mutually revamping support for one another. The heuristic-based improvised approach in this direction stands out as a convincing solution for addressing these issues. Innovative approaches regarding load balancing tactics and by configuring more adaptive applications could be instrumental for qualitative enhancement of the standards of networking in future. For this a dedicated research-oriented effort has to be pursued relentlessly for securing better solutions by virtue of improvisation and up gradation of the existing infrastructure along with software modelling.

Acknowledgements I am thankful to all concerned with this endeavour.

References

1. Badger, L., Grance, T., Patt-Corner, R., Voas, J.: "Cloud computing synopsis and recommendations", National Institute of Standards and Technology, Special Publication 800–146, 2012, Available Online at: http://csrc.nist.gov/publications/nistpubs/800-145/SP800-145.pdf
2. Irwin, N.: "These 12 technologies will drive our economic future", Available Online at: http://www.washingtonpost.com/blogs/wonkblog/wp/2013/05/24/these-12-technologies-will-drive-our-economic-future/ (May 2013)
3. Buyya, R., Yeo, C.S., Venugopal, S., Broberg, J., Brandic, I.: Cloud computing and emerging it platforms: vision, hype, and reality for delivering computing as the 5th utility. Futur. Gener. Comput. Syst. **25**(6), 599–616 (2009)
4. Reiss, C., Tumanov, A., Ganger, G.R., Katz, R.H., Kozuch, M.A.: "Heterogeneity and Dynamicity of Clouds at Scale: Google Trace Analysis", SOCC (2012)

5. Bodik, P., Fox, A., Franklin, M.J., Jordan, M.I., Patterson, D.A.: "Characterizing, modeling, and generating workload spikes for stateful services", SOCC., pp. 241–252 (2010)
6. Biggs, T., Hatch, P.: "Banks, websites down as wild weather knocks out Amazon Web Services" (2016). Available Online at: http://www.afr.com/technology/banks-websites-down-as-wild-weather-knocks-out-amazonweb-services-20160605-gpc8ob
7. . Guo, Z, McDirmid, S., Yang, M., Zhuang, L., Zhang, P., Luo, Y., Bergan, T., Bodik, P., Musuvathi, M., Zhang, Z., Zhou, L.: "Failure recovery: when the cure is worse than the disease", HotOS., pp. 8–14 (2013)
8. Nagappan, M., Peeler, A., Vouk, M.: "Modeling cloud failure data: a case study of the virtual computing lab". SECLOUD, pp. 8–14 (2011)
9. Mars, J., Tang, L., Hundt, R., Skadron, K., Soffa, M.L.: "Bubble-Up: increasing utilization in modern warehouse scale computers via sensible co locations", MICRO., pp. 248–259 (2011)
10. Zhang, Y., Laurenzano, M.A., Mars, J., Tang, L.: "Smite: precise QoS prediction on real-system smt processors to improve utilization in warehouse scale computers". Proc. of 47th Annual IEEE/ACM International Symposium on Microarchitecture, pp. 406–418 (2014)
11. Cheng, D., Rao, J., Jiang, C., Zhou, X.: Elastic power-aware resource provisioning of heterogeneous workloads in self sustainable datacenters. IEEE Trans. Comput. **65**(2), 508–521 (2016)
12. Salehie, M., Tahvildari, L.: Self-adaptive software: landscape and research challenges. ACM Trans. Auton. Adapt. Syst. **4**, 2 (2009)
13. Klein, C., Maggio, M., Arzen, K.E., Hernandez-Rodriguez, F.: "Brownout: Building more Robust Cloud Applications", ICSE (May 2014)
14. Maggio, M., Klein, C., Arzen, K.E.: "Control strategies for predictable brownout in cloud computing", IFAC WC (Aug 2014)
15. Wang, L, Pai, V., Peterson, L.: "The effectiveness of requestredirection on CDN robustness", OSDI (2002)
16. Barroso, L.A., H¨olzle, U.: "The datacenter as a computer: an introduction to the design of warehouse-scale machines". Synthesis Lectures on Computer Architecture, Morgan and Claypool (2009)
17. Hamilton, J.: "On designing and deploying internet-scale services", LISA (2007)
18. Wolf, J.L., Yu, P.S.: On balancing the load in a clustered web farm. ACM Trans. Internet Technol. **1**, 2 (2001)
19. Stankovic, J.A.: An application of bayesian decision theory to decentralized control of job scheduling. IEEE Trans. Comput. **34**, 2 (1985)
20. Diao, Y., Wu, C.W., Hellerstein, J., Storm, A., Surenda, M., Lightstone, S., Parekh, S., Garcia Arellano, C., Carroll, M., Chu, L., Colaco, J.: "Comparative studies of load balancing with control and optimization techniques", ACC (2005)
21. Patterson, R.H., Gibson, G.A., Ginting, E., Stodolsky, D., Zelenka, J.: "Informed Prefetching and Caching", SOSP (1995)
22. Bahi, J.M., Contassot-Vivier, S., Couturier, R.: Dynamic load balancing and efficient load estimators for asynchronous iterative algorithms. IEEE Trans. Parallel Distrib. Syst. **16**, 4 (2005)
23. Cardellini, V., Colajanni, M., Yu, P.S.: Request redirection algorithms for distributed web systems. IEEE Trans. Parallel Distrib. Syst. **14**, 4 (2003)
24. Manfredi, S., Oliviero, F., Romano, S.: A distributed control law for load balancing in content delivery networks. IEEE/ACM Trans. Netw. **21**, 1 (2013)
25. Huang, C., Abdelzaher, T.: Bounded-latency content distribution feasibility and evaluation. IEEE Trans. Comput. **54**, 11 (2005)
26. Andreolini, M., Casolari, S., Colajanni, M.: "Autonomic Request Management Algorithms for Geographically Distributed Internet-Based Systems", SASO (2008)
27. Ranjan, S., Karrer, R., Knightly, E.: "Wide area redirection of dynamic content by Internet data centers", INFOCOM (2004)
28. Doyle, J., Shorten, R., O'Mahony, D.: Stratus: load balancing the cloud for carbon emissions control. Cloud Comput. IEEE Trans. **1**, 1 (2013)

29. Lin, M., Liu, Z., Wierman, A., Andrew, L.L.H.: "Online algorithms for geographical load balancing", IGCC (2012)
30. Lu, Y., Xie, Q., Kliot, G., Geller, A., Larus, J.R., Greenberg, A.: Join-Idle-Queue: a novel load balancing algorithm for dynamically scalable web services. Perform. Eval. **68**, 11 (2011)
31. Ni, L., Hwang, K.: Optimal load balancing in a multiple processor system with many job classes. IEEE Trans. Softw. Eng. **11**, 5 (1985)
32. Tantawi, A.N., Towsley, D.: Optimal static load balancing in distributed computer systems. J. ACM **32**, 2 (1985)
33. Mitzenmacher, M.: The power of two choices in randomized load balancing. IEEE Trans. Parallel Distrib. Syst. **12**, 10 (2001)
34. Pao, T.L., Chen, J.B.: "The Scalability of Heterogeneous Dispatcher-Based Web Server Load Balancing Architecture", PDCAT (2006)
35. BIG-IP Local Traffic Manager, Available Online at: http://www.f5.com/products/big-ip/big-ip-local-traffic-manager/
36. Garcıa, D.F., Garcıa, J.: TPC-W E-commerce benchmark evaluation. Computer **36**(2), 42–48 (2003)
37. Alomari, F., Menasce, D.: Efficient response time approximations for multiclass fork and join queues in open and closed queuing networks. Parallel Distribut. Syst. IEEE Trans. **99**, 1–6 (2013)
38. Schroeder, B., Wierman, A., Harchol-Balter. M.: "Open versus closed: a cautionary tale", NSDI (2006)
39. Tomas, L., Tordsson, J.: "Improving Cloud Infrastructure Utilization Through Overbooking", CAC., pp. 1–10 (2013)

Blockchain-Based Location Sharing in 5G Open RAN Infrastructure for Sustainable Communities

Mojtaba Enayati, S. Sree Lekshmi, Tom Toby, Makarand Prabhu, K. P. Rahul, S. Parvathy, and Seshaiah Ponnekanti

Abstract Current trends indicate that 5G network deployment is gaining traction around the world; this paradigm shift is expected to deliver gigabit connectivity and wide-ranging services to its subscribers. These benefits, however, are currently restricted to users in metropolitan cities. Expansion of these services to rural areas is hindered by the lack of economic incentives for providers, forcing remote populations to use legacy pre-5G technologies. The key features of 5G include small cells, low-cost radio access points, energy-efficient data centers, and low footprint dedicated networks. The impact of these features in an urban context has been widely discussed, but they also have immense potential to serve the unique and differentiated needs of rural communities, servicing quality-of-life improvements through precision agriculture, augmented reality telemedicine, and reliable disaster support.

These authors contributed equally to the work (Joint first authors).

M. Enayati (✉)
School for Sustainable Development, Amrita Vishwa Vidyapeetham, Amritapuri, India
e-mail: amiddids20031@am.students.amrita.edu

S. S. Lekshmi (✉) · T. Toby · M. Prabhu · K. P. Rahul · S. Parvathy · S. Ponnekanti
Center for Wireless Networks and Applications (Amrita WNA), Amrita Vishwa Vidyapeetham, Amritapuri, India
e-mail: sslekshmi@am.amrita.edu

T. Toby
e-mail: amenp2wna20013@am.students.amrita.edu

M. Prabhu
e-mail: amenp2wna20006@am.students.amrita.edu

K. P. Rahul
e-mail: amenp2wna20012@am.students.amrita.edu

S. Parvathy
e-mail: amenp2wna20009@am.students.amrita.edu

S. Ponnekanti
e-mail: seshaiahp@am.amrita.edu

© The Author(s), under exclusive license to Springer Nature Singapore Pte Ltd. 2022
A. K. Nagar et al. (eds.), *Intelligent Sustainable Systems*, Lecture Notes in Networks and Systems 333, https://doi.org/10.1007/978-981-16-6309-3_54

These critical services can be realized through dedicated spectrum allocation by regulating authorities, and the regionalization lowers costs, allowing for rapid deployment of 5G local networks. In this scenario, the interplay between local community networks and public mobile networks is of paramount importance. The roles of reference design, system architecture, and effective smart contracts in the creation of sustainable communication infrastructure have been outlined in this paper. Disaggregated radio access network (Open RAN) and blockchain have been integrated into the framework, allowing rapid, end-to-end secure data communication. The use of network slicing and effective co-operation between local community networks and mobile network operators, together with the elucidated framework, enable efficient, privacy-conscious networking which can be relied upon for the delivery of critical services.

Keywords Sustainability · LCN · 5G · Open RAN · Network slicing · Sustainable development goals · Blockchain · Smart contract · Data privacy.

1 Introduction

During the COVID-19 pandemic, the demand for connectivity rapidly increased, with employment, education, and social interaction shifted online. This trend is noticeable in all sectors like health care, banking, e-governance, e-commerce, and education. The world has been witnessing advancements in the wireless communication industry, bringing industrialization in urban communities. However, the situation is different in rural communities. There exists a digital divide between urban and rural areas, resulting in largely unconnected rural communities. This is due to the lack of interest of Mobile Network Operators (MNOs) to upgrade their network in such low investment rural areas. As a result, alternative communication infrastructure solutions are considered by various stakeholders like Local Community Networks (LCNs)/government councils/Non-Government Organizations (NGOs)/enterprises/utility companies [1]. There are ongoing initiatives for "connecting the unconnected" communities to offer Internet access and connectivity to unserved rural communities, realizing the Sustainable Development Goals (SDGs). The key components involve connecting rural people to multiple opportunities, businesses, and life-enhancing services. One of these solutions, offering affordable connectivity for fishermen communities, has already deployed as a part of the OceanNet framework [1]. An alternative program, the Amrita Multi-Modal Applications and Human–Computer Interaction (AMMACHI) Labs [1], has taken up nearly 101 Indian villages and continues research projects for sustainable development of the rural communities.

2 5G Framework and Technology Enablers in LCNs

Recent trends in the 5G and Beyond 5G (B5G) technologies can be explored to design and deliver secure, sustainable, and affordable connectivity solutions for underprivileged users. In particular, 5G LCNs can be rapidly deployed as dedicated coverage solution. One of the key enablers that brings a unique blend of low-cost, energy-efficient, battery backup operated sustainable community infrastructure is the Open RAN. In Open RAN, multiple vendors lead to cheap and flexible deployments. Open RAN separates the radio network into multiple functional modules connected by standard interfaces and operating on Common Off-the-Shelf (COTS) hardware. The Open RAN with the key elements of ETSI NFVI/Multi-Access Edge Computing (MEC) [2] frameworks leads to the next generation of micro-edge Data Centers (μDCs) which are easily deployable by LCNs. The primary aim of sustainable framework is to support diverse services to cater the societal needs of today's unconnected communities in the developing world. Network slicing technology in 5G can enable distinct services to the communities. The role of Open RAN and slicing in realizing sustainable infrastructure will be discussed in detail in Sects. 2.2 and 2.3. The following section describes various services in rural communities.

2.1 Location-Based Services in Sustainable Communities

With the availability of cheap smartphones, the usage of mobile apps has been increasing. However, user communities lack awareness of the privacy. Network providers are yet to provide comprehensive safeguards to preserve the privacy of the users. Recent initiatives to develop local/regional language applications have been emerging to enhance the societal engagements. GAIUS networks [3] have outlined an application platform named "blur app". These types of applications can facilitate health care, emergency services, and local marketplaces or enterprise zones to improve efficiency of small traders in rural areas. These context-aware applications require location information to dynamically adapt behavior as per the user's current location data. In future, sustainable communities relying on advanced 5G technologies will be open to security threats. The security threats that have been witnessed may apply to the apps used by sustainable communities. More details on how subscribers can be protected from upstream attacks can be found in [4]. Unintentionally, at present communities are not recognizing the value of keeping safe location data. Table 1 shows the types of rural services where sharing of location data becomes critical to protect the privacy.

The paper proposes a new blockchain-based approach using a novel consensus algorithm to utilize smart contracts and decentralized applications between MNOs and third-party applications. This will enable secure sharing of the critical information (location) to deliver services under the users' control. The paper is organized as

Table 1 Rural location-based critical services

Rural services	Criticality of location data
Social care-Web-based monitoring of old-age homes/childcare	To avoid kidnaps/attacks
Health care/telemedicine	To monitor patients
Retail outlets-local supermarkets/farm outlet	Home delivery to eliminate middlemen
Financial-Internet banking/insurance	To avoid illegal transactions
Transportation-bus/cars/ferry	App-based riding
Emergency-public safety	To transport victims/patients
Postal delivery-local/international	Tracking the status and integrate with banks
Recreation-fitness/sport centers/tourism	App-based assistance/experiences

follows. Section 1 highlights the digital divide between urban and rural communities. Section 2 describes the 5G framework design and technology enablers for secure sharing of location data between MNOs and third-party applications. Sections 3 and 4 explain the significance of blockchain for sharing critical information (location) between MNOs and third-party applications. Section 5 discusses the preliminary validation of localization approach with super-resolution Direction-of-Arrival (DoA) technique with MUSIC-SVD [5]. Section 6 concludes the paper.

2.2 *Key Components to Enable 5G Sustainable Infrastructure*

Figure 1 shows the system design framework highlighting architectural advancements in 5G such as Open RAN, intelligent edge Data Centers (DCs), and network slicing to deliver diverse services. The 5G architecture popularly known as Service-Based Architecture (SBA) [6] is diverse and composable that changes based on use case model of the network. To change from the legacy single networking to a flexible, on-demand network architecture, 5G network slicing technology plays a significant role. Network slicing helps to create dynamic network instances resulting in multiple logical networks. The end-to-end connection (from end devices, over the RAN, edge, transport, and packet core to the application) allocates a dedicated set of physical and virtualized network resources between the end-user and the application.

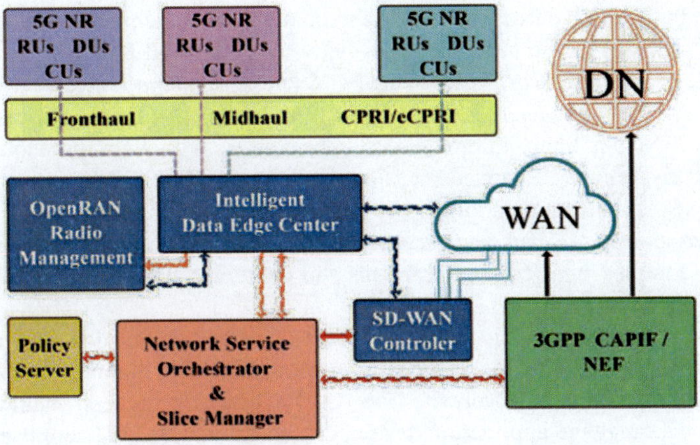

Fig. 1 Technology components in the 5G sustainable infrastructure [6]

In a typical service workflow, a slice request will be initiated from the vertical stakeholders via 3GPP Common API Framework (CAPIF)/Network Exposure Function (NEF) in the SBA. Thereafter, Network Service Orchestrator (NSO) will coordinate with the Policy Control Function (PCF) in the 5G SBA to enforce the policies related to the slice. In the Open RAN [6] paradigm, the RAN architecture has been disaggregated based on the split of the control and user plane protocols, and the network elements have been separated into Radio Units (RU), Distributed Unit (DU), and Centralized Unit (CU). The intelligent μDCs enable application controller to locally utilize the location and context analytics for intelligent topology selection (the combination of the nodes RU/DU/CU) at the edge. Software-Defined Wide-Area Network (SD-WAN) controllers oversee the transport domain. Open RAN manager supports the distributed multi-vendor virtual DU/CU and hardware RUs.

2.3 Overall System Design for Location Sharing Between MNOs and Third-Party Applications

The concept of smart contracts has been used widely in several critical areas such as financial sectors, and it is now recognized that smart contracts can deliver real revenues and cost savings in wide-ranging applications including telecom networks. Decentralized Autonomous Organizations (DAO) emerged during the last years by utilizing smart contracts where all the processes in a company are executed by smart contracts. 5G network has been evolving to embrace the network-of-networks paradigm, and affordable connectivity together with critical and sustainable communication services for rural communities dictates interplay between several business entities. The framework delineated in this paper maps onto the base of the pyramid as shown in Fig. 2 which corresponds to the manual contract that facilitates the roving (seamless moving) of the network slices between LCNs and MNOs. The network slice has been assumed to originate in the LCN and subsequently traverse the MNOs.

In the next step, however, Radio Access Network (RAN) sharing between with LCNs and MNOs is likely to morph into hybrid slices with a simultaneous mix of public and private slices. Recent discussions in the 3GPP, IETF, ETSI, and other international forums indicate technology evolution toward autonomous edge computing also termed as MEC. Smart contracts are expected to offer compelling benefits to reflect the complex inter-edge interactions and the underlying sharing mechanisms between LCNs and MNOs.

Finally, at the apex of the pyramid, smart contracts could play a pivotal role to facilitate easy and flexible business processes between mobile network operators, IT/cloud providers, Over-The-Top (OTT) content providers, managed SPs, local/regional language application developers, and neutral hosts (campuses, malls, exhibition sites, and smart cities). Figure 3 illustrates how LCN subscribers could be connected by distributed architecture enabling seamless and sustainable connectivity solutions. The key idea is to exchange private and public slice between LCN operators and MNOs to support sustainable slicing, assuring service continuity. This section outlines the main procedures for location sharing between MNOs and third-party

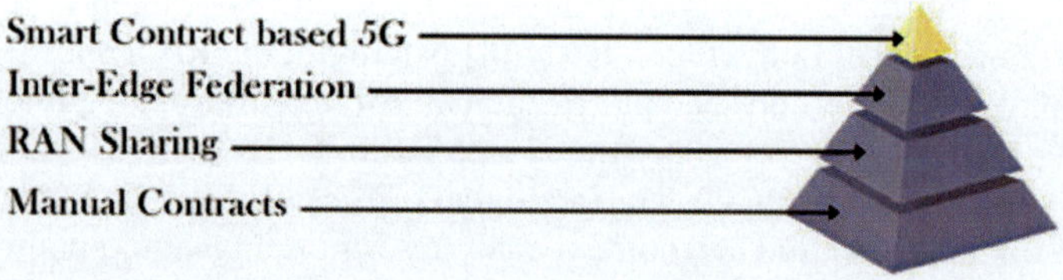

Fig. 2 Scope of smart contracts in 5G service provider (SP) networks

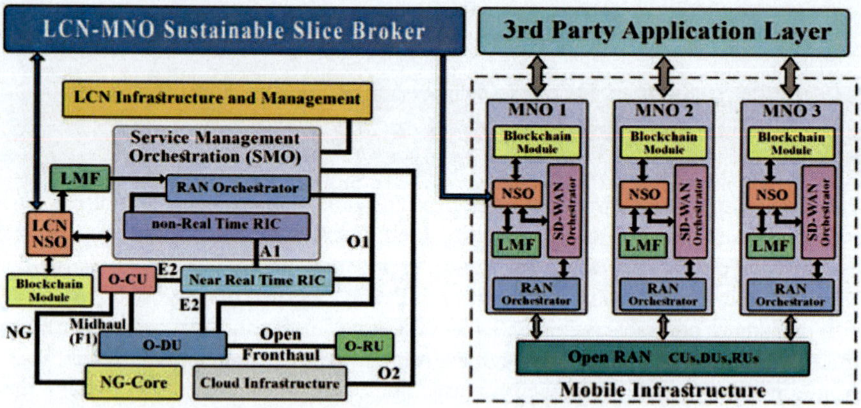

Fig. 3 Depiction of the location sharing workflow between MNOs and third-party applications

applications wherein the subscriber management and service activation expected to be undertaken by the LCN operators.

Slice requires coupling between different orchestrators at every stage to realize sustainable slicing. Sustainable slice needs to traverse between LCN and MNO Orchestrators. When trigger comes from third-party applications for location data, the trigger will pass through the blockchain module collocated with the management systems. The orchestrators will interwork with Location Management Function (LMF) to collect the location data from the RAN utilizing key localization techniques such as Enhanced Cell ID/Timing Advance (TA) Positioning, Observed Time Difference of Arrival (OTDoA). In LCN infrastructure, the LCN NSO will interact with Service Management and Orchestration (SMO) [7], which will connect to and orchestrate the RAN intelligence controller (RICs), cloud infrastructure, O-CU, and O-DU via O1, A1, and O2 interfaces.

When subscriber moves from LCN deployments to MNOs, NSO of MNOs will pass the information via slice broker with respective orchestrators to manage the required quality for the slice. NSO will absorb the LCN database and abstract the subscriber's profile that indicates the sustainable slice requirements. NSO further interacts with SD-WAN and RAN orchestrator and instantiates the Open RAN components (RUs/DUs and CUs) to enable the end-to-end flow. The key module blockchain will securely transfer the location data required for the third-party applications. The significance of blockchain in sharing the location data will be discussed in detail in the following Sects. 3 and 4.

3 Introduction to the Blockchain Technology

The blockchain technology, which was first introduced by the Bitcoin in 2008 [8], at the first glance may seem to be very disruptive. However, when designed wisely and implemented carefully, it has the potential of solving some of the most challenging problems of the new digital era. Blockchain can bring sustainability [9], trust, clarity, reliability, and automation, offering more security and better data privacy. It can facilitate public and private networks with an automated framework for sharing their infrastructure to realize the 5G network.

3.1 Evolution of the Blockchain Technology

Blockchain networks can be one of the three types: public, private, or consortium blockchain [10]. The public blockchain lets everyone join the network. A private blockchain, however, as its name suggests normally belongs to a specific organization and does not let entities from outside of the organization participate in the network. The consortium blockchain is like a private blockchain but comprises more than one organization. In such a system, nodes should be authorized before joining the network. Bitcoin's blockchain was the first generation of blockchains, and during the time, it was realized that it can be used in several other fields too. It could be designed to serve sustainability, data privacy, and energy efficiency. The second generation introduced the smart contract and decentralized application (DApp) concepts [11]. Utilizing smart contracts lets us decouple the mobile application from centralized servers. The third generation is going to be more sustainable, and it definitely will be much more energy efficient. Blockchain technology is a great enabler to realize the 5G network, by creating an automated leasing system for MNOs and LCNs to share their processing powers and networking infrastructure while keeping their data secure.

4 The Proposed Blockchain-Based Framework

The proposed solution is a consortium blockchain where the nodes of the system are MNOs and LCNs. The consensus algorithm used in the proposed solution is a novel BFT consensus called Proof_of_Connection (PoC), which uses the benefits of both Proof_of_Authority (PoA) and Proof_of_Stake (PoS) consensus algorithms for a higher performance. All nodes before participating in the system should be authenticated and stake a predefined amount of local digital assets to guarantee their honest behavior. Subsequently, they will install required mining software on their network management system (NMS) to perform required functions. Participants monitor each other's behavior and prevent the system from being disturbed by probable malicious activities, therefore, leading the system to a sustainable condition.

In a consortium blockchain, the communication between nodes will be faster and more secure. The system will be much more scalable, the block size setting can be much more flexible, and the block time will be much lesser in comparison with a public blockchain as shown in Table 2. Blockchain fork formation will not be an issue and platform updates happen easier as it is a private blockchain. The

Table 2 Suggested blockchain in comparison with public blockchain

Comparison factor	Proposed blockchain	Public blockchain
Sustainability	Designed to be sustainable	Depends on the system
Communication speed	High speed transactions	Vary in different systems
Scalability	Highly scalable	Can be a big challenge
Block time	Small amount of time needed	10 min in bitcoin
Forking probability	Not possible	Possible
Transaction fee	Almost free	Can be very high
Energy consumption	Energy efficient	Energy consuming in proof of work consensus
Decentralization	More centralized	More decentralized

transactions in the system can be much cheaper than in a public blockchain. Moreover, as there will be no need to compete on creating new blocks between the nodes, the system will be very energy efficient and much more sustainable than other networks. The aforementioned attributes lead to a high performance and sustainable system to support the 5G network for ultra-reliable and low latency communication, and connect the unconnected communities, especially in rural areas.

Full nodes of the blockchain will keep the ledger and will be responsible for running the smart contracts and hosting the decentralized applications. To join the network, any new node has to use a DApp to buy a minimum amount of local digital assets to be able to run its smart contracts. It also helps to stake the required digital assets for monitoring purposes. Moreover, small LCNs can choose to become a light node to only keep the block headers to run the system.

4.1 Workflow of the Handover Process Between LCNs and MNOs

To illustrate how the system works, imagine a subscriber tends to move toward its MNO's coverage zone border. When the user equipment (UE) measures the low Received Signal Received Power (RSRP), will try to latch to another network. Moreover, the subscriber needs to have a specific DApp called "SliceConnect" installed on the belonging smartphone. This helps the smart contract in monitoring the Quality of Service (QoS), to inform the subscriber using appropriate notifications, and to have a reputation system regarding the MNOs service delivery.

The subscriber's equipment searches for available networks, and the SliceConnect DApp will commence preparation phase to preserve sustainability in subscribers' connection experience. The SliceConnect DApp performs a technical evaluation of all neighbor MNOs by measuring their service provision capability based on the related factors and excludes incompatible MNOs. The result of the process will be a list of possible MNOs capable of serving the subscriber as per its need. The SliceConnect then starts a commercial bidding process between MNOs in the list based on their service pricing. The winner of the bidding process will be selected as target cell.

The selected visited network will create a new block on the blockchain containing the subscriber's identification data to request for authentication from its home SP which will be followed by a new block approving or rejecting the request. If the request is approved, the handover process gets started immediately after initiating a smart contract between the subscriber's visited network and the home network. All the block creations and smart contract initiations on the network takes place after applying digital signatures by the participants using their private keys. The SliceConnect DApp helps the smart contract to ensure that the delivered service meets the terms of the agreement based on the subscriber's need. An appropriate rating shall be applied to the SP for the subsequent requests. Depending on the

```
if (new subscriber requested for handover) then (create a new block for subscriber authentication)
    if (permission granted)
        while (subscriber is connected) do (initiate appropriate smart contract)
            if (any new request) then initiate new smart contract)
        end
    else (drop the request) end
end
```

Fig. 4 Handover and smart contract initiation process in MNO network management system

subscriber's request, home SP will be charged with proper amount of local digital assets.

As illustrated in Fig. 4, the service charging procedure is handled by a smart contract, and it is expected to take place before the service delivery. Upon any specific request, a predetermined service charge will be applied by the smart contract. If the traffic volume consumed by the user exceeds the predetermined traffic quota, a new smart contract will be initiated. If the pre-allocated quota was not fully utilized by the subscriber, the smart contract will reconcile with the billing and the accounting system at the Policy and Charging Rules Function (PCRF). The charging mechanism will take into consideration the data received from the SliceConnect DApp and releases only equivalent amount of digital assets to the SP's wallet, and the rest will be returned to the home SP's wallet.

4.2 *Workflow of the Location Sharing Process*

The amount of data transferred on the Internet is rapidly growing due to technological and mobile app penetration. End users share private data during online activities through the insecure Internet medium. This raises concerns about personal data privacy. The proposed solution can bring much more than a safe, reliable, and automated leasing framework for the MNOs to share their infrastructures to realize the slicing technique in 5G network. It can also offer a secure and trustworthy way for sharing subscriber's private data whenever and with whomever the user decides.

Currently one of the major issues with the current mobile applications is that users should grant various permissions at the time of application installation. This lets the applications collect data from user's device and misuse the data without explicit consent of the user [12]. There will be no way to prevent the installed application from accessing the data but uninstalling it. To mitigate the data misuse, utilizing Open RAN architecture allows MNOs promote trusted third-party applications offering specific services to the subscribers registered with the MNO. The proposed solution allows the MNO share its subscribers' private data in a protected manner, whenever the user requests for sharing it. Like all the subscribers who need to use the "SliceConnect" DApp which acts as an oracle to collect monitoring data for smart contracts, there will be a DApp at any third party's server to receive subscribers' private data from MNO via specific smart contract.

The following section illustrates the logical process of sharing users' location as an example of private data, giving the ability to receive a location-based service from a third party through its specific application. This makes the solution more robust and useful to serve the MNOs and subscribers at the same time. First, the third-party application needs to be verified by the MNO before being recommended to any subscriber for installation. When the user opens the application, which works based on the location, the request for user's current location will be sent to the SliceConnect DApp running on subscriber's smartphone. As a result, the SliceConnect DApp will inform the MNO about the request by sending the following message to the network: *Req(MNO_id, User_id, App_id, Requested_Data_Type).*

The aforementioned request will initiate a smart contract between the MNO and the specific third-party, whose identification number has been mentioned in the request from the user side DApp. All entities in the proposal will be having a digital wallet holding their local digital asset as a token or cryptocurrency and showing the history of their activities. Any smart contract will be granted access to the contract correspondents' digital wallets to reduce or deposit the service-related charges. The third party will pay the MNO for the user's location data through the smart contract. The smart contract will deduct the specific amount of digital assets from the service receiver's wallet and holds onto it until the agreement terms have been met.

After smart contract initiation as shown in Fig. 5, the MNO encrypts the user's location data with the receiver's public key and digitally signs it with its own private key. Asymmetric encryption will make sure that no one would be able to read the data on the block except the right receiver, and the digital signature will assure the receiver that the sender is who it claims to be. Upon receiving the expected data from the MNO, the DApp in the third-party side will let the smart contract know that the terms of the agreement have been met. Therefore, the respected digital asset will be deposited to MNO's digital wallet. The whole process of running DApps either on the user's side or the third party's side, which will be lying under the application layer, will be handled by the nodes on the network, lying under the blockchain layer

```
if (MNO1_digitalSignature=Valid && 3rdParty1_digitalSignature=Valid)
  if (3rdParty1Acc_balance >= ServiceFee)
  then (Deduct ServiceFee from 3rdParty1Acc and Order MNO1 to share the data
        and Set Timer as Service_deadline)
      while (Timer=Valid) do (wait for 3rdParty1DApp Signal)
            if ( Confirm Signal received) then (release ServiceFee -> MNO1)
            else if ( error signal received)
            then (release ServiceFee -> 3rdParty1 and set negative point for MNO1 reputation)
            end
      end
  end
end
```

Fig. 5 Smart contract to share location between MNOs and third parties of the network. The MNO's creating and holding the consortium blockchain will help each other to manage and maintain a secure, private, and reliable network to elevate their subscribers' convenience.

Transparency in the system lets the users be able to see where their location has been used and to whom it has been shared. Users will be having full control over their location sharing process as the user is the initiator of the data sharing process. The protocol can be used to control the sharing process of any type of private data like users' current location, based on the services offered by third party applications and verified by MNOs. Any communication needed during the process may be encrypted with Transport Layer Security protocol TLS 1.3. TLS provides the system with end-to-end security by running a mixed method of symmetric and asymmetric cryptography.

5 Application of Location Estimation Using DoA Techniques

Though location information is available on the network, SPs depend on third-party applications for location data. Presently, devices with a global positioning system (GPS) are used by over-the-top (OTT) providers to collect user location data. GPS is not an accurate location technology for dense 5G scenarios. In such a GPS denied environment, UE-assisted location tracking is preferred in which SPs utilize network intelligence to assist the OTTs. In 5G, the network topology will transform to dynamic beam-based cells that gather and sustain the precise location of users at the network edge. Consequently, 3GPP brings development in 5G positioning as a module of next-generation cellular technology to meet the diverse needs of different verticals.

In the first phase of the realization of the framework, a location module suitable for extracting the location has been designed in this paper. The module utilizes a modified form of the well-known direction of arrival estimation called Multiple Signal Classification (MUSIC). Further, the numerical stability of the algorithm has been enhanced by adopting the singular value decomposition (SVD). For the numerical evaluations, a suitable graphical user interface (GUI) has been developed in MATLAB as shown in Figs. 6 and 7 to validate the accuracy of the localization approach with modified super-resolution DoA estimation technique using MUSIC-SVD.

The accuracy of the localization technique has been evaluated in MATLAB, assuming static simulation scenarios (considering predetermined fixed angles of arrival). To reproduce the multi-path or dynamic environment, the performance analysis was undertaken to examine the sensitivity of the key degrees of freedom (DoFs) like the number of snapshots, the number of antenna elements, and signal-to-noise ratio (SNR). The DoA estimation GUI includes provisions for dynamic behavior based on the number of sources or users. The GUI has options to dynamically adjust

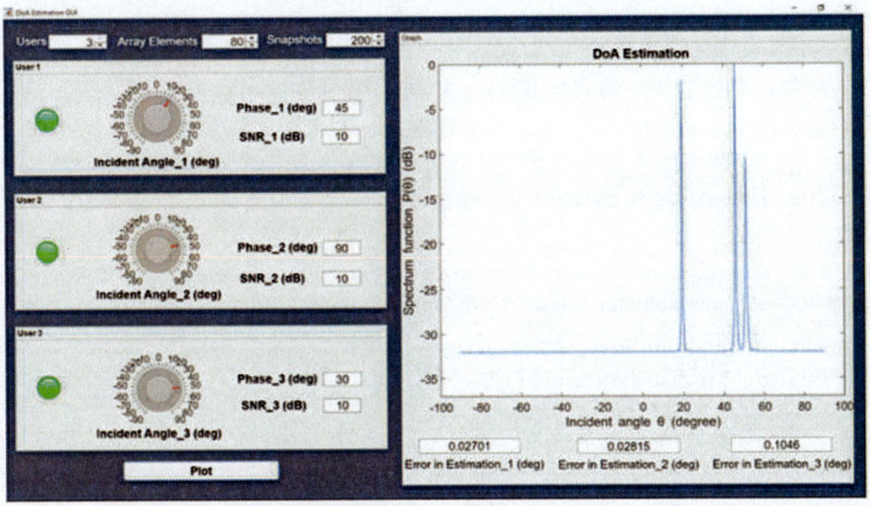

Fig. 6 DoA estimation with MUSIC-SVD for non-coherent sources

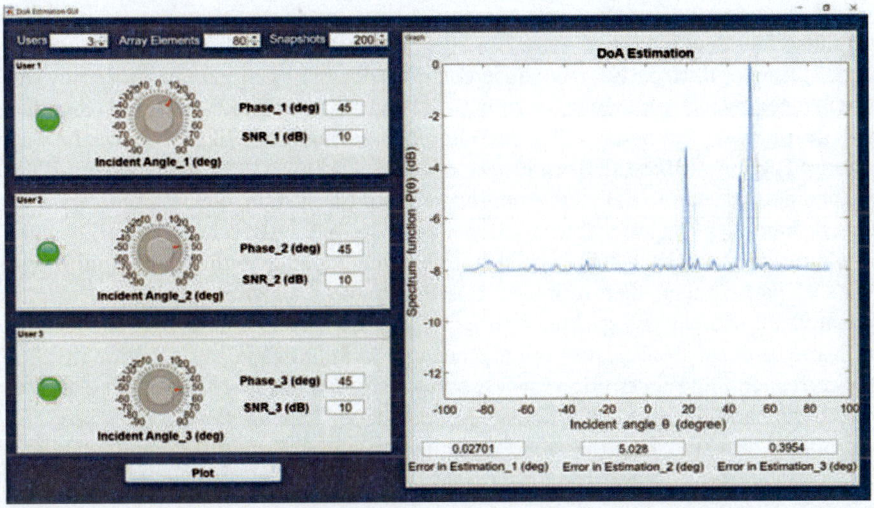

Fig. 7 DoA estimation with MUSIC-SVD for coherent sources

DoF values. The output includes a visual plot between spectrum function and incident angle. The GUI also shows the absolute error of estimation in degrees. Different scenarios were tested by varying DoF for coherent and non-coherent sources. Figure 6 shows the simulation considering the incident angle of arrivals as 20, 40, and 60 degrees for non-coherent signal sources with phases 45, 90, and 30 degrees, with SNR = 10, the number of snapshots as 200, and the number of antenna elements 80.

Figure 7 shows the simulation considering the incident angle of arrivals as 20, 45, and 50 degrees for coherent signal sources with phases 45 degrees, with SNR = 10, the number of snapshots as 200, and the number of antenna elements 80.

6 Conclusion and Future Scope

Based on the reference design framework proposed in the paper, the following elements will be considered for the testbed development of the LCN infrastructure.

- Small cell operating in the 5G New Radio (NR)-U (unlicensed).
- Virtualized 5G distributed unit (5G vDU).
- Open Source Front Haul Gateway (OFHG).
- Virtualized 5G Controlled Unit (5G vCU).
- Stand-alone 5G Core (5G SA) in the AWS.
- μONOS Near Real-Time RAN Intelligence Controller (nRT-RIC).
- μONOS non-RT RAN Intelligence Controller (NRT-RIC).

Going further, to utilize the blockchain to its full potential, security and deep learning are likely to play a key role in the next phase of the research. To serve the subscribers even further, there is a chance to utilize a secure blockchain-based marketplace for third parties from different fields of activity. Moreover, efficiency can be enhanced by using machine learning (ML) and artificial intelligence (AI) together with the proposed framework. These technology advances are likely to pave the way to formulate innovative and flexible service-level agreements (SLAs) between all the stakeholders to cater to the wider-ranging needs of the sustainable communities. The current framework is amenable to extensions to realize private Messenger DApp to be offered to the subscribers. The system has the capability of accepting different types of participants; therefore, one possible way of expanding the system is to incentivize different enterprises such as factory owners to share their networking infrastructures. Moreover, new approaches to address diverse consumption profiles and corresponding subscription models between the MNOs and LCNs can be used to extend the handover process articulated in this paper. Finally, the charging scenario and billing process using the smart contract can be developed further for the best performance of overall the system proposed in the paper.

Acknowledgements This project has been funded by the E4LIFE International PhD Fellowship Program offered by Amrita Vishwa Vidyapeetham. I extend my gratitude to the Amrita Live-in-Labs® academic program for providing all the support.

References

1. Enayati, M., Sree Lekshmi, S., Poluparthi, B.P., Rajagopal, L., Ponnekanti, S.: "Integrating Blockchain, open RAN And IoT platforms into rural community networks". Proceedings of the International Conference on Sustainable Development in Technology for 4th Industrial Revolution 2021 (ICSDTIR-2021) 12–13 March, 2021, Chattogram, Bangladesh
2. Hou, X., Ren, Z., Yang, K., Chen, C., Zhang, H., Xiao, Y.: IIoT-MEC: a novel mobile edge computing framework for 5G-enabled IIoT. IEEE Wireless Commun. Netw. Conf. (WCNC) **2019**, 1–7 (2019). https://doi.org/10.1109/WCNC.2019.8885703
3. GAIUS Networks, https://flipped.gaiusnetworks.com/. Last accessed 13 Jun 21
4. Sree Lekshmi, S., Bandodkar, S.S., Vippalapalli, V., Susarla, A., Ponnekanti, S.: Data optimization-based security enhancement in 5G edge deployments. In: Raj J.S., Iliyasu A.M., Bestak R., Baig Z.A. (eds.) Innovative Data Communication Technologies and Application. Lecture Notes on Data Engineering and Communications Technologies, vol. 59. Springer, Singapore, (2021). https://doi.org/10.1007/978-981-15-9651-3_30
5. Lekshmi, S.S., Sai Shibu, N.B., Nithin, P.S., Ponnekanti, S.: "Platform to effectively disseminate location information in 5G for supporting mission critical services," 2019 International Conference on Wireless Communications Signal Processing and Networking (WISPNET), pp. 517–522 (2019). https://doi.org/10.1109/WiSPNET45539.2019.9032854
6. Das, R.M., Sree, L.S., Ponnekanti, S., Paunovic, M.: "Key enablers to deliver latency-as-a-service in 5G networks," 2019 27th Telecommunications Forum (TELFOR), Belgrade, Serbia, pp. 1–4 (2019). https://doi.org/10.1109/TELFOR48224.2019.8971253
7. Next-generation Cloud RAN management and orchestration, Ericcson, https://www.ericsson.com/en/blog/2020/11/next-generation-cloud-ran-management-and-orchestration. Last accessed 13 Jun 21
8. Nakamoto, S.: Bitcoin: a peer-to-peer electronic cash system (2008). http://bitcoin.org/bitcoin.pdf. Last accessed 07 Jun 21
9. Lund, E.H., Jaccheri, L., Li, J., Cico, O., Bai, X.: Blockchain and sustainability: a systematic mapping study. IEEE/ACM 2nd International Workshop on Emerging Trends in Software Engineering for Blockchain (WETSEB), pp. 16–23 (2019). https://doi.org/10.1109/WETSEB.2019.00009
10. Ma, Y., Sun, Y., Lei, Y., et al.: A survey of blockchain technology on security, privacy, and trust in crowdsourcing services. World Wide Web **23**, 393–419 (2020)
11. Buterin, V.: Ethereum: a next-generation smart contract and decentralized application platform (2013). https://ethereum.org/en. Last accessed 14 Jun 21
12. Nugraha, R.A.: Analisis Yuridis Mengenai Perlindungan Data Pribadidalam Cloud Computing System Ditinjaudari UU Informasidan Transaksi Elektronik. Jakarta, Universitas Indonesia (2012)

Carbon Footprint Analysis Using Knowledge Graph

Sonam Sharma, Meghna Roy Chowdhury, and Sumukh Sirmokadam

Abstract Climate change is a global issue which needs to be addressed. Identifying and analyzing the data of one of the root elements, i.e., carbon footprint, and hence implementing required solutions serves as a good way to solve this issue. Although there is a lot of data available, there exists an information gap because of which we are not able to derive solutions. Knowledge graph is a possible solution where one can access information related to carbon footprint and thereby take actions to address the problem. In this paper, we discuss a knowledge graph approach where we discuss how can we connect scattered information that is present everywhere on the Web and create one knowledge graph which can talk about problem as well as the solution. We further complete this knowledge graph with node embedding and cluster the nodes. This further helps in getting the understanding on discussion of the problem and solution if present or can be derived. We can further query the graph and try and understand what policies can be build or measures can be taken to reduce carbon footprint.

Keywords Graph database · Knowledge graph · Carbon footprint · Climate change · NLP · Sustainability

1 Introduction

The twenty-first century is one with various developments and technical advancements. For the next generations to be able to exploit technology for the best, one must keep track of how the technical advancements affect the earth, an in turn its inhabitants. For instance, in 2020, India had a score of 35.6 in the Global Innovation

S. Sharma (✉) · M. Roy Chowdhury · S. Sirmokadam
Tata Consultancy Services, Mumbai, India
e-mail: sonam.sharma2@tcs.com

M. Roy Chowdhury
e-mail: meghna.r@tcs.com

S. Sirmokadam
e-mail: sumukh.sirmokadam@tcs.com

© The Author(s), under exclusive license to Springer Nature Singapore Pte Ltd. 2022
A. K. Nagar et al. (eds.), *Intelligent Sustainable Systems*, Lecture Notes in Networks and Systems 333, https://doi.org/10.1007/978-981-16-6309-3_55

Index (GII). Also, India ranks 45th on the innovation output and 57th on the innovation input subindices. [1]. At the same time, a survey dated in the year 2021 mentions that India ranks 3rd in the world for carbon emissions with about 2.46 billion metric tonnes of carbon dioxide produced year, which accounts for 6.8% of the total global emissions. Every Indian contributes 0.56 tonnes of carbon dioxide to the atmosphere annually on average, with 0.19 tonnes per capita being emitted by the poor, and 1.32 tonnes by the rich. [2]. The United Nations has set up 17 Sustainable Development Goals which it aims to achieve by 2030 with the help of nations around the world. Action in one area does impact the outcomes in other areas, hence one must ensure that advancements and development take into consideration social, economic, and environmental concerns [3]. In the context of global sustainability, carbon footprint research is one of the core topics for research. By definition, carbon footprint is the "total greenhouse gas (GHG) emissions caused by an individual, event, organization, service, place or product, expressed as carbon dioxide equivalent" [4]. Research of carbon footprint, integrated with latest technologies, can provide a good estimate of CO_2 usage in various industries, and hence the proper preventive measures in industry as well as domestic level can be taken. This way, the government can have a good estimate of economic factors like GDP and be environmentally sound at the same time.

In this generation of WWW, information is abundant. There are numerous Web pages online which give us an estimate of the carbon footprint in detail. However, assessing information from various sources according to the queries of end users is a tedious task. Artificial intelligence and machine learning have proved to be very useful in this regard. Although there is a good advancement in using AI for "good" (AI4Good), there must be more development toward addressing the sustainability of using AI in other global issues. Vinuessa et al. have presented an extensive study of the role of AI in achieving the sustainable development goals set up by UN. [5]. Strubell et al. demonstrate that the training of one deep learning, natural language processing (NLP) model results in approximately 600,000 lb of carbon dioxide emissions [6]. Following Google's introduction of Knowledge Graphs in 2012, this concept has attracted a great deal of research attention over the years. This state-of-the-art technology has been further developed by various companies like Airbnb, Amazon, Facebook, IBM, Microsoft, etc. A knowledge graph is a graph of data intended to accumulate and convey knowledge of the real world, whose nodes represent entities of interest and whose edges represent relations between these entities [7]. Open knowledge graph and enterprise knowledge graph are the two categories of knowledge graphs which are typically used. They have been extensively used in various applications like information retrieval, cybersecurity, fake news detection, medical applications, etc. [8].

In this paper, we present a methodology developed using specific knowledge graph using information about climate change crawled from the Internet using our in-house crawler which can further be used for policies, strategies, and planning. The task is listed as one of the goals and targets in UNs 2030 sustainability agenda. One of the climate change measures that we focus on is reduction in carbon dioxide emissions and carbon footprint. The purpose of this paper is to highlight the transformational potential AI has in climate change and analyzing carbon footprint and helps with the aim for their reduction.

2 Methodology

In this section, we discuss our approach in creation of purpose specific knowledge graph. Embeddings are created among graph nodes which helps in knowledge completion and understanding. This further helps in querying the graph and extract facts which helps in deriving policies, understand what measures needs to be taken by the population and multiple government bodies to work together as one goal and reduce carbon footprint.

2.1 *Dataset*

Using the in house developed crawler, using predefined keywords, we gather various articles/documents related to climate change, environmental change, and carbon footprint as shown in Fig. 1. The articles/documents are parsed to extract entities related to carbon footprint only. Extracted entities are further used to build the knowledge graph. Knowledge graph helps us to democratize the scattered/siloed data and enables easy access for the same. Below are the keywords we use in getting the articles from Web. Following are the keywords:

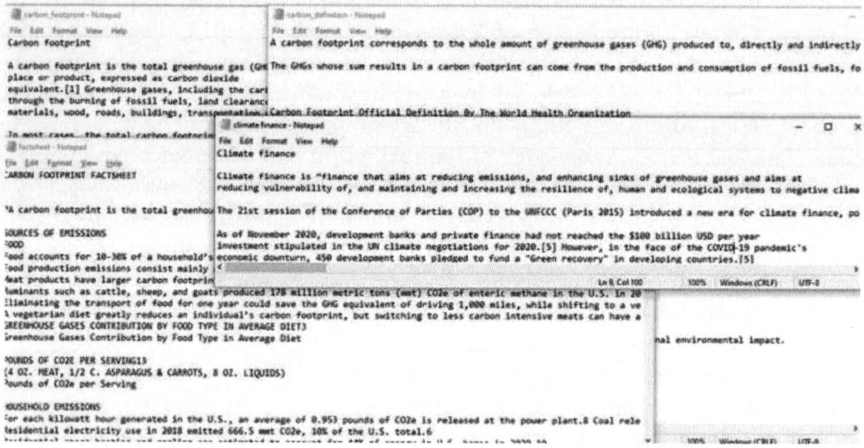

Fig. 1 Screenshot displaying some of the crawled information from the Web

climate action, carbon footprint, carbon foot print, climate adaptation, climate change, IPCC, carbon finance, climate effect, climate feedback, climate forcing, climate impact, climate mitigation, climate model, climate models, climate modeling, climate modelling, climate policy, climate policies, climate risk, climate risks, climate services, climate service, climate prediction, climate predictions, climate signal, climate signals, climate tipping point, climate variation, climate variations, Eco climatology, eco-climatology, regional climate, regional climates, urban climate, urban climates, climate adaptation, adaptive management, biodivers, bioeconomy, biological production, carbon, conservation, carbon emission, Anthropocene, atmosphere, clean development mechanism, glacier retreat, warming, greenhouse, ice-ocean interaction, ice-ocean interactions, nitrogen cycle, nitrogen cycles, ocean acidification, radiative forcing, sea ice, sea level, sea levels, thermal expansion, UNFCCC, ozone.

Additionally, data on CO2 and greenhouse gas emissions by Our World in Data is also leveraged. CO2 and greenhouse gas emissions dataset is a collection of key metrics maintained by Our World in Data [9]. This dataset, which includes information on CO2 emissions like annual, per capita, cumulative and consumption-based, other greenhouse gases, energy mix, etc., is kept up-to-date. The CSV file is in the format where one row describes a location/country with year and other related metrics. The population figures in the dataset are taken from UN world population prospects.

2.2 Preprocessing

As a first step we process the unstructured documents crawled from various Web sites/sources. We first link the documents with the related topic based on its contents. This helps us in building the taxonomy for the documents. Rapid Automatic Keyword Extraction (RAKE) is used for understanding documents and extracting keywords [10]. The extracted keywords act as topic which is addition to the number of topics, we have taken to crawl these articles (See Fig. 2) With the increase in number of articles we are able expand our taxonomy further.

With the raw text available with us from the crawled articles, we pass it to our *Text Distiller* module using which we clean the content and remove junk from the text which is further passed to the NLP pipeline. NLP pipeline has two modules named *Triples Extraction* and *Topic Cooccurrence. Triples Extraction* module is used to extract triples meaning in *Subject-Verb-Object* (SVO) form from the unstructured documents. *Topic Cooccurrence* module is then used to extract topics or keywords from documents and find their cooccurrence. If any two topics appear together in single document, we link them together in our graph. Figure 4 depicts various topics generated from documents.

	topic	topic_text
0	carbon definition	Carbon Footprint Definition\n\n\nA carbon foot...
1	carbon footpront	Carbon footprint\n\nA carbon footprint is the ...
2	clean accessible energy	3 lessons from India for achieving clean and a...
3	climate finance	Climate finance\n\nClimate finance is "finance...
4	climate action tracker	The COVID-19 pandemic has brought significant ...
5	Dhruba	Dhruba Purkayastha is CPI's India Director, wh...

Fig. 2 Extracted content from Web and link topics/keywords available with the article content

2.3 *Knowledge Graph*

First step in creating knowledge graph is to understand the graph data model, i.e., what will be the nodes and edges in graph database and how they will be connected. Here for our usecase, we have nodes and edges ***Subject–VERB–Object (SVO)*** as is a sentence structure where the *subject* comes first, the *verb* second, and the *object* third [11]. Apart from this, we have additional nodes such as *Topic* which explains number of topics this knowledge base has and each topic is connected to its relevant article(s) with relation CONTAINS_TOPIC and a *Topic* can be connect to another *Topic* if they occur in same article together, i.e., relation TOPIC_COOCCURANCE. *Article* node contains the unstructured text information and its raw form and its source from where it got extracted. Each article will have number of triples SVOs available which are connected via relation CONTAINS_SUBJECT and CONTAINS OBJECT. *Country* node explains the number of countries we have taken in our analysis and have come up in parsing the documents while preprocessing and entity extraction stage and *CO_2_Emissions_Knowledge* node have available dataset from [12, 13], where all the information is aggregated via country and year. In same way, we have linked the dataset with *country* node that has all information on available country and each CO2 emission-related knowledge is linked via YEAR relation of that specific country. After data understanding and modeling for preparation of data for graph database population, we populate out data to graph database, i.e., Neo4j with the available graph model understanding (see Fig. 3).

Data is preprocessed as per Neo4j requirement and using Cypher Query Language we populate Neo4j graph database. Neo4j supports Lucene indexing too [14, 15]. We index each node for it to be available for search. All the knowledge graph creation steps are explained in Fig. 4. Finally, we have knowledge graph available with us which can be used for querying and applying algorithms and acts as a baseline for

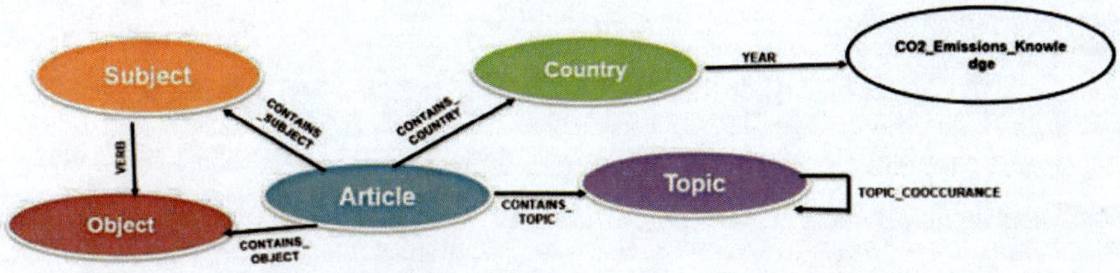

Fig. 3 Various topics available in Graph DB (Neo4J)

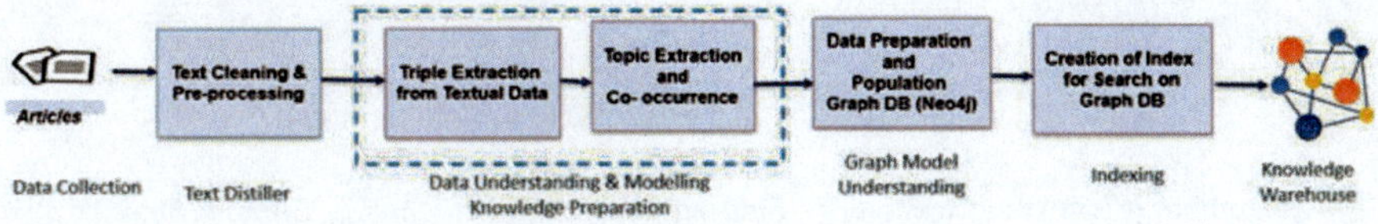

Fig. 4 Knowledge graph creation steps from unstructured textual content

further analysis. Subsequently, country details extracted from the article (if applicable) are saved in *Country* node and for each year starting from 1858 to 2019 we merge CO2 emissions data in *CO_2_Emissions_Knowledge* node.

This final graph represents the *"Knowledge World"* where all information related to climate impact and carbon footprint are available which have also integrated information from structured sources to complete the graph. This information is further enhanced with adding node embedding in the graph. We have used *node2vec* algorithm for embedding of nodes which is similar to *word2vec* on difference is that instead of using a list of sentences as input, we use a list of random walks [16, 17].

3 Results and Discussion

Our methodology aims at distilling knowledge from text articles which can further be used directly by multiple end user applications such as search engine which returns results in structured form (e.g., "Find all information containing Climate Finance as a Topic along with all related Topics of climate finance"). We first find all *Topics* related to "climate finance" within *node walk* distance of 2 and extract triples related to this specific topic. This natural language query is first converted to cypher query understandable form *"Climate Finance"* by NLP engine. The output to this query is shown in Fig. 5.

Another query is "Give me all the related topics which are common in at least 3 articles?". With related topic, we understand we only need to query on the topics node and to get the related topic we query the graph relations with *node walk* distance of (number of documents) *2, i.e., 6 in our case (see Fig. 6).

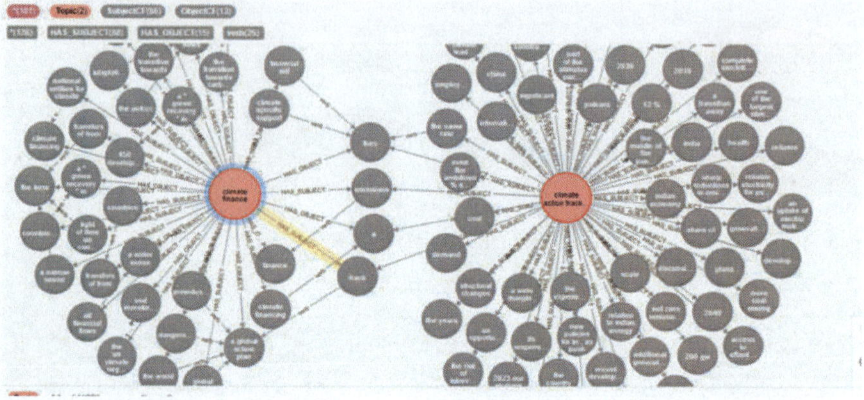

Fig. 5 Resultant graph result for query *"Find all information containing Climate Finance as a Topic along with all related Topics of climate finance"*

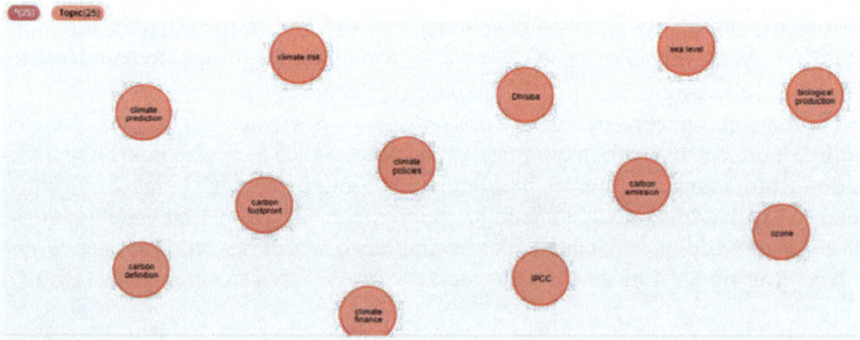

Fig. 6 Resultant graph result for query *"Give me all the related topics which are common in at least 3 articles?"*

Below are the embedding results of the most related topics: *[("climate adaptation", 0.7236759424209595), ("climate action", 0.6639043068885803), ("climate footprint", 0.5999650955200195), ("climate change", 0.5930663347244263), ("IPCC", 0.5898399353027344)].*

Finally, a query which we focus on is "*which industry contributes to maximum carbon footprint?*". Since, our graph nodes are not segregated the node for industry-wise comparison, this query is answered based on nodes embedding available with our nodes. Initially, as first step we try to find the most related terms related to the topic "*industry*".

We get following keywords: *industry, deforestation, energy, agriculture, fuel, greenhouse gas.* We add this to our query then try to extract facts from the graph.

Which can provide fact, i.e., related triple for the queried question. Following are the triples we get as an output:

- (fuel combustion, contributes, 8.6%)
- (such sector road transport, impacts, 10.5% to global warming)
- (1.2%, emission, Aluminum and non-ferrous metals production)
- (landfill of waste, takes, 1.5% carbon footprint).

4 Conclusion

After performing analysis of various sources of information related to carbon footprint and climate change using knowledge graphs, we conclude that leveraging knowledge graph gives significant benefits over traditional text-based search. First, information and data are crawled from various Web pages using specific keywords. Then a knowledge graph is formed by linking the entities with their respective topic. This has given us an insight about the statistics of carbon footprint from various industrial and domestic activities. Our solution can enable the not only the government and policymakers, but also common people to assess data from various sources by putting in their query, in one platform and make changes to reduce carbon footprint, therefore work toward fulfilling the sustainable goals set by the UN. AI applications which will leverage this knowledge graph to generate Insights can be built. Furthermore, the crawler can use a ranking mechanism for Web pages based on the veracity of information they contain. For a knowledge graph improve over time, it is essential to enhance extraction mechanism, quality assessment, and refinement methods.

References

1. India Brand Equity Foundation Homepage, http://www.ibef.org/industry/science-and-technology.aspx. Last accessed 14 Jun 2021
2. Study Finds India's Rich Emit 7x More Emissions Than the Poor—The Wire Science, https://science.thewire.in/environment/india-carbon-emissions-rich-poor-households/. Last accessed 1 Jun 2021
3. Sustainable Development Goals | United Nations Development Programme , https://www.undp.org/sustainable-development-goals. Last accessed 14 Jun 2021
4. Carbon Footprint, https://www.semtrio.com/en/carbon-footprint. Last accessed 14 Jun 2021
5. Vinuesa, R., Azizpour, H., Leite, I., et al.: The role of artificial intelligence in achieving the Sustainable Development Goals. Nat Commun **11**, 233 (2020). https://doi.org/10.1038/s41467-019-14108-y
6. van Wynsberghe, A.: Sustainable AI: AI for sustainability and the sustainability of AI. AI Ethics (2021). https://doi.org/10.1007/s43681-021-00043-6
7. Hogan, A., Blomqvist, E., Cochez, M., d'Amato, C., Melo, G.D., Gutiérrez, C., Gayo, J.E., Kirrane, S., Neumaier, S., Polleres, A., Navigli, R., Ngomo, A.N., Rashid, S.M., Rula, A., Schmelzeisen, L., Sequeda, J., Staab, S., Zimmermann, A.: Knowledge graphs. Commun. ACM **64**, 96–104 (2021)

8. Zou, X.: A survey on application of knowledge graph. J. Phys: Conf. Ser. **1487**, 012016 (2020). https://doi.org/10.1088/1742-6596/1487/1/012016
9. Our World in Data, https://ourworldindata.org/. Last accessed 14 Jun 2021
10. Rose, S., Engel, D., Cramer, N., Cowley, W.: Automatic keyword extraction from individual documents. In Text Mining, Berry, M.W., Kogan, J. (eds.) (2010). https://doi.org/10.1002/9780470689646.ch1
11. Dryer, M.: Order of subject, object, and verb. The World Atlas of Language Structures (2005)
12. Ritchie, H., Roser, M.: "CO_2 and greenhouse gas emissions". Published online at OurWorldInData.org (2020). Retrieved from: 'https://ourworldindata.org/co2-and-other-greenhouse-gas-emissions' [Online Resource]
13. Data on CO2 and greenhouse gas emissions by Our World in Data, https://github.com/owid/co2-data, Last accessed 14 Jun 2021
14. Cypher Query Language—Developer Guides, https://neo4j.com/developer/cypher/
15. Searching and Indexing with Apache Lucene - DZone Database, https://dzone.com/articles/apache-lucene-a-high-performance-and-full-featured. Last accessed 14 Jun 2021
16. Grover, A., Leskovec, J.: Node2vec: scalable feature learning for networks. In Proceedings of the 22nd ACM SIGKDD International Conference on Knowledge Discovery and Data Mining (KDD '16). Association for Computing Machinery, New York, NY, USA, 855–864 (2016). https://doi.org/10.1145/2939672.2939754
17. Mikolov, T., Chen, K., Corrado, G., Dean, J.: Efficient estimation of word representations in vector space. arXiv preprint arXiv:1301.3781 (2013)

Automatic Traffic Light Detection for Self-Driving Cars Using Transfer Learning

S. Gautam and A. Kumar

Abstract Self-driving vehicles are being tested to make them more road-ready and safer for a real traffic environment. Automobile giants: like Tesla, Waymo, Toyota, etc., are working exhaustively to cater to the needs of futuristic smart vehicles using deep learning methodologies. For a self-driving car to avoid collision situation, it must be able to accurately detect and classify traffic lights. After performing exhaustive experiments, we chose to compare the feature extraction capabilities of various pretrained CNN-based transfer learning models like VGG16, ResNet50, AlexNet, DenseNet121, InceptionV3, and Xception on freely available Lara & Lisa traffic light datasets. We segregated the Lisa traffic light dataset into day and night subsets and then manually separated the images into various traffic light classes like dayRed, dayYellow, nightYellow, nightGreen, road and traffic lights, LeftGreenArrow, and RightGreenArrow etc. We used the random forest classifier to identify the color of the detected traffic light. DenseNet121 achieved a top accuracy of 100% on the Lara traffic light dataset and the maximum accuracy of 99.88% on the Lisa Day-light dataset and 98.89% accuracy on the Lisa Night-light dataset.

Keywords Feature extraction · Traffic lights · CNN · Transfer learning · DenseNet121 · Random forest classifier

1 Introduction

Urbanization has highlighted life-threatening risks on roads. However, with the advent of self-driving technology, transportation and mobility have increased. Self-driving cars need to detect traffic lights accurately and act accordingly to make roads safer. To detect traffic lights, we considered features like color and shape. These features are extracted using CNN-based transfer learning models and then input to the random forest classifier [1]. Finally, the name of the traffic light color (red, yellow,

S. Gautam (✉) · A. Kumar
Department of Computer Science and Applications, Panjab University, Chandigarh, India

A. Kumar
e-mail: Anuj_gupta@pu.ac.in

© The Author(s), under exclusive license to Springer Nature Singapore Pte Ltd. 2022

A. K. Nagar et al. (eds.), *Intelligent Sustainable Systems*, Lecture Notes in Networks and Systems 333, https://doi.org/10.1007/978-981-16-6309-3_56

green, etc.) is provided as output. Some of the major classifiers are support vector machine (SVM) [2], XGBoost [3], etc. The CNN-based transfer learning models, like VGG16 [4], ResNet50 [5], DenseNet121 [6], etc., require an extensive amount of input data to learn more by seeing different examples, and for making our lives easier, they have been pretrained on ImageNet dataset[7]. In transfer learning, the network has already learned the various features of input images on which it has been trained and can be reused to train and test other images. Using transfer learning, the need for handcrafting the feature extraction methods gets eliminated. In this work, first, the dataset is segregated into day and night sequences, then the images were separated into traffic light classes manually. After that, the dataset is divided into train and test modules, and one-hot encoding is performed; the features extraction models, like VGG16, DenseNet121, InceptionV3, etc., have been applied, and finally, the extracted features have been input to random forest classifier to obtain the targeted class. Section 2 illustrates the related work done in the literature. Section 3 illustrates the layers used in CNN-based transfer learning models. Section 4 illustrates the proposed system design with the experiments performed and the results achieved and the limitations of our study; Sect. 5 shows limitation of our study. And in Sect. 6, Conclusion is generated along with the future forecast.

2 Related Work

Self-driving cars are the most rapidly growing transportation technology around the globe. Many successful attempts like Tesla's Autopilots, Google's Waymo have been made in this area. The success of a self-driving car certainly lies in accurate analysis of traffic lights. Various authors have proposed different methods for the accurate detection of traffic lights.

Authors in [8] applied You Only Look Once (YOLO) on Lisa traffic light dataset and achieved an Area Under Curve (AUC) of 90.49% for their daySequence1. Authors in [9] proposed a multi-feature fusion-based traffic light recognition system. They used vanishing lines to determine the ROIs and applied color segmentation and blob detection for feature extraction. By fusing the results, they achieved an accuracy of 96.07%. Authors in [10] integrated prior map information with (YOLOv4)-based detectors to recognize traffic lights in predefined routes. In [11], authors created a prospective assistance system to detect dynamic objects on roads like humans and changing traffic lights. They combined SVM with the hidden Markov model (HMM) for classification and state estimation of normal and blinking traffic lights. Authors in [12] presented a visual attention model-based framework to detect smaller objects in streets. Their experiments shows that the attention model-based system works better than the faster R-CNN model. They obtained average precision of 90% without tuning. In [13] the authors proposed a deep CNN-based high dynamic imaging and deep learning system to recognize lights. They proposed a nonparametric multicolor saliency model to extract lights with varying colors. Authors in [14] tested model-based detection using three different methodologies thresholding, back projection, and spotlight detection based on HSV, LUV, and white-top-hat morphological operation. Their system achieved an AUC of 51.4%. Tai et al. proposed a YOLOv4-based traffic light localization method on two South Korean datasets to detect all possible traffic light positions in the current scene and identify the actual ones from them. The authors focused on generalizing driver assistance systems (DAS) by capturing horizontal traffic light formats popular in South Korea, North America, and most European countries. Their system achieved a 95% accuracy at 30FPS [15]. Dennis et al. [16] developed a CNN-based traffic light detection model using SSD MobileNetV1 and faster RCNN InceptionV2. The authors trained their models on Bosch small traffic light dataset (BSTLD) and DriveU traffic light dataset (DTLD). They achieved the best performance on filtered BSTLD with mAP (of 56.84% using SSD MobileNetV1, while SSD MobileNetV1 gave lesser accuracy of 55.17% on DTLD. The best performance on DTLD was achieved using faster RCNN InceptionV2 with a mAP of 43.97%.

3 CNN Architecture

Due to advantages like pretrained models, we chose to apply CNNs in our study. CNN can take 3D inputs and generate 3D outputs. CNNs can generate the best output for computer vision or image analysis tasks. The basic CNNs are composed of six layers. The basic architecture of CNN is described in the subsequent section.

(1) **Convolutional layer** The convolution operation is performed in this layer by sliding an MxM filter on an input image.
(2) **Pooling** This layer increases the efficiency of CNN, by decreasing the size of convolutional feature maps thus reducing computational costs.
(3) **Fully connected layer** This layer serves as a connection between neurons of two different layers.
(4) **Activation functions** Activation functions learn and approximate continuous and complex relationships between variables.
(5) **Dropout layer** The dropout layer is used to overcome the problem of overfitting.
(6) **Loss layer** It attains truth labels from predicted values.

After performing exhaustive experiments on transfer learning models we chose the following six models which gave better accuracy on the selected dataset. These models are explained in the subsequent sections.

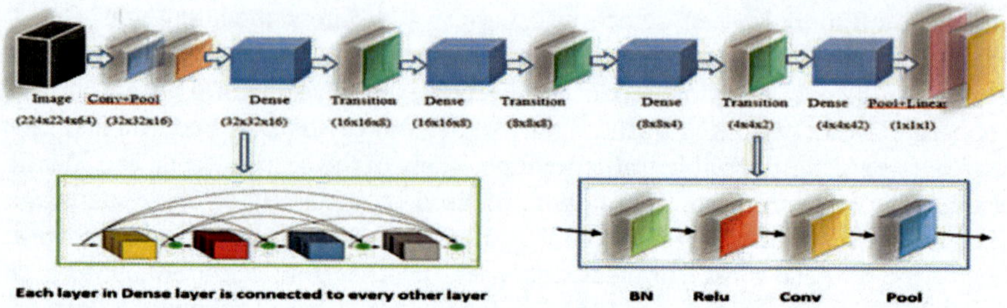

Fig. 1 Architecture of DenseNet is shown here

(1) **VGG16** This transfer learning model was proposed by [4] in 2014. VGG16 achieved top-1 accuracy of 71.3% and top-5 accuracy of 90.1%. It has 138 million parameters and a depth of 23 levels. It has a total of 16 convolution layers with 3 × 3 receptive fields, five maxpool layers of 2 × 2 size, three fully connected/dense layers followed by a softmax classifier.

(2) **ResNet50** ResNet[5] (ResiDual network) The power of ResNet lies in the fact that it solves the "Vanishing Gradient" problem in the backpropagation process. Its top-1 accuracy is 74.9% and top-5 accuracy is 92.1% on the ImageNet dataset.

(3) **AlexNet** The AlexNet architecture achieved top-1 accuracy of 63.30% and top-5 accuracy of 84.60%. It has 62 million parameters from five convolutional layers having 3 × 3 receptive fields with 3 × 3 receptive fields and two strides, three maxpooling layers, and two fully connected layers.

(4) **InceptionV3** It has over 23 million parameters with a depth of 159 having top-1 accuracy of 77.90% and top-5 accuracy as 93.7% from six convolution layers of 3 × 3 receptive field, and it has three inception layers and two maxpooling layers along with a softmax classifier[22].

(5) **Xception** It is a successor of the InceptionV3 model. The model achieved top-1 accuracy of 79.0% and top-5 accuracy of 94.5%. The model goes 126 layers deeper and works on over 22 million parameters [17].

(6) **DenseNet121** The compelling advantages of a DenseNet121 are reduction of vanishing gradient problem, feature propagation strengthening, and parameter reduction. It has more than 8 million parameters and achieved top-1 accuracy of 75.0% and top-5 accuracy of 92.3% from one convolution layer and a single 2 × 2 maxpool layer along with four dense layers and three transitional layers [6]. The DenseNet architecture is shown in (Fig. 1).

4 Proposed System Design

We divided the experiments into two parts by taking the entire Lisa traffic light dataset [18] and dividing it into two categories namely day and night. The datasets were augmented using scaling, padding, shearing, and horizontal flip. The pretrained

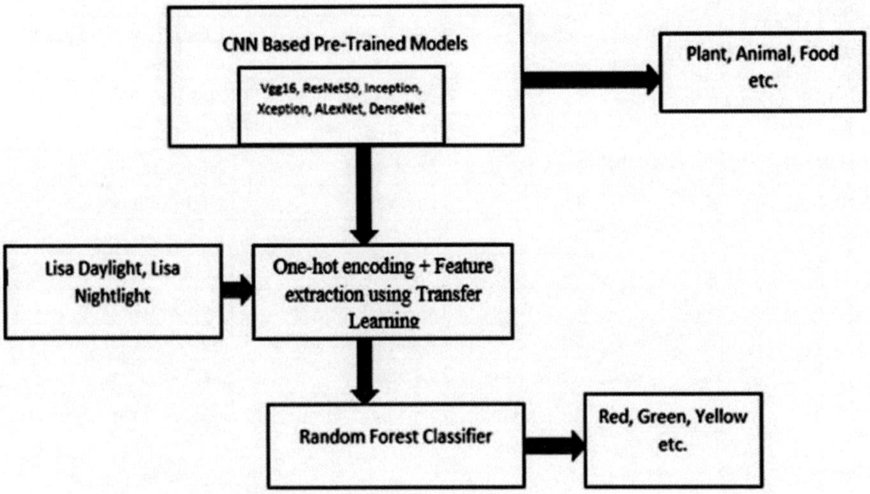

Fig. 2 Proposed system architecture

transfer learning models like VGG16, ResNet50, InceptionV3, Xception, AlexNet, and DenseNet were applied to both categories of the datasets. We made use of the features extracted by these pretrained models and reused them as input to random forest classifier. Accuracy was used as the evaluation metrics. The proposed system design is shown in (Fig. 2).

4.1 Datasets

In our study, we used two standard state of arts publically available datasets namely the Lara traffic light dataset and Lisa traffic light dataset. The Lara traffic light dataset contained 11,179 frames of a total of 8 min and 49 s of video sequence captured at 25 fps, shot on the roads of Los Angeles having 640 × 480 images. The Lisa traffic light dataset contains 23 min and 25s of video frames captured during the daytime and nighttime containing a total of 43,007 frames and 113,888 annotated images. The dataset was shot on the roads of San Diego, generating 1280 × 960 images. We manually separated both the datasets into traffic light classes having 8 daytime/nighttime classes as described in Table 1. We used a total of 37,352 images from the original dataset and further segregated 24,988 images for daytime and 12,364 images for nighttime.

Table 1 Shows the manual segregation of traffic light classes for daytime and nighttime datasets done in our study

Serial	Lisa daytime dataset classes	Lisa nighttime dataset classes
1	NoLight	NoLight
2	DayRed	NightRed
3	DayGreen	NightGreen
4	DayYellow	NightYellow
5	RedGreen	RedGreen
6	YellowGreen	YellowGreen
7	LeftGreenArrow	LeftGreenArrow
8	RaodVehicleLights	RaodVehicleLights
Total images	**24,988**	**12,364**

The bold indicates the total number of images used for day dataset and night dataset

4.2 *One Hot Encoding*

One hot encoding helps to generate **one Boolean column for each category/class** in the dataset. In other words, only one of the category columns can take a value as "**1**" for each sample.

Example: dayRed is given the numerical value as **1** and all the other classes are given values as "**0**". So, no other columns can attain the same numerical value thus helping ML algorithm to understand better.

4.3 *Feature Extraction Using Transfer Learning*

Transfer learning refers to the process of reusing feature extraction capabilities of pretrained models to train some new data and extract useful features from that. With the help of feature extraction methods, we can intensify the detection accuracy and curb overfitting. Table 2 shows transfer learning models with the number of parameters used and the number of extracted features.

4.4 *Classification of Traffic Lights*

After extracting features using transfer learning, we finally classified the traffic lights using random forest classifier. In the random forest classifier, we used hyper-parameters grid having parameters like n_estimators, random_state, max_features, max_depth, min_samples_split, min_sample_leaf, bootstrap, etc., and tuned them using RandomizedSearchCV with threefold cross-validation. After fitting the data

Table 2 List of pretrained transfer learning models

Model	Image size	Depth	Parameters	FLOPs	Features extracted
VGG16	224 × 224	23	138,357,544	19.6B	4096
InceptionV3	224 × 224	159	23,851,784	2B	1024
Xception	224 × 224	126	22,910,480	–	2048
AlexNet	227 × 227	96	62,378,344	1.5B	1280
ResNet50	224 × 224	50	25,636,712	11B	2048
DenseNet121	224 × 224	121	8,062,504	–	1024

Image size, depth, number of parameters, FLOPs, and finally the features extracted from each model

to the classifier, we predicted the accuracy of original labels against the predicted labels.

4.5 Experiments and Results

We conducted exhaustive experiments on the mentioned dataset using six different transfer learning models and achieved exemplary results. Initially, the datasets were loaded and images were augmented to feed to the transfer learning models. After extracting the features, results were input to the random forest classifier and the predicted accuracies were compared against original accuracies. The explanation of the dataset and its accuracies are provided below.

4.6 Evaluation on Lisa Day and Night Light Dataset

Lisa traffic light dataset is the most widely researched dataset in the literature. This dataset initially contained 13 daytime clips with categories having names dayClip1, dayClip2, etc. Since no proper categorization was done on the original dataset, we manually separated images into 8 day and 8 night light classes like dayRed, dayGreen, nightRed, nightYellow, etc., as mentioned in Table 1. Six different transfer learning models were applied, and hyperparameters were tuned to obtain shown accuracies. The transfer learning models along with their accuracies are shown in Table 3 for the day datasets and night datasets and an infographics is provided for better understanding.

These exhaustive experiments lead to the conclusion that the transfer learning model "DenseNet121" architecture outperformed VGG16, InceptionV3, Xception, AlexNet, and ResNet50 models by generating 99.88% accuracy on the daytime dataset and 98.89% accuracy on the night dataset. Table 4 shows a comparative analysis of state-of-art techniques with the proposed technique.

Table 3 Comparison evaluation of our study with the literature on day and night dataset in a tabular and infographics form

Sr	CNN-based transfer learning model	Day accuracy (%)	Night accuracy (%)
1	AlexNet	97.23	94.26
2	Vgg16	99.86	93.26
3	InceptionV3	97.51	89.39
4	Xception	99.16	91.08
5	ResNet50	96.30	94.89
6	DenseNet121	99.88	98.89

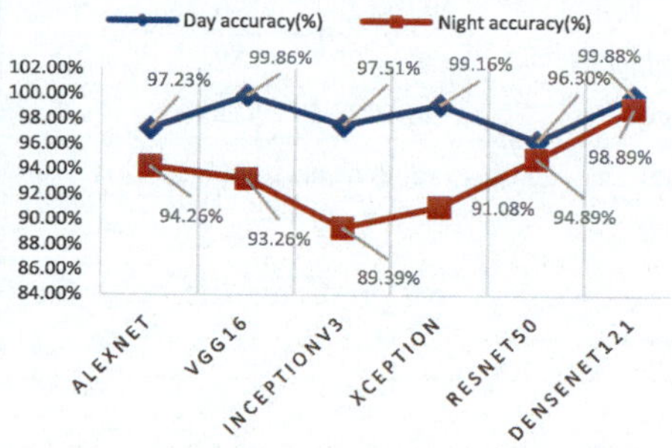

5 Limitations

The only limitations of the proposed system lie in the fact that it was tested only on readily available datasets after some modifications. A similar approach will be followed in the future to evaluate the traffic light detection system's performance on our own local dataset captured using a moderate quality camera.

6 Conclusion

In our study, we performed exhaustive experiments on six different pretrained transfer learning models. DenseNet121 in combination with random forest outperformed all the other transfer learning models. DenseNet121 solves the problem of vanishing and exploding gradient. From Table 4, we can observe that DenseNet121 combined with random forest classifier outperformed all other transfer learning models by achieving 99.88% accuracy on Lisa daytime dataset and 98.89% accuracy on Lisa nighttime dataset and a 100% on Lara traffic light dataset. Automatic traffic light detection system will serve as the pillar of safe transportation in the future by precisely detecting traffic lights and making the system act accordingly.

Table 4 Evaluation of state-of-the-art techniques in comparison to our study

References	Methodology	Dataset	Total accuracy	Day accuracy	Night accuracy
Our Study	DenseNet with random forest	Lara traffic light dataset	**100%**	–	–
Our Study	DenseNet with random forest	Lisa day and night Dataset	–	**99.88%**	**98.89%**
[8]	YOLO	Lisa traffic light day dataset	–	90.49%	–
[15]	YOLOv4	Two South Korean datasets	95%	–	–
[16]	SSD MobileNetV1 and Faster-RCNN InceptionV2	Bosch small traffic light dataset (BSTLD) and DriveU traffic light dataset (DTLD)	56.84% mAP using SSD MobileNet V1 on BSTLD mAP of 43.97% using faster RCNN InceptionV2 on DriveU traffic light dataset	–	–
[19]	SVM with polynomial kernel	Random day and night data from search engines	–	98.5%	98.8%
[20]	Random forest	UK road sign database	94.2%	–	–
[21]	CNNs	GTSRB	>95%	–	–
[22]	Recurrent-CNN	Firsthand Indian traffic sign dataset	99.19%	–	–

The bold indicates that the results achieved by DenseNet are the best among all other CNN based transfer learning models

References

1. Breiman, L.: Random forests. Mach. Learning **45**(1), 5–32 (2001)
2. Schuldt, C., Laptev, I., Caputo, B.: Recognizing human actions: a local SVM approach. In Proceedings of the 17th International Conference on Pattern Recognition, 2004. ICPR 2004, vol. 3, pp. 32–36, IEEE, (Aug 2004)
3. Chen, T., Guestrin, C.: Xgboost: a scalable tree boosting system. In Proceedings of the 22nd ACM Sigkdd International Conference on Knowledge Discovery and Data Mining, pp. 785–794, (Aug 2016)
4. Simonyan, K., Zisserman, A.: Very deep convolutional networks for large-scale image recognition. *arXiv preprint* arXiv:1409.1556 (2014)
5. He, K., Zhang, X., Ren, S., Sun, J.: Deep residual learning for image recognition. In Proceedings of the IEEE Conference on Computer Vision and Pattern Recognition, pp. 770–778, (2016)

6. Huang, G., Liu, Z., Van Der Maaten, L., Weinberger, K.Q.: Densely connected convolutional networks. In Proceedings of the IEEE Conference on Computer Vision and Pattern Recognition, pp. 4700–4708 (2017)
7. http://image-net.org/
8. Jensen, M.B., Nasrollahi, K., Moeslund, T.B.: Evaluating state-of-the-art object detector on challenging traffic light data. In Proceedings of the IEEE Conference on Computer Vision and Pattern Recognition Workshops, pp. 9–15 (2017)
9. Zhang, Y., Xue, J., Zhang, G., Zhang, Y., Zheng, N.: A multi-feature fusion based traffic light recognition algorithm for intelligent vehicles. In Proceedings of the 33rd Chinese Control Conference, pp. 4924–4929, IEEE (Jul 2014)
10. Possatti, L.C., Guidolini, R., Cardoso, V.B., Berriel, R.F., Paixão, T.M., Badue, C., Oliveira-Santos, T.: Traffic light recognition using deep learning and prior maps for autonomous cars. In 2019 International Joint Conference on Neural Networks (IJCNN), pp. 1–8, IEEE (Jul 2019)
11. Nienhüser, D., Drescher, M., Zöllner, J.M.: Visual state estimation of traffic lights using hidden Markov models. In 13th International IEEE Conference on Intelligent Transportation Systems, pp. 1705–1710, IEEE, (Sep 2010)
12. Lu, Y., Lu, J., Zhang, S., Hall, P.: Traffic signal detection and classification in street views using an attention model. Comput. Visual Media **4**(3), 253–266 (2018)
13. Wang, J.G., Zhou, L.B.: Traffic light recognition with high dynamic range imaging and deep learning. IEEE Trans. Intell. Transp. Syst. **20**(4), 1341–1352 (2018)
14. Jensen, M.B., Philipsen, M.P., Bahnsen, C., Møgelmose, A., Moeslund, T. B., Trivedi, M.M.: Traffic light detection at night: comparison of a learning-based detector and three model-based detectors. In International Symposium on Visual Computing, pp. 774–783, Springer, Cham (Dec 2015)
15. Tran, T.H.P., Jeon, J.W.: Accurate real-time traffic light detection using YOLOv4. In 2020 IEEE International Conference on Consumer Electronics-Asia (ICCE-Asia), pp. 1–4, IEEE (Nov 2020)
16. Hein, D.: Traffic Light Detection with Convolutional Neural Networks and 2D Camera Data, Doctoral dissertation, fu-berlin, (2020)
17. Chollet, F.: Xception: deep learning with depthwise separable convolutions. Proc 30th IEEE Conf. Comput. Vis. Pattern Recognition, CVPR 2017 2017-Janua 1800–7 (2017)
18. http://cvrr.ucsd.edu/LISA/lisa-traffic-sign-dataset.html
19. Widyantoro, D.H.: Circle and arrow traffic light recognition. In 2017 IEEE International Conference on Cybernetics and Computational Intelligence (CyberneticsCom), pp. 34–38, IEEE (Nov 2017)
20. Greenhalgh, J., Mirmehdi, M.: Traffic sign recognition using MSER and random forests. In 2012 Proceedings of the 20th European Signal Processing Conference (EUSIPCO), pp. 1935–1939, IEEE (Aug 2012)
21. Stallkamp, J., Schlipsing, M., Salmen, J., Igel, C.: Man vs. computer: benchmarking machine learning algorithms for traffic sign recognition. Neural Netw. **32**, 323–332 (2012)
22. Lin, C., Li, L., Luo, W., Wang, K.C., Guo, J.: Transfer learning based traffic sign recognition using inception-v3 model. Period. Polytech. Transp. Eng. **47**(3), 242–250 (2019)

An Exploratory Study to Evaluate the Quality of Research Publication by the Indian Researchers

Saptasindhu Panda, Kajal Panda, Sharmistha Pradhan, Sanjib Kumar Mahanta, and Sunil Dhal

Abstract In the recent days, huge number of papers has been published by the researchers in the name of Innovation and development. In this context, the authors are interested to make a study to evaluate the quality of research publication by the Indian researchers. For this, they have chosen a tridimensional methodology to evaluate the research article through primary study (qualitative and quantitative) and secondary study. A number of research parameters are taken into consideration as citation, journal, impact factor, number of reads, etc. for evaluating an article. The study shows that the Indian research articles have not made significant work in this field. A study has done through primary survey method for quantitative and qualitative data. Both the study shows that the Indian research articles have not significant impact in the field of research and innovation.

Keywords Impact factor · Citation · Read · Journals · Indian research · Research publication · Indian article

1 Introduction

In the epoch of emerging technologies, different demographic societies are exploring new ways to solve socio-economic problem, where Research Publications and Research Problems play a vital role to mitigate the issues. Researchers who have

S. Panda (✉) · K. Panda · S. Pradhan · S. K. Mahanta · S. Dhal
Sri Sri University, Cuttack, India
e-mail: saptasindhu.p2019mba@srisriuniversity.edu.in

K. Panda
e-mail: kajal.p2020mba@srisriuniversity.edu.in

S. Pradhan
e-mail: sharmistha.p2020mba@srisriuniversity.edu.in

S. K. Mahanta
e-mail: sanjiv.m2020mba@srisriuniversity.edu.in

S. Dhal
e-mail: sunildhal@srisriuniversity.edu.in

© The Author(s), under exclusive license to Springer Nature Singapore Pte Ltd. 2022

A. K. Nagar et al. (eds.), *Intelligent Sustainable Systems*, Lecture Notes in Networks and Systems 333, https://doi.org/10.1007/978-981-16-6309-3_57

just started their journey of research these articles would help them to give a path to resolve research problems. On the other hand, a number of research publications are available which does not give a clear visionary idea to the new researches what to be followed. Here the Author puts stress on the quality and standards of Indian Research Publications.

"India Todays" present report shows India was the third biggest maker of scientific research articles. China was at the top and USA was runners up, where China and USA distributed 20.67% and 16.54% articles respectively. In last one decade, the publication of scientific articles has been expanded. In 2008, it was observed that 48,998 science and engineering articles had been distributed but in 2018 the number increased to 1, 35, 778. In case of China, it was 2, 49, 049 in 2008 but the number escalated to 5, 28, 263 which is 7.81% rate of escalation, and it is also observed that it was developed at a pace of 0.71 percent annually from 3,93,979 in 2008 to 4,22,808 in 2018.

The utmost articles published worldwide are Germany, Japan, UK, Russia, Italy, South Korea, France as 104,396, 98,793, 97,681, 81,579, 71,240 respectively. The US and European Union research papers generally affect the majority of the citizens. The USA, the European Union and Japan had proficiency in health science, China and India worked in publishing engineering science journal articles, and conference papers. As mentioned in this report top 15 publishers, nations with higher-than-normal development incorporate, Brazil (five percent), South Korea (four percent), Russia (10%), India (11%) and Iran (11%), China (eight percent).

In the Nature Index table 2020, India is put twelfth over the sea in science research output. As per Nature Index in 2019, Top five in this ranking are USA, China, Germany, Joined Realm and Japan. India is after Spain and ahead of Italy. The Indian Institute of Science, Bangalore was first in 2019 among every single Indian institution. Homi Bhabha National Institute, Mumbai was in runner up, IIT Bombay was in third spot, IIT Madras was in fourth spot and In Kolkata (cultural capital of India)" Indian Institute Of Science Education and Research" stood fifth.

India is twelfth in the country-wise positioning in Nature Index publications and, from 2018 to 2019, it has improved its offer score by 4.7%. Just three Indian foundations include in the rundown of top 100 rising scholarly establishments around the world and those are - IISER Kolkata at 57; IIT Madras at 77; and IIT Guwahati at 100.IISc has improved its changed offer by two percent and had an include of 221 of every 2019, the highest among all Indian academic institutions. Homi Bhabha National Institute has improved by 19%; IIT Bombay by 7.4%; IIT Madras by 0.9%; lastly, IISER Kolkata has seen its adjusted share ascent by 69%.

2 Literature Review

The inceptive period of research clarified exercise of this study would lead positive and effective consequences for the quality of science which was extracted from Indian science and technology publication output by Buchandiran (2011). The research objective also aimed to improve decision-making and management processes in science policy. As investigated in an article named impact factor by giving more emphasis on the peer-review process: Measure of quality of research publication by Jain (2011). As per their study for ranking and evaluating the journals and judging the quality of Individual research academic publication, "Impact Factor" is being used as an index of quality and prestige and this key indicator is used more and more.

As per the survey operation by Quality improvement and quantity enhancement of Indian journals by [4], it was observed that Indian Journal failed miserably in quality. In order to assess research practice and quality for evaluating research a multidisciplinary approach is used: by Martensson, Uno Fors; Zander; and Nilson (2014) framed a model to create understanding of the generic dimensions of the quality of research practice though a multidisciplinary collaboration.

In addition to that, in trait of Research, qualifications and Research Publications in India- Is Quality of Research and Education Declining in India? by Peraman et al. [10] cited that from 2006 to 2009, the research competence of India has developed to 8th rank from 10th rank. With reference to reports by British council India (2014), four major challenges have been coded, against the progress of higher education and research in India, viz. (a) bad quality of teaching and learning, (b) high market interest hole, (c) lopsided development and admittance to a promising circumstance and (d) imperatives on research limit and advancement. Report by T.V Padma India Science Test from "Nature" (2015) stated that Indian examination is hampered by smothering organization, low quality instruction at most colleges and lacking financing" and "The thickness of researchers and specialists in India is one of the least on the planet, and there are a lot of significant regions where we can't do explore.

India was at the 13th place for its high-quality scientific publications according to Nature Index analysis 2014. Despite this achievement, in terms of predatory open access journals quality is very poor, several studies have revealed in India's scientific publication in predatory journals: need for regulating quality of Indian science and education by Seethapathy et al. [2].

Presently it has been investigated, by Indian space technologists by Sridhar (2017) in challenge of an investigation of correlation of use, citation and publishing of journal papers that the examples of the publishing of papers in journals by Indian. Which would prompt a superior connection as recommended here in a more controlled and strong examination with a bigger and homogeneous example.

3 Research Design

The research work has divided into three segments, i.e., PART A: Secondary data, PART B: Qualitative and PART C: Quantitative-based research.

PART A: Consists of the correlation between the factors those are significantly associated with the Quality of the Indian Research Journals. Here we have to collect some data regarding Indian Research Journal and observe whether they are positively correlated or not if they are then what's our conclusion if not then what. It is a

Secondary process through which we are going to observe statistically whether the journal published in India is quality Journals or not.

PART B: Quantitative Approach: A structured questionnaire of ten question has designed to collect the data from different segment of researcher spread across different age group. The question answer is in five-point scale, i.e., 1-Strongly Disagree, 2-Disagree, 3-Neutral, 4-Agree and 5-Strongly Agree. A form has designed to collect the data from different respondent through email received from different sources. The simple random technique is used to collect the data from different source. Before analyzing the data, it will be verified by the Cronbach's alpha bretel test to validate the data and the data adequacy will be tested through KMO test. Further the data will be analyzed through different analytics technique, i.e., correlation, multiple regression and ANOVA.

PART C: Qualitative Research: since techniques are frequently more flexible and subjective, it's essential to reflect on the methodology we took and clarify the decisions we made. A structured open-ended questionnaire will be prepared and survey will be conducted through a focused group. We planned to design seven number of groups for discussion. In each group, we will invite five numbers of experts in the field of research and academician. All the group discussion will be recorded, and further, it will be analyzed some qualitative technique.

4 Research Analysis

PART A: Here in this part, we are going to collect some data which are directly or indirectly correlated with the Quality of the Indian Published Journals. By using Pearson Correlation Coefficient method, we can wind-up with a solution that the Journals published by Indian Researchers are of good quality or not.

We collect the data on SJR Ranking, H-index, Total number of Document Published in the year 2019, Total number of Reference obtained in the year 2019, Number of documents cited within 2 years and number of Document referred in the year 2019 of different Indian Research Journals to observe whether the Publishing Journals are of good quality or not. We pile up the data based on above factors from the source "UGC Approved Listed Indian Journals in 2019". To decide whether the Journals published every year in Indian Research Journals are of good quality or not we have to calculate the correlation coefficient between the factors we have drawn from the collected data.

The following formula is used to calculate the simple correlation coefficient:

$$r = \Sigma(x_i - \overline{x})(y_i - \overline{y})\sqrt{}\Sigma((x_i - \overline{x})^2\Sigma(y_i - \overline{y})^2 \quad (1)$$

Here if we take the upper triangular matrix values of the table that signifies there is two negative Pearson Correlation Coefficient between Total Document published in 2019 and SJR ranking, also in between Total Reference in 2019 and SJR ranking having Significance value (2-tailed) 0.646 & 0.728. Here for this situation, the correlation is significant at the degree of 0.01 which is 99% (having ** demonstrates the level is 99%). Aside from this specific condition additionally there are a few situations where the correlation is significant at the level of 0.05 which is 95% (having * demonstrates the level is 95%).

From the above calculation, there is a negative correlation between Total number of Document Published and SJR ranking, Total Document Published and Citation per Document in 2 years, Total no of Reference in the year 2019 and Citation per Document in 2 years. This negative correlation indicates that if one factor increases other decreases. Logically, if we see when SJR ranking increases, the total number of documents should also increases, if it is doing so, then we can say the quality of the Journal is good. So, we here from the observation we got that when one parameter increases other one decreases which indicates the quality of the Journals are not that good. But also, if we focus on positive side then there is a strong positive correlation between Citation per document in 2 years and SJR ranking which indicates the respective journal have good quality of Cited documents. This implies the quality of the Journal is somehow good.

From the above fact, our finding is that some of the Indian research articles are significantly contributed on innovation and development but the overall of research publication by the Indian researchers are under scrutiny.

PART B: As per the background study of this research, we have considered ten variables, for our analysis in regard to the quality of paper. By utilizing the one sample t-test, independent sample t-test, and ANOVA; the author applied one sample t-test to verify the parameters impact on the quality of research. The result has taken from the SPSS software and it was found that for all the parameters P value, is less than 0.05 which indicates that the parameters give the negative remarks on the quality of research. This result shows the researchers must focus on those parameters for improvisation.

For applying ANOVA test, we have taken overall quality parameter as dependent and ten independent parameters (Review of problem, Literature Review, Review of techniques, Objective of research, Research Hypothesis, Research Design, Research Analysis, Conclusion of Research, Research solve the current research problem, References).

The compiled result of ANOVA and the Descriptive Statistics of independent parameter reflects as.

Hypothesis (Mean = 3.79, S.E = 0.118, S.D = 1.331, F = 1.285, P = 0.205), Conclusion (Mean = 3.69, S.E = 0.129, S.D = 1.450, F = 1.595 P = 0.067), Research Solve (Mean = 3.70, S.E = 0.128, S.D = 1.438, F = 1.547, P = 0.081). The result shows Research Hypothesis has been clearly defined by the Indian researchers at 5% level of significance. Further Research Conclusion and Research Solve the current research problem at 10% level of significance.

The other remaining parameters; Review (Mean = 3.93, S.E = 0.137, S.D = 1.539, F = 1.753, P = 0.036), Review Techniques (Mean = 3.77, S.E = 0.121, S.D = 1.364, F = 1.834, P = 0.026), Objective (Mean = 3.57, S.E = 0.127, S.D = 1.434, F = 2.596, P = 0.001),, Literature Review (Mean = 3.79, S.E = 0.123, S.D

= 1.389, F = 1.818, P = 0.027), Reference (Mean = 3.74, S.E = 0.129, S.D = 1.449, F = 1.954, P = 0.015), Research Design (Mean = 3.80, S.E = 0.124, S.D = 1.399, F = 2.029, P = 0.011), Research Analysis (Mean = 3.72, S.E = 0.126, S.D = 1.419, F = 2.682, P = 0.001) indicates that the parameters are not clearly defined in their research articles. Through the independent sample t-test we have tested that the segment of gender, age group and profession have the same opinion regarding the above fact.

From the above analysis, our finding is that the Indian research articles are not significant in many of the parameters for support the quality of research.

PART C: We have done a survey through tele-calling, where we have asked the target respondents some questions related to our research. We got 49 responses from which 11 are female and 38 are male. The response we have got, in which most of the researchers prefer to write mainly in software & IT sector, finance sector, material chemistry, porous polymer science, mechanical science, HR, OB (organizational culture and sub-culture), automobile, stock market, banking & macro-cycles areas. According to the survey, 57.1% people agree that according to their research topic, the information they find from previous research papers are enough for their research. Among our target respondents, 85–90% of the researchers prefer Indian article because language and writing pattern is easy and understandable. The scenarios are more relevant to them & 100% of them prefer foreign articles because the foreign articles give new perspective, it helps to identify the gap in Indian context and to upgrade with recent developments in tools and methods, the quality of research tends to be better and chances of the work being empirical is higher, the articles provide more insight and foreign articles have high visibility in scientific domain. According to our survey, 42.9% people are satisfied with the content Indian articles have.

Through qualitative research analysis, we find that the Indian research articles are simple, they have great level of citation, research area and they cover Indian scenarios which is easy to co-relate with as well as they explain well and easy to understand, however the articles have been lacking on high visibility, updated methodology and tools, true and proper guidance of experts, true and proper guidance of experts. Also, they do not cover global parameters. Some of them are hypothetical. Chance of the work being paraphrased or copied is higher, and chances of data manipulation are higher with very few substantial proofs of data actually being collected. From the above analysis, our finding is that Indian research articles are not up to the global standard.

5 Conclusion

Research and Innovation are the two mutually inclusive events plays a vital role in current scenario. In this context, the author had taken initiates to pursue a research work to evaluate the quality standard of the Indian research article. The research was carried out in three different dimensions. The first dimension undertakes the secondary dataset to carry forward the research and the finding was some of the Indian research articles are significantly contributed on innovation and development but the overall of research publication by the Indian researchers are under scrutiny. The second dimension focuses on the primary survey on quantitative parameters from the different researchers and the finding was the Indian research articles are not significant in many of the parameters for supporting the quality of research. Further in the third dimension, the research focuses on the primary survey on qualitative parameters and the finding was Indian research articles are not up to the global standard. Finally, we conclude that the researchers must put more effort on their research work and research publications.

Table 1 Pearson correlation between the variables that mentioned below

Pearson correlations

		SJR	V3	V5	V7	V9	V11
SJR	Coefficient	1	0.432**	−0.067	−0.05	0.964**	0.319*
	Significance		0.002	0.646	0.728	0	0.024
V3	Coefficient	0.432*	1	0.126	0.205	0.376**	0.291*
	Significance	0.002		0.384	0.154	0.007	0.04
V5	Coefficient	−0.067	0.126	1	0.942**	−0.041	0.023
	Significance	0.646	0.384		0	0.775	0.871
V7	Coefficient	−0.05	0.205	0.942**	1	−0.021	0.112
	Significance	0.728	0.154	0		0.883	0.438
V9	Coefficient	0.964**	0.376*	−0.041	−0.021	1	0.349*
	Significance	0	0.007	0.775	0.883		0.013
V11	Coefficient	0.319*	0.291*	0.023	0.112	0.349*	1
	Significance	0.024	0.04	0.871	0.438	0.013	

Where SJR—SC Imago Journal Rank (SJR) indicator of Indian Research Journals; V3—H-index of the Indian Research Journals; V5—Total number of Document published in the year 2019 in various Indian Research Journals; V7- Total number of Reference obtained by the respective Journals in the year 2019; V9- Number of Documents Cited from respective Indian Research Journals within 2 years; V11—Number of Documents referred from respective Indian Research Journals in 2019; N—no of observation.

Table 2. ANOVA one way analysis with taking all the parameters

ANOVA					
		Sum of Squares	Mean Square	F	Sig.
Review is essential for Indian research	Between Groups	74.160	3.708	1.753	0.036
	Within Groups	224.202	2.115		
	Total	298.362			
Review techniques are defined in the Indian article	Between Groups	60.242	3.012	1.834	0.026
	Within Groups	174.136	1.643		
	Total	234.378			
Objective of your research match with existing Indian research	Between Groups	85.170	4.259	2.596	0.001
	Within Groups	173.869	1.640		
	Total	259.039			
Hypothesis with research match with your study	Between Groups	43.577	2.179	1.285	0.205
	Within Groups	179.683	1.695		
	Total	223.260			
Conclusion defined in the Indian articles	Between Groups	61.310	3.066	1.595	0.067
	Within Groups	203.714	1.922		
	Total	265.024			
Background study mentioned properly in the article	Between Groups	62.136	3.107	1.818	0.027
	Within Groups	181.124	1.709		
	Total	243.260			
Reference of the study mentioned properly in the Indian articles	Between Groups	71.237	3.562	1.954	0.015
	Within Groups	193.188	1.823		
	Total	264.425			
Research design will be properly defined	Between Groups	68.282	3.414	2.029	0.011
	Within Groups	178.395	1.683		
	Total	246.677			
Research analysis properly defined in Indian Publications	Between Groups	85.278	4.264	2.682	0.001
	Within Groups	168.517	1.590		
	Total	253.795			
Research solve the current problem	Between Groups	58.885	2.944	1.547	0.081
	Within Groups	201.745	1.903		
	Total	260.630			

References

1. Makkar, S.: Problem of research quality in India: an analysis and suggestions for sensitisation. Int. J. Multidis. Res. Develop. **2**(8), 313–316 (2015)

2. Seethapathy, G.S., Santhosh Kumar, J.U., Hareesha, A.S.: India's scientific publication in predatory journals: need for regulating quality of Indian science and education. Curr. Sci. **111**(11), 1759–1764 (2016)
3. Gupta, B.M., Dhawan, S.M.: A Scientometric analysis of s&t publications output by India during 1985–2002. DESIDOC J. Library Inf. Technol. **28**(2), 73–85 (2010)
4. Jeevan, V.K.J.: Quality improvement and quantity enhancement of Indian LIS journals. Ann. Libr. Inform. Stud. **61**(03), 217–226 (2014)
5. Mahesh, G., Wadhwa, N.K.: Web of Science based ranking of Indian library and information science journals. Collnet J. Scientometrics Inf. Manage. **6**(2), 263–272 (2012)
6. Sridhar, M.S.: A study of correlation of use, citation and publishing of journal papers by Indian space technologists. Collect. Manag. **12**(02), 147–152 (1990)
7. Deshmukh, R.K., Dr. Taksande, P.G.: "Impact of correlation on research productivity". Int. J. Sci. Res. Comput. Sci. Eng. Inf. Technol. **3**(1), 1830–1835 (2018)
8. Pechlaner, H., Zehrer, A., Abfalter, D.: How can scientific journal quality be assessed? An exploratory study of tourism and hospitality journals. Tourism (Zagreb) **50**(4), 395–399 (2002)
9. Burkhardt, H., Schoenfeld, A.H.: Improving educational research: Toward a more useful, more influential, and better—funded enterprise. Educ. Res. **32**(9), 3–14 (2003)
10. Peraman, R., Kuppusamy, R., Muthukrishnan, R., Satyanarayana, M.: Quality of research, professional degree and research publications in India–is quality of research and education declining in India? PTB Rep. **1**(3), 96–97 (2015)
11. Basu, A., Aggarwal, R.: Indian scientific literature in science citation index: a report. information today and tomorrow. Scientometrics **20**(4), 3–8 (2001)
12. Beall, J.: "Unintended consequences: the rise of predatory publishers and the future of scholarly publishing." Editorial Office News **2**, 4–6 (2014)
13. Patra, S.K., Chand, P.: Library and information science research in India: a bibliometric study. Ann. Libr. Inform. Stud. **53**(2), 219–223 (2006)
14. Pichappan, P., Buchandiran, G.: Peer reviewing in Indian S&T journals. Curr. Sci. **90**(5), 615 (2006)
15. Konwar, B.K.: Research perspective in institutions of higher learning. University News **51**(17), 39–43 (2013)

Data Center Simulation for Resource Utilization Analysis at Extended Loads

Mathew Koshy Karunattu, S. Sayanth, Akash Vijayan, Ansamma John, and Manu J Pillai

Abstract Applications and services are widely and easily accessible to anyone at any time, creating the need for the availability of these services and applications crucial. Any cloud service or application can be down due to many reasons and a significant factor for such an outage is heavy traffic. This makes the testing phase and the resource requirement analysis of the applications and services very important for its reliable operation post deployment under heavy load. Testing by deploying applications using actual resources may not be very effective and cost efficient. Testing conditions like the real-world scenarios are necessary for the analysis of the application's performance under heavy traffic. This creates a need for a simulated environment where no actual execution of processes is required and hence minimize resource utilization, time, and money. This paper addresses such a simulation tool where a user can simulate applications with multiple processes and analyze the actual resource utilization of these applications once deployed in real life. The user can configure both the data center as well as the applications according to their needs and run a simulation which can produce resource requirement data for the application to run under any specified load and a graphical representation for further analysis of the generated data.

Keywords Data center · Simulation · Cloud failure · System resources · Seasonality · Load balancing · Resource utilization

1 Introduction

The demand for information and services is at its peak, ever since the introduction of the Internet services. This makes data management of utmost importance regarding an organization or business. Hence, the availability and accessibility of these data or services prove critical for any company to cater the needs of their customers as well as to improve their business agility. With access to smartphones and other

M. K. Karunattu (✉) · S. Sayanth · A. Vijayan · A. John · M. J. Pillai
TKM College of Engineering, Kollam Kerala 691005, India

© The Author(s), under exclusive license to Springer Nature Singapore Pte Ltd. 2022

A. K. Nagar et al. (eds.), *Intelligent Sustainable Systems*, Lecture Notes in Networks and Systems 333, https://doi.org/10.1007/978-981-16-6309-3_58

devices easily available for almost everyone, the need for data management is one of the core foundations for any services that a company has to offer. Therefore, one should make sure that any services or data provided should be reliable under any conditions. This can only be ensured by testing the services under various real-life scenarios. The main objective is to eliminate any downtime that could happen and affect the performance and availability of the service provided. But the testing of these applications and services on the actual cloud itself can prove to be risky and will be very expensive. The main two reasons that cause availability issues are due to failures at the data centers or the application because of poor implementation or architecture. The failure at the data center could be due to the data center being partially or completely down. This failure can be caused because of power outages, overheating, network issues, low storage, human errors, or failure in detecting issues. Most of the availability issues are due to failure on the application side which can be caused due to improper design, architecture, testing, race conditions, or unexpected situations that trigger unusual code paths. The above-mentioned problems create a need for an environment where one can safely and efficiently test the applications or services at low cost that can reproduce real-life scenarios.

2 Data Centers

To meet the ever growing demands of Web applications and online services, centralized facilities are necessary to serve the requests coming from computing devices located from every part of the world without any interruptions all the time. These facilities, called the data centers, run servers which are basically computing machines together with resources that run the application or service programs to serve the requests from the computing devices. Data centers are run and maintained by cloud service providers like Amazon Web Services, Microsoft Azure, and Google Cloud, and they provide a large number of different configurations for creating a server instance.

Amazon Web services provide server instances optimized for memory, computing, storage, and general purpose instances that provide balanced optimization between all the resources. The general purpose instances provide a range of instances with capacities starting from 1GB memory, 1 CPU core, and no storage to a configuration with 96 CPU cores, 384 GB of memory, and a 4×900 NVMe SSD storage with a network bandwidth of 100 Gbps. The CPU cores can be various chips from Intel, Apple, AMD, etc. And the operating systems can be Windows, Linux, UNIX, and so on. A maximum of 256 GB GPU memory can also be included in some of the instances [1]. This is the configuration of a single server instance, and in reality, a data center can house a large number of instances making them to accommodate a huge number of physical machines running together in enormous buildings with server management staff working all the time to keep the server up. In addition to this, data centers consume a huge amount of electricity which goes up to 12.4 Terawatts every year for Google data centers [2].

In terms of the software running in the server machines, every computing machine in the data center runs an operating system, device drivers, and various services. Besides that, they run the Web applications or the services they provide to the clients. Together they form a collection of different process instances, where each consumes different amounts of resources, and the number of process instances varies according to the load on the server.

3 Cloud Outages

It is possible that any cloud services can fail and cause outages or service unavailability due to many reasons. Almost half of the cloud service experience up to three outages a year on average apart from planned maintenance and failures of non-essential operations which can be categorized into four such as hardware failures, software errors, human errors, and unmanageable load [3].

Hardware failure can be anything affecting the physical devices in the data centers or power failures. Common hardware failures occur at network switches to cause network failures. Power outages can cause data center operations to be completely disrupted, necessitating traffic rerouting to backup data centers. Such a scenario occurred when the Silicon Valley NA14 instance of Salesforce went offline on May 9, 2016, causing an outage that lasted more than 24 h [4]. Extensive damage to business happened and the customers lost many hours of service and data. This was due to a power outage caused by a power breaker failure. The service was restored temporarily by switching from their primary data center in Washington to their secondary data center located at Chicago [3]. This compelled Salesforce to use Amazon Web Services for hosting their servers.

Outages due to software errors can be caused due to software upgrades, misconfigurations, or generally due to the bugs in the implementation of the application or supporting services. Storage failures due to corrupted databases can also cause outages [3]. On December 25, 2012, Netflix experienced a cloud failure, rendering thousands of subscribers unable to access any of the Netflix platform's contents for nearly 15 h. Errors in the Amazon Web Services' Elastic load-balancing API calls in the US-EAST-1 region led to the major service interruption.

Human errors cause failures due to misconfigurations and improper procedures followed during software upgrades [3]. Such an event was reported on November 18, 2014, when Microsoft Azure cloud service experienced interruptions in service impacting Azure storage and other services including its virtual machines in many regions [5]. This issue was primarily caused by human error, as the incremental deployment of changes across the small slices was not followed, and the configuration switch was incorrectly enabled for Azure Blob storage front ends.

Another cause of cloud application or service interruptions is heavy traffic. Cloud services are usually provisioned resources based on expected traffic, and unexpected surges in loads can cause server resources to run out causing the client requests to become unserviceable which renders the server unavailable. Unmanageable load is responsible for 9% of all the cloud outages and the reasons for the heavy traffic for primarily classified into four like user requests, code upgrades, misconfigurations, and flawed recovery [3]. Service unavailability can be caused due to a huge number of user requests that often happens on special occasions, days, or events such as Christmas eve [6], New Year eve [7], the President's inauguration day [8], the World Cup's opening [9], company acquisitions like WhatsApp's $19 billion acquisition by Facebook which led to a surge of sign ups [10]. Even though main requests can be monitored, and the service can be scaled elastically, non-monitored requests like database index file access, and authentication requests can cause huge cryptographic resource consumption. Code upgrades cause service failures by creating load on the server in some scenarios where a system upgrade demanding account migration requires a huge number of credential requests, or an upgrade that introduce a new code segment to keep data geographically close to the user accidentally overload a data center. Misconfigurations can also increase the load on servers when they trigger client re-queries multiple times, wrong traffic redirections which can cause huge loads going to a small set of servers, miscalculation of resource utilization of a server causing normal traffic on the server to be labeled as unserviceable. Flawed failure recovery can cause huge loads called recovery storms. These can occur due to the positive feedback loop problem where flawed recovery causes extra traffic that congests existing resources which again triggers more recovery traffic. For example, an automated recovery from misconfiguration causes clients to send requests causing an overwhelmed server, and then it causes the clients to retry the requests worsening the situation. Also, simultaneous reboot of many virtual machines can overload storage servers. The case of a group of dead servers rerouting traffic to a group of healthy servers can cause high traffic on the healthy servers.

4 Simulation

To test the services and applications in the most effective and efficient way while also considering the cost, the best way to go is with simulation of these applications and the services. Testing by the actual execution of all the processes within these applications and services under various real-life conditions is very time consuming as well as resource intensive. If an application or a service needs to be tested for a duration of one month, then actually executing the processes for this test would mean that it would take one month to complete the testing. Instead, if we just simulate the execution of the processes within these applications and services, then the above-mentioned test can be completed in just within a few minutes. Also, in case of simulation the resource utilization would be a whole lot less compared to the actual execution. This makes the simulation of the processes an ideal tool for testing purposes before the actual deployment of the products for the intended customers.

Creation of a simulation tool where the user can configure applications as well as each of the individual processes within these applications is the need of the hour.

The simulation tool can be built such that multiple applications can be simulated at once with the implementation of load balancers and chaos engineering to reflect real-world scenarios.

5 Methodology

The simulation of a real-life data center requires defining the various components in it. The component that primarily undergoes a simulation is a process instance which corresponds to a process instance of a program being executed on a server machine. The process instances belong to a parent process defined by parameters that determine the range of system resources the process instance would use during its execution. Various resources like memory, CPU, disk IO, network bandwidth, etc., can be considered. The processes collectively represent an application which runs independently in an instance of the server which has been allocated a fixed quota of system resources. Additional parameters are defined in the process model to make it more like the real-world data centers. Parameters like memory leak that represents the memory that will be left allocated even after process instance is deallocated, and a proportion parameter to maintain the proportion of the number of a particular type of process instance in the whole system is defined.

Additionally, the character of variation in the number of requests hitting the server over time called the seasonality can also be defined in the application definition. This seasonality component simulates the application execution over a time with the number of requests coming to the server changing continuously over time with peak times at certain time of a day and a low load at some other times. This seasonality could be extended to weeks, months, and years.

Since the load on the server varies over time, it is necessary to manage the load by distributing the load to different instances of the application in situations of high load. This process of balancing the load requires a single application to be able to run on multiple server instances. Hence, the tool supports the simulation of an application separately on different instances of server with each instance allocated a fixed quota of system resources. The requests coming to the application must be distributed among the different instances of the application in a manner, so the overall resource utilization of the application is maintained within a threshold.

Finally, the utilization of the various system resources by different applications and their instances must be tracked to draw a conclusion about the resource requirements of the applications under different loads. This requires the generation of resource utilization metrics at regular intervals during the simulation and its storage in some time series based databases. The generated resource utilization metrics must be visualized for easier representation and understandability of the generated data. The following sections describe the configuration of the simulation environment with various applications.

5.1 Configuration

The simulation tool has a mechanism to configure the environment in which the applications are to be simulated and to configure the applications itself to be simulated. The configurations are classified as system configuration and application configuration. System configuration has the definition of a compute system in the data center. It also defines the duration of the simulation, the interval in which the resource metric is generated, number of applications in the system, the time quantum for the round robin scheduling algorithm, total RAM, secondary storage size, maximum network bandwidth, maximum disk IO speed, and a base percentage usage of CPU. Application configuration defines the applications that are simulated. The data defined in the application config is used to make process instances which are defined in the architecture as app instances. They have a specific amount of memory, CPU, disk IO, and bandwidth usage, this data are taken randomly from the range of values specified in the application config. The seasonality and resource utilization dependency of each of the applications is defined in the application config. In addition to the modules specified above, it is necessary to consider another component which can start and coordinate the activities of other modules, called the Initiator.

5.2 Initiator

The starting point of the simulator is the initiator whose job is to invoke all the necessary functions for the simulation to start in a predefined order. The first thing the initiator does is to invoke the method that loads the system config and application config files into main memory. Then invokes the method to generate the app instances. Following this, it will create another module called load balancers and pass the corresponding application it should manage. When all the simulations are over, the initiator will upload the generated metrics to a Victoria Metrics [11], if available. The metrics data will always be saved in a file which will have the simulation start time as the name.

5.3 Load Balancer

Load balancer module simulates the working of a load balancer in a data center. It can scale up or down simulation instances according to the load-balancing policy defined in the application config. The simulation instances of an application are generated and managed by its load balancer, and it can create additional traffic if the load in any of the active simulation instances is less than the desired load at that time. The simulator module along with a system resource module that can handle the simulation of an application instance are generated by the load balancer.

5.4 Simulator

Simulator module simulates a compute system in a data center. This can simulate the running of application instances that have been generated. These instances of application are given to a scheduling algorithm which will simulate the execution and thus generates the resource utilization metrics data. Each of the simulation instances has a system resource manager that manages the resource allocation and deallocation of all the resources that a compute system has. This module keeps track of the compute instance's system clock, and the metrics generator module generates metrics data at the appropriate intervals. The metrics generated are passed on to the file handling module by the simulator at the end of the simulation to save the generated metrics. The file handling module will write the metrics data onto the output file in JSON line structure.

5.5 Visualizer

The representation and analysis of the resource utilization metrics generated during the simulation will be a Herculean task that can be simplified with the use of a graphical visualizer. The visualizer consists of two parts: a time series database (TSDB) and assssss visualization tool. The initiator will push the data to the TSDB once all the simulations are completed. The simulation's output is then viewed using a visualization tool such as Grafana [12] (Fig. 1).

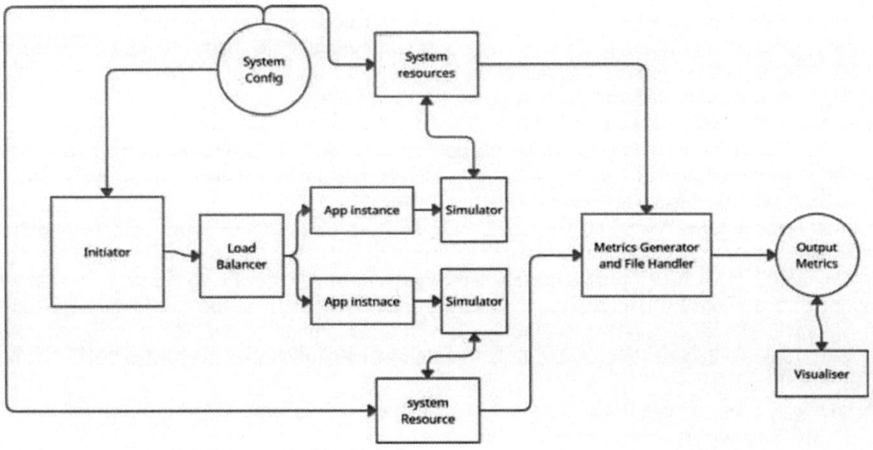

Fig. 1 Architecture of the simulation tool

6 Conclusion

Cloud outages happen to almost all major cloud services, making the services unavailable from time ranging from minutes to hours and sometimes even days, which causes significant loss of business and trust for the companies. Many factors can cause the outage, and out of that heavy traffic causes a significant share of the failures. The unexpected huge load on the server depletes the server resources, rendering the server unable to service the client requests and causing the application/service unavailability. Analyzing the resource requirements by deploying the application on a real server can be expensive and conditions of heavy traffic cannot be attained in a test environment. This paper proposes the design of a simulation tool that can simulate a data centre under different loads to analyze the resource utilization of the different simulated conditions and predict the resource requirements beforehand. The simulation tool can be used to analyze resource requirements for special events when considerable traffic is expected, so that more resources can be assigned ahead of time to deal with the extra traffic. This can help companies predict the resource requirements for every possible condition for a long period of time, so that heavy traffic do not deplete the server resources and cause failures. With comparatively fewer resources and effort, an efficient simulation mechanism may create a resource demand analysis for an average server, for an average period, and an efficient representation of the generated metrics can make the analysis easy and straightforward.

References

1. AWS: AWS Instance types (2021). https://aws.amazon.com/ec2/instance-types/
2. Bryce, R.: How Google powers its monopoly with enough electricity for entire countries (2020). https://www.forbes.com/sites/robertbryce/2020/10/21/googles-dominance-is-fueled-by-zambia-size-amounts-of-electricity/?sh=6f806b0468c9
3. Gunawi, H.S., Hao, M., Suminto, R.O., Laksono, A., Satria, A.D., Adityatama, J-f., Eliazar, K.J.: Why does the cloud stop computing? Lessons from hundreds of service outages. In: Seventh ACM Symposium on Cloud Computing, pp.1-16. Association for Computing Machinery, Santa Clara, CA, USA (2016) https://doi.org/10.1145/2987550.2987583
4. RIQ News desk: 6 Computing failures that shocked the world (2020). https://www.readitquik.com/articles/cloud-3/6-cloud-computing-failures-that-shocked-the-world/
5. Zander, J.: Final Root Cause Analysis and Improvement Areas: Nov 18 Azure Storage Service Interruption (2014) https://azure.microsoft.com/en-us/blog/final-root-cause-analysis-and-improvement-areas-nov-18-azure-storage-service-interruption/
6. Salesforce Trailblazer Community: RCM for NA14 Disruptions of service—May 2016 (2016). https://help.salesforce.com/articleView?id=000315819
7. White, C.: Twitter Site Back After 90 Minute Outage (2011). https://mashable.com/2011/12/31/twitter-is-down-2/
8. Grandoni, D.: Twitter Goes Down on Inauguration Day 2013 For Some Users (2013). https://www.huffpost.com/entry/twitter-down-inauguration-day_n_2522196
9. Roettgers, J.: Is the Wolrd Cup Bringing Down Twitter? (2010). https://gigaom.com/2010/06/11/is-the-world-cup-bringing-down-twitter/

10. Josh Constine: WhatsApp is Down, Confirms Server Issues (2014). https://techcrunch.com/2014/02/22/whatsapp-is-down-facebooks-new-acquisition-confirms/
11. Victoria Metrics https://victoriametrics.com/
12. Grafana. https://grafana.com/

Detection of Alzheimer's Disease Through Speech Features and Machine Learning Classifiers

Ajay Sankar Gulapalli and Vinay Kumar Mittal

Abstract The most common neurodegenerative disease identified in elderly people is Alzheimer's disease (AD). In general, identification of Alzheimer's is carried out based on medical images of the brain. The other non-invasive signal processing techniques for AD detection are reviewed. Possibly, these reviewed techniques can also provide an early diagnosis of AD, which might help delay the development of the disease. These models are based on the spontaneous speech (SS) features extracted, such as vocal, linguistic, acoustic, prosodic, and features. Various classifiers and machine learning algorithms that are feasible in AD detection are described. These are discussed and compared based on their performance accuracies of classification results. Few models developed on repository corpuses and others based on speech data collected from persons of different age groups, affected with AD, having mild cognitive impairment (MCI) and/or healthy control (HC) are presented. The models reviewed may help further research toward the development of a reliable machine assistive technology (MAT) for providing health care to elderly people and exploring cures or solutions to the diseases like AD, PD, vascular dementia, down syndrome, frontotemporal dementia, etc.

Keywords Speech recognition · Alzheimer's disease (AD) · Spontaneous speech (SS) · Mild cognitive impairment (MCI) · Healthy control (HC) · Machine assistive technology (MAT)

1 Introduction

Alzheimer's disease (AD) is a type of dementia. The human brain begins to shrink [1], also the nerve fibers get reduced. In this kind of dementia, the mental [2], cognitive, and linguistic functions are disturbed [3]. AD is a disease that is widely observed in elderly people, and that the severity may lead to death.

A. S. Gulapalli (✉)
Aditya Engineering College, Surampalem, India

V. K. Mittal
KL Deemed To Be University, Vijayawada, India

© The Author(s), under exclusive license to Springer Nature Singapore Pte Ltd. 2022
A. K. Nagar et al. (eds.), *Intelligent Sustainable Systems*, Lecture Notes in Networks and Systems 333, https://doi.org/10.1007/978-981-16-6309-3_59

The patient's spontaneous speech [4] and extracted features might be used for the early detection and progression of the diseases like Alzheimer's, Huntington's, and autism. In [5, 6], characterization of speech in children with autism spectrum disorder (ASD) using acoustic features is demonstrated. For AD speech, characterization through speech features is discussed in [7].

Earl-stage detection of disease may help in the development of new treatments, provides monitoring of the response to the interventions and arranged as per the need and care of the patient [8, 9].

Detection of AD is based on electroencephalographic (EEG) and MRI analysis of the gray matter (GM). Most of these techniques use machine learning methods to classify brain images (MRI or fMRI) [10]. And this kind of analysis might be costlier. So, a low-cost non-invasive technique is required [11]. If the technique can provide self-assessment of the disease, then it would help the patient in a better way. Computational linguistics is a powerful tool to show that fully automated analyses distinguish patients with Alzheimer's from healthy controls based on speech.

Figure 1 describes the architecture of AD detection through ML classifiers. The process of classifying the AD subject from a healthy control starts with speech data acquisition, which can be carried out by various devices like acoustic cardioid (AC) and smartphone (SP). In preprocessing step segregation of voiced data (vocal cords vibrate) and unvoiced data (no vibration of vocal cords), standardization, application of randomization, normalization, and removal of outliers is carried out. The voice activity detection in a signal is described in [12]. Then various signal processing methods are used for extracting the crucial features through which the impairment in speech is marked [13, 14].

A training dataset with samples collected from subjects of HC and AD is defined. Classifier selection plays a key role in classifying an individual between the HC and AD. But when a classifier is applied alone, discrimination accuracy might be very low. So, the classifiers are trained along with speech features for better accuracy. To develop machine assistive technology (MAT) [15] for diseases like AD and ASD, a study on patient's speech might be helpful. Since these kinds of diseases primarily affect the voice [16]. The MAT may reduce the effort in assisting the patient with AD (like the smart homes for AD patients and elderly people) [17–20].

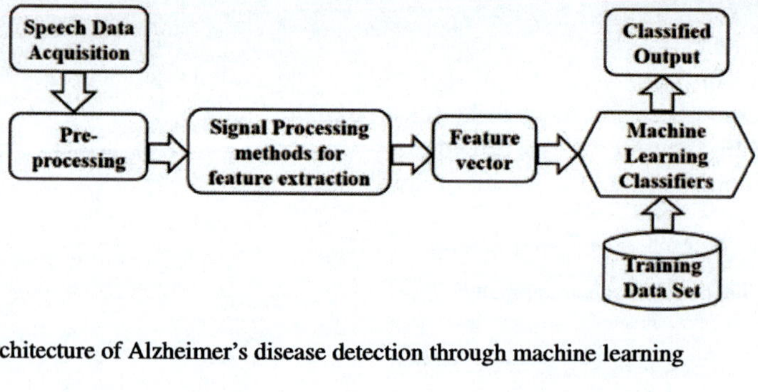

Fig. 1 Architecture of Alzheimer's disease detection through machine learning

The paper is organized as Chap. 2 reviews various database corpuses used. Chapter 3 gives a brief idea of signal processing methods used for preprocessing/feature extraction. Chapter 4 describes various features used in the literature. Chapter 5 deals with classifiers. Accuracy comparisons can be seen in the tabular forms given in each chapter. The mechanism involves speech data collection, feature extraction, classifier selection, and stating a decision-making algorithm.

2 Databases

The databases are developed by several health communities and associations, and their repositories are made available for experimental evaluation. The current chapter gives a literature review regarding the speech data collected from repository corpuses and speech data collected from different age groups with AD, MCI, and HC. A similar kind of data collection and processing for the detection of ASD is described in [7].

Dem@care project, BEA Hungarian Spoken Language Database, Penn Treebank Switchboard corpus, DementiaBank are the prominent repositories made available by various clinical institutions and research organizations. The datasets can be in various regional accents and languages. Automatic regional accent classification is illustrated in [21, 22]. The repository and customized speech datasets are summarized in Table 1.

3 Signal Processing Methods

The speech data derived from the databases are further applied with signal processing methods for preprocessing and feature extraction. Features extracted can be further used in developing, training, and testing the algorithm for the detection of AD. Signal processing methods/tools for the detection of AD are listed in Table 2. The signal processing methods/tools can be used in extracting the features either manually or automatically. Nonparametric Mann–Whitney U-test is a signal processing method to compare the differences between two independent groups [23, 27]. One-way ANOVA finds statistically significant distinguishment between independent groups [24]. The spectrogram is a function in MATLAB that returns short-time Fourier transform, also it uses windows to divide the signal into segments [36].

Mel-frequency cepstral coefficients (MFCC): MFCC reads the normalized speech signal, then finds the energy level and zero-crossing signals, if they are equal to unity, then signal values are extracted. MFCC then applies pre-emphasis, framing/segmentation, windowing, FFT, later mel-spectrum and mel-cepstrum are extracted with a desired cepstral coefficient [38].

Voice Activity Detection (VAD): An audio stream is separated into time intervals containing no speech data and time intervals with speech activity. VAD algorithm is useful in the selection of features that distinguish between speech and noise

Table 1 Summary of databases related to Alzheimer's disease

(a) S. no	(b) Database	(c) Place/institution	(d) Size	(e) Voice tasks	(f) Refs.
1	Dem@care project	The Memory Clinic in Nice, France	15 HC, MCI 23, AD 26	The subjects performed the tasks like counting backward task, a sentence repeating task, an image description task, and a verbal fluency task	[23]
2	BEA Hungarian Spoken Language Database	G´osy, Research Institute for Linguistics, Hungarian Academy of Sciences, Budapest, Hungary	75 speakers and 225 recordings, seven hours of speech data	Subjects performed tasks like 1. Need to talk on a one-minute animated film (immediate recall) 2. second day, second film presented, later, needs to talk on previous day 3. third day needs to talk on second film (delay recall)	[24]
3	Penn Treebank Switchboard corpus	Linguistic Data Consortium	72 members (46 HC, 26 AD)	Subjects need to describe the picture	[25]
4	DementiaBank	University of Pittsburgh School of Medicine	Total 473 samples, 233 HC, 240 AD	Subjects performed Boston Cookie Theft picture description task	[26]
5	*Dataset-1*	Cohort of Gipuzkoa-Alzheimer Project (PGA) of the CITA-Alzheimer Foundation	187 HC and 38 MCI	A categorical verbal fluency task, a picture description task and a spontaneous speech task are performed by the subjects	[27]
6	*Dataset-2*	Broca Hospital in Paris	48 people divided into 3 groups AD, MCI and HC with 16 people in each group,	Subjects involved in reading a sentence from the book "Le petit prince," a well-known French novel by Saint-Exupéry:	[28]

(continued)

Table 1 (continued)

(a) S. no	(b) Database	(c) Place/institution	(d) Size	(e) Voice tasks	(f) Refs.
7	*Dataset-3*	–	212 samples; with 100 AD and 112 HC	Subjects performed picture description task	[29]
8	*Dataset-4*	CREA in Salamanca, Spain For control group University of Salamanca	70 Subjects The group with AD (n = 35,) The control group (n = 35, men 80.6% and female 19.4%)	Several tasks were performed by the subjects: SVF; PVF; naming by definition; object naming; verbal fluency for proper names; immediate recall of a word list; deferred recall of a word list; delayed recall of data about a person	[30]
9	*Dataset-5*	Alzheimer Disease Research Centre of the University of Szeged, Hungary	A group of 38 HC and 48 MCI patients	Subjects need to recall the content of 2 short black and white films (one direct, one delayed) and by answering one question spontaneous speech	[31]
10	*Dataset-6*	–	50 healthy and 20 AD patients with subsets AZTIAHO, AZTIAHORE, AZTITXIKI	Subjects involved in friendly conversations and interactions on pleasant stories	[32, 33]
11	*Dataset-7*	Healthcare facilities at Istanbul	20 AD patients and 20 healthy subjects	Conversational speech of subjects recorded	[34]
12	*Dataset-8*	-	Total 60 participants: 30 HC and 30 AD	Recordings of friendly conversations and interactions on pleasant personal stories	[35]

Note (a) S. no., (b) corpuses or customized databases available, (c) place/institution at which databases are created, (d) number of participants, (e) voice tasks performed by the participants, (f) references

Table 2 Signal processing methods/tools for Alzheimer's disease

(a) S. no	(b) Signal processing method/tool	(c) Reference
1	Nonparametric Mann–Whitney U-test	[23, 27]
2	One-way ANOVA	[24]
3	Spectrogram function	[37]
4	Mel-frequency cepstral coefficients and filter banks energies (MFCC-FBE)	[25, 26, 28]
5	Voice activity detection (VAD)	[26, 29]
6	t-SNE algorithm	[29]
7	Praat voice analysis software	[30, 31, 33]
8	Temporal segmentation	[32]
9	Bispectrum estimation	[35]

properties. VAD also provides a detection scheme applied using features extracted [37].

t-Distributed Stochastic Neighbor Embedding Algorithm (t-SNE): This algorithm is useful in reducing the dimensionality for a high-dimensional dataset. t-SNE calculated the similarity points by considering conditional probability [29].

Temporal segmentation: This is designed based on temporal trajectories of short-term energy to segment and fundamental frequency [32].

Higher-order/third-order spectra/Bispectrum: It is used for extracting the variation from Gaussian parametric signals and phase relation of harmonic components [35].

4 Feature Sets

Spontaneous speech [39, 40] is identified to be one of the best tools for the identification of AD in elderly people in a non-invasive method. The features like vocal, acoustic [41, 42], linguistic [43], demographic [44], time domain, prosodic, temporal, etc., [45] are extracted. The feature extraction and disease detection process for autism in children is exemplified in [46].

The emotional response features and emotional temperature and autoregressive-based features (AR-B) achieved a considerable accuracy (~97.7%) [32]. Next to it, the features extracted from Dataset-1, i.e., classical features, time-domain features, classic perceptual features, and advanced perceptual features achieved better performance [27].

Classic features: Spectral-domain features like harmonicity, noise-to-harmonic ratio (NHR), pitch, harmonic-to-noise ratio (HNR), amplitude perturbation quotient (APQ), jitter, shimmer, and spectrum are classic features.

Time domain: Voiced/unvoiced segments, zero-crossing rate (ZCR), breaks.

Classic perceptual: Features like mel-frequency cepstrum coefficients (MFCC), energy, and short-time energy are classic perceptual. Advanced perceptual features are like voice quality, perception, adaptation: modulation spectra coefficients (MSC), linear predictive cepstral coefficients (LPCC), perceptual linear predictive coefficients (PLP), linear predictive cosine transform coefficients (LPCT), inferior colliculus coefficients (ICC), and adaptive component weighted coefficients (ACW).

Acoustic features: Further, the acoustic feature extracted from DementiaBank in [26] and a combined feature set of spontaneous speech fluency (SSF), HFD, maximum HFD, minimum HFD, variance HFD and standard deviation HFD, emotional response features and emotional temperature extracted from Dataset-6 in [32] shown a better accuracy. Emotion state classification and shout analysis is explained in [36, [47, 48].

Vocal features: This might include vocal reaction time (in seconds), relative length (patient sentence duration), amount of silence, number of insertions have shown better performance in classifying between HC and MCI, HC and AD, MCI and AD [23].

Higher-order spectral features: Features like FFT-based (FFT-B) and autoregressive-based features (AR-B) are also used in the disease detection and discrimination pro [35].

5 Classifiers

The extracted feature sets are used to train the algorithms designed using machine learning. [49] Explains the classification of ASD affected from normal children. The performance of classifiers might vary based on the datasets and extracted features.

Table 4 illustrated various available classifiers and their accuracies for the detection of AD. It was identified that KNN classier has performed with the highest accuracy with Dataset-8, extracted with higher-order spectral features (i.e., AR-B). Multilayer perceptron (MLP), deep learning by means of convolutional neural networks (CNN) (biologically inspired variants of MLPs), generalized regression neural network (GRNN) oracle ensemble classifiers have shown 95% of accuracy in classifying the subject from groups like healthy control, MCI, and AD patients.

6 Summary and Conclusions

The individual blocks in speech signal-based machine learning algorithm development were described. A review of speech signal databases is demonstrated. The features like vocal, acoustic, linguistic, demographic, time domain, prosodic, temporal, etc., extracted from the speech data from the datasets are reviewed.

The extracted features alone might not classify the subject, so the support of a classifier is needed for training the datasets. Various classifiers that can be used for the

Table 3 Features extracted and attained accuracy in Alzheimer's disease detection

(a) S. no	(b) Features	(c) Signal processing methods/tools	(d) Accuracy	(e) Database	(f) Ref
1	Vocal features	Nonparametric Mann–Whitney U-test	79% [HC & MCI], 87% [HC & AD], 80% [MCI & AD]	Dem@care project	[23]
2	Acoustic features	One-way ANOVA, MFCC-FBE, VAD, Praat voice analysis software	74–82%, 94.71%, 78.8%	BEA Hungarian Spoken Language Database, DementiaBank, Dataset-5	[24], [26], [31]
3	Linguistic features	One-way ANOVA	74–82%	BEA Hungarian Spoken Language Database	[24]
4	Acoustic and linguistic features	One-way ANOVA	80–86%	BEA Hungarian Spoken Language Database	[24]
5	Demographic, linguistic, and acoustic features	MFCC-FBE	93.1% [manual transcripts], 91.7% [ASR]	Penn Treebank Switchboard corpus	[25]
6	Demographic and acoustic features	MFCC-FBE	83.3%	Penn Treebank Switchboard corpus	[25]
7	Classic time domain, classic perceptual, advanced perceptual features	Nonparametric Mann–Whitney U-test	95%	Dataset-1	[27]
8	Prosodic, temporal parameters, and quality measures	MFCC-FBE	85%	Dataset-2	[28]
9	Speech rate and articulation rate	Praat voice analysis software	80%	Dataset-4	[30]
10	SSF	Temporal Seg	75.2%	Dataset-6	[32]
11	SSF + FD1	Temporal Seg	76.7%	Dataset-6	[32]
12	SSF + FD2	Temporal Seg	86.1%	Dataset-6	[32]
13	EF	Temporal Seg	90.7%	Dataset-6	[32]
14	EF + TE	Temporal Seg	97.7%	Dataset-6	[32]
15	SSF + EF	Temporal Seg	92.2%	Dataset-6	[32]
16	SSF + FD2 + EF	Temporal Seg	94.6%	Dataset-6	[32]

(continued)

Table 3 (continued)

(a) S. no	(b) Features	(c) Signal processing methods/tools	(d) Accuracy	(e) Database	(f) Ref
17	SSF + FD2 + EF + TE	Temporal Seg	94.6%	Dataset-6	[32]
18	Confusion rate, filler rate, word entropy, phone entropy, and rate of pauses in utterances	Praat voice analysis software	90%	Dataset-7	[34]
19	FFT-B	Bispectrum estimation	95.42%	Dataset-8	[35]
20	AR-B	Bispectrum estimation	97.71%	Dataset-8	[35]

Note-1 **HC:** Healthy controls,**AD**: Alzheimer's disease,**MCI**: Mild cognitive impairment, **SSF:** Spontaneous speech fluency,**EF:** Emotional response features,**FD1**: Higuchi fractal dimension (HFD),**FD2**: HFD, max HFD, min HFD, variance and standard deviation of HFD, **ET**: Emotional temperature

Note-2 (a) S. no., (b) Features used for the disease detection, (c) Signal processing methods/tools used for feature extraction and discrimination process, (d) Accuracy attained in disease detection, (e) Database from which features are extracted, (f) References.

detection of AD are demonstrated. The accuracy comparisons for various extracted features and classifiers are explained. These accuracies are for classifying a person between AD or MCI or HC. Several classifiers with the best accuracies (>90%) are observed. A combination of accurate features and the best classifier is crucial for the design of a reliable algorithm.

The scope of the work can be further extended for the detection of other types of dementia and psychoacoustic diseases.

Table 4 Classifiers and attained accuracy in Alzheimer's disease detection

(a) S. no	(b) Classifiers	(c) Accuracy	(d) Database	(e) Refs.
1	Support vector machine (SVM)	87% ± 3%, 80–86%, 78.8%, 93.79%, 90% 96.46%	Dem@care, BEA Hungarian Spoken Language Database, Dataset-5, Dataset-6, Dataset-7, Dataset-8	[23, 24, 31, 33–35]
2	AD/NL (Two neural network)	70.8%	Penn Treebank Switchboard corpus	[25]
3	Greedy approach (MLP)	94.4%	Penn Treebank Switchboard corpus	[25]
4	Bayesian networks (BN)	93.66%	DementiaBank	[24]
5	Meta-bagging (MB)	92.38%	DementiaBank	[24]
6	Random forest (RF)	90.90%, 78.8%	DementiaBank, Dataset-5	[26, 31]
7	Multilayer perceptron (MLP)	95%, 93.02	Dataset-1, Dataset-6	[27, 33]
8	Deep learning by means of CNN (biologically inspired variants of MLPs)	95%	Dataset-1	[27]
9	K-nearest neighbors (KNN)	85%, 87.59%, 97.71%	Dataset-2, Dataset-6, Dataset-8	[28, 33, 35]
10	GRNN oracle ensemble classifier	95%	Dataset-3	[29]
11	Decision tree (DT)	91.47%, 90%, 95.42%	Dataset-6, Dataset-7, Dataset-8	[33–35]
12	Naïve Bayes	87.59%, 95.94%	Dataset-6, Dataset-8	[33, 35]

Note (a) S. no., (b) Machine learning classifiers used for disease detection, (c) Accuracy attained in disease detection, (d) Database used for training the algorithm, (e) References

References

1. Honma, S.M., Gorno-Tempini, M.L., Miller, B.L., Vossel, K.A., Nagarajan, S.S., Houde, J.F.: Abnormal vocal behavior predicts executive and memory deficits in Alzheimer's disease. Neurobiol. Aging **52**, 71–80 (Apr 2017)
2. Fraser, K.C., Rudzicz, F., Hirst, G.: "Detecting late-life depression in Alzheimer's disease through analysis of speech and language". Linguistic Signal to Clinical Reality, pp. 1–11, San Diego, California (16 Jun 2016)
3. Greta, S., Ildiko, H., Veronika, V., Janos, K., Magdolna, P.: Speaking in Alzheimer's disease, is that an early sign? importance of changes in language abilities in Alzheimer's disease. Frontiers Aging Neurosci. **7**, 195 (2015)

4. Zhang, J., Pan, Z., Gui, C., Xue, T., Lin, Y., Zhu, J., Cui, D.: Analysis on speech signal features of manic patients. J. Psychiatric Res. **98**, 59–63 (2018)
5. Mohanta, A., Mittal, V.K.: "Analyzing autism speech of children in english vowels regions by analysis of changes in production features", in Proc 15th International Conference on Natural Language Processing (ICON 2018), Punjabi University, Patiala, India (Dec 2018)
6. Mohanta, A., Mittal, V.K.: "Acoustic Features for characterizing speech of children affected with ASD", in Proc. 16th International IEEE India Conference (INDICON 2019), Marwadi University, Rajkot, Gujrat, India (13–15 Dec 2019)
7. Zhou, L., Fraser, K.C., Rudzicz, F.: "Speech recognition in Alzheimer's disease and in its assessment", Interspeech 2016 September 8–12, 2016, San Francisco, USA
8. Laske, C., Sohrabi, H.R., Frost, S.M., López-de-Ipiña, K., Garrard, P., Buscema, M., Dauwels, J., Soekadar, S.R., Mueller, S., Linnemann, C., Bridenbaugh, S.A., Kanagasingam, Y., Martins, R.N., O'Bryant, S.E.: "Innovative diagnostic tools for early detection of Alzheimer's disease". Alzheimer's Dementia, pp. 1–18 (2014)
9. Alberdi, A., Aztiria, A., Basarab, A.: On the early diagnosis of Alzheimer's disease from multimodal signals: a survey. Artif. Intell. Med. **71**, 1–29 (July 2016)
10. Baldas, V., Lampiris, C., Capsalis, C., Koutsouris, D.: "Early Diagnosis of Alzheimer's type dementia using continuous speech recognition". In: Lin, J.C., Nikita, K.S. (eds.) Wireless Mobile Communication and Healthcare. MobiHealth 2010. Lecture Notes of the Institute for Computer Sciences, Social Informatics and Telecommunications Engineering, vol. 55. Springer, Berlin, Heidelberg
11. Weiner, J., Engelbart, M., Schultz, T.: "Manual and automatic transcriptions in dementia detection from speech". Interspeech 2017 August, Stockholm, Sweden (2017)
12. Graf, S., Herbig, T., Buck, M., et al.: Features for voice activity detection: a comparative analysis. EURASIP J. Adv. Signal Process. **2015**, 91 (2015). https://doi.org/10.1186/s13634-015-0277-z
13. Juan, J.G.M., Martínez-Sánchez, F., Carro, J., Sánchez, J.A., Perez, E.: "Acoustic markers associated with impairment in language processing in Alzheimer's disease". Spanish J. Psychol. **15**(2), 487–494 (2012)
14. Hardy, C.J.D, Marshall, C.R., Bond, R.L., et al.: Retained capacity for perceptual learning of degraded speech in primary progressive aphasia and Alzheimer's disease. Alzheimer's Res Therapy **10**, 70 (2018)
15. Rudzicz, F., Wang, R., Begum, M., Mihailidis, A.: "Speech interaction with personal assistive robots supporting aging at home for individuals with Alzheimer's disease". ACM Trans. Accessible Comput.
16. Martínez-Sánchez, F., Meilán, J.J.G., Carro, J., Ivanova, O.: "A prototype for the voice analysis diagnosis of Alzheimer's disease". J. Alzheimer's Disease **64**(2):473–481 (2018)
17. Lapointe, J., Bouchard, B., Bouchard, J., Potvin, A., Bouzouane, A.: Smart homes for people with Alzheimer's disease: adapting prompting strategies to the patient's cognitive profile. In Proceedings of the 5th International Conference on PErvasive Technologies Related to Assistive Environments (PETRA '12). Association for Computing Machinery, New York, NY, USA, Article 30, pp. 1–8 (2012)
18. Raei, P., Bouchachia, A.: A literature review on the design of smart homes for people with dementia using a user-centred design approach. In Proceedings of the 30th International BCS Human Computer Interaction Conference: Fusion! (HCI '16).BCS Learning and Development Ltd., Swindon, GBR, Article 51, pp. 1–8 (2016)
19. Pulido, M.L.B., Hernández, J.B.A., Ballester, M.Á.F., González, C.M. T., Mekyska, J., Smékal, Z.: "Alzheimer's disease and automatic speech analysis: a review". Expert Syst. Appl. **150**, 113213 (15 Jul 2020)
20. Konig, A., Satt, A., Sorin, A., Hoory, R., Derreumaux, A., David, R., Robert, P.H.: "Use of speech analyses within a mobile application for the assessment of cognitive impairment in elderly people". Curr. Alzheimer Res. **15**(Number 2), 120–129(10) (2018)
21. Krishna, G.R., Krishnan, R., Mittal, V.K.:"A system for automatic regional accent classification," 2020 IEEE 17th India Council International Conference (INDICON), pp. 1–5 (2020). https://doi.org/10.1109/INDICON49873.2020.9342577

22. Krishna, G.R., Krishnan, R., Mittal, V.K.: "Foreign accent recognition with South Indian spoken english," 2020 IEEE 17th India Council International Conference (INDICON), pp. 1–5 (2020). https://doi.org/10.1109/INDICON49873.2020.9342471
23. König, A., Satt, A., Sorin, A., Hoory, R., Toledo-Ronen, O., Derreumaux, A., Manera, V., Verhey, F., Aalten, P., Robert, P.H., David, R.: "Automatic speech analysis for the assessment of patients with predementia and Alzheimer's disease", Alzheimer's and Dementia: Diagnosis, Assessment and Disease Monitoring, vol. 1, Issue 1, pp. 112–124 (Mar 2015)
24. Gosztolya, G., Vincze, V., Tóth, L., Pákáski, M., Kálmán, J., Hoffmann, I.: "Identifying mild cognitive impairment and mild Alzheimer's disease based on spontaneous speech using ASR and linguistic features". Comput. Speech Lang. **53**, 181–197 (Jan 2019)
25. Sadeghian, R., David Schaffer, J., Zahorian, S.A.: "Speech Processing Approach for Diagnosing Dementia in an Early Stage", Interspeech 2017, Sweden (Aug 2017)
26. Al-Hameed, S., Benaissa, M., Christensen, H.: "Simple and robust audio-based detection of biomarkers for Alzheimer's disease". SLPAT 2016 Workshop on Speech and Language Processing for Assistive Technologies, San Francisco, USA (13 Sept 2016)
27. Lopez-de-Ipina, K., Martinez-de-Lizarduy, U., Calvo, P.M., Mekyska, J., Beitia, B., Barroso, N., Estanga, A., Tainta, M., Ecay-Torres, M.: "Advances on automatic speech analysis for early detection of Alzheimer disease: a non-linear multi-task approach". Curr. Alzheimer Res. **15**(Number 2), 139–148(10) (2018)
28. Mirzaei, S., ElYacoubi, M., Garcia-Salicetti, S., Boudy, J., Kahindo, C., Cristancho-Lacroix, V., Kerhervé, H., Rigaud, A.-S.: Two-stage feature selection of voice parameters for early Alzheimer's disease prediction. IRBM **39**(6), 430–435 (2018)
29. David Schaffer, J., Land, W.H.: "Predicting with confidence: classifiers that know wthat they don't know". Proc. Comput. Sci. **114**, 200–207 (2017)
30. Martínez-Sánchez, F., Meilán, J.J.G., García-Sevilla, J., Carro, J., Arana, J.M.: "Oral reading fluency analysis in patients with Alzheimer disease and asymptomatic control subjects". Neurología (English Edition) **28**(6), 325–331 (July–August 2013)
31. Tóth, L., Hoffmann, I., Gosztolya, G., Vincze, V., Szatlóczki, G., Bánréti, Z., Pákáski, M., Kálmán, J.: "A speech recognition-based solution for the automatic detection of mild cognitive impairment from spontaneous speech". Curr. Alzheimer Res. **15**, 130–138 (2018)
32. López-de-Ipiña, K., Alonso, J.B., Travieso, C.M., Solé-Casals, J., Egiraun, H., Faundez-Zanuy, M., Ezeiza, A., Barroso, N., Ecay-Torres, M., Martinez-Lage, P., Lizardui.: "On the selection of non-invasive methods based on speech analysis oriented to automatic Alzheimer disease diagnosis". Sensors **13**, 6730–6745 (2013)
33. Lo´pez-de-Ipin˜a, K., Alonso, J.B., Sole´-Casals, J., Barroso, N., Henriquez, P., Faundez-Zanuy, C.M. Travieso, M. Ecay-Torres, P. Martı´nez-Lage, H. Eguiraun.: On automatic diagnosis of Alzheimer's disease based on spontaneous speech analysis and emotional temperature. Cognitive Comput. **7**, 44–55 (2015)
34. Khodabakhsh, A., Kuşxuoğlu, S., Demiroğlu, C.:"Natural language features for detection of Alzheimer's disease in conversational speech," IEEE-EMBS International Conference on Biomedical and Health Informatics (BHI), Valencia, pp. 581–584 (2014)
35. Nasrolahzadeh, Mohammadpoory, Z., Haddadnia, J.: "Higher-order spectral analysis of spontaneous speech signals in Alzheimer's disease". CognNeurodyn **12**, 583–596 (2018)
36. Mohanta, Mittal, V.K.: "Human emotional states classification based upon changes in speech production features in vowel regions," 2017 2nd International Conference on Telecommunication and Networks (TEL-NET), pp. 1–6 (2017). https://doi.org/10.1109/TEL-NET.2017
37. Liu, L., Zhao, S., Chen, H., Wang, A.: A new machine learning method for identifying Alzheimer's disease. Simulation Modell. Practice Theory **99**, 102023 (2020)
38. On, C.K., Pandiyan, P.M., Yaacob, S., Saudi, A.: "Mel-frequency cepstral coefficient analysis in speech recognition," 2006 International Conference on Computing & Informatics, Kuala Lumpur, Malaysia, pp. 1–5 (2006). https://doi.org/10.1109/ICOCI.2006.5276486
39. López-de-Ipiña, K., Solé-Casals, J., Eguiraun, H., Alonso, J.B., Travieso, C.M., Ezeiza, A., Barroso, N., Ecay-Torres, M., Martinez-Lage, P., Beitia, B.: "Feature selection for spontaneous

speech analysis to aid in alzheimer's disease diagnosis: a fractal dimension approach". Comput. Speech Lang. **30**(1), 43–60 (Mar 2015)
40. Luz, S.: "Longitudinal monitoring and detection of Alzheimer's type dementia from spontaneous speech data," 2017 IEEE 30th International Symposium on Computer-Based Medical Systems (CBMS), Thessaloniki, pp. 45–46 (2017)
41. Haider, F., de la Fuente, S., Luz, S.: An Assessment of paralinguistic acoustic features for detection of Alzheimer's dementia in spontaneous speech. IEEE J. Select. Topics Sig. Process. **14**(2), 272–281 (Feb. 2020)
42. Mohanta, A., Mittal, V.K.: "Autism speech analysis using acoustic features", in Proc. 16th International Conference on Natural Language Processing (ICON 2019), IIIT Hyderabad, India (18–21 Dec 2019)
43. Fraser, Kathleen, C., Meltzer, J.A., Rudzicz, F., Fraser, K.C., Meltzer, J.A., Rudzicz, F.: "Linguistic features identify Alzheimer's disease innarrative speech". J. Alzheimer's Disease **49**(2), 407–422 (2016)
44. Land, W.H., Schaffer, J.D.: "A machine intelligence designed bayesian network applied to Alzheimer's detection using demographics and speech data".Proc. Comput. Sci. **95**, 168–174 (2016)
45. Nasrolahzadeh, M., Mohammadpoori, Z., Haddadnia, J.: Analysis of mean square error surface and its corresponding contour plots of spontaneous speech signals in Alzheimer's disease with adaptive wiener filter. Comput. Hum. Behav. **61**, 364–371 (2016)
46. Mohanta, A., Mukherjee, P., Mittal, V.K.: "Prosody features characterization of autism speech for automated detection and classification", in Proc 26th National Conference on Communications (NCC 2020), IIT Kharagpur, India (21–23 Feb 2020)
47. Ramdinmawii, E., Mohanta, A., Mittal, V.K.: Emotion recognition from speech signal," TENCON 2017–2017 IEEE Region 10 Conference, pp. 1562–1567 (2017). https://doi.org/10.1109/TENCON.2017.8228105
48. Mittal, V.K., Vuppala, A.K.: Changes in shout features in automatically detected vowel regions. Int. Conf. Signal Process. Commun. (SPCOM) **2016**, 1–5 (2016). https://doi.org/10.1109/SPCOM.2016.7746601
49. Mohanta, A., Mittal, V.K.: "Classifying speech of ASD affected and normal children using acoustic features", in Proc 26th National Conference on Communications (NCC 2020), IIT Kharagpur, India (21–23 Feb 2020)

An Approach to Paperless Administration of Offices in India Using Blockchain

Atharva Paliwal, Chaitanya Kapre, Yash Roy, and Dilipkumar A. Borikar

Abstract The growth of blockchain technology has attracted the attention of many industries. At present, the Indian government offices are being overloaded by the huge amount of hardcopies that they have to store at the offices. People on the other hand have to carry hardcopies at various government offices. For example—for opening a bank account, the user has to submit print copies of documents such as Aadhaar card (UIDAI), power bill, and PAN card. The banks have to store these documents in files. There have been many reported instances of fraud through unauthorized access to these documents. A lot of paper in the process adds to the physical storage space, personal verification, and problems in tracking the validity of the document. Environmental concerns necessitate minimization of causes of pollution, overuse of energy, and minimization of paper use for a greener world. Blockchain technology can help offices by resolving the problem of hardcopies and validation of the document and leading a world toward a paperless revolution. This paper compares the conventional procedure followed in the government offices and proposes a blockchain-based solution which focuses on the reducing of paper usage, proper authentication of the document using the digital signatures, and preventing people from carrying the bulky files at the offices by providing them a digital wallet.

Keywords Blockchain · Paperless offices · Digital signatures · Digital wallets

A. Paliwal (✉) · C. Kapre · Y. Roy · D. A. Borikar
Shri Ramdeobaba College of Engineering and Management, Katol Road, Gittikhadan, Nagpur, Maharashtra 440013, India
e-mail: paliwalap@rknec.edu

C. Kapre
e-mail: kaprecp@rknec.edu

Y. Roy
e-mail: royya@rknec.edu

D. A. Borikar
e-mail: borikarda@rknec.edu

© The Author(s), under exclusive license to Springer Nature Singapore Pte Ltd. 2022
A. K. Nagar et al. (eds.), *Intelligent Sustainable Systems*, Lecture Notes in Networks and Systems 333, https://doi.org/10.1007/978-981-16-6309-3_60

1 Introduction

Today, the concern is not only to store the document securely but also access the document effectively. Furthermore, paper-based administrations are facing many difficulties because of air pollution, plenty of water used in paper mills, and fire being the major threat. An A4 paper requires about 10 L of water per sheet for production. This advocates paperless administration and conservation of natural resources [1]. There are various steps taken toward transforming the offices into paperless offices or e-offices. But with the improvement of technology every day, digital printing and scanning have become a lot easier leading to criminals producing fake documents. Therefore, authenticating documents has become a most important work. It would be a difficult job for an organization to verify every document as quickly as possible who deals with a large number of documents. Hence, there is a need for a system which can be accessible as well as secured [2].

A blockchain is a decentralized ledger which can be used to store documents, verify the digital signature, and avoid fraudulent activities. There are many forging techniques available in paper-based systems, but with the help of blockchain technology, nobody can manipulate the data as it is immutable (after storing once).

2 Understanding the Current Paper Usage in Offices

Currently, paper is primarily used in almost 90% of the works that government offices do. From writing a small note to writing official documentation, paper is used everywhere. When a document is submitted, one has to submit the photocopies of the documents which lie in the offices as the proof, and the files created stay untouched for long. In offices, stamped paper has been widely used to collect taxes on documents requiring stamping such as leases, agreements, receipts, court documents, and many others. Furthermore, these documents are far-reaching, and storing them physically is very difficult because of lack of space and accessibility. This current paper-based administration compromises the security as it has continuous risk of fire, tampering, etc. Marinating papers in files and folders leads to accessibility issues [3].

3 The Blockchain Concept

A blockchain is the sequence of blocks that are linked together using a cryptographic chain. Each block in the blockchain consists of a hash of the previous block, data, and timestamp. It is a distributed public ledger that is used by many cryptocurrencies. Blockchain provides proper authentication, immutability of data, and integrity of transactions. The blocks in blockchain record a number of transactions that are maintained by a data structure called Merkle tree. When a transaction initiated, the hash of the previous block, public key of receiver, timestamp, and digital signature of sender is stored in the block and stored on blockchain. Only digital signatures are used for authentication, and the whole transaction is handled by blockchain software [4].

3.1 Properties of Blockchain [4]

1. Immutable—Transactions which are verified and are valid then the block is added to the blockchain. Every block in this chain of blocks holds the cryptographic hash of the previous block in the chain.
2. Decentralized—A blockchain network runs on the principle of peer-to-peer network, where no central authority is present to regulate the network.
3. Verification—A blockchain involves the method of digital signatures being signed by the peer members using the concept of public–private key cryptography.

3.2 Types of Blockchain

- Public blockchain—It represents a network without restrictions where participants can join any time. All participants in a public blockchain have equal rights, and there is no single entity which will run the network.
- Private blockchain—It is referred as *permissioned blockchain* which works on access controls that restrict the people to join the blockchain network.

4 Smart Contract Design and Development Using Solana Blockchain

4.1 Solana

Solana is an open-source decentralized Web-scale blockchain that provides a decentralized protocol for distributing computing and building decentralized applications. It has achieved commendable results in scaling the transactions, i.e., considering current system, it supports nearly 55,000 transactions per second (TPS) and 400 ms block times. Solana is most favorable for the DeFi applications because Solana's transaction fee is very small—$0.00001. The native currency of Solana's ecosystem is SOL token ($SOL). Having a maximum supply of 489 million SOL, currently there is a circulating supply of 26 million SOL. The working of Solana is presented in Fig. 1.

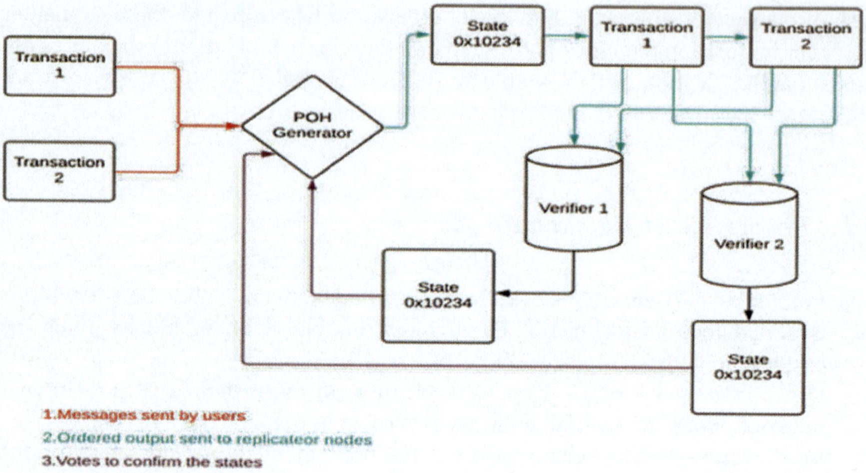

Fig. 1 Working diagram of Solana blockchain

4.2 *Working of Solana*

1. In the beginning, transactions are sent to the leader as input.
2. Then, the leader will carry out the messages and instruct them adeptly so that it can be processed by other nodes.
3. Transactions are executed by the leader which are later in RAM.
4. Leader then conceals the transactions and signature of the final state to verifiers.
5. The work assigned to verifiers is to execute the same transactions on their copies of the state and publish their signatures of the state when it receives confirmation.
6. Confirmations will act as votes for the consensus algorithm.

4.3 *Smart Contract*

A smart contract is a computer program which is written to automatically execute the events according to the contracts [7]. The main aim of the smart contract is to reduce the trust of any central authority. The working of the smart contract is done by deploying the smart contract into the blockchain which is done by sending a transaction from a wallet. The transaction is included in the block and added into the blockchain. Anchor is an emerging framework which has a similar role in Solana as that of truffle in Ethereum, and it is a savings protocol which aims toward producing a simple and convenient savings product. In Solana, on-chain programs can be written using the Rust programming language.

4.4 IPFS

Interplanetary file system (IPFS) is a peer-to-peer hypermedia protocol. IPFS is a peer-to-peer, open-source distributed file system to store data and files with high integrity and resiliency [4]. IPFS is a protocol designed for storing versioned file data in a decentralized version. IPFS uses a distributed hash table approach which maps the keys to values and provides the access to the key-value store. The nodes in IPFS have a unique node ID which resembles the hash of a public key. Whenever a file is added in the IPFS network, the file is divided into chunks of 256 KB. This chunks of file have object data and links which are placed in the Merkle DAG. The hash value is returned to the receiver, and that hash is used to retrieve the file from IPFS.

5 Literature Survey

Tyhulu and others have discussed the effect of a paperless working environment on the Master of the High Court [8]. Selvi and others have developed a project for storage and retrieval of the scanned documents. They incorporated Oracle database and application server in their product that was deployed on three-tier architecture [9]. Rehiel and others have discussed the problem caused by using the paper and its document. They have proposed a five action plan to develop a paperless office [10].

Warasart and Kuacharoen have proposed an approach for document verification wherein the integrity of the text message and author of the document is verified using the QR code and digital signature. But, the method is not convenient for the users when dealing with a large number of documents [11]. A comparative analysis with future directions for secure document storage solutions in health care based on IPFS and blockchain has been discussed in [4].

6 Methodology

6.1 Case-I: The User is to Apply and Acquire any Government Issued Document

In this era of digitalization, many of the documents that used to be made offline can now be made online by registering on the office Web site and the hard copy of their document is being delivered to them. However, many of the users who are not used to the online resources have to make their documents offline. In the platform, users can register through the resources that they feel comfortable with.

If a user wishes to create a document, she will either visit the office, submit all the necessary documents or can fill in all the information online as presented in

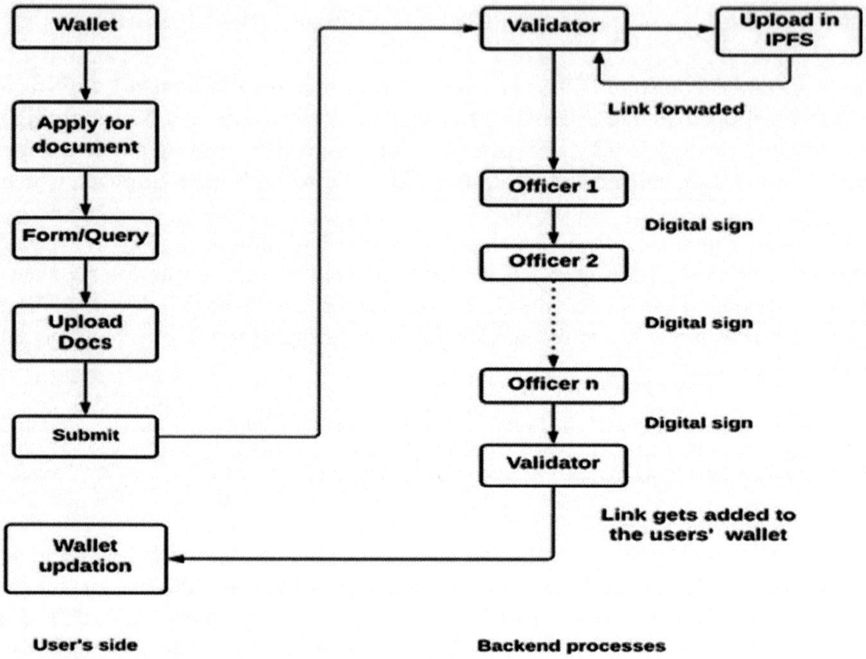

Fig. 2 Workflow of the proposed system

Fig. 2. For creation of a document, there are some prerequisites of documents that are required for the verification. For instance—if a user desires to make an Aadhaar card, then the user can fill the particular forms prior and submit the copy of necessary documents online. Though for verifying fingerprints and photocopy, users have to visit the offline government center for further procedure. In this way, the proposed application is saving time, travel cost, and neglecting the use of paper in application procedure.

Creation and Uploading of Document on Platform and Link Generation

After the creation of the document and validation of those prerequisite documents, the e-copy of the document will be generated and sent for verification to the authorities. The document generated is now uploaded on the platform by the validator who will provide a proper description for the document (to whom the document belongs to). Thereafter, the document will be uploaded on the IPFS blockchain that will reveal information related to the uploaded document, i.e., the description, uploading timestamp, the type of documenttext, and PDF or URL such as cryptographic hash that act as a fingerprint of the document. The validator now can see the link generated and verify details of the document before sending it for further verification to the senior authorities.

Digital Signatures and Verification of the Document (Hierarchy in Office)

After the verification of the document by the validator and checking all the information printed in the document, the document is now sent to the senior authorities for the digital signatures, so as to ensure that the document is validated completely. The senior officers will provide the digital signatures to the document, and if there are any discrepancies, then reverting back the document to the validator by stating the discrepancies would be viable. All the updates and verifications of the document are being stored on the blockchain, so as to keep track of the changes that were being made in the document while it was made. The senior authorities will be providing their digital signatures by the hierarchy present in the office (e.g., A junior authority will forward the document to the senior authority after verifying digital signature). All this is done on the private blockchain of the office and will only be visible to the people who are present on the blockchain network. So for each user, a proper hierarchy is being followed by the office, and the document will now finally be available to the final validator who will now be assigned to send these links to the wallet provided to the user.

Uploading Document to User's Wallet

The validator who has been assigned the work of distributing the document to the user's wallet will now send the document to the user's wallet. The wallet will be assigned to the person whenever registration is made in the offices for the first time. For example, when a child is born, parents have to visit government offices to make a birth certificate. This being the first document of the child, the validator at the office will issue a digital wallet in the name of the child. This wallet will contain all the documents that child is going to make in future. In this way, the wallet for a user will be generated. The wallet will be secured and can only be opened by a secret pin that would be given to the user at the time of registration.

The validator will now transfer the document to the wallet that was issued to the person when the user registered for the first document. The link which defines the user's document will, however, be not visible to the user, the user will only be able to see the document, and the link verification will take place in the backend. In this way, the user can avail the work done in a very less time and without the need of visiting the offices too frequently as the document is directly getting updated to the user's wallet. The text message will be sent to the user as soon as the document is being added to the wallet on the registered mobile number by the user.

6.2 Case-II: When the User Shares the Document

After updatin from the wallet if required by any of the other offices as seen in Fig. 3.

For example, if a user desires to make an bank account, required document would be Aadhaar card, and user now instead of using a traditional mechanism of submitting hard copies to the banks can directly send user's respective document via digital

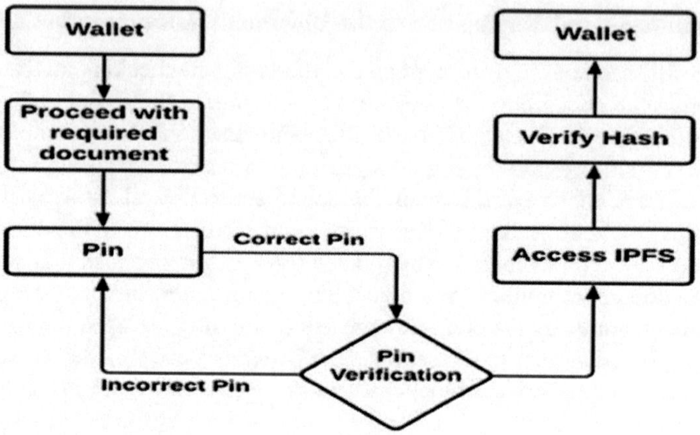

Fig. 3 Workflow of system (for Case-II scenario)

wallet to the bank's wallet from where the validator at the bank will be verifying the document by the link which is being processed in the backend. The user now will visit the bank for security purposes like fingerprint verification, etc.

6.3 Case-III: Document Updating

If a user finds that some incorrect information has been entered in the document or to update the address information, in that scenario, the user will have to visit the document publisher office and ask the validator to update information as shown in Fig. 4.

For example—if a user wishes to update an address on the Aadhaar card, then the user has to visit the Aadhaar card center. Later, the validator updates the information and will generate a new document for the user with the new address which will be uploaded on the platform for the link generation process. The hierarchy system of the office will be followed for the document validation by digital signatures of the authorities.

7 Implementation

The proposed solution has been implemented on Ethereum blockchain with Matic network, and the part of the solution that is implemented is the platform that will be used by the offices where the validator will upload the documents to generate the link from IPFS as seen in Fig. 5. The issue with Ethereum is that it is not scalable as it allows only 20 transactions per second [12], this would make the process of

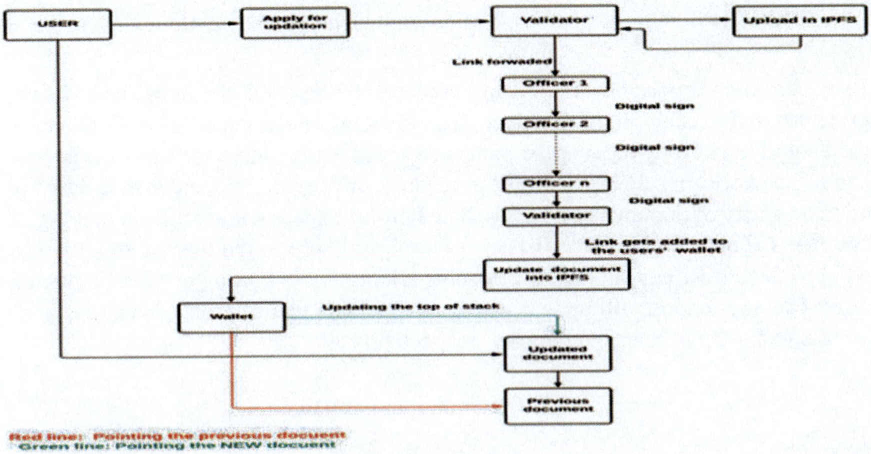

Fig. 4 Workflow of system (for Case-III scenario)

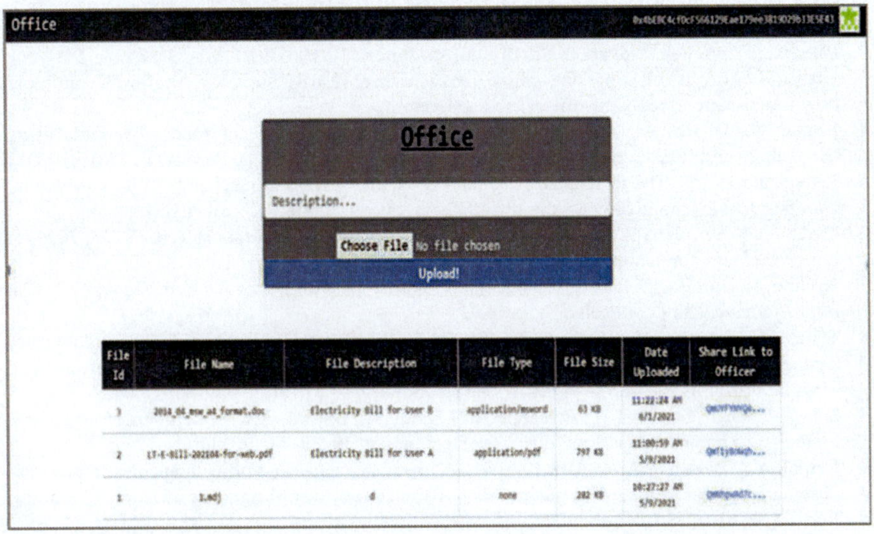

Fig. 5 Validator side snapshot

generating documents in the government offices very slow. This paper, therefore, proposes a Solana blockchain-based solution to handle the issue of scalability.

For implementing, the proposed blockchain system at different offices will necessitate the provisioning of distinct private blockchains for each office. This will need an extra cost and account for additional computational powers to carry out such huge load of transactions. It would also demand for regular maintenance in dealing with heavy network traffic.

8 Conclusion

The goal of the proposed solution is to reduce the paper used in offices and to incorporate use of digital medium to reduce dependencies on the paper. This blockchain-based solution not only reduces the paper usage but also reduces the time that people have to invest during the application procedure. It provides a greater transparency and traceability of documents which will be helpful to detect the frauds in making of documents. This smart contract-based solution will help in the idea of smart world and aims to address environmental concerns arising out of cutting of trees in making paper. The application will open a whole new perspective of creation and usage of documents by the offices and the user, respectively.

References

1. Shah, I.A., Amjed, S., Alkathiri, N.A.: The economics of paper consumption in offices. J. Bus. Econ. Manag. **20**(1), 43–62 (2019). https://doi.org/10.3846/jbem.2019.6809
2. Pal, S.K.: Changing Technological Trends for E-Governance. In: E-Governance in India. Palgrave Macmillan, Singapore (2019). https://doi.org/10.1007/978-981-13-8852-1_5
3. Glaser, K.: How much paper does office use each year? (2018). https://www.gflesch.com/blog/how-much-paper-does-your-office-use-each-year
4. Kumar, S., Bharti, A., Amin, R.: Decentralized secure storage of medical records using Blockchain and IPFS: a comparative analysis with future directions, Published: 27 April 2021
5. Jayachandran, P.: "The difference between public and private blockchain". https://www.ibm.com/blogs/blockchain/2017/05/the-difference-between-public-and-privateblockchain/
6. Hyde, A.: Why solana is the 'World Computer' Blockchain Developers Need. (5 Aug 2019) at https://medium.com/solana-labs/
7. Bachini, J., Building Solana Smart Contracts and dApps with James Bachini. (28 Apr 2021) at https://jamesbachini.com/solana-tutorial/
8. Tyhulu, L., Sibande, B., Zilwa, N., Langa, S., Hollis-Turner, S., Bruwer, J.: The influence of a paperless office environment on the sustainability of the Master of the High Court in Cape Town. Environ. Econ. **7**, 33–44 (2016). https://doi.org/10.21511/ee.07(3).2016.04.
9. Selvi, S., Khan, S., Rani, U., Prasad, B.V.N., Paul, A., Biswal, A.K.: Document management system—go green 'a paperless office' for steel plants. **35**, 39–41 (2011)
10. Rehiel, A., Shittu, A., Al-Matari, E., E.M., Alanazi, A.: Paperless office management: a feasibility analysis for saudi arabia government offices: case study in ministry of labor. J. Manage. Res. **6** (2014). https://doi.org/10.5296/jmr.v6i3.5700
11. Warasart, M., Kuacharoen, P.: Paper-based document authentication using digital signature and QR code. 2012 4th International Conference on Computer Engineering and Technology (ICCET 2012)
12. Mechkaroska, D., Dimitrova, V., Popovska-Mitrovikj, A.: Analysis of the Possibilities for Improvement of BlockChain Technology. 1–4 (2018). https://doi.org/10.1109/TELFOR.2018.8612034.

Key Technology Adoption Indicators for Smart Universities: A Preliminary Proposal

Dewar Rico-Bautista, César D. Guerrero, César A. Collazos, Gina Maestre-Gongora, María Camila Sánchez-Velásquez, Yurley Medina-Cárdenas, Diana Teresa Parra-Sánchez, Antón Garcia Barreto, and Jose Swaminathan

Abstract Technology management in universities impacts the quality of their administrative and academic processes. Smart technologies, such as big data, artificial intelligence, the Internet of Things, and cloud computing, continue to be disruptive and highly relevant. The factors for adopting smart technology and the key indicators will contribute to the knowledge of the transformation processes of current universities in Smart Universities. An evolutionary approach for a traditional university to progress by levels toward a smart university can be seen in a maturity model that integrates such factors and indicators. Although maturity models are numerous and widely applicable, there is little documentation on how to develop them. So far, there

D. Rico-Bautista (✉) · M. C. Sánchez-Velásquez · Y. Medina-Cárdenas · A. G. Barreto
Universidad Francisco de Paula Santander, Ocaña, Colombia
e-mail: dwricob@ufpso.edu.co

M. C. Sánchez-Velásquez
e-mail: mcsanchezv@ufpso.edu.co

Y. Medina-Cárdenas
e-mail: ycmedinac@ufpso.edu.co

A. G. Barreto
e-mail: anton@ufpso.cdu.co

C. D. Guerrero · D. T. Parra-Sánchez
Universidad Autónoma de Bucaramanga, Bucaramanga, Colombia
e-mail: cguerrer@unab.edu.co

D. T. Parra-Sánchez
e-mail: dparra486@unab.edu.co

C. A. Collazos
Universidad del Cauca, Popayán, Colombia
e-mail: ccollazo@unicauca.edu.co

G. Maestre-Gongora
Universidad Cooperativa de Colombia, Medellín, Colombia
e-mail: gina.maestre@campusucc.edu.co

J. Swaminathan
Vellore Institute of Technology, Vellore, India
e-mail: jose.s@vit.ac.in

© The Author(s), under exclusive license to Springer Nature Singapore Pte Ltd. 2022
A. K. Nagar et al. (eds.), *Intelligent Sustainable Systems*, Lecture Notes in Networks and Systems 333, https://doi.org/10.1007/978-981-16-6309-3_61

is a lack of useful tools to support the practical adoption of the developed models in universities. This article presents 52 key indicators associated with the proposed smart technology adoption factors that are the basis for the maturity model.

Keywords ICT adoption · ICT indicators · Maturity model · Smart technologies · Smart university · Technology adoption

1 Introduction

The adaptation of cities provides more comfortable, more useful, and more efficient environments by generating sustainable services in the long term that will facilitate and improve citizens' quality of life [47]. Furthermore, the arrival of the Internet has driven the growing use of digital tools, in all environments, especially in education, becoming a growing challenge for all institutions, which are focused on permanently establishing high performance by using tools that are at the forefront.

Maturity models are tools oriented to the improvement of processes in organizations, outlining evolution paths. They are reference models that evaluate an organization's current situation to identify and prioritize improvement measures. These models indicate predictable patterns in the evolution of organizations distributed in evolutionary stages, thus forming a roadmap that facilitates decision-making. Each stage has a higher degree of superiority and complexity concerning the immediate previous one. Thus, the initial stage represents that an organization has few capabilities in the domain considered. The highest stage of maturity represents full maturity. The intermediate stages represent a path of continuous progression of the organization's capabilities [9, 22]. Within each stage, a set of variables provide the criteria, characteristics, and conditions that must be met to advance in each of them and subsequently reach the level of maturity internally established in the organization. In this sense, it is vital to analyze the context in which the organization is located to derive and prioritize recommendations to achieve higher levels of maturity as the fulfillment of each of the stages' criteria is given [4].

It is important to emphasize that the maturity models provide a plan of action that organizations use to implement good practices for one or more areas within their processes, starting from a state where practices are poorly defined and inconsistent to a level of continuous innovation. This procedure is effective through a wide application in industrial and governmental organizations. Organizational performance can be measured by comparing institutional practices with the essential practices obtained from the models to establish, implement, evaluate and improve processes within an organization [11].

Adopting digital technologies in the organization requires improving connectivity, access to information, and response time in the processes. However, academic training is framed in a dynamic context, adapted to the changes brought about by globalization. In particular, the model's design incorporates the target audience's

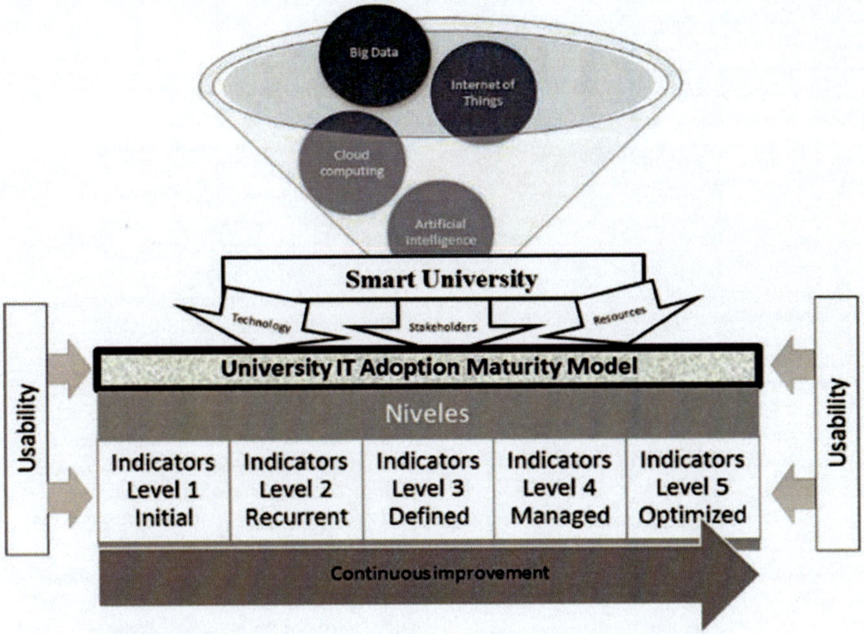

Fig. 1 Projection of IT adoption maturity model. *Source* Adapted from [15]

needs and how those needs will be met [5, 6]. The importance of a standard development framework is highlighted when considering the purpose for which a model can be applied, including whether the resulting maturity assessment is *descriptive, prescriptive,* or *comparative* (see Fig. 1).

We organize this article as follows. Section 2 presents the method. Section 3 shows the key indicators of smart technology adoption. Finally, Sect. 4 presents our conclusions.

2 Method

This article has presented a summary of the process conducted so far in Table 1.

Table 1 Summary of the article's methodology

Phase	Summary
1. *A conceptual framework* [1, 43]	The smart university concept has evolved as follows: • From smart city to smart university, published in [19, 29] • Smart campus vs smart university, published in [7, 44] • IoT as an emerging technology, published in[13, 29, 36] • Smart Technologies, published in [38] • Conceptual framework [33, 34, 43] • Strategic map [10, 40, 46]
2. *Characterization*	*Smart technologies* [38], which shows the concept of a smart university defined from Smart Technologies: *Cloud Computing; Big Data; Artificial intelligence and Internet of Things.* The key technologies for a smart university are Artificial intelligence and the Internet of Things [18, 45]
3. *Proposal of models for the adoption of smart technology* [37, 39, 41]	The methodology was adapted from Petersen [30], further developed in Revelo et al. [31]. The analysis of the results of Dybå et al. [20] was taken into account, which provides a list of important databases in Computer and Engineering Sciences. They were classified according to smart technologies: Artificial Intelligence, Big Data, Cloud Computing, and IoT [8, 16, 17, 21, 38]
4. *Integration* [32]	The integration of the four proposed models of smart technology adoption is generated
5. *User-centered smart technology* [35]	Smart technologies have become more powerful, so it is necessary to have tools that can measure their adoption by institutions. The objective of measuring adoption is to focus on the users of the processes. A User-Centered Smart University Model is the goal of this transformation process, based on an individual's perception, safety and risks, and organizational support
6. *Key Technology Adoption Indicators for Smart University*	The product is generated in this document. After the generation of adoption factors by smart technology, indicators associated with each technology are presented
7. *Maturity Model*	The main product is integrating adoption factors, indicators, and levels on a Smart University concept approach

3 Key Indicators of Smart Technology Adoption: First Approach

Maturity models have been applied for long periods in the field of information systems, obtaining good results. However, at the time, maturity approaches originated from quality management and were later oriented to information technologies (IT) to manage software processes. These approaches were also used to increase organizational development and achieve objectives from an initial basic state [26, 27, 42]. With the emergence of new information technologies, maturity models have been restructured to amplify their focus areas. For example, one of these areas is cloud data governance. The proposed models constitute a tool to assess the organization's vision and capability to implement data governance for cloud computing services [1]. To understand and address data governance in the cloud, models have emerged that classify by levels the states in which an organization related to this field of action can be in the following levels [1].

Maturity in cloud applications is extremely different from application to application. For example, many software applications exposed to consumers from the cloud are traditional software products with rudimentary security and access controls added to them and have been released as SaaS application [28]. Similarly, organizations must consider that the migration of core or non-core services has different impacts and associated risks. For example, when the migration of services is performed in the cloud, specific criteria must be used to improve these migration and management processes [23–25]. These are aspects that arise from the analysis before the implementation of a maturity model. Thoroughly knowing the technology and its application within the organization is an advantage in implementing the models.

A fundamental premise for any organization to generate value from IT is to align the IT capabilities and resources it possesses or acquires with the current and future business's strategies, objectives, and needs. The alignment mentioned above can be achieved by applying maturity models, which establish scales to quantify the organization's process. The scale ranges from 0 (no maturity) to 5 (optimal state of maturity). With the application of these models, it is possible to obtain a general balance of the current situation and how it could be in the future, as a strategy that promotes the growth of strategies that contribute to the fulfillment of the objectives [12].

The maturity models applied to educational environments represent a methodology used to design, develop, and continuously improve the main functions of an educational institution, such as education, teaching, learning, research, services, management, administration, control, and security. In other words, it can be seen as an opportunity for improvement, starting from a traditional university, adding intelligence to each internal process executed in the institution, resulting in an intelligent education using fields, classrooms, pedagogy, and intelligent tools. When discussing maturity levels in intelligent environments, reference is made to the following levels: initial, repeatable, defined, managed, and optimized. The latter becomes the highest degree of maturity, which involves using intelligent devices within the classroom [14].

Education is a fundamental pillar in the intellectual development of each individual. It is a process that in many institutions is taught traditionally, with one methodology applied to a group of students. With the arrival of the technological era, a series of tools have been proposed to tailor this process according to each individual's needs, establishing a specific and independent educational pace for each student. Smart education is one of the main elements to describe the new learning process in the information age. Promoting the use of technology in education has begun providing adaptive education. Students learn in different ways and have personalized services, resulting in learning from anywhere and anytime [3].

Adopting technology in tertiary institutions is important to guarantee good results in the future, so we propose models that seek to understand how to measure this adoption. The key indicators presented in Table 2 relate to the smart technology adoption

Table 2 Factors and indicators by smart technology for its adoption

Factors	Questions to keep in mind	Indicator
	Internet of Things	
Volunteering	• Are the faculty and administration interested in the use and adoption of IoT?	• Degree of interest in the use of IoT, measured via the staff and faculty of the institution
IT knowledge implemented in the institution	• Is the educational institution prepared to use IoT?	• Quantity of training before and after IoT implementation (Staff and faculty Training) Percentage of productivity achieved after adoption Efficient information analysis
Experience of its use	• Are there staff and faculty who have interacted with IoT in other situations?	• The number of people who have used IoT • Increase in the number of people trained in the use of IoT
Privacy	• Is the information protected?	• Tools that provide information privacy and can be adopted together with IoT
Cost of devices	• What is the value of implementing IoT within an academic institution?	• Value of IoT implementation in the institution
Data Security	• Are there tools to protect information?	• Tools that provide security
Price–Cost	• Do profits warrant the price of IoT adoption?	• Price vs. Perceived profits
Compatibility with other systems implemented in the institution	• Is the equipment that the institution currently has compatible with IoT?	• Compatibility testing

(continued)

Table 2 (continued)

Factors	Questions to keep in mind	Indicator
Infrastructure requirements	• What are the requirements to implement IoT within an educational institution?	• Percentage of sensor networks installed and devices in operation • Availability of hardware for IoT use • Quality of the current infrastructure
Relevance in educational processes and environments	• Does IoT improve educational processes?	• Tools needed for IoT adoption
Resistance to change	• Are staff and faculty willing to use IoT if it is adopted?	• Test of resistance to change in ICT environments • Meetings to express doubts regarding the adoption of IoT
	Cloud Computing	
How technology handles information	• How does Cloud Computing manage information? • Are there improvements in the outcome of processes when using Cloud Computing?	• Efficiency and effectiveness in the execution of tasks in set times • Security, privacy
Knowledge of technology	• Is the educational institution prepared to use Cloud Computing?	• Quantity of training given before IoT implementation (Staff and faculty training) • Percentage of productivity level
Experience	• Are there staff and faculty who have interacted with Cloud Computing in other situations?	• The number of faculty and/or administrative staff who have used Cloud Computing • Increase in the number of people trained in the use of Cloud Computing
Volunteering	• Do the faculty and administration have an interest in the use and adoption of Cloud Computing?	• Faculty and/or administrative staff who support the adoption of technology within the institution
Social Influence	• What are the institution's references to review the social influence of Cloud Computing?	• Cloud Computing Usage Ranking Ranking of cloud computing adoption results in similar environments

(continued)

Table 2 (continued)

Factors	Questions to keep in mind	Indicator
Performance expectations in educational settings	• What is the percentage of improvement in educational processes when using Cloud Computing?	• Level of advancement in educational processes using technology • Performance in task execution • Flexibility in information acquisition
Competitiveness	• Are there any higher education institutions making use of Cloud Computing? • Does the purchase value of the technology merit the results to be obtained?	• Innovation • Evolution in the streamlining of processes when using technology • Scalability
Resistance to wear and tear	• Are faculty willing to use Cloud Computing if it is adopted?	• Percentage of faculty and/or administrators who oppose the use of Cloud Computing
Information Availability	• Will the information be available at the time it is required?	• The number of times the system failed
Infrastructure requirements	• What kind of infrastructure will the higher education institution adapt to?	• Results obtained vs. expected results
	Big data	
Perception of the institution	• What is the opinion of the faculty and/or management about the technology? • Will the use of Big Data improve results in educational processes?	• Degree of interest in the use of Big Data, as measured by institution staff • Results in executing processes • Costs vs. Profit • Ranking of Big Data use in educational environments • Use of technological tools in executing academic activities
Safety and risks	• Are there tools to protect information? • Is the information available at locations other than the institution?	• Efficient information analysis • Tools that guarantee the security and privacy of information
Institutional Support	• What benefits does the institution seek with the adoption of Big Data?	• Degree of usefulness and performance about the execution of tasks
	Artificial intelligence	
Expectations of the educational institution regarding performance	• What results are expected from the adoption of Artificial Intelligence?	• Percentage of teachers and/or administrators interested in using artificial intelligence in the development of educational processes

(continued)

Table 2 (continued)

Factors	Questions to keep in mind	Indicator
Degree of Efficiency of Artificial Intelligence in the institution	• Can Artificial Intelligence help in the development of educational processes?	• Degree of compliance with planned activities with the use of Artificial Intelligence
Ease of use of artificial intelligence	• Are there training sessions planned on the use of Artificial Intelligence at educational environments?	• Percentage of personnel familiar with the technology
Knowledge of teachers and administrators in the use of artificial intelligence	• Is the educational institution prepared to use Artificial Intelligence?	• Number of training sessions provided
Social Influence	• What are my references to check the social influence of Artificial Intelligence?	• Ranking of use of Artificial Intelligence • Ranking of results of the adoption of Artificial Intelligence in similar environments
Technological innovation within the institution	• What benefits can I obtain with the use of Artificial Intelligence?	• Resources to implement Artificial Intelligence
Security and Trust	• Is the information safe if Artificial Intelligence is used?	• Number of tools available to keep information secure
Environment	• Are physical modifications required to implement Artificial Intelligence?	• Devices, Tools

factors generated in previous documents. The factors with their indicators will be fundamental pillars to propose the maturity levels of smart technology adoption [2].

This list of indicators associated with the factors is the first approximation to the generation of maturity levels of adoption for smart technology in universities.

4 Conclusion

Each technology brings special characteristics that adjust to the requirements of the context. In each smart technology, some elements are essential for their implementation. When we talk about IoT, the weight falls on the utility that the institution perceives at its use, what improvements it can bring, and which tools are favorable.

For the smart technology cloud computing, the central value is in handling the information. Therefore, the universities require elements that improve, make agile, and safeguard the information and count on it at the moment in which it is necessary.

Implementing this technology anticipates these needs and gives an added value of keeping all information centralized. For example, a university can create its private cloud within its infrastructure and thus manage to meet all the requirements. Therefore, it is important to emphasize the institution's orientation when implementing this new technology that brings many benefits. However, in turn, it requires staff and faculty who know how it works.

This technology brings many benefits to higher education environments. We can talk about smart tutors, which, based on algorithms, can create a pattern where the requirements of each student are established. It allows us to create personalized teaching with ongoing support; to this, each area teacher can be aware of developing the work with their students, achieving improvements in terms of learning techniques.

For Big Data, three main factors revolve around its adoption. First, the institution's perception before adopting technology is important because the institution must have a good perception value to have good results in the implementation of technology. The second and more important aspect must do with risks and security. This technology is based on the handling and administration of information. For this reason, this section represents a fundamental value. It must be guaranteed that a university trusts that using this technology could result in a good analysis of its information. Moreover, institutional support, which is related to the functionality of the technology already rolled out at the institution; for example, a university expects a high performance that must do with the level of utility and at the same time with the speeding up of processes.

For the model validation process, the following processes could be conducted:

- Factors identified during the initial stages of model development.
- Definition of a validation tool.
- Selecting a panel of experts.
- Releasing the results of the validation tool.
- Conveying the results.
- Discussing the results.

References

1. Al-Ruithe, M., Benkhelifa, E.: Cloud data governance maturity model. In: Proceedings of the Second International Conference on Internet of things, Data and Cloud Computing. pp. 1–10, ACM, New York, NY, USA (2017). https://doi.org/10.1145/3018896.3036394
2. Alrashed, S.: Key performance indicators for Smart Campus and Microgrid. Sustain. Cities Soc. **60**(May), 102264 (2020). https://doi.org/10.1016/j.scs.2020.102264
3. Bajaj, R., Sharma, V.: Smart Education with artificial intelligence based determination of learning styles. Proc. Comput. Sci. **132**, 834–842 (2018). https://doi.org/10.1016/j.procs.2018.05.095
4. Becker, J. et al.: Maturity models in IS research. 18th Eur. Conf. Inf. Syst. ECIS 2010 (2010)
5. Bertolli, M.P. et al.: Revisión de modelos de madurez en la medición del desempeño. Inge Cuc. **13**(1), 70–83 (2017). https://doi.org/10.17981/ingecuc.13.1.2017.07
6. Bruin, T. De et al.: Understanding the main phases of developing a maturity assessment model. Australas. Chapter Assoc. Inf. Syst. 8–19 (2005). https://doi.org/10.1108/14637151211225225

7. Cerdeira Ferreira, F.H., Mendes de Araujo, R.: Campus inteligentes: conceitos, aplicações, tecnologias e desafios. Relatórios Técnicos do DIA/UNIRIO. **11**(1), 4–19 (2018)
8. Choix, R.A., et al.: Factores determinantes en la adopción de tecnologías de información (TI) en las pymes. VinculaTégica EFAN. **1**(1), 806–828 (2015)
9. Van Dijk, F.W. et al.: Van Dijk—Cloud maturity models (2017)
10. Dong, Z.Y., et al.: Smart campus: definition, framework, technologies, and services. IET Smart Cities. **2**(1), 43–54 (2020). https://doi.org/10.1049/iet-smc.2019.0072
11. Duarte, D., Martins, P.V.: A maturity model for higher education institutions. CEUR Workshop Proc. **731**, 25–45 (2011)
12. Fernández Martínez, A., Llorens Largo, F.: Gobierno de las TI para universidades (2016)
13. Guerrero, C.D., Rico-Bautista, D.: Center for excellence and internet acquisition of things: a commitment to competitiveness from alliances between government, academia and productive sector [Centro de excelencia y apropiación en internet de las cosas: Una apuesta a la competitividad desde. RISTI—Rev. Iber. Sist. e Tecnol. Inf. 2020, E28, 615–628 (2020)
14. Heinemann, C., Uskov, V.L.: Smart university: literature review and creative analysis. In: Smart Universities. pp. 11–46, Springer International Publishing, Germany (2018). https://doi.org/10.1007/978-3-319-59454-5_2
15. Hung, C.N., et al.: Building a maturity model of information security governance for technological colleges and universities in Taiwan. Appl. Mech. Mater. **284–287**, 3657–3661 (2013). https://doi.org/10.4028/www.scientific.net/amm.284-287.3657
16. López Cabrera, M.V. et al.: Factors that enable the adoption of educational technology in medical schools. Educ. Medica. 20(xx), 3–9 (2019). https://doi.org/10.1016/j.edumed.2017.07.006.
17. Luzardo Briceño, M. et al.: Factores que influyen en la adopción de las Tecnologías de Información y comunicación por parte de las universidades. Dimensión Enseñanza-Aprendizaje. Educere **21**(68), 143–153 (2017)
18. Maestre-Gongora, G. et al.: Mapping concept and challenges for smart technologies: A systematic study approach. RISTI—Rev. Iber. Sist. e Tecnol. Inf. 2020, E32, 28–40 (2020)
19. Maestre Góngora, G.P.: Revisión de literatura sobre ciudades inteligentes: una perspectiva centrada en las TIC. INGENIARE. **19**(137) (2015). https://doi.org/10.18041/1909-2458/ingeniare.19.531
20. Malheiros, V. et al.: A visual text mining approach for systematic reviews. In: First International Symposium on Empirical Software Engineering and Measurement (ESEM 2007). pp. 245–254 IEEE (2007). https://doi.org/10.1109/ESEM.2007.21
21. Martín García, A.V. et al.: Factores determinantes de adopción de blended learning en educación superior. Adaptación del modelo UTAUT*. Educ. XX1. 17, 2, (2014). https://doi.org/10.5944/educxx1.17.2.11489
22. Medina-Cárdenas, Y. et al.: Mejoramiento continuo en la educación superior. Autoevaluación: perspectiva del estudiante. Hojas y Habl. 52–60 (2015)
23. Medina-Cárdenas, Y., Rico- Bautista, D.: Modelo de gestión basado en el ciclo de vida del servicio de la Biblioteca de Infraestructura de tecnologías de Información (ITIL). Rev. Virtual Univ. Católica del Norte. **27**, 1–21 (2009)
24. Medina-Cárdenas, Y., Rico-Bautista, D.: Mejores prácticas de gestión para la calidad de los servicios en tecnologías de información. Gerenc. Tecnológica Informática. **11**(29), 47–58 (2012)
25. Medina-Cárdenas, Y., Rico-Bautista, D.: Model of administration of services for the universidad of Pamplona: ITIL. Sci. Tech. Sci. Tech. Año XIV. **14**(39), 314–319 (2008)
26. Medina-Cárdenas, Y.C., Rico-Bautista, D.: Strategic alignment under a technology management organizational approach: ITIL and ISO 20000. Rev. Tecnura. **20**(1), 82–94 (2016). https://doi.org/10.14483/22487638.11681
27. Medina Cárdenas, Y.C. et al.: Modelo estratégico para la gestión tecnológica en la organización: plan táctico de la calidad (ITIL & ISO 20000). Instituto Tecnológico Metropolitano (2016). https://doi.org/10.22430/9789585414006.

28. Mitchell, I., Isherwood, S.: Cloud adoption The definitive guide to a business technology revolution. (2014). https://doi.org/10.1145/2554850.2555067
29. Parra Valencia, J.A. et al.: IOT: una aproximación desde ciudad inteligente a universidad inteligente. Rev. Ingenio. **13**(1), 9–20 (2017). https://doi.org/10.22463/2011642X.2128
30. Petersen, K. et al.: Systematic mapping studies in software engineering. In: 12th International Conference on Evaluation and Assessment in Software Engineering, EASE 2008. (2008). https://doi.org/10.14236/ewic/EASE2008.8
31. Revelo Sánchez, O. et al.: La gamificación como estrategia didáctica para la enseñanza/aprendizaje de la programación: un mapeo sistemático de literatura. Lámpsakos. **1**(19), 31–46 (2018). https://doi.org/10.21501/21454086.2347
32. Rico-bautista, D. et al.: Smart university: a vision of technology adoption Universidad inteligente: una visión de la adopción de la tecnología. Rev. Colomb. Comput. **22**(1), 44–55 (2021). https://doi.org/10.29375/25392115.4153
33. Rico-Bautista, D. et al.: Analysis of the potential value of technology: Case of universidad francisco de paula santander Ocaña [Análisis del valor potencial de la tecnología: Caso universidad francisco de paula santander Ocaña]. RISTI—Rev. Iber. Sist. e Tecnol. Inf. E17, 756–774 (2019)
34. Rico-Bautista, D.: Conceptual framework for smart university. J. Phys. Conf. Ser. (2019). https://doi.org/10.1088/1742-6596/1409/1/012009
35. Rico-Bautista, D. et al.: Latin American Smart University: Key Factors for a User-Centered Smart Technology Adoption Model. In: Sustainable Intelligent Systems. pp. 161–173 (2021). https://doi.org/10.1007/978-981-33-4901-8_10
36. Rico-Bautista, D. et al.: Smart university: a review from the educational and technological view of internet of things. In: M., P. et al. (eds.) International Conference on Information Technology and Systems, ICITS 2019. pp. 427–440 Springer Verlag, Systems and Informatics Department, Universidad Francisco de Paula Santander Ocaña, Algodonal Campus Vía Acolsure, Ocaña, 546551, Colombia (2019). https://doi.org/10.1007/978-3-030-11890-7_42
37. Rico-Bautista, D. et al.: Smart university: big data adoption model. In: 2020 9th International Conference On Software Process Improvement (CIMPS). pp. 52–60 IEEE (2020). https://doi.org/10.1109/CIMPS52057.2020.9390151
38. Rico-Bautista, D. et al.: Smart university: Characterization of the current situation of intelligent technologies, based on two case studies [Caracterización de la situación actual de las tecnologías inteligentes para una universidad inteligente en Colombia/latinoamérica]. RISTI—Rev. Iber. Sist. e Tecnol. Inf. 2020, E27, 484–501 (2020)
39. Rico-Bautista, D. et al.: Smart University: Key Factors for an Artificial Intelligence Adoption Model. In: García, M. V. et al. (eds.) Advances and Applications in Computer Science, Electronics and Industrial Engineering. pp. 153–166 Springer Singapore, Singapore (2021). https://doi.org/10.1007/978-981-33-4565-2_10
40. Rico-Bautista, D. et al.: Smart university: Strategic map since the adoption of technology [Universidad inteligente: Mapa estratégico desde la adopción de tecnología]. RISTI—Rev. Iber. Sist. e Tecnol. Inf. 2020, E28, 711–724 (2020)
41. Rico-Bautista, D. et al.: Smart university:IoT adoption model. In: 2020 Fourth World Conference on Smart Trends in Systems, Security and Sustainability (WorldS4). pp. 821–826 IEEE (2020). https://doi.org/10.1109/WorldS450073.2020.9210369
42. Rico-Bautista, D. et al.: Strategic management appropriation: A question of organizational skills. FACE Rev. la Fac. Ciencias Económicas y Empres. **15**(2), 71–80 (2015)
43. Rico-Bautista, D.W.: Conceptual framework for smart university. J. Phys. Conf. Ser. **1409**, 012009 (2019). https://doi.org/10.1088/1742-6596/1409/1/012009
44. Sánchez-Torres, B. et al.: Smart Campus: Trends in cybersecurity and future development. Rev. Fac. Ing. **27**(47), 93–101 (2018). https://doi.org/10.19053/01211129.v27.n47.2018.7807
45. Shvetsova, O.A.: Smart education in high school: new perspectives in global world. In: 2017 International Conference "Quality Management, Transport and Information Security, Information Technologies" (IT&QM&IS). pp. 688–691 (2017). https://doi.org/10.1109/ITMQIS.2017.8085917

46. Viñán-Ludeña, M.S. et al.: Smart university: an architecture proposal for information management using open data for research projects. Adv. Intell. Syst. Comput. 1137 AISC, March, 172–178 (2020). https://doi.org/10.1007/978-3-030-40690-5_17
47. Zapata-ros, M.: La universidad inteligente La transición de los LMS a los Sistemas Inteligentes de Aprendizaje en Educación Superior The smart university. **57**(10), 1–43 (2018)

Smart Farm Water Requirement Analysis Using Machine Learning

R. Akshay Sunny, R. Jyosthna, R. Melvin Raj, and M. Manoj

Abstract Smart farming includes incorporating information and communication technology into machinery, equipment, and sensors for use in agricultural production systems. More robotics and artificial intelligence are expected to debut into farming as a result of increased technologies, such as the Internet of Things and machine learning.Thus as a result, this article has two objectives: (i) to forecast the weather, soil moisture, and temperature based on available data. (ii) as well as to water the plants appropriately.Smart farming plays an essential part in smart cities and is a necessary aspect. Furthermore, the need to reduce water waste has triggered new research directions in the quest for water-saving solutions. In this article, a solution for intelligent irrigation of any kind of crop has been proposed, taking into account emerging technology such as the Internet of Things and machine learning.An application of this solution has been presented to demonstrate the viability of the suggested solution.

Keywords Machine learning · Regression analysis · Data in agriculture · Smart farming and agriculture

1 Introduction

In recent years, technological advancements such as the use of computer systems and data transfer have brought about significant improvements in the farming-working environment. These modifications necessitate new data from manufacturing processes, industries, and agents engaged in development in order to include decision-making data for production as well as technical and administrative problems. Smart farming, which is focused on the integration of information and communication technology into agricultural machinery, machines, and sensors, allows for the generation of a large amount of data and statistics with the gradual introduction of automation into the process. To enable smart farming, data transmission and data concentration

R. Akshay Sunny (✉) · R. Jyosthna · R. Melvin Raj · M. Manoj
School of Electronics and Communication Engineering, REVA University, Bangalore, India

© The Author(s), under exclusive license to Springer Nature Singapore Pte Ltd. 2022
A. K. Nagar et al. (eds.), *Intelligent Sustainable Systems*, Lecture Notes in Networks and Systems 333, https://doi.org/10.1007/978-981-16-6309-3_62

Fig. 1 Data analyzing

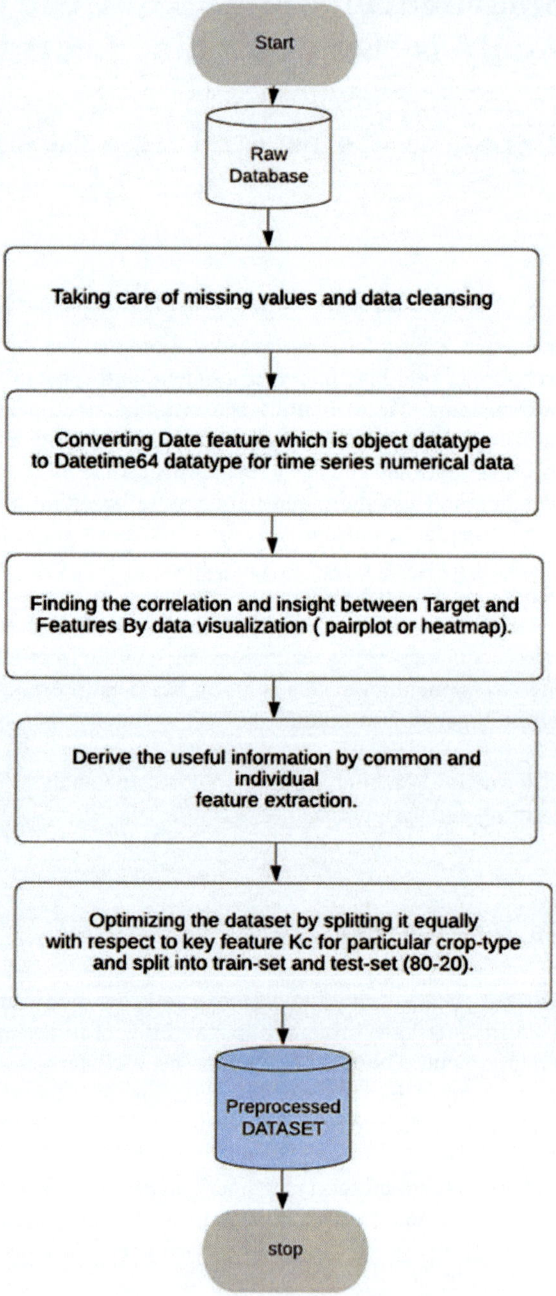

in remote storage systems are required [1]. Smart farming and smart agriculture literature is relatively new. In the scientific literature, there is no consensus on the definition of smart farming or the terms used to illustrate it. The phenomenon of "smart farming" is being propelled by remarkable progress in the Internet of Things (IoT) and machine learning [2]. Meteorological patterns such as the temperature, wind speed, sun, humidity, and precipitation are the foundation for progress in this field also including convergence of Internet technology and future-oriented technologies for use as smart objects.

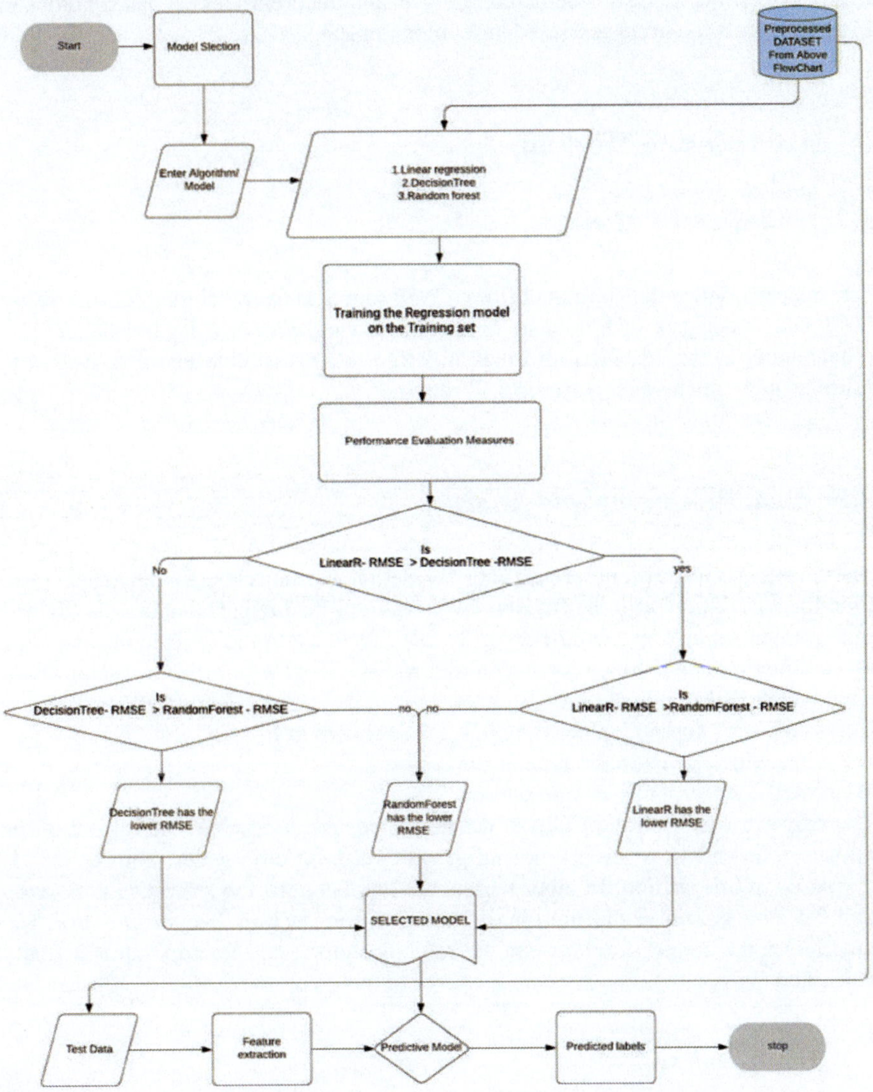

Fig. 2 Model selection and prediction

2 Methodology

We make predictions using regression analysis. A compilation of mathematical tools for estimating relationships between a dependent variable and one or more independent variables is known as regression analysis [3, 4].

For coding, we utilize the Python programming language. We begin by gathering raw data, then do data analysis and run machine learning models on the preprocessed dataset. The data analysis is depicted in Fig. 1. The selection and prediction of machine learning models is depicted in Fig. 2, and the preprocessed dataset block in Fig. 2 connects the preprocessed dataset block in Fig. 1.

3 Materials and Working

3.1 Dataset and Model

The meteorological data obtained from a Web source is from February 2021 to May 2021 which consists of humidity, temperature, wind speed, solar irradiance, and others as input factors. This information is fed into a machine learning system to estimate how much water the crop will require.

3.2 Data Processing and Analysis

The dataset should be preprocessed data. By making the data clean for better accuracy. Find the correlation with the features by pairplot or heatmap.The dataset is divided into training set and test set (70-30 or 80-20) . We have to remove the missing data and outliner of data points. As we see water need is linear with Eto, conclusion water need = coff * Eto
Coff is taken as kc with values 0.4 , 0.8 ,1.2 as shown in Fig. 5.
Kc varies with change in the type of the crop.
We have the kc of maize crop in dataset.
The crop evapotranspiration ETo of the reference crop used gives the impact of the climate. The impact of crop type and growth stage on crop water requirements is discussed in this section. In other words, the link between the reference grass crop and the crop grown in the field is discussed in this section. The crop factor, Kc, determines the connection between the reference crop and the crop that is actually cultivated [5].

$$\text{ETo} * \text{Kc} = \text{ETc} \tag{1}$$

where ETc = The amount of water required by the crop. Kc = Crop factor which is dependent on type of crop, the growth stage, and the climate. ETo = Evatranspiration occurred. Both ETc and ETo are expressed with the help of same unit, i.e., mm per day (Fig. 4).

3.3 Regression Analysis

Regression analysis is a set of statistical methods used for the estimation of relationships between a dependent variable and one or more independent variable. The model selection of regression analysis is using dataset.
The dataset is split equally with respect to key feature Kc (0.40, 0.80, 1.20) into train set and test set. So that our machine learning model will understand the underlying logic [6].
The input features has min temperature, max temperature, humidity, wind speed, solar irradiance, sun, and Kc (in order) as shown in Fig. 3 (Fig. 4, 5 and 6).

3.4 The Model Selection

The model is selected according to the data fitting, by running the dataset on different machine learning algorithms as shown in Fig. 2. We come into result by performance evaluation measures (RMSE and Rsquared/accuracy).

	date	year	month	day	Min_Temp	Max_Temp	Humidity	Wind_Speed	Solar_Irradiance	Sun	Kc	ETc	ETo	Water_Need_1day
7	2021-02-25	2021	2	25	17.90	34.900	13.0	7.70	10.20	22.3	0.40	12.91	5.16	20.656
21	2021-03-11	2021	3	11	15.00	35.700	60.0	4.90	12.20	27.2	0.80	7.89	6.31	25.248
39	2021-03-29	2021	3	29	17.38	33.210	67.5	6.51	10.50	24.3	0.80	6.94	6.82	27.280
16	2021-03-06	2021	3	6	11.70	39.200	52.0	4.80	15.00	31.3	0.80	9.70	7.76	31.040
77	2021-05-06	2021	5	6	17.90	38.000	69.0	5.50	12.90	29.3	1.15	8.40	9.66	38.640

Fig. 3 Dataset

Fig. 4 Water need correlation

```
Water_Need_1day      1.000000
ETo                  0.999146
Max_Temp             0.679560
Sun                  0.669282
Kc                   0.636462
Solar_Irradiance     0.578666
ETc                  0.553597
month                0.448735
Min_Temp             0.266192
day                 -0.081631
Wind_Speed          -0.193729
Humidity            -0.263644
year                      NaN
Name: Water_Need_1day, dtype: float64
```

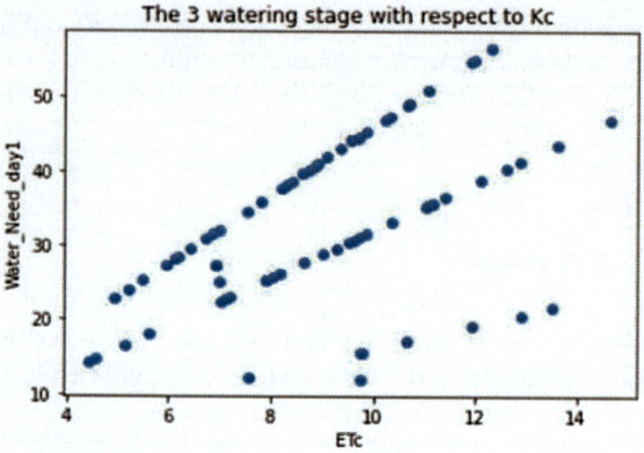

Fig. 5 Growth stage of crop (Kc)

```
[[ 2.    25.    17.9   34.9   13.     7.7   10.2   22.3    0.4  ]
 [ 3.    11.    15.    35.7   60.     4.9   12.2   27.2    0.8  ]
 [ 3.    29.    17.38  33.21  67.5    6.51  10.5   24.3    0.8  ]
 [ 3.     6.    11.7   39.2   52.     4.8   15.    31.3    0.8  ]
 [ 5.     6.    17.9   38.    69.     5.5   12.9   29.3    1.15 ]
 [ 4.    25.    16.6   41.1   59.     5.5   14.4   31.6    1.15 ]
```

Fig. 6 Training set

3.5 *Performance Evaluation Measures*

The evaluation measure is to indicate the performance of the different model. The four evaluation measures are:
Mean Absolute Error (MAE):

$$\mathrm{MAE} = \frac{1}{N}\sum_{i=1}^{N} \mid Yi - \widehat{Y}i \mid \tag{2}$$

Root Mean Squared Error(RMSE):

$$\mathrm{RMSE} = \sqrt{\sum_{i=1}^{n} \frac{(\widehat{y}i - yi)^2}{n}} \tag{3}$$

Mean Squared Error(MSE):

$$MSE = \frac{1}{n}\sum_{i=1}^{n}(yi - \widehat{y}i)^2 \tag{4}$$

R Squared:

$$R^2 = 1 - \frac{\sum(y - \widehat{y})^2}{\sum(y - \overline{y})^2} \tag{5}$$

4 Applications

This unit can water plants without the need for human interference. This low-cost, high-impact project uses precision and machine learning to ensure that the plants are watered according to their needs while also taking care of the crop dependent on meteorological conditions such as temperature, humidity, wind speed, and soil data [7].

5 Advantages

It is inexpensive and hence practical to mount around multiple crop fields, allowing farmers to relax during both sunny and stormy-cloudy days. The system ensures that the crop is watered focusing on the climatological factors described earlier.
The additional advantages are as follows: Analyzing production quality and outcomes in relation to treatment might help farmers improve product quality by adjusting procedures. All conservation initiatives, such as improved water consumption and productivity per land unit, have a direct impact on the environment.

6 Disadvantages

The machine learning approach for smart agriculture/farming that has been proposed in this paper is solely for maize. For different crops, the model must be adjusted based on their Kc, Etc, and Eto values.

7 Results

The outcome of predicted data with actual data can be shown (Figs. 7 and 8).

The below fig shows us the prediction difference using time series graph and comparison with actual data.
Comparison Graph (Table 1).

```
[[31.22 30.75]      [[30.4  30.75]      [[31.75 30.75]
 [23.59 22.76]       [25.25 22.76]       [25.88 22.76]
 [26.04 20.66]       [35.23 20.66]       [30.41 20.66]
 [24.38 23.92]       [22.77 23.92]       [23.75 23.92]
 [46.87 46.97]       [44.62 46.97]       [43.43 46.97]
 [33.18 34.59]       [27.42 34.59]       [26.15 34.59]
 [45.93 49.04]       [56.58 49.04]       [49.44 49.04]
 [31.7  28.15]       [28.2  28.15]       [26.38 28.15]
 [15.   16.51]       [25.   16.51]       [21.14 16.51]
 [23.48 22.76]       [25.25 22.76]       [25.51 22.76]
 [30.9  32.15]       [23.07 32.15]       [26.52 32.15]
 [33.14 31.55]       [25.   31.55]       [28.06 31.55]
 [40.42 39.77]]      [42.06 39.77]]      [38.55 39.77]]
```

Fig. 7 Array of comparison, left hand side has predicted value and right side has actual value (from left—linear, decision tree, and random forest)

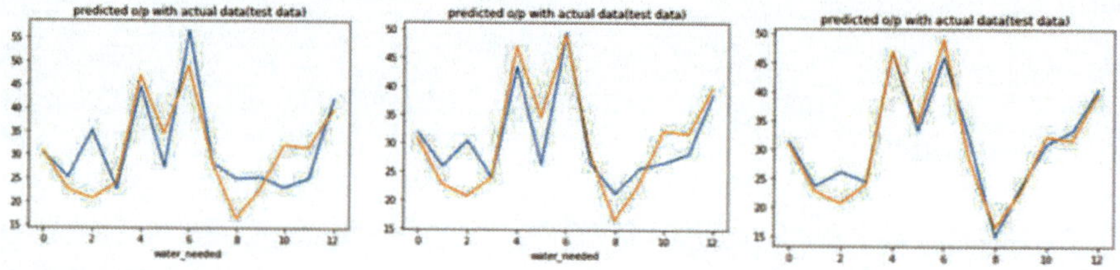

Fig. 8 Comparison between three different models, i.e., decision tree versus random forest versus linear (from left)

Table 1 Accuracy with linear regression model is 94%

Model	*R* square	RMSE
LR	0.94	2.17
Decision tree	0.25	6.46
Random forest	0.80	4.53

8 Conclusion

The suggested idea provides a low-cost, low-power, and highly accurate method for predicting the amount of water necessary for maize, as well as reducing the human workload of watering plants/crops under harsh climates such as scorching summers or bitter winters.

Smart farming would be more effective when we know the expense and the market value of water and plants. This would also be economical as the water is used based on the minimal and effective amount predicted by the model.

The project's future potential is enormous and can be scaled up in the future. As it is extremely adaptable in terms of growth, the project may be upgraded in the near future as and when the need arises. Given the fast growth of the population and the emergence of several new technologies throughout the years, it is critical that the agricultural sector shall soon follow.

This idea might be better developed, and it could be linked to a mobile application, so that farmers may complete their work even while they are not at home. Drones may also be used to collect real-time data and monitor the progress of the crops. It will be exciting to witness innovation in the agriculture sector.

References

1. Alves, R.G., Souza, G., Maia, R.F., Tran A.L.H., et al.: A digital twin for smart farming. In: 2019 IEEE Global Humanitarian Technology Conference (GHTC) (2019)
2. Nayyar, A., Puri, E.V.: Smart Farming: IoT based smart sensors agriculture stick for live temperature and moisture monitoring using Arduino, cloud computing and solar technology. In: Conference: The International Conference on Communication and Computing Systems (ICCCS-2016) (2016, November)
3. Liakos, K., Busato, P., Moshou, D., Pearson, S., Bochtis, D.: Machine learning in agriculture: a review. Sensors **18**, 2674 (2018). https://doi.org/10.3390/s18082674
4. Sidhu, R.K., Kumar, R., Rana, P.S.: Machine learning based crop water demand forecasting using minimum climatological data. Multimedia Tools Appl. **79**, 13109–13124 (2020). https://doi.org/10.1007/s11042-019-08533-w
5. Gupta, A., Nagda, D., Nikhare, P., Sandbhor, A.: Smart Crop Prediction using IoT and machine learning. Int. J. Eng. Res. Tech. (IJERT) NTASU—2020 **9**(3) (2021)
6. Treboux, J., Genoud, D.: Improved machine learning methodology for high precision agriculture. Glob. Internet Things Summit (GIoTS) **2018**, 1–6 (2018). https://doi.org/10.1109/GIOTS.2018.8534558
7. Mishra, S., Mishra, D., Santra, G.: Applications of machine learning techniques in agricultural crop production: a review paper. Indian J. Sci. Technol. **9**. 10.17485/ijst/2016/v9i38/95032

Abnormal Activity Detection Using Deep Learning

A. Dhanush Kumar, P. Shushruth Reddy, Kriti C. Parikh, C. Meghana Sarvani, and P. Loel Maansi

Abstract Identifying abnormal activity is a laborious job, and this has led to the advancement in the domain of deep learning for surveillance which assured performance gain. Abnormal detection plays a pivotal role in the field of research and application systems. Detecting real-world problems such as accidents, burglary, explosion, fighting, robbery and other critical events is crucial, so we developed a deep learning-based algorithm to reduce manual work and time. To provide an optimal solution for the same, we have designed a model that detects abnormal events. To achieve this objective, we have implemented two approaches. The first approach is the model that we have trained (auto-encoder), and the second approach is the pre-trained model (C3D feature extraction). In approach 1, we have introduced a spatiotemporal auto-encoder, which is based on a 3D convolution neural network. After training the model with auto-encoder, our model has reconstructed the frames with a minimum loss of 8% and a maximum accuracy of 71%. To improve accuracy and to reduce loss, we have exercised approach 2 (C3D extraction) that computes features from the fully connected layers of the C3D which resulted in increased accuracy. To detect the abnormal activity in banks, roads and many more crowded areas, this can be implemented through CCTV cameras to automate and simplify manual work.

Keywords Spatiotemporal auto-encoder · C3D feature extraction · 3D convolution neural network · Long short-term memory (LSTM)

1 Introduction

Abnormal event detection is the vital application in building pervasive and smart environment applications. Event recognition is based upon learning and analyzing the patterns of regular activities of the individuals that can be applied to different domains varying from the smart healthcare systems to the intelligent building systems for security. Rapid advancements in deep learning techniques provide optimal solutions

A. Dhanush Kumar (✉) · P. Shushruth Reddy · K. C. Parikh · C. Meghana Sarvani · P. Loel Maansi
Mahatma Gandhi Institute of Technology, Hyderabad 500075, India

© The Author(s), under exclusive license to Springer Nature Singapore Pte Ltd. 2022

A. K. Nagar et al. (eds.), *Intelligent Sustainable Systems*, Lecture Notes in Networks and Systems 333, https://doi.org/10.1007/978-981-16-6309-3_63

to diverse problems such as video processing, robotics, text processing and online advertising that has led to significant growth in various applications for event recognition. Existing anomalous event detectors primely rely on motion data acquired from several frames to detect anomalous activity. However, motion data may not be able to apprehend all kinds of anomalous activities. Hence, we implement surveillance videos that apprehend a variety of anomalous activities. Our model learns abnormal activities from dataset containing normal and abnormal activities. We consider each video as a bag and video segment in a bag representing an each instance. In addition to this, the model also learns to predict high anomaly score automatically for anomalous video instances. We also introduced sparsity and temporal smoothness constraints to improvise anomaly during training, and it is replaced by leveraging weakly labeled training videos. In this project, we mainly focus on the problem of anomalous activity detection automatically without much human intervention. Our proposed work includes multiple feature extraction and dominant behavior analysis.

2 Proposed Methodology

2.1 Avenue Dataset

Avenue dataset consists of 30,652 (15,324 testing, 15,328 training) frames in total which is split into 16 training videos and 21 unseen videos. The training videos consists of only normal situations, whereas unseen videos comprise both normal and anomalous events. Our evaluation of the video is not only restricted to the frame level but also includes spatial location. Additionally, abnormal events are labeled using rectangles. The challenges faced while using avenue dataset are as follows.

Normal patterns seldom appear in the training dataset. Slight camera shake results in partial distortion (in testing video, frame 1051–1100). Few outliers (data points that differ from other observations) are also incorporated in the training data.

2.2 UCF Crime Dataset

Due to the limitations observed from the avenue dataset, we have opted for a large-scale dataset (UCF) for evaluating our method. UCF dataset contains long, untrimmed surveillance videos which covers a total of 13 real-world anomalies, a few of which includes fighting, abuse, shooting, accident, robbery, etc. These are specifically chosen as they have a considerable impact on public safety. The dataset has been divided into two parts, i.e., the training data and the testing data. The training data consists of 800 normal and 810 anomalous videos. The testing data includes 150 normal and 140 anomalous videos. Both training and testing datasets accommodate different anomalies at various temporal locations in the videos (Table 1).

Table 1 Major differences between avenue dataset and UCF crime dataset

Avenue dataset	UCF crime dataset
Length of the dataset is 30 min	Length of the dataset is 128 h
It consists of 37 videos	It consists of 1900 videos
It consists of 839 average frames	It consists of 7247 average frames
Examples: run, throw, etc.	Examples: arrest, assault, accident, fighting, etc.

3 Data Preprocessing

3.1 Approach 1

- In this procedure, we have processed each video footage into different segments split over five frames per second with each video folder consisting of distinct data.
- Secondly, each frame which is in the form of an image with a resolution of [227 × 227 × 1] is resized by using interpolation as inter-area (a method that resamples the images using area pixel relation).
- Now, these images are converted into grayscale to reduce dimensionality, and later, normalization is done to unit variance.
- Furthermore, we store the features of images/frames of the footage in the form of an array and then save these in a NumPy file (.npy extension) which aids us in training the model.
- To evaluate, the array of an image functions as an input to the model which helps in classifying the videos for abnormal detection.

3.2 Approach 2

- Preprocessing the features into fully connected layers results in increased accuracy. Hence, preprocessing is a major aspect for the model to learn the features and predict the respective classes.
- Moreover, we used the pre-trained C3D network to process the video and then extract the elements from the video. Now, these images are converted into grayscale to reduce dimensionality, and later, normalization is done to unit variance.
- Therefore, we extract the C3D features for every 16 frame clips, resizing the video frames to [240 × 320] pixels which are split over 30 frames per second (features for each video segment are obtained by averaging 16 frames clip which is considered within the specific segment).

4 Network Architecture and Model Training

The performance of a classifier in activity recognition is primarily dependent on the features that are extracted from raw data.

4.1 Approach 1

Auto-encoders consist of two parts, namely encoding and decoding which aims to pare dimensionality by limiting the count of encoder outputs compared to that of inputs. Likewise, training of the auto-encoder is achieved by the method of back-propagation in an unsupervised approach. Therefore, the resultant output is obtained by reducing the loss of reconstruction from original inputs. By adding nonlinearity, the auto-encoders will pull out more prominent features compared to linear transformations like principal component analysis (PCA) and many more (Fig. 1).

Spatiotemporal Auto-encoder

- Firstly, to learn the features of the data, we introduced a 3D convolutional spatiotemporal auto-encoder.
- Our architecture includes spatial encoder and decoder as well as temporal encoder and decoder.
- Each spatial encoder and decoder consists of two layers, namely 3D convolutional and 3D de-convolutional.
- However, temporal encoder and decoder consist of three-layered convolutional long short-term memory (LSTM) model.
- Here, spatial encoder learns the spatial features of the input using small squares, whereas, the temporal model learns the patterns of the encoded spatial features.
- It is necessary to define the specifications such as the number of filters, filter dimensions and the number of layers to the convolution model for training each layer.

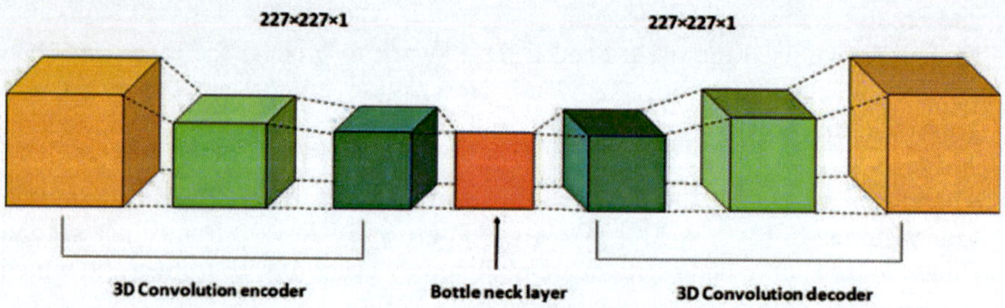

Fig. 1 Network architecture of spatiotemporal auto-encoder

- Generally, if filters are more, then a more number of features are extracted which in turn help the model for better prediction.
- To ensure all the input sequence data are of the same length, we implement padding to truncate.
- Tanh is the activation function, which restricts the value between -1 and 1 to ensure the symmetry of encoding and decoding.
- However, large parameters affect the computational time and memory, so, balancing the parameters is also the key factor.
- To learn the temporal dependencies between inputs, long sequences are the key for the model, and to acquire long sequences information, LSTM—variant of RNN—is used.
- LSTM is a variant of RNN which consists of a feedback gate or recurrent gate called as forget gate.
- It prevents backpropagated errors from vanishing/exploding.

Therefore, it can be used on long sequences and to derive information better. Once the architecture is developed, the model is trained on the preprocessed data by using the optimizer "Adam," which integrates both the AdaGrad and RMSProp algorithms to develop a suitable model. This can handle sparse gradients on noisy problems and has been evaluated on testing data (unseen data), to check whether the model is capable of detecting the video as normal or abnormal (using thresholding). Loss is calculated in terms of reconstruction error, which is defined as the difference between the input frame and the reconstructed frame.

Thresholding is the value that represents the sensitivity of the model. Based on thresholding, the data that has been fed to the model is classified as either a normal event or an abnormal event (Fig. 2).

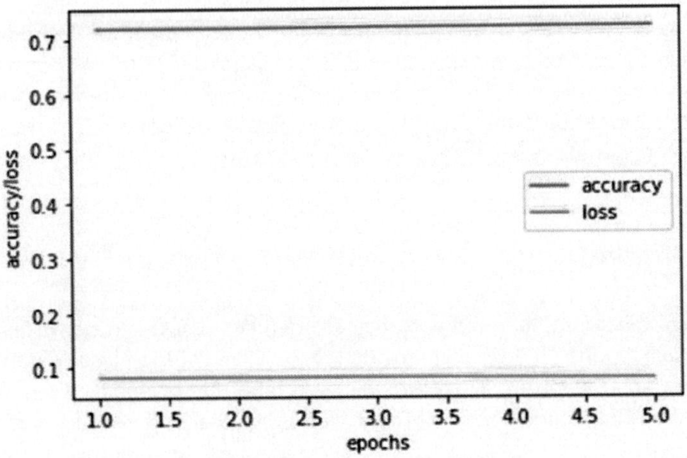

Fig. 2 Plot of accuracy and loss of auto-encoder

By thresholding to 0.0005, our model attained a reconstruction loss of 7% with an accuracy of 72% which is quite low and is not acceptable. Henceforth, we have implemented the pre-trained model C3D network to compute features of these videos to reduce the loss and improve the accuracy.

4.2 Approach 2

In approach 1, we used preprocessed videos and trained them using the auto-encoder. On contrary, in approach 2 we used a pre-trained C3D model to compute the features. These features are then stored in the text format followed by dense layers to reduce to a single dimension.

- Initially, we have a dataset (videos) that is split into 32 temporal segments. Furthermore, these temporal segments are divided into positive (abnormal) and negative (normal) bags. Each segment in the bag is termed a bag instance (video segment).
- These instances now undergo the C3D feature extraction process for each video segment. Each video consists of 32 sets of features, and based on the videos selected, it is processed as an input to the model.
- This model consists of five layers (three dense layers and two dropout layers).
- Dense layer—This is a fully connected layer, i.e., all the neurons in a layer are connected to those in the succeeding layer.
- Dropout layer—This works by randomly positioning the outgoing edges of hidden units (neurons that make up hidden layers) thus preventing overfitting.
- In the output layer of the model, we employed softmax as an activation function that predicts whether the activity is normal or abnormal. Instance score gives information regarding each segment's acquired value.
- Based on the anomaly score and the threshold value, we get the desired result. In approach 1, we used preprocessed videos and trained it using auto-encoder.
- 3D convolution network has the same architecture throughout the network. C3D extracts information from video in spatiotemporal at different levels of abstraction and detects minute changes (Figs. 3, 4, 5 and 6).

5 Conclusion

In abnormal detection, loss and accuracy are two key aspects of performance. The first approach resulted in less accuracy due to the usage of the biased dataset, which is not an optimal solution. Therefore, to overcome this problem, we have used a pre-trained model, which consists of real-world data that enhances its performance. We have analyzed the video in real-time and identified the abnormal activities like violence or theft by building a system that automates the task of video surveillance. In this paper, we have shown, observed and contrasted the efficiency of two models (training from scratch and pre-trained).

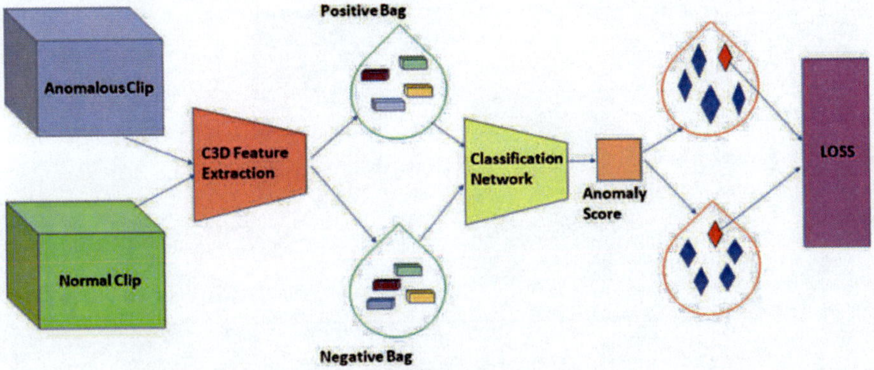

Fig. 3 Block diagram of approach 2 (using C3D feature extractor)

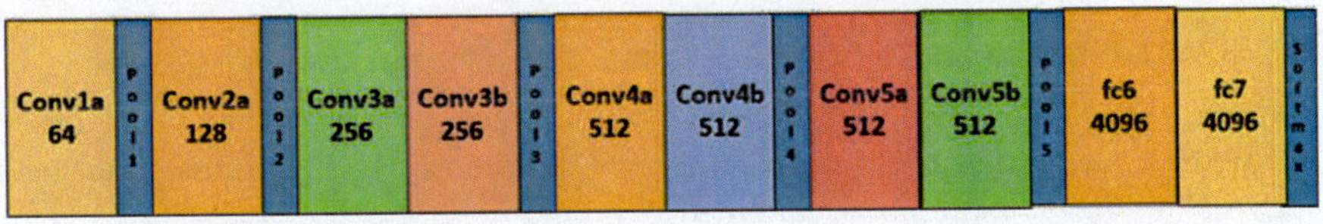

Fig. 4 Network architecture of C3D feature extractor

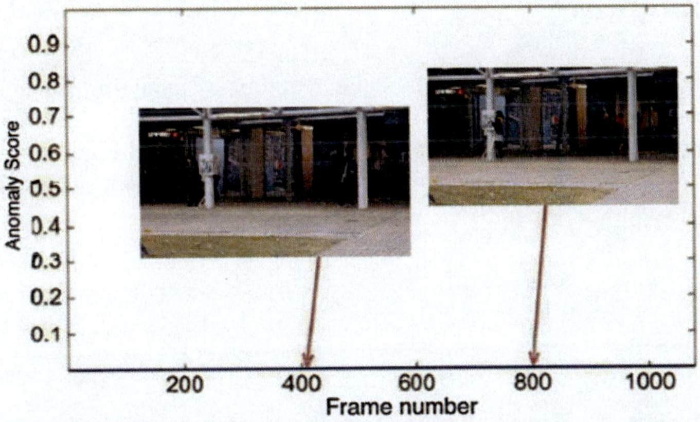

Fig. 5 Observation for normal event detected

5.1 *Advantages*

- It automates the process of deep surveillance in real time.
- It has the ability to monitor wide areas.
- It also helps in classifying different videos.

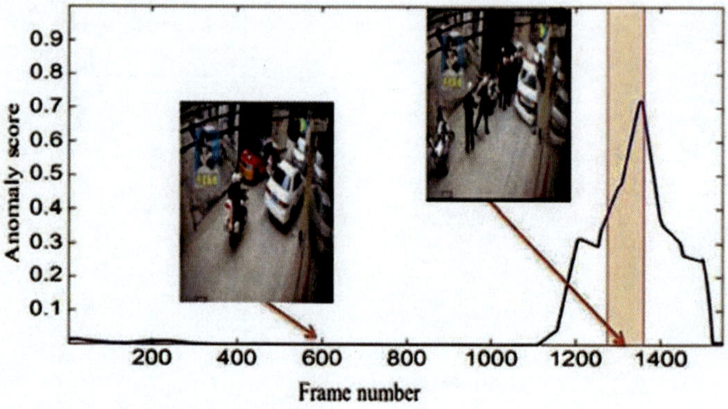

Fig. 6 Observation for abnormal event detected

5.2 *Drawbacks*

Our model concentrates on specific activities (either normal or abnormal), furthermore making its applications streamlined.

5.3 *Future Scope*

- In future, we can implement our model in real time by including more classes for specifically detecting changes.
- Additionally, multiple class detection will be taken into consideration.
- Furthermore, we can implement this model in the control room and also in drones to ease the work of the police.

References

1. Sultani, W., Chen, C., Shah, M.: RealWorld anomaly detection in surveillance videos. In: 2018 IEEE/CVF Conference on Computer Vision and Pattern Recognition (2018)
2. Lu, C., Shi, J., Jia, J.: Abnormal event detection at 150 FPS in Matlab. In: International Conference on Computer Vision. In: ICCV (2013)
3. Gul, M.A., Yousaf, M.H., Nawaz, S., Rehman, Z.U., Kim, H.W.: Patient monitoring by abnormal human activity recognition based on CNN architecture. Article published in MDPI (2020)
4. Cui, X., Liu, Q., Gao, M., Metaxas, D.N.: Abnormal detection using interaction energy potentials. In: CVPR (2011)
5. Chong, Y.S., Tay, Y.H.: Chapter 23 Abnormal Event Detection in Videos Using Spatiotemporal Autoencoder. Springer (2017)

Noisy Dense Label Prediction with Noise-Robust Loss Functions

Redwan Hasif Alvi and Rashedur M. Rahman

Abstract Image segmentation is an actively research field in the domain of deep learning and computer vision. Due to its use in a wide range of applications such as medical imaging problems, automated disease detection, self-driving vehicles and so on, image segmentation is widely used in dealing with building extraction tasks. Although deep learning models such as U-Net performs well with building extraction tasks that require semantic segmentation, presence of noisy data can still affect the performance of these models. Moreover, it is not unusual to have noisy labels present in data sets that are available online or collected automatically via the Internet. As such, we commonly come across label noise present in data sets used for building extraction tasks that can decrease the performance of a learning task. In this paper, we implement multiple well-known loss functions used for image segmentation on an existing U-Net model for learning building extraction and compare the results of each of the loss functions' performance. Furthermore, we implement some of the widely used loss functions such as cross entropy (CE), Dice loss (DL) and Tversky loss (TL) on a noise-robust intermediate false positive–false negative matrix model to evaluate the performances of these loss functions when dealing with noisy data.

Keywords Image segmentation · Automated building extraction · Dense label · Noise label · Noise-robust · Loss function

1 Introduction

A multitude of sectors stretching from software to medical imaging has been transformed with the advent of deep learning. Deep learning models have achieved tremendous results when dealing with computer vision tasks such as image classification,

---

R. H. Alvi · R. M. Rahman (✉)
Department of Electrical and Computer Engineering, North South University, Dhaka 1229, Bangladesh
e-mail: rashedur.rahman@northsouth.edu

R. H. Alvi
e-mail: redwan.alvi@northsouth.edu

© The Author(s), under exclusive license to Springer Nature Singapore Pte Ltd. 2022

A. K. Nagar et al. (eds.), *Intelligent Sustainable Systems*, Lecture Notes in Networks and Systems 333, https://doi.org/10.1007/978-981-16-6309-3_64

image segmentation, object detection and applications such as (but not limited to) cybersecurity, melanoma classification and bioinformatics. One of the significant contributions of deep learning for tasks such as building extraction is image segmentation. With the help of segmentation, we are able to not only tell apart one building from another, but also pinpoint exactly where that particular building or landmark exists. This in turn has dramatically assisted in developing algorithms to detect buildings and landmarks in various building maps and extraction tasks, which help with urban planning and land management.

We can define image segmentation as a dense prediction task (pixel-level classification). The various pixels in an image represent different elements in the image when grouped together. Via semantic segmentation method, deep learning models can classify these pixels into the elements in an image. Here, one crucial factor when developing complex segmentation-based architectures is the selection of loss function of the models. This is because the loss function plays a key role in the training stage of a model and launches the learning process for the deep learning model. It dictates how good the predications of the model would be and affects the adjustment of coefficients (i.e. weights and biases) for backpropagation. Hence, the loss function choice has a direct influence on the model performance.

In addition to the loss function, the data set also plays a major role in the performance of a deep learning model during segmentation. Unfortunately, large data sets require labelling process which can be intensive and time consuming. This leads to the production of cheaper and less labour-intensive data sets through online surveys and crowdsourcing. In majority of the building extraction tasks, the labels for any data can be obtained freely from the available vector data such as those from OpenStreetMap. These data commonly have incorrect or noisy labels due to annotators prioritizing speed over accuracy and having difference in opinions when it comes to labelling a complex image. Thus, the choice of loss functions plays important when dealing with data sets that have significant noise present.

In this paper, we use an existing label noise-robust U-Net model and apply various popular custom segmentation loss functions that include cross entropy (CE), balanced cross entropy (BCE), Dice loss (DL), combination of Dice loss and mean absolute error (Dice-MAE), Tversky loss (TL), as well as implement a custom loss that combines the DL and CE to get a combination loss function (COMB). The data set used for this study contains aerial images of Kutupalong mega camp procured by the United Nations International Organization for Migration. The data set contains artificially added noise, and we compare the differences in performance between the loss functions when dealing with the clean data and the noisy data.

2 Related Works

Deep learning has revolutionized the way we approach image segmentation. Plenty of work has been done in the last decade regarding segmentation that includes the application of different loss functions, dealing with noisy data and testing custom/robust loss functions in order to improve the performance of models for segmentation.

Jadon [1] presents a survey of loss functions that are used for semantic segmentation. In her work, the author summarized fifteen popular loss functions for segmentation. The loss functions include binary cross entropy, weighted binary cross entropy, balanced cross entropy, focal loss, Dice loss, Tversky loss, focal Tversky loss, sensitivity specificity loss, shape-aware loss, combo loss, exponential logarithmic loss, distance map-derived loss, Hausdorff distance loss, correlation maximized structural similarity loss and a custom log-cosh Dice loss. The author compared the results of each loss functions on NBFS skull stripping data set. The study concluded that focal Tversky and log-cosh Dice loss performed the best, achieving a Dice coefficient of 0.98, while the other functions came very close to the optimal result.

Cordeiro [2] presents a survey on how to train a model when dealing with noisy labels. The study classified the main techniques as robust losses, sample weighting, sample selection, meta-learning and combined approaches. Cordeiro surveys multiple data sets and their results with each technique. DivideMix, which is a combined approach, performed best across all data sets with an average accuracy of 77%.

Furtado [3] takes three different problems related to medical image segmentation, namely MRI, CT and EFI scans, and quantifies loss function, variations and segmentation scores of different targets. This study achieved the best result with Dice loss across all data sets and achieved 1 to 6 percentage points (pp) higher with Dice loss when compared to cross entropy.

3 Model Framework

The deep learning models are the current state-of-the-art system for automated building extraction tasks, and it has proven to give excellent results in the field, i.e. in cases where there are clean data, noisy data or both. For our work, we use two existing U-Net architectures implemented by Ahmed [4] for automated building extraction tasks and implement custom loss functions to these models in order to compare the performance of the models with these loss functions.

The first is a novel U-Net segmentation model (cleaning network) used to evaluate the training and classification performance of the model on the clean, noise-free data samples. In the model, each of the intermediate network contains eight blocks for downsampling and upsampling (four blocks for each). Each block contains two convolutional layers followed by a single dropout layer. In the case of the downsampling block, max pooling is carried out on the dropout layer in order to downsample it equal to half the output row and column size. On the other hand, the upsampling block dropout layer is interpolated to have double the output row and column size.

For the second model, false positive (FP) and false negative (FN) error matrices undergo approximation separately as an intermediate step of estimating building versus non-building prediction. The base network learns the noisy observed label approximated as the mean distribution prediction between the given noisy label and its corresponding true clean label. The FP error network learns the FP error matrix, whereas the FN error network learns the FN error matrix. Ultimately, we combine the FP and FN error matrix model outputs, as well as the observed label model, in the cleaning network in order to learn noise-free label approximation. The authors [4] achieved this through transfer learning by feeding the outputs from the base network and both the FP FN error matrices to the cleaning network. We train the cleaning network in the first step by setting the base network and error networks as non-trainable. In the second step of the training, we train the full network in an end-to-end fashion against clean labels.

3.1 Data Source

In this work, we have used the same data set used by Ahmed in his study [5]. The data set contains 258 images samples of size 512 $\times$ 512 from Kutupalong mega camp curated by the United Nations International Organization for Migration on 17 September 2018. We get the observed noisy labels from OpenStreetMap. We obtain the true clean labels from the relabelling carried out by Ahmed in his work [5]. The images are ultra-high-resolution aerial RGB images, meaning that the images have three channels and are not greyscale. The red, green and blue channels have a spatial resolution of 10 cm. As stated in their study, this data set is unique in the sense that we can obtain ground truths for the error matrices since we have access to both the observed noisy labels and the true clean labels. Additionally, we require the error matrices only for pre-training the dense label noise prediction model. The model directly outputs building maps during testing corrected by error matrices. The data set is very much imbalanced in that it has a more number of negative samples, i.e. not infrastructures.

3.2 Loss Functions

Loss functions are the mathematical representation of objectives. We must ensure that our loss functions are capable of dealing with even edge cases to make sure that our deep learning models are able to learn any objective fast and accurately. Since our objective deals with dense pixel semantic segmentation, the number of classes at pixel level is two. We have implemented multiple noise-robust loss functions in order to evaluate which performs the best for our FP FN error matrix model and data set. More details of this error matrix model could be found in the previous work [4].

3.2.1 Binary Cross Entropy

For any given random variable or case, the difference between two probability distributions is defined by binary cross entropy [6]. It is most popular and widely used loss function for classification tasks. Since segmentation is dense pixel classification, cross entropy achieves good results overall. We can define binary cross entropy as follows:

$$L_{\mathrm{BCE}}(y, y') = -(y\log(y') + (1-y)\log(1-y'))$$

where y' is the predicted value of the model and y is the ground truth label.

3.2.2 Balanced Cross Entropy

A variant of binary cross entropy—in balanced cross entropy, [7] we add weights to both the positive and negative examples with a certain coefficient beta β (for our model, we used beta = 0.3). It is popular for cases where we have skewed data. We define balanced cross entropy by

$$L_{\mathrm{BalCE}}(y, y') = -(\beta * y\log(y') + (1-\beta) * (1-y)\log(1-y'))$$

3.2.3 Dice Loss

The Dice coefficient is the go-to metric to measure and compare the similarity between two images. This coefficient has been adapted into the Dice loss [8], and the loss function is naturally noise-robust. It is because the Dice loss implicitly establishes a balance between foreground and background classes, which can be defined as follows:

$$\mathrm{DL}(y, y') = 1 - \frac{2yy' + 1}{y + y' + 1}$$

Here, in order to ensure that the function does not stay undefined in edge cases (i.e. $y = y' = 0$), 1 is added to both the numerator and the denominator.

Wang [9] proposed adding an additional metric in the form of mean absolute error to the Dice loss in order to make it more noise-robust. The mean absolute error can be defined as a loss by

$$L_{\text{MAE}}(y, y') = \frac{\sum_i^N |y_i - y_i'|}{N}$$

By combining the two functions, we get

$$L_{\text{DL-MAE}} = \text{DL}(y, y') + L_{\text{MAE}}(y, y')$$

3.2.4 Tversky Loss

Tversky loss [10] uses the Tversky index which can also be considered as a generalization of the Dice coefficient similar to the Dice loss. The difference is that with the Tversky index, we add a weight to the false positives and false negatives by using the beta coefficient [11]:

$$\text{TI}(y, y') = \frac{yy'}{yy' + \beta(1 - y)y' + (1 - \beta)y(1 - y')}$$

where when beta = 0.5, it can be solved into regular Dice coefficient. Hence, we can define the Tversky loss as follows:

$$\text{TL}(y, y') = 1 - \frac{yy' + 1}{yy' + \beta(1 - y)y' + (1 - \beta)y(1 - y') + 1}$$

3.2.5 Combination Loss

It is also possible to combine multiple loss functions. The combination loss (comb loss) is the weighted sum of Dice loss and a modified cross entropy [12]. The purpose here is to use the advantage of Dice loss' flexibility in dealing with class imbalances and use cross entropy for smoothing the curve. The combination loss is defined as follows:

$$L_{m-\text{bce}} = -\frac{1}{N}\sum_i \beta(y - \log(y')) + (1 - \beta)(1 - y)\log(1 - y')$$

$$\text{CL}(y, y') = \alpha L_{m-\text{bce}} - (1 - \alpha)\text{DL}(y, \hat{y})$$

where α is a constant.

4 Results and Discussion

The model has to run predictions on approximately 33 million pixels of testing data. In order to evaluate the performance of the model, we measure the total true positives (TP), true negatives (TN), false positives (FP) and false negative (FN) predictions. Thus, the parameters we calculate are the precision, recall, F1-score and intersection over union (IoU).

$$\text{Precision} = \frac{\text{TP}}{\text{TP} + \text{FP}}$$

$$\text{Recall} = \frac{\text{TP}}{\text{TP} + \text{FN}}$$

$$F1 - \text{score} = \frac{2\text{TP}}{2\text{TP} + \text{FP} + \text{FN}}$$

$$\text{IoU} = \frac{\text{TP}}{\text{TP} + \text{FP} + \text{FN}}$$

Table 1 Result summary for clean model

	Threshold	Precision	Recall	F1	IoU
Bal CE	0.25	0.87869	0.76631	0.81866	0.69299
	0.5	0.88654	0.75985	0.81832	0.69251
	0.75	0.89407	0.7528	0.81738	0.69116
Binary CE	0.25	0.80171	0.90348	0.84956	0.73847
	0.5	0.89217	0.83019	0.86006	0.75448
	0.75	0.94247	0.72639	0.82044	0.69555
Combo Loss	0.25	0.91285	0.68643	0.78361	0.64421
	0.5	0.91648	0.68188	0.78196	0.64199
	0.75	0.91998	0.67711	0.78008	0.63945
Dice-MAE	0.25	0.80219	0.88052	0.83953	0.72344
	0.5	0.80506	0.87912	0.84047	0.72483
	0.75	0.80795	0.87767	0.84137	0.72618
Tversky	0.25	0.84257	0.72495	0.77935	0.63847
	0.5	0.84446	0.72388	0.77953	0.63872
	0.75	0.84634	0.72276	0.77969	0.63892
Dice loss	0.25	0.86113	0.77054	0.81332	0.68537
	0.5	0.86378	0.76846	0.81334	0.6854
	0.75	0.86641	0.76627	0.81327	0.6853

Table 1 shows the results of the vanilla U-Net for each loss function against three levels of noise represented by the threshold value (i) 0.25 (ii) 0.5 and (iii) 0.75.

From the table, we can see that binary cross entropy scores the highest F1-score with 0.86, with Dice-MAE Loss coming very close with 0.84 (both at threshold 0.5). Balanced cross entropy performed decently with a F1-score 0.81, whereas both the combination loss and Tversky loss performed the worst with F1-scores of sub 0.80.

Additionally, we can also see that both the cross entropy and Dice-MAE loss perform similarly when it comes to precision, recall and IoU scores. For both, the precision and recall values are always somewhere between the 0.80 to 0.90 mark. This shows the both Dice-MAE loss and binary cross entropy are almost equally adept at distinguishing positive and negative cases for our data set. On the other hand, balanced cross entropy scores significantly higher in precision (into the high 0.80 s) than recall (mid 0.70 s). It is the same for combination loss where it scores even higher in precision (above 0.90) than in recall (below 0.70). Finally, Tversky loss also does much better in precision than in recall, although relatively lower in terms of score when compared to either balanced cross entropy or combination loss. This shows that unlike binary cross entropy or Dice-MAE loss, these loss functions are stronger at identifying either positive or negative cases.

Figure 1 shows the noise and error matrices for an image sample for each loss function.

Table 2 shows the results of the intermediate FP-FN-error matrix model [4], taking the loss functions that perform best for this model with the intermediate step. Again, binary cross entropy performed the best with this case with F1-score of 0.87. However, unlike with the vanilla U-Net, the intermediate FP-FN-EM model preforms identically with the Dice-MAE loss as well as with Tversky loss, with both scoring 0.84 F1-score. One of the reasons for this is the fact the Tversky loss adds weights to the false positive and false negatives. Since the intermediate FP-FN-EM model combines the FP and FN error matrix model outputs in the cleaning network, we can see why Tversky loss suddenly performs on par with Dice-MAE loss here, because Tversky loss puts emphasis on the false positives and negatives by adding certain weights to those values. This shows that Tversky loss can perform identical to or even outperform the likes of Dice-MAE loss and binary cross entropy in the I-FP-

FN-EM model given better optimization when making the model. In summary, this loss function is weighted by the constants 'alpha' and 'beta' that penalize false positives and false negatives, respectively, to a higher degree in the loss function as their value is increased. The beta constant, in particular, has applications in situations where models can obtain misleadingly positive performance via highly conservative prediction. This also shows that Dice loss and Dice-MAE are noise-robust as having similar scores even with the presence of noisy labels.

Figure 2 shows the noise and error matrices for an image sample to see the difference between the Dice-MAE loss and the Tversky loss as evaluated by the I-FP-FN-EM model proposed in [4].

If we analyse these results, we can conclude that for our data set and models, binary cross entropy is the best loss function to use. This may seem counter-intuitive given how well Dice loss function performs in segmentation-related tasks [13]. However,

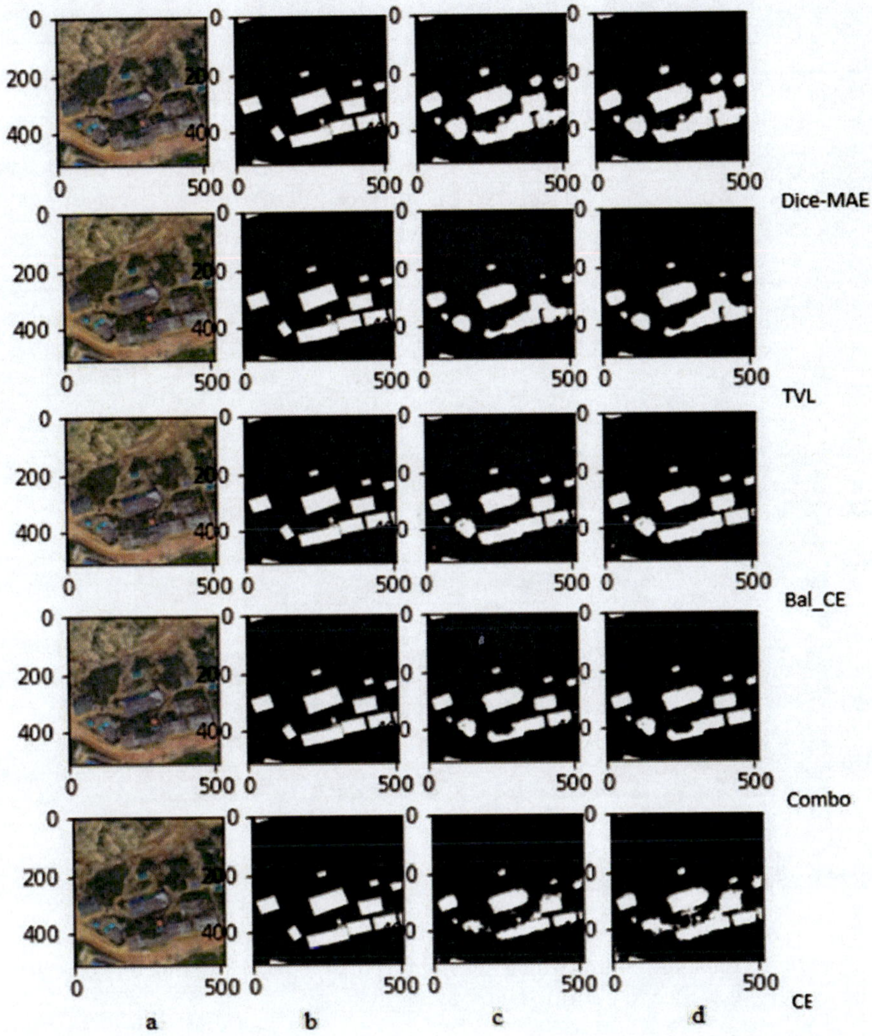

Fig. 1 Example of building prediction **a** image, **b** clean label **c** noisy label, **d** predicted label

we have to consider some key factors in order to understand why binary cross entropy performed the best in our case. In our building extraction task, we are dealing with dense pixel segmentation. This is where binary cross entropy works best in [14]. In other words, Dice loss usually performs best with segmentation tasks since it views the image as a whole. Therefore, it performs really well when comparing multiple image samples one against another (i.e. medical imaging cases such as pneumonia lesions). Since we are doing dense label prediction with our model, binary cross entropy performs better as we do not have to look at the entire picture. Thus, even

Table 2 Result summary for I-FP-FN-EM model

	Threshold	Precision	Recall	F1	IoU
Dice-MAE	0.25	0.81375	0.87333	0.84249	0.72784
	0.5	0.8171	0.87021	0.84282	0.72834
	0.75	0.82054	0.86675	0.84301	0.72862
Tversky	0.25	0.82973	0.85714	0.84321	0.72893
	0.5	0.83421	0.85301	0.84351	0.72936
	0.75	0.83878	0.84854	0.84363	0.72956
Binary CE	0.25	0.76109	0.94634	0.84366	0.7296
	0.5	0.86551	0.89424	0.87964	0.78514
	0.75	0.92819	0.80131	0.86009	0.75453

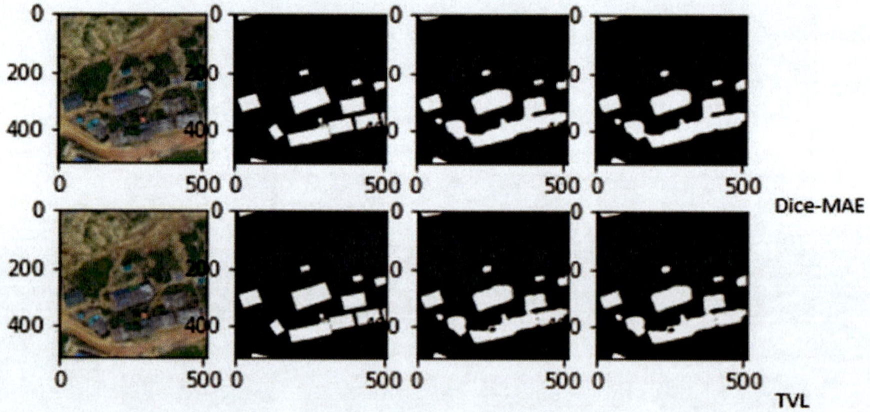

Fig. 2 Example of building prediction for I-FP-FN-EM model

after being an imbalanced set (which heavily favours the Dice loss function), binary cross entropy was able to perform better in our data set.

5 Conclusion and Future Work

This paper focuses mainly on the effect of different loss functions and their robustness when dealing with noisy label data in automated building extraction task. The six loss functions discussed in this work can be used for dealing with image segmentation, and these functions have proven to be noise-robust given that all the loss functions perform similar to each other when noisy dense labels are introduced. However, it is important to consider the type of task when choosing any of the loss functions. This is evident by the fact that even though Dice loss generally considered the most popular loss function in image segmentation, binary cross entropy can still outperform Dice loss because of its strength in pixel classification. Additionally, combining two functions may also lead to better robustness as evident with the combination of Dice and mean absolute error work better than the Dice loss. On the other hand, data set plays a huge role in determining how well we can train the model in automated building extraction, since clean label and its corresponding observed label are very rarely available in pairs. This is significant for the error matrices used for I-FP-FN-EM model used in this study, and hence, it limits the model applications.

Acknowledgements This work is supported by Faculty Research Grant (CTRG-20-SEPS-14), North South University, Bashundhara, Dhaka 1229, Bangladesh.

References

1. Jadon, S.: A survey of loss functions for semantic segmentation. 2020 IEEE Conf Comput Intell Bioinforma Comput Biol CIBCB 2020 (2020). https://doi.org/10.1109/CIBCB48159.2020.9277638
2. Cordeiro, F.R., Carneiro, G.: A survey on deep learning with noisy labels: how to train your model when you cannot trust on the annotations? Proc—2020 33rd sibgrapi conf graph patterns images. SIBGRAPI **2020**, 9–16 (2020)
3. Furtado, P.: Testing segmentation popular loss and variations in three multiclass medical imaging problems. J Imaging **7**, 16 (2021)
4. Ahmed, N., Rahman, R.M., Adnan, M.S.G., Ahmed, B.: Dense prediction of label noise for learning building extraction from aerial drone imagery. Int J Remote Sens. https://doi.org/10.1080/01431161.2021.1973685
5. Ahmed, N., Rahman, R.M.: Label noise tolerance of deep semantic segmentation networks for extracting buildings in ultra-high-resolution aerial images of semi-built environments. Geocarto Int. https://doi.org/10.1080/10106049.2021.1992022
6. De, M.Y., Liu, Q., Qian, Z.B.: Automated image segmentation using improved PCNN model based on cross-entropy. In: 2004 Int. Symp. Intell. Multimedia, Video Speech Process. ISIMP 2004. pp. 743–746 (2004)
7. Xie, S., Tu, Z.: Holistically-nested edge detection. Int J Comput Vis **125**, 3–18 (2015)
8. Sudre, C.H., Li, W., Vercauteren, T., Ourselin, S., Cardoso, M.J.: Generalised dice overlap as a deep learning loss function for highly unbalanced segmentations. Lect Notes Comput Sci (including Subser Lect Notes Artif Intell Lect Notes Bioinformatics) 10553 LNCS, pp. 240–248 (2017)
9. Wang, G., Liu, X., Li, C., Xu, Z., Ruan, J., Zhu, H., Meng, T., Li, K., Huang, N., Zhang, S.: A noise-robust framework for automatic segmentation of COVID-19 pneumonia lesions from CT images. IEEE Trans. Med. Imaging **39**, 2653–2663 (2020)
10. Salehi, S.S.M., Erdogmus, D., Gholipour, A.: Tversky loss function for image segmentation using 3D fully convolutional deep networks. In: Lect. Notes Comput. Sci. (including Subser. Lect. Notes Artif. Intell. Lect. Notes Bioinformatics). Springer, pp. 379–387, (2017)
11. Abraham, N., Khan, N.M.: A novel focal tversky loss function with improved attention U-net for lesion segmentation. Proc—Int Symp Biomed Imaging 2019 April, pp. 683–687 (2018)
12. Taghanaki, S.A., Zheng, Y., Zhou, S.K., Georgescu, B., Sharma, P., Xu, D., Comaniciu, D., Hamarneh, G.: Combo loss: handling input and output imbalance in multi-organ segmentation. Comput. Med. Imaging Graph **75**, 24–33 (2018)

13. Fidon, L., Li, W., Garcia-Peraza-Herrera, L.C., Ekanayake, J., Kitchen, N., Ourselin, S., Vercauteren, T.: Generalised wasserstein dice score for imbalanced multi-class segmentation using holistic convolutional networks. Lect. Notes. Comput. Sci. (including Subser Lect Notes Artif Intell Lect Notes Bioinformatics) 10670 LNCS, pp. 64–76 (2017)
14. Lguensat, R., Sun, M., Fablet, R., Mason, E., Tandeo, P., Chen, G.: EddyNet: a deep neural network for pixel-wise classification of oceanic eddies. Int. Geosci. Remote Sens. Symp., 2018-July, pp. 1764–1767 (2017)

A Systematic Review on Convolutional Neural Networks for Sentinel-2 Image Classification

Anjali R. Sharma and Suraj T. Sawant

Abstract The progression of various scientific domains, owing to the enhancement of computer engineering and related technologies, has created a competitive niche. Inter-domain coalescence of certain scientific areas emboldens and envisages a technologically advanced morrow. Deep learning and remote sensing exemplify one such relationship which has bloomed recently. The advancing intervention of deep learning methods has gracefully accelerated the field of remote sensing by outperforming the conventional approaches, thus enhancing it inside out. This paper presents a systematic review of one of the most dug segments of remote sensing, that is, satellite image classification, using a few chosen convolutional neural network frameworks and related approaches. The dataset focused is multispectral Sentinel-2 images which is an open-access source that requires minimal preprocessing. This study explores different architectural convolutional neural network models and their performances for the classification that can assist in various applications.

Keywords Convolutional neural networks · Remote sensing · Sentinel-2 · Image classification

1 Introduction

Remote sensing (RS), generally, is associated with gauged interactions between earth surface materials and the electromagnetic energies, which detects, recognizes, and classifies the objects on the earth's surface with the help of satellites. It is a significant source of data used for mapping of land cover and land use (LCLU); agricultural and environmental studies, natural resource management, weather forecasts, global change research, mineral exploration, etc. The evolution of deep learning (DL) has helped the development of state-of-the-art solutions in many research domains. It holds a promising future to fulfill the needs of challenges coming ahead in the province of RS image processing. With ever-advancing technologies that leverage

A. R. Sharma (✉) · S. T. Sawant
Department of Computer Science and IT, College of Engineering Pune, Pune, India

© The Author(s), under exclusive license to Springer Nature Singapore Pte Ltd. 2022

A. K. Nagar et al. (eds.), *Intelligent Sustainable Systems*, Lecture Notes in Networks and Systems 333, https://doi.org/10.1007/978-981-16-6309-3_65

the computing power of modern Graphical Processing Units (GPU) and Tensor Processing Units (TPU) to function and reason like humans and extract features of the input images.

Managing RS data has never been less of a challenge. The ever-expanding compilation and collection of data need to be administered well, so that significant information can be extracted from it and can be utilized judiciously. An easily accessible, feasible, and up-to-the-minute satellite dataset by Copernicus Missions Satellites-called Sentinel, is a constellation of six pairs of satellites. Sentinel-2 equipped with multispectral instruments (MSI) having high resolution is used for looking at changes on the earth's surface. Sentinel-2 satellite MSI has 13 spectral bands to give varied perspectives of the earth's land and vegetation. In RS, image classification is defined as a computer-assisted interpretation and distinction of remotely sensed images to assimilate the information about data characteristics such as structure, reference, and scene. There are two methods of image classification. In supervised classification, the system extracts features based on statistical characteristics from the training dataset, whereas in unsupervised classification, the software is used to cluster commonly occurring land scene types, simultaneously providing analysis of those scene types for later references. Gradual developments in satellite image classification have drawn immense attention of researchers to improve the approaches and to come up with solutions providing better accuracy. Image classification can prove to be an efficient process for information extraction, as it is one of the vastly explored and used segments of RS. An exhaustive study of different classification methods is often required to understand various classifiers suitable for focused applications. The classification process is harnessed by controlling the settings of parameters precisely, like the number of layers or intensity of hidden layers. Thus, it is surmised that DL presents quintessential approaches for the classification of RS images.

The following sections are harmoniously outlined to grasp the approaches efficiently. In section two, the researches contributed in RS domain is discussed. The third section selectively describes and distinguishes various convolutional neural network (CNN) models used in the recent past. In section four, detailed analysis and comparisons are carried out. The results obtained are used to enlist the disparities in the researches done in the fifth section. In the sixth section, a proposed methodology and certain significant improvements are suggested accounting for the specified domain collaboration. Lastly, the conclusion is drawn through the review work.

2 Related Work

The intervention of CNN- and DL-based concept in RS has flourished because of many significant research works in the past. CNN models proved to be very efficient even in their simplest form showing visible improvements over ML models. The training of a CNN is initiated using a huge number of imprecise data which are then later refined on a small set of precisely labeled data. For other machine learning (ML) methods (e.g., support vector machine (SVM)), accuracy is poorer than the CNN-based approaches which show significant advantages like removal of discontinuities due to patchy borders, improved accuracy with lesser parameters, and reduced execution time due to convolution operations. Merging certain ML algorithms, like SVM, along with fine-tuning of CNN always showed better performance. A meta-analysis [1] concluded that DL methods have surpassed the trending ML method. The new concept of transfer learning of finely-tuned pretrained CNN models, always performed better in terms of accuracy and efficiency. An ensembling approach [2] for image classification proposes initial targeting of a single optimized CNN and then further adding CNN to serve different factors of the experiment like weights, land types, augmentation values, requirements, etc., forming a group of CNNs for optimization.

A trained deep-CNN [3] is presented that performs multilabel image classification of high spatial resolution RS images. The transfer learning is implemented by using PlanetScope trained models on the Sentinel-2 dataset. The test for the transferability of the proposed model results in higher accuracy. The transferring of the model requires a lot of quantitative assessments. The paper [4] proposes a novel pipeline pixel-wise classification method with a deep neural network for high spatial resolution images. Simply implemented for high-resolution urban RS resulting in improved accuracy, highest compared to other CNN models. Feasibly implemented and embedded into an existing system, [5] has suggested a lightweight and deeper CNN model with fewer parameters which successfully extracts information and classifiers a large amount of non-preprocessed dataset. Another lightweight CNN [6] elucidating deep attention-oriented multiple branch fusion of features and deep features are extracted by synchronizing multiple convolutions and joining them with various branches to reduce the parameters. In a meta-analysis [7], researches related to every domain of DL in the RS field were studied and reviewed. The summarized sub-field of DL in RS are image fusion, object detection, LCLU classification, segmentation, scene classification, image registration, and their accuracy assessment. Some of the DL methods are not suitable or optimum for all the RS applications; thus, further researches are required in these sub-areas. Most of the research articles cited in this paper have not explicitly mentioned details of the frameworks used in their articles, so the details are unavailable.

3 Convolutional Neural Network Models for Classification

CNN is competent enough to extract numerous features from voluminous datasets generated in the field of RS. The following study discusses some CNN models and their results to understand the current progress in RS image classification specifically for the Sentinel-2 datasets. An optimal CNN framework is selected based on the optimum performance factors like efficiency, accuracy, time complexity, etc. One of the recent researches [8] in image classification determines the accuracy of conventional ML algorithms like random forest (RF), SVM, and k-nearest neighbors (KNN). The experiment is performed on the Sentinel-2 dataset by automatically identifying several heterogeneous landforms classes. An ensemble CNN approach [9] is modeled to maximize efficiency, accuracy, and performance for the classification of clouds, shadows, and land cover types. The output from the ensemble model is the scores combined from all the CNN models used for the multilabel and multimodal classification. Another research [10], proposes the step-by-step integration of multiple individual CNNs along with SVM for classification. The architecture comprises a group of parallel CNNs processing different sets of spectral bands for feature extraction, and the output is fed to SVM for classification which resulted in the highest classification accuracy.

In paper [11], a multi-attention-driven strategy for multilabel classification to address a major issue regarding limited information content of images that can be retrieved by the satellite. At first, a k-branch CNN model adds spatial resolution and spectral contents of the scenes to each branch. The scores are assigned to the scenes, and global descriptors are established on each score. The results are presented in terms of both classification (Recall (averaging sample)) and ranking (label ranking average precision (LRAP)). A multimodal approach [12] of classification throws light on a less perceived problem of cross-modality learning in RS by proposing a set of well-defined modules. This approach leverages the self-adversarial, interactive learning and label propagation notions in a DL-based CNN model for pixel classification. The framework is based on three stream neural networks consisting of a pair of CNNs, fully connected layers, and other layers to classify multimodal images. The paper [13] incorporates CNN and random forest (CNN-RF) in RS, which is a method enhanced by a newly devised optimal feature selection method. The approach is executed using feature extraction, feature selection, and finally classification using CNN-RF. In research [14], focus was made on crop classification using pixel-based recurrent neural network and CNN for the classification of multitemporal Sentinel-2 images.

Table 1 gives brief information about the models studied above. It can be easily interpreted that CNN models are achieving results way better than typical ML models.

Table 1 Comparison of contemporary classifier models

Classifier models	Accuracy (%)	Classes/ labels	Dataset	Year
(Conventional ML methods) [8] RF	84.00			
SVM	82.70	5	Sentinel-2	2019
KNN	79.60			
Ensemble CNN (DenseNet20, ResNet50, VGG10) [9]	80.00	17	Sentinel-2, Planetscope	2019
2 CNNs + RF [10]	81.32			
3 CNNs + SVM	84.32	6	Sentinel-2	2019
4 CNNs + SVM	89.43			
K-branch CNN [11]	Recall-68.50	43	BigEarth Data	2020
	LRAP- 85.90		(Sentinel-2)	
X-ModalNet [12]	88.23	15	Sentinel-2 (MSI)	2020
CNN + RF [13]	94.27	4	Sentinel-2	2020
Pixel R-CNN [14]	90.02	15	Sentinel-2	2021

The accuracy percentage depicted in the table is compared considering that all the models were implemented on Sentinel-2 datasets, though they have different numbers of classes and areas of study.

4 Comparative Analysis

It has been not more than a decade that the CNN models have portrayed their efficiency. Since then researchers have come up with many CNN models and their variations performing appreciably well. The traditional ML-based methods presented satisfactory results and performance. But, it is observed that some classes showed lower accuracy compared to others. The implementation of DL techniques was suggested for performance improvement and achievement of quality results [8]. The ensemble CNN model [9] comparably performed better on high-resolution images. The added advantage of this model over conventional ML methods is that it does not require typical RS image preprocessing steps. The implementation of CNN was defined only for feature extraction in this approach, yet yielding better results surpassing older methods in terms of speed and efficacy. Availability of limited data was the only issue faced [10]. In the K-branch CNN model [11], notable improvements in multilabel image classification is seen compared to other contemporary attention-driven multilabel RS methods. The reduced parameters result in computationally efficient solutions and reduced complexity. The outstanding advantage of X-ModalNet model [12] is that the multiple modes of RS data are accessed and overlapped to increase the accuracy rate. This intervention of multiple nodes and DL-based methods provided dramatic results for image classification. The hybrid CNN-RF model [13] presents highly accurate crop and land cover classification with a lower execution time compared to other typical hybrid models. Pixel R-CNN model [14] is contributed by two DL algorithms—RNN and CNN, for land and crop classification. This model is capable to extract features automatically by discovering certain significant correlations among multiple images. The research asserts to have achieved a significant improvement in terms of accuracy compared to other non-DL mainstream classification methods.

Some of the widely available datasets for classification using DL methods are Eurosat (27,000 images), SAT-4 (500,000), SAT-6 (405,000), etc. Out of all the datasets, Eurosat [15] is highly used as it is obtained from Sentinel-2 images, and it is considered to be a benchmark dataset for CNN frameworks. The observation has led to the conclusion that the Sentinel-2 dataset and DL models are optimal pairing for superior results. The researchers who are interested to dive deep into the classification (LULC) of satellite imagery using DL approaches can make use of the DL models mentioned above. Throughout the review, it has been identified that the accuracy of classification using DL models is directly proportional to size of the dataset. This is in line with the research gaps mentioned in Sect. 5. The researchers who are interested in LULC classification may also think of the creation of a standard dataset.

5 Identification of Research Gaps

Coming across several papers and reviews, having varying goals and producing varying results, many issues are recognized during the study. One of the major shortcomings in this area of study is the limited availability of open-access datasets. As known, the quality aspect of CNN models depends upon a larger number of datasets. The more the data is available for training, the better and accurate the model performs. This insufficiency restrains the development of the RS domain. Further, with multiple datasets procurable to train the CNN models for image classification, the drawback observed was the availability of limited categories and labels for these datasets. Augmenting the data is one of the solutions, but it adds up to the cost in the process. It can be inferred that most of the researches focuses on deriving higher accuracy, whereas time efficiency is focused on only by a few. The transfer learning method is modeled to overcome this drawback. But, more such time-efficient approaches are needed to be discovered. Another limitation that cannot be ignored is the restricted accessibility and availability of high-end technologies. The researchers seek enhancements in the GPU technologies and other application software. These developments are needed to generate and deal with the larger amount of training and testing data from RS. Generally, while listing the obstacles faced during the research, the individual factors are contemplated. But, what most of the researchers stumble upon is the "imbalance issue" of the problems. A solution competently balancing all the issues like time efficiency, better accuracy, data availability and accessibility, refinement of data, multi-labeling, etc., is required. Most of the research works are motivated to find answers to individual problems, unable to provide solutions to this disproportion. Several such research gaps and hindrances were discovered in the review. But only a few critical difficulties have been specified that need to be addressed and resolved sooner.

6 Outline of Proposed Methodology

A proposed methodology is outlined consisting of a deeper study followed by the implementation of a DL model and improvement of this model is outlined in Fig. 1. A basic CNN model has at least one layer having convolution, multiplication layers, and a few other layers. A striking advantage of CNN model is that there is minimal or no need of preprocessing the input images. The process begins with feature extraction where images/ vectors are input to the CNN model, several trained filters having specific hyperparameters scan them and generate feature maps at convolutional layers. A typical convolutional layer consists of a kernel, that is, a matrix of learnable weights with a convolution function. Some activation (ReLU) and loss functions are applied to the feature maps, and the filters extract significant features from them. The feature maps generated at previous layers can be used as input for subsequent convolution operations, simultaneously generating more refined feature

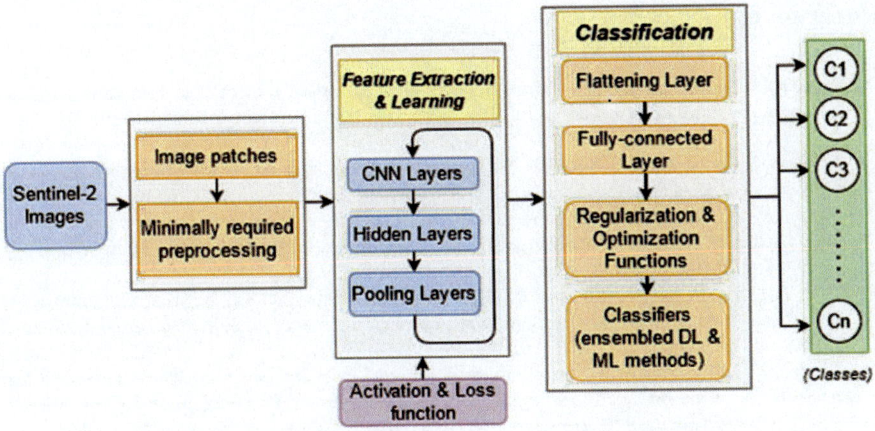

Fig. 1 Outline of proposed methodology

maps. The pooling layers downsize the space of these feature maps by deducing a random, average, or maximum value from the representative values in given window size without affecting the number of the generated feature maps. Thus, CNN models are trained to be robust, avoid problems like over-fitting, and reduce computational costs considerably. The feature maps obtained after feature extraction is then processed for classification. The flattening layer brings all the feature maps into a single vector passing it to fully connected layer without loosing any information. The fully connected layer linearly separates and maps the features, synchronizing with the output layer. A regularization method and optimization algorithms (Adam optimizer) can be added as required. Finally, by giving the resulting feature maps to the DL/ML based classifiers, the classification result is produced. The objective of these advances has always been to dive deeper into the aforementioned domain collaboration and discover its true potentiality.

7 Conclusion

The review highlights many unresolved issues aiming to draw readers' attention toward the budding phase of DL in RS. It is figured out that DL-based approaches like CNN in RS are studied by very few researchers, and only recently this field has shown a significant boom. Such extensive and systematic literature surveys and reviews help to reveal overviews of contemporary researches and advancements. Many types of research were performed under specified constraints, and they claimed their models to be better than other coexisting models. These models have evidently outperformed others, but even for the coherent comparisons, there are fewer competing researches. This clearly shows that more such works in the RS domain need to be encouraged.

References

1. Heydari, S.S., Mountrakis, G.: Meta-analysis of deep neural networks in remote sensing: a comparative study of mono-temporal classification to support vector machines. ISPRS J. Photogrammetry Remote Sens. **152**, 192–210 (2019)
2. Minetto, R., Segundo, M.P., Sarkar, S.: Hydra: an ensemble of convolutional neural networks for geospatial land classification. IEEE Trans. Geosci. Remote Sens. **579**, 6530–6541 (2019)
3. Pomente, A., Picchiani, M., Del Frate, F.: Sentinel-2 change detection based on deep features. In: IGARSS 2018-2018 IEEE International Geoscience and Remote Sensing Symposium, pp. 6859–6862. IEEE (2018)
4. Guo, R., Liu, J., Li, N., Shibin Liu, F., Chen, B.C., Duan, J., Li, X., Ma, C.: Pixel-wise classification method for high resolution remote sensing imagery using deep neural networks. ISPRS Int. J. Geo-Inf. **7**(3), 110 (2018)
5. Patowary, G., Agarwalla, M., Agarwal, S., Sarma, M.P.: A lightweight CNN architecture for land classification on satellite images. In: 2020 International Conference on Computational Performance Evaluation (ComPE), pp. 362–366. IEEE (2020)
6. Shi, C., Zhao, X., Wang, L.: A multi-branch feature fusion strategy based on an attention mechanism for remote sensing image scene classification. Remote Sens. **13**(10), 1950 (2021)
7. Ma, L., Liu, Y., Zhang, X., Ye, Y., Yin, G., Johnson, B.A.: Deep learning in remote sensing applications: a meta-analysis and review. ISPRS J. Photogrammetry Remote Sens. **152**, 166–177 (2019)
8. Baamonde, S., Cabana, M., Sillero, N., Penedo, M.G., Naveira, H., Novo, J.: Fully automatic multi-temporal land cover classification using Sentinel-2 image data. Procedia Comput. Sci. **159**, 650–657 (2019)
9. Shendryk, Y., Rist, Y., Ticehurst, C., Thorburn, P.: Deep learning for multi-modal classification of cloud, shadow and land cover scenes in PlanetScope and Sentinel-2 imagery. ISPRS J. Photogrammetry Remote Sens. **157**, 124–136 (2019)
10. Nijhawan, R., Joshi, D., Narang, N., Mittal, A., Mittal, A.: A futuristic deep learning framework approach for land use-land cover classification using remote sensing imagery. In: Advanced Computing and Communication Technologies, pp. 87-96. Springer, Singapore (2019)
11. Sumbul, G., Demir, B.: A deep multi-attention driven approach for multi-label remote sensing image classification. IEEE Access **8**, 95934–95946 (2020)
12. Hong, D., Yokoya, N., Xia, G.-S., Chanussot, J., Zhu, X.X.: X-ModalNet: a semi-supervised deep cross-modal network for classification of remote sensing data. ISPRS J. Photogrammetry Remote Sens. **167**, 12–23 (2020)
13. Yang, S., Gu, L., Li, X., Tao, J., Ren, R.: Crop classification method based on optimal feature selection and hybrid CNN-RF networks for multi-temporal remote sensing imagery. Remote Sensing **12**(19), 3119 (2020)
14. Mazzia, V., Khaliq, A., Chiaberge, M.: Improvement in land cover and crop classification based on temporal features learning from Sentinel-2 data using recurrent-convolutional neural network (R-CNN). App. Sci. **10**(1), 238 (2020)
15. Helber, P., Bischke, B., Dengel, A., Borth, D.: Eurosat: a novel dataset and deep learning benchmark for land use and land cover classification. IEEE J. Sel. Top. Appl. Earth Obs. Remote Sens. **12**(7), 2217–2226 (2019)

Categorical Analysis of Code Smell Detection Using Machine Learning Algorithms

Ankita Bansal, Utkarsh Jayant, and Abha Jain

Abstract In software development, a code smell is some surface indication that possibly corresponds to a deeper problem in the programme. Whether a piece of code has a code smell or not is very subjective and it varies mostly by the language and developer, but also the developmental technologies. The term code smell was popularised by Kent in the late 1990s. Since then several machine learning algorithms have been employed in order to detect code smells. Code smells have been categorised into several categories. With the lack of studies done in respect to smells and their relation to their respective classes, it is imperative to look into it for a deeper understanding of smells. In this paper, we have tried to analyse and study the detection accuracies of these smells categorically for unambiguous understanding of how smells work for efficient refactoring.

Keywords Code smells · Bloaters · Change preventers · Couplers · Dispensables · Object-oriented tool abusers · Code smell detection · Machine learning · Support vector machines · J-48 algorithm · Random forest · Naive Bayes · Logistic regression · God Class · Feature envy · Data class · Switch statements · Long method · Long parameters list

1 Introduction

Most of the code is a mess. Most of the new requirements change the existing code. The unfortunate part is that much of our work involves altering imperfect code. However, the fortunate part is that every big mess consists of many small ones. Certain small problems occur so frequently that they have been given names and are

A. Bansal (✉)
Netaji Subhas University of Technology, New Delhi, India
e-mail: lncs@springer.com

U. Jayant
Netaji Subhas Institute of Technology, New Delhi, India

A. Jain
Deen Dayal Upadhyaya College, University of Delhi, Delhi, India

© The Author(s), under exclusive license to Springer Nature Singapore Pte Ltd. 2022

A. K. Nagar et al. (eds.), *Intelligent Sustainable Systems*, Lecture Notes in Networks and Systems 333, https://doi.org/10.1007/978-981-16-6309-3_66

collectively known as 'Code Smells'. In software development, a code smell is some surface indication that possibly corresponds to a deeper problem in a programme. Whether a piece of code has a code smell or not, is very subjective and it varies mostly not only by the language and developer, but also varies due to the developmental technologies. Kent [1] had popularised the term code smell in the late 1990s and the usage of the said term drastically increased after it got featured in the book Refactoring: *Improving the Design of Existing Code* in 1999 by Fowler [2]. One major misconception that people have is that they think of code smells as bugs, they are not bugs of the programme. Code smells do not prevent the programme from functioning and are not technically incorrect; although code smells do indicate some weakness in design that may, in future, result in the slowing down of development or increase the risk of failure in the future. Code smells could lead to risk of failure as well as slowing down development of the programme of source code in the future.

Code refactoring is a process in which we change the internal structure of our source code whilst maintaining the external behaviour or functionality of the programme. Refactoring is not fixing bugs or re-writing code or improving the observable interface of the code. It majorly involves restructuring the existing code in small fragments which have a strong cumulative effect. Refactoring is done on the source code to get rid of code smells. Some of the major types of code smells used in study are data class, long parameters list, feature envy, god class, long method, switch statements, etc. [3]. These smells are classified into five categories namely bloaters, change preventers, couplers, dispensable and object-oriented tool abusers [4]. Each smell and category has its own refactoring technique.

Usually, the term 'bad' is associated with code smells, but really, if we look at the subtle points in the definition, we can infer that the code smells are easy to spot and their presence *might* indicate a problem. It's only when an issue is diagnosed that we can help find a cure to resolve the issue. Although code smells are mostly subjective in determining whether some code belongs to smelly or non-smelly instances, machine learning algorithms have shown to have high accuracies in code smell detection [5–7]. Each smell and its respective category has their own refactoring method and techniques. Once we determine the presence of smells in a given piece of code, we can begin to refactor the code. In order to perform refactoring effectively, it is imperative for us to first detect the smells present in the source code of the programme effectively and then begin working towards refactoring them.

This study is aimed with two objectives: First of all, although the detection of whether a source code contains smelly or non-smelly instances is partly subjective, we deduce the accuracies of several machine learning algorithms in detecting some code smells. Secondly, code smells and their respective categories have their own refactoring techniques and it is imperative to prioritise the refactoring process so as to not waste time and resources on futile smells(which have minimal effect on the source code) in order to carry out the task efficiently. In this study, we have selected the code smells which could have the maximum impact on the functionality in the long run and we try to find out which machine learning algorithms best determine the presence of smelly and non-smelly instances and further we try to determine which smells and their respective categories have the highest accuracies when it comes to detection.

The paper structure is as follows: Sect. 2 talks about the various code smell categories and the smells they contain. Section 3 discuses the approach we have taken to carry out our study. In the next section, we discuss the results and observations. Finally, the work is concluded in Sect. 5.

2 Basic Terminologies

"Once you give a complex idea a name, if we could all learn what that name stood for; it means that we can talk to each other in an unambiguous way without having miscommunications"-Sandi Metz.

Common code smells are grouped into five categories namely bloaters, change preventers, object-oriented tool abusers, dispensables and couplers. Each of these categories has been explained in this section.

2.1 Bloaters

Bloaters are usually pieces of code that have grown to such humongous sizes that it becomes difficult to work with them and maintain in the long run. Bloaters are not visible from the very beginning, they develop over time with the development of the software. It takes conscious effort to make note of them and remove them and not ignore them forever. Some examples of bloaters are data clumps, large class, long parameters list, long methods, primitive obsession, etc.

Long Method. If a method contains too many lines of code it belongs to this smelly class where the methods are responsible for multiple tasks and are disproportionately larger. It is always recommended to have short methods as opposed to having large methods as it is believed that the long methods can contain a duplicate code and if we divide this long method into smaller short methods this duplication can be avoided so that the short methods share logic among each other. The benefit of having short methods is that it helps make the understanding of the code easy and the extension and maintenance becomes easy across the software lifecycle as well.

Long Parameter Lists. These smells surface when a method has more than four parameters passed to it. Ideally, a method should not have more than four parameters passed on to it. This smell may be a result of a method carrying out a lot of processing. With the growth of parameter lists the maintenance as well as the understanding becomes difficult. This leads to painful usage.

2.2 Tool Abusers or Object-Oriented Abusers

Tool abusers are the smells which are inappropriate, incomplete or at times incorrect applications of principles of object-oriented programming which is a result of the developer unnecessarily applying his knowledge to places where it could be uncertain or risky.

Switch Statements. These smells surface when the next step is carried out by a complex set of conditionals such as switch or if. Instead of using beautiful concepts of object-oriented programming like polymorphism if procedural thinking is applied, it leads to this reaction. If the same conditional statements are duplicated across the system, when there is change in one switch statement in order to maintain the consistency others need to be changed as well.

2.3 Change Preventers

Change preventers are the pieces of code which prevent the developers from making further changes to the given source code. It basically says that if there is a need to make changes to one part of the code then it would lead to the developer making massive changes to other parts of the code making it highly complicated and expensive. Some examples of change preventer smells are divergent change, parallel inheritance hierarchies, shotgun surgery, God Class, etc.

God Class. When an object knows too much or in some cases does too much that part of smell is known as God Class or God Object. This increases the challenges in the maintainability of code and the tight coupling.

2.4 Dispensables

Dispensables are usually the smells which are futile to the functionality of the code. Once these dispensable pieces are removed the code becomes straightforward, efficient and cleaner which also ultimately enhances the readability of the code. Some common examples of dispensables are comments, data class, duplicate code, lazy class, speculative generality, etc.

Data Class. These smells are the classes that contain data fields and corresponding settings and getting methods to modify or read that data. They are also called crude methods and are used as containers of data which are required by other classes. These classes are extensively manipulated by other classes.

2.5 Couplers

Coupler smells suggest the excessive attraction between two classes. These smells may occur from one class to the other when there happens to be a disproportionate delegation. Some common coupler smells are feature envy, inappropriate intimacy, incomplete library class, middle man, message chains, etc.

Feature Envy. The functionality works on the data and the data exist together in the same object in object-oriented programming. If an object makes a lot of calls to other classes in order to achieve any outcome or to even obtain data, it becomes a case of feature envy.

3 Methodology

To carry out this study, we have taken six smells which have shown to have one of the most common occurrences and impact on the code in the long run. These smells are namely god class, data class, feature envy, long method, switch statements and long parameters list. The dataset was collected from QualitusCorporus Software systems http://essere.disco.unimib.it/reverse/MLCSD.html [8, 9].

The study was carried out in multiple steps. Firstly, correlation-based feature selection was applied to each dataset. Each dataset consisted of 420 instances. For each machine learning algorithm, the train test split was taken to be 70:30. Here, in this study, we have employed 5 of the most widely used machine learning algorithms for the detection of code smells namely random forest, support vector machines, J-48, logistic regression and Naive Bayes. The performance of these algorithms have been measured using the following performance metrics: precision, recall, area under the curve and accuracy.

For now, we have the results for six code smells, namely god class, feature envy, long method, long parameters list, switch statements and data class.

These aforementioned machine learning algorithms were trained for each smell, one smell at a time.

The detection process that was followed is as below:

1. Training dataset is Clj where j values from 1 until n and the set of classes that were obtained from the object-oriented systems is Clj.
2. For all the values of j, Cj is marked whether it represents a smelly instance or not.
3. The decision dataset (DDS) represents the set of classes that were extracted from object-oriented systems from which we aim to detect the smelly instance.

In order to detect the code smell from the decision dataset, the steps below were applied:

1. Computation of metric: The algorithm takes the training dataset as an input. Each class of the training dataset has object-oriented metrics being computed.

2. Classifier's training: Using the first step and training dataset, the algorithm is trained.
3. In the final step, the testing of the decision dataset takes place and the detection of the existence of the code smell was done with the help of the Weka tool [10].

Once all the algorithms were applied to each smell, we determined which machine learning algorithm suited best for each smell and subsequently the category it belongs. Once the best algorithm was selected, it was compared against the smells to determine the order of accuracies shown by each class to deduce the results [11, 12].

4 Result Analysis

Tables 4.1, 4.2, 4.3, 4.4, 4.5 and 4.6 show the performance of random forest, support vector machines, J-48, logistic regression and naive Bayes for detection of code smells experimented on our dataset.

The detection of smells by the aforementioned machine learning algorithms was compared using classification metrics: accuracy, precision, recall, area under the curve and f-measure. Random Forest algorithm showed to have the highest accuracies in detecting the smelly and the non-smelly instances followed by J-48 algorithm, Support Vector Machines, Logistic Regression and Naive Bayes algorithm respectively. However, in Table 4.5, we can see that support vector machines showed the highest accuracies followed by the order as described above. Feature Envy (Table

Table 4.1 Data class results

Machine learning algorithms	Accuracy	Recall	Precision	Area under the curve	F-measure
Random forest	97.62	94.74	97.3	0.999	0.976
Naive Bayes	72.22	94.74	52.17	0.928	0.733
Support vector	92.86	89.47	87.18	0.919	0.929
J48	96.03	97.37	90.24	0.981	0.961
Logistic regression	84.13	60.53	82.14	0.943	0.839

Table 4.2 Feature envy results

Machine learning algorithms	Accuracy	Recall	Precision	Area under the curve	F-measure
Random forest	95.24	91.89	91.89	0.980	0.952
Naive Bayes	84.92	67.57	78.13	0.899	0.846
Support vector	92.06	78.38	93.55	0.881	0.918
J48	96.03	91.89	94.44	0.947	0.960
Logistic regression	88.1	75.68	82.35	0.945	0.879

Table 4.3 God class results

Machine learning algorithm	Accuracy	Recall	Precision	Area under the curve	F-measure
Random forest	96.83	97.44	92.68	0.995	0.968
Naive Bayes	91.27	92.31	81.82	0.978	0.914
Support vector	93.65	87.18	91.89	0.919	0.936
J48	96.03	94.87	92.5	0.957	0.960
Logistic regression	93.65	87.18	91.89	0.984	0.936

Table 4.4 Long method results

Machine learning algorithms	Accuracy	Recall	Precision	Area under the curve	F-measure
Random forest	99.21	100	97.44	1.000	0.992
Naive Bayes	92.86	97.37	82.22	0.936	0.930
Support vector	97.62	97.37	94.87	0.975	0.976
J48	96.83	97.37	92.5	0.970	0.968
Logistic regression	93.65	86.84	91.67	0.947	0.948

Table 4.5 Long parameters list results

Machine learning algorithms	Accuracy	Recall	Precision	Area Under the curve	F-measure
Random forest	92.86	86.11	88.57	0.987	0.928
Naive Bayes	78.57	77.78	59.57	0.866	0.793
Support vector	96.03	97.22	89.74	0.964	0.961
J48	90.48	86.11	81.58	0.915	0.906
Logistic regression	88.89	91.67	75	0.952	0.892

Table 4.6 Switch statements results

Machine learning algorithms	Accuracy	Recall	Precision	Area under the curve	F-measure
Random forest	88.89	68.42	92.86	0.929	0.883
Naive Bayes	81.75	78.95	66.67	0.893	0.821
Support vector	79.37	52.63	71.43	0.718	0.784
J48	86.51	86.84	73.33	0.866	0.868
Logistic regression	84.92	52.63	95.24	0.844	0.834

Table 4.7 Combined analysis of random forest algorithm's results

Code smell	Accuracy	Recall	Precision	Area under the curve	F-measure
Data class	97.62	94.74	97.3	0.999	0.976
Feature envy	95.24	91.89	91.89	0.980	0.952
God class	96.83	97.44	92.68	0.995	0.968
Long method	99.21	100	97.44	1.000	0.992
Long parameters list	92.86	86.11	88.57	0.987	0.928
Switch statements	88.89	68.42	92.86	0.929	0.883

4.2) had better detection for smelly instances by J-48 algorithm. Furthermore, in Table 4.7, we can observe that Support Vector Machines showed the least accuracy for detection of smelly instances. Tree-based classifiers i.e. Random Forest and J-48 algorithm achieved the best performance among the pool of algorithms applied in this paper.

4.1 Combined Analysis

Among the several machine learning algorithms that we applied, random forest algorithm has shown to have the best detection of the smelly instances across the various smells taken in our study. Table 4.7 shows the combined analysis of random forest results applied through various smells. From Table 4.7, we can see that the detection of long method was carried out most effectively followed by data class, god class, feature envy, long parameters list and switch statements in decreasing order of smell detection efficiencies.

However, it should be noted that the combined analysis of other algorithms must also be taken into consideration to get a more holistic idea of detection efficiencies. The order in which these smells appeared according to decreasing accuracies in detection from the examination of other algorithms were long method, god class, data class, feature envy, long parameters list and switch statements.

5 Conclusion

Different code smells have their respective different effect on the source code and it is suggestible that once the code smells are detected that we get rid of those smells and categories first which have the most adverse effect on the programme while it is futile to work extensively work on getting rid of those smells which have minimal effect, even in the long run.

In this work, we have presented the use of machine learning algorithms in the detection of code smells. Most of the algorithms taken for this study have shown high accuracy in the detection (>90%). However, in future, it studies the imbalanced data vis-à-vis the lower prevalence of smells needs to be addressed. Feature selection helped a lot in improving the performance of the machine learning algorithms for each code smell respectively.

The overall comparative accuracies of the applied algorithms were random forest, J-48 algorithms, support vector machines, logistic regression and naive Bayes respectively in decreasing order. Now, according to their precedence among the smells, categorically the order of smells which they detect are dispensables, change preventers, bloaters, couplers and object-oriented tool abusers.

In the long run, this study can act as a baseline reference for code smell detection, and ultimately code refactoring categorically. The current limitations of the research are the lesser availability of a variety of other smells dataset and ambiguity in the presence of code smell instances in a source code. The practical advantages are that once we start detecting smells categorically and refactor them accordingly it drastically increases performance of refactoring process as the group of smells with the most accurate detection and similar refactoring styles can be taken care of at a time making the whole refactoring process more structured, hassle free and quick.

While in reference studies, smells were detected, in our study we have categorically studied the detection accuracies of these machine learning algorithms to expand our insights into the smells and how they impact the code categorically. As a result, we believe that machine learning models can be effectively used to detect code smells and categorically refactoring presents a better scope towards cleaner codes in the future.

In future studies, more smells, along with algorithms, can be added in order to widen the spectrum for analysis to further aid in refactoring the prevalent code smells.

References

1. Fowler, M.: 9 Feb 2016 https://martinfowler.com/bliki/CodeSmell.html. Last accessed 5/6/2021 (2021)
2. Fowler, M.: Refactoring: Improving the Design of Existing Code. Addison-Wesley, Boston, MA, USA (1999)
3. Azeem, M.I., Palomba, F., Shi, L., Wang, Q.: Machine learning techniques for code smell detection: a systematic literature review and meta-analysis. Inf. Softw. Technol. (2019). https://doi.org/10.1016/j.infsof.2018.12.009
4. Refactoring Guru, Code Smells. https://refactoring.guru/refactoring/smells. Last accessed 6/62021 (2021)
5. Di Nucci, D., Palomba, F., Tamburri, D., Serebrenik, A., Lucia, A.: Detecting Code Smells using Machine Learning Techniques: Are We There Yet? https://doi.org/10.1109/SANER.2018.8330266 (2018)
6. Arcelli Fontana, F., Mäntylä, M.V., Zanoni, M., et al.: Comparing and experimenting machine learning techniques for code smell detection. EmpirSoftware Eng. **21**, 1143–1191 (2016). https://doi.org/10.1007/s10664-015-9378-4
7. Kaur, A., Jain, S., Goel, S.: A support vector machine based approach for code smell detection. In: 2017 International Conference on Machine Learning and Data Science (MLDS), pp. 9–14. https://doi.org/10.1109/MLDS.2017.8 (2017)

8. Tempero, E., Anslow, C., Dietrich, J., Han, T., Li, J., Lumpe, M., Melton, H., Noble, J.: The qualitas corpus: a curated collection of java code for empirical studies. In: 2010 17th Asia Pacific Software Engineering Conference (APSEC), IEEE, pp. 336–345 (2010)
9. Palomba, F., Di Nucci, D., Tufano, M., Bavota, G., Oliveto, R., Poshy-vanyk, D., De Lucia, A.: Landfill: an open dataset of code smells with public evaluation. In: 2015 IEEE/ACM 12th Working Conference on Mining Software Repositories (MSR), IEEE, pp. 482–485 (2015)
10. Azadi, U., Fontana, F.A., Zanoni, M.: Machine learning based code smell detection through WekaNose. In: Proceedings of the 40th International Conference on Software Engineering: Companion Proceedings (ICSE '18). ACM, New York, NY, USA, 288–289 (2018). https://doi.org/10.1145/3183440.3194974
11. Fontana, F.A., Mäntylä, M.V., Zanoni, M.: Comparing and experimenting machine learning techniques for code smell detection. Empirical Softw. Eng. 1–49 (2015, June). https://doi.org/10.1007/s10664-015-9378-4
12. Fontana, F.A., Zanoni, M.: Code smell severity classification using machine learning techniques. Knowl. Based Syst. **128** (2017). https://doi.org/10.1016/j.knosys.2017.04.014

Foreign Accent Recognition Using a Combination of Native and Non-native Speech

Radha Krishna Guntur, Krishnan Ramakrishnan, and Vinay Kumar Mittal

Abstract A system for identifying speakers with geographically neighbouring native languages from a short utterance of their foreign accented speech is particularly challenging. The present research is aimed at studying the Mother Tongue Influence (MTI) of a speaker on a second language. It deals with automatically classifying between three south Indian English accents. An efficient method of classifying the speakers according to the regional accent has been verified. Identifying the regional accent of a speaker from non-native English speech has been carried out and a reference classification performance has been provided. The language partitioning experiments proved that the class of speakers in both native, non-native speech sets, and also their combination showed clear separation. Foreign accent recognition and classification tests yielded accuracies of nearly 94%, and 89%, respectively.

Keywords Regional accent · No-native speech · Cepstral features

1 Introduction

Accent is referred to as a distinct pattern of pronunciation of speakers belonging to a geographical location or a particular nation. Problem of inferring the native language of a speaker from his or her foreign accented speech is an accent classification task [1]. Most of the native listeners possess the knowledge pertaining to regional accent of the speaker in addition to the geographical location within a particular region [2]. Identification of speakers, classification of their dialectal zones is important in a multilingual country like India. However, there is very little research on the question of accuracy with which accent features can be used to identify a speaker's regional, eth-

R. K. Guntur (✉)
Department of ECE, VNRVJIET, Hyderabad, India

K. Ramakrishnan
Centre for CYS, Amritha University, Coimbatore, India

V. K. Mittal
Department of ECE, KLEF, Vijayawada, India

© The Author(s), under exclusive license to Springer Nature Singapore Pte Ltd. 2022
A. K. Nagar et al. (eds.), *Intelligent Sustainable Systems*, Lecture Notes in Networks and Systems 333, https://doi.org/10.1007/978-981-16-6309-3_67

nic origin [4]. A novel framework leveraging native language characteristics of South Indian speakers that gets reflected in their non-native English accents is explored in the present study related to Native Language Identification (NLI). The aim is to train the model in one language and testing is made in another. Basic structure of a typical NLI system includes the use of GMM, and SVM classifiers used as an LID system, which can produce good recognition scores [3].

Analysis and classification of speech collected from particular groups of second language learners is mainly dealt by the speech processing application of Native Language Identification (NLI) [5, 6]. The model training can be in native language and testing is made in non-native language [7, 8]. The activities involved in experimentally developing a *foreign accent recognition system* based on regional accent are described in the subsequent sections. In the training phase, acoustic model parameters were estimated from the training data in each language. Samples of English speech specifically collected from the native South Indian speakers of the languages *Telugu* (TEL), *Tamil* (TAM), and *Kannada* (KAN) are tested for recognition.

Validation of the proposed i-vector based system utilizing combined native language (L_1) and second language (L_2) data to train models for NLI task is the main objective of the present work. In the present study, the regional origin is considered as generalization of speaker identity, and the English speaker as a special sub-type of language. Three native language models were developed using cepstral based systems utilizing MFCCs for achieving the desired task. Speaker and accent partitioning have been clearly demonstrated for L_1 data, L_2 data, and also for their combination.

The organization of this paper is as follows. The collection of the database is described in Sect. 2. Description of an Automated Recognition system is given in Sect. 3. Experiments related to validation of partitioning of accented speech are presented in Sect. 4. Detection of accent classification using English speech is presented in Sect. 5. Discussion of the results obtained is given in Sect. 6, and conclusions drawn, and future scope are discussed in Sect. 7.

2 Database Collection and Recording Protocols

The proposed work is to examine data from people speaking Indian English. *Foreign accent*, which manifests itself as a specific articulatory behaviour is considered in the multilingual Indian scenario for the present study. Speech data has been collected from South Indian speakers speaking in native language, and also when speaking in English influenced by the corresponding regional language. Identifying the person is the goal, rather than what is being spoken. An application specific corpus has been developed. Commonly specified options in speech analysis task, which are addressed in building up of the speech corpus are: (a) Type of speech (b) Number of speakers (c) Types of speakers (d) Speaking environment (d) Transmission system (e) Type and amount of system training (f) Spoken input format.

Template of the data collection process is shown in Table 1. The speakers are divided into two separate groups: training set and testing set, which consisted of the

Table 1 Template used during data collection showing the file name and related demographic details

Native language	Name	Age/sex	File name

Table 2 Summary of data used for the NLI task of regional accent identification: (a) attributes (b) values for *training set* and (c) values for *testing set*

(a) Speaker attributes	(b) Native speech set	(c) Non-native speech set
Total number of speakers	60	75
Speakers per native language group. (Kannada, Tamil, Telugu)	20	25
Speech duration per speaker	300 s	120 s

native and non-native speech corpus, respectively. Native speech samples of twenty speakers from each of the language groups mentioned, namely: KAN, TAM, and TEL, with each speaker speaking for 300 s, formed the training set (Native speech). A total of 3600 s of native speech has been collected for each language group. A totally different subset of speakers belonging to each of the three language groups, who have the proficiency to speak English apart from their mother tongue, is selected as the testing set (Non-native speech). English utterances of speakers belonging to three native language groups formed the testing set. The test samples, each for a duration of sixty seconds have been recorded from speakers numbering 25 from each of the three native language groups. These speakers are capable of reading English. The TIMIT corpus, consisting of Standard English language speech samples was also utilized for the validation studies of speaker and accent space partitioning [9]. The details are given in Table 2.

3 Automated Recognition of Foreign Accent

Research in the field of *foreign accent recognition* is being undertaken mostly as a sub task of *Accent and Dialect Identification* [10]. Automatic detection of Foreign Accent identification has assumed importance in Speech Technology. Modeling approaches were inspired by two extensively studied related tasks namely Automatic Speaker Verification/Identification [11]. Also, nativeness is considered as a generalization of Speaker Identification and similar binary classification techniques can be applied. Non-native English speaker is seen as a special subtype of language. Here, features and modelling techniques used in LID like MLLR, prosody and word N-gram systems were employed. Model combination yielded EER of 8.6%. Similar to Speaker Recognition, Language Identification approaches are categorized by the level of linguistic information which is applied [12]. Practical approaches are based on machine learning techniques.

For native language identification in text-independent mode, usual practice is to create models representing the sequential statistics of more basic units in each of the languages. Approaches that model the static distribution of acoustic feature vectors and also utilizing the patterns of change of these vectors over substantially long periods are generally used. Non-native speakers, while speaking a learned language, substitute phonemes from their first language instead of learning new phonemes. It is possible to make a comparison of the acoustic properties that may differ between native and non-native speakers of a language. The GMM representation of feature distributions in a speech based system is dependant on the distribution of acoustic features. The component densities of these distributions model some hidden classes underlying the native language. Development of algorithms and techniques for computers, to obtain information from data automatically by computational and statistical methods, is a pattern learning problem.

Statistical models like SVM and GMM classifiers are trained on data labelled with information derived from the corresponding native languages. The possible features that cab be derived from learners speech for the any NLI task include the frame level MFCCs, or and lexical features like errors related to language use. The later needs speech recognizer [13]. The signal processing front-end is meant for converting the speech waveform to desired parameters for further analysis. The Gaussian Mixture Models (GMMs), which are generally used for representation of probability distribution of features has been developed, and particularly the i-vector approach, considered to be the state-of-the-art has also been employed.

Support Vector Machine (SVM) classifiers are known to be a very good fit for NLI [14], and is one of the effective supervised classifiers. SVM searches for a maximum-margin hyper plane for data separation [12, 15]. Accuracies of foreign accent classification were investigated by using the SVM classifier. The speech signal is first processed to extract attributes relevant to the foreign accent [16]. The i-vector representation, which is successfully applied to LID, can similarly be employed in accent recognition, where differences between classes are much finer [17]. The i-vector method [18] that is extensively used in accent, speaker, and language recognition, helps in getting a low dimensional representation which can preserve speaker specific information.

4 Experiments Validating Partitioning of Language Related Accent

4.1 Extraction of Cepstral Features

Cepstral features are closely related to the linguistic content of speech. The shaping of articulatory behaviour by an individual's mother tongue into distinct patterns that are audible in all the second languages that the speaker learns after critical period constitutes one aspect of foreign accent. The shape of the vocal tract is manifested in the envelope of the short-time power spectrum.

Information regarding the phonetic content of the utterance is captured by the MFCC and LPCC features [5] The particular configuration of the vocal tract corresponding to the phoneme that is being uttered is obtained from the MFCC parametrization. The LPCCs which are known to capture extra information from speech that discriminates different languages and the PLPs taking advantage of psychoacoustic principles and robust against noise have also been studied, but found less efficient. Most speech processing applications utilize certain properties of speech signals in accomplishing their tasks. These properties of speech and their extraction from a speech signal S(n) which provides effective representation of foreign accent identification is discussed below.

4.2 Language Related Accent Partitioning Using Combined Native and Non-native Speech

Language-specific differences in phonological development might be related to differences in phoneme and phoneme sequence frequency across languages [19, 20]. To validate the assumption that the language accent of different native languages is separable, experiments using a combined set of native, and non-native speech data were conducted. The distribution of spectral features representing the vocal tract features, which are influenced by the linguistic content, may be unique for speakers of a language group. The spectral features like MFCCs which encodes information regarding vocal tract size and shape, are not only useful for speaker, and language modelling, but also for accent group modelling. Data corresponding to each linguistic class contained the concatenation of MFCC features corresponding to 10 native speech samples of different speakers and 10 non-native speech samples from exclusive set of speakers belonging to the same class. Results of SVM classification have shown an accuracy 86.1%.

4.3 Non-native Accent Partitioning on English Speech Set

The non-native English speech data has also been tested using an SVM classifier. The Text-Independent(TI) database described in Table 2, consisting of non-native English utterances have been used for these experiments. The tools available in MATLAB for Classification were used for these studies. The comparative analysis of the MFCC distribution is presented in the form of scatter plots as shown in Figs. 1, and 2. Non-native accent classification accuracy of 89.2% is noted using SVM classifier during these studies. From these results, it is evident that the performance of a typical classifier using text-independent data can result in good Native Language Identification accuracies.

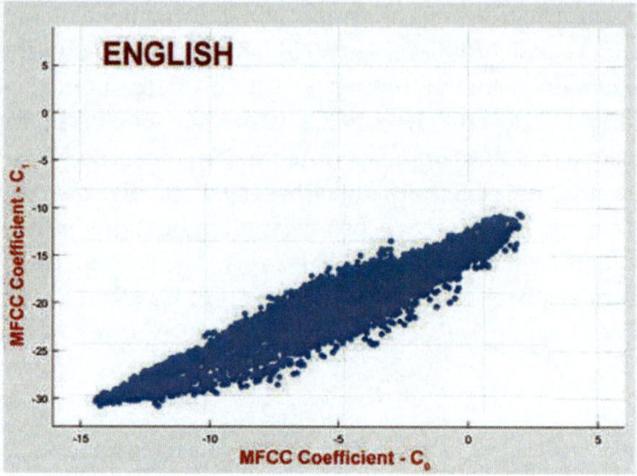

Fig. 1 Distribution of MFCC coefficients for native English speakers when speaking English

5 Detection of South Indian English Speakers Using Native Language Identification Approaches

A generic method is proposed for automatic classification of speakers based upon their foreign accent. Text-independent data shown in Table 2 containing native speech, and non-native speech has been utilized for the purpose of training and testing, respectively. The approach followed is to train language class specific GMMs using all the data for the native language class. The likelihood of the input feature stream is computed in the recognition stage for each of the native language models, and the best scoring model is selected. Native Language Identification task has been implemented by testing the unknown speaker with one of the developed models for determining the mother tongue. English utterances were collected from speakers belonging to the three languages Kannada (KAN), Tamil (TAM) and Telugu (TEL) for testing. Series of experiments using plain GMM technique, and GMM-UBM system were carried out. The i-vector based classifier produced the highest classification accuracy of 93.94%. Table 3 shows a consolidated list of all the classification and detection experiments.

6 Discussion of Results

A framework which can handle any set of L1 and L2 systems and detection of native language based on individual non-native speech has been addressed. The performance of the proposed system for foreign accent recognition system is on par with the best performing systems reported for NLI. It improves prediction performance by (a) reducing the data needed for learning (b) improving the predictive accuracy of algorithms (c) increasing the comprehension of constructed models (d) provides

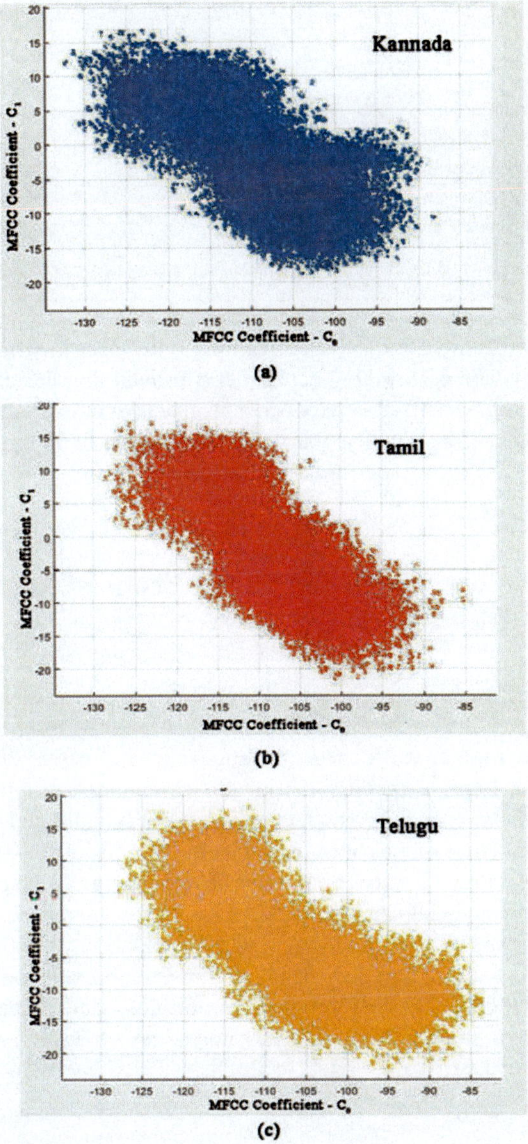

Fig. 2 Distribution of MFCC coefficients for Non-native speakers of South Indian languages **a** Kannada **b** Tamil, and **c** Telugu when speaking English

Table 3 Recognition and classification accuracies of native and non-native English speech by South Indian speakers employing different classification techniques

Type of data employed	Classifier used	Recognition accuracy
Native speech	$GMM-UBM$	93.3%
Combined native and non-native speech	SVM	86.1%
Non-native speech	SVM	89%
Native speech for training, and Non-native speech for testing	$i-vector$	93.9%

opportunity for data visualization (e) Reduces the training and utilization times. The language partitioning experiments proved that the training classes have clear separation. For different experiments on accent partitioning and classification accuracies above 85%. Finally, the validation testing of the proposed system employing the English speech has been successful.

7 Conclusions and Future Scope of Regional Accent Identification

A novel method is proposed for identifying the mother tongue of speakers belonging to regionally close language groups. Identification of speakers of Kannada, Tamil and Telugu, while speaking English, a non-native language is experimentally verified. Using a specifically developed database, employing cepstral features and the i-Vector classifier, a novel method has been validated for Native Language Identification. This task has been implemented by considering regional origin as generalization of speaker identity and the English speaker as a special sub-type of language. Future investigations can be planned for a variety of speech styles by varying the composition and characteristics of speaker population, and also including multi-disciplinary information. Not only negative transfers, but also positive transfers are informative for identifying the foreign accent. Imminent need exists for inclusions of more number of languages, particularly languages of Eastern and central India. Use of pitch contours to train models of English speech involving Indians await future investigations.

References

1. Jiao, Y, Tu, M., Berisha, V., Liss, J.: Accent identification by combining deep neural networks and recurrent neural networks trained on long and short term features. In: Interspeech, vol. 2016, pp. 2388–2392. San Fransico (2016)

2. Hanani, A., Russel, M.J., Carey, M.J.: Human and computer recognition of regional accents and ethnic groups from British English speech. Comput. Speech Lang. **47**, 59–74 (2013)
3. Radha Krishna, G., Krishnan , R., Mittal, V.K.: Native Language Identification from South Indian English Speech. In: Workshop on Machine Learning in Speech and Language Processing, September 7th, 2018, Organised by the SIG on Machine Learning in Speech and Language Processing of ISCA Along with INTERSPEECH 2018, Hyderabad (2018)
4. Harper, M.P., Maxwell, M.: Spoken Language Characterization. Springer Handbook on Speech Processing and Speech Communication, pp. 1–15 (2006)
5. Nisioi, S.: Feature analysis for native language identification. In: 16th International Conference on Intelligent Text Processing and Computational Linguistics, pp. 1–15 (2015)
6. Radha Krishna, G., Krishnan, R.: Native language identification based on English accent. In: International Conference on Natural Language Processing (ICON), pp. 69–74 (2014)
7. Luengo, I., Navas, E., Sainz, I., Saratxaga, I., Sanchez, J., Odriozola, I., Hernaez, I.: Text independent speaker identification in multilingual environments. In: Language Resources and Evaluation Conference (LREC) (2008)
8. Radha Krishna, G., Krishnan, R., Mittal, V.K.: An automated system for regional nativity identification of Indian speakers from English speech. In: 16th IEEE India Council International Conference INDICON 2019 (2019, December)
9. Radha Krishna, G., Krishnan, R., Mittal, V.K.: Identification of Indian English by speakers of multiple native languages. In: Sixth International Symposium on Signal Processing and Intelligent Recognition Systems (SIRS'20) (2020, October)
10. Shriberg, E., Ferrer, Ł., Kajarekar, S., Scheffer, N., Stolcke, A., Akbacak, M.: Detecting non-native speech using speaker recognition approaches. In: Proceedings IEEE Odyssey-08 Speaker and Language Recognition Workshop. Stellenbosch, South Africa (2008)
11. Koreman, J.: Category similarity in multilingual pronunciation training. In: Interspeech, vol. 27, pp. 2578–2582 (2018)
12. Wu, T., Duchateau, J., Martens, J.P., Van Compernolle, D.: Feature subset selection for improved native accent identification. Speech Commun. **52**(2), 83–98 (2010)
13. Malmasi, S., Evanini, K., Cahill, A., Tetreault, J., Pugh, R., Hamill, C., Napolitano, D., Qian, Y.: A report on the 2017 native language identification shared task. Assoc. Comput. Linguist. **2017**, 62–75 (2017)
14. Zampieri, M., Ciobanu, A.M., Dinu, L.P.: Native language identification on text and speech. In: Proceedings of the 12th Workshop on Innovative Use of NLP for Building Educational Applications, ACL, pp. 398–404 (2017)
15. Bahari, M.H., Saeidi, R., Van Hamme, H., Van Leeuwen, D.: Accent recognition using i-vector, Gaussian mean supervector and Gaussian posterior probability supervector for spontaneous telephone speech. In: ICASSP, pp. 7344–7248 (2013)
16. Moustroufas, N., Digalakis, V.: Automatic pronunciation evaluation of foreign speakers using unknown text. Comput. Speech Lang. **21**, 219–230 (2007)
17. De Marco , A., Cox, S.J.: Native accent classification via I-vectors and speaker compensation fusion. In: Interspeech, pp. 1472-1476 (2013, May)
18. Dehak, N., Kenny, P.K., Dehak, R., Dumouchel, P., Ouellet, P.: Front-end factor analysis for speaker verification. IEEE Trans. Audio Speech Lang. Process. **19**(4), 788–798 (2011)
19. Graham, C., Post, B.: Second language acquisition of intonation: peak alignment in American English. J. Phonetics **66**(2018), 1–14 (2018)
20. Ikeno, A., Hansen, J.H.L.: Recognition cues in native English accent variation: "listener accent, perceived accent, and comprehension". In: 2006 IEEE International Conference on Acoustics Speed and Signal Processing Proceedings, vol. 1, pp. I–401–I–404

A Novel Upgraded Uniform Embedding Technique for JPEG Steganography

Savina Bansal, R. K. Bansal, and Rakesh Kumar

Abstract JPEG steganography finds extensive applications in the area of concealing the intelligent data in images. Minimal distortion embedding-based steganographic technique provides a precise distortion measure with feasible coding. This paper presents a new Upgraded Uniform Embedding Distortion (U^2ED) for JPEG steganography that tries to maintain the statistical imperceptibility in both spatial and DCT domains. A new distortion measure is proposed for both AC modes and DCT blocks in the proposed U^2ED scheme that aims to restrict the secret data into coarse and noisy regions, thus making it tough to be detected by the steganalyzer. Syndrome trellis codes are utilized to accomplish minimum distortion due to embedding of secret message. Effectiveness of the proposed scheme is analysed using BOSS base image database as used in literature using GFR steganalyzer combined with ensemble classifier. Experimental result based on 100 testing images indicates the goodness of the proposed U^2ED against Uniform Embedding Revisited Distortion (UERD) at different payloads in terms of probability of error. Also, the proposed scheme competes well with JPEG Universal Wavelet Relative Distortion (J-UNIWARD) for quality factor 75 and tends to outperform it for quality factor 95.

Keywords JPEG steganography · Uniform embedding · Syndrome trellis code

1 Introduction

Steganography aims towards achieving maximum statistical imperceptibility while hiding confidential information into a cover medium usually digital image. There is always a trade-off between statistical imperceptibility and sufficient payload capacity

---

S. Bansal (✉) · R. K. Bansal · R. Kumar
Giani Zail Singh Campus College of Engineering and Technology, Maharaja Ranjit Singh Punjab Technical University Bathinda, Punjab 151001, India
e-mail: savinabansal@mrsptu.ac.in

R. K. Bansal
e-mail: rkbansal@mrsptu.ac.in

© The Author(s), under exclusive license to Springer Nature Singapore Pte Ltd. 2022
A. K. Nagar et al. (eds.), *Intelligent Sustainable Systems*, Lecture Notes in Networks and Systems 333, https://doi.org/10.1007/978-981-16-6309-3_68

[14]. Therefore, the target of a steganographer is to better the statistical imperceptibility for a fixed payload. This can be attained by adaptively utilizing the location of the cover image pixels or coefficients for data embedding [16].

Currently, the minimal distortion embedding strategy is considered as the most secure approach for JPEG image steganography. This technique involves a well-designed cost function and a realistic coding scheme to lessen the distortion. The total cost for an additive cost function is computed as the summation of modification cost assigned to each cover elements. For content-based image steganography, large costs are allotted where the detection rate is expected to be higher (flat regions such as background) and smaller costs are allotted for textured or rough regions where the embedding changes will be tough to detect.

A practical coding approach to implement minimal distortion embedding is syndrome trellis coding [1]. It includes an easy to deal with optimizable cost function and a non-binary embedding method for coding the secret message while minimizing the additive cost. Syndrome trellis codes (STCs) have embedding rate near the rate-distortion limit, so the work left for steganographer is to design an effective distortion function.

2 Related Work

Uniform embedding as introduced by Guo et al. [2] for non-side informed JPEG steganography was later extended [3] for side informed JPEG steganography. Their scheme uses non-zero AC DCT coefficients for modification and uniformly distribute the embedding modification over quantized DCT coefficients by incorporating STCs. Holub et al. [4] presented a novel method called UNIWARD for content adaptive JPEG steganography. This method uses wavelet domain to derive distortion function for spatial (S-UNIWARD), side informed JPEG (SI-UNIWARD) and non-side informed JPEG (J-UNIWARD) domain and use all DCT coefficients as possible cover elements for modification. Their experimental results showed good security performance but constraint computational complexity. A major improvement over [3] was presented in Guo et al. [5] by considering the relative variations in coefficients of cover images and named this JPEG steganography technique as Uniform Embedding Revisited Distortion. DCT coefficients (including DC, zero AC, and non-zero AC) are utilized for embedding modification and uniformly spread the embedding changes over the cover elements. Their additive distortion function helps in achieving high embedding security and reduced computational time. Pan et al. [6] presented an improved distortion function for UERD [5] as IUERD. In their scheme, the distortion function efficiently utilized the mutual correlation between DCT blocks, which leads to more statistical undetectability and its effectiveness of securely embedding payload is checked using JRM steganalyzer.

Wang et al. [7] presented hybrid distortion function steganography (HDS) which exploit spatial and DCT domain to resist multidomain steganalysis. Their distortion function calculates block distortion as fluctuation in block and DCT coefficients changes as quantization step. In progression to exploit spatial and DCT domain for distortion calculation, Su et al. [8] decomposed a distortion measure known as GUED using an exponential model. Their results claim to have improved embedding security and reduced computation cost over UERD and UNIWARD. Further, Wei et al. [9] presented Residual Block Value method to measure the embedding cost due to modification in DCT block and quantization step for selection channel embedding risk. Hu et al. [10] presented a block entropy transformation scheme (BET) based on block entropy embedding (BEE). Their hybrid distortion function is designed using statistics of decompressed cover images and DCT coefficients.

3 Proposed Work

This work is an upgrade to an earlier work named Uniform Embedding Revisited Distortion (UERD) for non-side informed JPEG images [5], wherein the concept of generalized uniform embedding has been exploited. As this work is an upgradation of UERD so, its brief overview is given below for better understanding the work proposed here.

3.1 UERD Scheme

In UERD, the cost of modifying a coefficient depends on coefficient of variation parameter (CV is the fraction of standard deviation to the mean). The larger the CV, more the probability of changing coefficient and vice versa.

Let $x_{i,j}$ denotes a DCT coefficient at location (a, b) in the c DCT block. The cost function for $x_{i,j}$ is given as:

$$\rho_{i,j} = \rho(a, b), \text{(mode)} * \rho(m, n), \text{(block)} \tag{1}$$

where $\rho(a, b)$, (mode) and $\rho(m, n)$, (block) denotes the cost measures for the corresponding AC modes and DCT blocks, respectively. Specifically, $\rho(a, b)$, (mode) $=$ $q_{a,b}$ where $q_{a,b}$ indicates the quantization step for $x_{i,j}$, whereas $\rho(m, n)$, (block) is the function of energies of the $(m, n)^{\text{th}}$ block and its adjacent blocks. The block energy, say $D_{m,n}$, is defined as

$$D_{m,n} = \sum_{0}^{7}\sum_{0}^{7}|x_{a,b}| * q_{a,b} \tag{2}$$

here, $x_{a,b}$ specifies quantized DCT coefficient in $(m, n)^{\text{th}}$ DCT block. Thus, UERD only consider the DCT coefficients in build of distortion function $\rho_{i,j}$.

3.2 Proposed U²ED Scheme

Here, new cost measures for AC modes and their corresponding DCT blocks are constructed depending on the statistics of both spatial domain and DCT domain. For DC mode, the distortion is calculated as the average of adjoining AC coefficients residing in the same DCT block. Inspired by J-UNIWARD [4], for $\rho(a, b)$, (mode), we compute the impact of unit modification on mode(a, b) calculated in wavelet domain as an alternative to quantization step $q_{a,b}$ in UERD. Let $W(a, b)$ denote the embedding modifications in spatial domain during modification of mode(a, b) by unity, and it can be obtained by utilizing two-dimensional inverse discrete cosine transformation (2D-IDCT) as:

$$W(a, b) = q_{a,b}.f(u, v; a, b) \tag{3}$$

where $q_{a,b}$ indicates quantization step for mode(a, b) and $f(u, v; a, b)$ is 2D-IDCT with index (u, v), i.e. $f(u, v; a, b) = \frac{Z_a.Z_b}{4}\cos\frac{\pi(2u+1)a}{16}\cos\frac{\pi(2v+1)b}{16}$, $Z_0 = 1/\sqrt{2}$, $Z_r = 1(r > 0), 0 \le u, v \le 7, 0 \le a, b \le 7$.

The relative changes in wavelet domain are calculated using8-tap Daubechies wavelet directional filter bank, $K = \{k_1, k_2, k_3\}$ where $k_1 = l.h'$,$k_2 = h.l'$, $k_3 = h.h'$. Here, l is low-pass and h is high-pass one-dimensional decomposition filter. The relative changes in wavelet coefficients at mode(a, b), are represented as the convolution ('⊙') between spatial modifications $W(a, b)$ and directional filter K:

$$\rho(a, b), (\text{mode}) = \sum_{k=1}^{3}\sum \text{abs}(W(a, b) \odot K_k) \tag{4}$$

The distortion function for DCT block is dependent on image content (i.e. the embedding changes are restricted in complicated textured regions) because of the fact that the Warden cannot distinguish content and noise from highly coarse regions. Also, from literature it is evident that most of the steganalyzer (Discrete Cosine Transform Residuals, Gabor Filter Residuals) are constructed based on the noise residuals model, features set of steganalysis and machine learning. Therefore, statistical properties of image model are exploited to design a steganalysis resistive distortion function. From this, we can deduce that residuals of image in spatial domain are a beneficial measure to calculate the distortion of DCT blocks. For building the distortion function $\rho(m, n)$, (block), we explore the Wiener Filter residuals (Wiener filter is a low-pass adaptive denoising filter) in spatial domain because of its better performance at lower computation cost, followed by a variance estimator as per [11].

$$\hat{R} = \hat{\sigma}_n^2 = \frac{r_n - \hat{r}_n^2}{p^2 - q} \tag{5}$$

where r_n is residuals value, $\hat{r}_n$ is least square estimation of residuals and $p^2 \times q$ is the size of matrix for parametric models. For numerical stability, a lower limit of 0.01 is set for estimated variance. To epitomize the cost measure for $(m, n)^{\text{th}}$ DCT block, $\hat{R}$ is divided into 8 by 8 block in arrangement with the other blocks in the same domain.

$$E_{m,n} = \sum_{a=0}^{7}\sum_{b=0}^{7} \hat{R}_{m,n}(a, b) \tag{6}$$

The cost measure for $(m, n)^{\text{th}}$ DCT block $\rho(m, n)$, (block) considers the cost measure $E_{m,n}$ and its eight neighbouring blocks as well, for all quantized DCT coefficient $x_{i,j}$ in the $(m, n)^{\text{th}}$ block:

$$\rho(m, n), \text{(block)} = \left(E_{m,n}\right)^{-1} + 0.25 * \sum_{d \in \bar{d}} (E_d)^{-1} \tag{7}$$

where $\bar{d} = \{(m-1, n-1), (m-1, n), (m-1, n+1), (m, n-1),$
$(m, n+1), (m+1, n-1), (m+1, n), (m+1, n+1)\}$.

In succinct, we propose a new distortion function $\rho_{i,j}$ for U[2]ED as shown in Eq. (8). The embedding costs for both DC and AC coefficients are calculated as follow.

$$\rho_{i,j} = \begin{cases} 0.5 * (\rho_{i+1,j} + \rho_{i,j+1}), & \text{if}(a, b) = (0, 0) \\ \rho(a, b), \text{(mode)} * \rho(m, n), \text{(block)}, & \text{otherwise} \end{cases} \tag{8}$$

4 Experimental Results and Performance Analysis

The effectiveness of the offered scheme is analysed from experimental results. The U[2]ED scheme is compared with the most closely resembling UERD and J-UNIWARD schemes.

4.1 Experimental Setup

Image sets from BOSS base ver. 1.01 [12] database is used to implement all experiments. The database contains 10,000 images in Portable Gray Map (PGM) format with each image having dimensionality of 512 by 512 pixels. All images are then JPEG compressed for quality factor 75 and quality factor 95 using MATLAB command imwrite. In our experiments, 100 images (50 for training and 50 for testing) are randomly selected from database. The embedding payload ranges from 0.1 to

0.5 bpnzac having a step size of 0.1 bpnzac. GFR-17000D [15] JPEG steganalyzer is employed to evaluate the pragmatic security performance of the proposed U^{2}ED JPEG steganographic scheme with UERD and J-UNIWARD. The ensemble classifier [13] with Fisher linear discriminant (FLD) base learner is utilized for classification of cover and stego images in our experiments. The total set of selected images are divided into two parts. One half of the images are used for training of ensemble classifier and the other half for testing of trained classifier. To evaluate the pragmatic security performance the minimum total probability of error P_e parameter is used. The P_e is the calculated as average of false alarm and missed detection of the ensemble classifier, under equal probability of occurring a cover or stego image, as is used in [5]

$$P_e = \min \frac{P_{\mathrm{fa}} + P_{\mathrm{md}}}{2} \tag{9}$$

where P_{fa} represents probability of false positive (indicating the cover image as stego image),P_{md} is probability of false negative (indicating a missed detection). The performance is evaluated using mean P_e over ten random tests with respect to relative payload. The higher value of P_e reflects strong empirical security.

4.2 *Performance of U^{2}ED*

We compare the proposed U^{2}ED scheme for non-side informed images with the UERD and J-UNIWARD schemes for JPEG steganography. Figure 1 and Fig. 2 show the security performance of the involved schemes against the GFR-17000D steganalyzer, for quality factor 75 and 95 respectively. As gathered from Fig. 1 for quality factor 75, U^{2}ED outperforms UERD by a sizeable margin of 8, 12 and 11% for

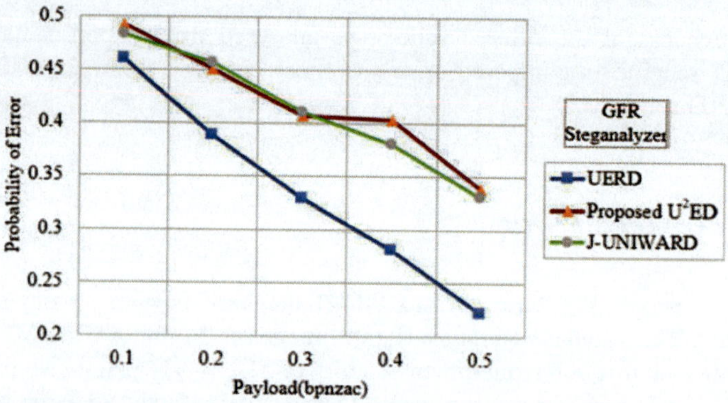

Fig. 1 Payload versus probability of error for JPEG quality factor 75

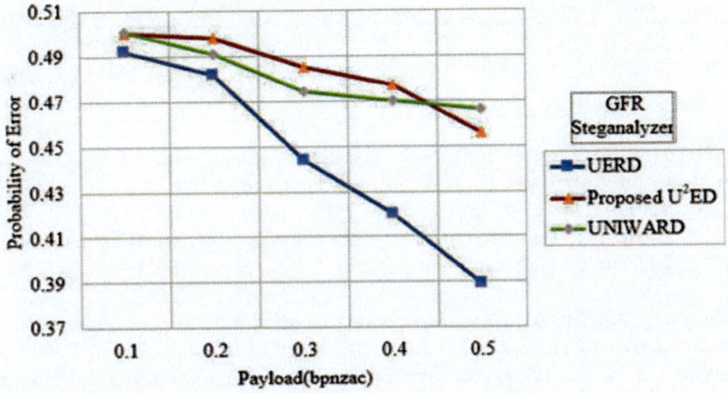

Fig. 2 Payload versus probability of error for JPEG quality factor 95

payload 0.3, 0.4, 0.5 bit per non-zero AC coefficients (bpnzac) respectively. Also, U^2ED has comparable security performance with J-UNIWARD for payloads less than 0.3 bpnzac and outperforms it for payloads greater than 0.3 bpnzac in all the tested JPEG images.

For quality factor 95, as gathered from Fig. 2 U^2ED outperforms UERD with a margin of 4%, 5%, and 7% for payloads 0.3, 0.4, and 0.5 bpnzac respectively. And, it outperforms J-UNIWARD for payload less than 0.5 bpnzac.

5 Conclusion

This paper presents a new distortion function named U^2ED for non-side informed JPEG steganography. This function utilizes the statistics of both spatial and DCT domains. The proposed distortion measure aims to maintain the statistical imperceptibility in both spatial and DCT domains by restricting the embedding changes in hard-to-detect regions. Experimental results indicate the superiority of U^2ED (in terms of surviving steganalysis attack) over UERD and J-UNIWARD for given set of training and testing data. More exhaustive results shall be presented in our future work.

References

1. Filler, T., Judas, J., Fridrich, J.: Minimizing additive distortion in steganography using syndrome-trellis codes. IEEE Trans. Inf. Forensics Secur. **6**(3), 920–935 (Sept. 2011). https://doi.org/10.1109/TIFS.2011.2134094
2. Guo, L., Ni, J., Shi, Y.Q.: An efficient JPEG steganographic scheme using uniform embedding. IEEE Int. Workshop Inf. Forensics Security (WIFS) **2012**, 169–174 (2012). https://doi.org/10.1109/WIFS.2012.6412644

3. Guo, L., Ni, J., Shi, Y.Q.: Uniform embedding for efficient JPEG steganography. IEEE Trans. Inf. Forensics Secur. **9**(5), 814–825 (May 2014). https://doi.org/10.1109/TIFS.2014.2312817
4. Holub, V., Fridrich, J., Denemark, T.: Universal distortion function for steganography in an arbitrary domain. EURASIP J. on Info. Security **2014**, 1 (2014). https://doi.org/10.1186/1687-417X-2014-1
5. Guo, L., Ni, J., Su, W., Tang, C., Shi, Y.: Using statistical image model for JPEG steganography: uniform embedding revisited. IEEE Trans. Inf. Forensics Secur. **10**(12), 2669–2680 (Dec. 2015). https://doi.org/10.1109/TIFS.2015.2473815
6. Pan, Y., Ni, J., Su, W.: "Improved uniform embedding for efficient JPEG steganography," in Proc. Int. Conf. Cloud Computing and Security Cham, Switzerland: Springer, pp. 125–133 (2016)
7. Wang, Z., Zhang, X., Yin, Z.: "Hybrid distortion function for JPEG steganography." J. Electron. Imaging **25**(5), 050501 (6 Sept 2016). https://doi.org/10.1117/1.JEI.25.5.050501
8. Su, W., Ni, J., Li, X., Shi, Y.: A new distortion function design for JPEG steganography using the generalized uniform embedding strategy. IEEE Trans. Circuits Syst. Video Technol. **28**(12), 3545–3549 (2018). https://doi.org/10.1109/TCSVT.2018.2865537
9. Wei, Q., Yin, Z., Wang, Z., et al.: Distortion function based on residual blocks for JPEG steganography. Multimed Tools Appl **77**, 17875–17888 (2018). https://doi.org/10.1007/s11042-017-5053-7
10. Hu, X., Ni, J., Shi, Y.: Efficient JPEG steganography using domain transformation of embedding entropy. IEEE Signal Process. Lett. **25**(6), 773–777 (June 2018). https://doi.org/10.1109/LSP.2018.2818674
11. Fridrich, J., Kodovský, J.: "Multivariate gaussian model for designing additive distortion for steganography." 2013 IEEE International Conference on Acoustics, Speech and Signal Processing, pp. 2949–2953 (2013). https://doi.org/10.1109/ICASSP.2013.6638198
12. Bas, P., Filler, T., Pevný, T.: "Break our steganographic system": the ins and outs of organizing BOSS. In: Filler, T., Pevný, T., Craver, S., Ker, A. (eds.) Information Hiding. IH 2011. Lecture Notes in Computer Science, vol. 6958. Springer, Berlin, Heidelberg (2011). https://doi.org/10.1007/978-3-642-24178-9_5
13. Kodovsky, J., Fridrich, J., Holub, V.: Ensemble classifiers for steganalysis of digital media. IEEE Trans. Inf. Forensics Secur. **7**(2), 432–444 (2012). https://doi.org/10.1109/TIFS.2011.2175919
14. Kaur, S., Bansal, S., Bansal, R.K.: Steganography and classification of image steganography techniques. Int. Conf. Comput. Sustain. Global Develop. (INDIACom) **2014**, 870–875 (2014). https://doi.org/10.1109/IndiaCom.2014.6828087
15. Song, X., Liu, F., Yang, C., Luo, X., Zhang, Y.: "Steganalysis of adaptive JPEG steganography using 2D Gabor filters." In: Proc. 3rd ACM Workshop Inf. Hiding Multimedia Security, pp. 15–23 (2015)
16. Kaur, S., Bansal, S., Bansal, R.K.: An efficient adaptive data hiding scheme for image steganography. In: Satapathy, S., Bhatt, Y., Joshi, A., Mishra, D. (eds.) Proceedings of the International Congress on Information and Communication Technology. Advances in Intelligent Systems and Computing, vol. 438. Springer, Singapore (2016). https://doi.org/10.1007/978-981-10-0767-5_40

Insights into Developers' Abandonment in FLOSS Projects

Rajdeep Kaur and Kuljit Kaur

Abstract The development of successful software relies on the efforts of volunteer contributors that are globally dispersed and co-operate with each other through the Internet. The success of an OSS (Open Source Software) project depends on the attraction of new members for participation and retention of old developers. The crew members do not receive financial aid and are not bound by any employment contract. Therefore, they have freedom to join or leave the project as per their convenience. After continuous participation in the project, some developers left the project or join some other project that provides more opportunities and career growth. The major cause behind developers' dropout is that they face many problems during early contribution. Therefore, developer dropout is the main issue in the development of an effective project. The existing literature examined the joining patterns, process of joining, motivation/retention, and hitches of newcomers/developers. But the area of developer abandonment has not achieved enough maturity. In order to predict developers' turnover in OSS projects and overcome this issue, it is important to analyze the various factors that force developers to leave the project and influence their permanence. The primary goal of the current study is to survey the literature to analyze developers' abandonment and several factors reported in the existing literature that impact the dropout of contributors in OSS projects. Our study results confirmed that most of the studies explored the different factors that lead to dropout of developers and few studies investigated the impact of abandonment on the project. The current study will help OSS project administrators to comprehend the predictive factors in abandonment. Based on this knowledge, they can make decisions in advance to tackle this problem.

Keywords Open source software · Abandonment · Influencing factors · Software developer

R. Kaur (✉) · K. Kaur
Guru Nanak Dev University, Amritsar, India
e-mail: rajdeep.rsh@gndu.ac.in

K. Kaur
e-mail: kuljitchahal.cse@gndu.ac.in

© The Author(s), under exclusive license to Springer Nature Singapore Pte Ltd. 2022
A. K. Nagar et al. (eds.), *Intelligent Sustainable Systems*, Lecture Notes in Networks and Systems 333, https://doi.org/10.1007/978-981-16-6309-3_69

1 Introduction

OSS is distributed under license along with its source code and users of OSS can study, modify and redistribute software [1], even though OSS development practices have been in use from last ten years. OSS has become more popular among developers, businessmen, government, and researchers nowadays [2]. Apache Web Server and Linux Operating System are very popular OSS projects [3]. OSS development practices are used by private households and several organizations[4]. As specified by Gartner Inc. more than 2/3 of applications by organizations depend on OSS [5]. IBM claims that approximately 90% of IT professionals rely on OSS for future application development [6]. Instead of their widespread acceptance, most of OSS projects disappear. Chengalur Smith predicts that around 80% of the projects dwindle [7, 8]. The survival of an OSS project depends upon retaining old members and attracting new participants [9]. Motivation and retention of new developer are necessary for sustainability of an OSS project [10]. Although several studies have examined the joining process of developers, obstacles of novices during early contribution, motivation of developers behind participation, and retention but few researchers studied developer abandonment. This motivated us to conduct this study. The current study presents the survey of existing literature and provides understanding about OSS developers' abandonment as well as various factors that influence developers' dropout.

The disengagement of team members is a serious and costly issue in OSS projects. Abandonment occurs frequently that degrade the performance of a project. It is the rate at which developers withdraw a project and their movement inside and outside a project [11]. In order to better understand why this phenomenon occurs in OSS projects and how developers' abandonment can be reduced. It is necessary to investigate the various influencing factors that lead to developers' withdrawal in OSS projects.

The remainder of the paper is structured as follows: Sect. 2 contains some related work. Section 3 discusses developer abandonment, presents the analysis of studies on abandonment, and discussed the influencing factors investigated in the existing literature. Section 4 provides the distribution of research studies with respect to the year. Conclusions and future work is presented in Sect. 5 followed by acknowledgment and references used in the study.

2 Related Word

This section summarized some existing work that explored the abandonment of OSS developers.

In the year 2019, Miller et al. [12] have identified the causes and predictive factors behind the disengagement of established contributors through conducting survey among recently disengage contributors. In the year 2017, Lin et al. [13] investigated the effect of time of joining, action type, frequency of file maintenance, and sort of job on developers' turnover intention. This study identifies a sub-group of developers who have a high possibility to leave a project. This study does not investigate the effect of turnover on OSS projects. In the year 2015, Steinmacher et al. [14] examined the difficulties faced by novices when they start their initial participation. This study identified barriers and classified barriers according to the barrier model. In the year 2014, Schilling et al. [4] surveyed the literature on retention, commitment, and attraction of OSS developers. This study revealed that the majority of research examined the developers' commitment in comparison with attraction/retention. This work does not explore developers' abandonment. In the year 2013, Steinmacher et al. [15] investigated the impact of lack of response, type of response like as answer civility, usefulness, and the author of response on newcomers' desertion. This study considers only newcomers who left the project. The abandonment of core members and short-term contributors is not considered by this work.

Complementing all the aforementioned studies, Our work analyzes developer abandonment and provides understanding about various influencing factors discussed in the existing literature that influence abandonment of OSS developers.

3 Analysis of Research Studies on Abandonment

In OSS projects abandonment of developers is a critical and costly issue that influences the sustainability of the project. It takes place when a member takes a decision to leave a project due to some reasons, moves to some other high-interest project, withdraws a project temporarily or permanently that is under development [16]. The rate at which developers withdraw a project is known as abandonment and it includes all changes that occurred in the development teams of a project. These changes can occur on the external level when a member either joins or leave a project and at the internal level, changes take place when the role of member changes in a project [11].

Many researchers examined the abandonment of developers and presented various factors that influence developers' abandonment. Table 1 presents the detail of abandonment studies and various influencing factors investigated by these studies.

Ferreira et al. [17] in the year 2020, investigated the turnover of core developers in 174 FLOSS projects and examined the impact of the size of team, scripting language, and owner of the project on core members' turnover intention. In the year 2019, Miller et al. [12] surveyed established contributors that disengage open-source contribution recently and identified predictive factors and causes of established contributors' disengagement using the survival modelling technique. In the same year, Iaffaldano et al. [18] interviewed active developers to investigated the inactivity and abandonment of developers. This study discovered the sleeping and dead states that represent temporary and permanent cessations taken by developers from contribution. In 2019, Avelino et al. [19] presented an investigation of TF developers' abandonment in OSS projects. In the year 2017 Lin et al. [13] analyzed the five industrial OSS projects WikiMedia, OpenStack, GlusterFS, Xen project, and Cloud Stack to study the impact of joining time, rate of maintaining files, sort of activity, and job type on developers'

Table 1 Studies on abandonment

S.No	Author(s), Year of Publication	Title	Studied Factors	Reference
1.	Ferreira et al. (2020)	"Turnover in open-source projects: The case of core developers."	Size of team, coding language, owner of the project, project age	[17]
2.	Miller et al. (2019)	"Why Do People Give Up FLOSSing? A Study of Contributor Disengagement in Open Source."	Activity level, working hours and support, organizational affiliation, team size, project popularity, job transition	[12]
3.	Iaffaldano etal. (2019)	"Why do developers take breaks from contributing to OSS projects? A preliminary analysis."		[18]
4.	Avelino et al. (2019)	"On the abandonment and survival of open source projects: An empirical investigation."		[20]
5.	Constantinou and Mens, (2017)	"An empirical comparison of developer retention in the Ruby Gems and npm software ecosystems."	Intensity as well as the frequency of social and technical activity	[22]
6.	Lin et al. (2017)	"Developer turnover in global, industrial open source projects: Insights from applying survival analysis."	Joining time, frequency of file maintenance, sort of activity, and job type	[13]
7.	Foucault et al. (2015)	"Impact of developer turnover on quality in open-source software."	Software quality	[11]
8.	Vasilescu et al. (2015)	"Gender and Tenure Diversity in GitHub Teams."	Gender diversity, tenure diversity	[23]
9.	Steinmacher et al. (2013)	"Why do newcomers abandon open source software projects."	Lack of answer, type of answer, and author of the answer	[15]
10.	Khondhu et al. (2013)	"Is it all lost? A study of inactive open source projects."		[25]

(continued)

Table 1 (continued)

S.No	Author(s), Year of Publication	Title	Studied Factors	Reference
11.	Zanetti et al. (2013)	"The rise and fall of a central contributor: Dynamics of social organization and performance in GENTOO community."	Social organization and bug handling performance	[27]
12.	Garcia et al. (2013)	"The role of emotions in contributors activity: A case study of the GENTOO community."	Emotions	[24]
13.	Yu et al. (2012)	"An empirical study of volunteer members' perceived turnover in open source software projects."	Hygiene factors	[26]
14.	Sharma et al. (2012)	"Examining turnover in open source software projects using a logistic hierarchical linear modelling approach."	Developer-level factors (past activity, position of developer, number of projects involved) project-level factors (project tenure, size, and age)	[28]
15.	Hynninen et al. (2010)	"Off-site commitment and voluntary turnover in GSD projects."	Commitment	[21]
16.	Izquierdo-Cortazar et al. (2009)	"Using software archaeology to measure knowledge loss in software projects due to developer turnover."		[19]
17.	Hall et al. (2008)	"The impact of staff turnover on software projects: the importance of understanding what makes software practitioners trick."	Project success	[29]
18.	Robles et al. (2006)	"Contributor turnover in libre software projects."		[30]

turnover by applying survival analysis technique. This study identifies a sub-group of developers who have a high probability to abandon a project. The results revealed that early joiners have higher probability to stay in the project, the developers who deal with file modification have high survival rates as compared to developers who deal with file creation, and developers whose main job is coding have high likelihood to stay in the project in comparison with developers whose primary job is documentation. In the same year, Constantinou and Men's [20] examined the impact of intensity and frequency social and technical activity of developer on their continuance in an ecosystem through analyzing two software ecosystems RubyGems and npm and identify that the developers that don't communicate with other developers, whose social and technical intensity is not very strong, whose commit or communication frequency is less, and do not communicate or commit for long time period have high likelihood to withdraw project. In year 2015, Foucault et al. [11] studied five large OSS projects Ansible, Rails, Jenkins, Angular J S, and J Query and examined the effect of developers' disengagement on quality of software. This study also proposes metrics to measure turnover. Study revealed that external turnover negatively impact software quality and internal turnover of developers does not have any effect on software quality. In the same year, Vasilescu et al. [21] examined the association of tenure and gender diversity with turnover. This study observed that existence of developers with different duration of engagement in GitHub increases the probability of turnover. In the year 2013, Garcia et al. [22] analyzed the association between emotions and contributors' activity based on the case study of the GENTOO project. This work revealed that turnover causes negative emotions in the project. Steinmacher et al. [15] in the same year examined the data of mailing list, issue manager (Jira), and users' mailing list of Hadoop common project to analyze the effect of lack of reply, nature of reply like answer civility and usefulness, and author of response on the desertion of the novice. They surveyed leavers to understand the reason behind the dropout. The authors found that author of response and type of reply received by novice influence their choice to abandon a project. But the absence of reply does not impact the decision to stay or leave a project. Survey results revealed that newcomer withdrawal in a project influence by negative responses of the author and when they do not get the required support. In the same year, Khondhu et al. [23] analyzed the sourceforge.net projects and explain how to find out a project that has been abandoned. This study also examined the characteristics and attributes of disengaged projects. Zanetti et al. [24] evaluated the relation of GENTOO's community social organization and bug handling performance with central contributor activity and desertion. This study demonstrated that performance increase when the central contributor is active, but his unexpected dropout has a negative influence on the efficiency of bug handling and response time. In the year 2012, Yu et al. [25] surveyed Launchpad.net and sourceforge.net developers to investigate the association of hygiene factors with volunteer members' dissatisfaction and developers' turnover. The authors found that project rule and management is key hygiene factors for dissatisfaction, personal needs for software functionality and future rewards. In the same year, Sharma et al. [26] have studied the Sourceforge.net projects and 201 developers to analyze the impact of project-level factors and developer-level factors on turnover of developers in OSS projects. The authors found that previous activity, developers' position, size of the project, and age of the project influence developer turnover. The turnover rates are not similar among Sourceforge.net projects. In the year 2010, Hynninen et al. [27] surveyed the developers to examined the connection between commitment and voluntary turnover in off-site global software development projects. This study found

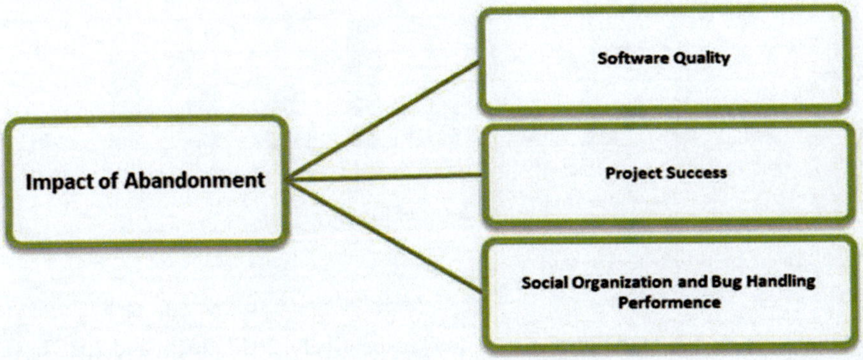

Fig. 1 Impact of abandonment

that low commitment is strongly associated with high turnover rates at off-site in GSD projects. In the year 2009, D. Izquierdo-Cortazar et al. [28] examined the influence of turnover on software projects on the basis of developers' contributed knowledge. In the year 2008, Hall et al. [29] have studied the correlation between high turnover and project success. The authors also examined the association among high staff turnover and low motivation of staff members through survey of 89 software professionals. The authors identify that staff turnover can be reduced by enhancing motivation of member that increase the chances of project success. In year 2006, Robles et al. [30] used activities data of source code management repositories, to examine the effect of developer joining and abandonment on liber software projects.

Figure 1 presents the impact of abandonment, and Table 1 illustrates the various factors identified in the literature that influence developer abandonment. As discussed above team size, programming language, project owner, activity level, working hours and support, organizational affiliation, project popularity, job transition, gender diversity, tenure diversity, emotions, commitment, type of answer, author of the answer, joining time, percentage of maintaining files, kind of job, sort of action, project rule, management, project-level factors such as size and age of the project, developer-level factors like previous activity and role of developers and intensity and frequency of social-technical activities of developers influence dropout of developers. Moreover, three researchers studied the influence of developers' turnover on the quality of software, project success, and performance of the social organization and bug handling.

4 Year-Wise Distribution of Research Studies

The year-wise distribution of studies is presented in Fig. 2. The yearly distribution of studies starts with only one publication in 2006, 2008, 2009, 2010, and 2020 contributed by Robles et al. [30], Hall et al. [29], Izquierdo-Cortazar et al. [28],

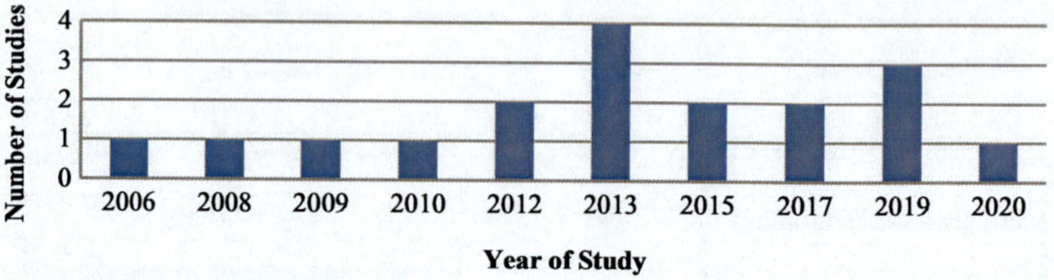

Fig. 2 Year-wise distribution of studies

Hynninen et al. [27], and Ferreira et al. [17] respectively. 2012, 2015, and 2017 have two studies respectively, conducted by Yu et al. [25], Sharma et al. [26], Foucault et al. [11], Vasilescu et al. [21], Constantinou and Mens[20], and Lin et al. [13]. The year 2013 has maximum publication. In this year, 4 studies authored by Steinmacher et al. [15], Khondhu et al. [23], Zanetti et al. [24], and Garcia et al. [22]. Finally, in year 2019, there were 3 studies contributed by Miller et al. [12], Iaffaldano et al. [18], and Avelino et al. [19].

5 Conclusions

In this research, we surveyed the literature to understand developers' abandonment in OSS projects and various predicting factors in abandonment. This study identified several factors evidenced in the existing literature that influence the dropout of contributors in OSS projects such as the size of the team, coding language, project owner, activity level, working hours and support, organizational affiliation, project popularity, job transition, gender diversity, tenure diversity, emotions, commitment, type of reply, author of the answer, joining time, amount of maintaining files, sort of activities, kind of work performed by developers, developers' previous activity, developer role, project size, project age, and hygiene factors like project regulation and administration influence the choice of the developer to stay or leave in the project. The intensity and frequency of the social-technical activity of developers also impact the developers' permanence in open-source software projects. Some studies investigated the impact of developer abandonment on OSS projects. The disengagement of central contributors influences the performance of social organization and bug handling. Furthermore, developers' dropout has a negative impact on software quality and success of the project. Through in this study, we observed that most of the studies investigate the various factors that influence the withdrawal of developers and identified a group of developers who have high probability to abandon the project. The aforementioned studies don't consider why abandonment occurs in the OSS project and how it can be reduced. Only three studies examined the impact of abandonment. There are many open issues related to developer abandonment that need to be resolved. Further research is required to examine the actual causes of abandonment, the impact of turnover on project performance, the effect of technical issues on volunteer member dissatisfaction, and the impact of turnover on the evolution of the project. It is necessary to develop strategies to reduce developers' abandonment and tackle this problem before occurrence.

Acknowledgements The authors of the paper will be appreciative of UGC (University Grants Commission), Govt. of India for providing economic aid under the scheme of Rajiv Gandhi National Fellowship to carry out this research work. The authors are also thankful for the infrastructural as well as academic support provided by Computer Science Department of Guru Nanak Dev University in Amritsar, Punjab.

References

1. Steinmacher, I., Wiese, I., Conte, T.U., Gerosa, M.A.: Increasing the self-efficacy of newcomers to open source software projects. Proc—29th Brazilian Symp. Softw. Eng. SBES 2015. pp. 160–169 (2015). https://doi.org/10.1109/SBES.2015.10
2. Stewart, K.J., Gosain, S.: The Impact of Ideology on Effectiveness in Open Source Software Development Teams. **30**, 291–314 (2013)
3. Colazo, J., Fang, Y.: Impact of license choice on open source software development activity. J. Am. Soc. Inf. Sci. Technol. 997–1011 (2009)
4. Schilling, A.: What do we know about floss developers' attraction, retention, and commitment? a literature review. Proc. Annu. Hawaii Int. Conf. Syst. Sci. 4003–4012 (2014). https://doi.org/10.1109/HICSS.2014.495
5. Gartner highlights key predictions for IT organisations and users in 2008 and beyond, http://www.gartner.com/it/page.jsp?id=593207
6. Open source technologies play a key role in future of application developement. (2011)
7. Chengalur-Smith, I., Sidorova, A., Daniel, S.: Sustainability of free/libre open source projects: a longitudinal study. J. Assoc. Inf. Syst. **11**, 657–683 (2010)
8. G, M., S, C.: F/oss research repositories & research infrastructures. In: NSF Workshop on Free/Open Source Software Repositories and Research Infrastructures. , Irvine (2008)
9. Zhou, M., Mockus, A.: Who will stay in the FLOSS community? modeling participant's initial behavior. IEEE Trans. Softw. Eng. **41**, 82–99 (2015). https://doi.org/10.1109/TSE.2014.2349496
10. Fang, Y., Neufeld, D.: Understanding sustained participation in open source software projects. J. Manag. Inf. Syst. **25**, 9–50 (2009). https://doi.org/10.2753/MIS0742-1222250401
11. Foucault, M., Palyart, M., Blanc, X., Murphy, G.C., Falleri, J.R.: Impact of developer turnover on quality in open-source software. Proc. 2015 10th Jt. Meet. Found. Softw. Eng.—ESEC/FSE 2015. 829–841 (2015). https://doi.org/10.1145/2786805.2786870
12. Miller, C., Widder, D.G., Kästner, C., Vasilescu, B.: Why do people give up FLOSSing? a study of contributor disengagement in open source. IFIP Adv. Inf. Commun. Technol. **556**, 116–129 (2019). https://doi.org/10.1007/978-3-030-20883-7_11
13. Lin, B., Robles, G., Serebrenik, A.: Developer turnover in global, industrial open source projects: Insights from applying survival analysis. Proc.—2017 IEEE 12th Int. Conf. Glob. Softw. Eng. ICGSE 2017. 66–75 (2017). https://doi.org/10.1109/ICGSE.2017.11
14. Steinmacher, I., Graciotto Silva, M.A., Gerosa, M.A., Redmiles, D.F.: A systematic literature review on the barriers faced by newcomers to open source software projects. Inf. Softw. Technol. **59**, 67–85 (2015). https://doi.org/10.1016/j.infsof.2014.11.001

15. Steinmacher, I., Wiese, I.: Why do newcomers abandon open source software projects?Hum. Asp. 25–32 (2013). https://doi.org/10.1109/CHASE.2013.6614728
16. K, E.-M., Z.H, P.: On information systems project abandonment: an exploratory study of organizational practices. MIS Q. 67–86 (1991)
17. Ferreira, F., Silva, L.L., Valente, M.T.: Turnover in Open-Source Projects: The Case of Core Developers, (2020). https://doi.org/10.1145/3422392.3422433
18. Iaffaldano, G., Steinmacher, I., Calefato, F., Gerosa, M., Lanubile, F.: Why do developers take breaks from contributing.pdf. In: In Proceedings of the 2nd International Workshop on Software Health. pp. 9–16 (2019)
19. Avelino, G., Constantinou, E., Valente, M.T., Serebrenik, A.: On the abandonment and survival of open source projects: An empirical investigation. Int. Symp. Empir. Softw. Eng. Meas. 2019-Septe, (2019). https://doi.org/10.1109/ESEM.2019.8870181.
20. Constantinou, E., Mens, T.: An empirical comparison of developer retention in the RubyGems and npm software ecosystems. Innov. Syst. Softw. Eng. **13**, 101–115 (2017). https://doi.org/10.1007/s11334-017-0303-4
21. Vasilescu, B., Posnett, D., Ray, B., Van Den Brand, M.G.J., Serebrenik, A., Devanbu, P., Filkov, V.: Gender and tenure diversity in github teams. Conf. Hum. Factors Comput. Syst. Proc. 2015-April, 3789–3798 (2015). https://doi.org/10.1145/2702123.2702549
22. Garcia, D., Zanetti, M.S., Schweitzer, F.: The role of emotions in contributors activity: A case study of the GENTOO community. Proc. 2013 IEEE 3rd Int. Conf. Cloud Green Comput. CGC 2013 2013 IEEE 3rd Int. Conf. Soc. Comput. Its Appl. SCA 2013. 410–417 (2013). https://doi.org/10.1109/CGC.2013.71
23. J, K., A, C., K. J, S.: Is it all lost? A study of inactive open source projects. In: IFIP International Conference on Open Source Systems. pp. 61–79
24. Zanetti, M.S., Scholtes, I., Tessone, C.J., Schweitzer, F.: The rise and fall of a central contributor: Dynamics of social organization and performance in the GENTOO community. 2013 6th Int. Work. Coop. Hum. Asp. Softw. Eng. CHASE 2013—Proc. 49–56 (2013). https://doi.org/10.1109/CHASE.2013.6614731
25. Yu, Y., Benlian, A., Hess, T.: An empirical study of volunteer members' perceived turnover in open source software projects. Proc. Annu. Hawaii Int. Conf. Syst. Sci. 3396–3405 (2012). https://doi.org/10.1109/HICSS.2012.97
26. Sharma, P.N., Hulland, J., Daniel, S.: Examining turnover in open source software projects using logistic hierarchical linear modeling approach. IFIP Adv. Inf. Commun. Technol. 378 AICT, 331–337 (2012). https://doi.org/10.1007/978-3-642-33442-9_30
27. Hynninen, P., Piri, A., Niinimäki, T.: Off-site commitment and voluntary turnover in GSD projects. Proc. 5th Int. Conf. Glob. Softw. Eng. ICGSE 2010. 145–154 (2010). https://doi.org/10.1109/ICGSE.2010.25
28. Izquierdo-Cortazar, D., Robles, G., Ortega, F., Gonzalez-Barahona, J.M.: Using software archaeology to measure knowledge loss in software projects due to developer turnover. Proc. 42nd Annu. Hawaii Int. Conf. Syst. Sci. HICSS. 1–10 (2009). https://doi.org/10.1109/HICSS.2009.498
29. Hall, T. (Brunel U., Beecham, S. (University of H., Verner, J. (National I.A.S., Wilson, David (University of Technology, S.: The Impact of Staff Turnover on Software Projects : The Importance of Understanding What Makes Software Practitioners Tick. Methodology. 30–39 (2008). https://doi.org/10.1145/1355238.1355245
30. Robles, G., Gonzalez-Barahona, J.M.: Contributor turnover in libre software projects. IFIP Int. Fed. Inf. Process. **203**, 273–286 (2006). https://doi.org/10.1007/0-387-34226-5_28

A Machine Learning Approach for Heart Attack Prediction

Sanjay Gour, Punita Panwar, Divya Dwivedi, and Chetan Mali

Abstract There are lot of activities in the human body which are unpredictable in general form. Heart attack is one of them, and it is so serious activity in human body which cause in death of human. But this action gives some sim-terms before it committed. Although it is not highly noticeable in normal conditions, it committed suddenly. So this is one of the highly unpredictable occurrences in human body. With the technology advancement some of the data mining algorithms were developed to predict the heart attack. In such continuation, different data mining algorithms, with machine learning approach are able to predict heart attack occurrence in human body. It is one of the typical diagnosis tasks, but it should be achieved accurately and efficiently with the help of machine learning. The present research paper is an attempt to model and solve the heart attack prediction problem. Different machine algorithms such as KNN, SVC, RFC, decision tree and logistic regression are taken here to form a model for the study. The forte of the projected model is sufficient and is competent enough to predict the occurrence of heart attack in the human body by the compilation of some sim-terms occurring in the body. So machine learning approach is a good approach to predict occurrence of heart attack timely and we may go for proper treatment and precaution for the same.

Keywords Machine learning · Data mining · Algorithm · Decision tree · KNN

S. Gour (✉) · P. Panwar
Department of Computer Science and Engineering, Jaipur Engineering College and Research Centre, Jaipur, India

D. Dwivedi
Department of Computer Science and Engineering, Bhupal Nobels University, Udaipur, India

C. Mali
Guest Faculty, Department of Computer Science& IT Mohanlal Sukhadia University, Udaipur, India

© The Author(s), under exclusive license to Springer Nature Singapore Pte Ltd. 2022
A. K. Nagar et al. (eds.), *Intelligent Sustainable Systems*, Lecture Notes in Networks and Systems 333, https://doi.org/10.1007/978-981-16-6309-3_70

1 Introduction

The role of predictive analytics is associated from machine learning under the domain of artificial intelligence where the descriptive analytics has no direct relation with machine learning. The important difference here is prediction of future casting or forecasting. The role of prediction is very much important when the result of prediction is associated with human life. In the present scenario, the role of machine learning for prediction is very much crucial. With such reference, in this study the role of machine learning is consider to predict the occurrence of heart attack in human body. So we can say that machine learning is a way of operating and predicting of hidden, formerly unidentified, identified and latent beneficial information. Machine learning approach comprises several classifiers of supervised learning, where unsupervised learning is utilised to predict and discover the accuracy of the provided dataset. To predict the occurrence of heart attack in human body, in the present study, some key approaches are taken in the consideration which are KNN, SVC, RFC, decision tree and logistic regression. These algorithms or approaches are taken here to form a model for the study. The aim of this research is to examine whether the person is likely to be committed the heart attack in near future. This prediction will be committed by the help of some selected attribute which are concern from the medical data. Actually, these attributes are holding sim-terms of heart attack in human body. Therefore, a dataset of sufficient attribute is taken in the consideration by which we can predict the said occurrence. Here, we take care that we utilised essential attributes for the prediction.

2 Review of Related Work

Fahd Saleh Alotaibi gives a scenario of implementation of machine learning model through the combination of five diverse algorithms. In the study, Rapid miner tool is used, which calculated high accuracy than other tool as MATLAB and Weka. The result of the study was very significant and highly accurate. The outcomes point out that the system can be useful in prevention of chances of heart attack [1].

A soundless work associated with the identification of Cardiovascular Heart disease by utilising machine learning approach has motivated in this work. The study contains a momentary survey and cardiovascular disease forecast has been prepared through algorithms including logistic regression, KNN, RFC, etc. [2].

A significant amount of work carries out in the direction of prediction of heart disease using machine learning. In the same direction Harshit Jindal et al. worked for the same under the title of heart disease prediction using machine learning algorithms. This paper produced a model for cardiovascular disease detection by utilising three machine learning approaches. The study predicts cardiovascular disease through the medical history of the patient.

Some of the machine learning approaches which are taken here to form the model were Logistic regression, Random Forest Classifier and KNN [3].

The model including IHDPS had the capability to compute the decision boundary by utilising the preceding and novel model of machine learning. It simplified the imperative and key knowledge just like the family history associated with the heart disease. The study also carried out in the direction of identification of risk factors related to coronary heart disease which is identified through approaches of neural network and was accurately capable to predict the suffering of the patient or not [4].

Parthiban, Latha and Subramanian et al. introduced the prediction of heart dieses through a numbers of attributes by utilising neural networks which was implemented in R programming. The model of the study was developed on the basis of deep neural network. There were output perceptron and it carries 120 hidden layers which are key technique of confirming precise consequence of heart disease [5].

Rajesh et al. worked to predict the heart disease by using various algorithms of machine learning. Naive Bayes algorithm is taken with dataset which consist data on risk factors. They also utilised decision trees and other grouping of algorithms for the prediction of heart disease with said dataset. In the final result, they found that small size dataset gives good result with naive Bayes algorithm and while the dataset is big than decision trees results are better [6].

Raihan M et al. depicted that supervised network has been recommend for the identification of heart diseases, while the assessment of the model was completed by the doctors with unfamiliar data. The projected model was trained with the previous learned data and predicted the result thus computing the precision of the model [7].

3 Dataset

For the study, a well-organised dataset has been taken with the necessary attributes required for the prediction of heart attack occurrence. In the present study, a dataset imported from the online facilities with kaggle.com and was saved in the comma separated (.csv) form. This dataset consists of key medical details of 303 diverse patient of diverse age groups. The attributes available in the dataset was sufficient to develop the model and predict the desire outcomes. The attributes which are available in the dataset are as age, sex, chest pain, blood pressure, serum cholesterol, fasting blood sugar, resting electrocardiograph, maximum health rate achieved, exercise induced angina, ST depression, slop, defect type, etc. Therefore, there were 13 attributes in the dataset. Four attributes were dropped as they are insignificant feature for the study. So finally 09 attributes are considered for modelling and prediction.

4 Methodology

The straightforward methodology used for the prediction of heart attack in human body. The methodology used in the study is shown by the diagram (Fig. 1). The methodology comprises various machine learning algorithm which are used to train the machine as well as test the dataset for final prediction.

At the stage-1, dataset for the study was imported in the prescribed software platform, and here we are utilising python programming language for the same.

At the stage-2, the operation of the removal of insignificance features/variables takes place. Here the un-wanted variables were removed from the imported dataset.

At the stage-3, the process of label encoding of categorical variable takes place, and it is one of the importance aspects for representing the numerical values.

At the stage-4, this stage is concern from the separation of the dataset for further process. Here, dataset separated for performing two major activities. First part of the split dataset is taken as training dataset, and rest of the dataset is known as the test dataset.

At the stage-5, dataset which is used to train the machine are processed with various machine learning algorithms, these are, namely K-nearest neighbours (KNN), support vector classifier, random forest classifiers, decision tree, and logistic regression.

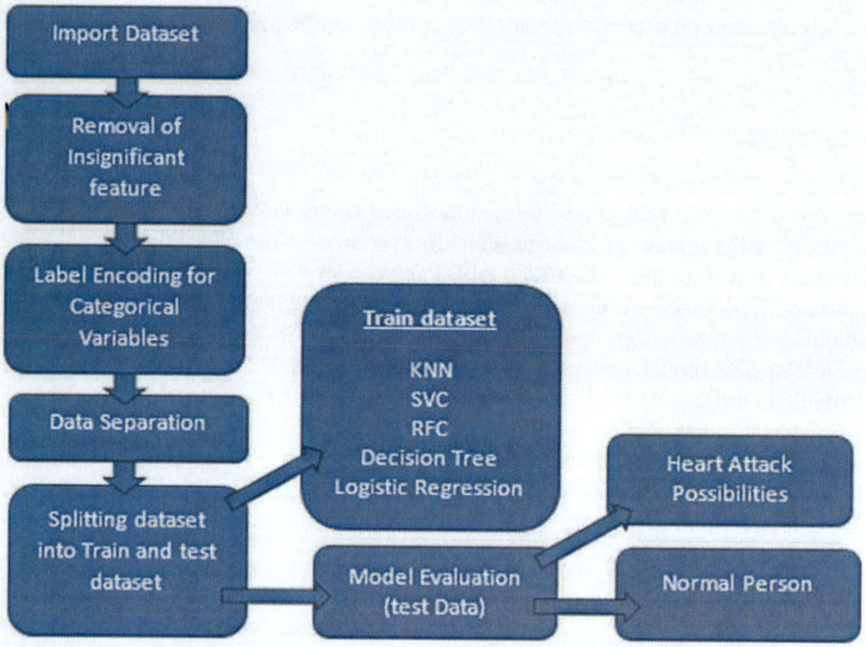

Fig. 1 Flow chart of methodology

At the stage-6, dataset is split as the test dataset taken in the account. Here, such dataset process for final outcome in association of model prepared with the help of machine learning algorithms which already trained the machine on the training dataset.

At the stage-7, the final consequence of the model evolution is recorded. The final output is available in the categorical form which shows either the possibilities of heart attack in the person or the normal person.

5 Experimental and Result

The experimental result of the study is depicted on the basis of the accuracy received from various algorithms. From these consequences, we can observe that though almost every scholar or expert of the machine learning field are utilising the algorithms like KNN, diagnosis and prediction of heart diseases and similar one but the result of these algorithms are different. The difference in the result obviously depends on the dataset, feature selection and modelling (Fig. 2).

Here in this study, the algorithms that we considered for modelling are accurate enough to solve the desire problem, and these are very cost-effective and efficient and provide solutions in the rapid time span. Python programming framework is used to complete the study and experimental (Table 1).

As per the experimental maximum accuracy received from SVC and logistic regression algorithms, both the algorithms give almost equal accuracy in the current modelling which is 80%. Similarly, KNN and random forest classifiers are also

Fig. 2 Code of accuracy output

```
x_train shape:  (212, 8)
x_test  shape:  (91, 8)
y_train shape:  (212,)
y_test  shape:  (91,)
KNeighborsClassifier: 0.7582417582417582
SVC Classifier: 0.8021978021978022
RandomForestClassifier: 0.7472527472527473
DecisionTreeClassifier: 0.6703296703296703
LogisticRegression: 0.7912087912087912
```

Table 1 Accuracy of various algorithms

SN	Algorithm	Accuracy (%)
1	SVC	80.21
2	Logistic regression	79.12
3	KNN	75.82
4	Random forest	74.72
5	Decision tree	67.03

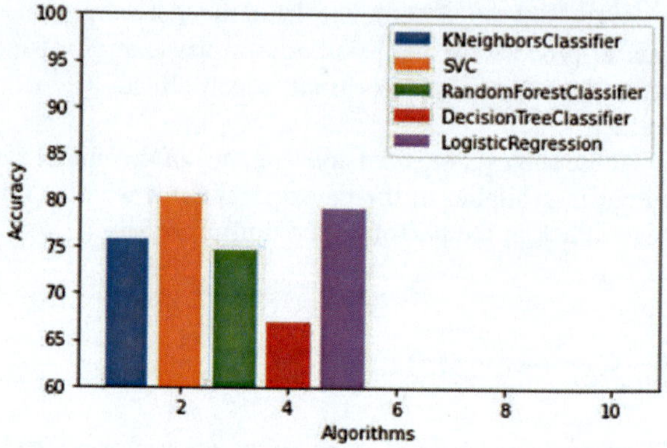

Fig. 3 Accuracy graph

performing well with almost 75% accuracy. Algorithm decision tree classifier is not performed as much as compared to the others, and here, the accuracy of decision tree classifier is near to 67% (Fig. 3).

As per the graph and result, it is found that the algorithm KNN and logistic regression are more efficient that the other algorithms considered in the model. The decision tree performance is not up to mark as the other in the present study.

6 Conclusion

In the present study what we found that with normal dataset, the decision trees algorithm does not give solution with good accuracy; while we looking towards the SVC and logistic regression, we are receiving more accurate results or accuracy with prospects of the entire other possibilities. Similarly, algorithm KNN and random forest classifiers are also performing well in the satisfactory mode. At the end, we can say that by this experiment that SVC is more accurate when the dataset size is near to 300 records and variables are 08–12. Similar performance will be aspect by the logistic regression algorithm. The study is able to find out prediction for heart attack occurrence timely by the help of other body sim-terms. So we can say that machine learning algorithms are capable to predict heart attack occurrence efficiently, effectively and accurately.

References

1. Alotaibi, F.S.: Implementation of machine learning model to predict heart failure disease. Int. J. Adv. Comput. Sci. Appl. **10**(6), 261–268 (2019). www.ijacsa.thesai.org

2. Ganna, A., Magnusson, P.K., Pedersen, N.L., de Faire, U., Reilly, M., Ärnlöv, J., Ingelsson, E.: Multilocus genetic risk scores for coronary heart disease prediction. Arterioscler. Thromb. Vasc. Biol. **33**(9), 2267–2272 (2013)
3. Jindal, H., Agrawal, S., Kheral, R., Jain, R., Nagrath, P.: Heart disease prediction using machine learning algorithms, IOP Conf. Series: Materials Science and Engineering 1022, 012072 (2021), IOP Publishing. https://doi.org/10.1088/1757-899X/1022/1/012072
4. Jabbar, M.A., Deekshatulu, B.L., Chandra, P.: Heart disease prediction using lazy associative classification. In 2013 International Mutli-Conference on Automation, Computing, Communication, Control and Compressed Sensing (iMac4s), pp. 40–6. IEEE, (Mar 2013)
5. Parthiban, L., Subramanian, R.: "Intelligent heart disease prediction systemusing CANFIS and genetic algorithm." Int. J. Biol. Biomed. Med. Sci. **3**(3) (2008)
6. Rajesh, N., Maneesha, T., Hafeez, S., Harikrishna: Prediction of heart disease using machine learning algorithms. Int. J. Eng. Technol. **7**(2.32), 363–366 (2018)
7. Raihan, M., Mondal, S., More, A., Sagor, M.O.F., Sikder, G., Majumder, M.A., Ghosh, K.: Smartphone based ischemic heart disease (heart attack) risk prediction using clinical data and data mining approaches, a prototype design. In 2016 19th International Conference on Computer and Information Technology (ICCIT), pp. 299–303, IEEE, (Dec 2016)

Bio-Inspired Framework for Security in IoT Devices

Dion Mariyanayagam, Pancham Shukla, and Bal S. Virdee

Abstract One of the many things COVID-19 has taught humanity is that the Internet is not just a commodity but a vital service integral to the modern world. As we become ever more connected, there is a growing need to secure data and communication streams. If data is valued, then it should be protected. Unfortunately, some of the least secure devices in modern electronic systems are the Internet of Things (IoT) devices—partly due to their low processing power and always—on functionality. Polymorphism is the notion of changing one's form. In biological organisms, polymorphic (mutating or changing) viruses trick the natural security mechanisms by changing their unique signatures (e.g. DNA or proteins). In computing, antivirus software systems are adapted to detect and remove constantly changing software viruses. However, polymorphism at the firmware level and over the wireless medium is neither well understood nor explored for IoT devices. This paper proposes a novel and bio-inspired framework for securing distributed IoT devices often assumed to be working at the intersection of engineering, computing, and cybersecurity domains. The proposed framework attempts to exploit the notion of polymorphism in resource-constrained (e.g. memory, power, bandwidth) IoT devices. The framework's core aim is to detect, reject, and block foreign agents individually or collaboratively and in real-time within a client and server model by changing the access credentials and encryption keys as soon as an unauthorised client is detected. The framework proposed for the bio-inspired framework for security in IoT devices is designed to remain operationally compartmentalised, functionally integrated, and objectively unified.

Keywords Internet of Things (IoT) · Polymorphism · Bio-inspired

---

D. Mariyanayagam (✉) · P. Shukla · B. S. Virdee
Centre for Communications Technology, London Metropolitan University London, London, UK

P. Shukla
e-mail: p.shukla@londonmet.ac.uk

B. S. Virdee
e-mail: b.virdee@londonmet.ac.uk

© The Author(s), under exclusive license to Springer Nature Singapore Pte Ltd. 2022
A. K. Nagar et al. (eds.), *Intelligent Sustainable Systems*, Lecture Notes in Networks and Systems 333, https://doi.org/10.1007/978-981-16-6309-3_71

1 Introduction

Internet of Things (IoT), which enables a wide variety of embedded devices, sensors and actuators (known as smart things) to interconnect and exchange data, is a promising network scenario for bridging the physical devices and virtual objects in the cyber world. However, such smart devices and sensitive data are vulnerable to security threats. Security is, therefore, the central area of focus for researchers in the field of IoT. Consequently, it is essential to develop cryptography technologies to secure the data from unauthorised access. Moreover, this can be achieved by transforming the data into an unrecognisable and unrelatable form. However, it is not easy to find one straightforward approach that will fit all IoT applications. There are various types of devices connected to an IoT network. Some devices can afford heavyweight and high-security methodologies, but most IoT devices are resource-constrained. They need a security solution that acts fast. Simultaneously, it needs to be simple in its complexity and versatile.

Last but not least important factor is trusted security. In general, the dedicated cryptographic algorithms need to be lightweight in terms of area, memory footprint, power and energy consumption. Therefore, the proposed research is pertinent since security and privacy in the IoT are not fully addressed.

Polymorphism is a very uncommon term and rarely researched topic within the traditional tech, cybersecurity and engineering industries. However, in the computer science discipline, one knows that a polymorphic code can change the initial code, but the code's function produces the same result; for example, $4 + 4$ and 2×4 both have the same results 8.

The nature of changing or mutating forms such as viruses (e.g. malware) that bypass the security system is more familiar to hackers than many security experts [1]. However, these mutating forms are well studied and documented in the human sciences, particularly biology. Thus, the notion of the proposed bio-inspired framework for security closely follows and resembles the biological functions of the immune system. In particular of a human white blood cell due to its ability to detect (intrusion detection), change (changing encryption keys and passwords) and fight (eject from the network) foreign cells inside the human body [2].

1.1 Framework Operations

The design and idea of the bio-inspired framework for IoT security, like many engineering solutions for real-world problems, takes inspiration from the most experienced engineer who is nature her self—in particular, looking at how the immune system (adaptive immune responses and innate immune responses [3][4]) in the human body. For example, how the immune system fights to combat new viruses and infections through vaccines.

This framework's fundamental nature is based on a server and client model whereby the server and client communication are encrypted. The encryption key is in the ledger (similar to blockchain technologies) and shared with authorised clients. Authorised clients are approved depending on their Universal Unique Identifier (UUID); either they are in the approved UUID address list or not. The server has a list of approved UUID addresses, and only authorised clients can connect to the server. Therefore, all authorised clients have the same ledger. This ledger contains two variables as objects which are inside an array. The first object is the Wi-Fi details (password), and the second object includes the encryption key to which the client and server communicate. The Wi-Fi details automatically change if the server detects any foreign clients trying to connect.

When any form of intrusion is detected, for example an unauthorised client successfully connects to the server, a signal (trigger) is sent between the server and client to successfully apply the change between one polymorphic form to another, for example like the transition from A - > B, where A is the original form and B is the new form.

The signal (trigger) is a message which contains a randomly chosen number that corresponds to an element in the array (ledger) that has the new credential and encryption key to use, therefore resulting in its new polymorphic form. More detailed information for the framework is in Chap. 3.

1.2 Framework Aims and Objectives

The bio-inspired framework for security in IoT devices needs to:

- Establish a wireless network with encrypted communication between the server and client devices.
- Have clients that read sensor data and send the data to the server successfully.
- Support a lightweight code to maximise the number of resources the device can utilise for sensors.
- Run multiple sensors from one IoT device without the polymorphic framework interfering and burdening the processing power with an ability to invoke the processor's sleep mode.
- Have the ability to detect unapproved clients.
- Have the means to eject unapproved clients.
- Have the function to change the security form by changing the server credentials and encryption key.
- Successfully reconnect all clients back to the server after changing forms and receive data.

2 Literature Review

Designing with a vision for the future helps devices be in service for longer as the device can be upgraded through software and modified by adding additional hardware. The paper "lattice-based cryptography for IoT in a quantum world: are we ready?"[5] shows how current FPGA's and other IoT platforms could implement lattice-based cryptography. In order to implement lattice-based cryptography in an IoT platform, there are a few challenges. One of the challenges is communication bandwidth, as most embedded processors are memory-constrained, therefore only suited for minor security parameters such as IoT applications with limited transmission bandwidth (through Wi-Fi). Security strength often balances performance and security as any brute-force cryptanalytic efforts require more computational resources increasingly for the required search on a block cypher such as AES-128 and other similar cyphers. However, the trade-off between performance and the required security is generally less desirable due to their associated overhead. This paper gives an excellent overview of what IoT devices are lacking. However, it does open a pandora box to a more significant impact on quantum computing and its ability to crack modern security systems with ease.

One of the challenging issues in IoT devices is the tampering of firmware, as it is challenging to detect and recover from the tampered firmware. The paper "ChainVeri: blockchain-based firmware verification system for IoT environment" [6] proposes a new blockchain-based firmware verification system that used a shared palette (ledger) to check the devices to see any firmware tampering have been made. The palette comprises the block header, which encompasses the block hash, block size, block version, previous block hash, time, difficulty, and nonce. The palette also has another module called the verification information, verifying the device model, firmware version, verifier, and identification. Having these functions in the palette allows the blockchain to know the block's version when the structure of the palette changes. It can also verify the device through a universally unique identifier and check and verify its firmware.

Due to IoT devices and storage's limited computational power, traditional off-the-shelf solutions are resource-containing on these devices and therefore not recommended for security implementation. The paper "Pseudo-Random Number Generator and Hash Function for Embedded Microprocessors"[7] proposes a solution to overcome this problem by implementing lightweight techniques for efficient Pseudo-Random Number Generator and Hash function to reduce memory consumption and accelerate performance. The reason behind using a pseudo-random number generator is to build and generate harder to break keys and other secret parameters on embedded processors. This paper does propose a relatively efficient way to harden existing security methods such as AES by implementing a pseudo-random number generator; therefore, not too much change needs to occur on the broader system for integration.

"Lightweight Cryptography Algorithms for resource-constrained IoT devices: A Review, Comparison and Research Opportunities" [8] is an excellent paper that provides a holistic view and compares various lightweight cryptography algorithms that are available in the market. The critical common challenges with conventional IoT-based cryptography are limited memory (RAM, ROM, and registers), limited computational power, the small surface area of the device, lower battery power (no battery power with RFID tags), and real-time operations. The way this paper compares the different cryptography algorithms is by three characteristics, physical, which is the physical area, memory and battery power. Performance determines the computing power (latency and throughput); lastly, the security characteristic measures minimum security strength in bits, attack models, and side-channel and fault-injection attacks. With these various parameters, the thesis can benefit by paying closer attention to what is valued in this paper's comparisons of different lightweight cryptography algorithms.

A comprehensive and detailed thought for what type of IoT devices should be implemented in an IoT platform in any IoT system development. Therefore, a study compares and "Reviews Low-End, Middle-End, High-End IoT devices" [9]. The way the IoT devices are classified are as follows: low-end devices (such as at ATTINY85) are classified as having less than 50kB of RAM, less than 250kB of Flash, devices that do not support an RTOS to devices with RTOS, communication protocols range from gateway communication, lightweight protocols such as Constrained Application Protocol (CoAP), and communication protocols such as HTTP. Lastly, security vulnerabilities whereby data is compromised, causing a medium to a high threat. Middle-End devices (like the ESP8266 and ESP32) provide more outstanding features and processing capabilities such as more RAM, Flash, higher clock speeds, and various communication protocols like Wi-Fi, Bluetooth and Bluetooth Low Energy (BLE). High-end devices (Raspberry Pi, PandaBoard, HummingBoard, and more) typically are single-board computers with powerful processing units and plenty of RAM to provide a graphical user interface or even the ability to run custom operating systems like Windows 10 IoT, Ubuntu, Linux, and Raspberry OS. The main reason why this study is essential for this IoT development is that it helps developers to decide which board to use for a particular functionality. Such as if there is a requirement to gather air pollution data in a given environment, there isn't a need to use a high-end device as they are more expensive for scaling up, instead of using a middle-end device would prove fruitful as they have enough processing power and a communication protocol to gather the data and send it to a server.

All the related studies have contributed in some way to the development of the framework. In fact, the proposed framework incorporates relevant features from the interdisciplinary research summarised in Sect. 2. For example, the use of a pseudo-random number generator and hash functions is a lightweight approach to develop harder to break keys and is an excellent use case for the trigger system for the proposed framework. Additionally, the labelling and classification of IoT devices into tiers according to their performance and features allows for various testing scenarios on the efficiency of the proposed framework.

3 Bio-Inspired (Polymorphic) Security Framework

Humans have an innate and adaptive immunity in terms of immunology. Innate immunity responds and recognises generic targets on foreign agents (pathogens), whereas adaptive immunity recognises specific targets using "randomly generated receptors that have a virtually unlimited recognition repertoire" [10]. There are two different types of adaptive immunity called humoral immunity and cell-mediated immunity, and please see the figure below for a visual representation. The properties of adaptive immune responses are Specificity, Diversity, Memory, Clonal expansion, Specialisation, Contraction and homeostasis, and nonreactivity of self [2]. Please see Table 1 for each feature and function for adaptive immunity properties.

a. An antigen is any toxin or foreign substance that induces an immune response in the body.
b. B Lymphocytes produce proteins called antibodies which mediates a humoral immune response.

Referring back to human biology, we can see that human immunology can detect foreign agents by using a known threat database, which, if those known threats are detected, an immune response is activated.

For programmers and engineers alike, adaptive immunity seems like an intrusion detection system with a database for known security risks. The detection of a foreign agent that correlates to a known security risk can then be subject to a specifically targeted response. It is also relatively common to block known incoming threats by either white listing specific sites, applications or communication streams or black-listing threats known to the host/organisation. Bio-Inspired framework for security in IoT devices is based around adaptive immunity, and a top-down overview of the framework is shown in Fig. 1.

Table 1 Properties of adaptive immune responses

Feature	Function
Specificity	To ensure the targeted foreign agents (antigens[*a]) gets the appropriate responses
Diversity	To have a large variety of responses for a large variety of antigens
Memory	To know which response was the most effective to protect against future exposure of the same foreign agent
Clonal expansion	Increases the number of antigen-specific lymphocytes[*b] from a small number of naive lymphocytes allows for a higher concentration-response as more cells are aware of the antigen
Specialisation	Specialisation gives the ability to generate a more specific defence against different microbes and antigens
Contraction and homeostasis	The ability to respond to an unknown or newly encountered antigen
Nonreactivity of self	Not to harm the host during the response to an antigen.

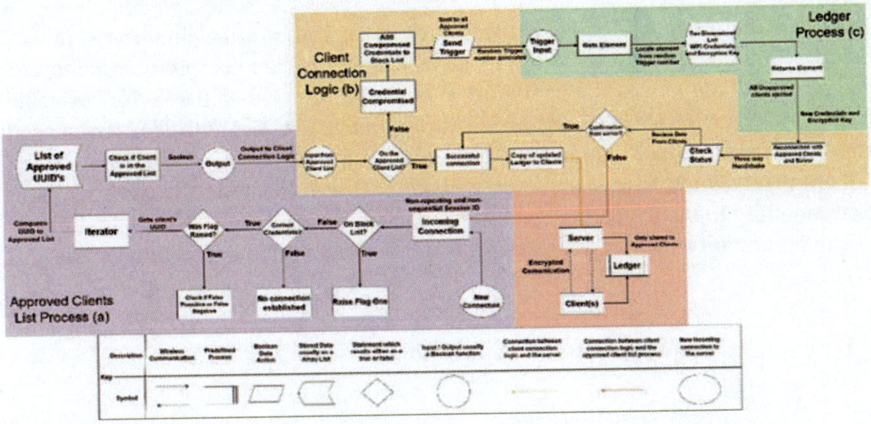

Fig. 1 A high-level overview of the bio-inspired framework for security in IoT devices

3.1 *Approved Clients List Process (ACLP)*

The beginning of the client journey starts with a new connection. The approved client list is the first line of defence against any unapproved clients. Additionally, the approved client list also acts as a RADIUS Server. The new client connection flows into the incoming connection function, where it is given a session ID that is non-repeating and non-sequential. Once given the session ID, the process then checks if the client has the correct credentials to connect to the server. The framework checks if the incoming client is on the block list and raises a flag if true. If the credentials are incorrect, then there is no connection established between the incoming client and the server. If the credentials are valid, the server will check if the flag for raised to see if there was a false positive or a false negative, then it moves on to request the client's UUID to compare using an iterator (a search function) with the approved UUID addresses. The iterator function will return and output either a "true" or "false" whether the UUID address is on the list or not. Referring back to the properties of adaptive immune responses, the ACLP deeply reflects features such as specificity, diversity, memory and specialisation.

3.2 *Client Connection Logic (CCL)*

The return output from the approved client list becomes an input variable to the client connection logic, as this function will react accordingly depending on the variable. If the variable is "true" (the connected client is on the UUID address list), this is considered a successful connection, and the server gives a copy of the ledger with the element location for the current password and encryption key. At this stage, the client and the server have successfully connected and can communicate with other clients and the server when necessary. Contrariwise, if the variable returns a "false", then the implication is that the current server credentials are compromised, and this unapproved client knows the password to connect. As soon as the "false" variable returns, the credentials used are added to the block list, and a trigger (composed from a random number generator) is sent to all approved connected clients. When a client receives the trigger value immediately, it will locate the ledger's element corresponding to the trigger value. The primary purpose of CCL reflects features of adaptive immunity such as nonreactive to self and contraction and homeostasis.

3.3 *Ledger Process(LP)*

The ledger only shared between approved clients contains a database comprised of a two-dimensional list, the password to connect to the server (Wi-Fi credentials) and the encryption key required to decypher the communication stream between client to client (node and peer to peer communication) and from client to server. The ledger's returned element allows the approved clients to reconnect to the server with the new credentials. As a result, all the unapproved clients get automatically rejected from the system as they do not have the newly updated credentials. When the clients attempt reconnection after the trigger is activated, a three-way handshake is performed, the server sends a synchronisation, the client syncs to the server synchronisation and acknowledges it then the server acknowledges the client's acknowledgement. After a three-way handshake is successful, the client gets a copy of the updated ledger with the credentials' new element location. However, if the three-way handshake is unsuccessful (the server did not acknowledge the client), then the client is reverted to an incoming connection to double-check if it is in the approved client list and if it is, it will get a copy of the ledger, and if it is not, it will be rejected. LP reflects adaptive immunity functions such as contraction and homeostasis, and clonal expansion.

The server and client's communication stream is encrypted using a block cypher to add additional security against any form of data theft and intrusion, such as a man in the middle attack listening or probing intrusions.

4 Conclusion and Future Work

In summation, we presented a novel bio-inspired framework for the security of IoT devices by an amalgamation of interdisciplinary concepts from immunology and computer systems security, which allows the framework to auto-detect foreign agents, reject unauthorised clients, and the polymorphic ability to change client credentials and encrypted communication streams.

The key features of the proposed framework are auto-detection of foreign agents, rejection of unauthorised clients, and demonstrate the notion of polymorphism by changing client credentials and encrypted communication streams.

Nonetheless, the next step is a system realisation of the bio-inspired framework using different categories of IoT devices. The future works aim to test the encumbrance on the framework for three categories of IoT devices: low-end, middle-end, and high-end. Authors envisage that the effectiveness of the system realisation of this framework will be governed by technological challenges such as onboard memory, processing speed, power consumption/management and fault-tolerant connectivity.

Acknowledgements This work is supported by the University's Rescaling Funds awarded to Dr Pancham Shukla for developing a PhD researcher. The first author, Dion, would like to thank his PhD supervisor Dr Pancham Shukla and Prof Bal Virdee for their guidance and support.

References

1. Li, X., Loh, P.K.K., Tan, F.: "Mechanisms of polymorphic and metamorphic viruses," Proc—2011 Eur. Intell. Secur. Informatics Conf. EISIC 2011, pp. 149–154 (2011). https://doi.org/10.1109/EISIC.2011.77
2. Abbas, A.K., Lichtman, A.H., Pillai, S.: Basic Immunology: Functions and of the Immune System (2016)
3. Janeway, C.A., Travers, P., Walport, M., Shlomchik, M.: Immunobiology—The Immune System in Health and Disease, 5th edn. Garland Publishing Library, New York (2007)
4. Murphy, K., Weaver, C., Weaver, M., Geha, R., Notarangelo, L.: Janeway's Immunobiology, 9th ed. New York and London, Garland Science (2017)
5. Khalid, A., McCarthy, S., O'Neill, M., Liu, W.: "Lattice-based Cryptography for IoT in A Quantum World: Are We Ready?," Proc—2019 8th Int. Work. Adv. Sensors Interfaces, IWASI 2019, pp. 194–199 (2019). https://doi.org/10.1109/IWASI.2019.8791343.
6. Lim, J., Kim, Y., Yoo, C.: "ChainVeri : Blockchain-based Firmware Verification System for IoT environment," pp. 1050–1056 (2018). https://doi.org/10.1109/Cybermatics.
7. Seo, H., Choi, J., Kem, H., Park, T., Kim, H.: "Pseudo random number generator and Hash function for embedded microprocessors. 2014 IEEE World Forum Internet Things WF-IoT **2014**, 37–40 (2014). https://doi.org/10.1109/WF-IoT.2014.6803113
8. Thakor, V.A., Razzaque, M.A., Khandaker, M.R.A.: "Lightweight Cryptography algorithms for resource-constrained IoT devices: a review. Comparison Research Opportunities," IEEE Access, pp. 1–17 (2021). https://doi.org/10.1109/ACCESS.2021.3052867
9. Ojo, M.O., Giordano, S., Procissi, G., Seitanidis, I.N.: A review of low-end, middle-end, and high-end Iot devices. IEEE Access **6**, 70528–70554 (2018). https://doi.org/10.1109/ACCESS.2018.2879615
10. Flower, D., Timmis, J.: In Silico Immunology, 1st edn., vol. 1(1). London: Springer (2006)

Reaction Capacity Identification Problem: Is There Any Way to Formalize Scientific Knowledge and Automate Reasoning so that Intelligent Systems Can Solve It?

Karina A. Gulyaeva and Irina L. Artemieva

Abstract The article describes the methodology to develop intelligent systems that are intended to address the reaction capacity identification problem, which can only be solved (at least partially) by professional chemists. The chosen approach can be called ontological since it starts with the ontology models. Utilizing the system approach, we start with the "Systems" meta-ontology model development. "Chemical compounds and their electron properties" ontology model is developed in the terms of "Systems" meta-ontology model. This ontology model has several modules at each system's level. These modules contain entities associated with each system's subsystem. These ontology models are at the heart of meta-ontology editor, "Chemical compounds and their electron properties" ontology editor, and knowledge editors. Knowledge engineers and chemists work together to fill the editors with relevant information. Since chemical compound structural formula input is the essential system interface, the JSME component has been introduced to the system. End users enter the chemical compound, they are interested in, into the intelligent system with the help of JSME. The intelligent system, in its turn, analyzes the parent compound structure, searches for the functional fragments that the parent compound possesses in its knowledge base, analyzes several characteristics found, and reasons on the reaction capacity of the parent compound. The prototype of the intelligent system is provided. It is written with the help of the Django framework.

Keywords Applied logic · Intelligent system · Ontology · Ontology model · Reaction capacity · Reaction capacity identification · Functional group

1 Introduction

Reaction capacity is an umbrella term for the chemical compound reactivity. The possibility, variety, and special features of chemical reactions are those aspects that are implied by the term "Chemical compound reaction capacity". Reaction capacity is often described with the help of scalar terms: {low, high}. Major factors that

K. A. Gulyaeva (✉) · I. L. Artemieva
Far Eastern Federal University, Vladivostok, Russia

© The Author(s), under exclusive license to Springer Nature Singapore Pte Ltd. 2022

A. K. Nagar et al. (eds.), *Intelligent Sustainable Systems*, Lecture Notes in Networks and Systems 333, https://doi.org/10.1007/978-981-16-6309-3_72

determine reaction capacity are the chemical compound structure (initial substance structure and reactive chemical structure) and reaction conditions. A contemporary view of reaction capacity is based on the assessment of the electron density redistribution in the initial substance stemming from the presence of the reactive chemical. For the most part, any chemist can predict the results of such redistribution taking into account the structure of the chemical compound under review. The more knowledge the scientist possesses, the more concise his or her predictions are. The domain expert acquires this kind of knowledge during dozens of years of studying and hands-on laboratory work in his or her domain area. Scientist brain "knowledge base" has hundreds of rules of various detail levels (generally, in the form of equalities, inequalities, or implications, e.g., "If the outermost electron shell is completely filled, a chemical bond cannot be formed") and cases (precedents, e.g., inert noble gases; and counter examples, e.g., excimers). Although humans do not have absolute knowledge about the microcosm laws of nature (and it is hardly likely that humans ever have such knowledge), we can reason about the chemical compound reaction capacity based on the incomplete information we possess. Almost any chemist can undertake the following thought experiment: Choose a chemical compound and mark the fragments of the compound (these fragments can have non-empty intersections with other fragments of the same compound) that have the structure he or she has already seen (one after the other). For each fragment marked in the previous step, the chemist searches his or her "knowledge base" to extract rules and cases relevant to the fragment. Evaluating the extracted information, the chemist can reason about the initial compound reaction capacity. However, one should be aware of the fact that a whole does not always possess the properties of its components. For instance, benzene does not participate in addition reactions that are typical for alkenes because of its uniform π-electron cloud. Chemists (especially—organic chemists) acquire such knowledge by studying the characteristics and typical reactions of the most frequently encountered functional groups—factual evidence that has been inquisitively collected by scientists for hundreds of years now. The concept of the functional group, as a subset of atoms organized in such a way that defines physical and chemical properties and reaction capacity of the chemical compound that is comprised of it, penetrates organic, medical chemistry, the study of toxins, spectroscopy in many of its aspects, and other branches of chemical science. Already classics S. Patai's "The Chemistry of Functional Groups", which describes many classes of organic chemistry functional groups, consists of more than a hundred volumes now (the first volume appeared in 1964) [14].

The design and development of the intelligent system intended to solve the reaction capacity identification problem—a problem originally meant solely for humans (specifically, domain experts)—contain many challenges, traps, and pitfalls. Since the DENDRAL project, intelligent system development has seemed a promising solution for many chemistry tasks that require both brute force algorithms and methods associated with artificial intelligence (AI) [12]. While the former methods (i.e., brute force algorithms) are intended to deal with exponentially growing datum volumes in chemistry (the number of databases filled with relevant "Chemical" information and database table records has been growing ever since, e.g., the number of CAS registry compounds exceeded 100 M in 2014 [2]), the latter—AI—methods are evolving rapidly now. Nevertheless, the best knowledge formalization technique is still an open issue. Knowledge-based systems differ from other information systems on the basis of separation of concerns: Information components (i.e., knowledge base) are detached from problem-solver components. Knowledge-based systems and efficient and effective methods to develop them have been elaborated by the following prominent scientists, some of which have received the Nobel Prize: Feigenbaum et al. [12], Newell and Simon et al. [13], Gruber [5], Pospelov et al., Kleschev et al. [9–11], and others.

This paper's major focus is the presentation of the overview of the ontological methodology to develop the intelligent system intended to solve the reaction capacity identification problem. System requirement analysis and design steps can be found in [6, 8].

2 Ontologies, Meta-Ontologies, and Their Mathematical Models

Ontology, tracing its origin to pre-Socratic Greek philosopher Parmenides of Elea, has also become an information technology concept. Defined by T. Gruber as "an explicit specification of a conceptualization" [5], ontology has gained its importance in application domain modeling. Frankly, the approaches to model application domain differ. We adhere to the methodology developed by A. Kleschev and his collaborators at the Institute of Automation and Control Processes (Far Eastern Branch of the Russian Academy of Sciences). Some special aspects of this approach can be found [1, 7, 11]. Ontology is defined similarly to Gruber's definition. It contains terms, term definitions, and term attributes. According to this approach, the ontology model is developed in mathematical terms with the help of the mathematical apparatus of algebraic systems. The ontology model is described using applied logics language. This language is declarative. It has a kernel, standard extension, and several specialized extensions ("Intervals", "Quantifiers", etc.). The language is defined in [11].

Specifically for our task, we have also defined two specialized extensions: "Complex numbers" and "Orders" (partial and linear). These extensions define the terms we need to develop two major ontology models: the "Systems" meta-ontology model and the "Chemical compounds and their electron properties" ontology model. The latter model is defined in terms of the former model (that is why "Systems" is called a meta-ontology).

Some ontology models can have recursive definitions. The following sample from the "Systems" meta-ontology model defines types, or sorts, "Systems" (Eq. 1), "Subsys", "DirectSubsys", "AllSubsys" (Eq. 2) and sets a restriction (Eq. 3), which means that any system cannot be its subsystem. The function "AllSubsys" is calculated recursively (Eq. 5) making use of the function "DirectSubsys" (Eq. 4). These examples are written in the declarative language of applied logics.

$$\textbf{\textit{Sort}}\ \text{Systems} \in (\{\}\,\mathrm{N}\,)\setminus\{\emptyset\} \tag{1}$$

$$\forall(\text{FName} \in \{\text{Subsys, DirectSubsys, AllSubsys}\})$$

$$\textbf{\textit{Sort}}\ \text{FName} \in \text{Systems} \rightarrow \{\}\text{Systems} \tag{2}$$

$$\underline{\text{Restriction}}\ \forall(\text{FName} \in \{\text{Subsys, DirectSubsys, AllSubsys}\})$$

$$\forall(\text{sys} \in \text{Systems})\text{sys} \notin \text{FName(sys)} \tag{3}$$

$$\text{DirectSubsys} \equiv (\lambda(s \in \text{Systems})\{\}\text{Subsys}(s)) \tag{4}$$

$$\text{AllSubsys} \equiv (\lambda(s \in \text{Systems})\text{DirectSubsys}(s)\cup$$

$$(\cup(\text{subs} \in \text{DirectSubsys}(s))\ \text{AllSubsys(subs)})) \tag{5}$$

"Chemical reaction" and "Chemical compound" are generally viewed by chemists as systems. The idea of representing chemical reaction as a system with several levels comes quite naturally. Figure 1 depicts chemical reaction as a five-level system and embodies all its subsystems.

Based on the "Systems" meta-ontology model, the "Chemical compounds and their electron properties" ontology model contains at least one module for each of the "Chemical reaction" system's subsystem. Each module has its terms, term definitions, and term attributes (e.g., "Chemical compound" subsystem contains modules: "Chemical bonds", "Molecule", "Diatomic molecules", "Molecule representation", and "Functional groups"). Some module terms represent entities of the subsystem this module belongs to (e.g., "Atoms" entity belongs to the "Atom" subsystem's module "Atom", "Molecules" entity belongs to the "Chemical compound" subsystem's module "Molecule", etc.).

3 Editors Based on the Ontology Models

After the above-mentioned ontologies and their mathematical models have been designed, the editors for "Systems" meta-ontology and "Chemical compounds and their electron properties" ontology can be developed. Afterwards, the editor for the knowledge base is developed. The hierarchical arrangement is the following: A meta-editor based on the "Systems" meta-ontology is developed first, then the editor for the "Chemical compounds and their electron properties" ontology is developed

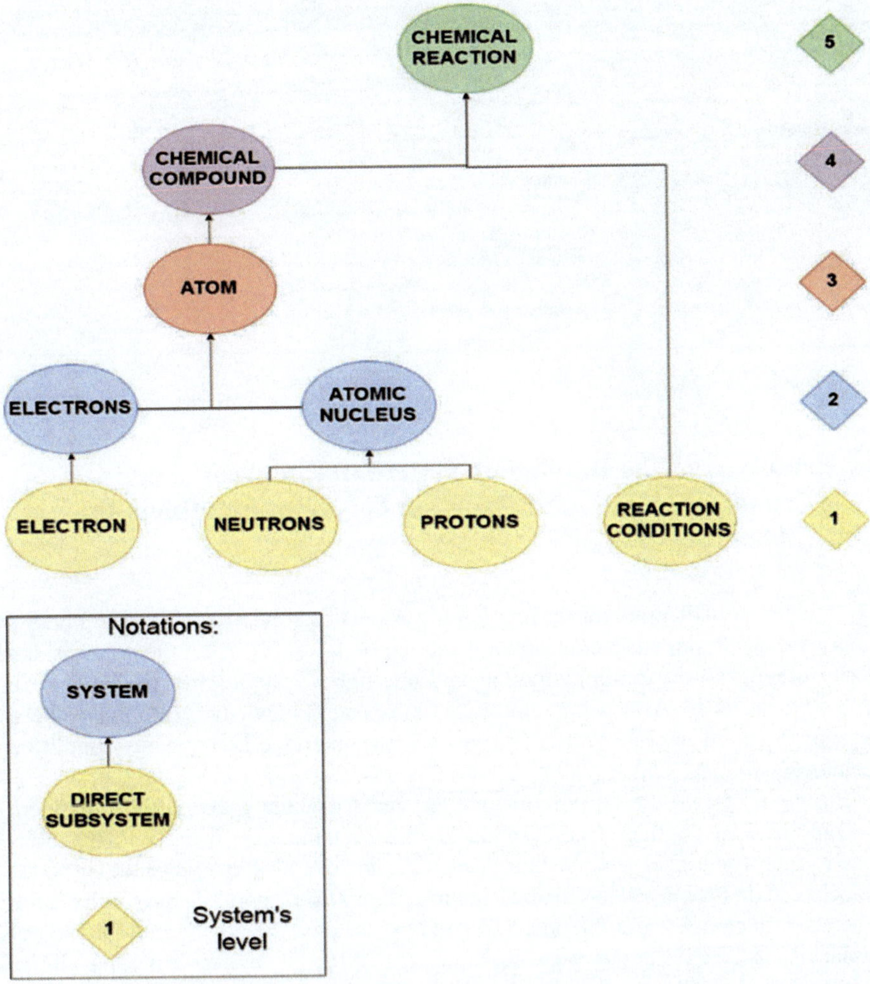

Fig. 1 Chemical reaction system and its each level subsystems

(utilizing the meta-editor), and then the editor for the knowledge base is developed (utilizing the "Chemical compounds and their electron properties" editor).

The stack of technologies used in the development process may differ. We have used the Django web framework. It is a Model-View-Template framework, which heavily utilizes the ORM concept. Moreover, the Admin panel is immensely customizable. Figure 2 demonstrates the work in the "Systems" meta-editor, which is developed with the help of Django Admin. Figure 2 shows that an engineer chooses the direct subsystem for the system provided ("Atom" is a direct subsystem of the "Chemical compound"). It is presumed that a knowledge engineer works with ontology editors, and a chemist works with knowledge editors.

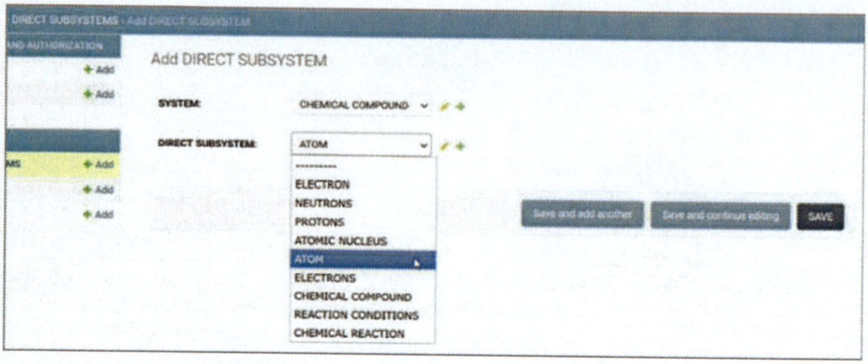

Fig. 2 "Systems" meta-editor

4 Prototype of the Intelligent System that Solves a Chemical Compound Reaction Capacity Identification Problem

The architectural diagram for the intelligent system that solves a chemical compound reaction capacity identification problem can be found in [8]. After the analysis and design stages, all the ontology models and intelligent system artifacts should have already been made. After all the ontology editors and knowledge editors have been developed (or in parallel with this step), all the interface components should be developed.

Structure representation components are indispensable parts of the intelligent system that works with molecular input. Chemists should be provided with the interface to enter functional groups, their properties, and chemical reactions these groups can engage in. Moreover, intelligent system primary users, who can have or have not education in chemistry or chemistry-related fields, should have the interface to put molecules in their structural form into the system. After any molecule has been introduced by the user ("parent" molecule), the intelligent system searches its knowledge base (substructure search is performed utilizing Ullmann's algorithm for subgraph isomorphism [15]). If any molecular fragment that the parent molecule possesses has already been entered into the system by the chemist, then the system will mark the atoms in the parent molecule that belong to the fragment found. Figure 4 depicts benzene, and Fig. 3 depicts the results of the search for the knowledge base fragments in the parent molecule. The system "understands" that the parent molecule contains three benzenes, and that the benzene is a uniform structure (and not a structure that contains alternating single-double bonds or alkenes).

The intelligent system described above has the opportunities to reason stronger as its knowledge base extends. Now it can make some predictions concerning the electron density shifts due to the presence of particular functional groups, compare atom electronegativities, etc.

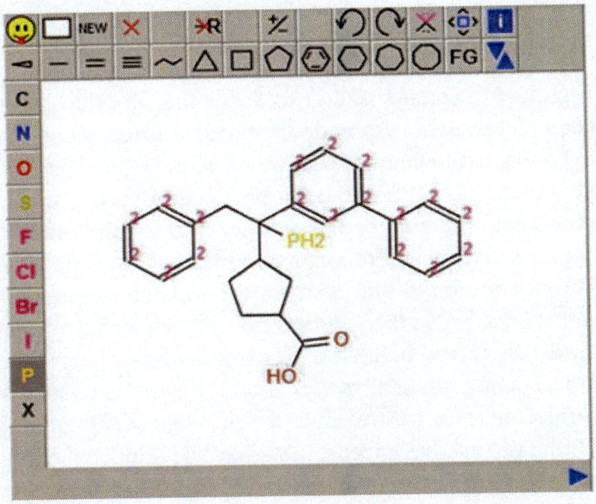

Fig. 3 Parent molecular structure (entered by the user) with three found benzenes (JSME [3, 4])

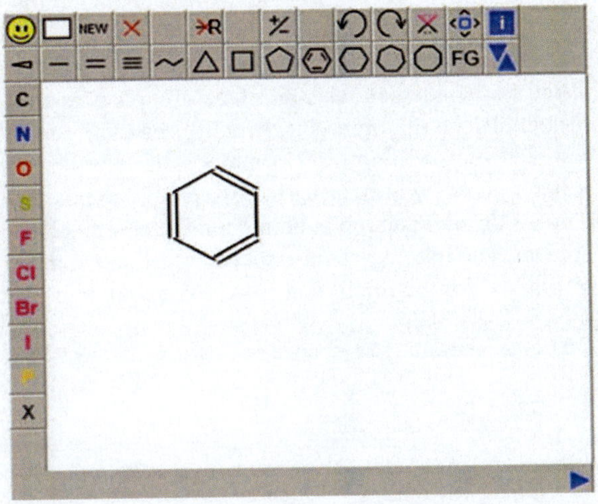

Fig. 4 Benzene from the knowledge base (JSME [3, 4])

Molecular input is made with the help of JSME, which is a lightweight JavaScript editor (JSME consists of a JavaScript bootstrap file and six HTML files weighing 257–271 KB depending upon the browser type) that can be easily embedded into the HTML page [3, 4]. It has been tested successfully in all the major browsers: Firefox, Chrome, Safari, Opera, and IE (including IE 6 and IE 7). Molecular editing functions fully perform on iPhone, iPad, and Android (small bug has been found during molecule rotation on Android). JSME input and output formats include SMILES and MOL files used by many cheminformatics libraries. It has a BSD license, which is very convenient [3, 4].

5 Conclusion

Intelligent systems intend to solve the problems originally meant for the human brain. Intelligent system development has many challenges if the application domain is connected to science. Chemists acquire their knowledge by dedicating dozens of years to studying and laboratory experiments. This helps them to predict chemical compound reaction capacity by examining the initial compound structure, reactive agent structure, and reaction conditions. Chemists examine functional groups of the parent compound and analyze their characteristics. The intelligent system that is intended to solve the reaction capacity identification problem should imitate professional chemists' behavior. To automate reasoning, the application domain ontology and its mathematical model should be analyzed and developed. This approach has turned out to be fruitful since the problem stated in this paper has been solved with it. Utilizing a system approach, we have developed the "Systems" meta-ontology model and the "Chemical compounds and their electron properties" ontology model, which is based on the "Systems" meta-ontology. Then the ontology and knowledge base editors are developed with the help of the Django framework (grounded on the models that have been developed previously). Knowledge engineers and chemists work with these editors to fill the knowledge base. The intelligent system prototype is provided in the current paper. The interface components that facilitate chemist compound structural formula input and overall user input are developed utilizing JSME. The complete methodology and its major steps have been developed and discussed in this paper in such a way that tries to be as precise as possible without getting the reader too bored. Further research is being conducted on collaborative knowledge base development. The ontologies introduced in this paper and the described methodology can be utilized for the intelligent system development in almost any domain of interest.

Acknowledgements The reported study was funded by RFBR, project number 19-37-90137.

References

1. Artemieva, I.L.: Ontology development for domains with complicated structures. In: Wolff, K.E., Palchunov, D.E., Zagoruiko, N.G., Andelfinger, U. (eds.) Knowledge Processing and Data Analysis. KPP 2007, KONT 2007. Lecture Notes in Computer Science, vol. 6581. Springer, Berlin, Heidelberg (2011)
2. Chemical Abstracts Service. 100-millionth-fun-facts, https://www.cas.org/support/documentation/chemical-substances/cas-registry-100-millionth-fun-facts. Last accessed 09 May 2019
3. Ertl, P., Bienfait, B.: JSME: a free molecule editor in JavaScript. In: J. of Cheminformatics.-2013.-Vol. 5, No. 24. P.1–6.
4. Ertl, P.: Molecular structure input on the web. J. Cheminform. **2**, 1 (2010). https://doi.org/10.1186/1758-2946-2-1
5. Gruber, T.: A translation approach to portable ontology specifications. Knowl. Acquisition. J. **5**(2), 199–220 (1993)
6. Gulyaeva, K. A., Artemieva, I. L.: Concept system formalization: ontology models in organic chemistry intelligent system design and development. 2019 International Multi-Conference

on Engineering, Computer and Information Sciences (SIBIRCON), pp. 0928–0933 (2019). https://doi.org/10.1109/SIBIRCON48586.2019.8958217
7. Gulyaeva, K.A., Artemieva, I.L.: Applied logics to develop ontology model of the complex-structured domains: organic chemistry and biochemistry. In: Tavares, J., Dey, N., Joshi, A. (eds.) Biomedical Engineering and Computational Intelligence. BIOCOM 2018. Lecture Notes in Computational Vision and Biomechanics, vol. 32. Springer, Cham (2020). https://doi.org/10.1007/978-3-030-21726-6_7
8. Gulyaeva, K.A., Artemieva, I.L.: Ontology models in intelligent system engineering: a case of the knowledge-intensive application domain. In: Osipov, G., Panov, A., Yakovlev, K. (eds.) Artificial Intelligence. Lecture Notes in Computer Science, vol. 11866. Springer, Cham (2019). https://doi.org/10.1007/978-3-030-33274-7_8
9. Kleschev, A.S., Chernyakhovskaya, M.Y., Shalfeeva, E.A.: The paradigm of an intellectual professional activity automation. Part 1 the Features of an Intellectual Professional Activity. Ontology Designing (Russ.), vol. 3 (9), pp. 53–69 (2005)
10. Kleschev, A.S., Shalfeeva, E.A.: An ontology of intellectual activity tasks. Ontol. Design. **5**, 179–205 (2015). https://doi.org/10.18287/2223-9537-2015-5-2-179-205
11. Kleshchev, A., Artemjeva, I.: A mathematical apparatus for domain ontology simulation. An extendable language of applied logic. In: Int. Journal "Information Theories and Applications", vol.12(num 2), 149–157 (2005)
12. Lindsay, R.K., Buchanan, B. G., Feigenbaum, E.A., Lederberg, J.: Applications of artificial intelligence for organic chemistry. The DENDRAL Project. McGraw-Hill, New York, USA (1980)
13. Newell, A., Shaw, J.C., Simon, H.A.: Report on a general problem-solving program. The RAND Corporation, Paper P-1584, December 30 (1958)
14. Patai, S., Rappoport Zvi, Z.: The Chemistry of Functional Groups. Wiley, UK (2006)
15. Ullmann, J.R.: An Algorithm for Subgraph Isomorphism. In: J. ACM. vol. 23(1), pp. 31–42 (1976)

Author Index

A

Aalam, Zunaid, 357
Abdel-khier, Somaya, 161
Acevedo-Duque, Angel, 61
Agarwal, Shalini, 557
Akhund, Tajim Md. Niamat Ullah, 73, 83, 247
Akshay Sunny, R., 665
Alieksieiev, Volodymyr, 335
Ali, K. M. Akkas, 247
Alves-Foss, Jim, 537
Alvi, Redwan Hasif, 683
Andrade, Roberto O., 33
Andrikov, D. A., 43
Anilkumar, Rahul, 527
Arévalo, Diana, 33
Armas, Rolando, 13
Arora, Sandhya, 499
Artemieva, Irina L., 759
Aruna Kumari, A., 515
Arun Kumar, A., 367
Awal, Gaganmeet Kaur, 547

B

Bannaka, Kenya, 139
Bansal, Ankita, 703
Bansal, Anureet, 547
Bansal, R. K., 723
Bansal, Savina, 723
Barreto, Antón Garcia, 651
Barros, Alistair, 83
Batra, Misha, 499
Batra, Reenu, 453
Bendre, Varsha, 287
Bhoyar, Pravin Kumar, 429
Biswas, Siddhartha Sankar, 453
Boby, Farjana Abedin, 247
Borikar, Dilipkumar A., 641
Brenes, Jose A., 103
Brézillon, Patrick, 277

C

Cachicatari-Vargas, Elena, 51, 61
Cai, Zhenhua, 173
Cazares, María Fernanda, 33
Cediel, Graciela Tristancho, 51
Chandra, Sagrika, 219
Chitrao, Pradnya, 429
Choudhary, Vandana, 219
Collazos, César A., 651
Cruz-Ayala, Katiusca, 61

D

Dabriwal, Mansi, 325
Dan, Ma, 123
Dargis, Roberts, 267
Deejungwiphat, Noppawee, 237
Dhal, Sunil, 607
Dhanush Kumar, A., 675
Dhanvijay, Sapana, 381
Díaz, Laura Sofía Bazán, 1
Divekar, Rajiv, 429
Doja, Faraz, 453
Dwivedi, Divya, 741

E

Ed-Doughmi, Younes, 131
El Ayachi, Rachid, 131
Enayati, Mojtaba, 571

© The Editor(s) (if applicable) and The Author(s), under exclusive license to Springer Nature Singapore Pte Ltd. 2022
A. K. Nagar et al. (eds.), *Intelligent Sustainable Systems*, Lecture Notes in Networks and Systems 333, https://doi.org/10.1007/978-981-16-6309-3

F
Fahim, Md. Monjourur Roshed, 73
Feng, Gang, 173
Fernando, Ma. Corazon G., 421
Fojcik, Marcin, 93
Fuertes, Walter, 33

G
Gang, Feng, 123
Garg, Sachin, 209
Garkaje, Ginta, 267
Gautam, S., 597
Ghalsasi, Prachi, 499
González-Díaz, Romel Ramón, 51, 61
Gosko, Didzis, 267
Gour, Sanjay, 741
Gruzitis, Normunds, 267
Guanilo-Gómez, Santos Lucio, 51, 61
Guerrero, César D., 651
Gulapalli, Ajay Sankar, 627
Gulyaeva, Karina A., 759
Guntur, Radha Krishna, 713
Gupta, Bhoomi, 209, 219, 229
Gupta, Manav, 219
Gupta, Piyush Kumar, 453
Gupta, Sachin, 209

H
Hassan, Ammar M., 161
Hassan, Loay, 161
Heiden, Bernhard, 335
Henge, Santosh Kumar, 515
Himel, Abdul Hadi, 247
Honglan, Zou, 123
Hu, Kaitang, 413

I
Islam, Md. Ariful, 73

J
Jain, Aabha, 499
Jain, Abha, 703
Jaisswal, Aaditree, 499
Jamadar, Mohammed Zayd, 507
Jani, Preetida Vinayakray, 295
Jayant, Utkarsh, 703
Jesmin, Rashida, 73
John, Ansamma, 617
Juanatas, Irish C., 421
Junzheng, Yang, 123
Jyosthna, R., 665

K
Kale, Shiv Pranathi, 287
Kapre, Chaitanya, 641
Karn, Rajneesh, 381
Karunattu, Mathew Koshy, 617
Kaur, Amritpal, 357
Kaur, Kuljit, 731
Kaur, Rajdeep, 731
Khatun, Tania, 73
Kiran, V., 313
Kone/Tapsoba, Lydie Simone, 23
Kozaki, Yasuhiro, 139
Kshirsagar, Akshada, 287
Kuarkamphun, Titiwat, 237
Kumar, A., 597
Kumar, Brajesh, 399
Kumar, Rakesh, 723
Kurbanov, S. V., 43

L
Lagman, Ace C., 421
Lakshmi Praveena, T., 473
Lam, Man-Lai, 257
Larco, Andrés, 13
Lasmanis, Viesturs Julijs, 267
Legaspi, John Benedict C., 421
Lekshmi, S. Sree, 571
Li, Ming, 413
Lin, Pei-Jung, 257
Lin, Yang-Cheng, 113
Liu, Bo, 113
Liu, Chengzhen, 173
Liu, Shijie, 173
Loel Maansi, P., 675

M
Maestre-Gongora, Gina, 651
Mahanta, Sanjib Kumar, 607
Mali, Chetan, 741
Malo, Sadouanouan, 23
Manocha, Prabhjot Singh, 357
Manoj, M., 665
Marín-Raventós, Gabriela, 103
Mariyanayagam, Dion, 749
Martos, Patricia Janet Uceda, 1
Mathur, Harsh, 399
Medina-Cárdenas, Yurley, 651
Meghana Sarvani, C., 675
Melvin Raj, R., 665

Mielnik, Pawel, 93
Mitsunari, Kenichiro, 139
Mittal, Vinay Kumar, 627, 713
Modi, Raghav, 465
MohanKumar, Harshith, 507
Montenegro, Carlos, 13
Murakami, Katsuhiko, 139
Muthu Lakshmi, N. V., 473

N
Nakata, Yasuo, 139
Natalia, Friska, 149
Natarajan, S., 507
Nayak, Deekshith, 313
Newaz, Nishat Tasnim, 83
Nieto-Chaupis, Huber, 183, 197
Nithin, H., 507
Nithin, R., 443
Niyomrat, Pannawat, 237
Noda, Ikuhiro, 139

O
Omer, Osama A., 161
Omori, Masato, 139

P
Padmapriya, K., 527
Paliwal, Atharva, 641
Panchal, Omkar, 347
Panda, Kajal, 607
Panda, Saptasindhu, 607
Pande, Sachin Suresh, 325
Pant, Janhvi, 429
Panwar, Punita, 741
Parikh, Kriti C., 675
Parra-Sánchez, Diana Teresa, 651
Parvathy, S., 571
Patel, Ronakkumar, 465
Patil, Jigisha, 287
Pawar, Mahima, 325
Phogat, Manu, 453
Pillai, Manu J, 617
Ponnekanti, Seshaiah, 571
Prabhu, Makarand, 571
Pradhan, Sharmistha, 607
Prasomphan, Sathit, 237
Priyalakshmi, G., 443

Q
Qianghua, Wang, 123

R
Rahman, Rashedur M., 683
Rahul, K. P., 571
Rahul Raj, D. N., 313
Ramakrishnan, Krishnan, 713
Rathi, Raj, 325
Ravindrababu, Sandeep Gogineni, 537
Rico-Bautista, Dewar, 651
Rocha, Gina Alexandra Albán, 489
Rodak, Zuzanna, 93
Roy Chowdhury, Meghna, 587
Roy, Yash, 641
Rustagi, Parth, 229

S
Sánchez-Rubio, Manuel, 33
Sánchez-Velásquez, María Camila, 651
Saba, Sabrina, 247
Salah, Mostafa, 161
Santana, Paulina Maricela Jácome, 489
Satale, Rushikesh, 347
Sawant, Suraj T., 695
Sayanth, S., 617
Shankar Lingam, M., 367
Shao, Suying, 413
Sharma, Anjali R., 695
Sharma, Sonam, 587
Sharma, Tushar, 547
Sheth, Siddhesh, 347
Shukla, Pancham, 749
Shushruth Reddy, P., 675
Singh, Ajay, 295
Singh, Sandeep, 357
Sinha, Priyanshu, 229
Sinha, Shrey Kumar, 325
Sirmokadam, Sumukh, 587
Solano Cueva, Elki Mimar, 1
Song, Wenrong, 173
Sroa, Rohit, 229
Sudirman, Sud, 149
Sultana, Atia, 83
Suny, Md. Faridul Islam, 73
Swaminathan, Jose, 651

T
Tahir, Hassane, 277
Tainsakukdej, Burapa, 237
Tajane, Kapil, 347
Takamatsu, Kunihiko, 139
Telkar, Prajwal S., 313
Tharanitaran, N. M., 443

Toby, Tom, 571
Tokarz, Krzysztof, 93
Tonino-Heiden, Bianca, 335
Traore, Yaya, 23
Tumbare, Tanmay, 347

V

Varaprasad, B. K. S. V. L., 527
Vatsal, Saumitra, 557
Vats, Prashant, 453
Vayas, Luis Gonzalo Maiza, 489
Vega, Wilmer Patricio Collaguazo, 489
Venugopal, Varsha, 537
Vijayan, Akash, 617
Virdee, Bal S., 749

W

Wang, Song, 173
Whaiduzzaman, Md, 83
Wu, You-Hsun, 113

X

Xuan, Tao, 173

Y

Yanez, Cesar, 13

Z

Zaman, Zahura, 73, 83